Edited by
Uwe Bovensiepen,
Hrvoje Petek, and
Martin Wolf

Dynamics at Solid State Surfaces and Interfaces

Related Titles

Kolasinski, K. W.

Surface Science

Foundations of Catalysis and Nanoscience

2008

ISBN: 978-0-470-03304-3

Breme, J., Kirkpatrick, C. J., Thull, R. (eds.)

Metallic Biomaterial Interfaces

2008

ISBN: 978-3-527-31860-5

Wetzig, K., Schneider, C. M. (eds.)

Metal Based Thin Films for Electronics

2006

ISBN: 978-3-527-40650-0

Lanzani, G. (ed.)

Photophysics of Molecular Materials

From Single Molecules to Single Crystals

2006

ISBN: 978-3-527-40456-8

Bordo, V. G., Rubahn, H.-G.

Optics and Spectroscopy at Surfaces and Interfaces

2005

ISBN: 978-3-527-40560-2

Butt, H.-J., Graf, K., Kappl, M.

Physics and Chemistry of Interfaces

2003

ISBN: 978-3-527-40413-1

Edited by
Uwe Bovensiepen, Hrvoje Petek, and Martin Wolf

Dynamics at Solid State Surfaces and Interfaces

Volume 1: Current Developments

WILEY-VCH Verlag GmbH & Co. KGaA

The Editors

Prof. Dr. Uwe Bovensiepen
Faculty of Physics
University of Duisburg-Essen
Germany
uwe.bovensiepen@uni-due.de

Prof. Hrvoje Petek
Department of Physics
University of Pittsburgh
USA
petek@pitt.edu

Prof. Dr. Martin Wolf
Department of Physics
Free University Berlin
Germany
wolf@physik.fu-berlin.de

Cover
The cover figure depicts (i) a time-resolved experiment at an interface using time-delayed pump and probe femtosecond laser pulses (left) and the detected response (right) being either reflected light or a photoemitted electron. In addition (ii) charge transfer from an excited resonance of an alkali atom to single crystal metal substrate is shown. The false color scale represents the wave packet propagation which was calculated by A. G. Borisov including the many-body response of the metal. The figure was designed and created by A. Winkelmann.

1st Reprint 2012

Library of Congress Card No.: applied for

British Library Cataloguing-in-Publication Data
A catalogue record for this book is available from the British Library.

Bibliographic information published by the Deutsche Nationalbibliothek
The Deutsche Nationalbibliothek lists this publication in the Deutsche Nationalbibliografie; detailed bibliographic data are available on the Internet at http://dnb.d-nb.de.

Composition Thomson Digital, Noida, India
Printing and Binding betz-druck GmbH, Darmstadt
Cover Design Adam Design, Weinheim

Printed in the Federal Republic of Germany
Printed on acid-free paper

ISBN: 978-3-527-40937-2

Contents

Dynamics at Solid State Surface and Interfaces Vol.1: Current Developments
Edited by Uwe Bovensiepen, Hrvoje Petek, and Martin Wolf

ISBN: 978-3-527-40937-2

Preface

The physical properties and functionality of solid-state materials are determined both by their geometric and electronic structures and by various elementary processes, such as electron–phonon coupling or collective excitations. A microscopic understanding of the functional properties requires a detailed insight into the dynamics of these elementary processes, which occur mostly on ultrafast (typically femtosecond) timescales. Femtosecond laser spectroscopy can directly access these timescales and has shown remarkable achievements during the past two decades with many successful applications to solid-state and surface dynamics. In particular, for solid interfaces, which play a key role in the function of solid-state materials and devices, a profound microscopic understanding has been developed through experiment and theory in recent years. This success is based on the availability and recent improvements of appropriate surface-sensitive ultrafast spectroscopy techniques, on the increasing sophistication in the preparation of well-defined model systems, and on the development of theoretical methods to describe surface and electronic structure dynamics.

This book intends to provide a comprehensive overview of the fundamental concepts, techniques, and current developments in the field of ultrafast dynamics of solid-state surfaces and interfaces. Our goal is to make these concepts and insights accessible also to nonexperts and younger researchers in this field. Volume 1 summarizes the present status of research on quasiparticle dynamics, collective excitations, heterogeneous electron transfer, photoinduced modification of materials, and applications of novel techniques to study the dynamics of solids and interfaces. Volume 2 discusses the fundamental concepts and provides introductory information on elementary processes, which should be valuable, in particular, for newcomers to the field including students and postdocs. We hope that this book will also help trigger new developments and future research on ultrafast dynamic processes of solids, their interfaces, and nanostructured materials.

Brijuni, August 2010

Uwe Bovensiepen
Hrvoje Petek
and *Martin Wolf*

Dynamics at Solid State Surface and Interfaces Vol.1: Current Developments
Edited by Uwe Bovensiepen, Hrvoje Petek, and Martin Wolf

ISBN: 978-3-527-40937-2

List of Contributors

Martin Aeschlimann
University Kaiserslautern
Department of Physics and Research
Centre OPTIMAS
Erwin-Schrödinger-Str. 46
67663 Kaiserslautern
Germany

Johannes V. Barth
Technische Universität München
Physikdepartment E20
85748 Garching
Germany

Michael Bauer
Christian-Albrechts Universität zu Kiel
Institut für Experimentelle und
Angewandte Physik
24118 Kiel
Germany

Francesco Bisio
CNR-SPIN
Corso Perrone 24
16152 Genova
Italy

Mischa Bonn
Philipps-Universität Marburg
Fachbereich Physik
Renthof 5
35032 Marburg
Germany

Uwe Bovensiepen
Universität Duisburg-Essen
Fakultät für Physik
Lotharstr. 1
47048 Duisburg
Germany

and

Freie Universität Berlin
Fachbereich Physik
Arnimallee 14
14195 Berlin
Germany

Adrian L. Cavalieri
Max Planck Institut für Quantenoptik
Hans-Kopfermann-Strasse 1
85748 Garching
Germany

Dynamics at Solid State Surface and Interfaces Vol.1: Current Developments
Edited by Uwe Bovensiepen, Hrvoje Petek, and Martin Wolf

ISBN: 978-3-527-40937-2

Cheng-Tien Chiang
Max-Planck-Institut für
Mikrostrukturphysik
Weinberg 2
06120 Halle
Germany

Evgueni V. Chulkov
Universidad del País Vasco
Facultad de Ciencias Químicas
Depto. de Física de Materiales and
Centro de Física de Materiales
CFM-MPC, Centro Mixto CSIC-UPV/
EHU
20080 San Sebastián
Basque Country
Spain

Markus Donath
Westfälische Wilhelms Universität
Münster
W.-Klemm-Str. 10
48149 Münster
Germany

Pedro M. Echenique
Universidad del País Vasco
Facultad de Ciencias Químicas
Depto. de Física de Materiales and
Centro de Física de Materiales
CFM-MPC, Centro Mixto CSIC-UPV/
EHU
20080 San Sebastián
Basque Country
Spain

Thomas Elsaesser
Max-Born-Institut für Nichtlineare
Optik und Kurzzeitspektroskopie
Max-Born-Straße 2 A
12489 Berlin
Germany

Ralph Ernstorfer
Max Planck Institut für Quantenoptik
Hans-Kopfermann-Strasse 1
85748 Garching
Germany

and

Technische Universität München
Physikdepartment E11
85747 Garching
Germany

Thomas Fauster
Universität Erlangen-Nürnberg
Lehrstuhl für Festkörperphysik
Staudtstr. 7
91058 Erlangen
Germany

Peter Feulner
Technische Universität München
Physikdepartment E20
85748 Garching
Germany

Alexander Föhlisch
Universität Hamburg
Institut für Experimentalphysik
Luruper Chaussee 149
22761 Hamburg
Germany

Martin E. Garcia
Universität Kassel
Fachbereich Mathematik und
Naturwissenschaften
Institut für Theoretische Physik
Heinrich-Plett-Str. 40
34132 Kassel
Germany

Sean Garrett-Roe
University of California
Department of Chemistry
Berkeley, CA 94720-1460
USA

Jens Güdde
Philipps-Universität Marburg
Fachbereich Physik
Renthof 5
35032 Marburg
Germany

Charles B. Harris
University of California
Department of Chemistry
Berkeley, CA 94720-1460
USA

Muneaki Hase
University of Tsukuba
Institute of Applied Physics
1-1-1 Tennodai
Tsukuba 305-8573
Japan

Rigoberto Hernandez
Georgia Institute of Technology
School of Chemistry & Biochemistry
Center for Computational Molecular
Sciences & Technology
Atlanta, GA 30332-0430
USA

Ulrich Höfer
Philipps-Universität Marburg
Fachbereich Physik
Renthof 5
35032 Marburg
Germany

Philip Hofmann
University of Aarhus
Institute for Storage Ring Facilities and
Interdisciplinary Nanoscience Center
8000 Aarhus C
Denmark

Rupert Huber
Universität Konstanz
Fachbereich Physik
Universitätsstr. 10
78464 Konstanz
Germany

Kunie Ishioka
National Institute for Materials Science
Advanced Nano-characterization Center
Sengen 1-2-1
Tsukuba 305-0047
Japan

James E. Johns
University of California
Department of Chemistry,
Berkeley, CA 94720-1460
USA

Henry C. Kapteyn
University of Colorado and NIST
JILA
Boulder, CO 80309-0440
USA

Reinhard Kienberger
Max Planck Institut für Quantenoptik
Hans-Kopfermann-Strasse 1
85748 Garching
Germany

and

Technische Universität München
Physikdepartment E11
85747 Garching
Germany

Patrick S. Kirchmann
Stanford Institute for Materials & Energy Science
McCullough Building 232
476 Lomita Mall
Stanford, CA 94305-4045
USA

Jürgen Kirschner
Max-Planck-Institut für Mikrostrukturphysik
Weinberg 2
06120 Halle
Germany

Tillmann Klamroth
Universität Potsdam
Institut für Chemie
Karl-Liebknecht-Str. 24-25
14476 Potsdam-Golm
Germany

Stephan W. Koch
Philipps-Universität Marburg
Fachbereich Physik
Renthof 5
35032 Marburg
Germany

Ferenc Krausz
Max Planck Institut für Quantenoptik
Hans-Kopfermann-Strasse 1
85748 Garching
Germany

Alfred Leitenstorfer
Universität Konstanz
Fachbereich Physik/LS Leitenstorfer
Fach M 696
78457 Konstanz
Germany

Wen-Chin Lin
Department of Physics
National Taiwan Normal University
Taipei 11677
Taiwan

Stefan Mathias
University of Kaiserslautern
Department of Physics and Research Center OPTIMAS
67663 Kaiserslautern
Germany

Yoshiyasu Matsumoto
Kyoto University
Graduate School of Science
Department of Chemistry
Kyoto 606-852
Japan

Luis Miaja-Avila
University of Colorado and NIST
JILA
Boulder, CO 80309-0440
USA

Torsten Meier
Philipps-Universität Marburg
Fachbereich Physik
Renthof 5
35032 Marburg
Germany

Alexey Melnikov
Freie Universität Berlin
Fachbereich Physik
Arnimallee 14
14195 Berlin
Germany

Dietrich Menzel
Technische Universität München
Physikdepartment E20
85748 Garching
Germany

and

Fritz Haber Institut der MPG
Faradayweg 4-6
14195 Berlin
Germany

Oleg V. Misochko
Russian Academy of Sciences
Institute of Solid State Physics
Chemogolovka
142432 Moscow Region
Russia

Eric Muller
University of California
Department of Chemistry,
Berkeley, CA 94720-1460
USA

Matthias Muntwiler
Paul Scherrer Institut
WSLA/122
5232 Villigen PSI
Switzerland

Margaret M. Murnane
University of Colorado and NIST
JILA
Boulder, CO 80309-0440
USA

Tadaaki Nagao
National Institute for Materials Science
WPI Research Center for Materials
Nanoarchitectonics
1-1 Namiki, Tsukuba-City
305-0044 Ibaraki
Japan

Luca Perfetti
Laboratoire des Solides Irradiés
Ecole Polytechnique
91128 Palaiseau Cedex
France

Hrvoje Petek
Department of Physics and Astronomy
University of Pittsburgh
G01 Allen Hall
3941 O'hara St.
Pittsburgh, PA 15213
USA

Walter Pfeiffer
Universität Bielefeld
Fakultät für Physik
Universitätsstr. 25
33615 Bielefeld
Germany

Martin Pickel
Westfälische Wilhelms-Universität
Münster
W.-Klemm-Str. 10
48149 Münster
Germany

and

Max-Born-Institut für Nichtlineare
Optik und Kurzzeitspektroskopie
Max-Born-Straße 2 A
12489 Berlin
Germany

M. Rohleder
Philipps-Universität Marburg
Fachbereich Physik
Renthof 5
35032 Marburg
Germany

Peter Saalfrank
Universität Potsdam
Institut für Chemie
Karl-Liebknecht-Str. 24-25
14476 Potsdam-Golm
Germany

Anke B. Schmidt
Westfälische Wilhelms-Universität
Münster
W.-Klemm-Str. 10
48149 Münster
Germany

and

Max-Born-Institut für Nichtlineare
Optik und Kurzzeitspektroskopie
Max-Born-Straße 2 A
12489 Berlin
Germany

Vyacheslav M. Silkin
Universidad del País Vasco
Facultad de Ciencias Químicas
Depto. de Física de Materiales
Apdo. 1072
20080 San Sebastián
Basque Country
Spain

and

IKER BASQUE
Basque Foundation for Science
48011 Bilbao
Basque Country
Spain

Irina Yu. Sklyadneva
Donostia Physics International Centre
(DIPC)
P. Manuel de Lardizabal 4
20018 San Sebastian
Basque Country
Spain

Julia Stähler
Freie Universität Berlin
Fachbereich Physik
Arnimallee 14
14195 Berlin
Germany

and

Fritz-Haber-Institut der Max-Planck-
Gesellschaft
Abteilung Physikalische Chemie
Faradayweg 4-6
14195 Berlin
Germany

Matthew L. Strader
University of California
Department of Chemistry
Berkeley, CA 94720-1460
USA

H. Ueba
Department of Electronics
Toyama University
930-8555 Toyama
Japan

Tijo Vazhappilly
Universität Potsdam
Institut für Chemie
Karl-Liebknecht-Str. 24-25
14476 Potsdam-Golm
Germany

Kazuya Watanabe
Kyoto University
Graduate School of Science
Department of Chemistry
Kyoto 606-8502
Japan

Martin Weinelt
Max-Born-Institut
für Nichtlineare Optik und
Kurzzeitspektroskopie
Max-Born-Straße 2 A
12489 Berlin
Germany

and

Freie Universität Berlin
Fachbereich Physik
Arnimallee 14
14195 Berlin
Germany

Stephan Wethekam
Humboldt Universität
Institut für Physik
Brook-Taylor-Str. 6
12489 Berlin
Germany

Aimo Winkelmann
Max-Planck-Institut für
Mikrostrukturphysik
Weinberg 2
06120 Halle
Germany

Helmut Winter
Humboldt Universität
Institut für Physik
Brook-Taylor-Str. 6
12489 Berlin
Germany

Michael Woerner
Max-Born-Institut für Nichtlineare
Optik und Kurzzeitspektroskopie
Max-Born-Straße 2 A
12489 Berlin
Germany

Martin Wolf
Freie Universität Berlin
Fachbereich Physik
Arnimallee 14
14195 Berlin
Germany

and

Fritz-Haber-Institut der Max-Planck-
Gesellschaft
Abteilung Physikalische Chemie
Faradayweg 4-6
14195 Berlin
Germany

Wilfried Wurth
Universität Hamburg
Institut für Experimentalphysik
Luruper Chaussee 149
22761 Hamburg
Germany

Xiaoyang -Y. Zhu
University of Minnesota
Department of Chemistry
Minneapolis, MN 55455
USA

and

University of Texas at Austin
1 University Station A5300
Austin, TX 78712-016
USA

Eeuwe S. Zijlstra
Universität Kassel
Fachbereich Mathematik und
Naturwissenschaften
Institut für Theoretische Physik
Heinrich-Plett-Str. 40
34132 Kassel
Germany

Part One
Quasiparticle Dynamics

Dynamics at Solid State Surface and Interfaces Vol.1: Current Developments
Edited by Uwe Bovensiepen, Hrvoje Petek, and Martin Wolf

ISBN: 978-3-527-40937-2

1
Nonlinear Terahertz Studies of Ultrafast Quasiparticle Dynamics in Semiconductors

Michael Woerner and Thomas Elsaesser

Quasiparticle concepts play a fundamental role in describing the linear and nonlinear responses of semiconductors to an external electric field [1, 2]. Basic optical excitations in the range of the fundamental bandgap are the Wannier exciton and the exciton-polariton, whereas the polaron, an electron coupled to the Coulomb-mediated distortion of a polar crystal lattice, is essential for charge transport. Such quasiparticles display nonequilibrium dynamics in the femto- to picosecond time domain, governed by microscopic couplings in the electronic system and between charge carriers and lattice excitations. Ultrafast optical spectroscopy [3] in combination with extensive theoretical work [1, 4, 5] has provided detailed insight into quantum coherent quasiparticle dynamics and into a hierarchy of relaxation phenomena, including decoherence, carrier thermalization, and carrier cooling as well as trapping and recombination.

In recent years, the generation of ultrashort electric field transients in the terahertz (THz), that is, far-infrared frequency range has made substantial progress [6]. In particular, THz field strengths of up to 1 MV/cm and (sub)picosecond time structures have been achieved with the help of THz sources driven by femtosecond laser pulses. Such transients open new ways for studying charge transport in semiconductors under high-field nonequilibrium conditions [7]. Both ballistic transport phenomena, where the strong interaction of the carriers with the external field leads to negligible friction of transport on a short timescale, and the regime of coherent quantum kinetic transport become accessible. In the latter, the quantum nature of quasiparticles and their coherent quantum phase are essential, requiring a theoretical description well beyond the traditional Boltzmann equation approach based on scattering times. New phenomena occur in the time range below such scattering times, for example, quantum coherent electron–phonon interactions [8] and extended real- and k-space motions of electron wave packets.

In this chapter, we review our recent work in this exciting new area of THz research. The main emphasis is on polaron dynamics in the polar semiconductor GaAs [9] and on coherent high-field transport of electrons in the femtosecond time domain. Following this introduction, we discuss a theoretical description of the static and dynamic properties of polarons (Section 1.1.1). We then present a summary of

Dynamics at Solid State Surface and Interfaces Vol.1: Current Developments
Edited by Uwe Bovensiepen, Hrvoje Petek, and Martin Wolf

ISBN: 978-3-527-40937-2

the experimental techniques in Section 1.1.2, followed by our results on quantum kinetic polaron dynamics (Section 1.1.3) and coherent high-field transport (Section 1.1.4). Conclusions are presented in Section 1.1.5.

1.1
Linear Optical Properties of Quasiparticles: The Polarization Cloud around a Charge Carrier

In a polar or ionic solid, a free electron distorts the crystal lattice, displacing the atoms from their equilibrium positions. One considers the electron together with its surrounding lattice distortion a quasiparticle [10, 11], the Fröhlich polaron [12, 13]. In thermal equilibrium, a Fröhlich polaron is characterized by a self-consistent attractive potential for the electron caused by a surrounding cloud of longitudinal optical (LO) phonons. In Figure 1.1, the polarized lattice (Figure 1.1a) and the potential energy as a function of the relative distance between the electron and the center of the LO phonon cloud (Figure 1.1b and c) are shown schematically. The electron–phonon coupling strength $\alpha = 0.067$ [14] determines the polaron

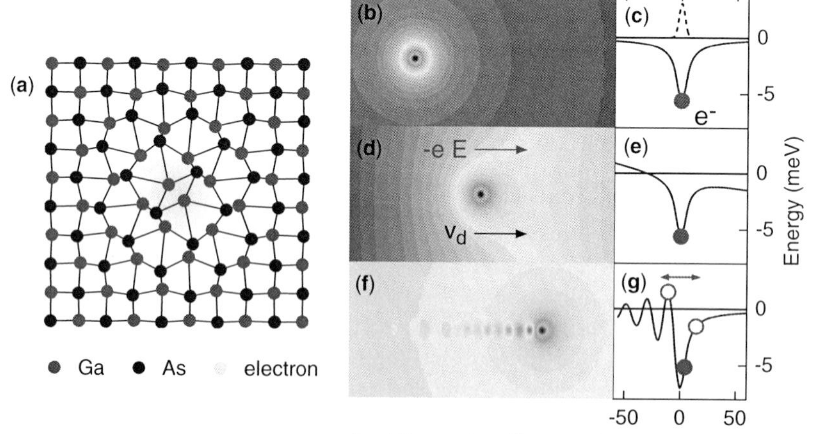

Figure 1.1 (a) Lattice distortion of the Fröhlich polaron in GaAs. Self-induced polaron potential (b, contour plot and solid line in (c)) and electron wave function (dashed line in (c)) of a polaron at rest. (d and e) Linear transport: For low applied electric fields, the total potential is the sum of the applied potential and the zero-field polaron potential. (f and g) Nonlinear transport: In a strong DC field (which has been subtracted from the potentials shown in (f) and (g)), the drifting electron (red dot) is displaced from the minimum of the LO phonon cloud and generates coherent phonon oscillations in its stern wave. As the amplitude of coherent LO phonons exceeds a certain threshold, the polaron potential eventually causes electron oscillations (shown as open circles) along the relative coordinate *r* on top of the drift motion of the entire quasiparticle. (Please find a color version of this figure on the color plates.)

binding energy of 5 meV, its radius of 2.7 nm (at room temperature), and its effective mass, which is slightly larger than the effective mass of the quasifree electron.

1.1.1 Theoretical Models Describing Static and Dynamic Properties of Polarons

In this section, we first give a brief overview of the theoretical literature on polaron physics, followed by a short description of the theoretical model we used for simulations of our nonlinear THz experiments on Fröhlich polarons in n-type GaAs.

Beginning with the late 1940s, polaron physics has been the subject of extensive theoretical literature. Pioneering work was performed by Lee *et al.* [13] and by Feynman [15] who introduced the path integral method – a standard tool being applied in a wide range of theoretical studies – for describing polaron behavior. Peeters and Devreese calculated the radius, the self-induced potential, and the number of virtual optical phonons of a polaron at rest [14]. More recently, the quasistationary high-field properties were investigated. Jensen and Sauls studied polarons near the Cerenkov velocity [16] and found that strong velocity (or momentum) fluctuations on top of the drift velocity cause the strong friction force around the threshold velocity for phonon emission. Janssen and Zwerger studied the nonlinear transport of polarons [17] with the important result that "... quantum effects become irrelevant for large fields or transport velocities ... ." This interesting fact establishes a link to the so-called classical polaron model [18] that has much in common with the treatment by Magnus and Schoenmaker [19] who calculated "an exact solution" of the time-dependent electron velocity in the linear regime within the Caldeira–Leggett model [20]. The results of Refs [19, 21] have clearly shown that memory effects in the electron–phonon interaction or energy nonconserving transitions (or collisional broadening) lead to interferences between the electron–electric field and the electron–phonon interaction prolonging in turn the ballistic transport on ultrafast timescales.

Based on such theoretical work, we developed a new approach to get a more specific insight into the microscopic nonlinear dynamics of polarons on ultrafast timescales. We performed calculations within a nonlinear and time-dependent extension of the linear model presented in Ref. [19]. We consider a single electron interacting with the local electric field in the x-direction and with the phonon modes of the crystal via different types of electron–phonon interactions. The quantum mechanical Hamiltonian [13] reads

$$\begin{aligned} H(t) &= \varepsilon(\vec{p}) + exE_{\mathrm{loc}}(t) + \sum_{b,\vec{q}} \frac{P_{b,\vec{q}}^2 + \omega_b^2(\vec{q})Q_{b,\vec{q}}^2}{2} \\ &\quad + \sum_{b,\vec{q}} M_b(\vec{q}) \times [P_{b,\vec{q}} \cos \vec{q}\vec{r} + \omega_b(\vec{q})Q_{b,\vec{q}} \sin \vec{q}\vec{r}], \end{aligned} \tag{1.1}$$

where $\vec{r} = (x, y, z)$ and $\vec{p} = (p_x, p_y, p_z)$ denote the position and momentum operators of the electron, respectively. The dispersive band structure of the lowest conduction band is described by $\varepsilon(\vec{p})$ that can be obtained from, for example, pseudopotential calculations [22–24]. For small excursions of the electron within the lowest minimum of the conduction band, the effective mass approximation $\varepsilon(\vec{p}) = \vec{p}^2/2m_{\text{eff}}$ is sufficient to describe the polaron correctly. The local electric field $E_{\text{loc}}(t)$ is the sum of the externally applied electric field and the field re-emitted by the coherent motion of all electrons. The latter contains the linear and nonlinear responses of the system and accounts for the radiative damping of the electron motion [25]. $Q_{b,\vec{q}}$ and $P_{b,\vec{q}}$ are the coordinate and the conjugate momentum of the phonon of branch b with the wave vector $\vec{q} = (q_x, q_y, q_z)$ and angular frequency $\omega_b(\vec{q})$. For simplicity, we limit our calculations to the polar coupling to longitudinal optical phonons ($b =$ LO) with a constant frequency $\omega_{\text{LO}}(\vec{q}) = \omega_{\text{LO}} = \text{const.}$

$$M_{\text{LO}}(\vec{q}) = \sqrt{\frac{e^2}{\varepsilon_0 V}\left(\frac{1}{\varepsilon_\infty} - \frac{1}{\varepsilon_S}\right)} \times \frac{1}{|\vec{q}|} \tag{1.2}$$

and coupling to acoustic phonons ($b =$ AC) via the deformation potential Ξ with an averaged sound velocity c_S,

$$M_{\text{AC}}(\vec{q}) = \sqrt{\frac{\Xi^2}{\varrho V c_S^2}}, \tag{1.3}$$

$$\omega_{\text{AC}}(\vec{q}) = \frac{2c_S q_{\text{zb}}}{\pi} \sin\left(\frac{\pi|\vec{q}|}{2q_{\text{zb}}}\right). \tag{1.4}$$

ε_S is the static relative dielectric constant and ε_∞ is the dielectric constant for frequencies well above the optical phonon frequency, but below electronic excitations. The difference $\varepsilon_\infty^{-1} - \varepsilon_S^{-1}$ is proportional to the polar electron–LO phonon coupling constant α [14]. V is the quantization volume that determines the discretization of the k- and q-spaces with the zone boundary $q_{\text{zb}}$and ϱ stands for the mass density of the crystal.

From Eq. (1.1), we derive the Heisenberg equations of motion for the expectation values of quantum mechanical operators like $\langle x \rangle$, $\langle p_x \rangle$, and so on. In this process, new quantum mechanical operators containing combinations of canonical variables, for example, $\langle P_{b,\vec{q}} \sin(\vec{q}\vec{r}) \rangle$, appear on the right-hand side of the equations of motion. Since we are interested only in the expectation values of the relevant observables and would like to close the infinite hierarchy of equations at some level, we expand and subsequently approximate those expectation values. In lowest order, one exactly obtains the equations of motion of the classical polaron [18, 26]. The classical polaron model predicts, however, an unrealistically high binding energy in the self-induced potential as classical particles correspond to infinitely small wave packets. To overcome this problem, one has to go one step further in the expansion of the expectation values of quantum mechanical operators and consider the finite size of the electron wave packet $\Delta x^2 = \langle x^2 \rangle - \langle x \rangle^2$. As shown in detail in Refs [27, 28], the dynamics of Δx^2 is inherently connected with the dynamics of both the variance of its conjugate momentum $\Delta p_x^2 = \langle p_x^2 \rangle - \langle p_x \rangle^2$ and $\Delta xp = \langle xp_x + p_x x \rangle/2 - \langle x \rangle\langle p_x \rangle$, which is the covariance of x and p. The main result of Refs [27, 28] is that under certain circumstances (which are fulfilled in our case), continuous position measurements of the electron caused by various fluctuating forces of the environment lead to decoherence phenomena in such a way that an initially Gaussian electron wave packet (in Wigner space) stays Gaussian in its further evolution and adjusts its size Δx^2 continuously to the respective momentum uncertainty according to $\Delta x^2 = \hbar^2/4\Delta p_x^2$ (cf. minimum of Heisenberg's uncertainty relation). Such continuous position measurements of the electron also lead to a small random walk in phase space, that is, diffusion of both the position and momentum of the particle. According to the arguments of the authors of Refs [27, 28], this diffusion is ineffectual in comparison to the wave packet localization and, thus, we completely neglect it in the following. The application of the approximations and arguments discussed above lead to the following system of equations of motion for the expectation values of the operators:

$$\frac{d\langle x\rangle}{dt} = \langle v_x\rangle, \quad \text{with the velocity operator } v_x = \frac{\partial \varepsilon(\vec{p})}{\partial p_x}, \tag{1.5}$$

$$\begin{aligned}\frac{d\langle p_x\rangle}{dt} &= eE_{\text{loc}}(t) + \sum_{b,\vec{q}} \exp\left[-\frac{1}{2}\vec{q}^{\,2}\Delta x^2\right] M_b(\vec{q}) \\ &\quad \times \left[\langle P_{b,\vec{q}}\rangle q_x \sin q_x\langle x\rangle - \omega_b(\vec{q})\langle Q_{b,\vec{q}}\rangle q_x \cos q_x\langle x\rangle\right],\end{aligned} \tag{1.6}$$

$$\frac{d\langle Q_{b,\vec{q}}\rangle}{dt} = \langle P_{b,\vec{q}}\rangle + M_b(\vec{q}) \cos q_x\langle x\rangle \exp\left[-\frac{1}{2}\vec{q}^{\,2}\Delta x^2\right], \tag{1.7}$$

$$\frac{d\langle P_{b,\vec{q}}\rangle}{dt} = -\omega_b^2(\vec{q})\langle Q_{b,\vec{q}}\rangle - M_b(\vec{q})\omega_b(\vec{q}) \sin q_x\langle x\rangle \exp\left[-\frac{1}{2}\vec{q}^{\,2}\Delta x^2\right]. \tag{1.8}$$

For simplicity, we use here spherical Gaussian wave packets with momentum $\langle p_x\rangle$ in the x-direction and isotropic momentum fluctuations $\Delta p_x^2 = \Delta p_y^2 = \Delta p_z^2$. Consequently, the expectation values of the kinetic energy $\langle \varepsilon(\vec{p})\rangle = \varepsilon_{\text{kin}}(\langle p_x\rangle, \Delta p_x^2)$ and the velocity operator $\langle v_x\rangle = V_x(\langle p_x\rangle, \Delta p_x^2)$ are functions of both $\langle p_x\rangle$ and Δp_x^2. Both two-dimensional functions have been derived from pseudopotential calculations [22–24]. The so far missing dynamical variable Δp_x^2 (in turn determining $\Delta x^2 = \Delta y^2 = \Delta z^2 = \hbar^2/4\Delta p_x^2$) can be inferred from an equation of motion of the kinetic energy

$$\frac{d\langle \varepsilon(\vec{p})\rangle}{dt} = \frac{\partial \varepsilon_{\text{kin}}(\langle p_x\rangle, \Delta p_x^2)}{\partial\langle p_x\rangle}\frac{d\langle p_x\rangle}{dt} + \frac{\partial \varepsilon_{\text{kin}}(\langle p_x\rangle, \Delta p_x^2)}{\partial \Delta p_x^2}\frac{d\,\Delta p_x^2}{dt} \tag{1.9}$$

using the following arguments. The temporal change of the total electron energy $d\langle E(\vec{p})\rangle/dt$ splits naturally into a ballistic coherent (first term) and an incoherent contribution (second term), the latter of which is connected to the velocity fluctuations of the electron (see also discussion of Eqs. (16) and (A4) of Ref. [16]). Since the acceleration of the electron in the external field does not change its momentum fluctuations, it exclusively contributes to the first term on the rhs of Eq. (1.9). In general, the friction force due to phonon scattering (second term rhs of Eq. (1.6)) will contribute to both terms in Eq. (1.9). In the typical situation, however, the energy relaxation time is distinctly longer than the momentum relaxation time (cf. Figure 13 of Ref. [29]). Thus, in good approximation, we assume that the friction force exclusively contributes to the incoherent contribution of electron energy change leading to the following implicit equation of motion for the expectation value of the momentum fluctuations Δp_x^2:

$$\begin{aligned}\frac{\partial\varepsilon_{\mathrm{kin}}(\langle p_x\rangle,\Delta p_x^2)}{\partial\Delta p_x^2}\frac{\mathrm{d}\Delta p_x^2}{\mathrm{d}t} &= V_x(\langle p_x\rangle,\Delta p_x^2)\sum_{b,\vec{q}}\exp\left[-\frac{1}{2}\vec{q}^2\Delta x^2\right]M_b(\vec{q}) \\ &\quad\times\left[\langle P_{b,\vec{q}}\rangle q_x \sin q_x\langle x\rangle-\omega_b(\vec{q})\langle Q_{b,\vec{q}}\rangle q_x\cos q_x\langle x\rangle\right] \\ &\quad+\Gamma_{\mathrm{loss}}(p_x,\Delta p_x^2,T_{\mathrm{L}}) \\ &\quad\times\left[\varepsilon_{\mathrm{kin}}(\langle p_x\rangle,m_{\mathrm{eff}}k_{\mathrm{B}}T_{\mathrm{L}})-\varepsilon_{\mathrm{kin}}(\langle p_x\rangle,\Delta p_x^2)\right].\end{aligned} \tag{1.10}$$

Emission and absorption of incoherent phonons are described by the energy relaxation rate $\Gamma_{\mathrm{loss}}(p_x,\Delta p_x^2,T_{\mathrm{L}})$, which is generally a “slow” process occurring on a timescale of several hundreds of femtoseconds (cf. Figure 13 of Ref. [29]). Thus, it can be well described by the Fermi’s golden rule (FGR) approach like in the semiclassical Boltzmann transport equation. In the absence of external electric fields, this term relaxes the wave packet size to its value at thermal equilibrium, that is, $\Delta p_x^2 = m_{\mathrm{eff}}k_{\mathrm{B}}T_{\mathrm{L}}$.

Equations (1.5)–(1.8) are similar to those of the classical polaron [18, 26]. Quantum mechanics, that is, Planck’s constant $\hbar$, enters this system of equations only through the bandwidth-limited wave packet size $\Delta x^2 = \hbar^2/4\Delta p_x^2$, the dynamics of which is determined by Eq. (1.10). We would like to stress the fact that the corresponding dynamics of the velocity (or momentum) fluctuations is determined by incoherent heating and cooling processes, both typically occurring on a timescale of several hundreds of femtoseconds. As a consequence, one expects negligible changes of Δx^2 on ultrafast timescales <200 fs.

Our main motivation for developing the nonlinear set of polaron equations of motion, (1.5)–(1.8) and (1.10), is their conceptual simplicity and the fact that they are tractable for arbitrary driving fields $E(t)$ on a standard personal computer. Before applying them to our nonlinear THz experiments, we calculate some linear and quasistationary properties of the polaron and compare the results with other theories and experiments.

1.1.2
Experimental Signatures of Linear and Quasistationary Polaron Properties

1.1.2.1 The Fröhlich Polaron at Rest

The radius, the self-induced potential, and the number of virtual optical phonons of a polaron at rest have been calculated by Peeters and Devreese [14]. In our model, the stationary ground state of the polaron can be simply determined by setting $\langle x\rangle$, $\langle p_x\rangle$, and all time derivatives of Eqs. (1.5)–(1.8) and (1.10) to zero and solving the implicit equations for Δp_x^2, $\langle Q_{b,\vec{q}} \rangle$, and $\langle P_{b,\vec{q}} \rangle$. The electric field created by the self-induced polaron potential acting on an electron at a position $\langle x \rangle \neq 0$ can be calculated with the help of the second term on the rhs of Eq. (1.6). The corresponding self-induced potential at $T_\mathrm{L} = 300$ K is shown in Figure 1.1b and c. We find an excellent agreement with the result of the path integral method in Ref. [14].

As sketched in Figure 1.1a, the electron (yellow cloud) polarizes the surrounding lattice of cations and anions via the Coulomb interaction, resulting in turn in a self-induced potential trap for the electron (solid line in Figure 1.1c). The size and depth of this polaron potential depends self-consistently on the electron–LO phonon interaction strength α and the temperature T_L of the coupled system. In addition to the thermodynamically averaged self-induced polaron potential, the electron experiences the fluctuating Coulomb potential of the surrounding ions caused by the quantum mechanical zero-point motion and thermal fluctuations of the latter. These fluctuating forces determine the spatial extension of the electron wave packet Δx^2 (dashed line in Figure 1.1c), which is called polaron radius in Ref. [14]. With increasing lattice temperature T_L, the coherence length of the electron shrinks leading to an enhanced interaction with LO phonons of higher q vectors and concomitantly to an increased average number of virtual phonons in the polaron cloud. This extended interaction with high-q LO phonons deepens and narrows the self-induced potential trap and reduces significantly the polaron radius. As already discussed in the conclusion of Ref. [14], this leads to the counterintuitive result that the polaron binding energy, that is, the depth of the self-induced trap, increases with the temperature.

Experimentally, it is very difficult to access any of the characteristic properties of the polaron at rest. For instance, strong magnetic fields have been applied to n-type semiconductor structures to measure the cyclotron resonance frequency [30]. One expects polaronic effects once the magnetic length meets the polaron radius or once the Landau level splitting meets the LO phonon energy. The experiments of Ref. [30] have shown that a separation of polaronic signatures from other effects, for example, band structure nonparabolicity, is extremely difficult.

1.1.2.2 **Frequency-Dependent Mobility of the Fröhlich Polaron**

Here, we briefly reconsider the linear response as discussed in Ref. [19]. In particular, we shall show that the linear limit of our theory, that is, $|q_x \langle x \rangle| \ll 1$ for all q_x, gives the quantitatively correct frequency- and temperature-dependent mobility of the polaron $\mu(\omega, T)$, ranging from the DC mobility up to the free carrier absorption (FCA) in the mid-infrared (MIR) spectral range.

The linearized versions of Eqs. (1.5)–(1.8) are those of linearly coupled harmonic oscillators. Similar to Eq. 16 in Ref. [19], the frequency-dependent mobility $\mu(\omega)$ is obtained from an integro-differential equation, which is solved by means of the Laplace transformation:

$$\frac{\mathrm{d}\langle v(t)\rangle}{\mathrm{d}t} = \frac{e}{m_\mathrm{eff}} E(t) - \int_{-\infty}^{t} \mathrm{d}t' \alpha(t-t') \langle v(t') \rangle. \tag{1.11}$$

This equation contains the memory kernel $\alpha(t)$ that in general allows describing the influence of quantum coherences on the transport behavior, that is, quantum kinetic phenomena. For a common frequency ω_{LO} for all LO phonon modes (the interaction of the electron with acoustic phonons via the deformation potential plays a minor role in the linear regime and, thus, is neglected), one has to introduce a damping mechanism to the memory kernel $\alpha(t)$ (cf. Eq. 14 in Ref. [19]) in order to ensure an irreversible energy loss to the lattice. In contrast to the standard exponential damping, we apply here the following memory kernel:

$$\alpha(t, T) = \Omega^2_{\text{trap}}(T)\frac{\tau \cos(\omega_{\text{LO}}t)}{\tau + t}, \tag{1.12}$$

$$\Omega^2_{\text{trap}}(T) = \frac{e^2}{\varepsilon_0 m_{\text{eff}} V}\left(\frac{1}{\varepsilon_\infty} - \frac{1}{\varepsilon_S}\right)\sum_{\vec{q}} \frac{q_x^2 \exp\left(-\vec{q}^{\,2}\Delta x^2(T)\right)}{|\vec{q}|^2}, \tag{1.13}$$

where $\tau = 300$ fs (for GaAs at $T = 300$ K) is the decoherence time of the memory in the electron–phonon interaction. The frequency- and temperature-dependent mobility $\mu(\omega, T)$ (solid line in Figure 1.2) calculated from Eq. (1.11) with the memory function Eq. (1.12) fits a broad range of experimental data in a quantitative way, from the DC mobility $\mu_{\text{DC}}(300\text{ K}) \approx 9000\text{ cm}^2/(\text{V s})$ up to the free carrier absorption in the mid-infrared spectral range: $\alpha_{\text{FCA}}(\lambda = 10\,\mu\text{m},\ T = 300\text{ K}) \approx 10\text{ cm}^{-1}$ for $N_e = 10^{17}\text{ cm}^{-3}$ [31]. The theoretical model predicts correctly the $\alpha_{\text{FCA}}(\omega, T) \propto \omega^{-3}$ dependence for mid-infrared frequencies ω above the LO phonon (cf. dotted line in Figure 1.2). The temperature dependence of both $\mu_{\text{DC}}(T)$ and $\alpha_{\text{FCA}}(\omega, T)$ is essentially determined by the temperature-dependent $\Omega^2_{\text{trap}}(T)$ (a comparison with the experiment shows that τ is almost insensitive to T).

The discussion of the linear mobility of the Fröhlich polaron in thermal equilibrium shows that the FCA in the mid-infrared spectral range is a sensitive probe of the

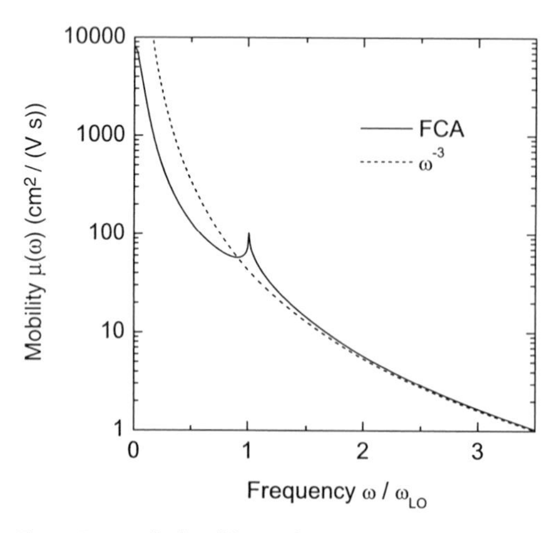

Figure 1.2 Calculated linear frequency-dependent mobility.

incoherent LO phonon population in the polaron cloud around the electron. Previously, this probe was exploited in mid-infrared pump–probe experiments on n-type InAs [32] to measure photoexcited incoherent hot, that is, nonequilibrium, phonon populations as a function of time via the transient enhancement of the free carrier absorption in the mid-infrared spectral range. So far, the theoretical description of such phenomena was based on second-order perturbation theory neglecting any coherence in the LO phonon density matrix and thus excluding possible interferences between different quantum mechanical pathways. The different theoretical approach discussed in this chapter, which is based on the quantum kinetic theory presented in Ref. [19], deals intrinsically with coherent phonons and, thus, is predetermined to describe novel FCA phenomena connected to *coherent nonequilibrium* LO phonons in the polaron cloud around the electron.

1.1.2.3 Quasistationary High-Field Transport of Polarons

In this section, we discuss the quasistationary high-field transport of polarons in GaAs. In particular, we shall show that the quasistationary drift velocity and energy of polarons are identical to results from a semiclassical Boltzmann transport equation approach. To this end, we compare the result of our polaron model with ensemble Monte Carlo simulations performed by M.V. Fischetti [29].

An external electric field E acting on the polaron (Figure 1.1d and e) induces charge transport, which is described by the drift velocity v_d. In the regime of linear response, the separation of the electron from the center of the polaron potential is negligible (Figure 1.1e) and the response to an external field is fully determined by the center-of-mass motion of the entire quasiparticle. A weak DC field (Figure 1.1d) induces a dissipative drift motion along the real space coordinate $x(t) = v_d t = \mu E t$ with a mobility $\mu = e\tau/m \approx 9000\,\text{cm}^2/(\text{V s})$ determined by the electronic charge e, the momentum relaxation time τ of the electron–phonon interaction, and the effective polaron mass m.

Above the electric field strength of ≈ 3 kV/cm, the drift velocity v_d starts to depend in a nonlinear way on the applied DC field E. In this regime, the quasistationary polaron potential is strongly distorted. In the frame of reference of the quasiparticle, a coherent standing wave of LO phonon oscillations appears as a stern wave of the moving electron (Figure 1.1f), similar to wake fields in plasmas [33, 34]. The enhanced generation of coherent nonequilibrium phonons in the stern wave of the quasiparticle creates the strong friction force in the nonlinear regime of polaron transport. In addition, the term with the drift velocity times the friction force in Eq. (1.10) drives the incoherent momentum fluctuations of the electron, which in turn provide access to interactions with phonons of larger q values. As a result, the wave packet size of the electron Δx^2 shrinks leading to an enhanced coupling to lattice degrees of freedom, in this way creating even higher friction forces.

The limit of quasistationary transport is achieved when we apply an electric field $E_{\text{loc}}(t)$ to Eqs. (1.5)–(1.8) and (1.10), which varies distinctly slower in time than the inverse of the incoherent energy loss rate $\Gamma_{\text{loss}}(p_x, \Delta p_x^2, T_L)$ (Eq. (1.10)) that is of the order of several hundreds of femtoseconds. The result of such a calculation is shown as symbols in Figure 1.3. In (b), we plot the drift velocity of polarons as a function of

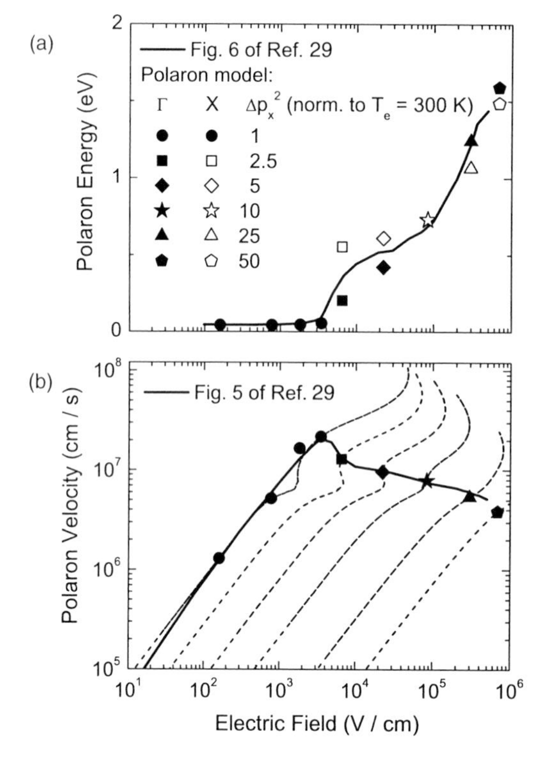

Figure 1.3 Calculated quasistationary high-field transport of polarons (symbols) in GaAs. For comparison, the black solid lines show the result of the semiclassical Boltzmann transport equation, that is, ensemble Monte Carlo simulations of M.V. Fischetti [29]. (b) Stationary drift velocity of polarons (symbols) as a function of the applied electric field. The dashed lines show drift transport of polarons with various fixed values of the wave packet size $\Delta x^2 = \hbar^2/4\Delta p_x^2$. (a) Corresponding energy of the polarons in both the Γ-valley and the X-valley of the crystal.

the applied electric field strength. If the electron–phonon matrix elements (Eqs. (1.2) and (1.3)) depend exclusively on the phonon wave vector $|\vec{q}|$, one gets a common drift velocity–friction force characteristics for all electrons independent of the conduction band valley they drift in. (It is worth to mention that an additional dependence of (1.2) and (1.3) on the electron momentum $\vec{p}$ could be easily incorporated into our model.) The dashed lines show drift transport of polarons with various *fixed* values of the wave packet size $\Delta x^2 = \hbar^2/4\Delta p_x^2$. Such a transport behavior is expected on a timescale, which is shorter than the respective energy relaxation time, but long enough to ensure transport in the drift limit, that is, outside the quantum kinetic regime. If this time window is large enough, one expects in this regime the well-known phenomenon of incoherent velocity overshoot, which is at the heart of the Gunn effect [35, 36]. In contrast, the polaron energy as a function of the applied field strength (Figure 1.3a) obviously depends on the electron valley. We show here the energy of the polarons in both the Γ-valley (solid symbols) and the X-valley (open symbols). The L-valley shows similar values (not shown).

Now, we compare our polaron model with the result (solid lines Figure 1.3) of the semiclassical Boltzmann transport equation calculated within the ensemble Monte Carlo approach by M.V. Fischetti [29]. The two models and experiments [37] agree excellently in the drift regime of high-field carrier transport. Thus, we have proven that our dynamic polaron model contains the results of the semiclassical Boltzmann transport equation [29] as a limiting case for a timescale on which the drift transport picture is still valid. On ultrafast timescales (i.e., $t < 300\,\text{fs}$), however, the semiclassical Boltzmann transport equation fails and predicts wrong scenarios of high-field transport phenomena, as will be demonstrated in the experiments discussed in the following. In contrast, the presented polaron model, that is, Eqs. (1.5)–(1.8) and (1.10), provides correct predictions in the ultrafast time domain as well.

1.2 Femtosecond Nonlinear Terahertz and Mid-Infrared Spectroscopy

Most of the THz experiments performed so far apply THz radiation as a linear probe. Studies of the nonlinear optical response and transport require THz fields of sufficient strength. In the mid-infrared spectral range, the generation and field-resolved detection of high-field transients [38] has allowed the field-resolved nonlinear experiments on intersubband transitions in n-type modulation-doped GaAs/AlGaAs quantum wells providing valuable information on both intersubband Rabi oscillations [39–41] and nonlinear radiative coupling phenomena between quantum wells [42, 43]. Recently, we developed a simple and reliable method to generate THz pulses with high electric field amplitudes in the spectral range below 5 THz [44]. In following we first present this THz source. We then discuss nonlinear THz experiments on n-type GaAs, providing new insight into quantum kinetic transport phenomena. Such results are in sharp contrast to the predictions of the semiclassical Boltzmann transport equation [29].

1.2.1 Generation of High-Field Terahertz Transients

Despite many advances in recent years, the generation, detection, and use of electromagnetic radiation [45, 46] in the frequency range of 0.1–10 THz are still far less developed than in other frequency ranges. Most THz generation schemes [6, 47, 48] provide small electric field amplitudes in the spectral range $\nu \approx 1\,\text{THz}$. The highest amplitudes (150 and 350 kV/cm) reported so far [49, 50] have been obtained using large-aperture photoconductors with bias voltages up to 45 kV. Here, we present a simple method to generate electric field amplitudes of more than 400 kV/cm. Using electro-optic sampling, we directly measure the electric field as a function of time. In this way, we fully characterize the electric fields in amplitude and phase, in contrast to interferometric methods requiring additional assumptions for field reconstruction.

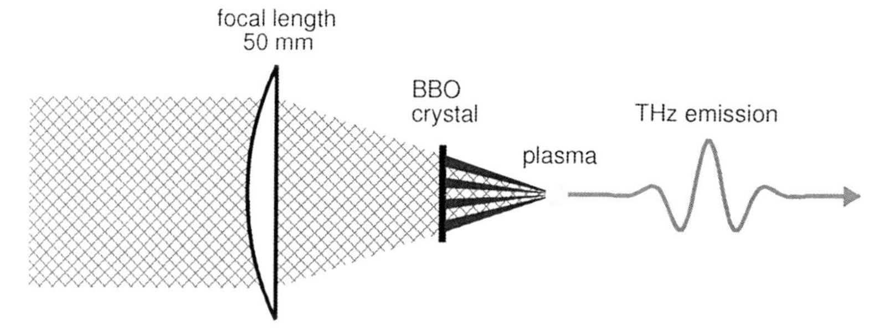

Figure 1.4 Schematic of the setup for THz generation. Incident on the lens is the output of a Ti:sapphire amplifier with pulse energies of up to 500 μJ and pulse lengths down to 25 fs. The BBO crystal of 0.1 mm thickness is cut for type-I phase-matched second-harmonic generation. In the focal region, the intensity is high enough to generate a plasma in nitrogen gas, which then acts as the source of THz radiation.

The nonlinear interaction of the fundamental (frequency ω) and the second harmonic (2ω) of femtosecond optical pulses in a laser-generated plasma in nitrogen gas is applied to generate intense THz transients. Compared to previous implementations of this method [47, 51, 52], we achieve much higher electric field amplitudes using tighter focusing and shorter pulses (both leading to higher intensities). The spectrum of the THz pulses generated extends to 7 THz, considerably higher than previously demonstrated.

Our setup for THz generation is shown schematically in Figure 1.4. A Ti:sapphire oscillator amplifier system generates pulses with a spectral width of 40 nm (corresponding to a bandwidth-limited pulse length of 22 fs) with pulse energies up to 0.5 mJ at a repetition rate of 1 kHz. Both the pulse energy and the chirp of these pulses (and thus the actual pulse length) can be varied by an acousto-optic pulse shaper [53, 54] between the oscillator and the amplifier. These pulses are focused by a fused silica lens with 50 mm focal length. A 0.1 mm thick BBO crystal cut for type-I SHG is inserted in the convergent beam about 5 mm before the focus. The THz radiation is generated in the plasma in the focal region. Both the focal length of the lens and the position of the BBO crystal are results of an optimization aimed at high electric field amplitudes. If one moves the BBO crystal closer to the focus, the intensity of the second harmonic and – as a result – the THz amplitudes become higher. This approach is, however, limited by damage in the crystal [55].

An undoped Si plate under Brewster's angle serves to separate the generated THz radiation from the remaining pump beam. Apart from its high transmission, Si has the further advantage of very low dispersion in the THz range, so it does not distort the electric field transients.

Using off-axis parabolic mirrors, the THz radiation is focused onto either a (110) ZnTe crystal or onto a (110) GaP crystal for electro-optic sampling (cf. Figure 1.7) [38]. To combine the THz radiation and the probe pulses, we use a second Si plate under Brewster's angle, which has high reflection for the s-polarized probe beam and high transmission for the p-polarized THz beam. The whole setup is enclosed and purged with dry nitrogen gas to prevent absorption from the rotational lines of water molecules.

Electric field transients measured with a 0.4 mm thick ZnTe electro-optic crystal are shown in Figure 1.5a. The THz detection range of ZnTe is limited [56, 57] by its optical phonon resonance (5.3 THz) to frequencies below about 4 THz. To check whether the spectrum (Figure 1.5c) of the THz transient generated with the 25 fs input pulse extends to higher frequencies, we have measured the same transient with a GaP (optical phonon frequency of 11 THz) electro-optic crystal (Figure 1.5b). The latter measurement gives a much broader spectrum and an even higher amplitude of the electric field of more than 400 kV/cm, corresponding to a THz pulse energy of about 30 nJ.

Apart from providing very high electric field amplitudes, the present method of THz generation has the additional advantage of being easily tunable by changing the pulse length of the input pulse, that is, by changing the chirp with the acousto-optic pulse shaper. Transients measured with the ZnTe electro-optic crystal for different lengths of the input pulse are shown in Figure 1.5a and the corresponding spectra in Figure 1.5c.

Another interesting property of our THz plasma source is its bandwidth. A transient measured with our setup using a 0.4 mm thick ZnTe electro-optic crystal is shown in Figure 1.6a. The THz detection range of ZnTe has a gap at its optical phonon resonance around 5.3 THz. This transient was optimized for a high-frequency cutoff of the THz radiation as shown in Figure 1.6b.

1.2.2 Electric Field-Resolved THz Pump–Mid-Infrared Probe Experiments

In this section, we present the first nonlinear THz pump–mid-infrared probe experiment, which shows an interesting quantum kinetic phenomenon of the electron–LO phonon dynamics of rapidly accelerated carriers in n-type GaAs. The experimental setup is shown in Figure 1.7. Both a high-field THz transient generated by four-wave mixing in a dry nitrogen plasma [44] and a synchronized mid-infrared transient generated by difference frequency mixing in GaSe [38] are focused collinearly onto the sample, a 500 nm thick layer of Si-doped (n-type) GaAs with a carrier concentration of $10^{17}\,\mathrm{cm}^{-3}$ (for details see Ref. [58]). After interaction with the sample, the time-dependent electric fields of the THz and MIR pulses are measured with electro-optic sampling in a thin ZnTe crystal [38]. Both the THz pump and the MIR probe beam are chopped with different frequencies allowing independent measurements of $E_{\mathrm{THz}}(t)$, $E_{\mathrm{MIR}}(t,\tau)$, and $E_{\mathrm{Both}}(t,\tau)$, the latter transient with both pulses interacting with the sample. τ is the delay between the THz and the MIR pulse and t is the real time. Figure 1.8 shows such transients for $\tau = 77$ fs. The nonlinear signal of interest is obtained by subtracting the two single-color measurements from $E_{\mathrm{Both}}(t,\tau)$: $E_{\mathrm{NL}}(t,\tau) = E_{\mathrm{Both}}(t,\tau) - E_{\mathrm{THz}}(t) - E_{\mathrm{MIR}}(t,\tau)$. The sample shows a coherent, nonlinear emission, which is for this particular τ in phase with the MIR pulse, demonstrating a THz field-induced MIR gain of the sample. The nonlinear transmission change is given by

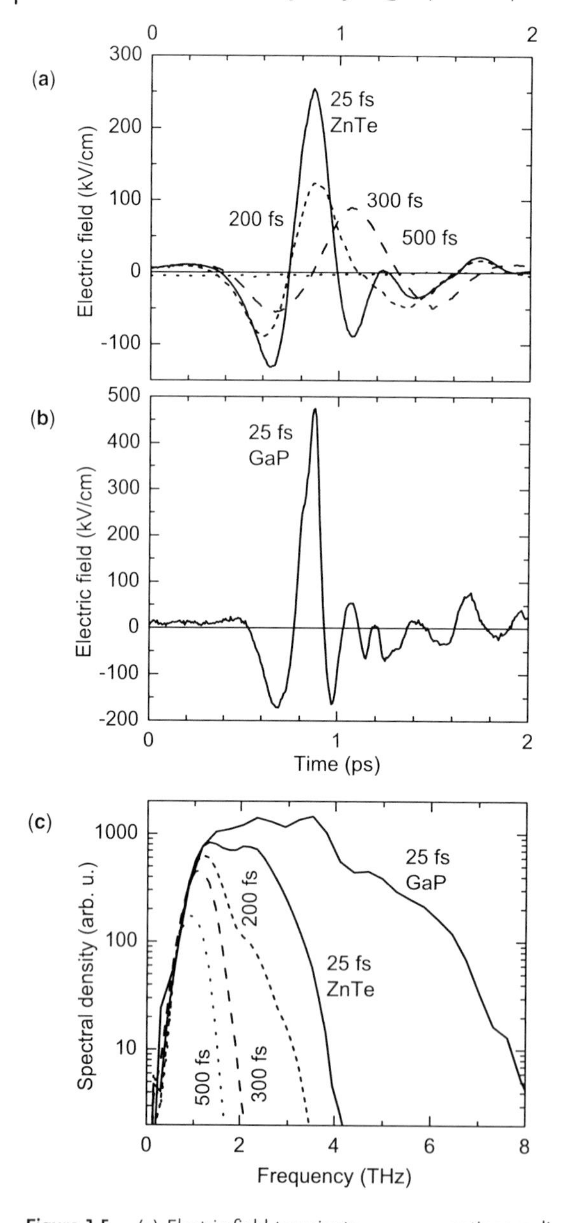

Figure 1.5 (a) Electric field transients measured by electro-optic sampling in ZnTe for different pulse lengths of the incoming pulse (pulse energy held constant at 0.5 mJ), varied by changing the amount of chirp. (b) Electric field transient for a 25 fs pulse measured by electro-optic sampling in a 0.1 mm thick GaP crystal. (c) Spectra obtained by Fourier transform of the transients. For the shortest pulse length of 25 fs, the spectrum measured with ZnTe as electro-optic crystal shows a high-frequency cutoff around 4 THz.

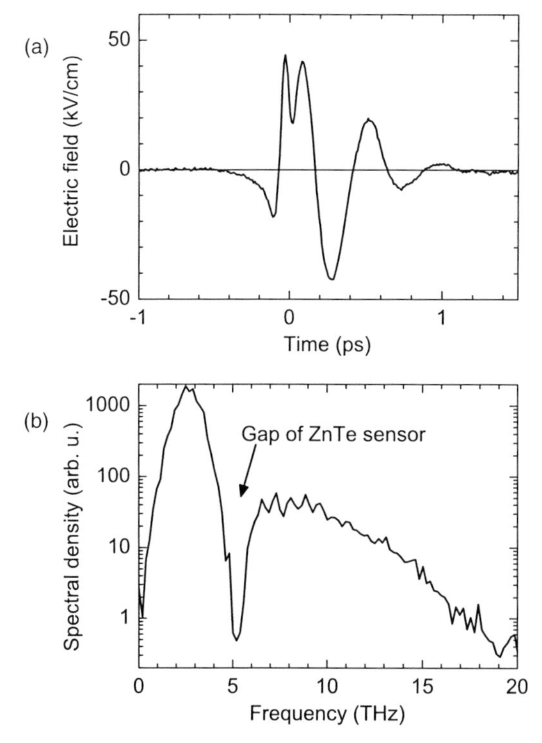

Figure 1.6 (a) Electric field transient measured by electro-optic sampling in a thin ZnTe crystal. (b) Spectrum of the pulse obtained by Fourier transform of the transient. For the shortest pump pulse length of 25 fs, we observe spectral components up to 20 THz.

$$\frac{\Delta T}{T_0}(t,\tau) = \int_{-\infty}^{t} E_{\mathrm{NL}}(t',\tau)E_{\mathrm{MIR}}(t',\tau)\mathrm{d}t' / \int_{-\infty}^{\infty} \left|E_{\mathrm{MIR}}(t',\tau)\right|^2 \mathrm{d}t', \tag{1.14}$$

which is for $t \to \infty$ identical to the usual pump–probe signal.

1.2.3 Nonlinear Terahertz Transmission Experiments

In most experiments, to study the time-resolved high-field transport, one uses a static electric field [59–61]. Time resolution is obtained by photogenerating charge carriers with a short visible or near-infrared pulse. The drawbacks of this scheme are that (i) one always has electrons *and* holes, making it difficult to extract the electron response, and (ii) the possible electric field strengths are limited by electrical breakdown. To overcome such problems, we apply a strong time-dependent electric field in the THz range on n-doped GaAs, so that only electrons contribute to the transport.

The sample investigated was grown by molecular beam epitaxy and consists of a 500 nm thick freestanding layer of Si-doped (donor concentration of $N_{\mathrm{D}} = 2\times 10^{16}\ \mathrm{cm}^{-3}$) GaAs clad between two 300 nm thick $Al_{0.4}Ga_{0.6}As$ layers [62]. A few cycle

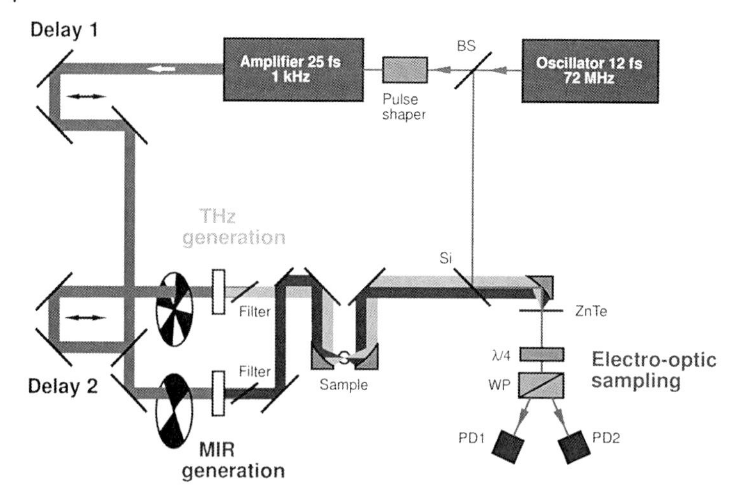

Figure 1.7 THz pump–MIR probe setup: Both terahertz $E_{THz}(t)$ and mid-infrared transients $E_{MIR}(t, \tau)$ propagate collinearly through a 500 nm thick n-type GaAs sample and are measured subsequently by electro-optic sampling in a thin ZnTe crystal. τ is the delay between the THz and the MIR field. Dual-frequency chopping of both incoming beams allow independent measurements of E_{THz}, E_{MIR}, and E_{Both} (both pulses are transmitted through the sample).

THz pulse with a center frequency of 2 THz and a field strength in the range up to 300 kV/cm was generated by optical rectification of 25 fs pulses from a Ti:sapphire oscillator–amplifier laser system and excites the sample placed in the focus of a parabolic mirror. The direction of the electric field is along the [100] direction of the sample. With a further pair of parabolic mirrors, the electric field of the transmitted THz pulse is transferred to a thin ZnTe crystal, where it is measured via electro-optic sampling [44, 63, 64]. The optics used ensures that, apart from a sign change, the electric field transients at the sample and at the electro-optic crystal are identical (Figure 1.9a). The entire optical path of the THz beam is placed in vacuum (for experimental details, see Refs [58, 62]). The electron current density [6]

$$j(t) = -env(t) = -2E_{em}(t)/(Z_0 d) \tag{1.15}$$

in the sample is proportional to the coherently emitted field $E_{em}(t) = E_{tr}(t) - E_{in}(t)$, which is given by the difference of $E_{tr}(t)$, the field transmitted through the sample, and $E_{in}(t)$, the field incident on the sample (n is the electron density in the sample and $Z_0 = \mu_0 c = 377\ \Omega$ is the impedance of free space). As the thickness of our sample $d = 500$ nm is much less than the THz wavelength $\lambda \approx 150\ \mu$m, all electrons in the sample experience the same driving field, which is identical to $E_{tr}(t)$ [25, 42, 62]. It should be noted that the detection scheme applied here is different from the frequently used setup where a large area of the sample is imaged as a small focal spot on the electro-optic crystal (Figure 1.9b). In the latter case, the electric field measured at the electro-optic crystal is proportional to the time derivative of the electric field at the sample.

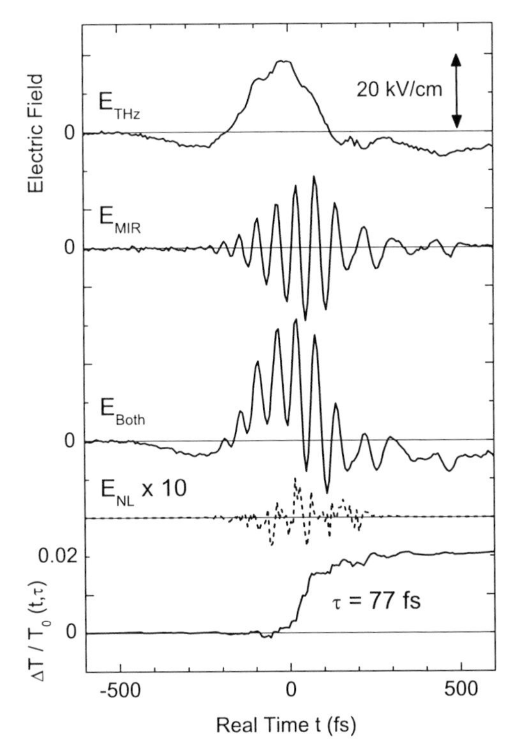

Figure 1.8 Measured transients for $\tau = 77$ fs. The curve at the bottom shows the buildup of the transmission change according to Eq. (1.14).

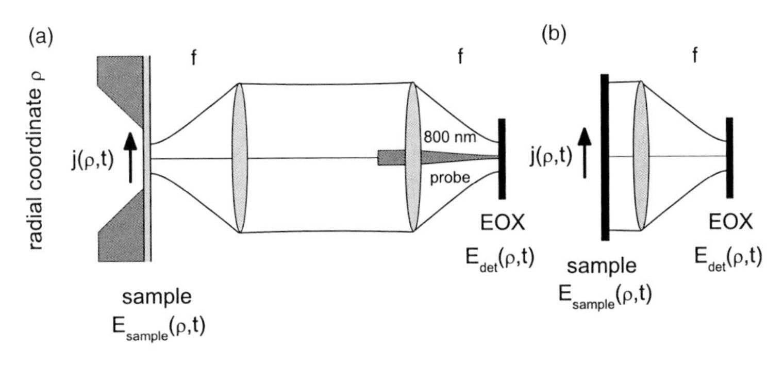

Figure 1.9 (a) Schematic of the optics used for transferring the electric field from the sample to the electro-optic crystal, EOX (in the actual setup, instead of lenses parabolic mirrors are used). Here, the on-axis ($\varrho = 0$) electric field at the sample is equal to minus the on-axis field at EOX, $E_{\text{sample}}(0,t) = -E_{\text{det}}(0,t)$. In contrast, in (b) $E_{\text{det}}(0,t) \propto d/dt E_{\text{sample}}(0,t)$.

1.3
Ultrafast Quantum Kinetics of Polarons in Bulk GaAs

In this section, we show that a high electric field in the terahertz range drives the polaron in a GaAs crystal into a highly nonlinear regime where – in addition to the drift motion – the electron is impulsively moved away from the center of the surrounding lattice distortion [65].

1.3.1
Experimental Results

The time-dependent electric field $E_{NL}(t, \tau)$ radiated from the nonlinear intraband polarization $E_{NL}(t, \tau) = E_{Both}(t, \tau) - E_{THz}(t) - E_{MIR}(t, \tau)$ is shown as the solid line in Figure 1.10a. The sample shows a coherent nonlinear emission, which is for this particular τ in phase with the MIR pulse, demonstrating a THz field-induced MIR gain of the sample. From $E_{NL}(t, \tau)$, we calculate the time-integrated mid-infrared transmission change $\Delta T/T_0(\tau)$ (circles in Figure 1.10b). This nonlinear signal shows an oscillatory behavior with a period of 120 fs, the period of the LO phonon in GaAs. Such oscillations correspond to a periodic switching between optical gain ($\Delta T/T_0 > 0$) and absorption ($\Delta T/T_0 < 0$) on the intraband transitions probed.

We observe the oscillatory behavior of intraband absorption and gain for THz driving fields of E_{THz} between 10 and 30 kV/cm, the maximum field applied in

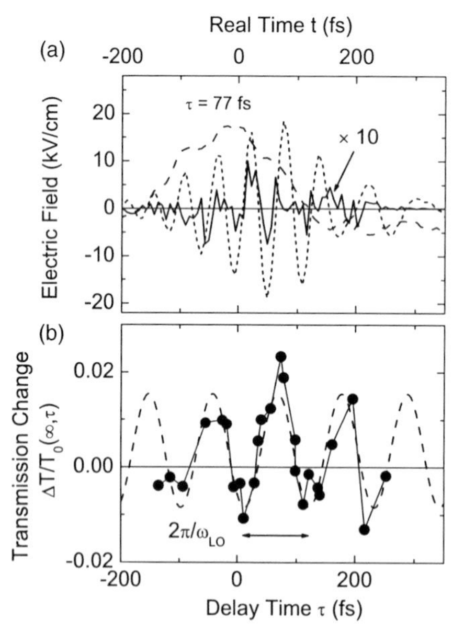

Figure 1.10 (a) $E_{THz}(t)$ (dashed line), $E_{MIR}(t, \tau)$ (dotted line), and $E_{NL}(t, \tau)$ (solid line) for $\tau = 77$ fs. (b) Transmission change $\Delta T/T_0(\infty, \tau)$ of the mid-infrared pulse as a function of τ (dots). Black dashed line: sine wave with the LO phonon frequency for comparison.

our measurements. In this range, the oscillation period does not depend on E_{THz} and is always identical to the LO phonon period. For $E_{THz} < 10\,kV/cm$, oscillations are absent. Experiments with a sample of five times lower doping density do not show any nonlinear THz response. Thus, the oscillatory pump–probe signal stems exclusively from the THz excitation of electrons present by n-type doping.

The oscillatory behavior of transmission observed here for the first time is a manifestation of the highly nonlinear response of polarons to a strong external field. We analyze our findings in a nonlinear transport picture discussed in the following and described from a theoretical point in the next section. The duration of the positive half-cycle of our THz pulse (Figure 1.8) is well below the energy relaxation time of the polaron. Thus, in the drift limit, an external electric field E acting on the polaron (Figure 1.1d and e) induces charge transport, which is described by the drift velocity v_d shown as the leftmost dashed line in Figure 1.3b.

As the electron approaches the characteristic velocity of $v_0 = \sqrt{2\hbar\omega_{LO}/m} = 435\,km/s$, that is, the threshold kinetic energy sufficient for emission of a LO phonon, the drift velocity v_d depends in a nonlinear way on the applied DC field E, as shown in Figures 1.3(b) (leftmost dashed line) and 1.11b and discussed in Refs [16–18, 26]. In this regime, the electron motion is described by the momentary electron velocity $v_e(t)$ and the differential mobility $\mu_{diff}(v_e) = [\partial E(v_e)/\partial v_e]^{-1}$. In our experimental scheme, we monitor such motion via the electric field $E_{NL}(t,\tau)$ (Eq. (1.14)) that is radiated from the moving charge interacting with both the driving THz field and the field $E_{MIR}(t,\tau)$ of the probe pulse. The resulting change of the mid-infrared transmission $\Delta T/T_0(\infty,\tau) \propto -(\mu_{diff}[v_e(\tau)])^{-1}$ is determined in sign and amplitude by the inverse differential mobility.

In our experiments, the polaron potential is strongly distorted by the femtosecond terahertz field. First, the strong external field accelerates the electron, leading to a finite distance of the electron from the center of the polaron (along the coordinate r shown in Figure 1.1g). This distance is generated impulsively, that is, on a short timescale compared to the LO phonon oscillation period. As soon as the kinetic energy of the electron reaches $\hbar\omega_{LO}$, the electron velocity saturates by transferring energy to the lattice. Due to the impulsive character of this transfer, coherent LO phonon oscillations appear as a stern wave of the moving electron (Figure 1.1f). With increasing strength of such oscillations, the related electric field (polarization) alters the motion of the electron so that electron oscillations occur along the coordinate r with a frequency ω_{LO} (Figure 1.1g). On top of the drift motion of the entire quasiparticle with $v_{polaron}(t) \approx v_0$, the electron oscillations along the internal coordinate $r(t)$ are connected to a periodic modulation of the momentary electron velocity $v_e(t) = dr(t)/dt + v_{polaron}(t)$. In this way, the electron explores velocity regions that are characterized by different $[\mu_{diff}(v_e)]^{-1}$ of *positive* and *negative* signs (Figure 1.11b), thereby modulating the transmission of the mid-infrared probe pulses in an oscillatory manner (circles in Figure 1.10b). The oscillatory internal motion of the polaron is exposed to the fluctuating interaction with thermally excited LO phonons and influenced by other scattering mechanisms. Such processes result in a dephasing of the oscillations on

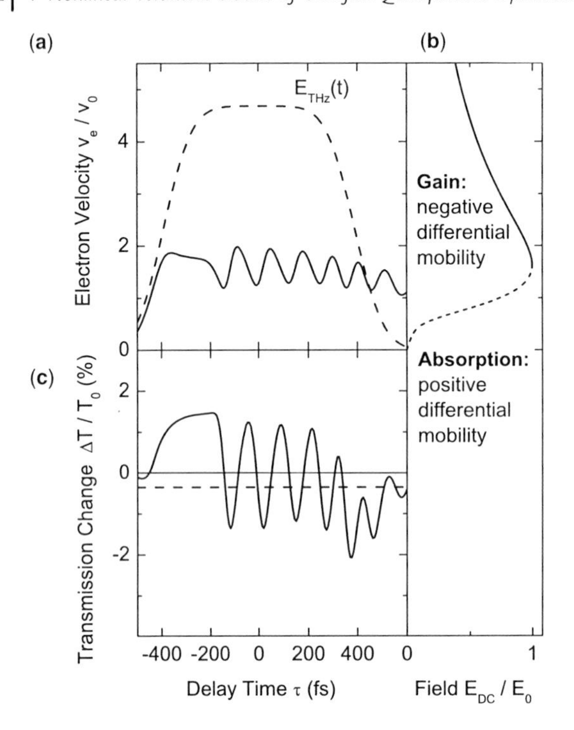

Figure 1.11 Results of model calculations. (a) Transient electron velocity (solid line) after nonlinear excitation with a strong THz field (dashed line). (b) DC drift velocity v_e as a function of the applied field E_{DC}. The transient modulation of the electron velocity in (a) and the mid-infrared transmission in (c) is caused by the periodic oscillation of v_e around the inflection point of the $v_e - E_{DC}$ characteristics changing periodically from a positive differential mobility, that is, absorption, to a negative differential mobility, that is, gain in the mid-infrared spectral range. (c) Calculated nonlinear transmission change $\Delta T/T_0(\tau)$ (solid line) of a short mid-infrared pulse as a function of τ. The transmission reduction due to linear intraband absorption is shown for comparison (dashed line).

a timescale of 0.5–1 ps, substantially longer than the time window studied in our experiments.

1.3.2
Discussion

The nonlinear response is analyzed quantitatively by considering the time-dependent self-consistent interaction potential between the electron and the LO phonon cloud already discussed in Section 1.1.1. The total local field acting on the electron is the sum of the externally applied fields [$E_{THz}(t)$ and $E_{MIR}(t, \tau)$], the field caused by the LO phonon cloud and the radiation reaction field leading to radiative damping. In Figure 1.11, the time-dependent electron velocity $v_e(\tau)$ (solid line in (a)) and the MIR transmission change (solid line in (c)) calculated for a 500 fs long terahertz pulse of 20 kV/cm amplitude (dashed line in (a)) are plotted as a function of delay time τ.

Similar to the data in Figure 1.10b, the nonlinear transmission change shows an oscillatory behavior changing periodically with the frequency of the LO phonon between gain and absorption. Even the strength of the nonlinear effect, which is approximately 10 times larger than the linear intraband absorption (dashed line in Figure 1.11c), is very well reproduced by the theory. A comparison of Figure 1.11a and c shows that – as expected and well reproduced by the theory – the oscillatory modulation of the mid-infrared transmission is directly connected to the transient motion of the electron along the internal coordinate *r*.

Our results highlight the quantum kinetic character of the nonlinear polaron response: the crystal lattice responds with coherent vibrations to the impulsive motion of electric charge. The timescale of such noninstantaneous process is inherently set by the LO phonon oscillation period and the picosecond decoherence of the LO phonon excitation. It is important to note that this nonlinear phenomenon occurs at comparably low electric field amplitudes of the order of $|E| = 10\,\text{kV/cm} = 0.1\,\text{V}/(100\,\text{nm})$. Thus, the quantum kinetic response plays a key role in high-frequency transport on nanometer length scales, in particular for highly polar materials such as GaN and II–VI semiconductors.

All theoretical results presented in this section were calculated on the basis of Eqs. (1.5)–(1.8) and (1.10) already discussed in Section 1.1.1. It is interesting to note that the size of the electron wave packet $\Delta x^2 = \hbar^2/4\Delta p_x^2$ remains almost constant on ultrafast timescales. There are two reasons for this behavior: (i) We start with a comparably large radius of Γ-valley polarons at room temperature, which, thanks to their high mobility, experience only a very weak friction force in the beginning. Thus, on ultrafast timescales, the first term on the rhs of Eq. (1.10) feeds the incoherent momentum fluctuations of the electron only weakly. (ii) In the quantum kinetic regime, the direction of the friction force strongly oscillates relative to that of the electron velocity leading in turn to canceling phenomena: a shrinkage and a *growth* of the wave packet size. This phenomenon is absent in the drift limit of carrier transport: here, the drift velocity and driving electric field are always in phase causing a strictly positive sign of the first term on the rhs of Eq. (1.10).

1.4 Coherent High-Field Transport in GaAs on Femtosecond Timescales

Eighty years ago, Felix Bloch showed that electron wave functions in the periodic Coulomb potential of the nuclei in a crystal are periodically modulated plane waves [66]. The spatially periodic modulation of the so-called Bloch functions restricts the allowed energies of the electrons, leading to a dispersive band structure $\varepsilon(\hbar\vec{k})$ (cf. first term on the rhs of Eq. (1.1)) containing both allowed (bands) and forbidden energy regions (gaps) [22]. Without scattering, an electron (charge $-e$) in an electric field $\vec{E}$ is expected to follow the dispersion of its band at a constant rate in momentum space [67]

$$\hbar \mathrm{d}\vec{k}/\mathrm{d}t = -e\vec{E}, \tag{1.16}$$

which is identical to Eq. (1.6) without the electron–phonon interaction term. A simple integration of Eq. (1.16) reads

$$\vec{k}(t) = \vec{k}(0) - e/\hbar \int_0^t \vec{E}(t')\mathrm{d}t'. \tag{1.17}$$

The corresponding velocity $\vec{v}$ in real space is given by

$$\vec{v} = \hbar^{-1}\nabla_{\vec{k}}\,\varepsilon(\hbar\vec{k}). \tag{1.18}$$

Thus, an electron moving in the periodic Coulomb potential of a crystal under the action of a constant external electric field is expected to undergo a coherent periodic oscillation both in real and momentum space. So far, such Bloch oscillations [66] have been observed only in artificial systems such as semiconductor superlattices [68–70], atoms and/or Bose–Einstein condensates in optical lattices [71], Josephson junction arrays [72], or optical waveguide arrays [73]. The absence of Bloch oscillations in electron transport through bulk crystals has been attributed to efficient scattering of electrons on a 100 fs timescale.

In the following, we demonstrate a novel regime of electron transport in bulk crystals driven by ultrashort high-field transients in the terahertz frequency range. Electrons in bulk n-type GaAs are subject to ultrashort electric field transients with very high field amplitudes of up to 300 kV/cm. The field transmitted through the sample is measured in amplitude and phase using the techniques described in Section 1.2.3. Under such conditions, electrons at room temperature perform a coherent ballistic motion within the lowest conduction band, in this way performing a partial Bloch oscillation. The coherent current observed at a crystal temperature of 300 K agrees well with the current expected for negligible scattering. This result, which is again in strong contrast to the prediction of the semiclassical Boltzmann transport equation [29], is fully confirmed by the dynamic polaron theory based on Eqs. (1.5)–(1.8) and (1.10).

1.4.1
Experimental Results

In Figure 1.12, we present experimental results at a sample temperature of 300 K for an incident THz pulse with an amplitude of 300 kV/cm. Figure 1.12 a and b shows the transients of the incident $E_{\mathrm{in}}(t)$ and of the transmitted $E_{\mathrm{tr}}(t)$ pulses. The difference between these transients yields the field $E_{\mathrm{em}}(t)$ emitted by the sample, as shown in Figure 1.12c. The velocity scale on the right-hand side is obtained from Eq. (1.15). Figure 1.12 d shows $-k(t)$ calculated according to Eq. (1.17) from $E_{\mathrm{tr}}(t)$. $k(0)$ was taken as zero, since the electrons initially occupy the conduction band minimum near $k = 0$ (Figure 1.12e). If one now plots $v(t)$ from (c) versus $-k(t)$ from (d), one obtains the dots in (f). To clarify the connection between the curves in (c) and (d) with the dots in (f), we have marked five moments t_1–t_5 (vertical lines in Figure 1.12a–d). Comparing these experimental results with the v versus k relationship from the conduction band structure (solid line in (f) calculated from Eq. (1.18)), one finds a good agreement, strongly pointing to ballistic transport across half the Brillouin zone.

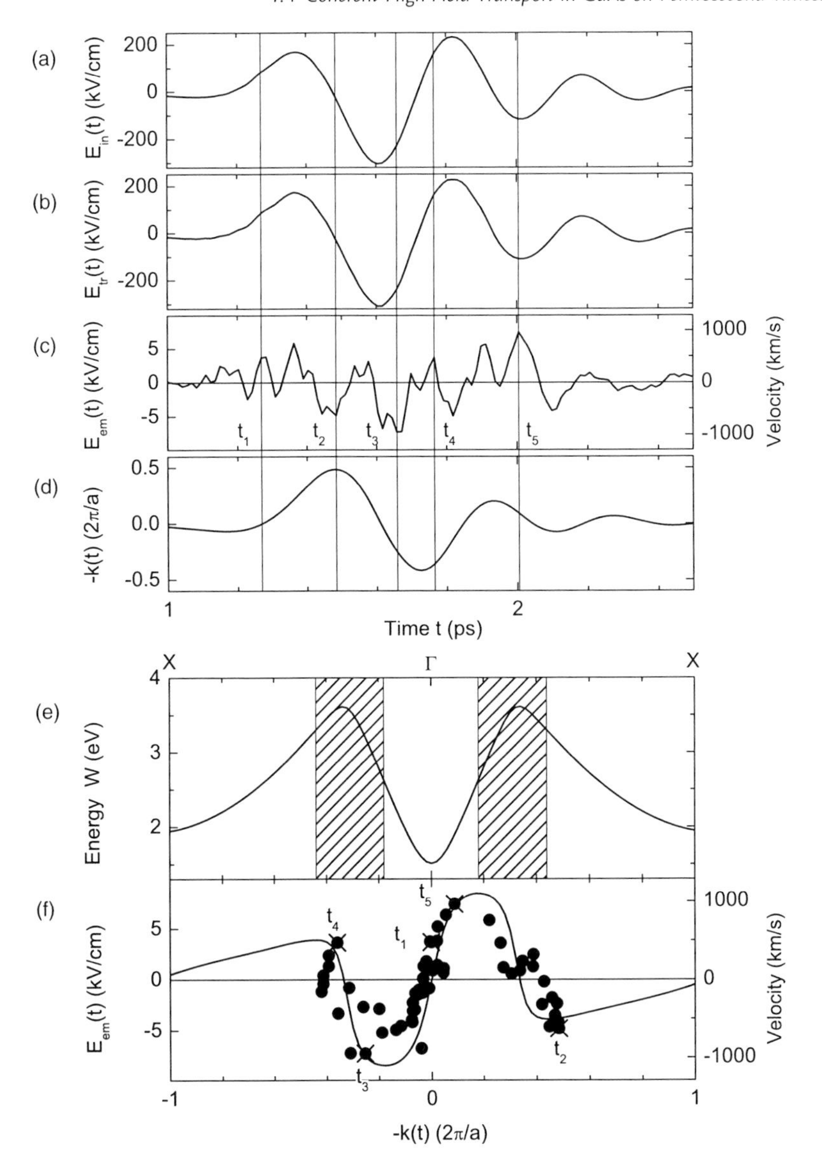

Figure 1.12 (a) Measured incident electric field as a function of time, $E_{in}(t)$. (b) Electric field transmitted through the sample, $E_{tr}(t)$. (c) Emitted electric field $E_{em}(t) = E_{tr}(t) - E_{in}(t)$. (d) $-k(t)$ calculated from the time integral of the electric field $E_{tr}(t)$ (Eq. (1.17)), in units of $2\pi/a$. (e) Lowest conduction band of GaAs in the [100] direction. The negative mass regions are hatched. (f) Dots denote $E_{em}(t)$ plotted versus $(e/\hbar)A(t)$. Crosses show the values at the times t_1–t_5, marked by vertical lines in (a)–(d). (e) Solid lines denote velocity v calculated from the conduction band $\varepsilon(k)$by $v = \hbar^{-1} d\varepsilon(\hbar k)/dk$.

To illustrate the effect of the band structure on the electron velocity, let us consider, for example, the time interval between t_2 and t_3. During this whole time, the electric field acting on the electron is negative (Figure 1.12b). The electron velocity is negative at t_2, then gets positive, and then gets negative again (Figure 1.12c). Thus, although the electric field has the same direction during the whole time between t_2 and t_3, there are times with positive and times with negative acceleration. This can be reconciled with equation of motion (Newton's law) only if the effective mass of the electron changes sign, which is exactly what happens. The effective mass of a band electron is given by $m_{\text{eff}} = \hbar^2[\mathrm{d}^2\varepsilon(\hbar k)/\mathrm{d}k^2]^{-1}$, that is, the sign of the effective mass is determined by the curvature of the band. In the conduction band of GaAs, the effective mass is positive around the Γ and the X points and negative around the band maxima (hatched areas in Figure 1.12e), explaining the change of the sign of the acceleration between t_2 and t_3. One should note that even for times as late as t_5, the data still agree with the velocity–momentum relationship expected for ballistic transport.

Our interpretation of the results is in agreement with the experimental data for all THz electric field amplitudes measured (not shown). For the two lowest applied field amplitudes (20–50 kV/cm), the emitted field $E_{\text{em}}(t)$ is approximately proportional to the time integral of the incident field, showing a linear Drude response. As we increase the field amplitude, we observe higher frequency components and a clipping of the emitted field amplitude around $|E_{\text{em}}(t)| < 7$ kV/cm. Since the emitted field is proportional to the electron velocity (Eq. (1.15)), this clipping is caused by the maximum velocity possible in the band structure. Figure 1.13 shows the emitted field (solid lines) for a particularly interesting field amplitude of 200 kV/cm of the incident field $E_{\text{in}}(t)$. At this field strength, a simulation with the traditional Boltzmann transport equation (dashed line in Figure 1.13b) completely fails, whereas our dynamic polaron model (dashed line Figure 1.13a) predicts correctly a quasiballistic high-field transport on ultrafast timescales.

1.4.2
Discussion

While our experimental results agree very well with the assumption of ballistic transport across half the Brillouin zone, the results do not agree with calculations based on the Boltzmann transport equation [29, 74]. While these calculations using Fermi's golden rule yield long scattering times ($\approx$200 fs) for electrons near the conduction band minimum, the scattering times decrease markedly for electrons with an energy enabling them to scatter into side valleys (L and X). Very short times (down to 3 fs) are obtained for electrons in the negative mass regions. With such short scattering times, it would be impossible on our timescale (100 fs) to have ballistic transport across these regions. Instead, one would expect that nearly all electrons are scattered into the side valleys before reaching the negative mass regions. Since electrons in the side valleys have rather low velocities (<200 km/s) [29, 37, 59], scattering into the side valleys would result in a drastic reduction of the electron velocity and thus of the emitted field. Since the return of electron into the Γ-valley takes quite long (>1 ps) [75], they would remain in the side valleys for the rest of the

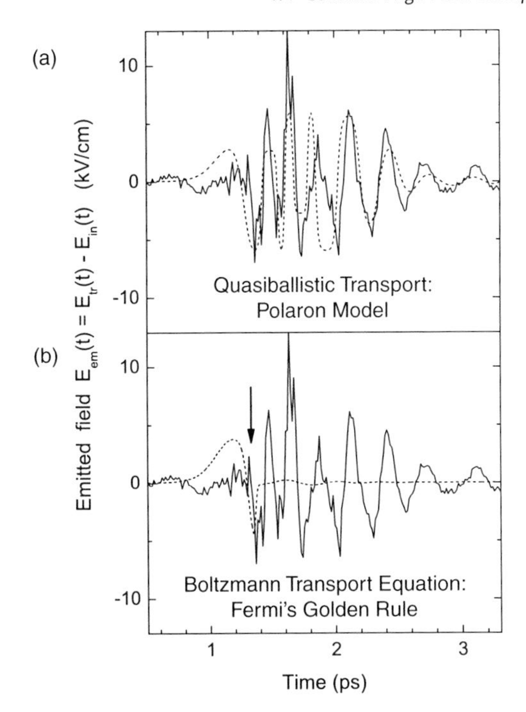

Figure 1.13 (a and b) Solid lines: emitted field transient $E_{em}(t)$ for an incident electric field $E_{in}(t)$ with an amplitude of 200 kV/cm. (a) Dashed line: result of the model calculation based on polaron model within the conduction band structure of GaAs. (b) Same experimental data as in (a), now compared to the results (dashed line) of a calculation assuming the intervalley scattering rates of Ref. [29]. At the time marked by the arrow, the electron energies are high enough for scattering into the side valleys.

pulse. Accordingly, one expects a strong signal $E_{em}(t)$ at the beginning of the pulse, but only very weak signals at later times, as shown by the dashed line in Figure 1.13b. This is in obvious disagreement with the experimental results.

To understand why calculations based on FGR do not agree with our experimental results, we have to consider the requirements for FGR to be valid. In our case, the main scattering mechanism for electrons is electron–phonon scattering, both deformation potential scattering with acoustic phonons and polar optical scattering with longitudinal optical phonons. For such periodic perturbations, FGR is only valid for times t large compared to the period of the perturbation [76], $t \gg \nu_{LO}^{-1} = 110$ fs [77]. Thus, FGR is not valid for the timescale of our experiment, in which the electrons are accelerated in one direction for about 250 fs. This is in contrast to most other experiments on high-field transport, which use a DC field for acceleration.

The dynamic polaron theory, which goes beyond Fermi's golden rule, predicts correctly a quasiballistic high-field transport in the quantum kinetic regime. The application of Eqs. (1.5)–(1.8) and (1.10), already discussed in Section 1.1.1, gives the result that the size of the electron wave packet $\Delta x^2 = \hbar^2/4\Delta p_x^2$ remains almost constant in our experiments on ultrafast timescales. In particular, a wave packet with large Δx^2 couples only weakly to phonons with large wave vectors, which would allow scattering into the side valleys. Thus, on ultrafast timescales, that is, $t < 200$ fs, intervalley scattering is effectively suppressed, which yields long momentum relaxation times allowing ballistic transport over several 100 fs. On longer timescales, Δx^2 decreases.

In summary, we have observed ballistic transport of electrons in GaAs across half the Brillouin zone by time-resolved high-field THz measurements. We present a model for high-field transport using polarons that agrees quantitatively with our experimental results on short timescales and yields the drift velocity on long timescales.

1.5 Conclusions and Outlook

The results presented here demonstrate the potential of nonlinear terahertz spectroscopy to unravel transport phenomena in the highly nonlinear and quantum kinetic regime. In contrast to studies on longer timescales, the ultrashort time structure of the electric field transients allows for driving fields with amplitudes up to MV/cm. While the work presented concentrates on transport in the polar semiconductor GaAs, there is a much wider range of systems to be studied by such techniques. Beyond bulk and nanostructured semiconductors and metals, the transfer and transport of charge in inorganic and organic molecular materials is readily accessible.

The present emphasis of nonlinear terahertz studies is on understanding the basic physical processes and interactions that govern charge transport and other field-driven phenomena. With the methods of coherent pulse shaping at hand, one may go beyond such analysis and steer charge transport by interaction with phase-tailored field transients or sequences of terahertz pulses. This may lead to new concepts for electronic and optoelectronic devices.

Acknowledgments

We would like to thank our former and present coworkers P. Gaal, W. Kuehn, T. Bartel, K. Reimann, and R. Hey for their important contributions to the work reviewed here. In part, the experiments were supported by the Deutsche Forschungsgemeinschaft.

References

1 Haug, H. and Koch, S.W. (1993) *Quantum Theory of the Optical and Electronic Properties of Semiconductors*, World Scientific, Singapore.
2 Mahan, G.D. (2000) *Many-Particle Physics*, 3rd edn, Kluwer, New York.
3 Shah, J. (1999) *Ultrafast Spectroscopy of Semiconductors and Semiconductor Nanostructures*, 2nd edn, Springer, Berlin.
4 Haug, H. and Jauho, A.P. (1996) *Quantum Kinetics in Transport and Optics of Semiconductors*, Springer, Berlin.

5 Kuhn, T. (1998) Density matrix theory of coherent ultrafast dynamics, in *Theory of Transport Properties of Semiconductor Nanostructures* (ed. E. Schöll), Chapman & Hall, London, pp. 173–214.
6 Reimann, K. (2007) Table-top sources of ultrashort THz pulses. *Rep. Prog. Phys.*, **70**, 1597–1632.
7 Ganichev, S.D. and Prettl, W. (2006) *Intense Terahertz Excitation of Semiconductors*, Oxford University Press, Oxford.
8 Betz, M., Göger, G., Laubereau, A., Gartner, P., Bányai, L., Haug, H., Ortner, K., Becker, C.R., and Leitenstorfer, A. (2001) Subthreshold carrier-LO phonon dynamics in semiconductors with intermediate polaron coupling: a purely quantum kinetic relaxation channel. *Phys. Rev. Lett.*, **86**, 4684–4687.
9 Gaal, P., Kuehn, W., Reimann, K., Woerner, M., Elsaesser, T., and Hey, R. (2007) Internal motions of a quasiparticle governing its ultrafast nonlinear response. *Nature*, **450**, 1210–1213.
10 Hase, M., Kitajima, M., Constantinescu, A.M., and Petek, H. (2003) The birth of a quasiparticle in silicon observed in time–frequency space. *Nature*, **426**, 51–54.
11 Huber, R., Tauser, F., Brodschelm, A., Bichler, M., Abstreiter, G., and Leitenstorfer, A. (2001) How many-particle interactions develop after ultrafast excitation of an electron–hole plasma. *Nature*, **414**, 286–289.
12 Fröhlich, H. (1954) Electrons in lattice fields. *Adv. Phys.*, **3**, 325–361.
13 Lee, T.D., Low, F.E., and Pines, D. (1953) The motion of slow electrons in a polar crystal. *Phys. Rev.*, **90**, 297–302.
14 Peeters, F.M. and Devreese, J.T. (1985) Radius, self-induced potential, and number of virtual optical phonons of a polaron. *Phys. Rev. B*, **31**, 4890–4899.
15 Feynman, R.P. (1955) Slow electrons in a polar crystal. *Phys. Rev.*, **97**, 660–665.
16 Jensen, J.H. and Sauls, J.A. (1988) Polarons near the Cerenkov velocity. *Phys. Rev. B*, **38**, 13387–13394.
17 Janssen, N. and Zwerger, W. (1995) Nonlinear transport of polarons. *Phys. Rev. B*, **52**, 9406–9417.
18 Bányai, L. (1993) Motion of a classical polaron in a dc electric field. *Phys. Rev. Lett.*, **70**, 1674–1677.
19 Magnus, W. and Schoenmaker, W. (1993) Dissipative motion of an electron–phonon system in a uniform electric field: an exact solution. *Phys. Rev. B*, **47**, 1276–1281.
20 Caldeira, A.O. and Leggett, A.J. (1983) Path integral approach to quantum Brownian motion. *Physica*, **121**, 587–616.
21 Rossi, F. and Jacoboni, C. (1992) Enhancement of drift-velocity overshoot in silicon due to the intracollisional field effect. *Semicond. Sci. Technol.*, **7**, B383–B385.
22 Chelikowsky, J.R. and Cohen, M.L. (1976) Nonlocal pseudopotential calculations for the electronic structure of eleven diamond and zinc-blende semiconductors. *Phys. Rev. B*, **14**, 556–582: erratum, 1984, 30, 4828.
23 Cohen M.L. and Bergstresser, T.K. (1966) Band structures and pseudopotential form factors for fourteen semiconductors of the diamond and zinc-blende structures. *Phys. Rev.*, **141**, 789–796.
24 Walter, J.P. and Cohen, M.L. (1969) Calculation of the reflectivity, modulated reflectivity, and band structure of GaAs, GaP, ZnSe, and ZnS. *Phys. Rev.*, **183**, 763–772: erratum, 1970, **B1**, 942.
25 Stroucken, T., Knorr, A., Thomas, P., and Koch, S.W. (1996) Coherent dynamics of radiatively coupled quantum-well excitons. *Phys. Rev. B*, **53**, 2026–2033.
26 Meinert, G., Bányai, L., and Gartner, P. (2001) Classical polarons in a constant electric field. *Phys. Rev. B*, **63**, 245203.
27 Bhattacharya, T., Habib, S., and Jacobs, K. (2002) The emergence of classical dynamics in a quantum world. *Los Alamos Sci.*, **27**, 110–125.
28 Bhattacharya, T., Habib, S., and Jacobs, K. (2003) Continuous quantum measurement and the quantum to classical transition. *Phys. Rev. A*, **67**, 042103.
29 Fischetti, M.V. (1991) Monte Carlo simulation of transport in technologically significant semiconductors of the diamond and zinc-blende structures. I. Homogeneous transport. *IEEE Trans. Electron Devices*, **38**, 634–649.

30 Sigg, H., Wyder, P., and Perenboom, J.A.A.J. (1985) Analysis of polaron effects in the cyclotron resonance of n-GaAs and AlGaAs–GaAs heterojunctions. *Phys. Rev. B*, **31**, 5253–5261.
31 Spitzer, W.G. and Whelan, J.M. (1959) Infrared absorption and electron effective mass in n-type gallium arsenide. *Phys. Rev.*, **114**, 59–63.
32 Elsaesser, T., Bäuerle, R.J., and Kaiser, W. (1989) Hot phonons in InAs observed via picosecond free-carrier absorption. *Phys. Rev. B*, **40**, 2976–2979.
33 Bingham, R. (2007) On the crest of a wake. *Nature*, **445**, 721–722.
34 Blumenfeld, I., Clayton, C.E., Decker, F.-J., Hogan, M.J., Huang, C., Ischebeck, R., Iverson, R., Joshi, C., Katsouleas, T., Kirby, N., Lu, W., Marsh, K.A., Mori, W.B., Muggli, P., Oz, E., Siemann, R.H., Walz, D., and Zhou, M. (2007) Energy doubling of 42GeV electrons in a metre-scale plasma wakefield accelerator. *Nature*, **445**, 741–744.
35 Gunn, J.B. (1963) Microwave oscillations of current in III–V semiconductors. *Solid State Commun.*, **1**, 88–91.
36 Ridley, B.K. and Watkins, T.B. (1961) The possibility of negative resistance effects in semiconductors. *Proc. Phys. Soc.*, **78**, 293–304.
37 Windhorn, T.H., Roth, T.J., Zinkiewicz, L.M., Gaddy, O.L., and Stillman, G.E. (1982) High field temperature dependent electron drift velocities in GaAs. *Appl. Phys. Lett.*, **40**, 513–515.
38 Reimann, K., Smith, R.P., Weiner, A.M., Elsaesser, T., and Woerner, M. (2003) Direct field-resolved detection of terahertz transients with amplitudes of megavolts per centimeter. *Opt. Lett.*, **28**, 471–473.
39 Luo, C., Reimann, K., Woerner, M., and Elsaesser, T. (2004) Nonlinear terahertz spectroscopy of semiconductor nanostructures. *Appl. Phys. A*, **78**, 435–440.
40 Luo, C.W., Reimann, K., Woerner, M., Elsaesser, T., Hey, R., and Ploog, K.H. (2004) Phase-resolved nonlinear response of a two-dimensional electron gas under femtosecond intersubband excitation. *Phys. Rev. Lett.*, **92**, 047402–1–04740-4.
41 Luo, C.W., Reimann, K., Woerner, M., Elsaesser, T., Hey, R., and Ploog, K.H. (2004) Rabi oscillations of intersubband transitions in GaAs/AlGaAs MQWs. *Semicond. Sci. Technol.*, **19**, S285–S286.
42 Shih, T., Reimann, K., Woerner, M., Elsaesser, T., Waldmüller, I., Knorr, A., Hey, R., and Ploog, K.H. (2005) Nonlinear response of radiatively coupled intersubband transitions of quasi-two-dimensional electrons. *Phys. Rev. B*, **72**, 195338-1–195338-8.
43 Shih, T., Reimann, K., Woerner, M., Elsaesser, T., Waldmüller, I., Knorr, A., Hey, R., and Ploog, K.H. (2006) Radiative coupling of intersubband transitions in GaAs/AlGaAs multiple quantum wells. *Physica E*, **32**, 262–265.
44 Bartel, T., Gaal, P., Reimann, K., Woerner, M., and Elsaesser, T. (2005) Generation of single-cycle THz transients with high electric-field amplitudes. *Opt. Lett.*, **30**, 2805–2807.
45 Dragoman, D. and Dragoman, M. (2004) Terahertz fields and applications. *Prog. Quantum Electron.*, **28**, 1–66.
46 Schmuttenmaer, C.A. (2004) Exploring dynamics in the far-infrared with terahertz spectroscopy. *Chem. Rev.*, **104**, 1759–1780.
47 Kress, M., Löffler, T., Eden, S., Thomson, M., and Roskos, H.G. (2004) Terahertz-pulse generation by photoionization of air with laser pulses composed of both fundamental and second-harmonic waves. *Opt. Lett.*, **29**, 1120–1122.
48 Löffler, T., Kreß, M., Thomson, M., Hahn, T., Hasegawa, N., and Roskos, H.G. (2005) Comparative performance of terahertz emitters in amplifier-laser-based systems. *Semicond. Sci. Technol.*, **20**, S134–S141.
49 Budiarto, E., Margolies, J., Jeong, S., Son, J., and Bokor, J. (1996) High-intensity terahertz pulses at 1-kHz repetition rate. *IEEE J. Quantum Electron.*, **32**, 1839–1846.
50 You, D., Jones, R.R., Bucksbaum, P.H., and Dykaar, D.R. (1993) Generation of high-power sub-single-cycle 500-fs electromagnetic pulses. *Opt. Lett.*, **18**, 290–292.
51 Cook, D.J. and Hochstrasser, R.M. (2000) Intense terahertz pulses by four-wave rectification in air. *Opt. Lett.*, **25**, 1210–1212.

52 Hamster, H., Sullivan, A., Gordon, S., White, W., and Falcone, R.W. (1993) Subpicosecond, electromagnetic pulses from intense laser–plasma interaction. *Phys. Rev. Lett.*, **71**, 2725–2728.

53 Tournois, P. (1997) Acousto-optic programmable dispersive filter for adaptive compensation of group delay time dispersion in laser systems. *Opt. Commun.*, **140**, 245–249.

54 Verluise, F., Laude, V., Cheng, Z., Spielmann, Ch., and Tournois, P. (2000) Amplitude and phase control of ultrashort pulses by use of an acousto-optic programmable dispersive filter: pulse compression and shaping. *Opt. Lett.*, **25**, 575–577.

55 Allenspacher, P., Baehnisch, R., and Riede, W. (2004) Multiple ultrashort pulse damage of AR-coated beta-barium borate. *Proc. SPIE*, **5273**, 17–22.

56 Leitenstorfer, A., Hunsche, S., Shah, J., Nuss, M.C., and Knox, W.H. (1999) Detectors and sources for ultrabroadband electro-optic sampling: experiment and theory. *Appl. Phys. Lett.*, **74**, 1516–1518.

57 Wu, Q. and Zhang, X.-C. (1997) 7 terahertz broadband GaP electro-optic sensor. *Appl. Phys. Lett.*, **70**, 1784–1786.

58 Gaal, P., Reimann, K., Woerner, M., Elsaesser, T., Hey, R., and Ploog, K.H. (2006) Nonlinear terahertz response of n-type GaAs. *Phys. Rev. Lett.*, **96**, 187402-1–187402-4.

59 Abe, M., Madhavi, S., Shimada, Y., Otsuka, Y., Hirakawa, K., and Tomizawa, K. (2002) Transient carrier velocities in bulk GaAs: quantitative comparison between terahertz data and ensemble Monte Carlo calculations. *Appl. Phys. Lett.*, **81**, 679–681.

60 Leitenstorfer, A., Hunsche, S., Shah, J., Nuss, M.C., and Knox, W.H. (2000) Femtosecond high-field transport in compound semiconductors. *Phys. Rev. B*, **61**, 16642–16652.

61 Schwanhäußer, A., Betz, M., Eckardt, M., Trumm, S., Robledo, L., Malzer, S., Leitenstorfer, A., and Döhler, G.H. (2004) Ultrafast transport of electrons in GaAs: direct observation of quasiballistic motion and side valley transfer. *Phys. Rev. B*, **70**, 085211.

62 Gaal, P., Kuehn, W., Reimann, K., Woerner, M., Elsaesser, T., Hey, R., Lee, J.S., and Schade, U. (2008) Carrier-wave Rabi flopping on radiatively coupled shallow donor transitions in n-type GaAs. *Phys. Rev. B*, **77**, 235204-1–235204-6.

63 Wu, Q., Litz, M., and Zhang, X.-C. (1996) Broadband detection capability of ZnTe electro-optic field detectors. *Appl. Phys. Lett.*, **68**, 2924–2926.

64 Wu, Q. and Zhang, X.-C. (1995) Free-space electro-optic sampling of terahertz beams. *Appl. Phys. Lett.*, **67**, 3523–3525.

65 Gaal, P., Reimann, K., Woerner, M., Elsaesser, T., Hey, R., and Ploog, K.H. (2007) Nonlinear THz spectroscopy of n-type GaAs, in *Ultrafast Phenomena XV* (eds P. Corkum, D. Jonas, D. Miller, and A.M. Weiner), Springer, Berlin, pp. 799–801.

66 Bloch, F. (1928) Über die Quantenmechanik der Elektronen in Kristallgittern. *Z. Phys.*, **52**, 555–600.

67 Ridley, B.K. (1993) *Quantum Processes in Semiconductors*, 3rd edn, Oxford University Press, Oxford.

68 Feldmann, J., Leo, K., Shah, J., Miller, D.A.B., Cunningham, J.E., Meier, T., von Plessen, G., Schulze, A., Thomas, P., and Schmitt-Rink, S. (1992) Optical investigation of Bloch oscillations in a semiconductor superlattice. *Phys. Rev. B*, **46**, 7252–7255.

69 Unterrainer, K., Keay, B.J., Wanke, M.C., Allen, S.J., Leonard, D., Medeiros-Ribeiro, G., Bhattacharya, U., and Rodwell, M.J.W. (1996) Inverse Bloch oscillator: strong terahertz-photocurrent resonances at the Bloch frequency. *Phys. Rev. Lett.*, **76**, 2973–2976.

70 Waschke, C., Roskos, H.G., Schwedler, R., Leo, K., Kurz, H., and Köhler, K. (1993) Coherent submillimeter-wave emission from Bloch oscillations in a semiconductor superlattice. *Phys. Rev. Lett.*, **70**, 3319–3322.

71 Bloch, I. (2008) Quantum coherence and entanglement with ultracold atoms in optical lattices. *Nature*, **453**, 1016–1022.

72 Delahaye, J., Hassel, J., Lindell, R., Sillanpää, M., Paalanen, M., Seppä, H., and Hakonen, P. (2003) Low-noise current amplifier based on mesoscopic Josephson junction. *Science*, **299**, 1045–1048.

73 Christodoulides, D.N., Lederer, F., and Silberberg, Y. (2003) Discretizing light behaviour in linear and nonlinear waveguide lattices. *Nature*, **424**, 817–823.

74 Littlejohn, M.A., Hauser, J.R., and Glisson, T.H. (1977) Velocity–field characteristics of GaAs with $\Gamma_6^c - L_6^c - X_6^c$ conduction-band ordering. *J. Appl. Phys.*, **48**, 4587–4590.

75 Tsuruoka, T., Hashimoto, H., and Ushioda, S. (2004) Real-space observation of electron transport in AlGaAs/GaAs quantum wells using a scanning tunneling microscope. *Thin Solid Films*, **464–465**, 469–472.

76 Messiah, A. (1964) *Mécanique Quantique*, vol. 2, Dunod, Paris.

77 Holtz, M., Seon, M., Brafman, O., Manor, R., and Fekete, D. (1996) Pressure dependence of the optic phonon energies in $Al_xGa_{1-x}As$. *Phys. Rev. B*, **54**, 8714–8720.

2
Higher Order Photoemission from Metal Surfaces

Aimo Winkelmann, Cheng-Tien Chiang, Francesco Bisio, Wen-Chin Lin, Jürgen Kirschner, and Hrvoje Petek

2.1
Introduction

Photoemission spectroscopy (PES) is a widely employed method for mapping the electronic structure of solids [1, 2]. In the simplest picture, the incident radiation induces transitions of electrons from occupied to unoccupied single-particle states separated by the photon energy. By invoking the laws of energy and momentum conservation, the photoemission process can be applied to map the occupied band structure of solids by detecting the photoelectrons in final states above the vacuum barrier.

A generalization of the linear one-photon photoelectric effect can be realized by nonlinear multiphoton processes induced by ultrashort laser pulses of high intensity, where the occupied initial states, the unoccupied intermediate states, and the coupling between them play a central role. The unique power of these effects stems from the fact that the dynamical processes in the intermediate excited states become directly accessible in the time domain when using delayed excitation pulses. Especially the technique of time-resolved two-photon photoemission (2PPE) has been used to gain information on the decay rates of electronic populations and their dephasing times [3–5].

Compared to 2PPE, higher order multiphoton photoemission processes (mPPE) in solids, which can extend the energy range of the studied unoccupied states and provide further information on photoexcitation processes, have received only little attention. This is mainly because their observation requires very intense optical fields and their considerably reduced photoelectron yields can be overwhelmed by space charge effects from the lower order processes [6–15]. Moreover, nonphotoelectric effect emission through the surface plasmon excitation and possibly tunneling, leading to ponderomotive acceleration of photoelectrons up to 0.4 keV energy [9, 16], has been reported with comparatively low external fields. Therefore, the mechanisms for high-order photoelectric excitations at solid surfaces, in particular to what extent they are governed by the band structure of metals, are largely unexplored.

Dynamics at Solid State Surface and Interfaces Vol. 1: Current Developments
Edited by Uwe Bovensiepen, Hrvoje Petek, and Martin Wolf

ISBN: 978-3-527-40937-2

As multiphoton photoemission experiments depend sensitively on the wave functions and energy levels of the states that are involved in the excitation, they are expected to provide band structure information about occupied and unoccupied states in solids with high precision [17]. Using procedures very similar to conventional angle-resolved PES, multiphoton resonances can be exploited to determine the separation of electronic bands at specific $\vec{\mathbf{k}}$-points by looking for resonances between the multiple states coupled by the incident radiation.

In this chapter, we summarize our observations of higher order photoemission at copper surfaces. The electronic structure of Cu(001) supports resonant three-photon photoemission with photon energies of $\sim$3 eV. We will show how coherent multiphoton photoemission carries information about the bulk band structure of copper via the resonant excitation of electronic bulk states through unoccupied intermediate surface states. In the 3PPE and 4PPE experiments, we can also observe electrons that have absorbed more energy than is necessary to overcome the work function of the material, and therefore undergo above-threshold photoemission (ATP). We will analyze an example of how momentum conservation restrictions can keep these electrons in quasibound states above the thermodynamic vacuum level. Finally, we will briefly discuss how the spin–orbit coupling (SOC) in the Cu d-bands can be exploited to excite and control the spin-polarized higher order photoemission using circularly polarized light for excitation and spin-dependent scattering for the spin-resolved detection of photoelectrons.

2.2 Observation of Higher Order Photoemission at Cu Surfaces

We investigated nonlinear photoemission spectra from Cu(001) and Cu(111) surfaces. The photoemission experiments were carried out in an ultrahigh vacuum system (pressure $<5 \times 10^{-11}$ mbar) [18, 19]. The Cu surfaces were prepared with standard sputtering and annealing procedures. All the experiments, unless specified, were carried out at 300 K. The electrons photoemitted along the surface normal were analyzed by a cylindrical sector analyzer (Focus CSA300) with parallel momentum ($k_{\|}$) resolution below ± 0.07 Å. The ultrashort excitation pulses were provided by the frequency-doubled output of a self-built Ti:sapphire oscillator operating at 81 MHz. The pulse central energy could be continuously varied in the range of $h\nu = 2.99$ -3.15 eV by tuning the phase-matching angle of the frequency-doubling β-BaB_2O_4 (BBO) crystal. At the energy of $h\nu = 3.07$ eV, the pulse length at the surface was $\leq$20 fs and the pulse energy $\sim$1 nJ. The peak power densities that were reached at the sample surface were in the order of 10 GW/cm^2, the measured sample current was in the order of 1 nA, corresponding to a total of less than 100 emitted electrons per pulse. Significant space charge effects beyond the experimental resolution could not be detected in the experiments reported below.

In Figure 2.1, we show multiphoton photoemission spectra recorded simultaneously for two-, three-, and four-photon excitation for Cu(001) and Cu(111) surfaces as a function of the final-state energy above the Fermi level E_F. The

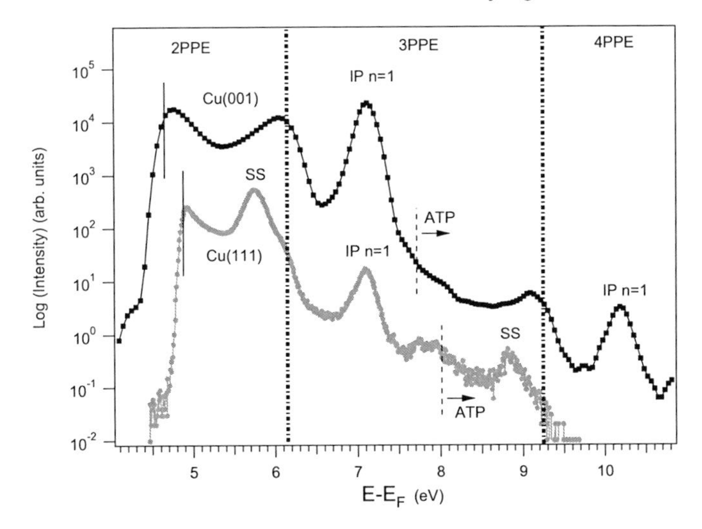

Figure 2.1 Nonlinear photoemission spectra from Cu(001) (black squares) and Cu(111) (gray circles), for emission perpendicular to the surface, single-photon energy of $h\nu = 3.07$ eV, and p-polarized light. Indicated are the energy regions that are reached from the occupied initial states by 2, 3, and 4 photons, respectively. Multiphoton replicas of the $n = 1$ image potential (IP) state peak and the Cu(111) Shockley surface state (SS) peak can be distinguished. Above-threshold processes are indicated by ATP. The energy resolution was $\approx$100 meV. The spectra are not normalized with respect to each other.

low-energy onsets of the spectra represent the work function thresholds of 4.6 eV for Cu(001) and 4.9 eV for Cu(111) [5]. The highest energies that can be reached by photoelectrons from the Fermi level by absorption of two and three photons are indicated by dashed–dotted vertical lines. These delimit the regions of 2PPE, 3PPE, and 4PPE.

In the 2PPE spectral region of Cu(111), we identify the well-known occupied Shockley surface state (SS) ($E_F - 0.4$ eV), while there are no distinctive features in the Cu(001) spectrum. In the 3PPE part of both surfaces, we can clearly see the $n = 1$ image potential (IP) state peak at a final-state energy of 7.1 eV. The binding energies of the $n = 1$ IP states measured with respect to the vacuum level are 0.59 eV for Cu(001) and 0.84 eV for Cu(111) [5]. Due to the difference of 0.3 eV between the work functions of the two surfaces, their IP states appear at approximately the same energy with respect to the Fermi level.

Above the IP states, we indicate by dashed lines the position of the vacuum levels plus one photon energy. Photoelectrons with energies above the indicated limits emerge with kinetic energies larger than the single-photon energy. The contribution of this above-threshold photoemission effect is seen especially clearly for the three-photon copy of the Cu(111) surface state near 8.8 eV: These electrons can however escape after absorbing only two photons (the 2PPE peak at 5.7 eV). A very similar process is apparent in the 4PPE IP state feature, which corresponds to absorption of two photons to reach the vacuum instead of only one.

The most striking effect is seen in the intensity of the 3PPE through the IP state at Cu(001) compared with Cu(111): The 3PPE signal from Cu(001) at the IP state can reach higher intensity than the 2PPE signal and is orders of magnitude stronger than the 3PPE signal through the IP state of Cu(111). In the next section, we will analyze how the Cu(001) electronic structure can support such an efficient 3PPE process.

2.2.1
Resonant 3PPE in the Cu(001) Electronic Band Structure

In order to interpret our experimental results on 3PPE from Cu(001), we show in Figure 2.2 the relativistic band structure of copper for the *k*-space line relevant to emission normal to the Cu(001) surface (Δ-line from Γ to X) according to the calculated band structure of Ref. [20]. We indicate a three-photon process for photon energies near 3 eV starting from the Cu d-bands via the unoccupied sp-band and the $n = 1$ image potential state, which has been demonstrated previously [18]. At sufficiently high laser fluence, this process leads to a 3PPE signal higher in intensity than the simultaneously observed nonresonant 2PPE signal.

As is well known, a theoretical band structure such as shown in Figure 2.2 usually does not describe experimental photoemission data quantitatively [22, 23]. Because of this, we assume in the following that the calculation still describes the *dispersion*

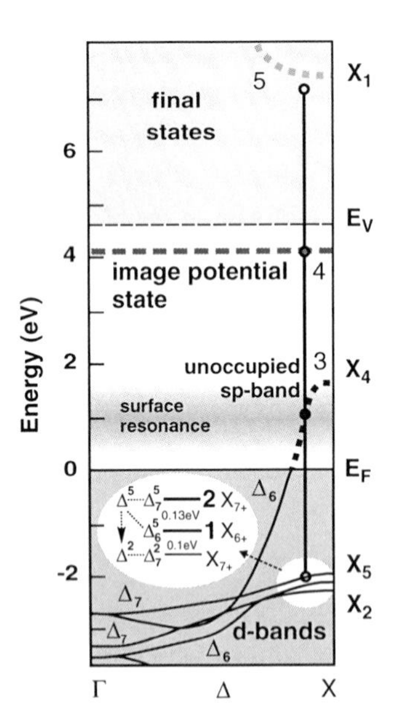

Figure 2.2 Relativistic bulk band structure [20] of Cu(001) with the proposed three-photon resonance for a photon energy near 3 eV. In addition, the $n = 1$ image potential state is shown at an energy of 4.04 eV [5]. A broadened unoccupied surface resonance is expected near 1 eV [21].

of the bands correctly, but we make the absolute values for the critical points adjustable to known experimental data. Then we try to find k-conserving resonances in the band structure as a function of photon energy. In a first approximation, this is realized by shifting the involved bands according to the number of photons needed to reach the final-state energy. A resonance is given when the energy levels cross (see also Ref. [17]). This imposes stringent conditions on the relative energy separation between the energy levels involved.

The image potential state is a surface state, which by definition does not disperse with momentum in the direction perpendicular to the surface and thus is drawn horizontally for the Γ–X direction of Figure 2.2, at an energy of 4.04 eV [5]. The observed spectral structure can be expected to be produced mainly by the IP state in combination with the initial d-bands that show a reduced $k_\perp$-dispersion near the X point. This is why we look for a two-photon resonance between the $n = 1$ image potential state and the initial d-bands. In addition, the dispersing unoccupied sp-band can supply states near simultaneous one-photon resonance to the d-bands and the image potential, respectively. Adjusting the critical points of the bulk band structure, the d-bands are assumed to have X_5 at −1.99 eV and X_2 at −2.18 eV [22]. Due to spin–orbit coupling, the band with Δ^5 spatial symmetry is split into Δ_7^5 and Δ_6^5 bands with reported experimental splittings between 100 [24] and 170 meV [25] in the region of interest near the X point. For comparison, the theoretical value is 160 meV [22] at the X point. The upper sp-band is assumed with X_1 at 7.67 eV [22] and thus cannot be reached resonantly with the photon energies in our experiment. In Figure 2.3, we show the bands after shifting up the IP state by one photon energy, the sp-band by two photon energies, and the d-bands by three photon energies to the final-state energy.

The results in Figure 2.3 strongly support a resonance mechanism in 3PPE from Cu(001). In the two panels of Figure 2.3, we can expect the near-crossing of three bands for a pair of photon energies approximately 110 meV apart, at 2.97 and 3.08 eV, which can be excited simultaneously with our 170 meV bandwidth laser pulse. The crossing of the three bands in Figure 2.3 means that not only we have a two-photon resonance between the d-bands and the IP state (which could be fulfilled for a range of photon energies due to the dispersion of the d-bands) but also the unoccupied sp-band is near one-photon resonance to the d-bands and the IP state. In this situation, the transitions involved are emphasized because energy conservation is fulfilled for the one-photon and two-photon transitions at the same time. The two separate resonances are labeled A and B in the higher resolution experimental data shown in Figure 2.3. In view of the very simple model, both the photon energies and the separation of the resonances are in good agreement with the measured values.

The 3PPE resonances discussed involve two initial states belonging to spin–orbit split Δ^5 d-bands. They both proceed at different k-values via the dispersing unoccupied portion of the sp-band states [26–28] to the image potential state. A possible contribution from an unoccupied surface resonance [21], which is expected near or higher than 1 eV above the Fermi level (Figure 2.2), cannot be excluded, but in our model it can hardly be responsible for the two simultaneous resonances

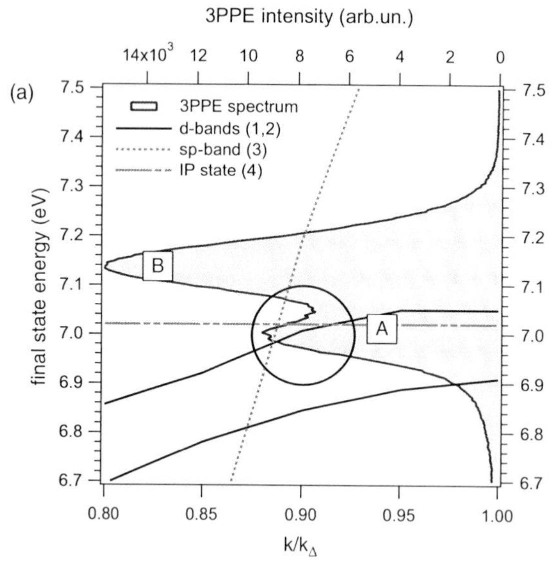

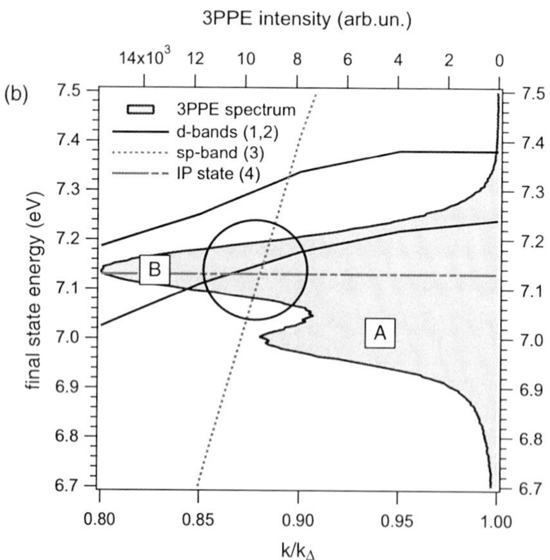

Figure 2.3 Analysis of the possible resonance conditions between the initial d-bands (solid black), the intermediate sp-band (dotted), and the image potential state (dash-dot-dotted). The bands are numbered according to Figure 2.2. The initial d-bands are shifted upward by the energy of three photons, the sp-band by two-photon energies, and the intermediate $n = 1$ image potential state is shifted by the energy of one photon to the final-state energy. The near crossing of three bands indicates simultaneous two-photon and one-photon resonances. Due to the broad pulse spectrum ($\approx$170 meV), the separate resonances for $h\nu = 2.97$ eV (a) and $h\nu = 3.08$ eV (b) connected to the two d-bands are simultaneously observed as features A and B in the experimental spectrum (solid gray, experimental data for 3.04 eV central photon energy of the laser pulse, energy resolution $\approx$50 meV).

observed. This is because of the missing $k_\perp$ dispersion of the surface resonance and the IP state that fixes their one-photon resonance at *one* specific photon energy for all $k_\perp$. The unoccupied surface resonance would not allow *two* photon energies to be in simultaneous two-photon and one-photon resonances with the IP state and the d-bands, that is, the crossing of three bands as shown in Figure 2.3 for two different photon energies would not be possible. In this respect, a dispersing intermediate-state band seems to be necessary to explain our observations (compare also Figure 2.6).

An interesting feature of the multiphoton resonances is that they proceed via the *same* image potential state, but still show a two-peak structure corresponding to the different initial states. This points to the influence of multiphoton excitation processes that cannot be pictured as sequential stepwise processes from one level to the next, in which the information (coherence) on the excitation pathway would be lost. The application of these coherent excitation processes for electronic structure mapping is discussed in the next section.

2.3 Electronic Structure Mapping Using Coherent Multiphoton Resonances

An analysis of the double-peak 3PPE structure A and B in Figure 2.3 shows that this part of the spectrum can be fitted by the sum of two Gaussian peaks with widths (FWHM) of 110 meV for (A) and 130 meV for (B) and a separation of 150 meV. Peak A is located at a final-state energy of 6.95 eV, while peak B is at 7.15 eV. Considering that the optical spectrum of the excitation pulses has a width of 170 meV, and the light field acts on the sample three times, it is striking that the resolution of the 3PPE spectrum is well below the excitation linewidth [29].

The mechanism behind the formation of the double-peak structure in Figure 2.3 can be analyzed by using the density matrix formalism, where the populations of the states involved are described by the diagonal elements ϱ_{aa} and the coherences between states by the off-diagonal elements ϱ_{ab} of the density matrix ϱ [5, 30–34].

We assume that the part of the Cu band structure that is relevant for the observed 3PPE process can be effectively simplified to a four-level system. The unperturbed Hamiltonian $\hat{H}_S$ of the system is written on the basis of four orthonormalized states ($\langle a|b\rangle = \delta_{ab}$) representing either one of the initially occupied d-band states $|i_{1,2}\rangle$, the intermediate unoccupied sp-band $|s\rangle$, the image potential state $|m\rangle$, and the final photoemitted state $|f\rangle$ at the respective energy levels $\varepsilon_{i,s,m,f}$. This simplification disregards the dispersion of the electronic states with crystal momentum $\mathbf{k}$ that is a major inhomogeneous contribution to nonresonant 2PPE spectra [34] and the various inter- and intraband relaxation mechanisms that would need to be part of a quantitative theory for a crystalline solid state system. This could be handled, for example, by introducing sets of k-dependent multilevel systems with appropriate couplings that represent the above effects [34, 35]. We also assume that the perturbation of the system by the laser pulse is weak. The 3PPE processes are assumed to proceed independently from either of the two different sets of initial d-band states $|i_{1,2}\rangle$.

The density operator $\hat{\varrho}$ of this simplified four-level system explicitly expands as.

$$\hat{\varrho} = \begin{pmatrix} \varrho_{ii} & |i\rangle\langle i| & \varrho_{is} & |i\rangle\langle s| & \varrho_{im} & |i\rangle\langle m| & \varrho_{if} & |i\rangle\langle f| \\ \varrho_{si} & |s\rangle\langle i| & \varrho_{ss} & |s\rangle\langle s| & \varrho_{sm} & |s\rangle\langle m| & \varrho_{sf} & |s\rangle\langle f| \\ \varrho_{mi} & |m\rangle\langle i| & \varrho_{ms} & |m\rangle\langle s| & \varrho_{mm} & |m\rangle\langle m| & \varrho_{mf} & |m\rangle\langle f| \\ \varrho_{fi} & |f\rangle\langle i| & \varrho_{fs} & |f\rangle\langle s| & \varrho_{fm} & |f\rangle\langle m| & \varrho_{ff} & |f\rangle\langle f| \end{pmatrix} \tag{2.1}$$
$$\varrho_{ii}(t = -\infty) = 1.$$

We have emphasized with a dark gray box the matrix element ϱ_{ii}, which is the only nonzero element in the ground state at time $t = -\infty$, representing the initial population of the relevant d-band state.

The other emphasized matrix element ϱ_{ff} (light gray box) determines the quantity we measure in the experiment: the very small population that is transferred to the final state by the ultrashort optical pulse. Initially, $\varrho_{ff} = 0$, and it can only become nonzero due to the coupling to the initial state via the intermediate states. This coupling is provided by the electric field $E(t)$ of the excitation pulse.

For simplicity, we assume here that the most relevant couplings are between those states that are nearly a photon energy $\hbar\omega_p$ apart in the level scheme:

$$|i\rangle \overset{\hbar\omega_p}{\longleftrightarrow} |s\rangle \overset{\hbar\omega_p}{\longleftrightarrow} |m\rangle \overset{\hbar\omega_p}{\longleftrightarrow} |f\rangle. \tag{2.2}$$

Starting at $t = -\infty$ from only $\varrho_{ii} \neq 0$, the interaction Hamiltonian $\hat{H}_{\text{int}}$ allows finite values to develop in the matrix elements during the time evolution of the density operator $\hat{\varrho}$. This time evolution is described by the equation of motion for the density matrix (Liouville – von Neumann equation) [5, 30, 32, 33, 36]:

$$\frac{d\varrho_{ab}}{dt} = -\frac{i}{\hbar}\langle a|\left[\hat{H}_S + \hat{H}_{\text{int}}, \hat{\varrho}\right]|b\rangle - \Gamma_{ab}\varrho_{ab}, \quad a, b = i, s, m, f. \tag{2.3}$$

The numerical treatment of Eq. (2.3) is in principle straightforward. With respect to the physical interpretation, however, a perturbative expansion of $\hat{\varrho}$ with respect to interactions with the electrical field is useful. A perturbative expansion allows to visualize the induction of nonzero matrix elements ϱ_{ab} as a sequence of steps in which the commutators $[H_{\text{int}}, \varrho_{ab}]$ in Eq. (2.3) are recursively expanded up to a specified order. This expansion can be symbolized by double-sided Feynman diagrams that keep track of both the time development of the population in the states involved as well as the coherences induced between these states that are described by the diagonal and off-diagonal elements of the density matrix, respectively [36]. Another related visualization allows to show the possible coupling of two generic elements ϱ_{ab} and ϱ_{cd} of the density matrix as Liouville space pathways [36].

These Liouville space pathways show that besides the sequential one-photon pumping of population from one state to the next (compare Eq. (2.2)):

$$\varrho_{ii} \rightarrow \varrho_{ss} \rightarrow \varrho_{mm} \rightarrow \varrho_{ff}, \tag{2.4}$$

there are contributions to ϱ_{ff} containing only coherences, schematically written as

$$\varrho_{ii} \rightarrow \varrho_{is} \rightarrow \varrho_{im} \rightarrow \varrho_{if} \rightarrow \varrho_{ff}, \tag{2.5}$$

which implies that these contributions produce no population in the intermediate states $|m\rangle$ (corresponding to the observable $N_m = \mathrm{Tr}[|m\rangle\langle m|\hat{\varrho}]$), due to the fact that the density matrix components ϱ_{mm} (as well as ϱ_{ss}) are not affected by these processes. It is emphasized that despite the fact that no population in the intermediate state is involved, these states nevertheless have a decisive role. They contribute via their coherent superposition with the other states. From our experimental data, we see that the coherent coupling of the d-band states to the image potential state via the sp-bands contributes to the initial-state peaks in the photoelectron spectrum ($\varrho_{ff}(E)$). We keep in mind that the pathways (2.4) and (2.5) are two particular pathways in a general coherent 3PPE process and occur together with all other pathways and should be added coherently in calculating $\varrho_{ff}(E)$.

Features A and B in Figure 2.3 are not a consequence of the presence of the IP state alone, but are inherently caused by a nonlinear resonance effect between initial, first intermediate states, and second intermediate (IP) states. Very qualitatively, a memory of its excitation pathway must be imparted on an electron because it starts at a specific energy level (one of the two d-bands), then goes through the *same* energy level of the IP state as the other electrons, and finally ends up at a specific energy again. In this sequential picture where only the actual population of the IP state is measured by the third ionizing photon, no initial-state peaks could show up in the photoelectron spectrum if the electrons from different initial states lose memory of their excitation pathway in the IP state before their final ionization. In this respect, the simple observation of two IP state related peaks (which do not move when the central pulse energy is tuned [29]) is a sign of specific nonlinear–optical pathways that cannot be pictured as sequential one-photon transitions, but that allow the electron to keep its initial-state information as it is emitted into the vacuum.

The resonance effect also provides an effective energy filter, which would explain why the widths of the observed peaks A and B are smaller than the optical spectrum of the excitation pulses. Strictly, energy conservation is imposed only on the overall coherent 3PPE process from an initial to a final state that are separated by an energy $\hbar\omega_{03}$. In principle, all three-photon processes that fulfill $\hbar\omega_{03} = \hbar\omega_{01} + \hbar\omega_{12} + \hbar\omega_{23}$ are allowed (with different single-photon energies $\hbar\omega_{01,12,23}$). However, if the excitation energy is additionally near one-photon resonances $\hbar\omega_{\mathrm{R}} \approx \hbar\omega_{01} \approx \hbar\omega_{12} \approx \hbar\omega_{23}$ between dispersing levels (compare also Figure 2.3), energy conservation is also approximately fulfilled for the intermediate levels. This favors the photon energies near the simultaneous one-photon, two-photon, and three-photon resonances $\hbar\omega_{03} = \hbar(\approx \omega_{\mathrm{R}}) + \hbar(\approx \omega_{\mathrm{R}}) + \hbar(\approx \omega_{\mathrm{R}})$ and thus provides an energy filtering mechanism.

The main physics of the coherent multiphoton photoemission effect is expected to be already present in three-level systems of initial, intermediate, and final states. For these systems, analytical solutions exist for single-frequency interactions in two-photon photoemission [37]. Assuming that the two initial d-band levels and the intermediate IP state play a dominant role in the mechanism we focus on here, we can simplify our four-level system (3PPE) to a three-level system (2PPE) by artificially moving the two initial d-band states to the original energy level of the first intermediate sp-bands, which are neglected in the following. We then apply the model of Wolf *et al.* [37] to our special simplified system that models 2PPE from two closely spaced, independent initial states via the same intermediate state.

For the intermediate state, we assumed a constant lifetime of 35 fs and a dephasing of 1 meV [5]. For the final state, we assumed infinite population lifetime and 1 meV dephasing rate [5]. The initial states are separated by 150 meV at energy levels that reproduce the experimentally observed final-state positions via the intermediate state at 4.04 eV, corresponding to the $n = 1$ IP state on Cu(001). We then calculated the resulting 2PPE spectra for a range of initial-state dephasing rates (assuming no population decay in the initial state). While for faster dephasing only a single peak is present, it turns out that – in order for the two-peak structure to appear – the dephasing time of the initial states must be in the order of 10–20 fs [29]. Although this value cannot be taken as a quantitative determination, nevertheless it compares well with direct time domain observations of Cu d-band hole dephasing times of $>$20 fs in the observed d-band region near the X point [38].

The result for a model 2PPE process shows how the dephasing rate between the initial and intermediate states contributes to the peak width of the transition. If the dephasing rate is high, the two independent transitions from the closely spaced initial states are broadened to an extent that they are not separable anymore and appear as a single peak. With decreasing dephasing, the peak widths of the two resonances become sharp enough to give two peaks in the final-state spectrum. This process can be expected to be of very general character, and a similar influence of dephasing is expected to be relevant in the actual 3PPE experiment. This means that a sufficiently low dephasing rate of the initial states with respect to the intermediate states will be required in any case to distinguish two initial-state peaks in a transition that proceeds via a common intermediate-state level, like the IP state considered here.

Our observations of the direct influence of coherent excitation pathways in 3PPE from Cu(001) are relevant for applications of nonlinear photoemission in electronic structure investigations, which can be carried out using high-intensity laboratory laser sources and accelerator-based free electron lasers. It has been predicted theoretically [17] that resonant two-photon photoemission could give the separation between occupied initial states and unoccupied intermediate *bulk* states within the fundamental limit imposed by the intrinsic energy and momentum widths of the coupled states. The important advantage in such two-photon measurements is that the optical transition from the initial state to the intermediate state would couple bulk states that are not influenced by the symmetry breaking due to the surface [39] and the perpendicular crystal momentum $k_{\perp}$ in such a transition would need to be conserved (at negligible photon momentum).

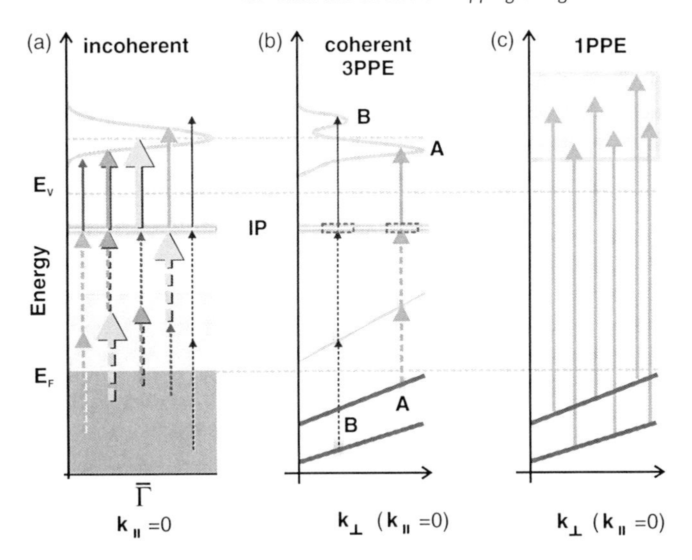

Figure 2.4 Principle of detecting coherent bulk band structure resonances in 3PPE via an image potential state on Cu(001). (a) Nonresonant incoherent excitation of the IP state from unspecific bulk bands at normal emission ($k_{\|} = 0$). The initial states are averaged over $k_{\perp}$ illustrating the loss of $k_{\perp}$ information in the sequential one-photon pathways due to the lack of a $k_{\perp}$ selection rule in the IP to final-state one-photon transition ($k_{\perp}$ is undefined for surface states). The colors of the arrows correspond to photon frequencies provided by the excitation pulse, their thickness corresponds to the respective spectral weight. The photons can act in any combination for sequential incoherent excitations. The peak position of the IP state is determined by the central photon energy of the pulse (thickest arrow) and the measured width of the IP state is determined by the width of the pulse spectrum. (b) Coherent 3PPE resonances in the bulk band structure coupled to the image potential state produce specific peaks A and B in the spectrum, providing $k_{\perp}$ information. A fixed photon energy defines each resonant pathway. The width of features A and B is not limited by the spectral pulse width. (c) The corresponding 1PPE process for the same level scheme using a single-photon energy three times as large as that in the corresponding 3PPE process is not k-selective if no resonance to a dispersing final-state band is present. (Please find a color version of this figure on the color plates.)

The extension of this idea to 3PPE is straightforward, and we illustrate in Figure 2.4 the qualitative difference between unspecific bulk excitations and coherent resonances that sensitively depend on the band dispersion. It can also be easily imagined that this approach can be generalized to multiple excitation frequencies that can be tuned to search for the simultaneously possible single-photon resonances in a multiphoton process. When surface states are involved in the multiphoton resonances, it has to be taken into account that the periodicity of the surface-state wave function may emphasize transitions near the bulk Brillouin zone boundaries [40, 41]. The use of multiphoton resonances corresponding to very well-defined k-space positions should also allow improved time domain investigations [3] of band-resolved electron dynamics. The demonstrated coherent coupling of bulk states to surface states through a two-photon resonance could also lead to improved possibilities of controlling electron motion at surfaces by interferometric techniques [42, 43].

2.4
Dynamical Trapping of Electrons in Quasibound States

In the previously discussed experiments shown in Figure 2.1, the IP states populated via two-photon excitation have been measured in normal emission, corresponding to $k_{||} = 0$. For $k_{||} > 0$, the energy of an electron in the IP states disperses with an effective mass m^* near the free electron mass m_e as $E_{IP}(k_{||}) = E_0 + \hbar^2 k_{||}^2/2m^*$. This dispersion can be seen in the measurements shown for Cu(001) in Figure 2.5. We detected electrons with $k_{||} > 0$ in off-normal emission by rotating the sample. Starting from the value at $k_{||} = 0$, the IP state disperses toward higher energies with increasing $k_{||}$. Interestingly, at one point, the kinetic energy of the IP electrons is more than one-photon energy larger than the energy of an electron that has overcome the vacuum barrier E_V. This means that electrons can stay long enough in the IP state to absorb an additional third photon although their total kinetic energy would be already sufficient to leave the sample after two-photon excitation. As we will discuss below, quasibound states can exist above the thermodynamic vacuum level on account of the vectorial nature of momentum conservation, in addition to quantitative energy conservation.

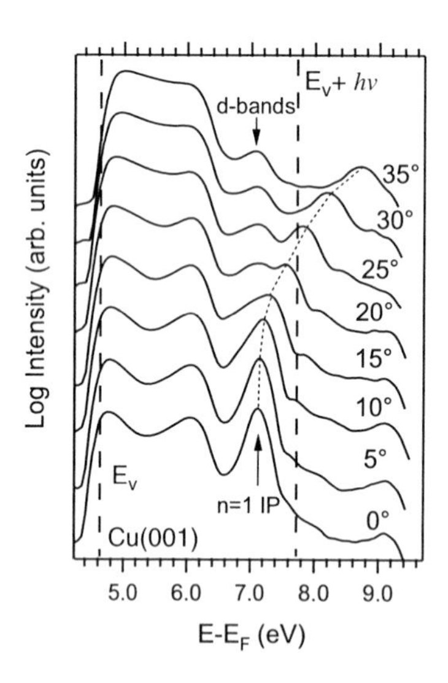

Figure 2.5 Dispersion of the $n = 1$ IP state on Cu(001) with $k_{||} \geq 0$. Above $E_V + h\nu$, the two-photon excited electron in the IP state already has higher energy than the work function $E_V - E_F$, but cannot leave the sample because its perpendicular momentum is too low to overcome the surface barrier. Only after absorbing a third photon, the electron is photoemitted.

If electrons are to be emitted from a sample, they have to overcome the potential barrier of the surface. In a classical picture, the perpendicular component of the electron momentum vector is reduced outside the sample due to the force acting along the surface normal direction when the electron overcomes the potential step at the surface. If the perpendicular momentum component $p_\perp$ of an electron, which travels along an inclined trajectory toward the surface barrier, is below the critical value $p_\perp^0$, then it is deflected back into the direction away from the surface, although its total kinetic energy might actually be sufficiently high to leave the sample. Quantum mechanically, the self-consistent reflection and transmission of the electron waves at the surface and at the bulk potential determine the formation of standing waves for the energies of surface states and resonances, which correspond to electrons trapped in a surface quantum well [44]. Under the simplest approximation, in this multiple scattering picture, only the projection of the electron wave vectors on the surface normal direction is relevant for scattering in a surface barrier potential that varies only in the dimension perpendicular to the surface.

This perpendicular momentum of the scattered electron waves at the surface is affected, for instance, by the exchange of reciprocal lattice vectors in low-energy electron diffraction (LEED). In LEED experiments, external electrons with total energies *above* the vacuum level can be trapped in the surface potential well by momentum restrictions. When entering the surface, the external electron beam is accelerated due to the lower inner potential of the sample and it can be diffracted into various beams, some of which can move nearly parallel to the surface. In this way, the momentum and energy in the surface parallel motion are increased for the diffracted beams and the respective electrons effectively lose "perpendicular energy." The effective perpendicular energy can drop below the vacuum level, and the electron waves are then scattered in a similar way as electrons in the bound states, that is, the IP states, of the one-dimensional perpendicular potential. This is seen near the emergence threshold of diffracted beams moving nearly parallel to the surface, where fine structure effects appear in the reflected electron intensity via the interference of waves reflected only at the bulk and diffracted waves that have been reflected at the bulk and surface barriers (possibly multiple times) [45]. Similar trapping effects can be seen in the scattering of He atoms at surfaces [46].

As we see in the measurements of Figure 2.5, multiphoton excitation can provide an alternative way to study electrons under restricted surface perpendicular momentum that are trapped in the IP states. According to the discussed model, the electrons in the IP state could, in principle, acquire any amount of energy if their motion is free electron-like in the direction parallel to the surface and as long as their perpendicular momentum stays low enough. If the surface-parallel velocity of the electrons in the IP state could be sufficiently increased by multiphoton excitation, this could give access to the dynamical screening properties of the electrons in the occupied states, which lead to a finite response time for the buildup of the image potential state and a velocity-dependent binding energy [47]. As the electron motion at the surface is not completely free, however, reflection by the periodic part of the lateral surface potential will influence the electron motion in the IP states and thereby impose limits on their velocity.

2.5
Above-Threshold Photoemission

The effect of above-threshold photoemission is well known in atomic physics [48]. It provides the possibility that an electron absorbs one or more photons than are actually necessary to ionize an atom or, in our case, to be photoemitted from a surface. We have seen in the previous section that momentum restrictions can lead to a situation in which an electron with sufficient energy to overcome the work function is trapped at a surface. However, the intrinsic ATP effect is seen even in the absence of similar restrictions.

This can be realized when looking at our experimental spectra in Figure 2.1. For Cu(111), we have two peaks that originate from the Shockley surface state ($E_B = -0.4$ eV): one at 5.7 eV, due to excitation by two photons. But clearly, there is a process in which the surface-state electron absorbs one additional photon, as we see by the three-photon copy of the SS peak near 8.8 eV. A very similar process gives rise to the $n = 1$ IP state feature in the 4PPE spectrum of Cu(001). This corresponds to photoemission of electrons from the IP state by two photons instead of only one, which would be sufficient enough to reach the vacuum. While multiphoton ATP effects from solids under high-intensity excitation can produce unwanted background effects that are largely unspecific with respect to the sample electronic structure, our experimental spectra of Figure 2.1 clearly display very specific signals from surface states and the Fermi edge from Cu(001) and Cu(111) in different nonlinear orders. Such information should be very useful for comparison with quantitative theories that predict the relative yields of the different orders of multiphoton excitation.

Compared to atomic physics, the nonlinear effects at solid surfaces show important differences. In this respect, one of the most obvious questions is how the complex electronic structure of a crystalline system influences multiphoton photoemission, and how the electronic structure of a solid is modified by the nonlinear interaction with the laser field [49]. Furthermore, the symmetry breaking at the surface introduces additional influences such as refraction and reflection of electrons and light at the solid–vacuum interface [50]. Correspondingly, the measurement of ATP effects from solids by ultrashort pulses can provide important information about the occupied and unoccupied electronic states and the relevant timescales of the elementary processes involved. It has been shown, for example, that multiphoton excitations can give access to final-state effects in photoemission from solids, like the process of photoexcited electrons leaving a sample [51]. The involved timescales can be sensed by comparing the signals of specific excited electronic states and the corresponding signals of higher order excitations induced by an additional laser pulse. The experimental scheme involves electrons that are photoemitted, for example, by attosecond XUV pulses. Access to the dynamics of the excited electrons is given by the time-resolved sideband structure in the photoelectron spectra, which is due to low-energy photon absorption and stimulated emission induced by the dressing field [13, 52] (see also chapters 21 and 22 in this book).

2.6
Spin-Polarized Multiphoton Photoemission

It is well known that photoexcited spin-polarized electrons can be produced at unpolarized targets, provided that the incident light has circular polarization. Spin–orbit coupling in combination with optical selection rules provides the mechanism by which electrons can be spin selectively excited. This forms the basis of various effects related to optical spin orientation [53]. For one-photon photoemission from metals, it has been demonstrated that significant spin–orbit coupling leads to strong spin-polarization effects of the photoelectrons even in a relatively low-Z material like copper [24]. The extension of the optical spin orientation effect to multiphoton transitions in solids would allow to transiently spin polarize excited intermediate states below the photoemission threshold that in principle cannot be sensed by one-photon photoemission. Such a multiphoton excitation scheme can then be directly exploited to study spin-dependent scattering of excited electrons in solids. In this respect, the Cu(001) surface provides an especially interesting system, since we have seen how the spin–orbit-coupled Cu d-bands can be coupled to the IP states. This is why we studied spin-resolved 3PPE under excitation of circularly polarized laser pulses.

The principle of the spin-polarized 3PPE from Cu(001) is shown in Figure 2.6a. The relativistic selection rules for normal emission from an fcc(001) surface dictate the final-state symmetry to be Δ_6^1. Furthermore, right circularly polarized (RCP) light couples spin-down (spin-up) electrons from a Δ_6^5 (Δ_7^5) symmetry initial state to a Δ_6^1 symmetry final state (and vice versa for left circular excitation) [54]. This provides

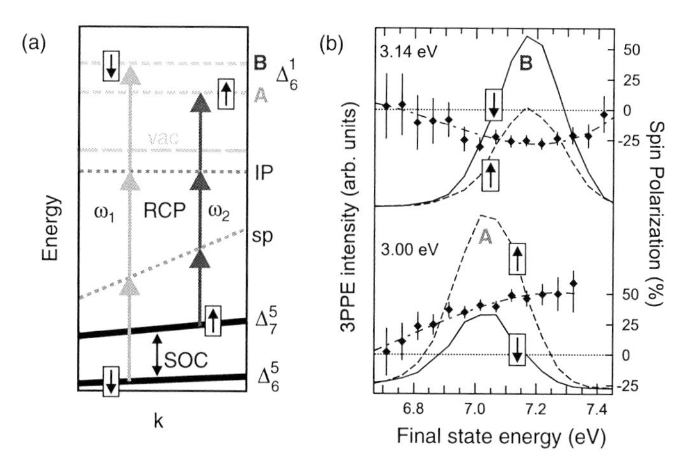

Figure 2.6 (a) Principle of spin-polarized three-photon photoemission from Cu(001). Spin-polarized electrons are selectively excited by the right circularly polarized (RCP) light due to the spin–orbit coupling in the Cu d-bands. By the two-photon resonance to the image potential state, spin-polarized electrons transiently occupy a surface state of nonmagnetic Cu. (b) Experimental data showing that oppositely spin-polarized electrons are excited by photons of 3.00 and 3.14 eV, respectively [19]. A and B symbolize the two possible resonances (Figure 2.3), which are not separately resolved due to the limited experimental resolution (100 meV) imposed by spin-resolved detection. Reprinted with permission from [17]. Copyright 2007 by the American Physical Society.

the basic mechanism by which oppositely spin-polarized photoelectrons are excited selectively from the two spin–orbit split bands of different symmetry to the unoccupied IP state.

The electrons photoemitted along the surface normal were analyzed by a cylindrical sector analyzer coupled with a newly developed spin detector. This detector performs spin analysis through very low-energy electron scattering on a magnetized ultrathin Fe/W(001) film [55].

In the experimental data shown in Figure 2.6b, we have demonstrated the creation of a large (40%) spin polarization of the electrons in the $n = 1$ IP state on the Cu(001) surface [19]. The spin polarization is achieved by resonantly coupling the spin–orbit split Cu d-bands with the IP state via a multiphoton transition excited by circularly polarized light. We see that both the sign and the magnitude of the IP state spin polarization can be tuned by balancing the resonant coupling to different spin–orbit split d-bands, which are selected by the photon energy in Figure 2.6. The spin polarization can also be interferometrically controlled by the delay between two ultrashort laser pulses, via selective interference of high- and low-frequency components of the excitation [56]. Using the additional degree of information provided by the spin, our excitation scheme can also provide a way to selectively study dynamics of electrons excited from d-bands of different double-group symmetry. These spin-resolved investigations may provide a basis for the coherent manipulation and probing of the spin dynamics of excited electrons at nonmagnetic surfaces via multiphoton photoemission. Such investigations can complement studies of spin-dependent electron dynamics at magnetic surfaces [57].

2.7 Summary and Outlook

In this chapter, we have shown how higher order photoemission can provide valuable information about the electronic structure of surfaces for the model system of Cu(001) excited by 3 eV laser pulses. Our findings suggest extensions to future studies of bulk band structure with tunable laboratory and free electron laser sources, exploiting multiphoton resonances for band mapping. If we note that when – in addition to the energy of the exciting radiation – the quantum mechanical phase relationships between the coupled states and the radiation also become a defining factor in the nonlinear excitation process [58], qualitatively different applications as compared to conventional band mapping with one-photon angle-resolved photoemission spectroscopy are expected by exploiting the unique properties of multiphoton transitions between occupied and unoccupied bulk states.

Acknowledgments

We thank Frank Helbig for invaluable technical support. H.P. is grateful for support by NSF grant CHE-0650756.

References

1 Hüfner, S. (1995) *Photoelectron Spectroscopy*, Springer-Verlag, New York.

2 Schattke, W. and Van Hove, M.A. (eds) (2003) *Solid-State Photoemission and Related Methods: Theory and Experiment*, Wiley-VCH Verlag GmbH.

3 Petek, H. and Ogawa, S. (1997) Femtosecond time-resolved two-photon photoemission studies of electron dynamics in metals. *Prog. Surf. Sci.*, **56**, 239.

4 Fauster, Th. (2003) Time resolved two-photon photoemission, in *Solid-State Photoemission and Related Methods: Theory and Experiment* (eds W. Schattke and M. van Hove), Wiley-VCH Verlag GmbH, Chapter 2, pp. 247–268.

5 Weinelt, M. (2002) Time-resolved two-photon photoemission from metal surfaces. *J. Phys. Condens. Matter*, **14**, R1099–R1141.

6 Banfi, F., Giannetti, C., Ferrini, G., Galimberti, G., Pagliara, S., Fausti, D., and Parmigiani, F. (2005) Experimental evidence of above-threshold photoemission in solids. *Phys. Rev. Lett.*, **94**, 037601.

7 Bisio, F., Winkelmann, A., Lin, W.-C., Chiang, C.-T., Nývlt, M., Petek, H., and Kirschner, J. (2009) Band structure effects in surface second harmonic generation: the case of Cu(001). *Phys. Rev. B*, **80**, 125432.

8 Fann, W.S., Storz, R., and Bokor, J. (1991) Observation of above-threshold multiphoton photoelectric emission from image-potential surface states. *Phys. Rev. B*, **44**, 10980–10982.

9 Irvine, S.E., Dechant, A., and Elezzabi, A.Y. (2004) Generation of 0.4-keV femtosecond electron pulses using impulsively excited surface plasmons. *Phys. Rev. Lett.*, **93**, 184801.

10 Kinoshita, I., Anazawa, T., and Matsumoto, Y. (1996) Surface and image-potential states on Pt(111) probed by two- and three-photon photoemission. *Chem. Phys. Lett.*, **259**, 445–450.

11 Lehmann, J., Merschdorf, M., Thon, A., Voll, S., and Pfeiffer, W. (1999) Properties and dynamics of the image potential states on graphite investigated by multiphoton photoemission spectroscopy. *Phys. Rev. B*, **60**, 17037–17045.

12 Luan, S., Hippler, R., Schwier, H., and Lutz, H.O. (1989) Electron emission from polycrystalline copper surfaces by multi-photon absorption. *Europhys. Lett.*, **9**, 489–494.

13 Miaja-Avila, L., Yin, J., Backus, S., Saathoff, G., Aeschlimann, M., Murnane, M.M., and Kapteyn, H.C. (2009) Ultrafast studies of electronic processes at surfaces using the laser-assisted photoelectric effect with long-wavelength dressing light. *Phys. Rev. A*, **79**, 030901.

14 Ogawa, S. and Petek, H. (1996) Two-photon photoemission spectroscopy at clean and oxidized Cu(110) and Cu(100) surfaces. *Surf. Sci.*, **363**, 313–320.

15 Schoenlein, R.W., Fujimoto, J.G., Eesley, G.L., and Capehart, T.W. (1988) Femtosecond studies of image-potential dynamics in metals. *Phys. Rev. Lett.*, **61**, 2596–2599.

16 Kupersztych, J., Monchicourt, P., and Raynaud, M. (2001) Ponderomotive acceleration of photoelectrons in surface-plasmon-assisted multiphoton photoelectric emission. *Phys. Rev. Lett.*, **86**, 5180–5183.

17 Schattke, W., Krasovskii, E.E., Díez Muiño, R., and Echenique, P.M. (2008) Direct resolution of unoccupied states in solids via two-photon photoemission. *Phys. Rev. B*, **78**, 155314.

18 Bisio, F., Nývlt, M., Franta, J., Petek, H., and Kirschner, J. (2006) Mechanisms of high-order perturbative photoemission from Cu(001). *Phys. Rev. Lett.*, **96**, 087601.

19 Winkelmann, A., Bisio, F., Ocana, R., Lin, W.-C., Nývlt, M., Petek, H., and Kirschner, J. (2007) Ultrafast optical spin injection into image-potential states of Cu(001). *Phys. Rev. Lett.*, **98**, 226601.

20 Eckardt, H., Fritsche, L., and Noffke, J. (1984) Self-consistent relativistic band structure of the noble metals. *J. Phys. F Met. Phys.*, **14**, 97–112.

21 Thörner, G., Borstel, G., Dose, V., and Rogozik, J. (1985) Unoccupied electronic

surface resonance at Cu(001). *Surf. Sci.*, **157**, L379–L383.

22 Strocov, V.N., Claessen, R., Aryasetiawan, F., Blaha, P., and Nilsson, P.O. (2002) Band- and *k*-dependent self-energy effects in the unoccupied and occupied quasiparticle band structure of Cu. *Phys. Rev. B*, **66**, 195104.

23 Strocov, V.N., Claessen, R., Nicolay, G., Hüfner, S., Kimura, A., Harasawa, A., Shin, S., Kakizaki, A., Starnberg, H.I., Nilsson, P.O., and Blaha, P. (2001) Three-dimensional band mapping by angle-dependent very-low-energy electron diffraction and photoemission: methodology and application to Cu. *Phys. Rev. B*, **63**, 205108.

24 Schneider, C.M., Demiguel, J.J., Bressler, P., Schuster, P., Miranda, R., and Kirschner, J. (1990) Spin-resolved and angle-resolved photoemission from single-crystals and epitaxial films using circularly polarized synchrotron radiation. *J. Electron Spectros. Relat. Phenomena*, **51**, 263–274.

25 Courths, R. and Hüfner, S. (1984) Photoemission experiments on copper. *Phys. Rep.*, **112**, 53–171.

26 Mugarza, A., Marini, A., Strasser, T., Schattke, W., Rubio, A., García de Abajo, F.J., Lobo, J., Michel, E.G., Kuntze, J., and Ortega, J.E. (2004) Accurate band mapping via photoemission from thin films. *Phys. Rev. B*, **69**, 115422.

27 Ortega, J.E., Himpsel, F.J., Mankey, G.J., and Willis, R.F. (1993) Quantum-well states and magnetic coupling between ferromagnets through a noble-metal layer. *Phys. Rev. B*, **47**, 1540–1552.

28 Wegehaupt, T., Rieger, D., and Steinmann, W. (1988) Observation of empty bulk states on Cu(100) by two-photon photoemission. *Phys. Rev. B*, **37**, 10086–10089.

29 Winkelmann, A., Lin, W.-C., Chiang, C.-T., Bisio, F., Petek, H., and Kirschner, J. (2009) Resonant coherent three-photon photoemission from Cu(001). *Phys. Rev. B*, **80**, 155128.

30 Knoesel, E., Hotzel, A., and Wolf, M. (1998) Temperature dependence of surface state lifetimes, dephasing rates and binding energies on Cu(111) studied with time-resolved photoemission. *J. Electron Spectros. Relat. Phenomena*, **88–91**, 577–584.

31 Pontius, N., Sametoglu, V., and Petek, H. (2005) Simulation of two-photon photoemission from the bulk sp-bands of Ag(111). *Phys. Rev. B*, **72**, 115105.

32 Ramakrishna, S. and Seideman, T. (2005) Coherence spectroscopy in dissipative media: a Liouville space pathway approach. *J. Chem. Phys.*, **122**, 084502.

33 Ueba, H. and Mii, T. (2000) Theory of energy- and time-resolved two-photon photoemission from metal surfaces influence of pulse duration and excitation condition. *Appl. Phys. A*, **71**, 537–545.

34 Weida, M.J., Ogawa, S., Nagano, H., and Petek, H. (2000) Ultrafast interferometric pump–probe correlation measurements in systems with broadened bands or continua. *J. Opt. Soc. Am. B*, **17**, 1443–1451.

35 Meier, T., Thomas, P., and Koch, S.W. (2007) *Coherent Semiconductor Optics*, Springer.

36 Mukamel, S. (1995) *Principles of Nonlinear Optical Spectroscopy*, Oxford University Press, New York.

37 Wolf, M., Hotzel, A., Knoesel, E., and Velic, D. (1999) Direct and indirect excitation mechanisms in two-photon photoemission spectroscopy of Cu(111) and CO/Cu(111). *Phys. Rev. B*, **59**, 5926–5935.

38 Petek, H., Nagano, H., and Ogawa, S. (1999) Hole decoherence of d bands in copper. *Phys. Rev. Lett.*, **83**, 832.

39 Strocov, V.N. (2003) Intrinsic accuracy in 3-dimensional photoemission band mapping. *J. Electron Spectros. Relat. Phenomena*, **130**, 65–78.

40 Goodwin, E.T. (1939) Electronic states at the surfaces of crystals I. The approximation of nearly free electrons. *Proc. Camb. Philol. Soc.*, **35**, 205–220.

41 Wallauer, W. and Fauster, Th. (1996) Exchange splitting of image states on Fe/Cu(100) and Co/Cu(100). *Phys. Rev. B*, **54**, 5086–5091.

42 Gauyacq, J.P. and Kazansky, A.K. (2007) Modelling of interferometric multiphoton photoemission. *Appl. Phys. A*, **89**, 517–523.

43 Güdde, J., Rohleder, M., Meier, T., Koch, S.W., and Höfer, U. (2007) Time-resolved investigation of coherently controlled electric currents at a metal surface. *Science*, **318**, 1287–1291.

44 Fauster, Th. (1994) Calculation of surface states using a one-dimensional scattering model. *Appl. Phys. A*, **59**, 639–643.

45 Jones, R.O. and Jennings, P.J. (1988) LEED fine structure: origins and applications. *Surf. Sci. Rep.*, **9**, 165–196.

46 Farías, D. and Rieder, K.-H. (1998) Atomic beam diffraction from solid surfaces. *Rep. Prog. Phys.*, **61**, 1575–1664.

47 Bausells, J. and Echenique, P.M. (1986) Velocity dependence of binding energies and lifetimes of image states at surfaces. *Phys. Rev. B*, **33**, 1471–1473.

48 Delone, N.B. and Krainov, V.P. (1999) *Multiphoton Processes in Atoms*, 2nd edn, Springer, Berlin.

49 Faisal, F.H.M. and Kamiński, J.Z. (1998) Floquet–Bloch theory of photoeffect in intense laser fields. *Phys. Rev. A*, **58**, R19–R22.

50 Faisal, F.H.M., Kamiński, J.Z., and Saczuk, E. (2005) Photoemission and high-order harmonic generation from solid surfaces in intense laser fields. *Phys. Rev. A*, **72**, 023412.

51 Cavalieri, A.L., Müller, N., Uphues, Th., Yakovlev, V.S., Baltuska, A., Horvath, B., Schmidt, B., Blümel, L., Holzwarth, R., Hendel, S., Drescher, M., Kleineberg, U., Echenique, P.M., Kienberger, R., Krausz, F., and Heinzmann, U. (2007) Attosecond spectroscopy in condensed matter. *Nature*, **449** (7165), 1029–1032.

52 Drescher, M., Hentschel, M., Kienberger, R., Uiberacker, M., Yakovlev, V., Scrinzi, A., Westerwalbesloh, Th., Kleineberg, U., Heinzmann, U., and Krausz, F. (2002) Time-resolved atomic inner-shell spectroscopy. *Nature*, **419** (6909), 803–807.

53 Meier, F. and Zakharchenya, B.P. (eds) (1984) *Optical Orientation*, North-Holland, Amsterdam.

54 Kuch, W. and Schneider, C.M. (2001) Magnetic dichroism in valence band photoemission. *Rep. Prog. Phys.*, **64**, 147–204.

55 Winkelmann, A., Hartung, D., Engelhard, H., Chiang, C.-T., and Kirschner, J. (2008) High efficiency electron spin polarization analyzer based on exchange scattering at Fe/W(001). *Rev. Sci. Instrum.*, **79**, 083303.

56 Winkelmann, A., Lin, W.-C., Bisio, F., Petek, H., and Kirschner, J. (2008) Interferometric control of spin-polarized electron populations at a metal surface observed by multiphoton photoemission. *Phys. Rev. Lett.*, **100**, 206601.

57 Schmidt, A.B., Pickel, M., Wiemhofer, M., Donath, M., and Weinelt, M. (2005) Spin-dependent electron dynamics in front of a ferromagnetic surface. *Phys. Rev. Lett.*, **95**, 107402.

58 Petek, H., Heberle, A.P., Nessler, W., Nagano, H., Kubota, S., Matsunami, S., Moriya, N., and Ogawa, S. (1997) Optical phase control of coherent electron dynamics in metals. *Phys. Rev. Lett.*, **79**, 4649–4652.

3
Electron Dynamics in Image Potential States at Metal Surfaces

Thomas Fauster

3.1
Scattering Processes

Electron dynamics describes the temporal evolution of an electronic state due to interactions with particles or quasiparticles. The ground state of the system is characterized by the stationary states whose time dependence is determined by the energy of the state. In the present context, we consider an excited electron at the surface, for example, in a surface or adsorbate state. On its relaxation path to the ground state (or thermal equilibrium) it is going to be scattered and the possible processes can be divided into the following categories:

- **Electron–electron scattering**: This most important process for energy relaxation leads to decay into bulk or surface states at lower energy with simultaneous creation of an electron–hole pair. The coupling to the electronic system can also include other quasiparticles such as excitons, plasmons, and magnons that are not covered in this chapter.
- **Electron–phonon scattering**: This scattering process mainly changes the direction of the electron motion and embraces both scattering to bulk bands and scattering within the surface state band. The energy loss is usually rather small compared to electron–electron scattering. Temperature is a convenient control of electron– phonon scattering.
- **Electron defect scattering**: Real surfaces contain always a nonnegligible amount of defects, such as steps or impurity atoms. The associated electron defect scattering mainly changes the electron momentum leaving the energy almost unchanged. In many aspects, it is similar to electron–phonon scattering. Electron defect scattering can be identified by controlling the defect density.

Image potential states [1] are a special class of surface states that exist on many metal surfaces (see chapter 3 of volume 2). The small binding energies relative to the vacuum level point to a weak coupling to the metal. Correspondingly, image potential states can have relatively long lifetimes (>10 fs) and can conveniently be studied with

Dynamics at Solid State Surface and Interfaces Vol. 1: Current Developments
Edited by Uwe Bovensiepen, Hrvoje Petek, and Martin Wolf

ISBN: 978-3-527-40937-2

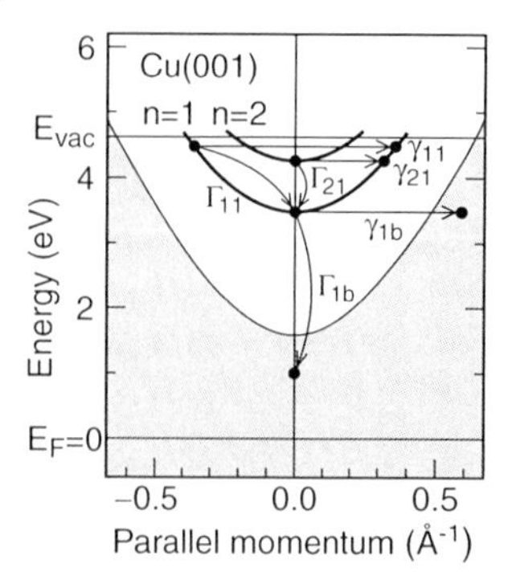

Figure 3.1 Projected bulk band structure (shaded area) for the Cu(001) surface with image potential bands ($n = 1, 2$). For clarity, the binding energy of the image potential states has been scaled by a factor of 2. Arrows indicate possible scattering processes.

time-resolved two-photon photoemission in considerable detail. Therefore, image potential states serve as a model system to investigate the various scattering processes. Figure 3.1 shows the parabolas of the lowest two image potential bands close to the vacuum level E_{vac} in the projected bulk band structure of Cu(001). Arrows illustrate possible scattering processes between surface states and bulk states.

Any scattering event will change energy and momentum of the electron. However, it is pragmatic to distinguish between processes with small energy loss (or gain) and ones with large energy loss. The latter processes are *inelastic* and imply the energy transfer to another particle (or quasiparticle) as illustrated in Figure 3.1 by arrows labeled Γ_{nm}. If the energy loss is small compared to the experimental resolution, the scattering is called *quasielastic* or just *elastic* for short. The corresponding horizontal arrows and are labeled γ_{nm}. Nevertheless, the momentum may be changed considerably in quasielastic scattering processes. An example is shown in Figure 3.1 by the arrow γ_{11} where the momentum is reversed in the $n = 1$ band by a scattering event.

An additional classification of scattering events is done according to whether the electron remains in its initial band or ends up in a different band. The former is termed *intraband* scattering as opposed to *interband* scattering. The processes Γ_{11} and γ_{11} are intraband scattering processes depicted in Figure 3.1. Intraband scattering leads to a redistribution of the population within the band. All other arrows indicate interband scattering. The most prevalent interband scattering is scattering to bulk bands that decrease the population in the image potential bands. For image potential states, interband scattering from the $n \geq 2$ bands to the $n = 1$ band is of particular interest. The processes are indicated by the arrows Γ_{21} and γ_{21} and are associated with a transfer of electrons from the $n = 2$ to the $n = 1$ band.

The distinction between the various scattering processes is summarized in Table 3.1 together with the notation used. Inelastic scattering rates are denoted by Γ and elastic scattering rates by γ. The indices indicate the initial and final state bands. Natural numbers are used for the image potential bands, whereas the index b represents bulk bands in Figure 3.1.

Table 3.1 Classification and notation of scattering processes.

	Intraband	Interband		
From band *n* to	band *n*	band *m*	bulk *b*	
Inelastic	Γ_{nn}	Γ_{nm}	Γ_{nb}	Energy loss
(Quasi-) elastic	γ_{nn}	γ_{nm}	γ_{nb}	Momentum change
Band population	redistributed	transferred	decreased	

3.2 Energies and Dispersion of Image Potential States

As evident from Figure 3.1, there are many scattering processes possible involving image potential bands and other states. In order to determine the associated scattering rates elaborate measurements are needed. As in conventional photoemission, the normal mode of two-photon photoemission (2PPE) is to count the number of emitted electrons as a function of the kinetic energy E_{kin}. An example is shown in Figure 3.2 for the Cu(001) surface[1] where the energy scale has been transformed to binding energy relative to E_{vac}. Three peaks can be identified for the lowest image

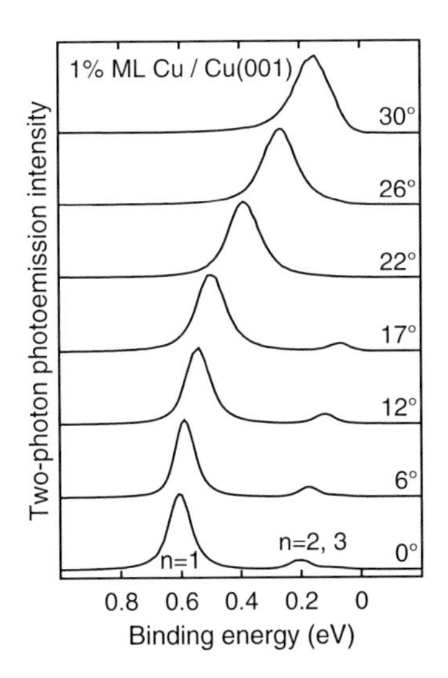

Figure 3.2 Two-photon photoemission spectra for the lowest image potential states on Cu(001) for different emission angles. The electrons are excited by photons with 4.62 eV energy into the image potential states and emitted by photons with an energy of 1.54 eV.

1) The small concentration of Cu adatoms is not relevant for the current discussion.

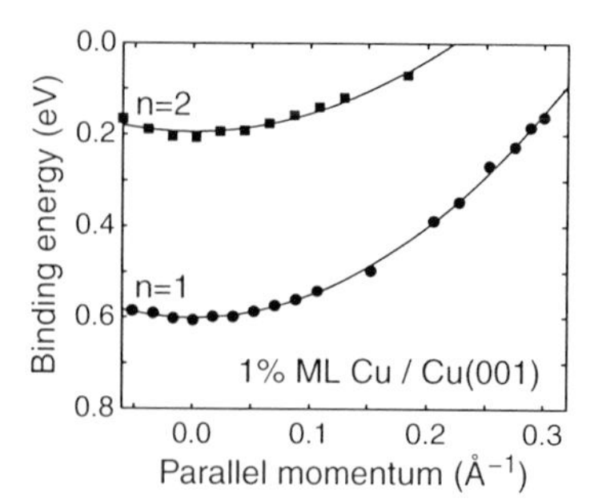

Figure 3.3 Dispersion of the lowest image potential states on Cu(001). Reprinted from Ref. [3]. With kind permission of Springer Science & Business Media.

potential states. For a discussion of the binding energies of image potential states, the reader may refer to the literature [2].

Spectra in Figure 3.2 were taken for different emission angles ϑ and the peaks show a parabolic dispersion. Using the expression for the parallel momentum $k_{\parallel} = \sqrt{2mE_{\text{kin}}}/\hbar \sin\vartheta$, the dispersion relation $E_n(k_{\parallel})$ is obtained as depicted in Figure 3.3. It can be described well by parabolas $E_n(k_{\parallel}) = E_n(0) + \hbar^2 k_{\parallel}^2/2m^*$ with effective masses m^* close to the free electron mass m.

The dispersion is derived from the energy of the peak maxima. Additional information provides the peak width that is determined by the energy resolution of the analyzer, the spectral bandwidth of the laser pulses, and the intrinsic linewidth of the electronic state. The last point will be discussed in more detail in Section 3.4.3. Figure 3.4 shows that the peak width increases with emission angle. The main reason is the limited angular resolution that translates into an energy width for strongly dispersing bands [5]. For the image potential bands with a parabolic dispersion characterized by an effective mass m^*, the kinetic energy varies with emission angle as described by the following expression:

$$E_{\text{kin}}(\vartheta) = \frac{E_{\text{kin}}(0)}{1 - \frac{m}{m^*}\sin^2(\vartheta)}. \tag{3.1}$$

If the electron energy analyzer has an angular resolution $\Delta\vartheta$, the following contribution $\Delta E_{\text{kin}}(\vartheta)$ has to be taken into account for the energetic peak width:

$$\Delta E_{\text{kin}}(\vartheta) = \frac{dE_{\text{kin}}(\vartheta)}{d\vartheta} \cdot \Delta\vartheta = \frac{E_{\text{kin}}(0) \cdot \frac{m}{m^*}\sin 2\vartheta}{\left(1 - \frac{m}{m^*}\sin^2\vartheta\right)^2} \cdot \Delta\vartheta. \tag{3.2}$$

Figure 3.4 shows the experimental results together with a fit to Eq. (3.2). The angular resolution $\Delta\vartheta$ is obtained as 1.6°. Other experimental contributions to the measured linewidth are the energy resolution of the analyzer and the spectral bandwidth of the pump and probe laser pulses. The decay and dephasing rates of the measured states may also contribute to the linewidth as discussed in Section 3.4.3. These last contributions may also depend on the energy and emission angle as shown in Section 3.3.2.

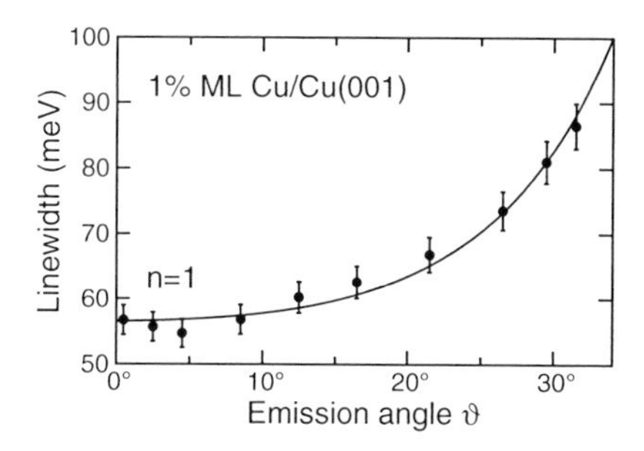

Figure 3.4 Peak width of the lowest image potential states on Cu(001) as function of emission angle. Adapted from Ref. [4].

3.3 Inelastic Scattering

In many spectroscopies, inelastic scattering is detected directly by the loss of energy and the particle appears with a reduced energy after the scattering event. In two-photon photoemission, the energy loss of an electron can be observed in two ways: (i) the electron appears after a scattering event at lower energy at a later time. (ii) The number of electrons in a particular state decreases with time. This decay of population can in many cases be attributed to inelastic scattering.

3.3.1 Lifetimes of Image Potential States

Two-photon photoemission has an additional experimental parameter compared to regular photoemission, that is, the time delay between the two laser pulses. Figure 3.5 shows an intensity map as function of pump–probe delay and binding energy. The data [6] were obtained on a Ru(0001) surface using a time-of-flight detector for

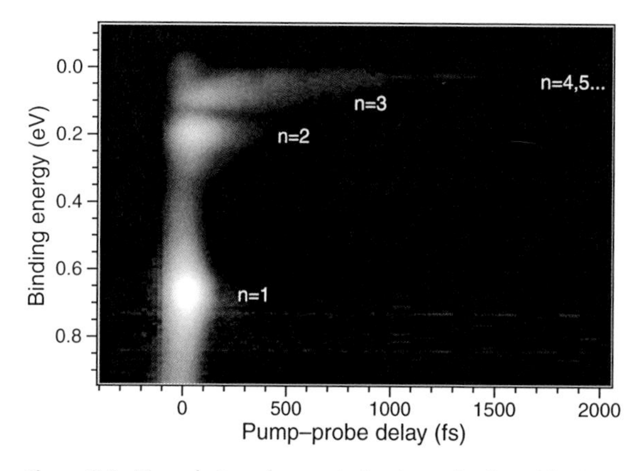

Figure 3.5 Two-photon photoemission intensity (logarithmic gray scale) for the lowest image potential states on Ru(0001) as a function of energy and pump–probe delay. Adapted from Ref. [6].

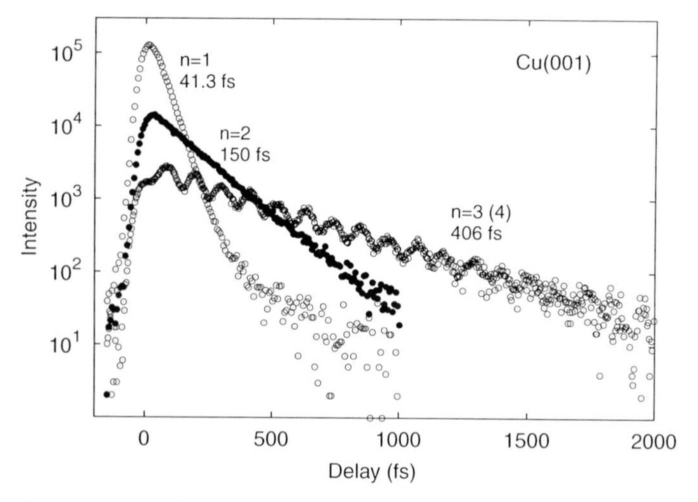

Figure 3.6 Two-photon photoemission signal for the lowest image potential states on Cu(001) ($k_{\|} = 0$) as a function of pump–probe delay. Reprinted from [23]. Copyright 2004, with permission from Elsevier.

electrons that analyzes arbitrary kinetic energies. The intensity maxima for the image potential states are clearly recognized. The energies of the peaks are independent of the time delay. For hemispherical analyzers that detect only a preset kinetic energy (or range of energies), it is convenient to measure the intensity of the peaks as a function of pump–probe delay. Such data are shown in Figure 3.6. for the lowest image potential states on the Cu(001) surface. Due the logarithmic ordinate axis, the exponential decay of the population appears as a linear decrease at large pump–probe delays. The lifetimes for the $n = 1$ and $n = 2$ states are $\tau_1 = 40$ and $\tau_2 = 150$ fs, respectively. Note that the curves cross at large delay times indicating that the $n = 2$ population exceeds eventually the $n = 1$ population. The trace for the $n = 3$ state shows regular oscillations on top of a linear slope corresponding to a lifetime $\tau_3 = 400$ fs. These quantum beats arise from a coherent excitation of the $n = 3$ and $n = 4$ image potential states by the short laser pulse with a spectral bandwidth comparable to the energy separation of the states [7]. From the oscillation period of $T = 117$ fs, the energy difference can be determined very accurately to $|E_3 - E_4| = h/T = 35$ meV.

The lifetimes τ_n shown in Figure 3.6 depend strongly on the quantum number n and consequently binding energy E_B. It is convenient to convert lifetimes to total decay rates $\Gamma_n = \hbar/\tau_n$ that are the sum of all scattering rates associated with a decrease in the population in the state n at a particular momentum or energy. Total decay rates Γ_n are plotted in Figure 3.7 as a function of binding energy E_B for several copper surfaces. For $n \geq 2$, an $E_B^{3/2}$ dependence indicated by the dashed lines is observed. It corresponds to the classical expectation of the round trip oscillation of the electron in the potential well formed by the image potential and the solid represented by a hard wall [8]. This picture assumes that decay processes occur predominantly when the electron hits the surface. Since the binding energy E_B is proportional to n^{-2}, it follows for the lifetimes $\tau_n \propto n^3$. The same result is obtained when the probability

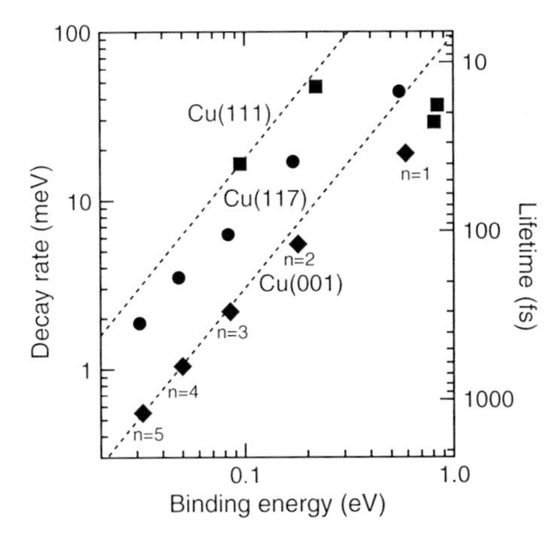

Figure 3.7 Decay rates for the image potential states as a function of binding energy for Cu(111) (squares), Cu(117) (circles), and Cu(001) (diamonds). The dashed lines indicate $E_B^{3/2}$ dependence. Reprinted from [23]. Copyright 2004, with permission from Elsevier.

to find the electron in the bulk is evaluated from the wave function of the image potential state. The overlap with bulk bands explains also the increase in the decay rate from Cu(001) to Cu(117) in Figure 3.7. The energies of the image potential states get closer to the effective bandgap and therefore the penetration of the wave function into the bulk becomes larger [9]. The limiting case is reached for Cu(111) where the states $n \geq 2$ are degenerate with bulk bands. The energy of the $n = 1$ state is still in the bandgap and has a decay rate smaller than the one of the $n = 2$ state [10].

Table 3.2 summarizes the available experimental data on lifetimes of image potential states on clean metal surfaces for normal emission ($k_{\parallel} = 0$). The main factors determining the lifetimes are (i) the bulk penetration of the wave function that

Table 3.2 Measured lifetimes in femtoseconds of image potential states on clean metal surfaces.

	τ_1	τ_2	τ_3	References
Cu(001)	40 ± 6	120 ± 15	300 ± 20	[7, 11]
Cu(1 1 15)	22 ± 3	55 ± 3	165 ± 10	[12]
Cu(1 1 11)	14 ± 2.5	45 ± 3	170 ± 10	[12]
Cu(119)	9 ± 2.5	39 ± 3	130 ± 10	[12]
Cu(117)	9 ± 2.5	39 ± 3	130 ± 10	[12]
Cu(115)	< 5	20 ± 3	65 ± 5	[12]
Cu(111)	18 ± 5	14 ± 3	40 ± 6	[10, 13, 14]
Cu(775)	18 ± 2			[15]
Ag(001)	55 ± 5	160 ± 10	360 ± 15	[11]
Ag(111)	36	18 ± 2		[16–18]
Ru(0001)	11 ± 2	57 ± 5	174 ± 10	[6, 19]
Ni(111)	7 ± 3			[20]
Pd(111)	25 ± 4			[21]
Pt(111)	26 ± 7	62 ± 7		[22]

strongly depends on the relative position in the bandgap of the projected bulk band structure [2]. This effect can be seen nicely for the copper surfaces where the effective bandgap depends on the surface orientation [12]. (ii) The density of available final states in particular close to the Fermi energy [23]. This influence can be pinpointed in Table 3.2 at the short lifetimes of the d-band metals compared to the noble metals. In fact, for many d-band metals the lifetimes are too short to be reliably measured in time-resolved 2PPE.

3.3.2
Momentum Dependence of Lifetimes

The overlap with bulk states is an important factor determining the lifetimes of surface or image potential states. For dispersing bands, the decay depends also on the parallel momentum. An example is shown in Figure 3.8 for the $n = 1$ image potential band on Cu(001) [24]. The data show that the lifetime decreases with increasing parallel momentum. Two effects contribute to the increase in the decay rate: (i) The energy increases with parallel momentum and the available phase space for inelastic decay increases. (ii) Inelastic scattering processes within the image potential band become possible for electrons with energies above the band bottom located at $k_{\|} = 0$.

Two experimental data sets for the decay rates $\hbar/\tau$ are shown in Figure 3.9 by filled [24] and open circles [25] that agree within the error bars. The linear dependence with energy (i.e., a quadratic dependence on parallel momentum) is characterized by a slope of 0.047 [24] compared to 0.035 [25]. The apparent discrepancy of the values for the slope arises from the larger error bars of the former data set that gives the values for small energies more weight.

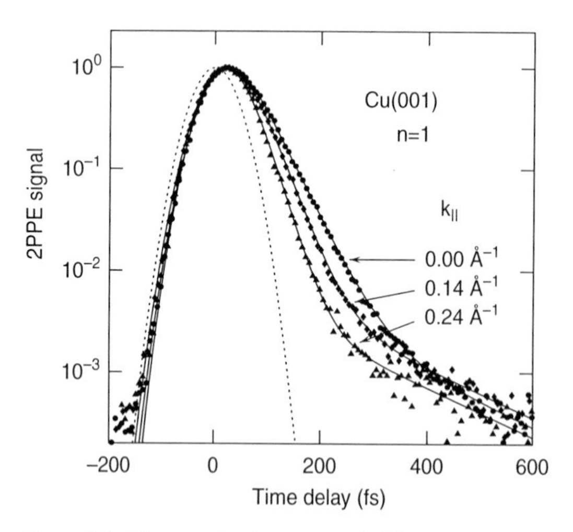

Figure 3.8 Time-resolved 2PPE signal of the $n = 1$ state of Cu(001) for three different values of the parallel momentum $k_{\|}$. The computed instrument function is shown as a dotted line. Reprinted with permission from [24]. Copyright 2002 by American Physical Society.

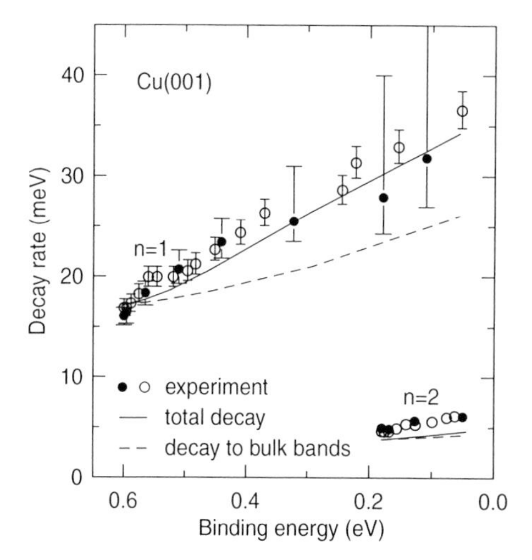

Figure 3.9 Experimental (dots: Ref. [24]; circles: Ref. [25]) decay rates for the $n = 1$ and $n = 2$ image potential states as a function binding energy. Solid and dashed lines show the calculated total decay rate and the contribution of decay to bulk bands [24].

The experimental data agree very well with the calculation (solid lines) [24]. Both inelastic intraband scattering processes and inelastic interband scattering processes to bulk bands (dashed lines) demonstrate a linear dependence on energy as shown in Figure 3.9. For the $n = 1$ band, both processes are calculated to contribute about the same amount to the increase in the decay rate [24].

3.3.3
Inelastic Intraband Scattering

In the previous section, inelastic intraband scattering was identified as one contribution to the momentum dependence of the decay rates. The other contributing factor is the energy dependence of the decay rate, and a quantitative assessment can be obtained from calculations [24]. A different approach is illustrated in Figure 3.10 that shows time-resolved 2PPE data for the $n = 1$ image potential state on Cu (001) [26]. They are taken at the bottom of the $n = 1$ band for $k_{\|} = 0$. The intensity is plotted on logarithmic scale and shows good statistics over four orders of magnitude. The lower curve of Figure 3.10 is a reference measurement that was taken with a photon energy of 4.43 eV that is below the threshold for excitation of the $n = 2$ state. It shows a simple linear decay in the semilogarithmic plot. For photon energies above this threshold, the data exhibit a biexponential decay. The short timescale corresponds to the lifetime of the $n = 1$ state, whereas the longer timescale is discussed in more detail in Sections 3.3.4 and 3.4.1.

The observed difference in the lifetimes of 36 fs for excitation with 4.43 eV photon energy and 40 fs wirh 4.65 eV photon energy is explained by cascade processes of inelastic intraband scattering (Γ_{11} in Figure 3.1) along the $n = 1$ band [24]. Electrons excited high into the image potential band decay downward in the band. The required

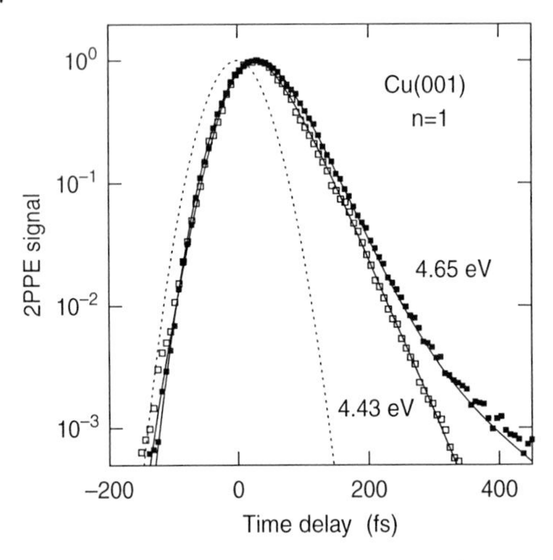

Figure 3.10 Logarithmic plot of pump–probe traces for various excitation conditions at the bottom of the $n = 1$ band on Cu(001). Results of numerical simulations described are displayed as thin lines. The computed instrument function is shown as a dotted line. Reprinted with permission from [24]. Copyright 2002 by American Physical Society.

time for the downward scattering is seen as an apparent longer lifetime. In order to avoid these cascade processes, pump pulses close to the excitation threshold should be used in order to obtain accurate lifetimes.

3.3.4
Inelastic Interband Scattering

Because the long-range image potential leads to a series of bands, scattering between these bands is possible. This provides a convenient model system to study the associated effects in detail, which are usually much harder to identify at other surfaces. Figure 3.6 shows a biexponential decay for the $n = 1$ state. The slower decay has within the statistical errors the same slope as the decay of the $n = 2$ state. This biexponential decay is also seen in Figure 3.8 for different parallel momenta. Figure 3.10 proves that the effect is due to the scattering from the $n = 2$ band to the $n = 1$ band because the slowly decaying component is absent for photon energies below the excitation threshold of the $n = 2$ band [24]. The relative intensity at which the number of electrons scattered from the $n = 2$ to the $n = 1$ band becomes comparable to the rapidly decaying population excited directly by the pump pulse into the $n = 1$ band is a measure for the strength of the interband scattering. More quantitative results are obtained from rate equation models [26] or fits using the optical Bloch equations [3]. Results for the clean Cu(001) surface are shown in Figure 3.11 by open symbols as a function of energy relative to the band bottom of the $n = 1$ band. The interband scattering rate is almost independent of energy (i.e., parallel momentum). For comparison, the total decay rate Γ_2 of the $n = 2$ band is plotted in the inset of Figure 3.11 that is on the clean Cu(001) surface more than one magnitude larger than the interband scattering rate Γ_{21}.

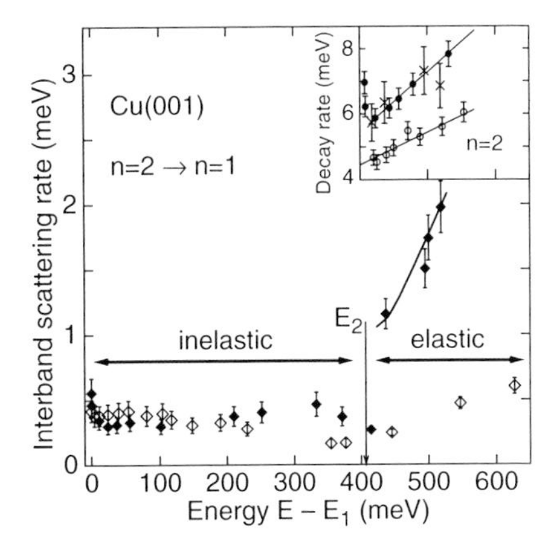

Figure 3.11 Interband scattering rates on the clean (open symbols) and covered (1% ML Cu, filled symbols) surface measured on the $n = 1$ band. The solid line shows the difference of the $n = 2$ decay rates between the covered and the clean surface from the inset. Reprinted in part from [12]. With kind permission of Springer Science & Business Media.

3.4 Quasielastic Scattering

Quasielastic scattering describes events with energy losses small compared to the experimental energy resolution. In the case of intraband scattering of image potential states, the electron just changes its direction or momentum. For perfect stepped surfaces, the momentum relative to the steps is reversed leading to asymmetries in the step-up or step-down directions. For point scatterers such as adatoms, the net population within the image potential band is not changed in intraband scattering. Consequently, it cannot be detected in time-resolved measurements and we have to resort to linewidth analysis as explained in Section 3.4.3.

Elastic interband scattering is observed from the appearance of the electron in a different image potential band at the same energy. A special case occurs when two image potential bands cross due to band umklapp. The strong coupling leads to a resonant interband scattering process. Elastic interband scattering to bulk bands (rate γ_{nb}) cannot be directly detected because the lifetime in bulk bands is quite short. It can be inferred indirectly from the difference of the total scattering rate Γ_n (measured by the lifetime) to the individual interband contributions $\Gamma_{nm} + \gamma_{nm}$.

3.4.1 Elastic Interband Scattering

Inelastic interband scattering from the $n = 2$ to the $n = 1$ band has been discussed in Section 3.3.4. There the $n = 2$ decay was observed at the bottom of the $n = 1$ band. For elastic interband scattering, the $n = 2$ decay has to be observed on the $n = 1$ band at the same energy. This implies measurements at finite parallel momentum. For the clean surface, the elastic interband scattering is quite small as can be seen in Figure 3.11 (open symbols). For 1% Cu adatoms, the interband scattering rate increases significantly for energies above the bottom of the $n = 2$ band [25]. The elastic character of the interband signal is proven by the fact that the decay rate measured for the $n = 2$ band (filled symbols in the inset of Figure 3.11) agrees within the experimental errors with the decay rate obtained for the slowly decaying component on the $n = 1$ band (crosses) at the same energy. Scattering of electrons by Cu adatoms changes the momentum with negligible change of energy. Scattering by other electrons is possible only on the clean surface that is always associated with an energy loss for the creation of an electron–hole pair. The scattering by the Cu lattice is incorporated via the crystal potential in the electronic band structure.

The total decay rate Γ_2 in the $n = 2$ band significantly increases by the addition of Cu adatoms as can be seen in the inset of Figure 3.11. This increase corresponds quite well to the elastic interband scattering rate as shown by the solid line in the main figure. From this good agreement, we conclude that elastic scattering from the $n = 2$ band to bulk bands by adatoms (γ_{2b}) is small [25].

3.4.2
Resonant Interband Scattering

Elastic interband scattering can be significantly enhanced on stepped surfaces. In addition, the interband scattering rate can show pronounced peaks as a function of energy in contrast to the smooth dependence observed for adatoms shown in Figure 3.11. Therefore, we use the term *resonant interband scattering* for stepped surfaces. The effect was discovered first on a clean Cu(001) surface [26]. Later results suggest that the sample used in the pioneering work must have had a nonnegligible number of steps [27].

Time-resolved two-photon photoemission measurements for the Cu(1 1 11) surface are shown in Figure 3.12 on a logarithmic intensity scale [12]. As indicated in the inset, the data are obtained on the $n = 1$ image potential band from the band bottom E_1 up to the vacuum energy. For energies below approximately 330 meV, a single exponential decay is observed. When the energy of the $n = 2$ band minimum is reached, a biexponential decay appears. For energies above the bottom of the $n = 2$ band, the interband scattering contribution becomes even stronger and a $n = 1$ decay is hardly visible in the spectra of Figure 3.12. The $n = 2$ contribution becomes weaker at even higher energies when a third component appears corresponding to the $n = 3$ decay with its typical quantum beat oscillations [7].

The relative contributions of the interband scattering processes can be obtained by fitting the experimental spectra of Figure 3.12 [28]. The results presented in Figure 3.13 show that the interband scattering from the $n = 2$ band can reach over 90% for Cu(1 1 11). The maximum contribution is observed for energies above the band minimum E_2 at energies depending on the terrace width [12]. These maxima coincide with the intersection of the $n = 1$ band with the backfolded $n = 2$ band (see inset in Figure 3.12 and Ref. [12]). The periodic step lattice provides an interaction that leads to mixing of the $n = 1$ and $n = 2$ bands. Therefore, the distinction between $n = 1$ and $n = 2$ bands disappears [29].

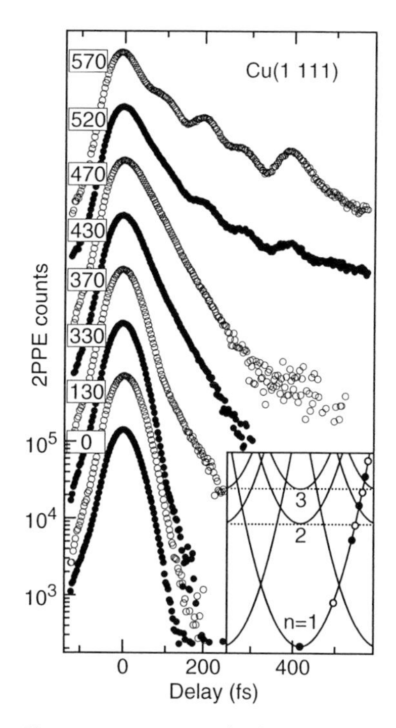

Figure 3.12 Time-resolved two-photon photoemission intensity measured on the $n = 1$ image potential band for the Cu(1 1 11) surface plotted on a logarithmic scale. The energy above the band bottom is indicated by the numbers (in meV) and marked in the band structure plot. Reprinted from [12]. With kind permission of Springer Science & Business Media.

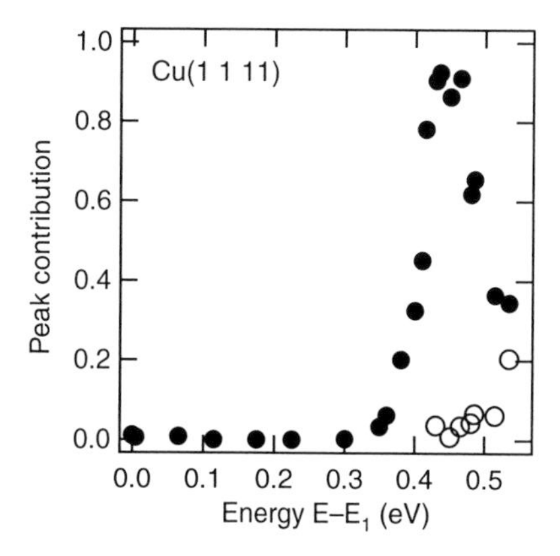

Figure 3.13 Interband scattering probability from $n = 2$ (filled circles) and $n = 3$ (open circles) to $n = 1$ image potential states as a function of energy above the band bottom E_1 for Cu(1 1 11). Reprinted from [12]. With kind permission of Springer Science & Business Media.

3.4.3
Elastic Intraband Scattering

The pump pulse in two-photon photoemission excites the image potential bands up to a maximum energy given by the photon energy. States with all possible parallel momenta according to the dispersion relation are populated. Because for low-index surfaces the dispersion shows no azimuthal dependence, elastic intraband scattering just redistributes the electron population in the band. This can be modeled quantitatively, for example, in a wave packet propagation approach [30, 31]. As a result, usual time-dependent measurements as a function of pump–probe delay at fixed energy do not yield information on elastic intraband scattering. However, the scattering event leads to a change in the phase of the wave function. This *dephasing* can be detected in the energetic linewidth.

A set of energy spectra from a Cu(001) surface for various pump–probe delays is plotted in Figure 3.14. The width of the peaks changes noticeably. The full width at half maximum of the peaks is shown in Figure 3.15 as a function of pump–probe delay. At negative time delay, the linewidth decreases linearly and attains a constant value for long positive delays. These surprising observations can be explained by describing the two-photon photoemission process in a density matrix formalism. The resulting optical Bloch equations [32] can be solved for special cases [33, 34]. The scattering processes are introduced by decay and dephasing rates Γ and Γ^*, respectively. The resulting line shape is represented in very good approximation by a convolution of a Lorentzian and a Gaussian function. The latter is determined by experimental parameters, that is, energy resolution of the analyzer and spectral bandwidth of the laser pulses. The Lorentzian is often referred to as the intrinsic linewidth. In 2PPE, it corresponds to twice the dephasing rate plus a delay-dependent term containing the decay rate that goes to zero for long delays [33]. In normal

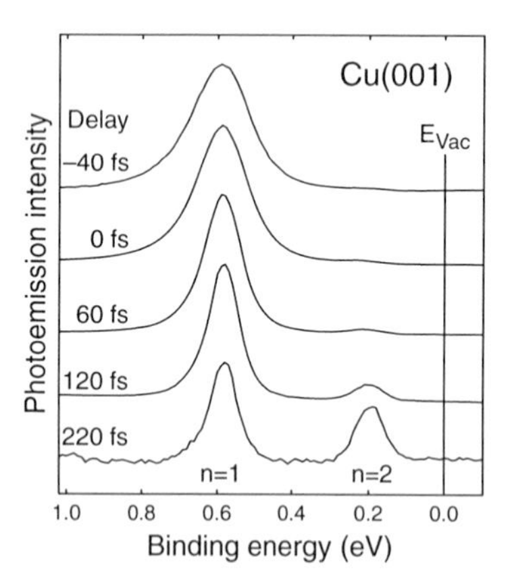

Figure 3.14 Two-photon photoemission spectra for the image potential states on Cu(001) for different pump–probe delays taken with $h\nu = 1.49$ eV probe and $3h\nu$ pump pulses. All spectra are normalized to the same height. Reprinted with permission from [24]. Copyright 2002 by the American Physical Society.

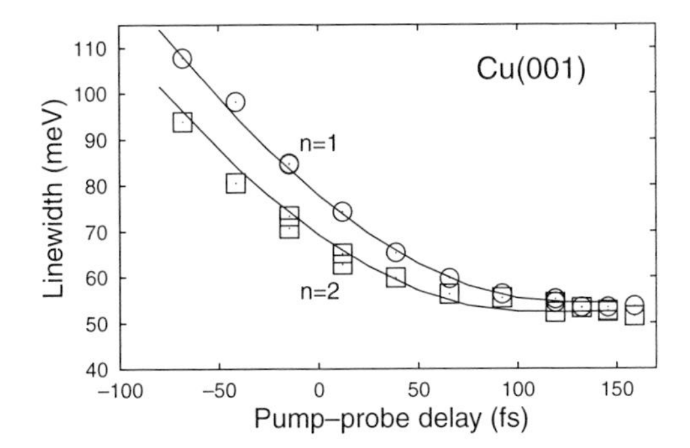

Figure 3.15 Linewidth for the $n = 1$ and $n = 2$ image potential states (circles and squares, respectively) on Cu(001) compared to the results of numerical calculations (solid lines). Reprinted with permission from [33]. Copyright 2002 by the American Physical Society.

photoemission, the intrinsic linewidth contains the sum of decay and dephasing rates, which cannot be disentangled.

The results of calculations are compared with the data in Figure 3.15. The decay and dephasing rates as well as the experimental parameters match the values obtained by independent measurements [33]. The almost identical linewidths for both image potential states at long delays (see Figure 3.15) indicates that the dephasing rates are the same for $n = 1$ and $n = 2$ image potential states. Taking into account the information about the experimental parameters, dephasing is found to be negligible on the clean Cu(001) surface [10].

Decay and dephasing can be illustrated nicely in the quantum beat pattern of Figure 3.6. The trace for the $n = 3(4)$ states shows an exponential decay and a superimposed oscillation. The decay rate is determined by the average decrease in the population. The temporal interference pattern depends on the relative phase of the time-dependent wave functions. The observation of the beating pattern with a constant amplitude on the logarithmic scale indicates that the scattering events change only the population and not the phase. Dephasing is therefore also negligible for the higher image potential states on clean Cu(001). Adsorbates can influence the decay or the dephasing in the quantum beat pattern depending on the associated scattering potential [35].

3.5 Electron–Phonon Scattering

Scattering by phonons is usually quasielastic because typical phonon energies at metal surfaces are smaller than the experimental resolution. An exception might occur for molecules [37]. The influence of phonons can conveniently be controlled by temperature. The number of phonons for sufficiently high temperatures is proportional to the absolute temperature T. The number of scattering events and the associated dephasing rate entering the linewidth are then expected to increase linearly with temperature. Figure 3.16 presents two-photon photoemission spectra

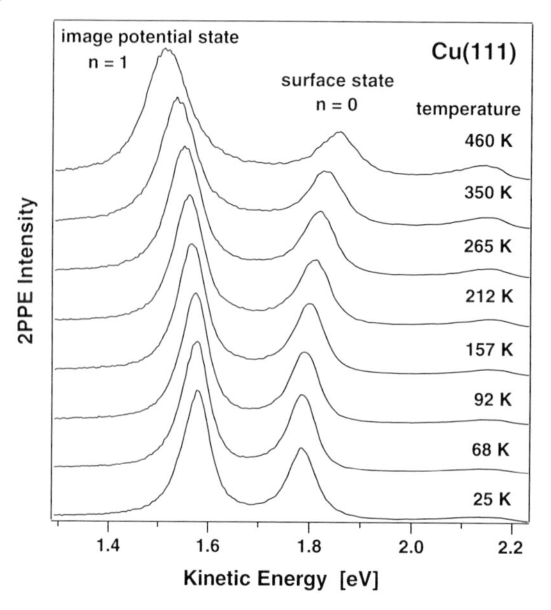

Figure 3.16 2PPE spectra from Cu(111) for various temperatures taken with $h\nu = 2.37$ eV probe and $2h\nu$ pump pulses. All spectra are normalized to the same height. Reprinted from [36]. Copyright 1998, with permission from Elsevier.

from Cu(111) for temperatures between 25 and 460 K [36]. The two peaks correspond to the occupied surface state ($n = 0$) and the $n = 1$ image potential state. The peak shifts with temperature are explained by the variation of the bandgap due to the thermal expansion of the lattice [36]. The linewidths also change with temperature

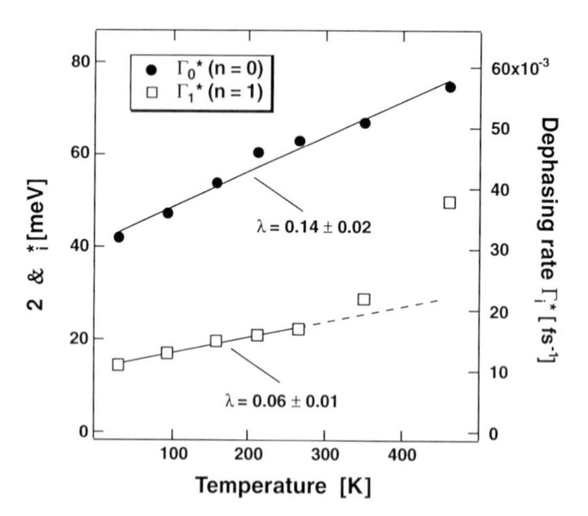

Figure 3.17 Pure dephasing rates versus temperature derived from the linewidths of the $n = 0$ and $n = 1$ states using the optical Bloch equations. The mass enhancement parameters λ are obtained from linear fits using the relation $\Gamma^{*}_{e-ph} = 2\pi\lambda k_B T$. Reprinted from [36]. Copyright 1998, with permission from Elsevier.

that allows to determine the dephasing rates plotted in Figure 3.17. For both states, a linear increase with temperature is found. For the $n = 1$ state, a deviation is found above 400 K because the energy of the state is outside the bandgap in this temperature range [36]. The associated larger penetration into the bulk increases the coupling to bulk phonons and accordingly the dephasing rate.

The linear increase in the dephasing rate with temperature has a slope relative to the electron–phonon mass enhancement parameter $\lambda = \Gamma^{*}_{\text{e-ph}}/2\pi k_B T$ [38]. For the $n = 0$ state, the value $\lambda = 0.14 \pm 0.02$ is in excellent agreement with the results of photoemission experiments [39–41]. For the $n = 1$ image potential state, $\lambda = 0.06 \pm 0.01$ is obtained. The smaller value for the $n = 1$ state is explained by the smaller overlap with bulk wave functions for the $n = 1$ state compared to the $n = 0$ state [36].

An extrapolation to $T = 0$ K yields dephasing rates of 40 and 14 meV for the $n = 0$ and $n = 1$ states, respectively. This indicates a significant amount of defects even on Cu(111) surfaces prepared according to the state of the art. For the occupied surface state, the value agrees with results from photoemission, scanning tunneling spectroscopy, and theoretical calculations [41]. Results for the $n = 1$ image potential state from other methods are not available. Values for the parameter $\lambda \leq 0.011$ have been calculated for $n = 1$ states on the (001) surfaces of Cu and Ag [42] in agreement with the small penetration of the wave function into the bulk for image potential states close to the center of the bandgap. No temperature dependence has been found experimentally for the Cu(001) surface in agreement with the negligible dephasing found in Section 3.4.3 [23]. For Pd(111), a strong temperature dependence of the linewidth is observed [43] in spite of the fact that the overlap with bulk bands should be similar to the case of Cu(001). A consistent interpretation of the data using $\lambda = 0$ is obtained when thermally created defects are taken into account [43].

3.6 Electron Defect Scattering

Surfaces with adatoms and steps have been discussed in previous sections to illustrate various scattering processes for electrons in image potential states. In the following section, the dependence on adatom concentration and step density will be discussed.

3.6.1 Scattering by Adatoms

The decay rates and interband scattering rates introduced in Figures 3.9 and 3.11 are plotted in Figure 3.18 as a function of Cu coverage. The top panel shows the average interband scattering rates for elastic ($E > E_2$) and inelastic ($E < E_2$) interband scattering. The elastic scattering rate γ_{21} increases strongly with Cu coverage confirming that this process is caused by the Cu adatoms. The linear increase is consistent with the occurrence of individual adatoms in the relevant coverage and temperature range [44, 45]. The inelastic interband scattering rate Γ_{21} shows no significant dependence on Cu coverage. The slightly larger rates for 3% ML Cu

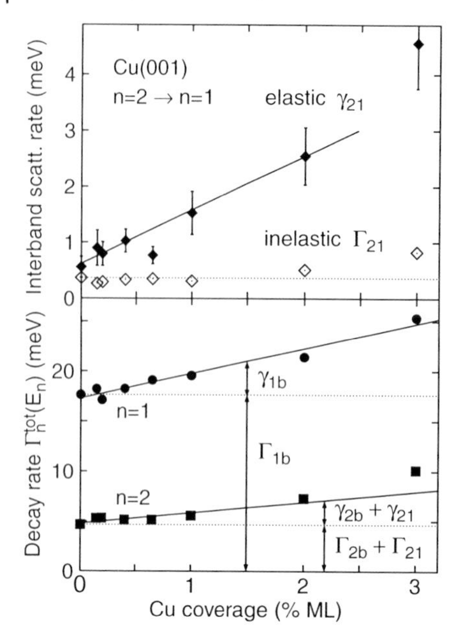

Figure 3.18 *Top*: Average elastic (filled diamonds) and inelastic (open diamonds) interband scattering rates versus Cu coverage. *Bottom*: Total scattering rates $\Gamma_n(E_n)$ of the $n = 1$ (circles) and 2 (squares) image potential states at the band bottom (filled symbols). Reprinted with permission from [25]. Copyright 2004 by the American Physical Society.

coverage in Figure 3.18 could indicate the breakdown of the assumption of well-separated adatoms. Γ_{21} does not increase with Cu coverage in contrast to γ_{21}. Therefore, electrons scattered elastically (γ_{21}) to the $n = 1$ band and then inelastically (Γ_{11}) down to the bottom of the $n = 1$ band do not contribute significantly to the long-lived component from which Γ_{21} is derived. Inelastic interband scattering Γ_{21} from the $n = 2$ to the $n = 1$ band must then be due to the interaction of the electrons in the $n = 2$ state with the underlying electron system in the bulk. These arguments are supported by the excellent agreement for Γ_{21} with theoretical calculations [46].

At the band bottom of the $n = 1$, band decay can proceed only via scattering to bulk band, and the total decay rate $\Gamma_1(E_1)$ is the sum of the inelastic decay rate Γ_{1b} and the elastic scattering rate γ_{1b}: $\Gamma_1(E_1) = \Gamma_{1b} + \gamma_{1b}$. $\Gamma_1(E_1)$ grows linearly with adatom coverage (bottom panel of Figure 3.18, filled circles). For the bottom of the $n = 2$ band, inelastic and elastic interband scattering terms have to be added: $\Gamma_2(E_2) = \Gamma_{2b} + \gamma_{2b} + \Gamma_{21} + \gamma_{21}$. Γ_{21} and γ_{21} are obtained from independent measurements of the scattered components on the $n = 1$ band (Figure 3.11). Figure 3.18 shows that Γ_{21} is independent of coverage and γ_{21} is proportional to the adatom concentration. If the bulk provides most of the final states for inelastic decay and the contribution of the adatoms is negligible, it is plausible that the inelastic decay rates Γ_{nb} are independent of coverage and the coverage dependence is carried by the elastic decay rates γ_{nb}. Note that inelastic and elastic processes are also described by different theoretical models [46, 47]. The results support the simple picture that scattering of electrons by a heavy adatom is elastic, whereas interaction with another electron of the same mass permits the exchange of energy leading to an inelastic scattering process. Both processes are usually accompanied by a change of momentum.

Data for scattering rates have been obtained for Cu and Co adatoms on the Cu(001) surface [25, 48]. The agreement for the inelastic scattering rates is excellent. The moderate agreement for the elastic scattering rates might be attributed to the fact that both experiment and theory are very sophisticated and challenging.

3.6.2
Scattering by Steps

In Section 3.4.3, it was shown that the linewidth for long delays contains information on the elastic intraband scattering rate. Such data are presented in Figure 3.19 for the $n = 1$ state at various stepped copper surfaces. The linewidth increases significantly with step density. The associated elastic intraband scattering rate is only slightly smaller than the decay rate [12]. The increase with step density is plausible because the step edges are the obvious scatterers. However, for a perfect step array the scattering by the steps should be incorporated in the band structure leading to small bandgaps at the intersection of backfolded bands [49]. The elastic intraband scattering should be negligible as demonstrated for the clean Cu(001) surface. The conclusion is that the observed increase in the linewidth with step density is primarily caused by disorder and fluctuation of steps.

Another interesting observation for stepped surfaces is the asymmetry of the resonant interband scattering rate for the upstair and downstair directions [12, 50]. This asymmetry can also be observed in scanning tunneling spectroscopy, and a good agreement with 2PPE measurements is obtained [49]. The asymmetric scattering rate can be explained only if some of the electrons are scattered into bulk bands.

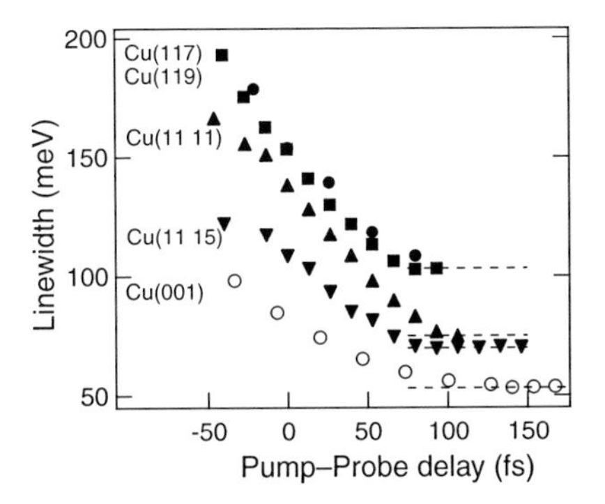

Figure 3.19 Linewidth of two-photon photoemission spectra for the $n = 1$ image potential state at $k_{\parallel} = 0$ as a function of pump–probe delay for stepped copper surfaces. Reprinted from [12]. Copyright 2007, with permission from Elsevier.

3.7
Summary and Outlook

Time- and angle-resolved two-photon photoemission of image potential states permits a detailed study of scattering processes at metal surfaces. Many scattering rates can be determined quantitatively in good agreement with theoretical calculations. This thorough understanding helps to identify and clarify scattering processes at other surfaces [51, 52]. However, a lot of interesting questions are still open for image potential states. The scattering behavior of adatoms has been studied in full detail only for Cu and Co on Cu(001). Other adsorbates such as small molecules might exhibit a different scattering behavior [14]. On stepped surfaces, the influence of the step regularity (terrace width distribution and frizziness) is not well understood. The study of surfaces with better ordered and straighter steps than vicinal copper surfaces would be worthwhile. Finally, both paths of research could join in the investigation of steps with adatom decoration.

Acknowledgments

This chapter is largely based on the excellent work of K. Boger and M. Roth and is a result of close cooperation with E. V. Chulkov, P. M. Echenique, U. Höfer, and M. Weinelt over many years. I express my sincere gratitude to all of them. I thank C. Gahl for providing Figure 3.5.

References

1 Echenique, P.M. and Pendry, J.B. (1978) *J. Phys. C*, **11**, 2065.
2 Fauster, T. and Steinmann, W. (1995) Chaptrer 8, in *Photonic Probes of Surfaces (Electromagnetic Waves: Recent Developments in Research)* (ed. P. Halevi), vol. 2, North-Holland, Amsterdam, p. 347.
3 Boger, K., Weinelt, M., Wang, J., and Fauster, T. (2004) *Appl. Phys. A*, **78**, 161.
4 Boger, Klaus. (2004) Ph.D. thesis, Universität Erlangen-Nürnberg.
5 Matzdorf, R. (1998) *Surf. Sci. Rep.*, **30**, 153.
6 Gahl, C. (2004) Ph.D. thesis, Free University Berlin.
7 Höfer, U., Shumay, I.L., Reuß, C., Thomann, U., Wallauer, W., and Fauster, T. (1997) *Science*, **277**, 1480.
8 Fauster, T., Reuß, C., Shumay, I.L., and Weinelt, M. (2000) *Chem. Phys.*, **251**, 111.
9 Roth, M., Pickel, M., Wang, J., Weinelt, M., and Fauster, T. (2002) *Appl. Phys. B*, **74**, 661.
10 Weinelt, M. (2002) *J. Phys.: Condens. Matter*, **14**, R1099.
11 Shumay, I.L., Höfer, U., Reuß, C., Thomann, U., Wallauer, W., and Fauster, T. (1998) *Phys. Rev. B*, **58**, 13974.
12 Roth, M., Fauster, T., and Weinelt, M. (2007) *Appl. Phys. A*, **88**, 497.
13 Weinelt, M. (2000) *Appl. Phys. A*, **71**, 493.
14 Kirchmann, P.S., Loukakos, P.A., Bovensiepen, U., and Wolf, M. (2005) *New J. Phys.*, **7**, 113.
15 Shen, X.J., Kwak, H., Mocuta, D., Radojevic, A.M., Smadici, S., and Osgood, R.M. (2002) *Chem. Phys. Lett.*, **351**, 1.
16 Lingle, R.L., Jr., Ge, N.-H., Jordan, R.E., McNeill, J.D., and Harris, C.B. (1996) *Chem. Phys.*, **205**, 191.

17 Gaffney, K.J., Wong, C.M., Liu, S.H., Miller, A.D., McNeill, J.D., and Harris, C.B. (2000) *Chem. Phys.*, **251**, 99.

18 Gundlach, L., Ernstorfer, R., Riedle, E., Eichberger, R., and Willig, F. (2005) *Appl. Phys. B*, **80**, 727.

19 Tegeder, P., Danckwerts, M., Hagen, S., Hotzel, A., and Wolf, M. (2005) *Surf. Sci.*, **585**, 177.

20 Link, S., Sievers, J., Dürr, H.A., and Eberhardt, W. (2001) *J. Electron Spectrosc. Relat. Phenom.*, **114–116**, 351.

21 Schäfer, A., Shumay, I.L., Wiets, M., Weinelt, M., Fauster, T., Chulkov, E.V., Silkin, V.M., and Echenique, P.M. (2000) *Phys. Rev. B*, **61**, 13159.

22 Link, S., Dürr, H.A., Bihlmayer, G., Blügel, S., Eberhardt, W., Chulkov, E.V., Silkin, V.M., and Echenique, P.M. (2001) *Phys. Rev. B*, **63**, 115420.

23 Echenique, P.M., Berndt, R., Chulkov, E.V., Fauster, T., Goldmann, A., and Höfer, U. (2004) *Surf. Sci. Rep.*, **52**, 219.

24 Berthold, W., Höfer, U., Feulner, P., Chulkov, E.V., Silkin, V.M., and Echenique, P.M. (2002) *Phys. Rev. Lett.*, **88**, 056805.

25 Boger, K., Weinelt, M., and Fauster, T. (2004) *Phys. Rev. Lett.*, **92**, 126803.

26 Berthold, W., Güdde, J., Feulner, P., and Höfer, U. (2001) *Appl. Phys. B*, **73**, 865.

27 Fauster, T., Weinelt, M., and Höfer, U. (2007) *Prog. Surf. Sci.*, **82**, 224.

28 Roth, M., Pickel, M., Weinelt, M., and Fauster, T. (2004) *Appl. Phys. A*, **78**, 149.

29 Lei, J., Sun, H., Yu, K.W., Louie, S.G., and Cohen, M.L. (2001) *Phys. Rev. B*, **63**, 045408.

30 Olsson, F.E., Borisov, A.G., Persson, M., Lorente, N., Kazansky, A.K., and Gauyacq, J.P. (2004) *Phys. Rev. B*, **70**, 205417.

31 Gumhalter, B., Šiber, A., Buljan, H., and Fauster, T. (2008) *Phys. Rev. B*, **78**, 155410.

32 Loudon, R. (1983) *The Quantum Theory of Light*, Oxford University Press, New York.

33 Boger, K., Roth, M., Weinelt, M., Fauster, T., and Reinhard, P.-G. (2002) *Phys. Rev. B*, **65**, 075104.

34 Wolf, M., Hotzel, A., Knoesel, E., and Velic, D. (1999) *Phys. Rev. B*, **59**, 5926.

35 Fauster, T. (2002) *Surf. Sci.*, **507–510**, 256.

36 Knoesel, E., Hotzel, A., and Wolf, M. (1998) *J. Electron Spectrosc. Relat. Phenom.*, **88–91**, 577.

37 Hotzel, A., Wolf, M., and Gauyacq, J.P. (2000) *J. Phys. Chem. B*, **104**, 8438.

38 Grimvall, G. (1981) *The Electron Phonon Interaction in Metals*, North Holland, Amsterdam.

39 McDougall, B.A., Balasubramanian, T., and Jensen, E. (1995) *Phys. Rev. B*, **51**, 13891.

40 Matzdorf, R., Meister, G., and Goldmann, A. (1996) *Phys. Rev. B*, **54**, 14807.

41 Eiguren, A., Hellsing, B., Reinert, F., Nicolay, G., Chulkov, E.V., Silkin, V.M., Hüfner, S., and Echenique, P.M. (2002) *Phys. Rev. Lett.*, **88**, 066805.

42 Eiguren, A., Hellsing, B., Chulkov, E.V., and Echenique, P.M. (2003) *J. Electron Spectrosc. Relat. Phenom.*, **129**, 111.

43 Sklyadneva, I.Y., Heid, R., Silkin, V.M., Melzer, A., Bohnen, K.P., Echenique, P.M., Fauster, T., and Chulkov, E.V. (2009) *Phys. Rev. B*, **80**, 045429.

44 Ernst, H.-J., Fabre, F., and Lapujoulade, J. (1992) *Phys. Rev. B*, **46**, 1929.

45 Dürr, H., Wendelken, J.F., and Zuo, J.-K. (1995) *Surf. Sci.*, **328**, L527.

46 Sarría, I., Osma, J., Chulkov, E.V., Pitarke, J.M., and Echenique, P.M. (1999) *Phys. Rev. B*, **60**, 11795.

47 Borisov, A.G., Gauyacq, J.P., and Kazansky, A.K. (2003) *Surf. Sci.*, **540**, 407.

48 Hirschmann, M. and Fauster, T. (2007) *Appl. Phys. A*, **88**, 547.

49 Roth, M., Weinelt, M., Fauster, T., Wahl, P., Schneider, M.A., Diekhöner, L., and Kern, K. (2004) *Appl. Phys. A*, **78**, 155.

50 Roth, M., Pickel, M., Wang, J., Weinelt, M., and Fauster, T. (2002) *Phys. Rev. Lett.*, **88**, 096802.

51 Weinelt, M., Kutschera, M., Fauster, T., and Rohlfing, M. (2004) *Phys. Rev. Lett.*, **92**, 126801.

52 Weinelt, M., Schmidt, A., Pickel, M., and Donath, M. (2007) *Prog. Surf. Sci.*, **82**, 388.

4
Relaxation Dynamics in Image Potential States at Solid Interfaces

James E. Johns, Eric Muller, Matthew L. Strader, Sean Garrett-Roe, and Charles B. Harris

The interface between metals and molecules remains an area of highly active research. The abrupt change in electronic environment between the electron gas of a metal and the covalent bonds of a molecule give rise to new surface states such as charge transfer excitons, metal-induced gap states (MIGS), and deep trap sites [1, 2]. The behavior of electrons at these interfaces dominates both charge injection and charge collection in the growing number of molecular electronic devices. We have studied the dynamics and energetics of these interfacial states, specifically focusing on two particular states, the image potential state (IPS) and molecular based lowest unoccupied molecular orbitals (LUMOs).

The image potential is the attractive potential that a charge feels near the surface of a conductor. An electron at the surface of metal induces a polarization in the metal, which exposes some of the underlying positive ionic cores. The actual shape of this polarization is quite complex inside the metal. Outside the metal, however, the potential is exactly that of a fictitious positive point charge placed inside the metal at a distance from the surface equal to that of the electron. This results in the classical $1/z$ distance-dependent form of the image potential. Quantum mechanically, this image potential can be solved analogously to a one-dimensional hydrogen atom, leading to a Rydberg-like series of bound states that converge to the vacuum level of the system.

Electrons from inside the metal can be either directly or indirectly excited into these states upon exposure to ultraviolet (UV) light. For bare metal surfaces, the average distance from the surface of an electron in the first IPS ($n = 1$, where n is the Rydberg quantum number) is on the order of 3 Å. Energetically, the states are bound by 0.9–0.5 eV below the vacuum level, and decay via direct tunneling in under 50 fs. In the presence of a nonmetallic adsorbate, the IPS is partially screened by electrons in the adsorbate. The IPS can also hybridize with other molecular states at the interface, altering the energy of the state and changing the overlap of the IPS with the metal surface. In systems with strong electron–nuclear coupling such as organic molecules, the presence of an excess external charge can also lead to an induced nuclear motion that can further change the wave function and energy of the IPS. This makes

Dynamics at Solid State Surface and Interfaces Vol. 1: Current Developments
Edited by Uwe Bovensiepen, Hrvoje Petek, and Martin Wolf

ISBN: 978-3-527-40937-2

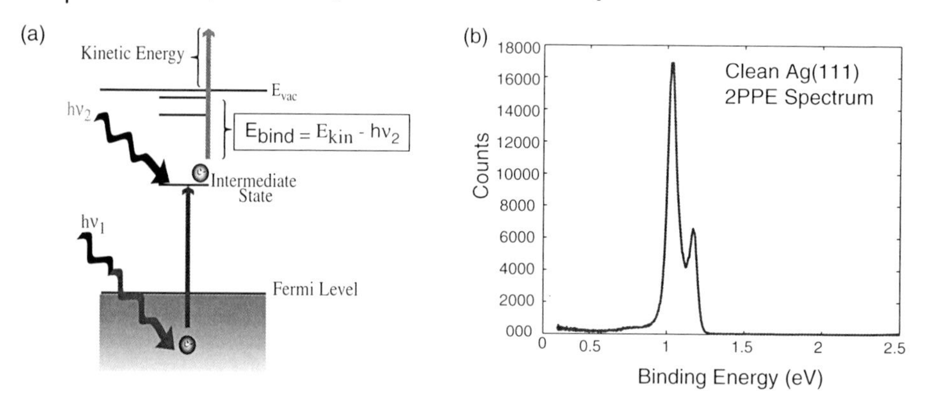

Figure 4.1 (a) Pictorial representation of the 2PPE process. (b) A representative 2PPE spectrum of the clean Ag(111) surface. The lower energy peak at −0.77 eV is the $n = 1$ image potential state, and smaller feature at −0.67 eV is the $n = 0$ surface state.

the image potential state a good experimental probe for the behavior of molecules at metal surfaces when exposed to excess charge.

In order to extract information from these systems, we use time- and angle-resolved two-photon photoemission (2PPE). This is an ultrafast pump-probe technique used to investigate excited state surface electrons, typically of metals or adlayers supported on metal substrates. A laser pump pulse with energy $h\nu_1$ excites an electron from below the Fermi level of the metal to an intermediate state. After some waiting time τ, a probe pulse with energy $h\nu_2$ photoemits excited electrons. The kinetic energy of these electrons is then measured either by measuring flight time to a detector or by using an electron energy analyzer. By subtracting the energy of the probe pulse from the measured kinetic energy spectrum, one can determine the absolute energy level of electronic states relative to the Fermi and vacuum levels. By changing the pump-probe delay time, changes in the energy of electronic states can be monitored. The basic scheme is shown in Figure 4.1 with a representative spectrum of clean Ag(111).

The ability to time-resolve photoemission of excited states is further improved by the ability to momentum resolve states (see Figure 4.2). At a crystal surface, quasi-particles bound in the z-direction (surface normal) still have a two-dimensional band structure, $E(k_x, k_y)$. The momentum of an electron parallel to a well-defined surface has been shown to be preserved during photoemission. Electrons of different momenta parallel to the surface are therefore photoemitted at different angles to the surface normal related by the following equation:

$$k_x = \sqrt{2m_e E_{kin}} \sin\theta \tag{4.1}$$

where k_x is the momentum wave vector, m_e is the mass of an electron, E_{kin} is the kinetic energy, and θ is the angle of the photoemitted electron relative to the surface normal vector [3]. Angle-resolved spectra can be taken either by varying the angle of the sample with respect to a fixed detector with a limited acceptance angle or by using

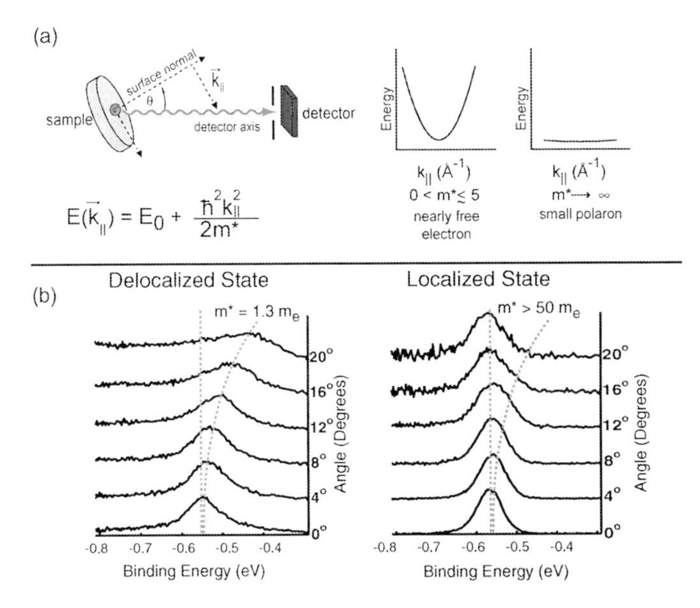

Figure 4.2 (a) As the angle of the sample is changed with respect to the detector, only specific momentum slices make it to the detector. Delocalized or nearly free electron-like particles have a parabolic dependence of the energy versus $k_{//}$. (b) Spectral representation of localized versus delocalized features. Both panels are collections of 2 ML toluene/Ag(111) 2PPE spectra taken at different angles. On the left-hand side, the IPS is delocalized, and its energy increases as the angle increases. On the right-hand side, time has elapsed since the initial excitation, and the IPS has collapsed into a small polaron. The energy of the feature then shows no angular dependence.

an array style detector. For spectra taken where the momentum in the *x*- and *y*-directions is the same, or laser spot size averages over many domains oriented in different directions, the momentum is often referred to as $k_{//}$ since a more specific direction is often experimentally impossible to define. From the measured surface band structure, we can determine the effective mass (m^*) from the equation:

$$m^* = \hbar^2 \left[\frac{\partial^2 E}{\partial k_{//}^2} \right]^{-1} \tag{4.2}$$

For metal-derived surface states, such as bare image potential states or Shockley surface states, the effective mass is approximately 1 and usually less than 5. For systems in which an electron is highly localized, because of either strong material interactions or a lack of crystallinity, the energy of the state does not depend on the momentum, that is, *k* is not a good quantum number, and the effective mass approaches infinity.

The first spectra of IPS electrons on bare metal surfaces were recorded using inverse photoemission in 1984–1985. Giesen *et al.* were the first to study them using 2PPE with nanosecond lasers [4–7], and Schoenlein *et al.* were the first to present time-resolved 2PPE [8], with Steinmann close behind [9]. During this time, IPS energies and lifetimes were interpreted using multiple reflection theory, first developed by Echenique and Pendry [10, 11]. In this theory, bound states are found by bouncing plane waves between the metal surface in a band gap and the image potential. When the accumulated phase shift from reflections off both boundaries equals 2π, a bound state occurs. Proper wave function matching is used to evaluate the phase shifts. The lifetime of the states is then proportional to the overlap of the calculated IPS wave function inside the metal. Once the systems examined involved adsorbates, such phases could no longer be easily calculated.

The next major model for understanding the electron dynamics at metal organic surfaces was the dielectric continuum model (DCM). Within the DCM, molecular adlayers are treated as structureless slabs that screen charges from the metal and build up surface charges at the metal/molecule and molecule/vacuum interfaces. This problem was first solved for the image potential by Cole [12], and the model was first applied to two-photon photoemission results by McNeil *et al.* [13]. This model takes only the low-frequency dielectric constant and electron affinity (here defined as the energy of the conduction band relative to the vacuum level) as inputs. This effective potential shown in Figure 4.3 is inserted into the time-independent Schrödinger equation that must be solved numerically. The lifetime of these states are determined by the penetration of the image potential wave functions into metal and the known lifetime of electrons in the metal at that energy. This model has been successfully used to explain the IPS energetic positioning and lifetime for several adsorbate systems including noble gasses [14–17], linear acenes, and alkanes [18–23].

This model, however, also has several inadequacies. For example, the DCM is a structureless model. As a result, it cannot account for materials with similar physical properties but with differing emergent behavior caused by differing morphological or crystalline properties. For example, our group has studied the dynamics of IPS

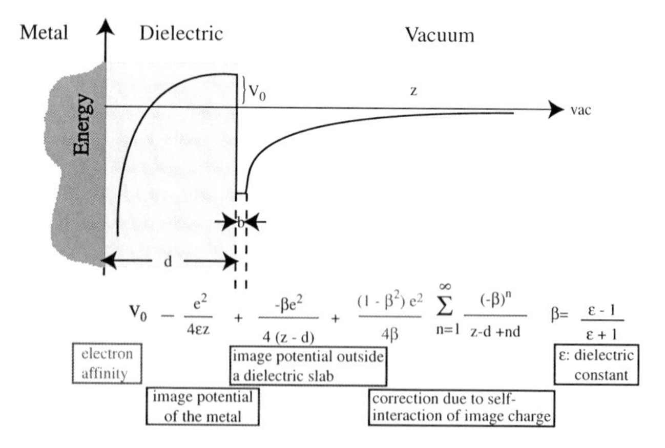

Figure 4.3 Dielectric continuum model potential for the IPS screened by nonconductive adsorbate as solved by Cole [12].

Table 4.1 IPS lifetimes for sample aromatic molecules with similar dielectric properties.

Molecule	Electron affinity	Dielectric constant	Trilayer exponential IPS lifetime (fs)
Benzene	0.5 eV	2.3	50
Toluene	0.4 eV	2.4	200
o-Xylene	0.4 eV	2.6	200
m-Xylene	0.38 eV	2.4	400
p-Xylene	0.5 eV	2.3	1000

electrons using two-photon photoemission of small aromatics physisorbed to Ag(111). Molecules such as benzene, toluene, and *para*-, *ortho*-, and *meta*-xylenes all have very similar electron affinity levels and nearly identical dielectric constants as shown in Table 4.1 [24, 25]. The dielectric continuum model can offer no insight into why the image potential states at these very similar molecular interfaces are so different. As all these molecules share a phenyl core with 0, 1, or 2 methyl substitutions, the difference in IPS lifetimes must be in some manner due to intermolecular forces at work caused by these groups. Furthermore, with the exception of benzene, the dynamics of the IPS in these states are not simple exponential decays as the model would propose. Thus, although it captures some of the basic physical elements for many systems, it remains insufficient as a general model for the interpretations of 2PPE spectra with aromatic adsorbates, and new methods were required.

The most common and current interpretive device for electronic spectra throughout all fields of chemistry and physics is density functional theory (DFT). Solving the well-known Kohn–Sham equations should yield the correct electronic ground state [26], and methods have been developed such as the Green's function approach [27] or time-dependent DFT to deal with excited states [28]. For a variety of reasons, DFT underestimates the electron spill out of the metallic surfaces due to the vanishing bandgap in the bulk and the superimposed surface bandgap. It also fails to accurately yield the correct functional form of the image potential. Progress has been made (and previously described therein) in the creation of one-dimensional potentials that effectively match the work function, image potential, and surface state to known experimental values [29]. Employing the GW approximation to LDA-DFT results in the accurate prediction of electronic energies and lifetimes, as well as breaking down these lifetimes into phonon contributions, electron–hole contributions, and electron–electron contributions. This task, however, is inherently computationally expensive. To date, the most complex system described in this level of detail is a monolayer of cesium over silver adlayers deposited on copper [30]. If, however, we consider a relatively simple system (from a chemist's perspective) such as linear alkanes deposited on a metal, the DFT is computationally far too expensive. One-dimensional models are no longer sufficient, as the lateral extent of the wave function collapses into a small polaron. Furthermore, if each unit cell contains only a single alkane, this still requires several metal atoms per unit cell, each of which drastically increases the computational time required. The computational formulism and power does not yet exist to study metal-induced excited states at surfaces in these systems.

In this chapter, we seek to review our current work that moves beyond the dielectric continuum model, into areas that are computationally too demanding to be tractable. We discuss two intermediate methods that we have recently used to understand and interpret 2PPE spectra of adsorbate films on Ag(111). The first model is a single-parameter phenomenological model for understanding the role of band structure on IPS decay and dynamics. This method was developed and applied for studying the IPS intraband relaxation (electronic cooling) of *p*-xylene/Ag(111) by Garrett-Roe *et al.* [24]. The second model we discuss is a practical model used to relate the dynamics of IPS solvation to the differential capacitance of molecules adsorbed to metal electrodes in electrochemistry, first published by Strader *et al.* [31].

4.1 Stochastic Interpretation of IPS Decay

The decay of electrons in the IPS of molecules adsorbed to metal surfaces has been studied in a wide variety of systems; these fall broadly into two general experimental cases. In the simplest cases, the adsorbate layers have a natural affinity for accommodating excess negative charge, and these are denoted as having a positive electron affinity. In these films, the wave function of the IPS is drawn in toward the metal surface, allowing for a high degree of bulk metal penetration. The electronic lifetime of the image potential state is then directly proportional to the lifetime of a bulk conduction band electron (typically a few femtoseconds) and the IPS shows rapid single exponential decay dynamics. For adsorbates that are not easily polarized and cannot accommodate extra charges such as small nonpolar hydrocarbons, the IPS electron resides on the vacuum side outside of the layer. The reduced overlap of the IPS with the metal resulting from the adsorbate allows relaxation processes that occur on relatively long timescales. Such processes include IPS scattering off of adsorbate phonons, adsorbate electrons, and electron–hole pairs created in the metal. Competition between these processes and tunneling back to the metal result in complex dynamics that strongly depend on the electrons' momentum parallel to the surface. Due to the difficulty in calculating the rates for each of these processes in any but the simplest of systems, we have searched for the simplest, phenomenological description of electronic relaxation of IPS electrons. Our results have led us to conclude that decay of these electrons can be modeled surprisingly well by the simple explanation of stochastic friction. A cursory overview of the model follows below. A more detailed description and explanation is given in Garrett-Roe *et al.* [24].

The basics of the model are quite simple. IPS electrons are treated as bound particles in the direction normal to the surface, and as nearly free electrons parallel to the surface, with definite momenta parallel to the surface given by $\hbar\mathbf{k}_{//}$. The band structure of the system is nearly free electron-like with a parabolic dispersion relation given by

$$E(\mathbf{k}_{//}) = \frac{\hbar^2 k^2}{2m^*} \tag{4.3}$$

where m^* is the standard effective mass of an electron in the band. In effect, this defines a parabolic potential energy surface in k-space. In the absence of any other forces, an electron would stay in its given k-state indefinitely. The IPS electron, however, is coupled to a bath of electronic and phonon modes of the adsorbate/metal system. The effect of this coupling is to slow the velocity of the electron parallel to the surface by dissipating energy to the bath. This deceleration of the electron in real space is equivalent to dragging the electron to the bottom of the band in k-space. Rather than attempting to calculate the individual mechanism of IPS/bath coupling, we will use the simplest model, stochastic friction. Coupling to the bath results in a random walk of the electron and gives rise to a linear frictional force

$$F_{\text{friction}} = -\Gamma_x \times \dot{x} \tag{4.4}$$

where Γ_x is the coefficient of friction in real space and $\dot{x}$ is the electron's velocity, as well as diffusional motion related to friction and the temperature of the ensemble:

$$D_x = \frac{k_b T}{\Gamma_x} \tag{4.5}$$

where D_x is the diffusion constant, k_b is Boltzmann's constant, and T is the absolute temperature of the bath. Simple physical considerations and unit analysis show that the friction in real space is inversely proportional to the friction in k-space:

$$\Gamma_k = \frac{\hbar^2}{\Gamma_x} \tag{4.6}$$

In our model then, the band structure of the IPS is like a harmonic potential energy, and the friction is like an overdamped restoring force. For a given initial distribution of overdamped oscillators, the population dynamics $p(k_x,t)$ has an analytic solution provided that the friction is Markovian (i.e., has no memory) [32]. Using the appropriate Green function for a momentum distribution,

$$G(k_x, t|k_{x0}, t_0) = \sqrt{\frac{1}{2\pi\sigma_{kx}^2(t)}}\, e^{\frac{-(k_x - \langle k_x\rangle(t))^2}{2\sigma_{kx}^2(t)}} \tag{4.7}$$

The population as a function of time and initial distribution is given by

$$p(k_x, t|k_0, t_0) = \int_{-\infty}^{\infty} G(k_x, t|k_{x0}, t_0) p(k_0, t_0) \mathrm{d}k_{x0} \tag{4.8}$$

where σ_{kx} is the time-dependent width of the temperature-dependent propagator and $\langle k_x \rangle(t)$ is the time-dependent average of the distribution. These can be written as

$$\sigma_{kx}^2(t) = \frac{D_{kx}(1-e^{-2B_{kx}(t-t_0)})}{B_{kx}} \tag{4.9}$$

$$\langle k_x \rangle(t) = k_{x0} e^{-Bk_x(t-t_0)} \tag{4.10}$$

Here, the D_{kx} is the diffusion constant in k-space of the electron and is given by

$$D_{kx} = \frac{k_b T}{\Gamma_{kx}} \tag{4.11}$$

and B_{kx} is simply a measure of the force dragging an electron to the bottom of the band and depends on the curvature of the band and the frictional constant as given below.

$$B_{kx} = \frac{\hbar^2}{m^* \Gamma_{kx}} \tag{4.12}$$

At this point, the physical picture of the solution can be adequately described. At time $t = 0$, a population of excited electronic states is created. This population can be experimentally determined via angle-resolved two-photon photoemission at pump-probe, temporal overlap. As time progresses, the high momentum IPS electrons exchange energy with the bath (electronic and phonon modes of the adsorbate and/or metal), dissipating energy and drifting to the bottom of the band. Due to the random nature of thermal fluctuations, however, the probability distribution does not collapse to $p(k_x, t = \infty) = \delta(0)$, but rather to a Boltzmann distribution of electron velocities. Extending this one-dimensional model to a 2D model in cylindrical coordinates as appropriate to electrons at a surface is an algebraic matter and the details are shown in the original publications [24, 32].

By construction, this model for intraband relaxation obeys the fluctuation dissipation theorem; however, as stated it does not allow actual population decay back into the metal. Electron tunneling dynamics can be simply accounted for by a first-order correction to the propagator, leading to exponential decay of the population in each state [33]. The resultant and final equation governing the kinetics is then

$$p(k_x, k_y, t | k_{x0}, k_{y0}, t_0) = \int \int_{-\infty}^{\infty} \mathrm{d}k_{x0}\, \mathrm{d}k_{y0} \left[G(k_x, k_y, t | k_{x0}, k_{y0}, t_0) - \varkappa_0(k_x, k_y) \right] p(k_{x0}, k_{y0}, t_0) \tag{4.13}$$

This equation can then be numerically propagated if the initial population distribution is known from two-photon photoemission measurements. In this formulism, the decay rate back to the metal is approximated as constant. Detailed work has shown that the tunneling rate is affected by the energetic proximity to the conduction band edge [17, 34]. At higher momenta, there is an increased probability of an electron scattering into the conduction band. Unfortunately, calculating this dependence for molecular adsorbates containing several atoms is computationally unfeasible at an atomistic level and gives incorrect values at the dielectric continuum model level. So, here we have neglected the dependence of the tunneling rate upon parallel momentum.

This model was first presented in the study of IPS decay at the p-xylene/Ag(111) interface. Shown below is the two-photon photoemission spectrum for several layers

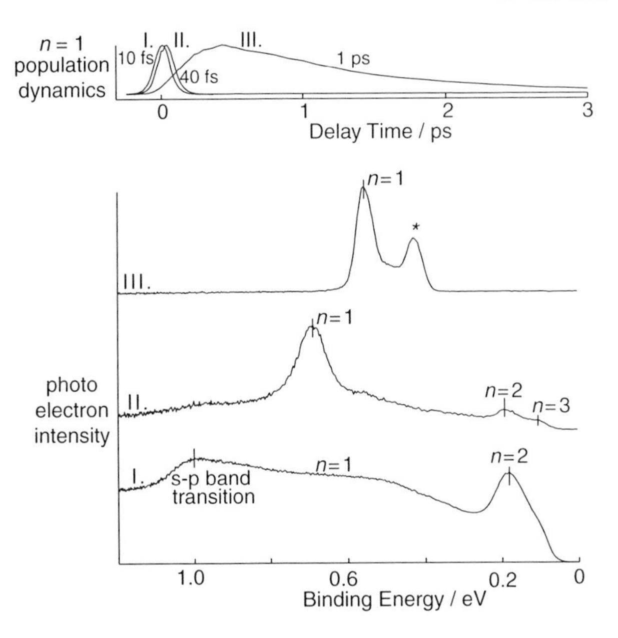

Figure 4.4 2PPE spectra of *p*-xylene/Ag(111) for increasing coverages. I, II, and III are one, two, and three monolayers, respectively. The asterisked feature for the thickest coverage is possibly a thicker coverage or different morphology, but is not dealt with in this chapter. The top panel shows the population dynamics for the $n = 1$ IPS state as a function of *p*-xylene coverage. Reprinted with permission from [24]. Copyright 2005 American Chemical Society.

of *p*-xylene on Ag(111) taken with a pump energy of 3.6 eV and a probe energy of 1.8 eV (Figure 4.4). All binding energies are referenced to the vacuum layer. Layer by layer growth of *p*-xylene/Ag(111) was achieved by careful control of the substrate temperature during dosing. Monolayer determinations were made by changes in the 2PPE spectrum. Upon formation of the first monolayer, there is a broad transition arising from the bulk s–p band in the metal [35], a broad $n = 1$ transition, and a sharper $n = 2$ feature. The breadth of these features is likely due to overlap of the $n = 1$ IPS state with the electron affinity of *p*-xylene [23]. The second monolayer shows a hydrogenic progression of image states for $n = 1$, 2, and 3 with a work function shift of −0.74 eV. Lifetimes for the $n = 1$ IPS are 10 and 40 fs for the first and second monolayers, respectively, and are near the lower end of times resolvable by our experimental apparatus. These times were obtained by fitting the intensity of the features to a single exponential decay, exponential rise, and convoluted with an instrument response function as obtained from spectra of clean silver.

After adsorption of additional *p*-xylene, however, the lifetime of the image potential state increases to 1.2 ps, nearly a 100-fold increase. Despite the positive (i.e., attractive) electron affinity of *p*-xylene, multilayers of *p*-xylene at the Ag(111) surface are acting as a dielectric barrier, pushing the $n = 1$ IPS wave function away from the surface and into the vacuum. This behavior is completely unexplained by the dielectric continuum model. Furthermore, the lifetime of the $n = 1$ state carries a strong momentum dependence. For low parallel momentum values, less than 100 pm^{-1}, there is a measurable rise time to the kinetics, starting at 300 fs for $k_{//} = 0\ pm^{-1}$, and decreasing to 0 with increasing momentum. Also, the net decay rate of the $n = 1$ signal decreases from 1200 fs at 0 pm^{-1} to 50 fs by 250 pm^{-1}. The full angle-resolved dynamics are presented in Figure 4.5.

This momentum-dependent relaxation is exactly the predicted behavior for intraband relaxation, and similar to the dynamics seen on N_2/Xe/Cu(111) [36]. Electrons at high parallel momenta scatter rapidly down the band to lower momentum states. Electrons pile up at low momenta leading to an observed rise time near the bottom of the band, and the decay at the well bottom is long and governed by the tunneling rate for electrons back into the metal.

In Figure 4.6, we present the fit of normalized angle-resolved dynamical populations to those predicted from our simple stochastic method at 100 K with a best fit friction coefficient of $\Gamma_x = 1200 \pm 500$ eV Å^2 fs, or 6×10^{-18} kg/s. The data are in excellent agreement with the simulations. The simulations are able to reproduce the rise time at low momenta, biexponential behavior around 80 pm^{-1} due to electronic cooling, and the short lifetimes of the higher momenta states. The higher momenta states, however, do show a discrepancy with the simulation. The experimental states still show a significant finite rise time, whereas the simulations show only decay. This can be explained by the size of the simulation. The simulation was performed only using a box size of 250 pm^{-1}. As a result, there are few high momenta electrons above 200 pm^{-1} in the simulation to scatter into these states. Nevertheless, the good agreement between the measured dynamics and the simulations suggest that random stochastic fluctuations based on a friction style formulism is a proper frame for thinking about electron decay rates in these complex systems.

One drawback of this method, however, is that it does not suggest any physical basis for the interpretation of the magnitude of friction coefficient. Echenique and coworkers have modeled electron decay rates for a variety of systems with atomistic detail using the GW and GW + T approximations in combination with plane wave density functional theory methods [37–46]. In their studies, they were able to break down the various decay pathways into their quasi-particle component parts. For bare surfaces and atomic adsorbate-covered surfaces, they deconstructed the inter- and intraband components into direct electron–hole scattering, electron–phonon scattering, and electron–electron scattering.

For systems with more complicated adsorbates, the work by Hotzel *et al.* remains the closest to this work. In their studies, they measured and modeled the decay of IPS electrons on the N_2/Xe/Cu(111) surface [36]. They too suggested an analogue to friction in their analysis of the experimental kinetics. In their modeling, however, they modeled only the dissipation and left no room for thermal fluctuations, resulting in a minimum dissipative energy of 6 ± 2 meV. This minimum dissipative energy led to the conclusion that librational motion of the N_2 molecules creates the frictional drag force on the IPS electrons. We suggest that the minimum dissipative energy of 6 ± 2 meV results from the finite thermal distribution of momenta caused by thermal fluctuations in the local electric field.

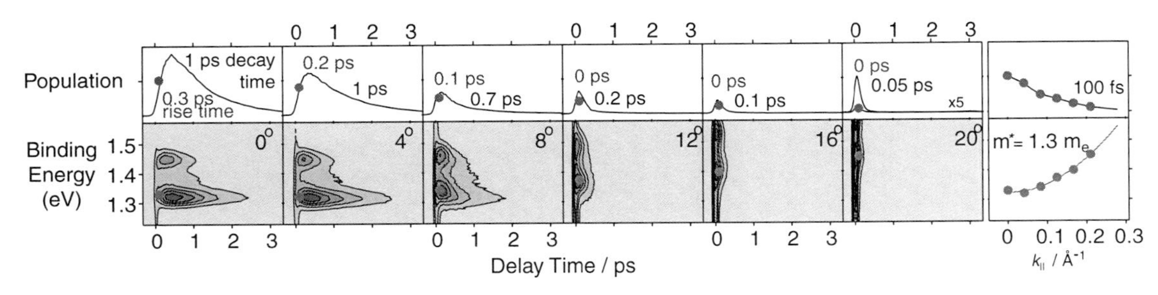

Figure 4.5 Angle-resolved dynamic scans of 3 ML of *p*-xylene/Ag(111). Dynamic population traces for each angle are given in the top panels, and contour plots of the kinetic energy versus time for each angle are given in the bottom panels. The bottom right panel shows the dispersion and effective mass that remain constant at all times, and the top right panel shows the 100 fs population distribution versus $k_{//}$. The red dots correspond to the main peak's center energy and the blue dots are the population after 100 fs. Reprinted with permission from [24]. Copyright 2005 American Chemical Society. (Please find a color version of this figure on the color plates.)

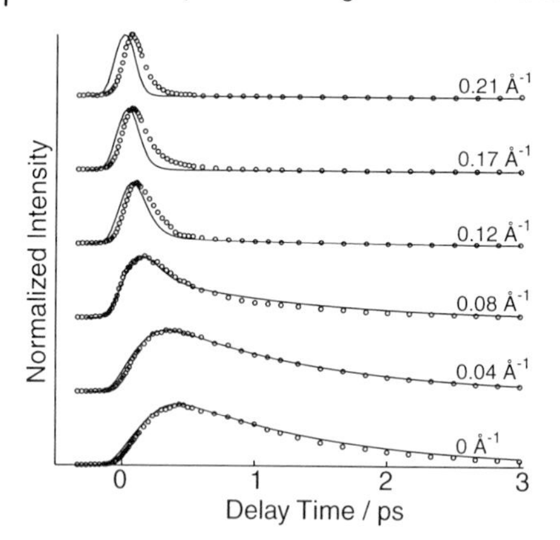

Figure 4.6 Normalized population for the $n=1$ IPS of 3 ML of *p*-xylene/Ag(111). Open circles are the experimental data, and the solid line is the best fit to the 2D friction model. Reprinted with permission from [24]. Copyright 2005 American Chemical Society.

In an attempt to explain the source of the frictional force, we compared the same two mechanisms for electronic friction as Hotzel *et al.* An electron traveling parallel to a metal surface can interact with collective surface excitations, that is, surface plasmons, provided that the surface plasmons are damped [47]. This is the same friction that a charge would experience in the metal due to the metal's finite conductivity [48–51]. The associated friction depends on the cube of the electrons' distance to the metal surface, z, and the conductivity of the metal at experimental temperatures, σ:

$$\Gamma_{\text{plasmon},x} = \frac{(Ze)^2}{4\pi\varepsilon_0 \cdot z^3 \cdot 16\pi\sigma} \tag{4.14}$$

Plugging in estimated values of $z=3\,\text{Å}$ and $\sigma=2.4\times10^6\ (\Omega\,\text{cm})^{-1}$ (at the experimental temperature of 100 K) [52] yields a frictional value of $1\times10^{-19}\,\text{kg/s}$. This is over an order of magnitude too low and represents an upper bound for plasmonic friction. The $z=3\,\text{Å}$ used is the expectation value for the position of $n=1$ on a bare metal surface. Since the *p*-xylene is acting as an electronic barrier, the actual value for z should be larger, making the friction due to surface plasmons multiple orders of magnitude too low to account for the intraband relaxation observed, as found for N_2/Xe/Cu(111).

The other metal-derived pathway for intraband relaxation is the creation of electron–hole pairs inside the metal. Inefficiently screened IPS electrons near the metal surface can lose energy by creating electron–hole pairs in the metal, and the substrate can donate energy by scattering thermally excited electron–hole pairs into the IPS electron. To estimate this, we examined the asymptotic limit of the friction due to interband single-electron excitations as reported by Ferrell and Nuñez [47, 48]:

$$\Gamma_{\text{electron-hole},x} = \frac{3m^*(Ze)^2\ln(k_{\text{TF}}z/1.4475)}{8\pi^2\varepsilon_0(k_{\text{TF}}z)^4} \tag{4.15}$$

where k_{TF} is the Thomas–Fermi screening wave vector. Using a value of k_{TF} appropriate for the Ag(111) surface (1.71 Å^{-1}) [53] and $z = 3$ Å results in a friction value of $\Gamma_{\text{electron–hole}, x} = 3.3 \times 10^{-17}$ kg/s. This is approximately five times larger than the measured friction. However, Thomas–Fermi screening requires a slowly varying potential over the length scale of several $1/k_{TF}$, which does not apply to an image potential electron at 3 Å. Integrating the equations provided by Nuñez *et al.* leads to a slightly larger value for the plasmonic friction and a slightly smaller value for the electron–hole contribution. By increasing the average distance from the surface to 5 Å, the integrated electron–hole friction drops to 4.9×10^{-18} kg/s, a number in surprisingly good agreement with the measured value of 6×10^{-18} kg/s given the simplicity of the models. This model can be applied to any nearly free electron-like system in which there is a strong separation of timescales for the net population decay and the fluctuating electric fields leading to intraband decay. The magnitude of the friction/diffusion constant can then be compared to theoretical mechanism for the intraband relaxations in cases where the electron–hole friction may not be dominant.

4.2 IPS Decay for Solvating Molecules

Solvation is the process whereby solvent molecules interact with a solute to energetically stabilize the system. The role of solvents and the solvation of excited states is crucial in many charge transfer and biological reactions. Despite the plethora of reactions that show a strong solvent dependence, the ability to monitor solvation in real time required the advent of ultrafast lasers. The first ultrafast experimental studies monitored the dynamic Stokes shift (the spectral difference between the onset of absorption and the onset of fluorescence) of a dye molecule in a solvent. These experiments observed the relaxation of the excited state molecular solute, which was dominated by reorientation of the surrounding first coordination shell of solvent molecules. The relaxation mechanism of these systems shed further insight into the behavior of liquid solvents in general via the fluctuation dissipation theorem. Excellent reviews on the subject can be found in the work of Fleming, Maroncelli, and Barbara [54–59].

Analogously to these studies, two-photon photoemission of the image potential state can show how adsorbed molecules on a metal surface react to excess charges, and a change in surface dipole. As an electron is promoted into the IPS, it creates an excited dipole between itself and the image charge, separated by a large distance. Any dynamic changes in the measured energy of the IPS electron must come from a molecular response of an adsorbate since the IPS electron has zero internal degrees of freedom. In this picture then, the IPS electron plays the role of a perfect solute and the adsorbate plays the role of a solvent that can reorient itself to lower the IPS electron's energy. Monitoring the change in energy as a function of time and temperature ($E(t,T)$) as the energy of the IPS is dissipated should then provide information on the thermal, fluctional motion of the molecules themselves.

Time-dependent IPS electronic energies were first reported by Ge *et al.* in the dynamic formation of a small polaron in bilayers of *n*-heptane on Ag(111) [60]. In this work, they discovered that although the IPS for 2 ML of *n*-heptane/Ag(111) is initially delocalized, the heptane molecules respond in a dynamic fashion to the excess charge and trap the electron in a localized final state. The energetic driving force for electron localization in this system is a stabilization of the small polaron over the delocalized state by approximately 10–20 meV. Ge *et al.* used the formal equivalency between small polaron formation and Marcus electron transfer theory to deduce that the 750 cm^{-1} methyl rocking motion of the alkane is the primary mode responsible for electron localization in these types of systems [60]. Similar dynamic polaron formation was observed for many straight-chained alkanes and cycloalkanes.

Although these systems have a small energy shift concurrent with electron localization, they do not represent the traditional, chemical view of solvation, meaning a large-scale reorientation of molecules to lower electrostatic energy. The first investigations into true solvation were published in 2002 by Gahl, Bovensiepen, and Wolf in a series of papers the solvation of electrons in various ice layers and by Miller, Liu, and Harris on the solvation of IPS electrons in various alcohols on silver [61–63]. In these studies, the kinetic energy of the electrons became truly dynamical quantities, not only changing during a localization event but also continuing to change for up to several picoseconds with solvation energies much greater than the thermal energy, exceeding 300 meV.

Liu *et al.* interpreted their results in terms of a locally changing work function. For an IPS with an alcohol adsorbate, it was concluded that rotation of the O−H bond is responsible for the observed solvation. As the bond rotates, it lowers the local work function resulting in a shift in the observed kinetic energy of electron. Bovensiepen and Wolf interpreted their data in terms of successive reorganization of neighboring hydrogen bond networks. Their work made a substantial contribution by discovering that localized electrons could have an observed slightly negative effective mass rather than the traditional flat band. A broad, Gaussian-shaped feature in the energy spectrum can be the result of strong inhomogeneous broadening of localized electrons in different sites. If different components of this distribution decay with different rates, a net apparent negative effective mass will grow over time. Furthermore, as they showed in H_2O/Cu(111), the same mechanism can lead to an apparent peak shift in energy that resembles solvation but is, in fact, inhomogeneous electron decay from a distribution of high and low energy electrons [62]. In order to be assured that solvation is occurring, the energy of a peak must have an energetic shift larger than the width of the original peak envelope.

In more recent solvation work, Zhu *et al.* collaborated with Wolf and Bovensiepen and discovered highly trapped localized electrons at specific bonding sites, which show continuous solvation dynamics on the timescales of minutes [64]. Sebastian *et al.* began their theoretical treatments on simultaneous solvation and localization in a deformable medium [65, 66]. Petek and coworkers made the first 2PPE studies of electron solvation in an alcohol on a semiconductor surface rather than a metallic one by studying MeOH/TiO_2 [67]. Below, we present a summary of our most recent work on electron solvation, a study of the differential capacitance of DMSO/Ag(111) with two-photon photoemission [31].

Dimethylsulfoxide (DMSO) is a small, highly polar organic molecule commonly used as an electrochemical solvent. In electrochemistry, as the applied potential on an immersed, metal electrode is varied, a polarization accumulates at the surface of the electrode to offset the potential. For small polar molecules, this results in a region surrounding the electrode with an excess of highly oriented dipoles. This potential-dependent accumulation of charge is termed the differential capacitance of the solvent, and is an important quantity for making precise electrochemical measurements. DMSO is known to have an abnormally low differential capacitance at noble metal electrodes compared to other small polar solvents such as acetonitrile [68]. Both Schröter *et al.* and Si *et al.* have proposed that low differential capacitance can be explained by the way DMSO binds to noble surfaces [69, 70]. Shown in Figure 4.7 is the proposed binding mechanism for DMSO to the Ag(111) surface. For the first monolayer, the DMSO is tethered to the surface by both the metal oxygen and metal sulfur bonds. Their hypothesis was that if DMSO is incapable of rotating to accommodate charges in the metal by these two strong bonds, then this would explain the lack of net dipole rotation and thus the low differential capacitance.

As shown in the preceding section, IPS electrons make excellent probes of a molecule's response to excess charge. If the monolayer and multilayers of DMSO

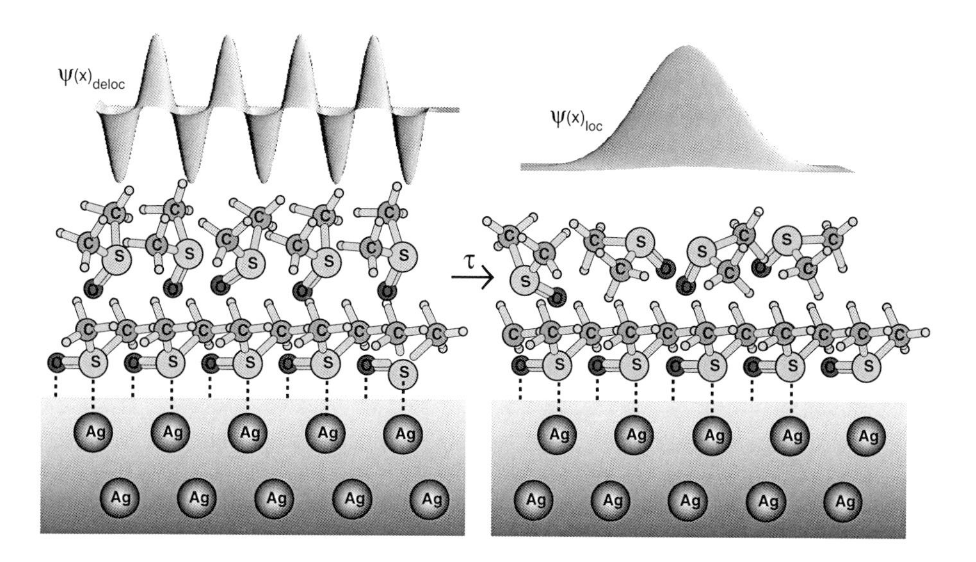

Figure 4.7 Proposed solvation and localization scheme for DMSO/Ag(111). The left panel shows the proposed equilibrium geometry of DMSO/Ag(111). The first monolayer of DMSO is tethered to the surface via bonding to the oxygen and sulfur. The initial IPS wave function is delocalized. After several hundred femtoseconds (right panel) the second layer of DMSO has rotated to expose its most positively charge areas to the IPS electron. This leads to a wave function collapse and localization.

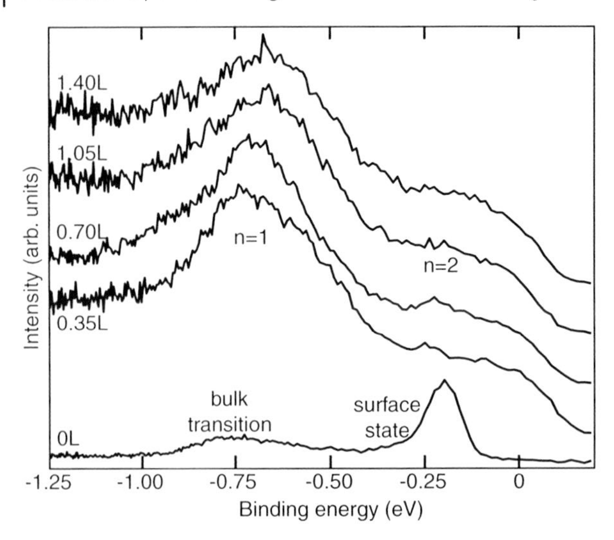

Figure 4.8 2PPE Spectra of DMSO/Ag(111) of increasing coverages. Reprinted with permission from [31]. Copyright 2008 American Chemical Society.

show remarkably different dynamic energy shifts, then this would lend support to the theory of reduced differential capacitance due to hindered rotation. The results for these experiments are presented in more detail in Ref. [31].

Mono- and multilayers of DMSO were successfully grown on a clean Ag(111) surface. Shown in Figure 4.8 are the two-photon photoemission spectra of increasing coverages of DMSO taken with a pump energy of 3.4 eV and a probe energy of 1.7 eV. This film was grown at 210 K and annealed to prevent the adsorption of multilayer peaks in accordance with other studies in the literature. Upon adsorption of DMSO, the bulk s–p transition and the Ag(111) Shockley-type surface state disappears, and the monolayer IPS $n = 1$ and $n = 2$ states appear at -0.7 eV and -0.2 eV, respectively, relative to the vacuum. This is accompanied by a large work function decrease of 1.2 ± 0.03 eV. This spectrum saturates after 0.7 Langmuir (L) of exposure at 210 K and was determined to be 1 ML. Increased exposure to DMSO below 200 K led to the formation of multilayer DMSO, with an average exposure of 1 ML per 0.5 L exposure as determined by two-photon photoemission. There was no evidence of layer-by-layer growth, so layer thickness should be considered as an average value.

In order to study the solvation response, we investigated the energy of the $n = 1$ IPS electrons as a function of angle, pump-probe delay time, and coverage. 2PPE spectra were taken at angles ranging from 0 to 24°, and at time delays from -333 fs to a few picoseconds. Shown in Figure 4.9 is a sample of the experimental data that allow us to draw several conclusions. In the first panel, the 1 ML, $n = 1$ IPS is visible with a binding energy of -0.7 eV that decreases slightly as a function of time. In panel two, we see the 2 ML, $n = 1$ IPS with a greatly enhanced lifetime and net solvation energy. In the third panel, the 1 ML, $n = 1$ spectra are shown collected at an angle of 24°. It is at a higher energy than the 0° spectrum, and indeed the 1 ML shows no evidence of any localization of the electron at any time point. There is a slight enhancement of the

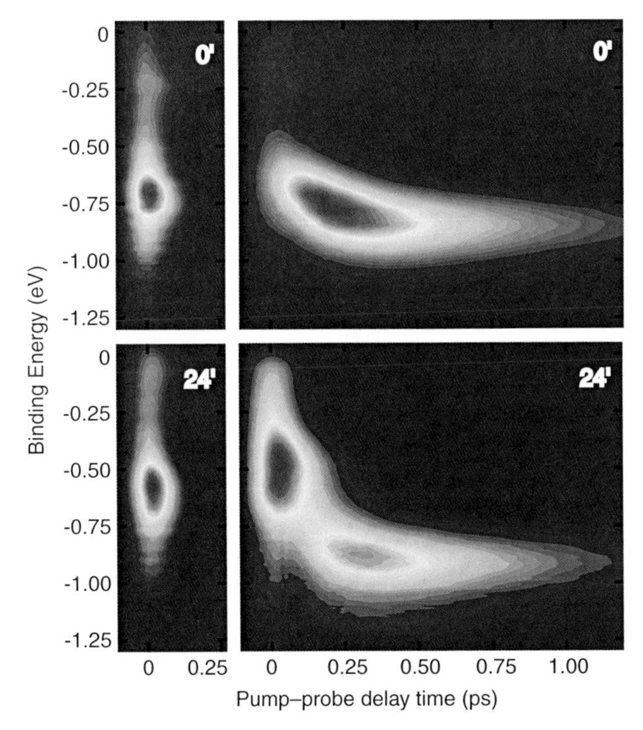

Figure 4.9 Dynamics of the IPS for mono- and multilayers of DMSO/Ag(111). The left half shows time-resolved spectra for the monolayer at 0 and 24′. The right half shows the spectra for multilayers at the same angles. Reprinted with permission from [31]. Copyright 2008 American Chemical Society.

effective mass, but the peak remains delocalized. In the fourth panel, the 24° spectrum of 2 ML is provided. The IPS electron is initially in a delocalized sate with an effective mass of 1.2 m_e. Within 120 fs, the IPS electron induces a response in the DMSO causing it to self-trap into the same state shown at long times at 0°. This state then continuously solvates for another 500 fs.

The lifetimes of the $n = 1$ IPS were measured for one, two, and three monolayers. For the first monolayer, the dynamics followed a simple exponential decay with a lifetime of 48 fs at 0°. Determining the lifetime of the second monolayer was slightly more difficult because there was no discernible difference between a localized and delocalized feature for collection angles less than 12°. In order to determine the lifetime, an exponential rise and decay time were fit for the localized feature at high angles. A single decay constant of 440 ± 40 fs was able to fit the decay of the long-lived feature at every angle, providing further evidence that this is a dynamically formed, localized state that comprises a superposition of k-states. The decay of electrons in the third monolayer IPS were biexponential, with one time constant of 440 ± 40 fs, and a secondary decay time of 5600 ± 500 fs. These increasing decay times indicate that the IPS electron is being pushed to the outside of the layer.

The solvation response of the IPS, $S(t)$, was measured as the change in energy of the IPS peak from its initial value at pump–probe overlap. The solvation response for

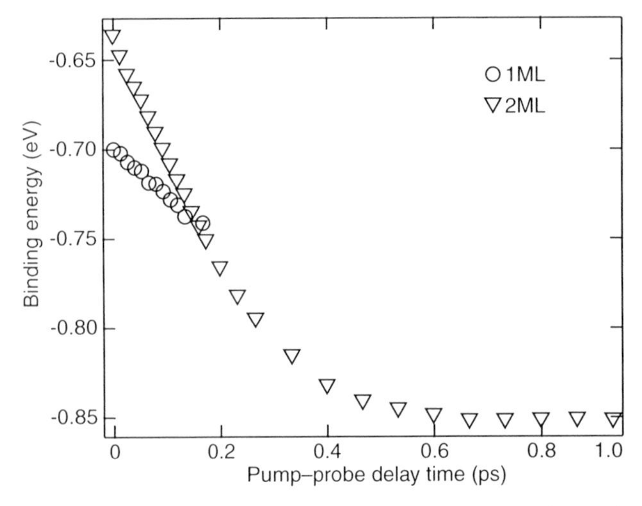

Figure 4.10 IPS peak energies over time for one and two monolayers of DMSO/Ag(111). Reprinted with permission from [31]. Copyright 2008 American Chemical Society.

one and two monolayers of DMSO is shown in Figure 4.10. The typical functional form of solvation in liquids is a half Gaussian representing the inertial solvation with exponential decays for each additional process that contributes to the net solvation. For the first monolayer of DMSO, the solvation is poorly characterized by this form, and a line was used to measure the change in IPS energy versus time. Since there was no localized peak for the first monolayer, this slope was measured for the delocalized peak. Values ranged from -0.25 eV/ps at 0° to -0.32 eV/ps at 24°. Despite the small mass enhancement this implies, the zero degree solvation is used for comparison to the 2 ML localized state because the wave function for the localized state is a superposition of k-states centered around $k_{//} = 0$ Å. The net solvation observed before tunneling back into the metal was 50 meV for the first monolayer at 0°.

The second ML showed a remarkable stronger solvation response. The solvation response of 2 ML DMSO fit remarkably well to an exponential relaxation time of 140 ± 20 fs. The data were also fit to the Gaussian plus exponential model, but the fit was indistinguishable within the error of the experiment. The total magnitude of the solvation at 0° was 220 ± 10 meV. A direct comparison between the total solvation between 1 and 2 ML of DMSO is nonsensical because of the significantly shorter lifetime of an IPS electron above a single monolayer. However, one can assign an upper bound to the solvation for 1 ML by noting that the solvation rate should decrease over time. By 420 fs, 95% of the solvation for 2 ML is finished. If we extend the initial linear solvation rate of 1 ML as an estimate of what the solvation would be, we can establish a 110 meV upper bound for the single-monolayer solvation response. That the solvation energy response is strongly suppressed in the monolayer over the multilayer supports the claims of Schröter and Si that the bonding of DMSO prevents molecular rotation in the first monolayer, thus limiting the ability of the DMSO to support an interfacial charge.

The impetus for this work was to understand a peculiarity in electrochemical capacitance measurements, but that in turn suggests a heuristic model for thinking about electron solvation by image potential states. An electron promoted from a metal surface into the image potential state creates a situation in which there is a positive charge (the image charge), a dielectric (either vacuum or molecular), and a negative charge in the IPS electron. If the electrostatics are such that the molecular medium repels the IPS electron to the vacuum interface, then the geometry is similar to that of a parallel plate capacitor. The capacitance, or charge stored per unit volume, can be cast as

$$C(t) = \varepsilon_0 \varepsilon(t) \frac{A}{d} \tag{4.16}$$

where C is the capacitance, ε_0 is the permittivity of free space, ε is the dielectric function of the adsorbate, A is the lateral spatial extent of the IPS electron and its image charge, and d is the separation of the IPS electron from the metal surface. For a delocalized electron, the area of an IPS electron has been estimated at approximately 7500 Å^2 using a disk of diameter equal to the decoherence length of a 4 eV surface electron of 100 Å [61]. For localized electrons, an estimation of the electron size can be made by Fourier transforming the intensity of the 2PPE spectrum as a function of $k_{//}$ as shown by Bezel *et al.* [71]. Since capacitance is the charge stored as a function of the electrical potential (here measured as the energy above the Fermi level of the IPS), and the charge is known, one can deduce the dielectric function $\varepsilon(t)$ by measuring the solvation energy over time. For this system, the area corrected molecular contribution to the dielectric constants for 1 and 2 ML were 0.034–0.091 and 26, respectively.

Obviously, this is pedagogical model rather than a quantitative one. For instance, in a parallel plate capacitor, the potential varies linearly from one plate to the next, whereas for the image potential state, the potential is known to vary as $1/z$. Nevertheless, the concept of the image potential capacitance provides a connection to other relevant studies. This dielectric function measured on an ultrathin film can then be compared with the dielectric relaxation measured by other ultrafast techniques such as time-resolved Stokes shift of fluorophores in various solvents. In a dynamic Stokes shift experiment, a rigid fluorophore is optically excited in liquid, creating a nonequilibrium dipole. As time evolves, the solvent fluctuations around the excited molecule lead to a reorganization of the solvent shell, which lowers the energy of the excited state relative to the ground state, creating a redshifting in the emission spectrum. The primary results of solvation theory is that the dynamical dielectric response is composed of two parts; the initial response of the solvent is composed of a reversible Gaussian decay as the first solvation shell responds, and finds a new equilibrium position, and a longer timescale exponential decay as molecules translate and undergo large-scale movement. The short timescale Gaussian, also called the inertial response, typically consists of small amplitude motions that reorient and rotate the molecular dipole of the first solvent shell in response to the new dipole of the fluorophore [56].

This response is exactly the proposed mechanism for DMSO solvation of the IPS, and the timescale for solvation should be similar to the inertial response of DMSO in the bulk liquid. For example, Horng *et al.* measured two time constants for the dielectric relaxation of coumarin 153, one at 214 fs, and one at 10.7 ps [54]. The shorter timescale corresponds closely to the two-photon photoemission solvation timescale of 140 ± 20 fs. Similar agreement has been found for studies of the IPS electron in multilayers of acetonitrile (MeCN) on Ag(111) with the dynamical Stokes shift of coumarin 153 in MeCN despite experimental differences in temperature, dimensionality, and solute. Barbara and coworkers [59] were the first to present fluorescence upconversion data on the dynamical Stokes shift of coumarin 311 and coumarin 102 in aprotic solvents such as MeCN. Singh *et al.* [72] reported more recently on the solvation of coumarin 151 in MeCN with faster time resolution. Singh measured the dynamic Stokes shift and fit it to a biexponential with time constants of 390 fs and 1100 fs. Szymanski and Harris have also measured the Gaussian component of IPS solvation for approximately 3 ML of MeCN/Ag(111) at 250 ± 20 fs. While the agreement is not exact, the similarity of the timescales suggest that the same basic mechanism is at work, namely, small reorientation of the adsorbate (solvent) to accommodate a change in dipole due to IPS (fluorophore) excitation.

It is important to note that the dynamic dielectric function measured in the two-photon photoemission is comprised only of molecular motion. The electronic dielectric component is too fast to monitor with this experiment, but it does play a role in the initial energy of the IPS state through electronic screening. Translation-derived components of the dielectric function, which can be important in electrochemistry and measured via other experiments, are here irresolvable both because they occur on slower timescales and because two-photon photoemission is usually performed at cryogenic temperatures to enhance spectral resolution. This solvation timescale corresponds to the ultrafast solvent inertial relaxation that is determined by the displacement of the equilibrium orientation of the system from the excited state.

Here, we have created a model to understand the dynamics of electron solvation at metal molecule interfaces, which elucidates the connection between 2PPE results with the broader communities of generalized ultrafast spectroscopy and electrochemistry. We have provided strong evidence that the solvation of the IPS at the interface between DMSO and silver is hindered by the molecule's inability to easily rotate its dipole in the monolayer regime. This corroborates the arguments of Schröter and Si that the electrochemical interfacial capacitance of DMSO at electrodes is low because of the tethered nature of its bond to the metal surface.

4.3 Conclusions

In this chapter, we have provided a brief review of the dynamics and solvation response of IPS electrons at metal/organic interfaces. Multiple reflection theory and the dielectric continuum model provided the first insights into the decay pathways for IPS electrons. Many of the underlying concepts are still crucial to how we think about these systems, including the round-trip time between the image potential and the metal surface, dielectric screening of the image potential leading to a change in energy, and wave function overlap with the metal. Nevertheless, these simple models are insufficient to explain the behavior of more complex systems that more closely resemble the applied world of electrochemistry or material and device chemistry/physics. We have presented here two systems in which the DCM has failed, and full DFT calculations remain intractable. In the toluene/Ag(111) system, the lifetime of the electrons tends in the opposite direction of that predicted by DCM theory with increasing coverage. We have constructed a generalized stochastic friction-based model to study the thermal redistribution of IPS electrons. Based on this model, we have assigned IPS scattering from electron–hole pairs created in the metal as the primary source of the electronic friction in this system. We have also presented the solvation of the IPS on DMSO/Ag(111). This small, aprotic solvent has an anomalously low interfacial capacitance. We are able to explain this on the basis of the different bonding mechanism for the monolayer versus multilayer. Furthermore, a model was proposed for comparing the solvation response of a dielectric to an IPS electron as a capacitive phenomenon, which allows comparison of 2PPE results to a broad range of inertial dielectric response literature on liquids.

References

1 Lang, N.D. and Kohn, W. (1973) *Phys. Rev. B*, **7**, 3541.

2 Smith, N.V. (1985) *Phys. Rev. B*, **32**, 3549.

3 Memmel, N. (1998) *Surf. Sci. Rep.*, **32**, 91.

4 Giesen, K., Hage, F., Himpsel, F.J., Riess, H.J., and Steinmann, W. (1985) *Phys. Rev. Lett.*, **55**, 300.

5 Giesen, K., Hage, F., Himpsel, F.J., Riess, H.J., and Steinmann, W. (1986) *Phys. Rev. B*, **33**, 5241.

6 Giesen, K., Hage, F., Himpsel, F.J., Riess, H.J., and Steinmann, W. (1987) *Phys. Rev. B*, **35**, 971.

7 Giesen, K., Hage, F., Himpsel, F.J., Riess, H.J., Steinmann, W., and Smith, N.V. (1987) *Phys. Rev. B*, **35**, 975.

8 Schoenlein, R.W., Fujimoto, J.G., Eesley, G.L., and Capehart, T.W. (1988) *Phys. Rev. Lett.*, **61**, 2596.

9 Steinmann, W. (1989) *Appl. Phys. A Mater.*, **49**, 365.

10 Echenique, P.M. and Pendry, J.B. (1978) *J. Phys. C Solid State Phys.*, **11**, 2065.

11 Echenique, P.M. and Pendry, J.B. (1989) *Prog. Surf. Sci.*, **32**, 111.

12 Cole, M.W. (1971) *Phys. Rev. B*, **3**, 4418.

13 McNeill, J.D., Lingle, R.L., Ge, N.H., Wong, C.M., Jordan, R.E., and Harris, C.B. (1997) *Phys. Rev. Lett.*, **79**, 4645.

14 McNeil, J.D., Lingle, R.L., Jordan, R.E., Padowitz, D.F., and Harris, C.B. (1996) *J. Chem. Phys.*, **105**, 3883.

15 Hotzel, A., Moos, G., Ishioka, K., Wolf, M., and Ertl, G. (1999) *App. Phys. B*, **68** (3), 615–622

16 Berthold, W., Höfer, U., Feulner, P., and Menzel, D. (2000) *Chem. Phys.*, **251** 123.

17 Wong, C.M., McNeill, J.D., Gaffney, K.J., Ge, N.H., Miller, A.D., Liu, S.H., and Harris, C.B. (1999) *J. Phys. Chem. B*, **103**, 282.

18 Velic, D., Hotzel, A., Wolf, M., and Ertl, G. (1998) *J. Chem. Phys.*, **109**, 9155.

19 Lindstrom, C.D., Quinn, D., and Zhu, X.Y. (2005) *J. Chem. Phys.*, **122**, 10.

20 Jørgensen, S., Ratner, M.A., and Mikkelsen, K.V. (2002) *Chem. Phys.*, **278**, 53.

21 Lingle, R.L., Ge, N.H., Jordan, R.E., McNeill, J.D., and Harris, C.B. (1996) *Chem. Phys.*, **205**, 191.

22 Gaffney, K.J., Wong, C.M., Liu, S.H., Miller, A.D., McNeill, J.D., and Harris, C.B. (2000) *Chem. Phys.*, **251**, 99.

23 Gaffney, K.J., Miller, A.D., Liu, S.H., and Harris, C.B. (2001) *J. Phys. Chem. B*, **105**, 9031.

24 Garrett-Roe, S., Shipman, S.T., Szymanski, P., Strader, M.L., Yang, A., and Harris, C.B. (2005) *J. Phys. Chem. B*, **109**, 20370.
25 Garrett-Roe, S. (2005) Ultrafast electron dynamics at dielectric/metal interfaces: intraband relaxation of image potential state electrons as friction and disorder induced electron localization, University of California, Berkeley.
26 Kohn, W. and Sham, L.J. (1965) *Phys. Rev.*, **140**, 1133.
27 Hedin, L. (1965) *Phys. Rev.*, **139**, A796.
28 Runge, E. and Gross, E.K.U. (1984) *Phys. Rev. Lett.*, **52**, 997.
29 Chulkov, E.V., Silkin, V.M., and Echenique, P.M. (1999) *Surf. Sci.*, **437**, 330.
30 Wiesenmayer, M., Bauer, M., Mathias, S., Wessendorf, M., Chulkov, E.V., Silkin, V.M., Borisov, A.G., Gauyacq, J.-P., Echenique, P.M., and Aeschlimann, M. (2008) *Phys. Rev. B*, **78**, 245410.
31 Strader, M.L., Garrett-Roe, S., Szymanski, P., Shipman, S.T., Johns, J.E., Yang, A., Muller, E., and Harris, C.B. (2008) *J. Phys. Chem. C*, **112**, 6880.
32 Bagchi, B. and Fleming, G.R. (1990) *J. Phys. Chem.*, **94**, 9.
33 Poornimadevi, C.S. and Bagchi, B. (1990) *Chem. Phys. Lett.*, **168**, 276.
34 Berthold, W., Hofer, U., Feulner, P., Chulkov, E.V., Silkin, V.M., and Echenique, P.M. (2002) *Phys. Rev. Lett.*, **88**, 4.
35 Miller, T., McMahon, W.E., and Chiang, T.C. (1996) *Phys. Rev. Lett.*, **77**, 1167.
36 Hotzel, A., Wolf, M., and Gauyacq, J.P. (2000) *J. Phys. Chem. B*, **104**, 8438.
37 Chulkov, E.V., Sarria, I., Silkin, V.M., Pitarke, J.M., and Echenique, P.M. (1998) *Phys. Rev. Lett.*, **80**, 4947.
38 Campillo, I., Pitarke, J.M., Rubio, A., Zarate, E., and Echenique, P.M. (1999) *Phys. Rev. Lett.*, **83**, 2230.
39 Osma, J., Sarria, I., Chulkov, E.V., Pitarke, J.M., and Echenique, P.M. (1999) *Phys. Rev. B*, **59**, 10591.
40 Campillo, I., Silkin, V.M., Pitarke, J.M., Chulkov, E.V., Rubio, A., and Echenique, P.M. (2000) *Phys. Rev. B*, **61**, 13484.
41 Chulkov, E.V., Kliewer, J., Berndt, R., Silkin, V.M., Hellsing, B., Crampin, S., and Echenique, P.M. (2003) *Phys. Rev. B*, **68**, p. 195422.
42 Machado, M., Berthold, W., Hofer, U., Chulkov, E., and Echenique, P.M. (2004) *Surf. Sci.*, **564**, 87.
43 Zhukov, V.P., Chulkov, E.V., and Echenique, P.M. (2005) *Phys. Rev. B*, **72**, p. 155109.
44 Vergniory, M.G., Pitarke, J.M., and Echenique, P.M. (2007) *Phys. Rev. B*, **76**, p. 245416.
45 Wiesenmayer, M., Bauer, M., Mathias, S., Wessendorf, M., Chulkov, E.V., Silkin, V.M., Borisov, A.G., Gauyacq, J.P., Echenique, P.M., and Aeschlimann, M. (2008) *Phys. Rev. B*, **78**, p. 245410.
46 Zhukov, V.P., Chulkov, E.V., and Echenique, P.M. (2008) *Physica Status Solidi A*, **205**, 1296.
47 Ferrell, T.L., Echenique, P.M., and Ritchie, R.H. (1979) *Solid State Commun.*, **32**, 419.
48 Nuñez, R., Echenique, P.M., and Ritchie, R.H. (1980) *J. Phys. C Solid State Phys.*, **13**, 4229.
49 Tomassone, M.S. and Widom, A. (1997) *Phys. Rev. B*, **56**, 4938.
50 Shier, J.S. (1968) *Am. J. Phys.*, **36**, 245.
51 Boyer, T.H. (1999) *Am. J. Phys.*, **67**, 954.
52 Lide, D.R. (ed.) (2009–2010) *CRC Handbook of Chemistry and Physics*, 90th edn, CRC Press.
53 Ashcroft, N.W. and Mermin, N.D. (1976) *Solid State Physics*, Sanders College Publishing.
54 Horng, M.L., Gardecki, J.A., Papazyan, A., and Maroncelli, M. (1995) *J. Phys. Chem.*, **99**, 17311.
55 Maroncelli, M. and Fleming, G.R. (1987) *J. Chem. Phys.*, **86**, 6221.
56 Castner, E.W., Maroncelli, M., and Fleming, G.R. (1987) *J. Chem. Phys.*, **86**, 1090.
57 Castner, E.W., Fleming, G.R., and Bagchi, B. (1988) *J. Chem. Phys.*, **89**, 3519.
58 Kimura, Y., Alfano, J.C., Walhout, P.K., and Barbara, P.F. (1994) *J. Phys. Chem.*, **98**, 3450.
59 Kahlow, M.A., Jarzeba, W., Kang, T.J., and Barbara, P.F. (1989) *J. Chem. Phys.*, **90**, 151.
60 Ge, N.H., Wong, C.M., Lingle, R.L., McNeill, J.D., Gaffney, K.J., and Harris, C.B. (1998) *Science*, **279**, 202.

61 Liu, S.H., Miller, A.D., Gaffney, K.J., Szymanski, P., Garrett-Roe, S., Bezel, I., and Harris, C.B. (2002) *J. Phys. Chem. B*, **106**, 12908.

62 Gahl, C., Bovensiepen, U., Frischkorn, C., and Wolf, M. (2002) *Phys. Rev. Lett.*, **89**, 4.

63 Gahl, C., Bovensiepen, U., Frischkorn, C., Morgenstern, K., Rieder, K.H., and Wolf, M. (2003) *Surf. Sci.*, **108** 532–535.

64 Bovensiepen, U., Gahl, C., Stahler, J., Bockstedte, A., Meyer, M., Baletto, F., Scandolo, S., Zhu, X.Y., Rubio, A., and Wolf, A. (2009) *J. Phys. Chem. C*, **113**, 979.

65 Sebastian, K.L., Chakraborty, A., and Tachiya, M. (2005) *J. Chem. Phys.*, **123**, p. 214704.

66 Sebastian, K.L., Chakraborty, A., and Tachiya, M. (2003) *J. Chem. Phys.*, **119**, 10350.

67 Li, B., Zhao, J., Onda, K., Jordan, K.D., Yang, J.L., and Petek, H. (2006) *Science*, **311**, 1436.

68 Roelfs, B., Schroter, C., and Solomun, T. (1997) *Ber. Bunsen. Phys. Chem.*, **101**, 1105.

69 Si, S.K. and Gewirth, A.A. (2000) *J. Phys. Chem. B*, **104**, 10775.

70 Schröter, C., Roelfs, B., and Solomun, T. (1997) *Surf. Sci.*, **380**, L441.

71 Bezel, I., Gaffney, K.J., Garrett-Roe, S., Liu, S.H., Miller, A.D., Szymanski, P., and Harris, C.B. (2004) *J. Chem. Phys.*, **120**, 845.

72 Singh, P.K., Nath, S., Kumbhakar, M., Bhasikuttan, A.C., and Pal, H. (2008) *J. Phys. Chem. A*, **112**, 5598.

5
Dynamics of Electronic States at Metal/Insulator Interfaces

Jens Güdde and Ulrich Höfer

5.1
Introduction

Modern materials and devices are frequently structured on a nanoscale. Their properties become increasingly dominated by interfaces between two solids that often have very different electronic structures. Not only the mechanisms of electron transfer processes across such interfaces, in particular, are important in terms of the performance of novel devices but also their microscopic understanding is a challenging fundamental problem. Unfortunately, there is a lack of suitable experimental methods that can reach the buried interfaces and provide interface-specific information. In addition, there is a lack of simple model systems that could allow us to study specific interface-related electronic properties under well-defined conditions.

On bare metal surfaces, image potential states [1, 2] have been proven to be an ideal model system for studying the dynamics of excited electrons. Time-resolved two-photon photoemission (2PPE) and accurate many-body calculations have led to a good understanding of many aspects of the electronic decay of these hydrogen-like states at metal surfaces [3–19].

Their model character for unoccupied surface states also made them a focus of the study of the influence of thin adsorbate layers on electron dynamics at surfaces [20–25]. For adsorbates that do not introduce unoccupied electronic levels around the vacuum energy, such as rare-gas layers, the image potential states will retain their simple character. Their coupling to the metal bulk, however, can be drastically modified due to a repelling of the image potential state wave function from the metal. This effect is particularly strong for adlayers with negative electron affinity, like Ar, where the bandgap of the insulating layer represents a tunneling barrier between the metal and the image-potential in the vacuum [26, 27].

We have recently shown that for thick Ar films on Cu(100) the image-potential within the insulating adlayer supports electronic states which are similar to image potential states, but located at the insulator/metal interface [28, 29]. In the limit of an

Dynamics at Solid State Surface and Interfaces Vol.1: Current Developments
Edited by Uwe Bovensiepen, Hrvoje Petek, and Martin Wolf

ISBN: 978-3-527-40937-2

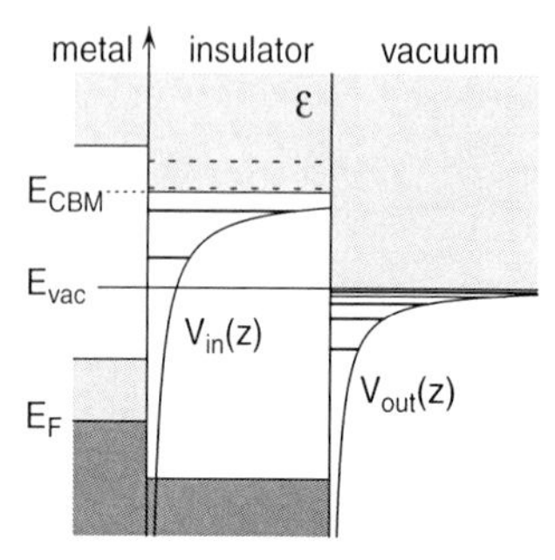

Figure 5.1 Schematic energy diagram of the sequence Cu–Ar–vacuum. Dark and light shades depict occupied and unoccupied bands, respectively. The finite insulating layer thickness leads to a screened image potential V_{in} of finite range, which supports a finite number of unoccupied quasibound states (solid lines) and additionally scattering resonances in the conduction band of Ar (dotted lines). The infinite range of the image potential V_{out} in front of the adlayer supports an infinite series of unoccupied bound image potential states.

insulating adlayer of infinite thickness, this long-range potential gives rise to an infinite series of electronic states with energies

$$E_{n'} = E_{\text{CBM}} - \frac{0.85\ \text{eV}}{(n'+a)^2} \times \frac{m_{\text{eff}}}{\varepsilon^2}, \quad n' = 1, 2, \ldots. \tag{5.1}$$

where n' denotes the quantum number, a the quantum defect ($0 \leq a \leq 0.5$ [30, 31]) and ε the dielectric constant of the insulator. Compared to the usual series of image potential states on bare metal surfaces, which converge towards the vacuum level E_{vac}, the reference level for the interface states is the affinity level of the insulator or the conduction band minimum E_{CBM} that represents the lowest energy of an excess electron in the bulk dielectric. For an adlayer with negative electron affinity $EA = E_{\text{vac}} - E_{\text{CBM}}$, such as Ar, the interface states are in principle not bound since $E_{n'}$ may be smaller than $-EA$. Decay into the vacuum, however, will be suppressed since the screened image potential forms a potential barrier on the vacuum side (Figure 5.1). Hence, we will call them "quasibound" states in the following discussion. The binding energy of these interface states with respect to E_{CBM} is lowered by a factor $m_{\text{eff}}/\varepsilon^2$ because the image potential is screened by the dielectric overlayer, and because the potential is periodically modulated for ordered adlayers, which leads to an effective electron mass $m_{\text{eff}} \cdot m_0$, where m_0 is the mass of the free electron. Owing to the same reason, the spatial spread of the wave function is enhanced by a factor of $\varepsilon/m_{\text{eff}}$. It can be estimated by the distance

$$z_{n'} = 4a_0(n'+a)^2 \times \frac{\varepsilon}{m_{\text{eff}}}, \tag{5.2}$$

where the probability density has its maximum. Here, a_0 is the Bohr radius [25][1)].

1) Note the typing error in Eq. (2) of Ref. [29].

For decreasing thickness of the insulating adlayer, the interface states shift to higher energies, and the number of quasibound states below E_{CBM} becomes finite like in a simple quantum well (Figure 5.1). The series of states will be continued by scattering resonances above E_{CBM} (dotted lines). Without the image potential, these resonances would correspond to quantum well states within the insulating layer, which form for increasing thickness the continuous conduction band of the dielectric solid. For large layer thicknesses, the potential V_{out} in front of the dielectric layer is given by the image potential of a dielectric solid [32], while it is dominated by the image potential of the metal for thin films [20, 33, 34].

In the following section, we will review experiments on interface states for Ar layers on Cu(100), which represents a simple model system to study the ultrafast electronic dynamics at buried interfaces. The main focus of this chapter will be on experiments using time-resolved two-photon photoemission spectroscopy, where the interface states are transiently populated as intermediate states. These states, however, can also be observed as enhancement of the final-state density in conventional photoelectron spectroscopy called the one-photon photoemission (1PPE) in the following discussion. Spectroscopically, 1PPE measurements have advantages in terms of a higher accessible energy range above the vacuum level, which makes it possible to determine the energy of the interface states as a function of Ar-layer thickness up to high quantum numbers n' [35]. The electron dynamics of the interface states could be investigated with 2PPE even if these states are buried under films as thick as 200 Å [28]. Relaxation of excited electrons in these states occurs on timescales between 40 and 200 fs via two distinct channels. Apart from inelastic decay into the metal, the population of the interface states decays by resonant tunneling through the Ar film into vacuum, which makes it possible to study the transfer of low-energy electrons through thin dielectric films.

5.2 Spectroscopy by One-Photon Photoemission

Owing to the negative electron affinity of Ar, the conduction band minimum E_{CBM} of the Ar layers on the Cu(100) surface is located 0.25 eV above the vacuum level, and the interface states are unbound with respect to decay into the vacuum even if this decay will be suppressed for thick Ar layers. On the one hand, this makes it possible to observe the unoccupied interface states also by conventional photoemission spectroscopy as final state resonances [35]. On the other hand, it was another reason why these states were not observed in earlier 2PPE experiments where one usually avoids pump photon energies above the work function [36] in order to suppress the one-photon photoemission background.

Figure 5.2 shows 1PPE spectra for different Ar coverages using a photon energy of 4.96 eV, which is 0.58 eV larger than the work function of the Ar-covered surface. Pronounced resonances of the photoemission intensity can be observed even for low Ar coverages. With increasing Ar thickness, the number of maxima increases, which shift to lower energies and converge to constant energies after crossing

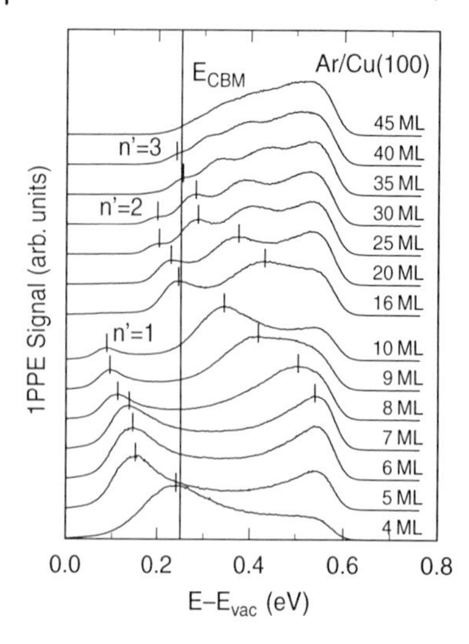

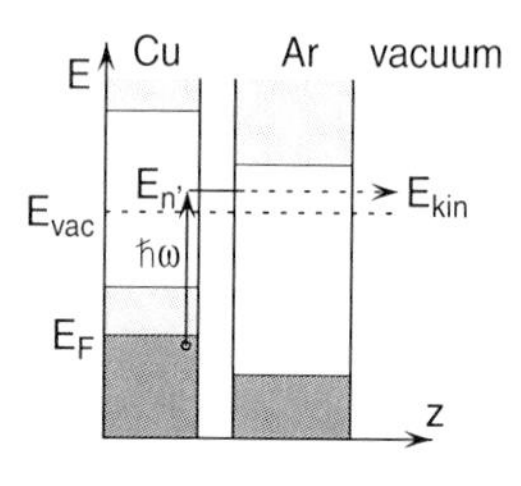

Figure 5.2 1PPE spectra of Ar films on Cu(100) for various thicknesses using a photon energy of 4.96 eV. The energy scheme on the right illustrates the excitation at the Ar/Cu interface and the transfer through the Ar layer by tunneling. Reprinted from [25]. Copyright 1998, with permission from Elsevier.

E_{CBM}, accompanied by a drop in intensity. These resonances have different character depending on their energy with respect to E_{CBM}. For a finite layer thickness, one has to distinguish between interface scattering resonances above E_{CBM} and quasibound interface states that are bound within the quantum well formed by the image potential inside the Ar layer but which can tunnel into the vacuum through the potential barrier at the surface.

The interface states of different character can be well reproduced by model calculations using a one-dimensional parameterized model potential that has been successfully used to describe the decoupling of image potential states by rare-gas layers on Cu(100) [24, 25, 35]. The model potential consists of three parts for the metal, the Ar-layer, and the vacuum. The metal is represented by a two-band model of nearly free electrons. The corresponding parameters of the potential have been adjusted to reproduce the size and location of the projected bandgap of the Cu(100) surface. The potential inside the Ar layer does account not only for the macroscopic properties of the insulator such as the dielectric constant and the electron affinity, that is, the location of the conduction band minimum E_{CBM}, but also for the corrugation of the potential due to the discrete nature of the single layers. The vacuum part is given by the screened image potential. Wave functions and energies have been computed by numerically solving the stationary Schrödinger equation for this model potential. While the stationary solutions below E_{vac} can be associated with the bound image potential states in front of the Ar layer, the quasibound interface states correspond to

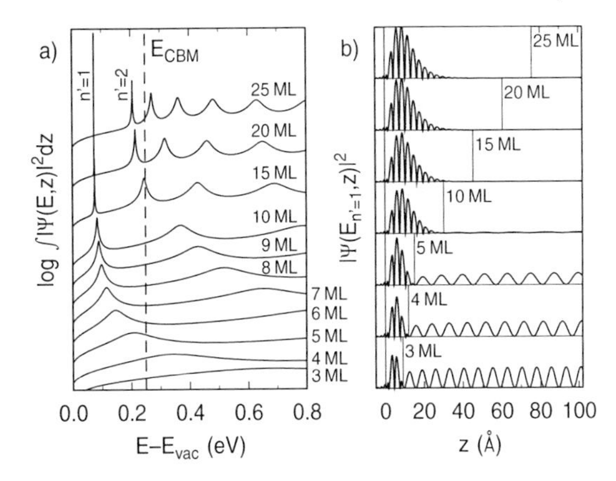

Figure 5.3 (a) Total probability density inside the Ar layer (logarithmic scale) as a function of energy above E_{vac} calculated by solving the one-dimensional Schrödinger equation within the atomic model. (b) Probability density of the $n' = 1$ interface state for different layer thicknesses. The Ar layers are depicted by the shaded areas. Reprinted from [25]. Copyright 1998, with permission from Elsevier.

resonances at energies $E > E_{vac}$, where one finds continuum solutions for all energies. Plotting the integrated probability density inside the Ar layer as a function of energy, as shown in Figure 5.3a, reveals resonant maxima that get sharper with increasing Ar thickness and decreasing energy. Figure 5.3b shows the corresponding probability densities of the lowest resonance that evolves into the $n' = 1$ interface state calculated at the resonance energy for various thicknesses. For small Ar coverages, the wave function is essentially a free wave, and only a scattering resonance appears. For increasing Ar thickness, the resonance energy shifts below E_{CBM} and the wave function gets confined to the Cu/Ar interface. Within the metal and the Ar layer, the probability density is modulated with the respective layer distance since the wave function within both regions corresponds to a Bloch wave in a periodic lattice. If the Ar thickness is larger than the spatial extent of the wave function, the binding energy becomes independent of thickness. In addition, the resonance gets sharper since the tunneling probability into the vacuum depends exponentially on the width of the barrier that increases with increasing film thickness. For photon energies larger than the work function but smaller than the bandgap of the Ar layer, the interface states can be occupied only by the excitation of electrons of the Cu substrate at the interface. In this case, the photoemission intensity depends on both the overlap of the interface state wave function with the occupied bulk state of the Cu substrate and the coupling to the vacuum where the electrons are detected. The overlap with the Cu bulk states converges rapidly for increasing film thickness leading to a coverage-independent excitation probability. The emission into the vacuum depends on the tunneling probability that is responsible for the extinction of the photoemission intensity for the lowest states with increasing Ar thickness.

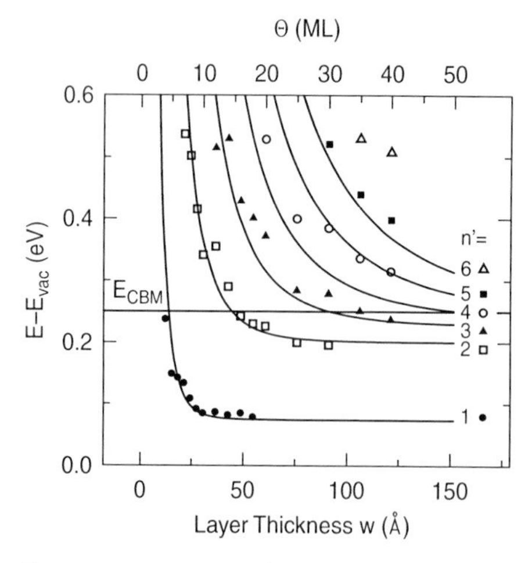

Figure 5.4 Experimental resonance energies obtained from the 1PPE spectra (data points) and calculated maxima of the probability density inside the Ar layer (solid lines) as a function of Ar coverage θ. Reprinted from [25]. Copyright 1998, with permission from Elsevier.

The agreement between experimental and calculated resonance energies is excellent for the lowest three interface states as shown in Figure 5.4 as a function of film thickness. Both the dependence on film thickness and the absolute values of the converged binding energies are correctly reproduced. This confirms the validity of the one-dimensional description of the interface states within the Ar layer using the same atomically corrugated potential as for the image potential states in front of the adlayer. However, the resonance energies of the states with quantum number $n' > 3$ are slightly larger than those calculated. On the one hand, these deviations could be caused by the film morphology at the surface, which gets probably less perfect for increasing thickness. First of all, this affects the higher members of the Rydberg series the spatial extent of which quadratically increases with n' resulting in an increased sensitivity to the ordering of the Ar layer at larger distances from the Cu/Ar interface. On the other hand, the one-dimensional potential, of course, cannot describe the full Ar band structure. The parameters of the atomic potential have been optimized for the description of the Ar conduction band near the $\bar{\Gamma}$-point, whereas the bandwidth has been overestimated [24]. Thus, larger deviations are expected for increasing energy above E_{CBM}.

5.3
Observation by Two-Photon Photoemission

Observation of the interface state dynamics in a time-resolved pump–probe 2PPE experiment requires a pump photon energy that is larger than the work function. However, there is no serious restriction for Ar/Cu, since the Ar bandgap acts as a tunnel barrier for the quasibound interface states. As long as $\hbar\omega_a < E_{n'} - E_F + \hbar\omega_b$,

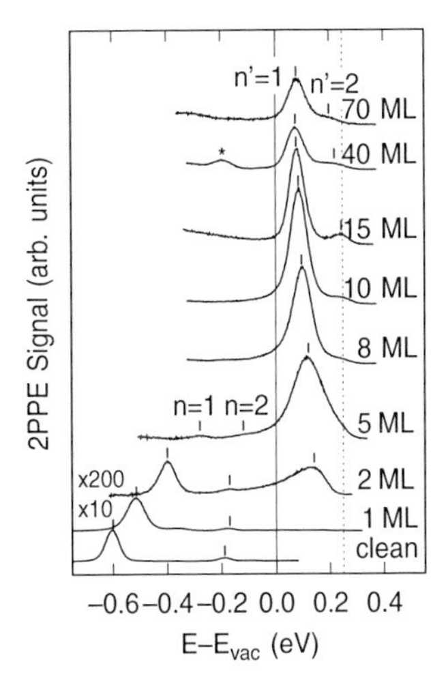

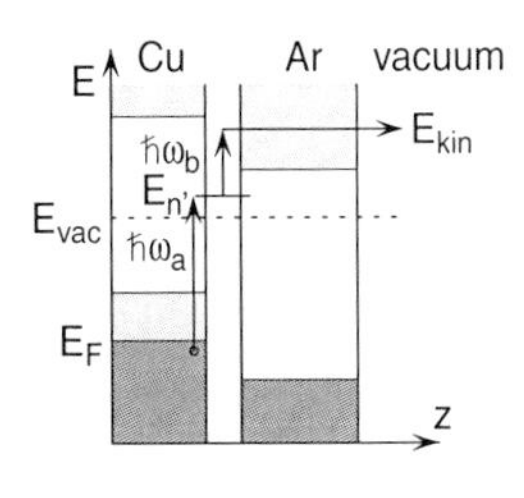

Figure 5.5 2PPE spectra of Ar/Cu(100) as a function of energy for increasing thickness of the Ar layer showing the $n' = 1, 2$ interface states above the vacuum level. The dotted line marks the conduction band minimum of bulk Ar. At the clean surface and at low thicknesses, the bound $n = 1, 2$ image potential states are visible. The asterisk in the 40-ML spectrum marks a defect state induced by the UV irradiation. (Reprinted with permission from [28]. Copyright 2005, the American Physical Society.) The energy scheme on the right illustrates the 2PPE process for the interface states.

with $\hbar\omega_a$ being the pump photon energy and $\hbar\omega_b$ the probe photon energy, interface states with energies $E_{n'} > E_{\text{vac}}$ can be observed without interfering with 1PPE background. Figure 5.5 shows energy-resolved 2PPE spectra for different Ar coverages recorded in normal emission. Here, the third harmonic of the Ti:sapphire oscillator was set to a pump photon energy of $\hbar\omega_a = 4.71$ eV, which is 0.34 eV higher than the work function of Ar/Cu(100). The fundamental pulses ($\hbar\omega_b = 1.57$ eV) served as probe pulses at zero delay with respect to the pump pulses. Below E_{vac}, the bound image potential states can be observed for the clean Cu(100) surface and for small Ar coverages. With increasing coverage the energies of these states shift upward, and their intensity drops rapidly due to the repulsion from the Cu interface, which decreases the overlap with the Cu bulk electrons and therewith the excitation probability. Starting at 2 ML, the $n' = 1$ interface state becomes visible as a broad feature above E_{vac}. It shifts downward, grows in intensity, and shrinks in linewidth for increasing thickness. Also for coverages larger than 10 ML, the $n' = 2$ state can be assigned. With further increase in the Ar coverage, the energies of the $n' = 1, 2$ interface states saturate at 0.076(5) and 0.199(15) eV above E_{vac} in excellent agreement with the calculated resonance positions (0.074 and 0.200 eV).

In contrast to the 1PPE spectra, the 2PPE intensity of the interface states decreases only slightly for high Ar coverages, and the $n' = 1, 2$ states can be clearly observed even for a coverage of 70 ML corresponding to a thickness of 213 Å. This can be understood in the following simple picture that is illustrated by the energy schemes in Figures 5.2 and 5.5: In order to be observable by 1PPE, the electron in the interface state has to pass the Ar layer by tunneling, whereas in the 2PPE scheme the probe pulse promotes the electron into the conduction band of the Ar layer where it can move ballistically and escape into the vacuum. Our spectroscopic results indicate that this transport occurs without appreciable loss of energy or momentum for layer thicknesses up to 100 Å where we still observe sharp peaks without attenuation of the maxima. For thicker Ar layers, however, the intensity decreases and the peaks broaden, most likely due to a reduction of the film quality at the surface.

5.4 Lifetimes

The possibility to observe the quasibound interface states by 2PPE over a wide range of layer thicknesses allows us to study in detail different relaxation processes contributing to the decay of the interface states. On the one hand, this is tunneling through the Ar layer, which strongly depends on layer thickness. This elastic electron transfer in the vacuum is accompanied by inelastic decay due to the creation of electron–hole pairs in the metal as for the image potential states on the clean surface. Since the latter depends on the overlap of the interface and Cu bulk states, which is almost independent of layer thickness, we could show that time-resolved 2PPE experiments as a function of coverage allow the separation of both processes [28].

The main panel of Figure 5.6a shows the lifetimes of the excited $n' = 1$ and $n' = 2$ states deduced from time-resolved 2PPE data as a function of layer thickness. The lifetime of the $n' = 1$ state increases strongly with increasing Ar coverage and reaches a saturation value of 105 ± 10 fs near 15 ML. The lifetime of the $n' = 2$ state is shorter for thin Ar films than that of the $n' = 1$ state but reaches a higher saturation value of 180 ± 20 fs near 30 ML. The large scatter in the $n' = 2$ lifetime above 30 ML probably originates from the contribution of energetically close-lying and long-living higher states $n' \geq 3$, which are experimentally difficult to separate.

Since thick layers effectively suppress the elastic decay by electron transmission, the saturation values of the measured lifetimes τ_{sat} can be identified with the inelastic decay channel due to e–h pair excitation. This relaxation mechanism is virtually independent of coverage, only confinement to extremely thin films may enhance the coupling of the wave functions to the metal. We can therefore obtain the thickness-dependent rates of elastic transfer Γ_{tr} by subtracting the nearly constant contribution of inelastic decay $\Gamma_{e-h} = \hbar/\tau_{sat}$ from the measured total decay rate $\hbar/\tau$ (inset, Figure 5.6a). The resulting transfer rates Γ_{tr} of both states exponentially decrease with layer thickness, the transfer through all films being faster for $n' = 2$ than for $n' = 1$. This dependence on layer thickness and electron energy is a characteristic of tunneling through the bandgap of Ar.

In a simplified but illustrating approximation, the transfer rates may be analyzed with the WKB formula for tunneling through a square barrier of height ΔE and width w, $\Gamma_{tr} = \Gamma_0 \exp[-2w\sqrt{2m\Delta E/\hbar^2}]$. From best fits to the data of Figure 5.6a, using

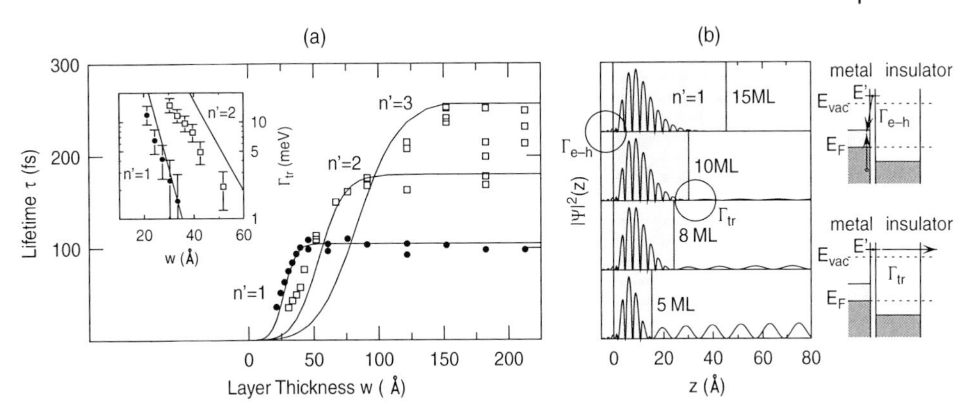

Figure 5.6 (a) Experimental lifetimes (symbols) of the $n' = 1, 2$ interface states as a function of Ar layer thickness (1 ML ≙ 3.04). Simulated curves were calculated as $\tau = (\Gamma_{tr}/\hbar + 1/\tau_{sat})^{-1}$ from the theoretical rates of elastic transfer Γ_{tr} and the experimental high-coverage lifetimes τ_{sat}. *Inset*: elastic rates Γ_{tr}. (b) Computed probability densities of the $n' = 1$ state for different Ar coverages. Reprinted with permission from [28]. Copyright 2005, the American Physical Society.

$m = 0.6\, m_e$, we obtain $\Delta E = 0.092$ eV (0.024 eV) for $n' = 1$ ($n' = 2$). These energies are smaller than the measured energy differences $E_{CBM} - E_{n'} = 0.17$ eV (0.05 eV) of $n' = 1$ ($n' = 2$). Obviously, the screened image potential increases more gradually than the rectangular potential used in the WKB formula. The attempt rates $\Gamma_0 = 0.55$ eV(0.27 eV) lie within the same order of magnitude, as expected from the classical oscillation time of an electron in the image potential [3].

For a more rigorous description of the tunneling through the Ar layers, we have determined the transfer rates $\Gamma_{tr}(w)$ from the calculated resonance spectra depicted in Figure 5.3. A resonance was identified as a maximum of probability density inside the layer. The shape of the resonances is nearly Lorentzian, in particular, when the width is small compared to the energetic distance of the resonances. However, we have determined their width by fitting Beutler-Fano profiles in order to account for the slight asymmetric line shapes due to contributions of interface states above the investigated energy range.

As shown in Figure 5.4, the resonance energies obtained in this way very accurately reproduce the experimental peak positions of Figure 5.2. The transfer rates $\Gamma_{tr}(w)$ correspond to the widths of the resonances, which are indicated by lines in the inset of Figure 5.6a.

The calculated values differ only by a factor of 2 from the experimental ones. Considering the approximation of the truly three-dimensional corrugation of the Ar potential by a one-dimensional model, this agreement is satisfactory and confirms our interpretation of this decay channel.

For illustration purposes, we compare in Figure 5.6b probability densities calculated at the resonance positions of $n' = 1$ for different layer thicknesses. For thin films, there is a considerable probability density in the vacuum region outside the layer. This leakage of the wave function is the origin of the elastic decay by electron transfer across the insulating Ar layer. It gets smaller when the adsorbate film grows and virtually vanishes for a thickness of 15 ML and higher. The probability densities of Figure 5.6b also show a small amplitude at $z < 0$ inside the metal. This penetration of the wave functions reflects the coupling to the substrate and is therefore a measure of the rate of inelastic decay by creation of e–h pairs at the metal surface.

Today, many-body theory can very accurately predict inelastic lifetimes of image potential states on clean noble-metal surfaces [13, 37]. In the case of Cu(100), one can divide the different contributions into a pure bulk contribution, a near-surface contribution, and a cross-term, all of which scale approximately to the amplitude of the wave function at the metal surface [37]. Dielectric screening in the Ar layer and the small effective mass of the interface states lead to a larger spread of the wave functions of the Ar/Cu interface states compared to the image potential states on clean Cu. One thus expects the interface states to interact less with the metal and have the longer lifetimes.

In fact, with $\tau_{n'=1} = 105$ fs the experimental lifetime of the first interface state is larger than that of the first image potential state $\tau_{n=1} = 40$ fs [3, 13]. However, the computed penetration values $p_{n'=1} = 0.82$, $p_{n=1} = 3.9$ differ by almost a factor of 5, while the ratio of the lifetimes is only 2.6. At this point, we can only speculate why the interface states decay faster than expected from penetration arguments. Enhanced screening of the Coulomb interaction inside the Ar layer should reduce rather than enhance the decay rate. Possibly, the strong oscillatory character of the interface states in Ar enhances the near-surface contributions to the decay compared to the smoother image potential states. However, only a full many-body calculation can probably resolve this issue.

5.5 Momentum-Resolved Dynamics

While we have so far discussed the inelastic dynamics at the $\bar{\Gamma}$-point due to interband transitions between the interface or image potential states and the Cu bulk states, further decay channels have to be considered for finite parallel momenta $k_{\parallel}$. On the one hand, it has been shown for image potential states on the clean Cu(100) surface that the interband decay has a $k_{\parallel}$-dependence due to the dispersion of the bulk bands, which results in a $k_{\parallel}$-dependence of the image-state wave function perpendicular to the surface [9]. On the other hand, it has been shown in the same work that intraband relaxation of electrons toward the band minimum has a comparable contribution to the observed decay rate. This process is particularly important for the hole decay in the occupied surface bands on the (111) noble metal surfaces, which is almost completely determined by intraband scattering [38]. The intraband scattering on the clean metal surfaces results from the same interaction with the bulk electrons as the interband scattering. Although the energy exchange is small in intraband decay, which results in a small Coulomb interaction, there is almost complete spatial overlap of image potential wave function for different $k_{\parallel}$. This makes the magnitude of intraband decay due to electron–electron interaction comparable to interband decay [9]. However, for dielectric covered metal surfaces, where intraband decay of image potential states was first observed, it has been suggested that adsorbate motion is mainly responsible for a redistribution in momentum space [39, 40]

In order to compare the momentum dependence of the inelastic decay between interface and image potential states, we have performed time-resolved 2PPE measurements as a function of parallel momentum for the $n' = 1$ interface state at an Ar coverage of 25 ML [29]. At this coverage, the $n' = 1$ interface state is completely confined to the Ar layers, and elastic decay due to tunneling can be neglected.

Angle-resolved 2PPE spectra have been obtained by employing a hemispherical electron analyzer equipped with an angle-resolved lens mode and a two-dimensional image-type detector that makes it possible to determine the dispersion $E(k_{\parallel})$ in a single measurement [29]. The 2PPE spectrum for 25 ML Ar/Cu(100) depicted in Figure 5.7 shows clearly the parabolic dispersion of the $n' = 1$ and even that of the $n' = 2$ can be resolved. The dispersion of the $n' = 1$ interface state reveals an effective mass of only $m_{\text{eff}} = 0.61 \pm 0.1$. For comparison, we measured the effective mass of the $n = 1$ image potential state of clean Cu(100) in a similar way and deduced 0.98 ± 0.1 in good agreement with earlier measurement [41].

The smaller effective mass of the $n' = 1$ interface state can be readily understood if one considers that these states are derived from the conduction band minimum of bulk Ar. There, an effective mass of 0.53 ± 0.01 has been measured using low-energy electron transmission spectroscopy [42]. This value is very close to our result. The small difference might have arisen from the energetic position of the $n' = 1$ state inside the bandgap of Ar, 170 meV below the conduction band minimum. Our

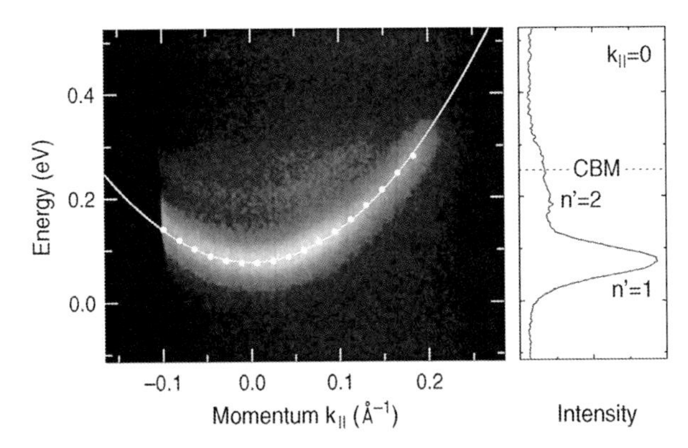

Figure 5.7 (a) $E(k_{\parallel})$ 2PPE spectrum for 25 ML Ar/Cu(100) acquired with a display-style electron analyzer. The dots along the dispersion parabola of the $n' = 1$ interface state mark the points used for evaluation of the time-resolved pump–probe traces. The faint parabola of the $n' = 2$ state is also visible. (b) 2PPE energy spectrum for $k_{\parallel} = 0$ obtained from a cut through the 2d spectrum with a width of about $0.018\,\text{Å}^{-1}$. Reprinted from [29]. Copyright 2006, IOP Publishing Ltd. and Deutsche Physikalische Gesellschaft. (Please find a color version of this figure on the color plates.)

result agrees also with recent theoretical work on resonance states in very thin Ar films, which gave an effective mass of 0.6 [43].

The momentum dependence of the inelastic lifetime of the $n' = 1$ state has been determined by recording 2D $E(k_{\|})$ spectra such as the one displayed in Figure 5.7 for different time delays between pump and probe pulses. The evaluation of the intensity decay along the dispersion curve revealed that the lifetime decreases from $\tau = 110 \pm 5$ fs at the band bottom ($k_{\|} = 0$, $E = E_0$) to $\tau = 84 \pm 5$ fs at $E - E_0 = 171$ meV, that is, at $k_{\|} = 0.18\ ^{-1}$.

Since independent relaxation processes are contributing additively to the total decay rate $\Gamma = \hbar/\tau$, it is more instructive to plot this property as a function of the energy of parallel motion $E_{\|} = \hbar^2 k_{\|}^2 / 2m_{\text{eff}} m_0$. These data are shown together with experimental data of the $n = 1$ image potential state on the clean surface in Figure 5.8. The solid line shows the theoretical results for the clean surface from Ref. [9] in excellent agreement with the experiment. Compared to the $n = 1$ state on the clean surface, the smaller decay rate of the $n' = 1$ interface state goes along with a weaker dependence on $k_{\|}$. This can be simply understood if the momentum-dependent relaxation dynamics of the $n' = 1$ state is governed by the same interaction with the bulk electrons as for the $n = 1$ state on the clean surface. Then, both the decay rate and its momentum dependence should scale with the same factor compared to the $n = 1$ state as confirmed by the dashed line in Figure 5.8. A similar relation has been found for the momentum-dependent decay rate of the image potential states on mono- and bilayers of Xe on Ag(111) [39]. There, the bilayer exhibits both a weaker

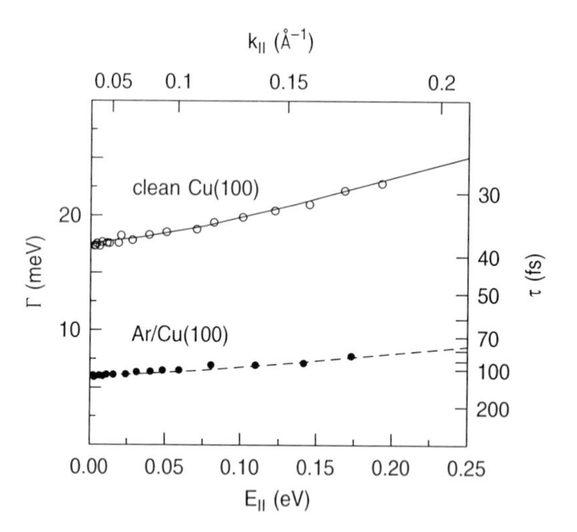

Figure 5.8 Decay rates of the $n = 1$ image potential state (open symbols) and the $n' = 1$ Ar interface state (closed symbols) on Cu(100) as function of the kinetic energy of parallel motion $E_{\|} = \hbar^2 k_{\|}^2 / 2m_{\text{eff}} m_0$. The solid curve indicates the theoretical values for the clean surface from Ref. [9]. The dashed curve represent the same curve but scaled by a factor of 0.34 (see text). Reprinted from [29]. Copyright 2006, IOP Publishing Ltd. and Deutsche Physikalische Gesellschaft.

dependence on $k_{\|}$ and a smaller decay rate at the band bottom compared to the monolayer.

The good agreement with the empirical scaled theoretical decay rate shows that the momentum-dependent dynamics of the interface states is most likely governed by the same mechanism as for the clean surface, namely, inter- and intraband scattering of electrons in the interface state with bulk electrons. Decay processes mediated by phonons in the Ar layer as proposed for the system N_2/Xe/Cu(111) [40] should lead to a stronger $k_{\|}$ dependence. However, scattering at defects is expected not to depend much on $k_{\|}$ [44], which could contribute to a constant offset of the experimental decay rate compared to the scaled theoretical results. On the other hand, the inelastic decay rate of the $n' = 1$ state is very reproducible and insensitive to film thickness as shown in Figure 5.6, which makes defect scattering due to imperfections of the Ar layer not very likely.

5.6 Summary

We have shown the existence of electronic states located at a metal–insulator interface that resemble image potential states but are modified due to screening of the image potential in the insulating layer. We have demonstrated that time-, energy-, and angle-resolved 2PPE experiments are possible despite the fact that these interface states are buried under Ar films as thick as 200 Å. The existence of such buried image potential states makes it possible to investigate electron transfer and decay processes, which are of general relevance to metal–insulator interfaces, in a simple and well-defined model system. In the case of Ar/Cu(100), distinctly different decay channels have been observed: elastic transfer through the Ar film and inelastic decay by e–h pair excitation in the metal. Similar experiments should be possible at other interfaces between metals or semiconductors and dielectric materials, where, in the case of chemisorbed adsorbates, also more strongly localized states will occur in addition to these image potential-type interface states.

Acknowledgments

We acknowledge funding by the Deutsche Forschungsgemeinschaft through grant No. HO 2295/1–4 and by the Center for Optodynamics, Philipps-University Marburg.

References

1 Echenique, P.M. and Pendry, J.B. (1978) The existence and detection of Rydberg states at surfaces. *J. Phys. C: Solid State Phys.*, **11**, 2065–2075.

2 Fauster, T. and Steinmann, W. (1995) Photonic probes of surfaces, in *Electromagnetic Waves: Recent Developments in Research*, vol. 2

(ed. P. Halevi), North-Holland, Amsterdam, pp. 347–411.

3 Höfer, U., Shumay, I.L., Reuß, C., Thomann, U., Wallauer, W., and Fauster, T. (1997) Time-resolved coherent photoelectron spectroscopy of quantized electronic states at metal surfaces. *Science*, **277**, 1480–1482.

4 Knoesel, E., Hotzel, A., and Wolf, M. (1998) Temperature dependence of surface state lifetimes, dephasing rates and binding energies on Cu(111) studied with time-resolved photoemission. *J. Electron Spectrosc. Relat. Phenom.*, **88**, 577–584.

5 Reuß, C., Shumay, I.L., Thomann, U., Kutschera, M., Weinelt, M., Fauster, T., and Höfer, U. (1999) Control of dephasing of image-potential states by adsorption of CO on Cu(100). *Phys. Rev. Lett.*, **82**, 153–156.

6 Sarria, I., Osma, J., Chulkov, E.V., Pitarke, J.M., and Echenique, P.M. (1999) Self-energy of image states on copper surfaces. *Phys. Rev. B*, **60**, 11795–11803.

7 Klamroth, T., Saalfrank, P., and Höfer, U. (2001) Open-system density matrix approach to image-potential dynamics of electrons at Cu(100): energy- and time-resolved two-photon photoemission spectra. *Phys. Rev. B*, **64**, 035420.

8 Shen, X.J., Kwak, H., Radojevic, A.M., Smadici, S., Mocuta, D., and Osgood, R.M. (2002) Momentum-resolved excited-electron lifetimes on stepped Cu(775). *Chem. Phys. Lett.*, **351**, 1–8.

9 Berthold, W., Höfer, U., Feulner, P., Chulkov, E.V., Silkin, V.M., and Echenique, P.M. (2002) Momentum-resolved lifetimes of image-potential states on Cu(100). *Phys. Rev. Lett.*, **88**, 056805.

10 García-Lekue, A., Pitarke, J.M., Chulkov, E.V., Liebsch, A., and Echenique, P.M. (2002) Role of surface plasmons in the decay of image-potential dtates on silver surfaces. *Phys. Rev. Lett.*, **89**, 096401.

11 Weinelt, M. (2002) Time-resolved two-photon photoemission from metal surfaces. *J. Phys. Condens. Matter*, **14**, R1099–R1141.

12 Rhie, H.S., Link, S., Dürr, H.A., Eberhardt, W., and Smith, N.V. (2003) Systematics of image-state lifetimes on d-band metal surfaces. *Phys. Rev. B*, **68**, 033410.

13 Echenique, P.M., Berndt, R., Chulkov, E.V., Fauster, T., Goldmann, A., and Höfer, U. (2004) Decay of electronic excitations at metal surfaces. *Surf. Sci. Rep.*, **52**, 219.

14 Schmidt, A.B., Pickel, M., Wiemhofer, M., Donath, M., and Weinelt, M. (2005) Spin-dependent electron dynamics in front of a ferromagnetic surface. *Phys. Rev. Lett.*, **95**, 107402.

15 Borisov, A.G., Chulkov, E.V., and Echenique, P.M. (2006) Lifetimes of the image-state resonances at metal surfaces. *Phys. Rev. B*, **73**, 073402.

16 Chulkov, E.V., Borisov, A.G., Gauyacq, J.P., Sanchez-Portal, D., Silkin, V.M., Zhukov, V.P., and Echenique, P.M. (2006) Electronic excitations in metals and at metal surfaces. *Chem. Rev.*, **106**, 4160–4206.

17 Winkelmann, A., Bisio, F., Ocana, R., Lin, W.C., Nyvlt, M., Petek, H., and Kirschner, J. (2007) Ultrafast optical spin injection into image-potential states of Cu(001). *Phys. Rev. Lett.*, **98**, 226601.

18 Fauster, T., Weinelt, M., and Höfer, U. (2007) Quasi-elastic scattering of electrons in image-potential states. *Prog. Surf. Sci.*, **82**, 224–243.

19 Güdde, J., Rohleder, M., Meier, T., Koch, S.W., and Höfer, U. (2007) Time-resolved investigation of coherently controlled electric currents at a metal surface. *Science*, **318**, 1287–1291.

20 Harris, C.B., Ge, N.H., Lingle, R.L., McNeill, J.D., and Wong, C.M. (1997) Femtosecond dynamics of electrons on surfaces and at interfaces. *Annu. Rev. Phys. Chem.*, **48**, 711–744.

21 Hotzel, A., Moos, G., Ishioka, K., Wolf, M., and Ertl, G. (1999) Femtosecond electron dynamics at adsorbate–metal interfaces and the dielectric continuum model. *Appl. Phys. B*, **68**, 615–622.

22 Zhu, X.Y. (2002) Electron transfer at molecule–metal interfaces: a two-photon photoemission study. *Annu. Rev. Phys. Chem.*, **53**, 221–247.

23 Machado, M., Chulkov, E.V., Silkin, V.M., Höfer, U., and Echenique,

P.M. (2003) Electron lifetimes in image-potential states at metal–dielectric interfaces. *Prog. Surf. Sci.*, **74**, 219.

24 Berthold, W., Rebentrost, F., Feulner, P., and Höfer, U. (2004) Influence of Ar, Kr, and Xe layers on the energies and lifetimes of image-potential states on Cu(100). *Appl. Phys. A*, **78**, 131.

25 Güdde, J. and Höfer, U. (2005) Femtosecond time-resolved studies of image-potential states at surfaces and interfaces of rare-gas adlayers. *Prog. Surf. Sci.*, **80** 49–91.

26 Berthold, W., Feulner, P., and Höfer, U. (2002) Decoupling of image-potential states by Ar mono- and multilayers. *Chem. Phys. Lett.*, **358**, 502–508.

27 Marinica, D.C., Ramseyer, C., Borisov, A.G., Teillet-Billy, D., Gauyacq, J.P., Berthold, W., Feulner, P., and Höfer, U. (2002) Effect of an atomically thin dielectric film on the surface electron dynamics: image-potential states in the Ar/Cu(100) system. *Phys. Rev. Lett.*, **89**, 046802.

28 Rohleder, M., Berthold, W., Güdde, J., and Höfer, U. (2005) Time-resolved two-photon-photoemission from buried interface states in Ar/Cu(100). *Phys. Rev. Lett.*, **94**, 017401.

29 Rohleder, M., Duncker, K., Berthold, W., Güdde, J., and Höfer, U. (2005) Momentum-resolved dynamics of Ar/Cu (100) interface states probed by time-resolved two-photon photoemission. *New J. Phys.*, **7**, 103.

30 Smith, N.V. (1985) Phase analysis of image states and surface states associated with nearly-free-electron band gaps. *Phys. Rev. B*, **32**, 3549–3555.

31 Echenique, P.M., Pitarke, J.M., Chulkov, E., and Silkin, V.M. (2002) Image-potential-induced states at metal surfaces. *J. Electron Spectrosc. Relat. Phenom.*, **126**, 163–175.

32 Milton Cole, E. and Morrel Cohen, H. (1969) Image-potential-induced surface bands in insulators. *Phys. Rev. Lett.*, **23**, 1238–1241.

33 Cole, M.W. (1971) Electronic surface states of a dielectric film on a metal substrate. *Phys. Rev. B*, **3**, 4418–4422.

34 McNeil, J.D., Lingle, R.L., Jordan, R.E., Padowitz, D.F., and Harris, C.B. (1996) Interfacial quantum well states of Xe and Kr adsorbed on Ag(111). *J. Chem. Phys.*, **105**, 3883–3891.

35 Rohleder, M., Duncker, K., Berthold, W., Güdde, J., and Höfer, U. (2007) Photoelectron spectroscopy of Ar/Cu(100) interface states. *Appl. Phys. A*, **88**, 527–534.

36 Haight, R. (1995) Electron dynamics at surfaces. *Surf. Sci. Rep.*, **21**, 277–325.

37 Echenique, P.M., Pitarke, J.M., Chulkov, E.V., and Rubio, A. (2000) Theory of inelastic lifetimes of low-energy electrons in metals. *Chem. Phys.*, **251**, 1–35.

38 Kliewer, J., Berndt, R., Chulkov, E.V., Silkin, V.M., Echenique, P.M., and Crampin, S. (2000) Dimensionality effects in the lifetime of surface states. *Science*, **288**, 1399–1402.

39 Wong, C.M., McNeill, J.D., Gaffney, K.J., Ge, N.H., Miller, A.D., Liu, S.H., and Harris, C.B. (1999) Femtosecond studies of electron dynamics at dielectric–metal interfaces. *J. Phys. Chem. B*, **103**, 282–292.

40 Hotzel, A., Wolf, M., and Gauyacq, J.P. (2000) Phonon-mediated intraband relaxation of image-state electrons in adsorbate overlayers: N_2/Xe/Cu(111). *J. Phys. Chem. B*, **104**, 8438–8455.

41 Giesen, K., Hage, F., Himpsel, F.J., Riess, H.J., Steinmann, W., and Smith, N.V. (1987) Effective mass of image-potential states. *Phys. Rev. B*, **35**, 975–978.

42 Perluzzo, G., Bader, G., Caron, L.G., and Sanche, L. (1985) Direct determination of electron band energies by transmission interference in thin films. *Phys. Rev. Lett.*, **55**, 545–548.

43 Marinica, D.C., Ramseyer, C., Borisov, A.G., Teillet-Billy, D., and Gauyacq, J.P. (2003) Image states on a free-electron metal surface covered by atomically thin insulator layer. *Surf. Sci.*, **528**, 78–83.

44 Boger, K., Weinelt, M., and Fauster, T. (2004) Scattering of hot electrons by adatoms at metal surfaces. *Phys. Rev. Lett.*, **92**, 126803.

6
Spin-Dependent Relaxation of Photoexcited Electrons at Surfaces of 3d Ferromagnets

Martin Weinelt, Anke B. Schmidt, Martin Pickel, and Markus Donath

6.1
Introduction

Ultrafast dynamics at surfaces and in thin films can be unraveled by femtosecond laser experiments. In particular, two-photon photoemission (2PPE) allows us to study the dynamics of photoexcited electrons directly in the time domain [1, 2]. Decay processes determine the lifetime of photoexcited carriers and are by nature inelastic. In a metal, an electron photoexcited to states above the Fermi level will decay into empty states below. To conserve energy and momentum, for example, a second electron out of the Fermi sea is excited. In contrast, in a quasielastic scattering process, only the momentum of the photoexcited carriers is changed. Momentum scattering usually does not alter the population, but is important in transport processes. Moreover, quasielastic scattering causes dephasing, that is, polarization decay upon optical excitation. Besides understanding these fundamental aspects [2–12], the driving power for this active research field are the impact of excited, hot electrons and holes in chemical reactions [13–16] as well as the importance of electron transport through semiconducting and ferromagnetic devices [1, 17]. A number of excellent review articles considering these topics have been published [1, 2, 18].

In this chapter, we describe some aspects of spin-dependent decay and dephasing processes at surfaces of 3d ferromagnets. Layered magnetic structures are ubiquitous in today's technical application, and transport processes in these devices will be affected by spin-dependent lifetimes of carriers and spin-dependent scattering at surfaces and interfaces.

In a ferromagnet, the band structure is exchange split with a spin-dependent density of states that leads to unequal occupation numbers for majority and minority spin carriers. Thus, relaxation processes become spin dependent and, in general, minority spin electrons have a shorter lifetime than their majority spin counterparts [10, 17, 19–22]. In addition, spin waves can be excited upon electron scattering. Low-energy magnons will cause dephasing while high-energy magnons contribute to inelastic decay of photoexcited carriers. We note that in the ground state of the

Dynamics at Solid State Surface and Interfaces Vol.1: Current Developments
Edited by Uwe Bovensiepen, Hrvoje Petek, and Martin Wolf

ISBN: 978-3-527-40937-2

ferromagnet, that is, for zero temperature $T = 0$ K, spin waves can only be emitted [20]. Conservation of orbital momentum requires that magnons are created by scattering of minority spin electrons. Therefore, magnon emission may lead to additional scattering channels for minority electrons and thus spin-dependent dynamics becomes a consequence of orbital momentum conservation.

In our studies, image potential states serve as well-defined model systems. Detailed investigations of these states over the past two decades have helped to develop a profound understanding of electron dynamics at metal surfaces [3, 12, 18, 23–25]. Two-photon photoemission allows us to address the energetics and dynamics of image potential states. They offer the unique possibility to study electron dynamics in a system, where momentum and energy of the excited electron are well defined and can be determined in angle-resolved two-photon photoemission. The lifetime of these states is unaffected by diffusion processes and refilling processes can be suppressed by a proper choice of the excitation energy. This allows us to follow momentum- and energy-dependent dynamics in time-resolved two-photon photoemission. Image potential states are introduced together with the spin-resolved extension of 2PPE in Section 6.2. Prior to our work, spin-resolved 2PPE studies investigated the spin-dependent dynamics of photoexcited bulk electrons in thin films of 3d metals [17, 26, 27]. A brief report on these results with regard to spin-dependent scattering processes is given at the outset of Section 6.3. Following this, we discuss in detail the spin-dependent decay and dephasing processes and give strong evidence for magnon contributions in both inelastic and quasielastic electron scattering in the 3d ferromagnet iron. The energetics of image potential states on ferromagnetic iron and cobalt surfaces is outlined in Section 6.4. The exchange splitting of these states reflects the exchange–split boundaries of the bulk bandgap. Magnetic linear dichroism in two-photon photoemission allows us to map spin–orbit hybridization points in the bulk band structure and elucidate the role of image potential states as sensors for the near-surface magnetization.

6.2
Spin-Resolved Two-Photon Photoemission on Image Potential States

6.2.1
Image Potential States at a Ferromagnetic Surface

Besides the electronic states derived from the bulk bands and intrinsic surface states such as Tamm or Shockley states [28, 29], image potential states occur at conductive surfaces [30, 31]. Extensive reviews on the physics of these states may be found in Refs [18, 23, 24, 32–34] and we refer the reader to chapters 3, 4 and 5 of this book for a detailed discussion.

Image potential states are a special class of electronic states at surfaces. An electron approaching a polarizable surface feels the attractive force of its own image potential, created by the polarization charge it induces in the surface region of the solid. If a gap in the surface-projected bulk band structure or a potential step prevents the electron from propagating into the solid, it may be transiently trapped several Ångstrom in front of the surface. Due to the long-range nature of the Coulomb potential, a Rydberg-like series of bound states is formed. These so-called image potential states appear within 0.85 eV below the vacuum level E_{vac}, to which they are pinned. In a ferromagnet, the electronic bands are exchange split, and the interaction of the image potential states with the surface of the ferromagnet lift the spin degeneracy [35]. In a multiple scattering approach, the two subsystems of majority and minority spin electrons can be treated separately, and the different positions of the spin-up and spin-down bulk band edges yield a spin-dependent crystal barrier. Hence, a spin-dependent quantum defect $a^{\uparrow\downarrow}$ can be introduced to describe the binding energies $E_n^{\uparrow\downarrow}$ of the image potential states in front of a ferromagnetic metal surface:

$$E_{vac} - E_n^{\uparrow\downarrow} = \frac{Ry}{16} \frac{1}{(n + a^{\uparrow\downarrow})^2}, \quad n = 1, 2, \ldots. \tag{6.1}$$

Accordingly, the spin splitting $\Delta E^{\uparrow\downarrow} = E_n^{\uparrow} - E_n^{\downarrow}$ is expected to scale with n^{-3} for large quantum number n. The image potential is induced by screening from the electrons near the Fermi level, where the ferromagnetic surface usually has a spin-dependent density of states [36]. Hence, based on the exchange interaction near the crystal surface, the image potential experienced by electrons outside the metal is also spin dependent [37]. For the Fe(110) surface, Nekovee *et al.* showed that, due to the negative polarization of the charge density in the surface layer, the contribution of the image potential barrier is, though comparatively small, opposite in sign to the contribution of the crystal barrier potential.

Image potential states are unoccupied and can be probed by inverse photoemission (IPE) [31, 38], that is, the detection of photons emitted by an incoming electron upon relaxation into an empty state, two-photon photoemission [39], and tunneling spectroscopy [40]. Spin-resolved IPE proved for the first time the magnetic exchange splitting of the $n = 1$ image potential state on Ni(111) and Fe(110) [19, 41]. Moreover, the linewidth analysis of spin-resolved IPE data suggests spin-dependent lifetimes of the first image potential state. Two-photon photoemission allows us to address not only the energy but also the time domain and its spin-resolved extension is described in the following section.

6.2.2 Spin-Resolved Two-Photon Photoemission

In 2PPE, a first laser pulse $\hbar\omega_a$ excites an electron from an occupied bulk or surface state below the Fermi level into an unoccupied intermediate state and from there a second laser pulse $\hbar\omega_b$ raises the excited electron above the vacuum level. The photoelectrons are then selected by energy and emission angle. In order to avoid direct photoemission, which can overwhelm the second-order signal of 2PPE, the photon energy of the pump pulse is usually kept below the work function $\Phi = E_{vac} - E_F$ of the sample, while chosen high enough to excite the states of interest from initial states below the Fermi level E_F.

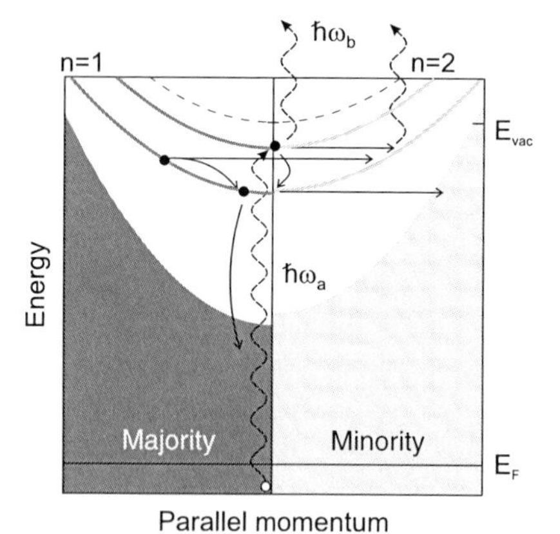

Figure 6.1 Two-photon photoemission scheme for a ferromagnetic surface. Within each spin subsystem, the ultraviolet pump pulse ($\hbar\omega_a$, UV) excites electrons from bulk states to the $n = 1$ and $n = 2$ image potential state parabolas. Arrows mark inelastic and quasielastic (horizontal arrows) scattering processes. They redistribute excited electrons among the image potential state bands and can include spin-flip scattering. The infrared pulse ($\hbar\omega_b$, IR) probes the intermediate-state population at a distinct time delay after the excitation with the UV pulse. In our experiment $\hbar\omega_a = 3 \times \hbar\omega_b$.

Pump and probe processes are illustrated in the schematics of the surface-projected bulk band structure of Figure 6.1. Photoexcitation occurs separately within each spin subsystem of the ferromagnet. State selectivity in 2PPE is obtained measuring the emission angle ϑ and kinetic energy E_{kin} of the outgoing photoelectron. Thus, we obtain the dispersion relation $E(k_{\|})$ with the parallel momentum:

$$k_{\|} = \sqrt{2m_e E_{\text{kin}}/\hbar^2}\,\sin\vartheta. \tag{6.2}$$

Herein m_e is the free electron mass and E_{kin} is the kinetic energy of the photoelectron measured with respect to the vacuum level of the sample surface. The latter is deduced from the position of the low-energy cutoff at normal emission $\vartheta = 0°$. To indicate E_{vac}, we have used a marker at $k_{\|} = 0$ in Figure 6.1, since the momentum parallel to the surface does not help to overcome the vacuum barrier. Therefore, the position of the low-energy cutoff disperses like a free electron parabola (dashed line). The electron trapped in front of the surface moves virtually free parallel to the surface and we observe parabolic image potential state bands at most low-index metal surfaces. Equation (6.1) defines the binding energy for $k_{\|} = 0$ at the bottom of the image potential state band.

To separate the 2PPE signals of majority spin electrons $I_{\uparrow}$ and minority spin electrons $I_{\downarrow}$, the spin polarization P of the photocurrent $I = I_{\uparrow} + I_{\downarrow}$ has to be determined:

$$I_{\uparrow\downarrow} = \frac{1 \pm P}{2} I, \tag{6.3}$$

$$P = \frac{I_\uparrow - I_\downarrow}{I}. \quad (6.4)$$

The polarization P of a spin-polarized current is measured in an electron scattering experiment, where spin–orbit coupling in a high-Z element causes a left–right asymmetry in the intensity of the scattered beams with count rates N_1 and N_2 (see Section 6.2.3) [42–44]. Alternatively, P can be determined by exchange scattering at the surface of a ferromagnet [45–47].

$$P = \frac{1}{S}\frac{N_1 - N_2}{N_1 + N_2} \quad (6.5)$$

The Sherman function S is a detector property that describes the spin sensitivity. S corresponds to the measured asymmetry of a fully spin-polarized beam with $P = 1$ and lies usually in the range of 0.1–0.3, but is difficult to determine exactly. Note that the small value of S implies comparable count rates N_1 and N_2, since $N_1 - N_2 \leq S \times (N_1 + N_2)$. Instrumental asymmetries have to be eliminated by repeating the experiment with reversed magnetization. In this case, the count rate N_i in Eq. (6.5) is replaced with the geometrical mean of the count rates N_i and $\overline{N}_i$ obtained from two measurements with opposite magnetization ($M_\uparrow$ and $M_\downarrow$) [48]:

$$P = \frac{1}{S}\frac{\sqrt{N_1\bar{N}_2} - \sqrt{N_2\bar{N}_1}}{\sqrt{N_1\bar{N}_2} + \sqrt{N_2\bar{N}_1}}. \quad (6.6)$$

By error propagation of

$$\Delta N_i = \frac{1}{2}\sqrt{\frac{N_i}{\bar{N}_j}(\Delta\bar{N}_j)^2 + \frac{\bar{N}_j}{N_i}(\Delta N_i)^2}, \quad i,j = 1,2 \wedge i \neq j, \quad (6.7)$$

the error of the spin polarization can be calculated as

$$\Delta P = \frac{2}{S}\frac{\sqrt{N_2^2\Delta N_1^2 + N_1^2\Delta N_2^2}}{(N_1 + N_2)^2}. \quad (6.8)$$

Finally, we obtain the random error of the 2PPE count rate of majority and minority spin electrons:

$$\Delta I_{\uparrow\downarrow} = \frac{1}{2S}\sqrt{(S+1)^2\Delta N_i^2 + (S-1)^2\Delta N_j^2}, \quad i,j = 1,2 \wedge i \neq j. \quad (6.9)$$

In the limit of large count rates, $\Delta N_i = \sqrt{N_i}$, since independent events are observed. As $\Delta N_i \approx \Delta N_j$ for small S, Eq. (6.9) yields $\Delta I \propto \Delta N_i / S$. Therefore, one defines the figure of merit of the spin detector as

$$\eta = S^2 \cdot \frac{N_1 + N_2}{N_0}, \quad (6.10)$$

where N_0 is the count rate without spin detector. Using the spin-polarized low-energy electron diffraction (SPLEED) for spin selection reduces the count rate by three to four orders of magnitude $(N_1 + N_2)/N_0 \approx 10^{-3}$–$10^{-4}$. Thus, a long data recording

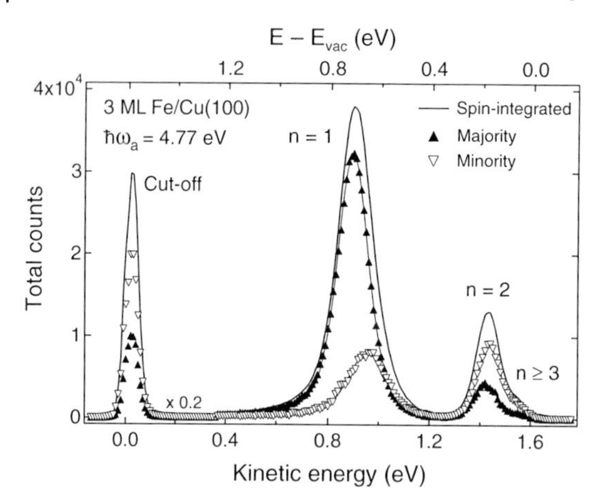

Figure 6.2 Spin- and energy-resolved 2PPE spectra of 3 ML Fe/Cu(100). Spectra are recorded in normal emission for a photon energy of $\hbar\omega_a = 3 \times \hbar\omega_b = 4.77\,\text{eV}$. The 3 ML iron film is magnetized out-of-plane. The exchange splitting $\Delta E_n^{\uparrow\downarrow}$ between majority and minority components decreases with n^{-3}. Note the different intensity and spin polarization of the first and the second image potential states $n = 1$ and 2.

time is in general required to obtain sufficient statistics [42, 48]. Furthermore, the Sherman function S adds a significant contribution to the statistical error in spin-sensitive experiments. In our experimental setup, we use a SPLEED detector with a Sherman function of $S = 0.24 \pm 0.02$. The uncertainty in the Sherman function of $\Delta S = \pm 0.02$ is a systematic error and is therefore not included in the statistical error of the displayed data. ΔS is, however, taken into account in the spin splitting of the image potential states as specified in our publications.

Figure 6.2 shows an example of a spin- and energy-resolved 2PPE spectrum of a three monolayer (ML) iron film on Cu(100). As just described, two spectra with opposite out-of-plane magnetization were successively recorded in normal emission with zero delay between pump and probe pulses. The maximum spin-resolved count rate in Figure 6.2 is $> 10^4$ counts/s and the error defined in Eq. (6.9) is within the symbol size. The kinetic energy scale is defined with respect to the sample, that is, the low-energy cutoff is set to zero kinetic energy. The kinetic energy corresponds to $E_{\text{kin}} = E_{\text{F}} - \Phi + \hbar\omega_a + \hbar\omega_b$ and thus the high-energy cutoff of the 2PPE spectrum to the Fermi level ($E_{\text{kin}} = 1.61 \pm 0.02\,\text{eV}$ for $\Phi = 4.75 \pm 0.02\,\text{eV}$). The binding energy scale on the upper ordinate refers to the vacuum level. Within 0.85 eV binding energy, we observe the first and second image potential states. The higher states $n \geq 3$ are not resolved. The exchange splitting $\Delta E_n^{\uparrow\downarrow}$ of the first and second image potential states of $\Delta E_1^{\uparrow\downarrow} = 56 \pm 10\,\text{meV}$ and $\Delta E_2^{\uparrow\downarrow} = 7 \pm 3\,\text{meV}$ scales with the penetration depth of the image potential state into the crystal, that is, with n^{-3} [10]. While for the $n = 1$ image potential state, the majority component dominates, intensities are reversed for the $n = 2$ state. As the exchange splitting of the image potential states is rather small, the wave function of minority and majority spin components is comparable. Thus, the spin-dependent intensities of the image potential states must mimic the

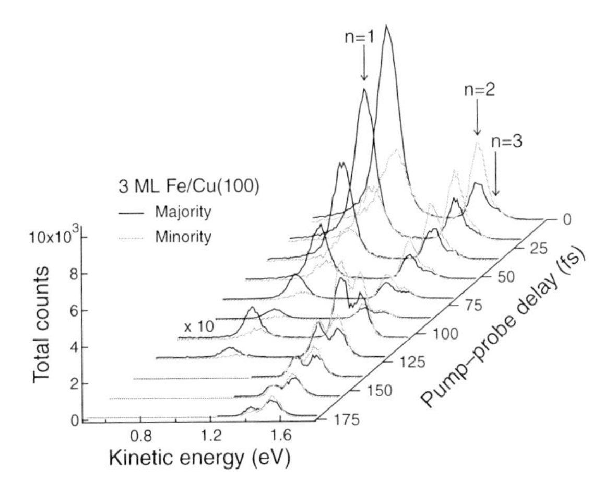

Figure 6.3 Spectra of energy-resolved 2PPE measurements of 3 ML Fe/Cu(100) for increasing pump–probe delay. The majority component is displayed in black, the minority component in gray.

spin-dependent density of the initial iron bulk states. As will be illustrated in Section 6.4, this allows us to access the spin-dependent bulk band structure and to follow the temperature dependence of the sample magnetization.

It is the power of 2PPE that we can not only observe the empty states of the electronic band structure but also access their dynamics. Figure 6.3 shows the temporal evolution of the image potential state peaks in the spin- and energy-resolved 2PPE spectrum of Figure 6.2. In all image potential state peaks, the intensity decreases exponentially with increasing pump–probe delay, fastest in the $n = 1$ state. The $n = 2$ peak decays on a slower timescale and after about 120 fs, the intensity of the $n \geq 3$ image potential states dominates. Moreover, the minority components decay faster than the majority components. This is most obvious for the second image potential state, where the intensity of the majority component overtakes the minority intensity after approximately 120 fs.

In a time-resolved measurement, we tune the analyzer to the peak maximum of the state of interest and record its intensity as a function of pump–probe delay. Such traces are shown in Figure 6.4 for the first image potential state on 3 ML iron on Cu (100). For each delay, we record four data points with the analyzer tuned to either of the two peak maxima exchange split by 56 meV and for the two reversed magnetization directions. This, and the dynamics in the count rate spanning over three orders of magnitude, allows us to determine rather precisely the spin dependence of the $n = 1$ lifetime. Since the intensity is plotted on a logarithmic scale, the exponential decay of the $n = 1$ population manifests as a straight line in the spectra at positive delay. Lifetimes have been evaluated solving optical Bloch equations (see Section 6.3.3) [49, 50]. The duration of the IR and UV pulses and delay zero were determined by autocorrelation and cross-correlation measurements [23]. From the simulation (solid lines in Figure 6.4), we obtain lifetimes of $\tau_1^{\uparrow} = 16 \pm 2$ and of $\tau_1^{\downarrow} = 11 \pm 2$ fs for the majority and the minority spin components, respectively.

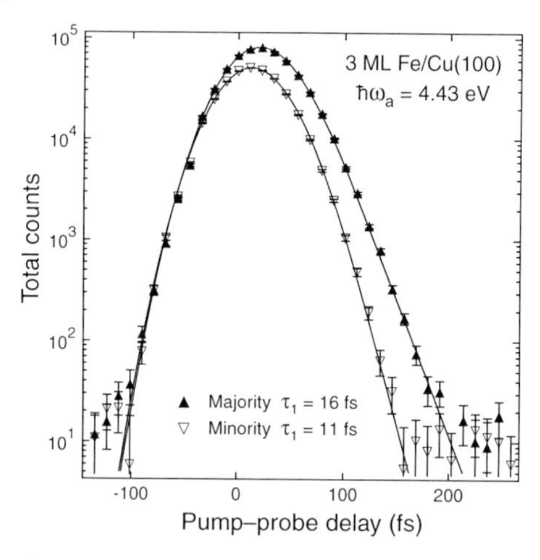

Figure 6.4 Time-resolved measurement of the $n = 1$ image potential state on 3 ML Fe/Cu(100). Spectra are recorded in normal emission with a photon energy of $\hbar\omega_a = 3 \times \hbar\omega_b = 4.43$ eV.

We now have a preliminary introduction to spin-resolved two-photon photoemission and a survey of some of the fascinating phenomena we encounter when knowing the electron's spin. The spin dependence of image potential state lifetimes will be further discussed in Section 6.3, while the use of their spin polarization and spin-dependent binding energies as a sensor of magnetism will be discussed in Section 6.4.

6.2.3
Key Aspects of the Experiment

The experimental setup sketched in Figure 6.5 for spin-, time-, angle-, and energy-resolved two-photon photoemission spectroscopy consists of two parts: (a) a laser system to generate femtosecond pump and probe pulses and (b) an ultrahigh vacuum system equipped with a photoelectron spectrometer combined with the SPLEED-type spin polarization detector. The key parameters of the setup are listed in Table 6.1.

a) The laser system is a Ti:sapphire oscillator pumped by a solid-state, diode-pumped, frequency-doubled Nd:vanadate laser (532 nm single frequency routinely run at 5.2 W). The design of the home-built oscillator is similar to the one described in Ref. [51]. The ultrashort laser pulses generated by our oscillator in the mode-locked state are frequency tripled via second harmonic generation (SHG) with an LBO crystal and subsequent sum-frequency generation (THG) in a BBO crystal. The third harmonic pulse $\hbar\omega_a$ is used as the pump pulse in the 2PPE experiment, while the fundamental frequency supplies the probe pulse $\hbar\omega_b$.
b) The home-built ultrahigh vacuum system with a base pressure below 3×10^{-11} mbar comprises components for sample preparation and characterization. The thickness of evaporated iron and cobalt films is monitored via medium-energy electron diffraction and a quartz microbalance. In a separate analysis chamber,

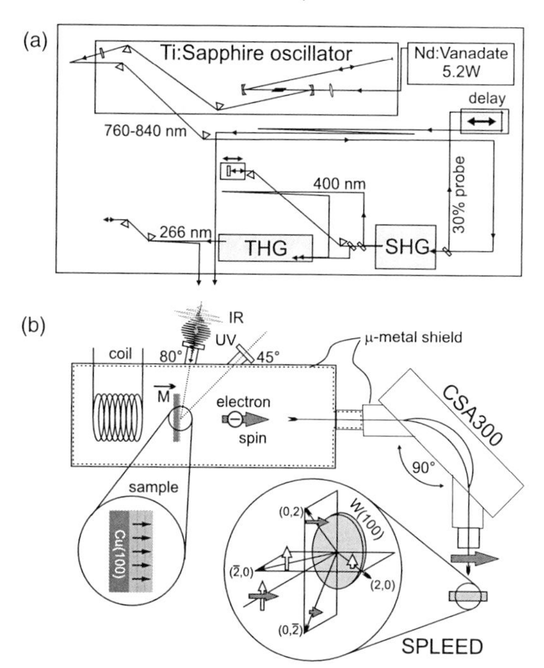

Figure 6.5 Experimental setup for time- and spin-resolved two-photon photoemission. (a) The laser source is a home-built 88 MHz Ti: sapphire oscillator with frequency tripling. (b) The electron analyzer is manufactured at FOCUS company and consists of a cylindrical-sector electron energy analyzer CSA300 in combination with a SPLEED-type spin polarization detector based on spin-polarized low-energy electron diffraction at a W(100) crystal surface. The setup shows a sample with perpendicular anisotropy. To magnetize cobalt films in-plane, the coil is rotated by 90° and centered below the sample. Sample preparation takes place in a separate chamber.

a CSA300 electron analyzer (Focus GmbH) is used for electron detection (see Figure 6.3). The analyzer is combined with a separately pumped spin polarization detector based on SPLEED [42]. The two-chamber system ensures that the tungsten crystal of the SPLEED detector can be cleaned without affecting the sample preparation and vice versa. Since the kinetic energy of the excited electrons is typically around 1 eV, the μ-metal shielding of the analyzer lens system has been elongated to cover the sample area. We reduced the entrance aperture of the analyzer by a factor of 9 to yield an angular resolution of $\pm 5°$.

In contrast to noble metals, investigating surfaces of ferromagnetic samples provides the experimentalists with several challenges: (i) *Growth of bulk single crystals without impurities*: Usually single crystals are grown from the melt. For example, in Fe, a structural phase transition at 1183 K prevents single crystal production. The phase transition can be avoided by adding a small percentage of Si, which, however, introduces a high impurity density and can modify the magnetic properties. In the case of Gd, single crystal growth from the melt inevitably comes with a high density of

Table 6.1 System parameters.

88 MHz Ti:sapphire oscillator and frequency tripling	
IR laser power	900 mW – IR at 5.2 W pump power
Tunability	725–850 nm
UV pump pulse	12 mW at 266 nm – 50 fs pulse duration
IR probe pulse	240 mW – 20–40 fs pulse duration
CSA300 electron energy analyzer and SPLEED detector	
Spin-resolved count rate	$\approx 10^4$ counts/s ($n = 1$ on Fe/Cu(100))
Sherman function	0.24 ± 0.02
Angular resolution	$\pm 5°$
Energy resolution	50 meV
Residual magnetic field	$\leq$2 mG between sample and analyzer
Combined energy resolution	70 meV

various impurities. (ii) *Defined magnetic structure*: In most cases, magnetic anisotropies prevent a single domain structure at the surface. A defined magnetic structure is, however, necessary for quantitative spin analysis in electron spectroscopy. (iii) *Magnetic fields*: The large stray field of a magnetic bulk sample poses a major obstacle for low-energy electrons. Likewise, a single domain state cannot be forced by external fields. A solution of all three problems is the preparation of ferromagnetic samples as thin films on a nonmagnetic single crystal. As the probing depth of photoelectrons is a few Ångstrom only, the electronic states, especially surface states, of thin films resemble the electronic structure of bulk crystal surfaces.

The thin films are in-plane or out-of-plane magnetized directly in the measurement position by an air coil mounted on an adjustable mechanical feedthrough. Reversing the magnetization direction without changing the sample position allows us to eliminate the experimental asymmetry. In the energy-resolved mode, we record spectra with both magnetization directions successively, while in the time-resolved mode we reverse the magnetization for each data point. The manipulator head holds two samples, a Cu(100) and a Cu(111) crystal. The latter is used to measure the cross-correlation of the occupied Shockley surface state.

6.3 Spin-Dependent Dynamics

6.3.1 Spin-Dependent Population Decay of Bulk Electrons

The spin dependence of the lifetime of carriers is a well-established characteristics of bulk 3d ferromagnets [19, 52–55]. First results in the time domain employing spin-resolved 2PPE were published in 1997 [17, 26]. In Ref. [17], Aeschlimann *et al.* studied hot electron lifetimes in fcc cobalt films on Cu(100). A suction voltage of 10 V was applied between the cesiated Co(100) thin film sample and the analyzer, which leads to an integration over part of the $\vec{k}$-space [56]. In fair agreement with Fermi liquid theory [57], spin-integrated lifetimes τ increase with energy as $(E - E_F)^{-2}$ when approaching the Fermi level [17]. In spin-resolved measurements, a ratio $\tau^\uparrow/\tau^\downarrow$ of ≈ 2 was established at about 1 eV above the Fermi level E_F, which decreases to ≈ 1.3 at $E - E_F = 0.6$ eV. Similar $\tau^\uparrow/\tau^\downarrow$ ratios were later found for iron and nickel thin films [27] and have been recently confirmed [58].

For a current review on the theoretical description of low-energy electron and hole excitations in metals, we refer to Ref. [22]. Qualitatively, the spin dependence of the electron lifetimes in the 3d ferromagnets can be explained by the exchange split density of d-states illustrated in Figure 6.6a. Decay rates of photoexcited electrons are mainly determined by inelastic electron–electron scattering, that is, the creation of electron–hole pairs sketched in Figure 6.6b. As pointed out in Ref. [19], this process depends on the number of available final states. Pure Coulomb scattering does not support a spin-flip process. Therefore, the larger density of unoccupied d-states in the minority channel leads to a shorter lifetime $\tau^\downarrow$ of minority spin electrons. To conserve energy and momentum, inelastic decay of the primary electron is accompanied by excitation of a secondary electron–hole pair in either the minority or the majority spin channel. The latter process is sketched in Figure 6.6b. The corresponding Feynman diagram in Figure 6.6c describes the propagation of the minority electron ($e^\downarrow$) and its change in momentum and energy upon Coulomb interaction with an electron–hole pair in the majority channel ($e^\uparrow - h^\uparrow$). The exchange of a virtual photon is described by the wiggly line.

In an experiment, where the excited electron leaves the sample, the minority spin electron and the majority spin hole left behind is a spin-flip excitation of the ferromagnet, the so-called Stoner excitation. In spin-polarized electron energy loss spectroscopy (SPEELS), such a "virtual spin-flip" process was found to be of similar strength as the direct scattering event without a spin-flip [59–62].

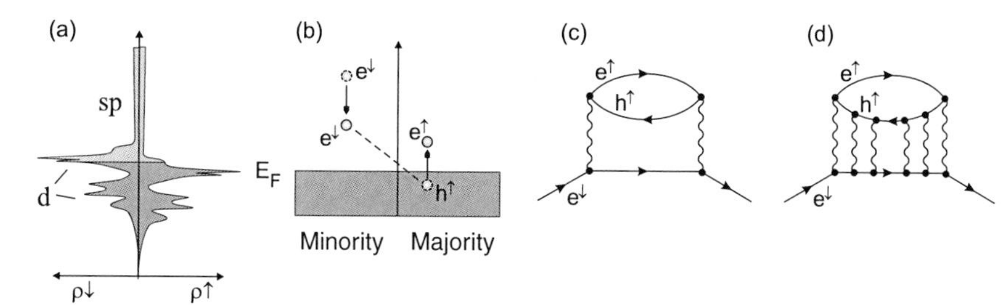

Figure 6.6 Schematic illustration of decay processes. (a) Density of states of an itinerant ferromagnet. The d-bands are significantly exchange split, while the sp-derived density of states ϱ shows only a weak spin dependence. (b) Inelastic electron–electron scattering. Depicted is an inelastic scattering process, where an electron decays in the minority channel, creating an electron–hole pair in the majority channel. (c) Feynman diagram for the contribution of electron–hole pair creation in the majority spin channel to the self-energy of a minority spin electron. (d) Feynman diagram of the final-state interaction depicted in (b) by the dashed line. The magnon emission corresponds to multiple scattering between the primary electron and the hole of an electron–hole pair in the opposite spin channel (Feynman diagrams after Refs [20] and [21]).

Besides the creation of Stoner pairs, spin-flip scattering of electrons can lead to excitation of collective modes, that is, magnons. Such spin waves have been found in SPEELS for low energies (≤ 10 eV) of the primary electron beam [63, 64]. The collective nature of a spin wave excitation is conceived in the ladder graph of Figure 6.6d [20, 21]. Here multiple scattering between the electron $e^{\downarrow}$ and the hole $h^{\uparrow}$ in opposite spin channels describes the *spread* of the spin-flip excitation via the creation of virtual electron–hole pairs. In the framework of many-body theory, the decay rate $\Gamma = \tau^{-1}$ of an excited electron is derived from the imaginary part of the self-energy Σ as

$$\Gamma = -\frac{2}{\hbar}\Im\Sigma. \tag{6.11}$$

The real part of Σ describes the renormalization of the single-particle energy, that is, the energy and dispersion of the dressed quasiparticle. The self energy Σ is usually calculated in the so-called GW approximation, evaluating the first term in a perturbation series of the screened Coulomb interaction W. To include Stoner and spin wave excitations, multiple scattering processes (*ee*, *hh*, and *eh*) are added in the calculation of the self-energy in the $GW + T$ extension [65, 66].

Note that independent of the source of the primary electron, the final state interaction $e^{\downarrow}-h^{\uparrow}$ describes the emission of a magnon in the bulk of the ferromagnet as indicated in Figure 6.6b by the dashed line. A hole in the majority spin channel and an electron in the minority spin channel correspond to the creation of a spin wave, which lowers the magnetization along the quantization direction by $2\mu_B$. In an inelastic electron scattering process, where a magnon is created, the total momentum, energy, and angular momentum must be conserved. At low temperatures, magnons are frozen out and only spin wave emission is possible. Even at room temperature, only magnons with small momentum are thermally excited. Considering magnons in the whole Brillouin zone, spin wave emission will dominate [67]. This dominance of magnon emission leads to an additional decrease of the lifetime of minority spin electrons, since inelastic decay of a majority spin electron can only be accompanied by magnon absorption. Moreover, if spin wave emission is dominant, the spin asymmetry in the lifetime ($\tau^{\downarrow} < \tau^{\uparrow}$) has its origin in the conservation of angular momentum rather than the details of the band structure.

Self-energy calculations by Hong and Mills, describing the 3d ferromagnets by a one-band Hubbard model, predict that spin wave emission is particularly important for inelastic electron scattering with low energy transfer [20]. In a nutshell, magnon emission in the 3d ferromagnets, even though a higher order term in inelastic electron–electron scattering, is predicted to be dominant at low excitation energies and expected to qualitatively vary with the amount of unoccupied minority spin d states. Thus, with the decrease of the exchange splitting from iron via cobalt to nickel, magnon emission upon inelastic decay of hot carriers should be less important [20].

This was corroborated by *ab initio* calculations of electron lifetimes in bcc iron and fcc nickel [21]. According to the calculations by Zhukov *et al.*, it is mainly magnon emission, which leads to the spin asymmetry of electron lifetimes in iron for energies up to 1.0 eV above E_F. For nickel, in contrast, decay rates are not affected by spin wave emission. Here, the asymmetry is a pure band structure effect.

Unfortunately, measured and calculated spin asymmetries differ significantly [17, 21, 27, 68]. In particular, the calculated lifetimes of majority spin carriers are much larger than experimental values and, thus, much higher spin asymmetries are predicted theoretically. Such large spin asymmetries have been indeed observed in transport experiments [69–71]. However, this *transport* lifetime of quasiparticles near the Fermi surface is of different nature than the lifetime of photoexcited carriers. Measurements of the electric current include spin-dependent electron velocities [72] and momentum scattering [20]. We should likewise mention that in the 2PPE experiment, measured lifetimes of photoexcited electrons in bulk states, low in energy above E_F, can be affected by both secondary electrons (cascade and Auger electrons) and transport processes. Secondary electrons repopulate the probed state, whereas ballistic transport and diffusion depopulate the state. These processes are again spin dependent and transport phenomena are made responsible for the difference in calculated and measured lifetimes of majority electrons [22]. It is fair to state that we have reached a qualitative understanding of spin-dependent electron lifetimes in bulk states of 3d ferromagnets, while a quantitative description still offers challenges.

6.3.2 Spin-Dependent Lifetimes of Image Potential States

At ferromagnetic surfaces, the image potential states provide a simple model system for a detailed investigation of spin-dependent electron dynamics in the time domain and the identification of spin-dependent scattering processes.

Image potential states posses energy and momentum, which can be unambiguously accessed in the 2PPE experiment. Their population decay is not affected by transport processes away from the surface, refilling processes can be controlled and determined, and due to the small overlap with bulk states, electron–phonon coupling is usually negligible or, where not, well accessible. Nevertheless, the ratio of majority and minority lifetimes $\tau^{\uparrow}/\tau^{\downarrow}$ mirrors the ratio of hot electron lifetimes in the bulk at comparable excitation energy above E_F.

In Figure 6.3, we have already depicted the measurement of the population decay of the $n = 1$ image potential state for 3 ML Fe on Cu(100). Significantly longer lifetimes have been observed for the $n = 2$ state (Figure 6.7). Here the spin dependence of the decay rate can be directly extracted from the slope of the time-resolved traces after the pump pulse is over, that is, after about 50 fs. The ratio $\tau^{\uparrow}/\tau^{\downarrow} = 54\,\text{fs}/43\,\text{fs}$ amounts to 1.26. We note that similar ratios have been observed for image potential states on Co/Cu(100) [25]. Thus, the lifetime ratios at the bottom of the image potential state band, for both iron and cobalt, are comparable to the lifetime of bulk electrons discussed in the previous section. Even in the high-energy region between 3 and 5 eV above E_F, the ratios predicted by theory for iron, cobalt, and nickel are much higher ($\tau^{\uparrow}/\tau^{\downarrow} \approx 3-5$) ([21, 68] Zhukov (2009) Unpublished). As already discussed for the time-resolved measurements of the image potential state lifetime, we can exclude carrier diffusion or refilling processes. Thus, we have to

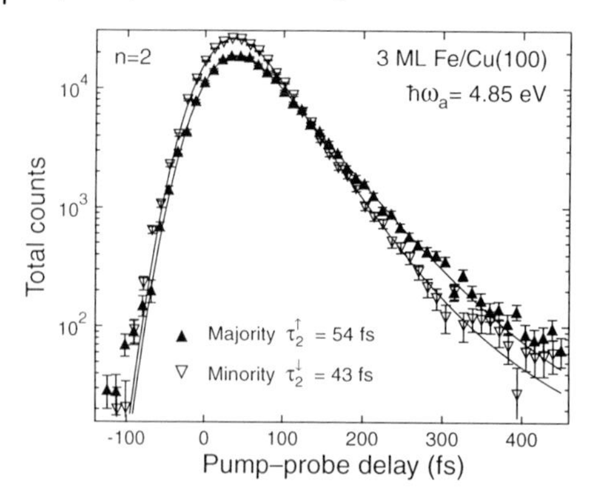

Figure 6.7 Time-resolved 2PPE measurement of the $n = 2$ image potential state on Fe for majority and minority electrons. The simulation (solid lines) was done using optical Bloch equations.

conclude that for large energy transfer in inelastic decay, the lifetime asymmetries are overestimated by *ab initio* theory.

6.3.3 Quasielastic Scattering: Spin-Dependent Dephasing

The lifetimes of states observed with 2PPE may be of the order of, or even shorter than, the duration of the laser pulses used in the experiment. Decay of the population and transitions between excited states occur during the excitation into the final state above the vacuum energy, while the pump pulse may still be active. We can therefore, in general, not treat 2PPE as a process of two consecutive and independent photoexcitation steps. However, the 2PPE process can be modeled with the Liouville–von Neumann equations in the framework of density matrix formalism [23, 49, 50, 73, 74]. The elements of the density matrix ϱ_{mn} can then be interpreted as follows: The diagonal elements ϱ_{nn} give the probability that the system is found in the energy eigenstate $|n\rangle$, or, in an ensemble, the average population of this state. The coherence between states $|m\rangle$ and $|n\rangle$ is given by the off-diagonal elements ϱ_{mn}. When a laser field induces a coherent superposition between states, the off-diagonal elements are proportional to the polarization, that is, the induced dipole moment in the system. The density matrix equation of motion with the phenomenological inclusion of dissipation is given by

$$\begin{aligned} \dot{\varrho}_{mn} &= -\frac{i}{\hbar}[H_0 + H_{\text{int}}, \varrho]_{mn} - \gamma_{mn}\varrho_{mn}, \quad m \neq n, \\ \dot{\varrho}_{nn} &= -\frac{i}{\hbar}[H_0 + H_{\text{int}}, \varrho]_{nn} + \sum_m \Gamma_{nm}\varrho_{mm} - \Gamma_n\varrho_{nn}, \end{aligned} \tag{6.12}$$

where H_0 is the unperturbed Hamiltonian of the system with $H_0|n\rangle = E_n|n\rangle$ and H_{int} describes the interaction with the laser field of the pump and probe pulses. The population ϱ_{nn} decays with the rate $\Gamma_n = 1/\tau_n$. As already discussed in Section 6.3.2, mainly inelastic scattering into the bulk and concomitant electron–hole pair creation governs this lifetime τ_n. Refilling of state $|n\rangle$ from other levels $|m\rangle$, that is, inelastic or quasielastic interband scattering, occurs with Γ_{nm}. The damping rate of the off-diagonal elements

$$\gamma_{mn} = \frac{1}{2}(\Gamma_m + \Gamma_n) + \Gamma^*_{mn} \tag{6.13}$$

accounts for the relaxation of the ϱ_{mn} coherence, that is, destroys the phase relation between the electron wave function and the laser field. Note that the population decay also causes decoherence, but the so-called *pure* dephasing rate Γ^*_{mn} leaves the population unchanged.

Generally, equations of motion (6.12) are solved numerically and the resulting simulation is then compared with the measured 2PPE traces as shown, for example, in Figure 6.4. Thus, with the help of the simulation, lifetimes considerably shorter than the laser pulse duration can be analyzed [3]. Of the parameters entering the simulations, as many as possible are collected from independent experiments. The pulse length of the probe pulse is determined by interferometric autocorrelation measurements. The cross-correlation trace obtained by two-photon photoemission from the occupied surface state on Cu(111) directly follows the time profile of the pump and probe pulses, providing a measure of the pump pulse length (for known probe pulse length), and also a precise measurement of the time zero between pump and probe pulses.

While conventional photoemission probes a time-constant population, in 2PPE, the probe pulse samples the strongly time-dependent transient population of the intermediate state. In both cases, decay and dephasing contribute to the linewidth of the measured state, but in 2PPE they contribute independently.

A look at Figure 6.8a shows that with increasing time delay between pump and probe pulses, the linewidth of the majority and minority spin $n = 1$ image potential state decreases. Linewidth versus pump–probe delay is plotted in Figure 6.8b. The line shape of the intermediate state in 2PPE is a convolution of a Gaussian function, which is determined by the experimental resolution (i.e., the energy resolution of the electron analyzer and the bandwidth of the probe pulse), with a Lorentzian curve, whose width depends on pump–probe delay [50]. For increasingly negative delay, where the probe pulse precedes the pump pulse, photoemission is mainly caused by the overlap of the pulse tails and the Lorentzian width increases linearly with decreasing delay. The pump and probe pulse durations determine the slope. An offset between two curves for large negative delay corresponds to the difference in decay rates. For a constant population of the intermediate state, ionization of the intermediate state by the probe pulse corresponds to conventional photoemission. But, while the pump pulse is active, population buildup and decay compete and the population of the intermediate state depends in a nontrivial manner on time. Therefore, as long as the pump pulse is active, both lifetime and dephasing contribute to the linewidth. In the limit of separated pump and probe pulses, that is, when the pump process is over and the population ϱ_{nn} decays exponentially $\varrho_{nn}(t) \propto \exp(-\Gamma_n t)$, the decay rate Γ_n does not contribute to the linewidth. It can be shown that for zero dephasing $\Gamma^* = 0$ and large delay between pump and probe pulses, the

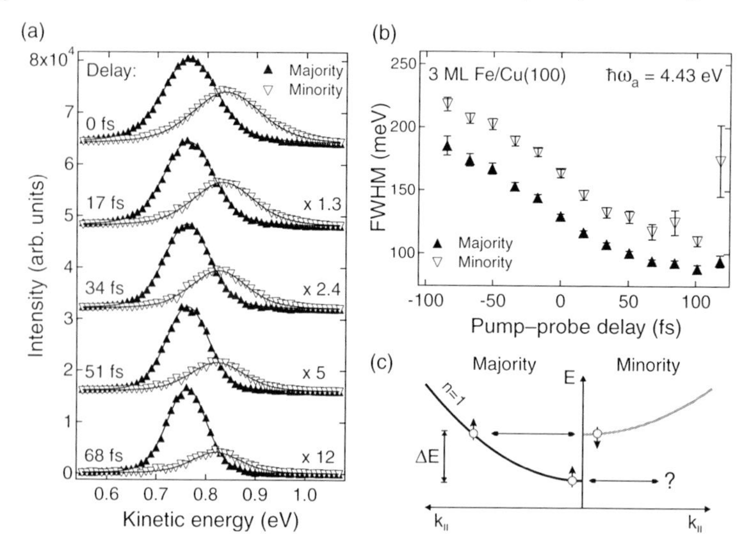

Figure 6.8 (a) Energy- and spin-resolved 2PPE spectra of the $n = 1$ image potential state for increasing pump–probe delay; the spectra are normalized on the majority spin channel at 0 fs, the numbers on the right indicate the multiplication factors of the intensity. (b) Full width at half maximum (FWHM) of the majority and minority components of the $n = 1$ image potential state on a 3 ML iron film as a function of pump–probe delay. (c) Schematic of the free electron-like dispersion parallel to the surface of the majority and minority $n = 1$ bands. A possible quasielastic (virtual) spin-flip process is indicated; not more than energy on the order of the exchange splitting can be transferred to a magnon.

photoemission signal equals a convolution of the Gaussian probe pulse with an exponential function describing the population decay [50]. This convolution yields a Gaussian, shifted but not broadened by the lifetime. Accordingly, for finite dephasing and at large delay, the dephasing rate $2\Gamma^*$ but not the lifetime τ contributes to the measured linewidth. Hence, we are provided with an excellent possibility to measure the dephasing rate.

As the probe pulse length and the experimental resolution are identical for majority and minority electrons, the difference in linewidths for a high pump–probe delay clearly suggests spin-dependent dephasing. Scattering events, which change the energy and/or momentum of an electron on a scale smaller than the experimental resolution, that is, leave the detectable population unchanged, but nevertheless destroy the phase coherence between intermediate and final state, are the origin of pure dephasing.

We interpret the spin-dependent dephasing in terms of electron–magnon scattering on the iron surface. The dispersion relation of magnons $\omega \propto q^2$ supports low-energy excitations and could easily lead to spin-dependent quasielastic scattering processes changing mainly the spin and momentum of the quasiparticle. An acoustic spin wave in iron is on the order of a few meV in energy for the necessary momentum transfer between the minority and majority bands, an energy scale that definitely qualifies as quasielastic for our energy resolution. This process is indicated schematically in Figure 6.8c by the horizontal double-headed arrow. Spin-flip scattering from minority to majority spin states corresponds to the emission of a magnon, while spin wave absorption satisfies angular momentum conservation when scattering occurs from majority spin into minority spin states. From the minimum of the minority spin band, an electron can easily scatter into the majority spin band via the emission of a magnon. Although at a sample temperature of 90 K, low-energy magnons may be already excited and, in principle, also a majority spin electron could absorb a magnon and scatter into a minority spin band, this is impossible at the minimum of the majority spin image potential band. Due to the exchange splitting, the majority spin electron simply lacks a corresponding minority spin image potential state to scatter into, as is indicated in Figure 6.8c. We have thus identified a definitely spin-dependent quasielastic scattering process, namely, magnon generation, as a source for the spin-dependent dephasing rate of the $n = 1$ image potential state in iron [10].

6.3.4 Inelastic Intraband Scattering: Magnon Emission

Besides quasielastic scattering, spin wave excitations may contribute to inelastic decay of photoexcited electrons. Theory predicts that magnon emission is particularly important at low energies, where spin wave excitations significantly increase the phase space for spin-flip decay [20, 21]. For image potential states, inelastic decay with small energy transfer is found in intraband scattering. Electrons excited to the image potential state with energies above the band bottom can decay via intraband scattering along the band. Furthermore, the electrons gain additional phase space for decay into bulk states. We already sketched these possible decay channels in Figure 6.1 for electrons photoexcited to the image potential state band. This increase in phase space leads to a nearly linear increase of the decay rate with energy above the band bottom. Due to the large overlap of initial- and final-state wave functions, intraband decay was shown to contribute equally strong as bulk decay to the increase in decay rate with energy, at least on the Cu(100) surface [75]. For this system, a linear increase of Γ with a slope of $\mathrm{d}\Gamma/\mathrm{d}E = 0.06 \pm 0.01\ (\mathrm{eVfs})^{-1}$ was found [23, 75].

Again for 3 ML iron on Cu(100), we measured the energy E_n as a function of the momentum parallel to the surface $k_{\parallel}$. The dispersion of the first and second image potential state bands is depicted in Figure 6.9. It can be described by two exchange split parabolic bands with $\Delta E_1^{\uparrow\downarrow} = 56$ meV and $\Delta E_2^{\uparrow\downarrow} = 7$ meV [10]. To follow the inelastic intraband scattering, we performed time-resolved measurements at fixed kinetic energy, parallel momentum, and spin along the image potential state parabola [76]. A cross-correlation was intermittently measured on Cu(111) to provide an independent measure of pulse durations and zero time delay. As shown in Figure 6.10, the increase in decay rate $\Gamma = 1/\tau$ with energy above the band bottom strongly depends on the spin: The minority spin slope of $\mathrm{d}\Gamma^{\downarrow}/\mathrm{d}E = 0.25 \pm 0.04\ (\mathrm{eVfs})^{-1}$ is twice as large as the slope $\mathrm{d}\Gamma^{\uparrow}/\mathrm{d}E = 0.12 \pm 0.02\ (\mathrm{eVfs})^{-1}$ for majority spin electrons.

We note that spin-dependent decay channels cannot explain this substantial difference. The iron 3d-bands, with their high density of states, lie within a ± 2.5 eV

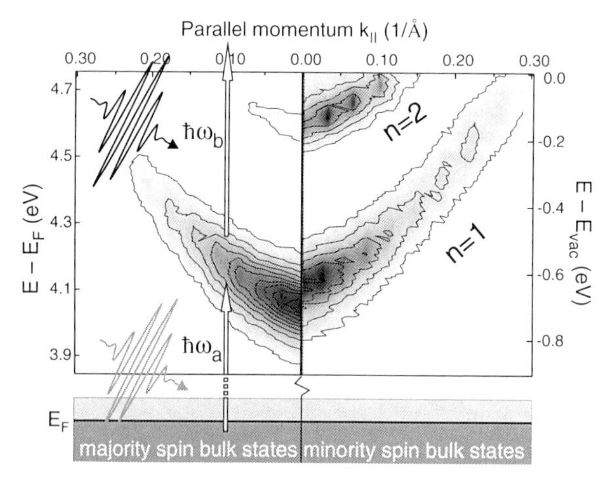

Figure 6.9 Dispersing first and second image potential state bands as a function of momentum parallel to the surface $k_{\|}$ of a 3 ML iron film on Cu(100). The energy scale on the left refers to the Fermi level E_F, the scale on the right to the vacuum level E_{vac} (work function $\Phi = E_{vac} - E_F = 4.75$ eV). The gray scale represents the intensity of the 2PPE signal in the majority (left-hand side) and the minority spin channels (right-hand side) with black indicating the maximum. The arrows depict the 2PPE process schematically.

interval around E_F. They determine the spin dependence of the decay rate at the bottom of the image potential state band but not the increase $d\Gamma/dE$ above. Neither the image potential state band (final states of intraband scattering) nor the sp bulk bands (unoccupied bulk states above the $n = 1$ band bottom at $E - E_F \geq 4.5$ eV) show a significantly spin-dependent density of states. In conclusion, the additional phase space for electron–hole pair creation is hardly spin dependent. Therefore, spin-flip processes must account for the twice as large slope $d\Gamma^{\downarrow}/dE$ in the minority channel.

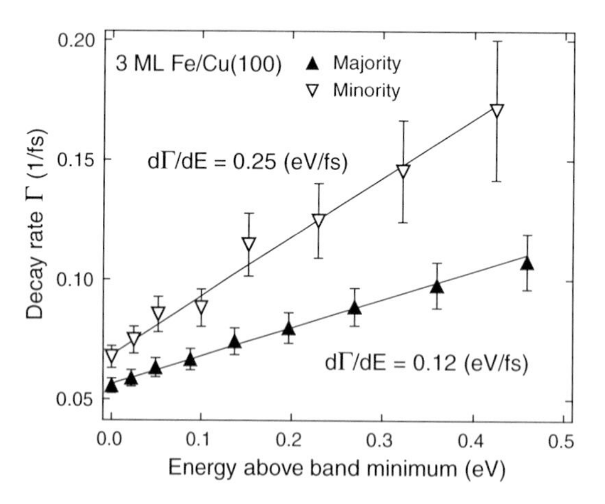

Figure 6.10 Decay rate $\Gamma_1 = 1/\tau_1$ of the first image potential state on the iron film in the majority and minority spin bands for increasing energy above the band minimum.

We have already emphasized that unless spin–orbit coupling is strong, Coulomb interaction does not mediate direct scattering of an electron from the minority spin into the majority spin band.

An electron, which previously resided in the minority spin band, can only appear in the majority spin image potential band via the process illustrated in Figure 6.6b. This leads likewise to the excitation of Stoner pairs in the bulk. Multiple scattering between this electron–hole pair, indicated by the dashed line in Figure 6.6b, is illustrated by Figure 6.6d. It describes the emission of a magnon and the net process can be viewed as spin-flip intraband scattering from the minority into the majority image potential state band.

Decay rates of bulk electrons, calculated via the GW and $GW + T$ approach at excitation energies corresponding to the energy losses in intraband decay along the image potential state parabola, confirm that magnon emission at low excitation energies yields a dominating contribution to the decay rate $\Gamma^{\downarrow}$ and provides a much higher slope of $\mathrm{d}\Gamma^{\downarrow}/\mathrm{d}E$ for minority spin electrons. Intraband decay creates electron–hole pairs in the bulk at energies around the Fermi level, where the joint density of empty minority spin states and filled majority spin states is large. Therefore, magnons play a dominant role in the lifetime of minority spin electrons and are responsible for the strongly spin-dependent increase of decay rates in iron. This is confirmed by calculations of the spectrum of magnetic excitations, that is, spin waves, Stoner excitations, and their hybridization. As described in Ref. [76], the three well-defined magnetic layers of the iron film lead to one acoustic and two optical spin wave branches. With increasing energy above the band bottom, excitation of optical magnons becomes possible and accounts for the large increase of the decay rate $\mathrm{d}\Gamma^{\downarrow}/\mathrm{d}E$ in the minority spin channel.

We have thus observed ultrafast magnon emission directly in the time domain, establishing the timescale of only a few femtoseconds for the generation of collective magnetic excitations. This conclusion further implies that the spin polarization of a hot electron population changes within femtoseconds after excitation.

6.4 Image Potential States: A Sensor for Surface Magnetization

In Section 6.2.2, we already pointed to the spin-dependent intensities and spin-dependent binding energies of the image potential states. These reflect the underlying bulk band structure of the ferromagnetic thin film, and we will illustrate in the following that image potential states can be used as a unique sensor of the surface magnetization.

6.4.1 Projecting the Spin Polarization of Bulk and Surface States

Figure 6.11 shows spin-resolved 2PPE results for 6 ML Co on Cu(100) for three-photon energies [77]. The spectra in panels (a) and (b) are for p and s polarization of

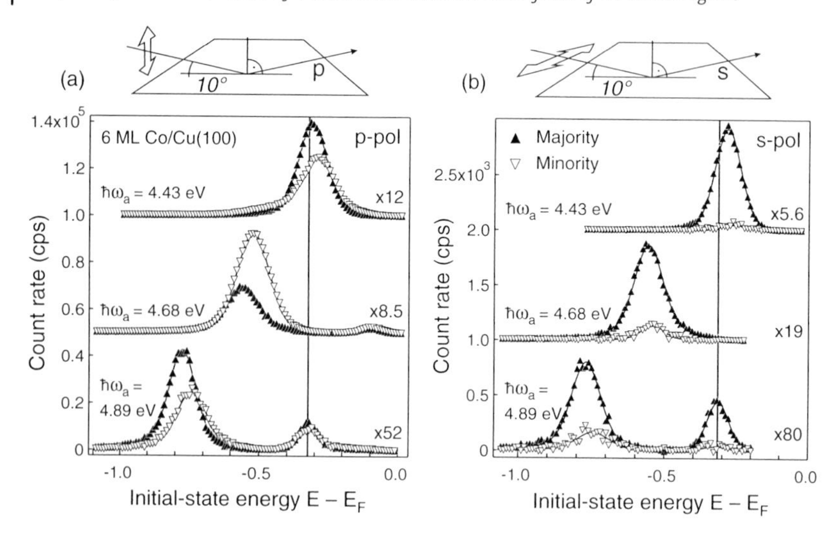

Figure 6.11 Pump pulse photon energy dependence of spin-resolved 2PPE spectra of 6 ML Co/Cu(100) for p-polarized pump pulses (a) and s-polarized pump pulses (b). The spectra are plotted versus the initial-state energy. The spin polarization is pinned to the initial-state energy. At a common initial-state energy, the spin polarization of the $n = 1$ (top spectra) and $n = 2$ (bottom spectra) image potential states is identical (indicated by the line). Note that the population excited with s-polarized pump pulses is considerably smaller. Reprinted from [77]. Copyright 2008 IOP Publishing Ltd.

the pump pulse, respectively. The dominating spectral feature represents the $n = 1$ state, while for higher photon energies the $n = 2$ state becomes populated as well. The energy scale is referred to the initial states, which means that the peaks of the image potential states shift to lower energy with increasing photon energy. The strongly spin-dependent intensities in the spectra reflect the spin-dependent initial states, because the spin is conserved in the dipole transition. In addition, the excitation probabilities depend on the wave function overlap between the bulk states and the image potential state and can therefore differ between $n = 1$ and $n = 2$ states [78]. In a first simplified approach, let us consider the nonrelativistic space group of the (100) surface with C_{4v} symmetry. For $k_{\parallel} = 0$, that is, for normal emission, we probe initial states along the ΓX direction of the Brillouin zone (cf. Figure 6.16). The image potential state wave function belongs to the Δ_1 representation. Due to dipole selection rules, for our experimental geometry, only electrons from states with Δ_1 and Δ_5 spatial symmetries can be excited into the image potential states with p-polarized light [79]. As the image potential states reside in the sp gap of the surface-projected bulk band structure, the excitation strength from sp-like Δ_1 states into the image potential states is much stronger [80], and hence the contribution from Δ_5 initial states in the spectra obtained with p-polarized light is negligible. Excitation with s-polarized light into image potential states is possible from Δ_5 states only.

As indicated by the vertical line, the $n = 1$ image potential state of the spectrum with $\hbar\omega_a = 4.43$ eV is populated from the same initial states as the $n = 2$ state of the

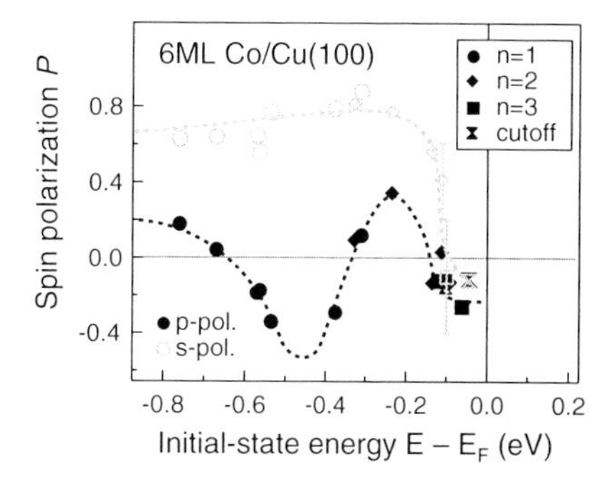

Figure 6.12 Spin polarization P of the initial states as reflected in the spin-dependent intensities of the image potential states $n = 1, 2, 3$ and the low-energy cutoff of the 2PPE spectra, measured with p- and s-polarized pump light. The dashed lines are a guide to the eye. Reprinted with permission from [83]. Copyright 2008 by the American Physical Society.

spectrum with $\hbar\omega_a = 4.89$ eV. The equivalent spin dependence of these peaks proves that it is completely determined by the initial states. By varying the photon energy, one can thus sample the spin polarization of the initial states over a broad range of energy.

The spin polarization as a function of initial-state energy is summarized in Figure 6.12. Data points stem from the $n = 1, 2, 3$ image potential states and the low-energy cutoff for p- and s-polarization, respectively. P reflects the spin polarization of the initial states at the respective energies. The change of sign in P for p-polarized light is due to an occupied surface state of minority spin character at 0.45 eV below E_F [77]. As the surface state does not alter the spin polarization for s-polarized pump pulses, it must be of Δ_1 symmetry. The surface state is also responsible for the resonant enhancement of the $n = 1$ minority intensity at the same initial-state energy in Figure 6.11a (p polarization). The sharp drop in the spin polarization close to the Fermi level marks the top of the majority d-band at the X point of the bulk Brillouin zone (see Figure 6.16). This shows that 6 ML Co on Cu(100) is a strong ferromagnet with the majority d-bands fully occupied.

6.4.2
Access to Spin–Orbit Coupling via Dichroism

Up to now we have based our discussion on a nonrelativistic band structure, where spin–orbit coupling is neglected. Spin–orbit coupling lowers the symmetry of a quantum mechanical system, thereby lifting degeneracies of electronic states. As a consequence, the symmetry character of the electronic states can no longer be described in terms of single-group representations accounting for spatial symmetries only, but has to incorporate the spin space. Now hybridization gaps occur at intersections between bands of the same double-group character, but different (spatial) single-group symmetry [81, 82].

A dependence of the photocurrent on the direction of the sample magnetization is known as magnetic dichroism in photoemission, in the case of linearly polarized

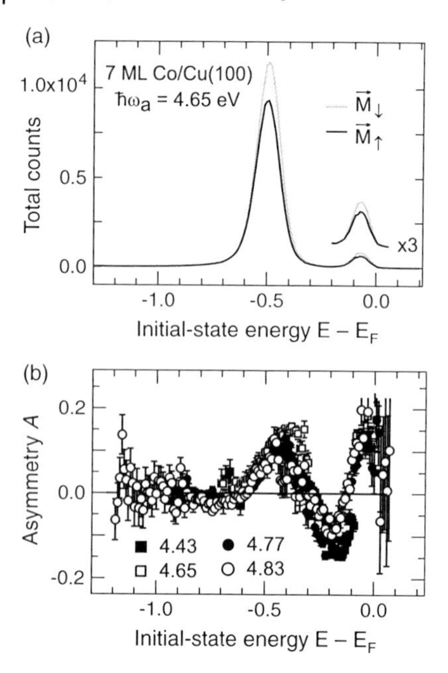

Figure 6.13 Energy-resolved, spin-integrated 2PPE spectrum of 7 ML Co/Cu(100) for antiparallel directions of the magnetization $M_\uparrow$ and $M_\downarrow$. The spectrum is plotted on an initial-state energy scale. The asymmetry A in (b) was extracted from spectra recorded at photon energies $\hbar\omega_a$ of 4.43, 4.65, 4.77, and 4.83 eV, respectively. Reprinted with permission from [83]. Copyright 2008 by the American Physical Society.

light also as magnetic linear dichroism. This means that in the presence of spin–orbit coupling, we expect different spectra for two different experimental geometries, if they are inequivalent with respect to the symmetry point group of the sample. The photon and electron momenta, $\vec{q}$ and $\vec{k}$, the polarization of the light $\vec{A}$, and the direction of the magnetization $\vec{M}$ enter into the characterization of an experimental geometry. In the Co samples, the in-plane magnetization lowers the rotational symmetry from the fourfold C_{4v} point group of the crystal surface to a twofold point group C_{2v}. In our experimental configuration, the two magnetization directions of the in-plane-magnetized Co sample are not equivalent, if we excite with p-polarized light.

The 2PPE spectra in Figure 6.13 reveal different intensities $I(\vec{M}_\downarrow)$ and $I(\vec{M}_\uparrow)$ for the two antiparallel directions of magnetization. To underline the dichroic effect, the asymmetry defined as

$$A = \frac{I(\vec{M}_\downarrow) - I(\vec{M}_\uparrow)}{I(\vec{M}_\downarrow) + I(\vec{M}_\uparrow)} \qquad (6.14)$$

recorded for various photon energies $\hbar\omega_a$ is collected in Figure 6.13b. The magnetic dichroism is a reliable sign for the existence of spin–orbit split regions in the band structure and the maxima and minima in A can be associated with the tops and bottoms of the spin–orbit hybridization gaps.

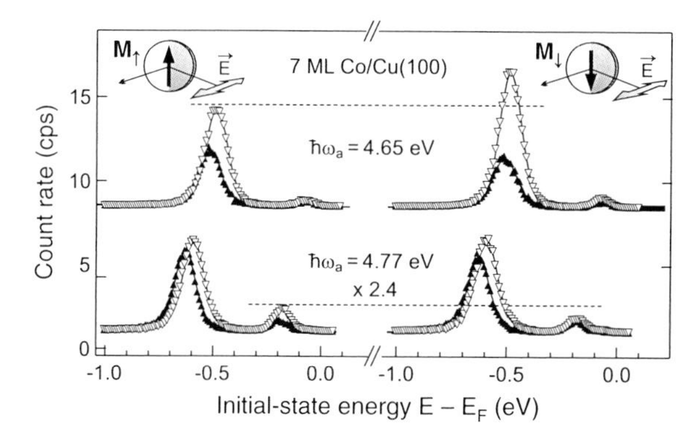

Figure 6.14 Spin- and energy-resolved 2PPE data. Filled up triangles and open down triangles denote majority and minority electrons, respectively. The left and right panels are recorded for magnetization up $M_\uparrow$ and down $M_\downarrow$. Reprinted with permission from [83]. Copyright 2008 by the American Physical Society.

For spin-polarized initial states, we expect also different intensities in each spin channel for different magnetization directions. The spin-resolved spectra of the Co sample in Figure 6.14 reveal that for $\hbar\omega_a = 4.65$ eV the minority component of the $n = 1$ state is strongly affected, while for $\hbar\omega_a = 4.77$ eV the $n = 2$ minority component changes intensity. Thus, the magnetic linear dichroism is caused by a change in count rate of the minority spin component.

In addition, this leads to apparently dichroic lifetimes of the image potential states. Observing the decay of the $n = 2$ population with spin-integrated 2PPE, the lifetime seems to depend on the magnetization direction (Figure 6.15). This is of course exclusively due to the dichroic nature of the excitation and the spin-dependent lifetimes.

Combining the spin information with the asymmetry A, we identify spin–orbit hybridization points in the band structure of the thin cobalt film and determine the double group character. We attribute the dichroic features centered at 70 and

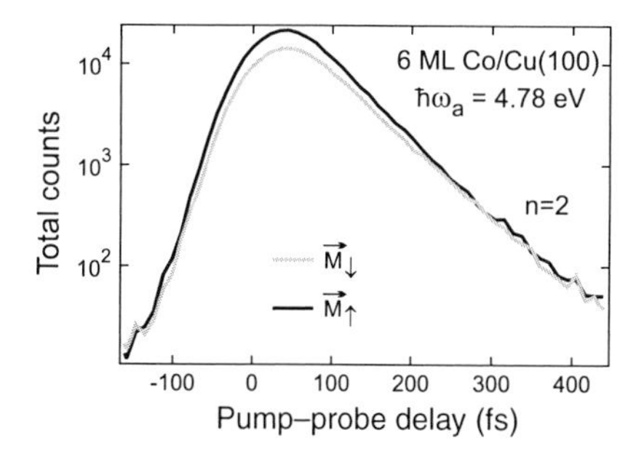

Figure 6.15 The time-resolved spectrum of the second image potential state shows that we indeed observe magnetic dichroism. The spin-dependent lifetimes of the image potential state electrons in combination with the spin polarization lead to different spin-averaged lifetimes for the two magnetization directions.

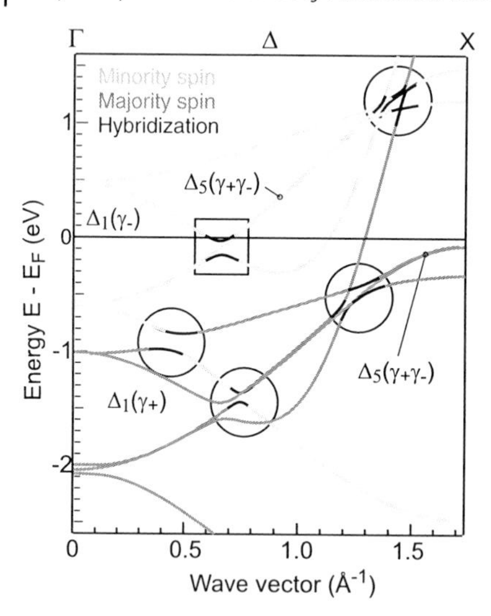

Figure 6.16 Spin-polarized fully relativistic band structure calculation of fcc cobalt along the ΓX direction. Bands are labeled according to the nonrelativistic space group (double-group character in parenthesis). Off the hybridization points (square and circles), majority and minority spin bands are plotted dark and light gray.

180 meV below E_F to the avoided crossing of the $\Delta_1^{\downarrow}$ and $\Delta_5^{\downarrow}$ minority bands (cf. square in Figure 6.16). The derived spin–orbit splitting of 110 ± 20 meV and the position close below E_F are in excellent agreement with the band structure calculation shown in Figure 6.16. The sign of the asymmetry agrees with the calculation of Ref. [84]. Moreover, we provide unequivocal experimental proof for the predicted minority spin and spatial character of the initial valence states.

The circles indicated in Figure 6.16 show that spin–orbit coupling can lead to a mixture of majority and minority spin bands, where two bands with the same double-group character but different spin character cross. Near such points, spin alone is no longer a good quantum number and, in particular, not necessarily conserved in a scattering event. This Elliott–Yafet spin relaxation mechanism, that is, the possibility that a scattering-induced transition includes a spin-flip, is a direct consequence of the mixing of electron states with opposite spin [85, 86].

In purely nonrelativistic electron–electron or electron–phonon scattering, spin-flip scattering is not possible because the Coulomb operator does not act on the spin part of the electron wave function. Only in the presence of spin–orbit coupling can electrons undergo a spin-flip and thus cause a net demagnetization [42]. It was conjectured that the spin-flip scattering probability increases significantly at hybridization points in the band structure. These so-called spin hot spots have been identified in *ab initio* calculations to overcome the discrepancy between measured spin lattice relaxation rates in aluminum and the much lower estimate based on purely atomic properties [87]. Avoided crossing and concomitant spin mixing of bands close to the Fermi level can likewise explain optically driven ultrafast demagnetization in the 3d ferromagnets, since excited electrons or holes relax toward the Fermi level in the first hundred femtoseconds of carrier thermalization via inelastic electron–electron scattering. Spin-resolved 2PPE in combination with linear magnetic dichroism allows us to pinpoint such spin–orbit hybridization points in the band structure of ferromagnets.

With the high count rates we achieve in spin-resolved 2PPE, even temperature-dependent studies of the image potential state exchange splitting become feasible. As the topic of magnetic phase transitions reaches beyond the scope of the present book, we here refrain from a further discussion and refer the interested reader to Refs [25, 88].

6.5 Summary

This chapter describes the spin-dependent relaxation of photoexcited electrons at surfaces of 3d ferromagnets. We use electrons in image potential states as rather unique probes to unravel the spin dependence of relaxation processes in iron and cobalt. Image potential states are well-defined electron states at surfaces, which form an almost two-dimensional electron gas with distinct energy and momentum. In addition, they are weakly coupled to the bulk, with the consequence that they mirror bulk properties but are almost decoupled from the bulk phonon system. Spin-resolved two-photon photoemission offers direct experimental access to these states and allows measurements with excellent energy, momentum, and time resolution.

We present numerous results that provide important insight into details of the electronic system as well as relaxation processes in ferromagnets. The image potential states are used to project the spin dependence of bulk and surface states. Moreover, we succeed in identifying spin–orbit hybridization gaps, which play an important role in explaining spin-flip scattering necessary to understand ultrafast demagnetization. This is made possible via linear magnetic dichroism.

The particular virtue of 2PPE is the option of measurements in the time domain. This is used not only to study the spin-dependent population decay of bulk electrons but also to measure the spin-dependent lifetimes of image potential state electrons. We are able to evaluate separately inelastic scattering and quasielastic scattering, including spin-dependent dephasing. By making use of momentum resolution, we could even prove magnon generation on the femtosecond timescale with a detailed study of intraband scattering.

The wealth of information obtained so far by energy-, momentum-, spin-, and time-resolved spectroscopy of image potential states promises further important results in the field of femtomagnetism in the future. Projects may include the extension to ferromagnetic materials with coupling mechanisms other than band ferromagnets, for example, ferromagnets with localized magnetic moments and indirect exchange coupling, like Gd. Worthwhile are also detailed spin-resolved studies of the electron system after short demagnetizing laser pulses. This kind of measurements will enable us to compare equilibrium and nonequilibrium magnetic phase transitions and the spin-dependent relaxation processes in these different situations.

Acknowledgment

This work was supported by the Deutsche Forschungsgemeinschaft through the priority program SPP1133 *Ultrafast Magnetization Processes.*

References

1 Haight, R. (1995) Electron dynamics at surfaces. *Surf. Sci. Rep.*, **21**, 275.
2 Petek, H.and Ogawa, S. (1997) Femtosecond time-resolved two-photon photoemission studies of electron dynamics in metals. *Prog. Surf. Sci.*, **56**, 239.
3 Hertel, T., Knoesel, E., Wolf, M., and Ertl, G. (1996) Ultrafast electron dynamics at Cu(111): response of an electron gas to optical excitation. *Phys. Rev. Lett.*, **76**, 535.
4 Höfer, U., Shumay, I.L., Reuß, Ch., Thomann, U., Wallauer, W., and Fauster, Th. (1997) Time-resolved coherent photoelectron spectroscopy of quantized electronic states on metal surfaces. *Science*, **277**, 1480.
5 Osgood, R.M. and Wang, X.Y. (1997) Image states on single-crystal metal surface, in *Solid State Physics* (eds H. Ehrenreichand F. Spaepen), vol. 51, Academic, San Diego, p. 1.
6 Ge, N.H., Wong, C.M., Lingle, R.L., Jr., McNeill, J.D., Gaffney, K.J., and Harris, C.B. (1998) Femtosecond dynamics of electron localization at interfaces. *Science*, **279**, 202.
7 Reuß, Ch., Shumay, I.L., Thomann, U., Kutschera, M., Weinelt, M., Fauster, Th., and Höfer, U. (1999) Control of the dephasing of image-potential states by CO adsorption on Cu(100). *Phys. Rev. Lett.*, **82**, 153.
8 Roth, M., Pickel, M., Wang, J., Weinelt, M., and Fauster, Th. (2002) Electron scattering at steps: image-potential states on Cu(119). *Phys. Rev. Lett.*, **88**, 096802.
9 Boger, K., Weinelt, M., and Fauster, Th. (2004) Scattering of hot electrons by adatoms at metal surfaces. *Phys. Rev. Lett.*, **92**, 126803.
10 Schmidt, A.B., Pickel, M., Wiemhöfer, M., Donath, M., and Weinelt, M. (2005) Spin-dependent electron dynamics in front of a ferromagnetic surface. *Phys. Rev. Lett.*, **95**, 107402.
11 Güdde, J., Rohleder, M., Meier, T., Koch, S.W., and Höfer, U. (2007) Time-resolved investigation of coherently controlled electric currents at a metal surface. *Science*, **318**, 1287.
12 Fauster, Th., Weinelt, M., and Höfer, U. (2007) Quasi-elastic scattering of electrons in image-potential states. *Prog. Surf. Sci.*, **82**, 224.
13 Dai, H.L.and Ho, W. (eds) (1995) *Laser Spectroscopy and Photochemistry on Metal Surface*, World Scientific, Singapore.
14 Bonn, M., Funk, S., Hess, C., Denzler, D.N., Stampfl, C., Scheffler, M., Wolf, M., and Ertl, G. (1999) Phonon- versus electron-mediated desorption and oxidation of CO on Ru(0001). *Science*, **285**, 1042.
15 Backus, E.H.G., Eichler, A., Kleyn, A.W., and Bonn, M. (2005) Real-time observation of molecular motion on a surface. *Science*, **310**, 1790.
16 Stépán, K., Güdde, J., and Höfer, U. (2005) Time-resolved measurement of surface diffusion induced by femtosecond laser pulses. *Phys. Rev. Lett.*, **94**, 236103.
17 Aeschlimann, M., Bauer, M., Pawlik, S., Weber, W., Burgermeister, R., Oberli, D., and Siegmann, H.C. (1997) Ultrafast spin-dependent electron dynamics in fcc Co. *Phys. Rev. Lett.*, **79**, 5158.
18 Echenique, P.M., Berndt, R., Chulkov, E.V., Fauster, Th., Goldmann, A., and Höfer, U. (2004) Decay of electronic excitations at metal surfaces. *Surf. Sci. Rep.*, **52**, 219.
19 Passek, F., Donath, M., Ertl, K., and Dose, V. (1995) Longer living majority than minority image state at Fe(110). *Phys. Rev. Lett.*, **75**, 2746.

20 Hong, J. and Mills, D.L. (1999) Theory of the spin dependence of the inelastic mean free path of electrons in ferromagnetic metals: a model study. *Phys. Rev. B*, **59**, 13840.

21 Zhukov, V.P., Chulkov, E.V., and Echenique, P.M. (2004) Lifetimes of excited electrons in Fe and Ni: first-principles GW and the T-matrix theory. *Phys. Rev. Lett.*, **93**, 096401.

22 Zhukov, V.P. and Chulkov, E.V. (2009) The femtosecond dynamics of electrons in metals. *Physics-Uspekhi*, **52**, 105.

23 Weinelt, M. (2002) Time-resolved two-photon photoemission from metal surfaces. *J. Phys. Condens. Matter*, **14**, R1099.

24 Fauster, Th. and Steinmann, W. (1995) Two-photon photoemission spectroscopy of image states, in *Photonic Probes of Surfaces, Electromagnetic Waves: Recent Developments in Research* (ed. P. Halevi), Electromagnetic Waves: Recent Developments in Research, vol. 2, North-Holland, Amsterdam, p. 347.

25 Weinelt, M., Schmidt, A.B., Pickel, M., and Donath, M. (2007) Spin-polarized image-potential-state electrons as ultrafast magnetic sensors in front of ferromagnetic surfaces. *Prog. Surf. Sci.*, **82**, 388.

26 Scholl, A., Baumgarten, L., Jacquemin, R., and Eberhardt, W. (1997) Ultrafast spin dynamics of ferromagnetic thin films observed by fs spin-resolved two-photon photoemission. *Phys. Rev. Lett.*, **79**, 5146.

27 Knorren, R., Bennemann, K.H., Burgermeister, R., and Aeschlimann, M. (2000) Dynamics of excited electrons in copper and ferromagnetic transition metals: theory and experiment. *Phys. Rev. B*, **61**, 9427.

28 Tamm, I. (1932) Über eine mögliche art der elektronenbindung an kristalloberflächen. *Phys. Z. Sowjetunion*, **1**, 733.

29 Shockley, W. (1939) On the surface states associated with a periodic potential. *Phys. Rev.*, **56**, 317.

30 Echenique, P.M. and Pendry, J.B. (1978) The existence and detection of Rydberg states at surfaces. *J. Phys. C*, **11**, 2065.

31 Dose, V., Altmann, W., Goldmann, A., Kolac, U., and Rogozik, J. (1984) Image-potential states observed by inverse photoemission. *Phys. Rev. Lett.*, **52**, 1919.

32 Echenique, P.M. and Pendry, J.B. (1990) Theory of image states at metal surfaces. *Prog. Surf. Sci.*, **32**, 111.

33 Dose, V. (1987) Image potential surface states. *Phys. Scr.*, **36**, 669.

34 Donath, M., Math, C., Pickel, M., Schmidt, A.B., and Weinelt, M. (2007) Realization of a spin-polarized two-dimensional electron gas via image-potential-induced surface states. *Surf. Sci.*, **601**, 5701.

35 Borstel, G. and Thörner, G. (1988) Inverse photoemission from solids: theoretical aspects and applications. *Surf. Sci. Rep.*, **8**, 1.

36 Wu, R. and Freeman, A.J. (1992) Spin density at the Fermi level for magnetic surfaces and overlayers. *Phys. Rev. Lett.*, **69**, 2867.

37 Nekovee, M., Crampin, S., and Inglesfield, J.E. (1993) Magnetic splitting of image states at Fe(110). *Phys. Rev. Lett.*, **70**, 3099.

38 Straub, D. and Himpsel, F.J. (1984) Identification of image-potential surface states on metals. *Phys. Rev. Lett.*, **52**, 1922.

39 Giesen, K., Hage, F., Himpsel, F.J., Riess, H.J., and Steinmann, W. (1985) Two-photon photoemission via image-potential states. *Phys. Rev. Lett.*, **55**, 300.

40 Binning, G., Frank, K.H., Fuchs, H., Garcia, N., Reihl, B., Rohrer, H., Salvan, F., and Williams, A.R. (1985) Tunneling spectroscopy and inverse photoemission: image and field states. *Phys. Rev. Lett.*, **55**, 991.

41 Passek, F. and Donath, M. (1992) Spin-split image-potential-induced surface state on Ni(111). *Phys. Rev. Lett.*, **69**, 1101.

42 Kirschner, J. and Feder, R. (1979) Spin polarization in double diffraction of low-energy electrons from W(001): experiment and theory. *Phys. Rev. Lett.*, **42**, 1008.

43 Gay, T.J. and Dunning, F.B. (1992) Mott electron polarimetry. *Rev. Sci. Instrum.*, **63**, 1635.

44 Yu, D.H., Math, C., Meier, M., Escher, M., Rangelov, G., and Donath, M. (2007) Characterization and application of a SPLEED-based spin polarisation analyser. *Surf. Sci.*, **601**, 5803.

45 Tillmann, D., Thiel, R., and Kisker, E. (1989) Very-low-energy spin-polarized electron diffraction from Fe(001). *Z. Phys. B*, **77**, 1.

46 Winkelmann, A., Hartung, D., Engelhard, H., Chiang, C.T., and Kirschner, J. (2008) High efficiency electron spin polarization analyzer based on exchange scattering at Fe/W (001). *Rev. Sci. Instrum.*, **79**, 083303.

47 Okuda, T., Takeichi, Y., Maeda, Y., Harasawa, A., Matsuda, I., Kinoshita, T., and Kakizaki, A. (2008) A new spin-polarized photoemission spectrometer with very high efficiency and energy resolution. *Rev. Sci. Instrum.*, **79**, 123117.

48 Kessler, J. (1985) *Polarized Electrons, Atoms and Plasmas*, 2nd edn, vol. 1, Spinger, Berlin.

49 Blum, K. (1981) *Density Matrix Theory and Applications*, Plenum, New York.

50 Boger, K., Roth, M., Weinelt, M., Fauster, Th., and Reinhard, P.G. (2002) Linewidths in energy-resolved two-photon photoemission spectroscopy. *Phys. Rev. B*, **65**, 75104.

51 Asaki, M.T., Huang, C.P., Garvey, D., Zhou, J., Kapteyn, H.C., and Murnane, M.M. (1993) Generation of 11-fs pulses from a self-mode-locked Ti:sapphire laser. *Opt. Lett.*, **18**, 977.

52 Pappas, D.P., Kämper, K.P., Miller, B.P., Hopster, H., Fowler, D.E., Brundle, C.R., Luntz, A.C., and Shen, Z.X. (1991) Spin-dependent attenuation by transmission through thin ferromagnetic films. *Phys. Rev. Lett.*, **66**, 501.

53 Schönhense, G. and Siegmann, H.C. (1993) Transmission of electrons through ferromagnetic material and applications to detection of electron spin polarization. *Ann. Phys.*, **2**, 465.

54 Getzlaff, M., Bansmann, J., and Schönhense, G. (1993) Spin-polarization effects for electrons passing through thin iron and cobalt films. *Solid State Commun.*, **87**, 467.

55 Passek, F., Donath, M., and Ertl, K. (1996) Spin-dependent electron attenuation lengths and influence on spectroscopic data. *J. Magn. Magn. Mater.*, **159**, 103.

56 Burgermeister, R. (1998) Spin- and time-resolved two-photon photoemission. PhD thesis. Eidgenössische Technische Hochschule Zürich.

57 Quinn, J.J. and Ferrell, R.A. (1958) Electron self-energy approach to correlation in a degenerate electron gas. *Phys. Rev.*, **112**, 812.

58 Goris, A., Doebrich, K., Panzer, I., Schmidt, A.B., Donath, M., and Weinelt, M. (2010) Hot electron lifetimes and virtual spin-flip scattering in cobalt, Submitted.

59 Vignale, G. and Singwi, K.S. (1985) Spin-flip electron-energy-loss spectroscopy in itinerant-electron ferromagnets: collective modes versus Stoner excitations. *Phys. Rev. B*, **32**, 2824.

60 Kirschner, J. (1985) Direct and exchange contributions in inelastic scattering of spin-polarized electrons from iron. *Phys. Rev. Lett.*, **55**, 973.

61 Modesti, S., Valle, F.D., Bocchetta, C.J., Tosatti, E., and Paolucci, G. (1987) Normal versus exchange inelastic electron scattering in metals: theory and experiment. *Phys. Rev. B*, **36**, 4503.

62 Abraham, D.L. and Hopster, H. (1989) Spin-polarized electron-energy-loss spectroscopy on Ni. *Phys. Rev. Lett.*, **62**, 1157.

63 Vollmer, R., Etzkorn, M., Anil Kumar, P.S., Ibach, H., and Kirschner, J. (2003) Spin-polarized electron energy loss spectroscopy of high energy, large wave vector spin waves in ultrathin fcc Co films on Cu(001). *Phys. Rev. Lett.*, **91**, 147201.

64 Tang, W.X., Zhang, Y., Tudosa, I., Prokop, J., Etzkorn, M., and Kirschner, J. (2007) Large wave vector spin waves and dispersion in two monolayer Fe on W (110). *Phys. Rev. Lett.*, **99**, 087202.

65 Zhukov, V.P., Chulkov, E.V., and Echenique, P.M. (2005) GW + T theory of excited electron lifetimes in metals. *Phys. Rev. B*, **72**, 155109.

66 Springer, M., Aryasetiawan, F., and Karlsson, K. (1998) First-principles

T-matrix theory with application to the 6eV satellite in Ni. *Phys. Rev. Lett.*, **80**, 2389.

67 Loong, C.K., Carpenter, J.M., Lynn, J.W., Robinson, R.A., and Mook, H.A. (2000) Neutron scattering study of the magnetic excitations in ferromagnetic iron at high energy transfer. *Phys. Rev. B*, **62**, 5589.

68 Grechnev, A., Marco, I.D., Katsnelson, M.I., Lichtenstein, A.I., Wills, J., and Eriksson, O. (2007) Theory of bulk and surface quasiparticle spectra for Fe, Co, and Ni. *Phys. Rev. B*, **76**, 035107.

69 Piraux, L., Dubois, S., Fert, A., and Belliard, L. (1998) The temperature dependence of the perpendicular giant magnetoresistance in Co/Cu multilayered nanowires. *Eur. Phys. J. B*, **4**, 413.

70 van Dijken, S., Jiang, X., and Parkin, S.S.P. (2002) Spin-dependent hot electron transport in $Ni_{81}Fe_{19}$ and $Co_{84}Fe_{16}$ films on GaAs(001). *Phys. Rev. B*, **66**, 094417.

71 Bass, J. and Pratt, W.P., Jr. (2007) Spin-diffusion lengths in metals and alloys, and spin-flipping at metal/metal interfaces: an experimentalist's critical review. *J. Phys. Condens. Matter*, **19**, 183201.

72 Zhukov, V.P., Chulkov, E.V., and Echenique, P.M. (2006) Lifetimes and inelastic mean free path of low-energy excited electrons in Fe, Ni, Pt and Au: Ab initio GW + T calculations. *Phys. Rev. B*, **73**, 125105.

73 Loudon, R. (1983) *The Quantum Theory of Light*, Clarendon, Oxford.

74 Boyd, R.W. (1992) *Nonlinear Optics*, Academic, San Diego.

75 Berthold, W., Höfer, U., Feulner, P., Chulkov, E.V., Silkin, V.M., and Echenique, P.M. (2002) Momentum-resolved lifetimes of image-potential states on Cu(100). *Phys. Rev. Lett.*, **88**, 056805.

76 Schmidt, A.B., Pickel, M., Donath, M., Buczek, P., Ernst, A., Zhukov, V.P., Echenique, P.M., Sandratskii, L.M., Chulkov, E.V., and Weinelt, M. (2010) Ultrafast generation of spin–wave excitations in an Fe film on Cu(100), submitted.

77 Schmidt, A.B., Pickel, M., Allmers, T., Budke, M., Braun, J., Weinelt, M., and Donath, M. (2008) Surface electronic structure of fcc Co films: a combined spin-resolved one- and two-photon photoemission study. *J. Phys. D*, **41**, 164003.

78 Pickel, M., Schmidt, A.B., Donath, M., and Weinelt, M. (2006) A two-photon photoemission study of spin-dependent electron dynamics. *Surf. Sci.*, **600**, 4176.

79 Eberhardt, W. and Himpsel, F.J. (1980) Dipole selection rules for optical transitions in the fcc and bcc lattices. *Phys. Rev. B*, **21**, 5572.

80 Wallauer, W. and Fauster, Th. (1996) Exchange splitting of image states on Fe/Cu(100) and Co/Cu(100). *Phys. Rev. B*, **54**, 5086.

81 Inui, T., Tanabe, Y., and Onodera, Y. (1990) *Group Theory and Its Applications in Physics*, Springer, Berlin.

82 Kuch, W. and Schneider, C.M. (2001) Magnetic dichroism in valence band photoemission. *Rep. Prog. Phys.*, **64**, 147.

83 Pickel, M., Schmidt, A.B., Giesen, F., Braun, J., Minár, J., Ebert, H., Donath, M., and Weinelt, M. (2008) Spin–orbit hybridization points in the face-centered-cubic cobalt band structure. *Phys. Rev. Lett.*, **101**, 066402.

84 Fanelsa, A., Kisker, E., Henk, J., and Feder, R. (1996) Magnetic dichroism in valence-band photoemission from Co/Cu(001): Experiment and theory. *Phys. Rev. B*, **54**, 2922.

85 Elliott, R.J. (1954) Theory of the effect of spin–orbit coupling on magnetic resonance in some semiconductors. *Phys. Rev.*, **96**, 266.

86 Yafet, Y. (1963) g factors and spin–lattice relaxation of conduction electrons, in *Solid State Physics* (eds F. Seitz and D. Turnbull), vol. 14, Academic, New York, p. 1.

87 Fabian, J. and Das Sarma, S. (1998) Spin relaxation of conduction electrons in polyvalent metals: theory and a realistic calculation. *Phys. Rev. Lett.*, **81**, 5624.

88 Pickel, M., Schmidt, A.B., Weinelt, M., and Donath, M. (2010) Magnetic exchange splitting in Fe above the Curie temperature. *Phys. Rev. Lett.* **104**, 237204.

7
Electron–Phonon Interaction at Interfaces

Philip Hofmann, Evgueni V. Chulkov, and Irina Yu. Sklyadneva

7.1
Introduction

Many-body effects and their interplay are at the heart of some of the most interesting problems in condensed matter physics, and frequently the simultaneous presence of different effects is found in complex materials. The electron–phonon (e–ph) interaction is one such effect that limits the lifetime of excited electrons (or holes) and has long been studied because of its role in many phenomena, from electrical conductivity to electronic heat capacity and BCS-type superconductivity. Several experimental techniques such as tunneling spectroscopy or heat capacity measurements have provided information on the electron–phonon coupling strength averaged over the bulk Fermi surface of metals [1].

Experimentally, recent advances in angle-resolved photoemission (ARPES) have opened the opportunity for a study of many-body effects in unprecedented detail. Most importantly, studies are not confined to averages over the Fermi surface but detailed information about the energy and k-dependence of the interaction has come within reach. This permits, for instance, to establish the symmetry of the superconducting gap in novel superconductors [2, 3].

The e–ph interaction stands out as a fundamental many-body process that can be tested by both experimental and theoretical methods. Much has been learned by studying the e–ph coupling on carefully chosen electronic surface states, for which good arguments can be made for the e–ph interaction to be the *only* many-body effect giving rise to a bosonic spectroscopic signature. Surface states have also played an important role because they have, as do the states in the cuprates, a merely two-dimensional dispersion, an essential prerequisite for the analysis of ARPES data.

In the most simple picture, the e–ph coupling changes the dispersion and the lifetime of the electronic states in a material. This situation is illustrated in Figure 7.1a. Very close to the Fermi level, within a typical phonon energy $\hbar\omega_D$, the dispersion is renormalized such that it is flatter at the Fermi energy. Consequently, the effective mass of the electrons at the Fermi level and the density of states are

Dynamics at Solid State Surface and Interfaces Vol.1: Current Developments
Edited by Uwe Bovensiepen, Hrvoje Petek, and Martin Wolf

ISBN: 978-3-527-40937-2

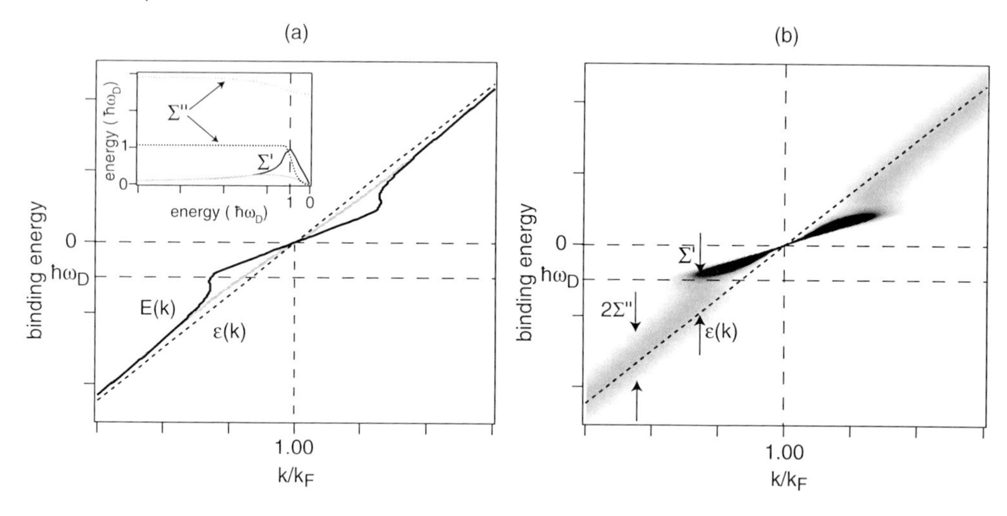

Figure 7.1 (a) Renormalization of the electronic dispersion close to the Fermi energy (schematic). The dashed line is the bare dispersion $\varepsilon(k)$; the solid line the renormalized dispersion $E(k)$ for a low temperature (black) and a higher temperature (gray). *Inset*: Real and imaginary parts of the complex self-energy for the electron–phonon coupling, Σ' and Σ'' for a low temperature (black) and a higher temperature (gray). (b) Spectral function $\mathcal{A}(\omega, \vec{k}, T)$ at a low temperature showing the sharpening of the quasiparticle peak near E_F. The arrows indicated how Σ' and Σ'' correspond to the renormalization of the dispersion and the finite width of the peak, respectively.

increased [1]. The increase in the effective mass is described by the electron–phonon mass enhancement parameter λ such that $m^* = m_0(1+\lambda)$, where m^* and m_0 are the effective masses with and without e–ph interaction, respectively.

The effect of the e–ph coupling on the dispersion and lifetime of the states can be expressed by the complex self-energy Σ, where the real part Σ' renormalizes the dispersion and the states acquire a finite lifetime τ through the imaginary part Σ''. In this context, both $\Sigma'' = \hbar/2\tau$ and the inverse lifetime $\Gamma = \hbar/\tau$ are frequently used. All the closely related quantities Σ', Σ'', Γ, and τ can be obtained from the spectral function $A(\omega, \vec{k}, T)$ that is defined later on in this chapter but for now can be taken to be proportional to the photoemission intensity in ARPES. Figure 7.1b shows a plot of the spectral function at low temperature and indicates how Σ' and Σ'' (or Γ) contribute to renormalization and broadening, respectively.

Typical results for Σ' and Σ'' (calculated in the Debye model) are given in the inset of Figure 7.1a. Σ' is small except for energies very close to the Fermi level. Σ'' is changing rapidly close to the Fermi level and is constant at higher energies. Σ' vanishes exactly at E_F such that the Fermi surface is not affected by the interaction. Σ'' only vanishes at E_F for zero temperature. Both are related by a Hilbert transformation, that is, from a spectroscopic point of view it is sufficient to determine either Σ' or Σ''.

The lifetime τ, inverse lifetime Γ, or the imaginary part of the self-energy Σ'' are all essentially the same quantity, describing the decay of excited electrons or holes. In this chapter, we are primarily interested in a decay that involves e–ph coupling, but we briefly discuss other scattering mechanisms as well, since they will eventually contribute to the total Γ and since we have to single out the e–ph contribution. In paramagnetic metals, Γ has three contributions, namely, e–ph, electron–electron (e–e) scattering, and electron–defect (e–df) interactions [4]. These contributions are additive such that

$$\Gamma = \hbar/\tau = \Gamma_{\text{e-df}} + \Gamma_{\text{e-e}} + \Gamma_{\text{e-ph}}. \tag{7.1}$$

$\Gamma_{\text{e-df}}$ takes into account elastic scattering processes by defects that limit the mean-free path of a carrier. $\Gamma_{\text{e-df}}$ is usually not strongly energy or temperature dependent and thus acts as a mere offset to Γ. Note, however, that while the defect scattering strength might not be temperature dependent, the number of defects is not: defects can be thermally excited at elevated temperatures and this can contribute to an increase in $\Gamma_{\text{e-df}}$ [5, 6]. Often, the defect scattering can be suppressed in experiments such as scanning tunneling spectroscopy measurements [7–9] or time-resolved two-photon photoemission [7, 10, 11]. It can also be strongly reduced in photoemission spectroscopy studies [7, 12].

$\Gamma_{\text{e-e}}$, the contribution from the predominantly inelastic e–e scattering, includes several decay channels related to charge density, spin density, singlet pair, and triplet pair fluctuations [7]. $\Gamma_{\text{e-e}}$ is energy dependent: it increases for higher binding energies because the phase space for inelastic e–e scattering is extended. The temperature dependence of $\Gamma_{\text{e-e}}$, on the other hand, is usually very small, in sharp contrast to $\Gamma_{\text{e-ph}}$ that increases at high temperatures because of the increased probability of phonon excitations. At sufficiently low temperatures, in the absence of defects and for large excitation energies, the e–e scattering is the most important process that limits the excitation lifetime. However, close to the Fermi level and in particular for high temperatures, $\Gamma_{\text{e-e}}$ can become smaller than $\Gamma_{\text{e-ph}}$ [4, 7, 13–17].

An important result from these considerations is that in many situations the e–ph is the only contribution to (7.1) with a significant temperature dependence and this can be exploited to experimentally single out the e–ph part from the other contributions.

In the following sections, we discuss how the e–ph can be described theoretically, how information about it can be extracted from experimental data, and we give a few selected examples. We do not attempt to present a complete overview on the current status of the field but rather focus on some historically and didactically valuable examples. Particular emphasis will be put on cases for which both experimental data and *ab initio* calculations are available.

7.2 Calculation of the Electron–Phonon Coupling Strength

A basic quantity of the e–ph interaction is the electron–phonon matrix that gives the probability of electron scattering from an initial electron state (i) with momentum $\mathbf{k}$ to a final electron state (f) by a phonon with momentum $\mathbf{q}$ and

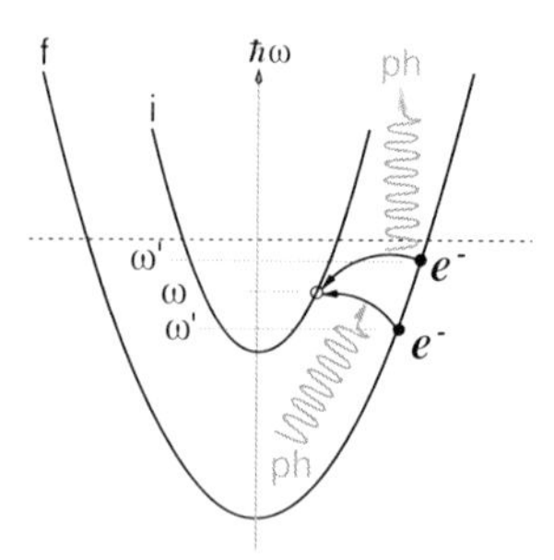

Figure 7.2 Phonon-mediated interband scattering from a final state of the hole to the initial state with energy ω by phonon emission and phonon absorption.

mode index ν:

$$g^{i,f}(\mathbf{k},\mathbf{q},\nu) = \sqrt{\frac{1}{2M\omega_{\mathbf{q},\nu}}}\langle\Psi_{\mathbf{k}i}|\hat{\varepsilon}_{\mathbf{q},\nu}\cdot\delta V^{SCF}_{\mathbf{q},\nu}|\Psi_{\mathbf{k}+\mathbf{q}f}\rangle. \quad (7.2)$$

Here, M is the atomic mass, $\Psi_{\mathbf{k}i}$ and $\Psi_{\mathbf{k}+\mathbf{q}f}$ are the electronic wave functions for the initial and final states, respectively. $\delta V^{SCF}_{\mathbf{q},\nu}$ gives the gradient of the self-consistent potential with respect to the atomic displacements induced by the phonon mode $(\mathbf{q},\nu)$ with frequency $\omega_{\mathbf{q},\nu}$ and phonon polarization vector $\hat{\varepsilon}_{\mathbf{q},\nu}$. Such phonon-mediated interband scattering is shown schematically in Figure 7.2.

The effectiveness of phonons with energy $\hbar\omega$ to scatter electrons is expressed in terms of the Eliashberg spectral function, $\alpha^2F(\omega)$. If the initial electron energy ε_i and momentum $\mathbf{k}$ are fixed, the corresponding state-dependent Eliashberg spectral function gives the e–ph coupling between the initial state and all other final states (ε_f) that differ in energy by $\hbar\omega$ due to the phonon emission (E) or absorption (A) processes:

$$\begin{aligned}\alpha^2F^{E(A)}(\varepsilon_i,\mathbf{k};\omega) &= \textstyle\sum_{\mathbf{q},\nu,f}\delta(\varepsilon_i-\varepsilon_f\mp\omega_{\mathbf{q},\nu})\\ &\quad\times|g^{i,f}(\mathbf{k},\mathbf{q},\nu)|^2\delta(\omega-\omega_{\mathbf{q},\nu}).\end{aligned} \quad (7.3)$$

The "−" and " + " signs in the delta function with electron energies correspond to a phonon emission and absorption, respectively. The sum is carried out over final electron states (f) and all phonon modes $(\mathbf{q},\nu)$. As one can see from Eq. (7.3), $\alpha^2F(\omega)$ is nothing else than the phonon density of states weighted by the e–ph coupling.

Figure 7.3 shows an example of the Eliashberg spectral function. It was calculated for a hole state in the $\overline{\Gamma}$ symmetry point on the Cu(111) surface state [14]. The figure also shows the calculated phonon dispersion, in which the surface-localized Rayleigh mode is clearly identified as split-off below the bulk continuum around the $\overline{M}$ point. This mode contributes significantly to the Eliashberg function. Its contribution has been singled out by the dashed line in $\alpha^2F(\omega)$.

While the electron–phonon mass enhancement parameter's original definition is related to the overall mass enhancement at the Fermi surface, a more spectroscopic interpretation of λ is to view it as a dimensionless parameter measuring the coupling

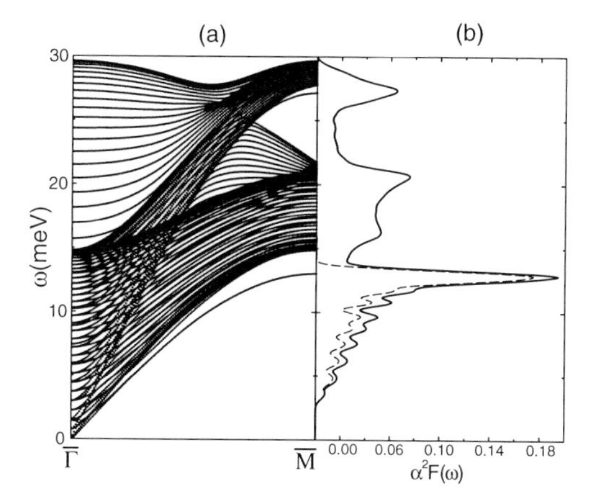

Figure 7.3 Example of a calculated Eliashberg function. (a) The phonon dispersion from a 31-layer slab calculation in the $\bar{\Gamma}\bar{M}$-direction of the surface Brillouin zone of Cu(111). (b) The Eliashberg function of the hole state at the $\bar{\Gamma}$-point (solid line) and the contribution from the Rayleigh mode to the Eliashberg function (dashed line). Reprinted with permission from [14]. Copyright 2002 by the American Physical Society.

strength for a hole of given energy ε_i and momentum $\mathbf{k}$.

$$\lambda(\varepsilon_i, \mathbf{k}) = \int_0^{\omega_{max}} \frac{\alpha^2 F^E(\varepsilon_i, \mathbf{k}; \omega) + \alpha^2 F^A(\varepsilon_i, \mathbf{k}; \omega)}{\omega} d\omega. \quad (7.4)$$

Here, ω_{max} is the maximum phonon frequency.

Very often, the energy change of the scattered electron due to the absorption or emission of a phonon is neglected because the phonon energies are much smaller than the electronic energy scale. While the typical phonon energy lies in the range of meV, the energies of electrons are of the order of eV. Thus, e–ph scattering changes mainly the direction of the electron motion (momentum) while the energy change is negligible. Therefore, one can assume that the initial and final electron energies coincide:

$$\delta\left(\varepsilon_i - \varepsilon_f \mp \omega_{q,\nu}\right) \approx \delta\left(\varepsilon_i - \varepsilon_f\right). \quad (7.5)$$

When this so-called quasielastic assumption is applied, the state-dependent Eliashberg spectral function and e–ph coupling parameter are

$$\alpha^2 F(\varepsilon_i, \mathbf{k}; \omega) = \sum_{q,\nu,f} \delta(\varepsilon_i - \varepsilon_f) |g^{i,f}(\mathbf{k}, \mathbf{q}, \nu)|^2 \delta(\omega - \omega_{q,\nu}), \quad (7.6)$$

and

$$\lambda\left(\varepsilon_i, \mathbf{k}\right) = 2 \int_0^{\omega_{max}} \frac{\alpha^2 F(\varepsilon_i, \mathbf{k}; \omega)}{\omega} d\omega. \quad (7.7)$$

This approximation allows us to use the same Eliashberg spectral function for both emission and absorption processes.

One can average $\alpha^2 F(\varepsilon_i, \mathbf{k}; \omega)$ over electron momentum $\mathbf{k}$ to obtain the energy-resolved spectral function. The latter is defined by the sum over all possible initial electron states with the same energy [18]. In particular, when the energies of initial and final electronic states coincide with the Fermi energy ($\varepsilon_i = \varepsilon_f = E_F$), we obtain the spectral function and the e–ph coupling parameter λ (following Eq. (7.7)) as the Fermi surface averaged quantities:

$$\begin{aligned}\alpha^2 F(E_F; \omega) = & \frac{1}{N(E_F)} \sum_{q,\nu} \delta(\omega - \omega_{q,\nu}) \sum_{k,i,f} |g^{i,f}(\mathbf{k}, \mathbf{q}, \nu)|^2 \\ & \times \delta(\varepsilon_i - E_F)\delta(\varepsilon_f - E_F),\end{aligned} \tag{7.8}$$

Here $N(E_F)$ is the electron density of states per atom and per spin at E_F.

The e–ph interaction introduces a shift in the dispersion of electronic states and changes their lifetime. The phonon-induced lifetime broadening of a hole (electron) state can be obtained from the imaginary part of the e–ph self-energy, Σ'', while the real part, Σ', allows to evaluate the shift in electronic energies. Both parts of the complex e–ph self-energy are fully determined by the Eliashberg function. The imaginary part of the electron–phonon self-energy is related to the Eliashberg function through the integral over all the scattering events that conserve energy and momentum [1]:

$$\begin{aligned}\Gamma_{e-ph}(\varepsilon_i, \mathbf{k}, T) &= 2\Sigma''(\varepsilon_i, \mathbf{k}; T) \\ &= 2\pi \int_0^{\omega_{max}} \{\alpha^2 F^E(\varepsilon_i, \mathbf{k}; \omega)[1 + n(\omega) - f(\varepsilon_i - \omega)] \\ &\quad + \alpha^2 F^A(\varepsilon_i, \mathbf{k}; \omega)[n(\omega) + f(\varepsilon_i + \omega)]\} d\omega.\end{aligned} \tag{7.9}$$

Here, f and n are the Fermi and Bose distribution functions, respectively. Note that the temperature dependence of Γ_{e-ph} is introduced exclusively by the Fermi and Bose distribution functions. The term in the first square brackets represents the phonon emission and the term in the second square brackets is associated with phonon absorption processes.

In the quasielastic approximation the contribution of phonons to a hole (electron) state linewidth is written as [1]

$$\begin{aligned}\Gamma_{e-ph}(\varepsilon_i, \mathbf{k}, T) &= 2\Sigma''(\varepsilon_i, \mathbf{k}; T) \\ &= 2\pi \int_0^{\omega_{max}} \alpha^2 F(\varepsilon_i, \mathbf{k}; \omega)[1 - f(\varepsilon_i - \omega) + f(\varepsilon_i + \omega) + 2n(\omega)] d\omega.\end{aligned} \tag{7.10}$$

Let us obtain the behavior of the electron–phonon linewidth in the limiting cases, $T \to 0$ and $T \gg \omega_{max}$. Note that at $T \to 0$, the Bose distribution function $n(\omega) \to 0$. Then, in the quasielastic approximation, we have

$$T \to 0, \ \Gamma_{e-ph}(\varepsilon_i, \mathbf{k}) = 2\pi \int_0^{\omega_{max}} \alpha^2 F(\varepsilon_i, \mathbf{k}; \omega) \, d\omega. \tag{7.11}$$

At $T = 0$, only phonon emission occurs. Since no electrons can scatter into a hole at the Fermi level, the linewidth for holes at E_F is equal to zero. Then, $\Gamma_{e-ph}(\varepsilon_i, \mathbf{k})$ increases monotonically up to a maximum value at $\omega = \omega_D$ (the maximum phonon energy) as more and more phonon modes become available (see the insert in Figure 7.1a). As the temperature increases, the linewidth increases for all electronic energies. This temperature dependence of the linewidth has often been used to extract the e–ph coupling parameter λ for electronic states with energies much larger than the maximum phonon energy. At elevated temperatures, when $k_B T$ is higher than the maximum phonon energy, the Tdependence of $\Gamma_{e-ph}(\varepsilon_i, \mathbf{k})$ becomes linear with a slope determined by the electron–phonon coupling parameter λ [1]:

$$\Gamma_{e-ph}(\varepsilon_i, \mathbf{k}) = 2\pi\lambda(\varepsilon_i, \mathbf{k})\, k_B \mathrm{T}. \tag{7.12}$$

As λ can be derived from the measurements of the lifetime broadening as a function of temperature, this expression enables us to compare experimental and theoretical results.

The real part of the self-energy, Σ', allows us to evaluate the renormalization of the electronic energy bands due to the interaction with the phonon modes of the lattice (Figure 7.1a). Using the electron–phonon self-energy representing the screening of the electrons by the lattice, one can obtain the renormalized band dispersion, $E(\mathbf{k})$:

$$E(\mathbf{k}) = \varepsilon_0(\mathbf{k}) + \Sigma'(\mathbf{k}, E) \tag{7.13}$$

Here, Σ' $(\mathbf{k}, E)$ is the real part of the self-energy:

$$\Sigma'(E, T) = \int_{-\infty}^{\infty} \mathrm{d}\nu \int_0^{\omega_{max}} \mathrm{d}\omega' \alpha^2 F\,(E, \omega')\, \frac{2\omega'}{\nu^2 - \omega'^2} f\,(\nu + E) \tag{7.14}$$

The technique commonly used to determine the mass enhancement factor λ at the Fermi energy is to calculate the slope of the Σ' at E_F because of the identity between the partial derivative of Σ' at the Fermi energy and λ (E_F):

$$\lambda = \frac{\mathrm{d}\Sigma'}{\mathrm{d}E}\Big|_{E_F, T=0K} \tag{7.15}$$

Thus, the theoretical evaluation of the e–ph interaction generally requires the knowledge of the low-energy electronic excitation spectrum, the complete vibrational spectrum, and the self-consistent response of the electronic system to lattice vibrations. A model approach for evaluating the e–ph interaction at surface states has been proposed in Refs [14, 16, 19]. The model combines three independent approximations: (1) one-electron wave functions and energies are calculated with a one-dimensional potential [20, 21]; (2) phonon frequencies and polarizations are obtained either from one-parameter force constant model [22] or from an embedded atom model [23]; and (3) the gradient of the one-electron potential is represented by the Ashcroft pseudopotential [24] screened within Thomas–Fermi approximation. A restriction of this model is that it can be applied only to s–p_z surface electronic states on simple and noble metal surfaces. All quantities that determine the e–ph coupling can also be obtained from *ab initio* calculations. An advantage of this approach is that all the three ingredients of the e−ph coupling matrix are precisely evaluated on the same footing irrespective of the surface state symmetry. To date, the most efficient method for calculating lattice dynamical properties of solids is linear response technique based on the solid-state Sternheimer theory [25, 26]. In this approach, atomic displacements are treated as perturbations and the electronic response to the perturbation is calculated self-consistently. This technique has been shown to be an efficient alternative to the frozen phonon method because this approach is not limited to commensurate phonon wave vectors $\mathbf{q}$, and arbitrary wave vector $\mathbf{q}$ can be treated without any supercell. Moreover, it does not require the knowledge of all unperturbed electronic states as the perturbative approach. This method has been implemented with different basis sets for representing electronic wave functions [27, 28].

7.3
Experimental Determination of the Electron–Phonon Coupling Strength

Angle-resolved photoemission is a unique experimental tool providing direct access to band structure and many-body effects in solids in general and on e−ph in particular. It is a firmly established experimental technique and many reviews are available, describing both its theoretical and it experimental fundamentals (see, for example, [29–33] or Chapter 3). In the following discussion, we focus on the essential points for the study of e−ph on surfaces states. We are concerned only with ARPES from nearly two-dimensional states using a spectrometer with infinitely high energy and k-resolution. We also note that the photoelectron wave vector parallel to the surface $\vec{k}_{\parallel}$ is conserved in the photoemission process and as we treat only quasitwo-dimensional (2D) states, this 2D wave vector is the only one of interest here. For brevity, we denote it as $\vec{k}$.

In this case, and under certain additional assumptions, the photoemission intensity is proportional to the hole spectral function of the sample times the Fermi distribution. The spectral function $\mathcal{A}$, in turn, is used to describe the electronic structure of a solid in the presence of many-body effects. $\mathcal{A}$ can be viewed as the probability of finding an electron with energy $\hbar\omega$ and momentum $\vec{k}$ at a given temperature T. The spectral function is determined by the unrenormalized dispersion $\varepsilon(\vec{k})$ and the self-energy Σ. It is usually assumed that Σ is independent of $\vec{k}$. Then, $\mathcal{A}$ has the form

$$\mathcal{A}(\omega, \vec{k}, T) = \frac{\pi^{-1}|\Sigma''(\omega, T)|}{[\hbar\omega - \varepsilon(\vec{k}) - \Sigma'(\omega, T)]^2 + \Sigma''(\omega, T)^2}. \tag{7.16}$$

A plot of a typical spectral function for the case of strong electron–phonon coupling is given in Figure 7.1b. Under the given assumption that $\mathcal{A}(\omega, \vec{k}, T)$ is proportional to the photoemission intensity, and taking into account that the measured kinetic energy of the photoelectron E_{kin} is merely the binding energy shifted by photon energy and work function, the remaining task is to extract Σ from the measured

$\mathcal{A}(\omega, \vec{k}, T)$ and hereby to gain the desired information about the e–ph coupling strength.

In a modern ARPES setup, the photoemission intensity can be measured for so many values $(\omega, \vec{k})$ that any cut through the spectral function can be extracted. However, we briefly relate the spectral function to the traditional measuring modes of ARPES, energy distribution curves (EDCs) and momentum distribution curves (MDCs).

An EDC is the photoemission intensity as a function of kinetic energy for a fixed photon energy and a fixed emission angle. The fact that the emission angle, not $\vec{k}$, is constant means that an EDC corresponds to a fairly complicated cut through the spectral function. Under certain conditions, for example, for normal emission or for a very small energy range, an EDC is taken at approximately constant $\vec{k}$. Even then an EDC calculated from (7.16) has a fairly complicated form. This is due to the energy dependence of Σ. A simple scenario arises when we assume that $\Sigma'(\omega, T) = 0$ and that $\Sigma''(\omega, T)$ does not depend on ω. Then, we get

$$\mathcal{A}(\omega, \vec{k}, T) = \frac{\pi^{-1}|\Sigma''(T)|}{[\hbar\omega - \varepsilon(\vec{k})]^2 + \Sigma''(T)^2}, \tag{7.17}$$

which is a Lorentzian with the maximum at $\varepsilon(\vec{k})$ and a full width at half maximum (FWHM) of $2|\Sigma''(T)|$. This approximation is relevant in the case of the e–ph as long as the binding energy of the peak is not too small (see Figure 7.3). However, care is necessary when an EDC linewidth is identified with $2|\Sigma''(T)|$ because of the above-mentioned problem that an EDC is strictly measured at a constant emission angle, not at a constant $\vec{k}$ [34–36].

The situation is simpler in case of MDCs because they are readily represented by (7.16). The maximum of an MDC is reached when $\hbar\omega - \varepsilon(\vec{k}) - \Sigma'(\omega, T) = 0$. Based on this, the renormalized dispersion is defined as the self-consistent solution of

$$E(\vec{k}) = \varepsilon(\vec{k}) + \Sigma'(E(\vec{k}), T). \tag{7.18}$$

Eq. (7.16) takes on a particularly simple form in the case of a linear dispersion. We consider only one direction in $\vec{k}$ space and write $\varepsilon(k) = vk$ such that the origin of the coordinates is at the Fermi level crossing. Then, it is easy to show that (7.16) is a Lorentzian line in k for a given ω with the maximum at

$$k_{\max} = (1/v)(\hbar\omega - \Sigma'(\omega)) \tag{7.19}$$

and

$$\mathrm{FWHM} = 2|\Sigma''(\omega)/v|. \tag{7.20}$$

We see that it is under certain condition possible to determine the real or imaginary part of the self-energy from the spectral function measured by ARPES. The next difficulty in the analysis is to relate this to the e–ph coupling strength. The most fundamental quantity for describing the e–ph is the Eliashberg function α^2F that cannot be directly extracted from the experiment. It is, however, closely related to

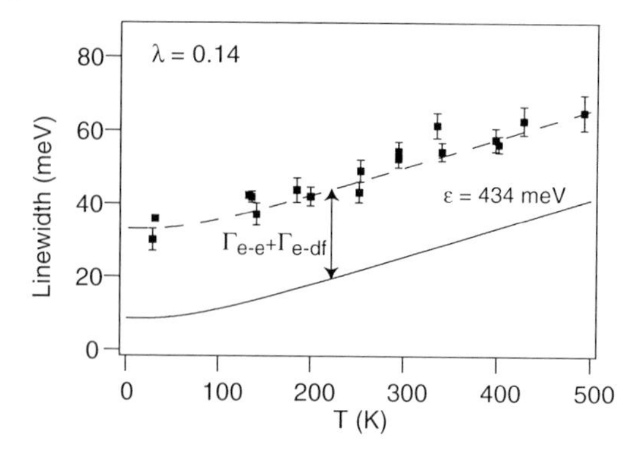

Figure 7.4 Temperature-dependent linewidth of the Cu(111) surface state at $\bar{\Gamma}$ (data points are taken from Ref. [44]). The solid line is the e–ph contribution to the linewidth calculated within a two-dimensional Debye model and assuming $\lambda = 0.14$. The dashed line is a rigid displacement of the solid line in order to take e–e and e–df scattering into account.

the Σ through (7.10) and (7.14). The difficulty is that there is no trivial inversion to these equations and that the e–ph coupling effects in Σ are temperature-dependent whereas $\alpha^2 F$ is not. In the following discussion, we briefly discuss and illustrate different approaches that have been proposed to extract information about the e–ph from photoemission data.

A simple approach used frequently (e.g., see Refs [37–46]) is to measure the temperature-dependent EDC linewidth of a state far away from E_F. In this case, we have seen from (7.17) that the linewidth is $2|\Sigma''(T)|$. Figure 7.4 shows the linewidth of the Cu(111) surface sate at the $\bar{\Gamma}$ point at a binding energy of 434 meV, as well as a calculation for the expected Γ_{e-ph} from (7.10), using a value of $\lambda = 0.14$. The data points are taken from a paper by McDougall *et al.* [44]. Evidently, the agreement between the calculation and the experiment is very good if the latter are rigidly shifted to higher energy. This is expected according to (7.1) because the measured linewidth does not only contain the e–ph contribution but also the e–e and e–df contributions that are assumed to be independent of temperature. In their original paper, McDougall *et al.* fitted the data points with a line, using (7.12) plus an offset, rather than the full expression (7.9). While this simplification is formally justified only for temperatures much higher than the Debye temperature Θ_D (343 K for Cu), it is, in practice, already quite useful for temperatures similar to Θ_D, as evident from the figure.

If the (surface) Debye temperature is too high for (7.12) to be a good approximation, this simple approach of data analysis becomes problematic. In order to extract information about the e–ph from temperature-dependent data, it is then necessary to use (7.9), but this requires a model for the Eliashberg function $\alpha^2 F$. Frequently, one employs a simple model for $\alpha^2 F$, such as the three-dimensional Debye model with

$$\alpha^2 F(\omega) = \lambda \left(\frac{\omega}{\omega_D} \right)^2, \quad \omega < \omega_D \tag{7.21}$$

and

$$\alpha^2 F(\omega) = 0, \omega > \omega_D,$$

which has also been used to calculate the solid curve in Figure 7.4. Alternative models are an Einstein model or a two-dimensional Debye model [7, 16]. Unfortunately, this introduces a certain degree of arbitrariness and it requires the precise knowledge of the surface Debye temperature.

Information about the e–ph strength near the Fermi level, and hence in the energy range most relevant for transport properties, can also be obtained by a direct observation of the band renormalization. However, the extraction of the self-energy near the Fermi level is not straightforward. The key problem is that the unrenormalized dispersion $\varepsilon(\vec{k})$ is not known. This is a familiar problem for high-temperature superconductors for which a strong band renormalization is found. Different solutions have been employed to solve this problem. One is to extrapolate $\varepsilon(\vec{k})$ from states at higher binding energy where the renormalization is negligible [47]. Another is to take $\varepsilon(\vec{k})$ from a calculation of the band structure that does not incorporate the many-body effects. The third is to obtain $\varepsilon(\vec{k})$ from a measurement of the dispersion at elevated temperatures where the renormalization due to electron–phonon coupling is negligible [32]. Finally, $\varepsilon(\vec{k})$ and the self-energy can be determined by a self-consistent fitting procedure [48].

A good illustration of the different possibilities is given in Figure 7.5 that shows surface-state dispersion for a Mo(110) surface state, the imaginary part of the self-energy, and the real part of the e–ph self-energy [47]. The dispersion, as determined

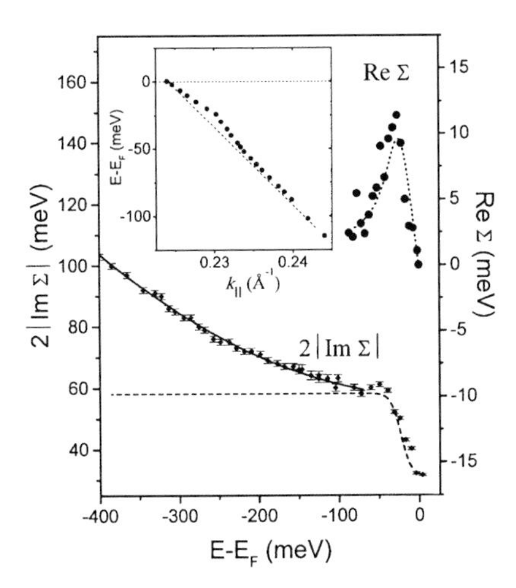

Figure 7.5 e–ph for a Mo(110) surface state after Valla *et al.* [47]. Σ' and Σ'' are plotted as a function of binding energy. Σ'' was obtained from the width of the quasiparticle peak. The dashed (dotted) line shows the calculated electron–phonon contribution to Σ'' (Σ'). The dashed line is shifted up by 26 meV. The inset shows the renormalized (points) and bare dispersion (dashed line) used to extract Σ'. Reprinted with permission from [47]. Copyright 1999 by the American Physical Society.

from MDCs, shows a clear kink close to the E_F that is caused by e–ph. Σ' was determined according to (7.18) from these data and the dispersion interpolated from higher binding energies, assuming that the position of the Fermi level crossing is not affected by the e–ph. The figure also shows the Σ'' determined from the EDC peak width, an approach that only works if the coupling is not too strong and the Fermi cutoff is taken into account. Close to E_F, Σ'' shows the typical signature of e–ph with a strong change in a small energy window, which is schematically shown in Figure 7.1. The dashed line shows a model calculation for the e–ph part of Σ'' and the dotted line shows the Kramers–Kronig transformation of this, which is in good agreement with Σ'. The calculated Σ'' stemming from a calculated bulk $\alpha^2 F$ and (7.9) agrees well with the data; and the surface Debye temperature is similar to the bulk value, suggesting a similar mass enhancement parameter λ. Interestingly, the measured Σ'' shows an increase at higher binding energies that cannot be accounted for by e–ph. This is ascribed to e–e. Σ'' is also 26 meV higher than the calculated value, which is ascribed to e–df scattering.

Figure 7.6 illustrates the case of stronger coupling that is found on the Be(0001) surface. Apart from the strong coupling, beryllium is a favorable material for the observation of e–ph because of the high phonon energies and Debye temperature that permit a detailed observation of the effect without the need of an exceedingly high-energy resolution. Figure 7.6 shows low-temperature high-resolution data from the work of Hengsberger *et al.* [49, 50]. EDCs from the Be(0001) surface are given as the dispersion approaches E_F. Close to the Fermi-level crossing, the EDCs clearly deviate from the Lorentzian line shape (7.17) because both conditions for the validity

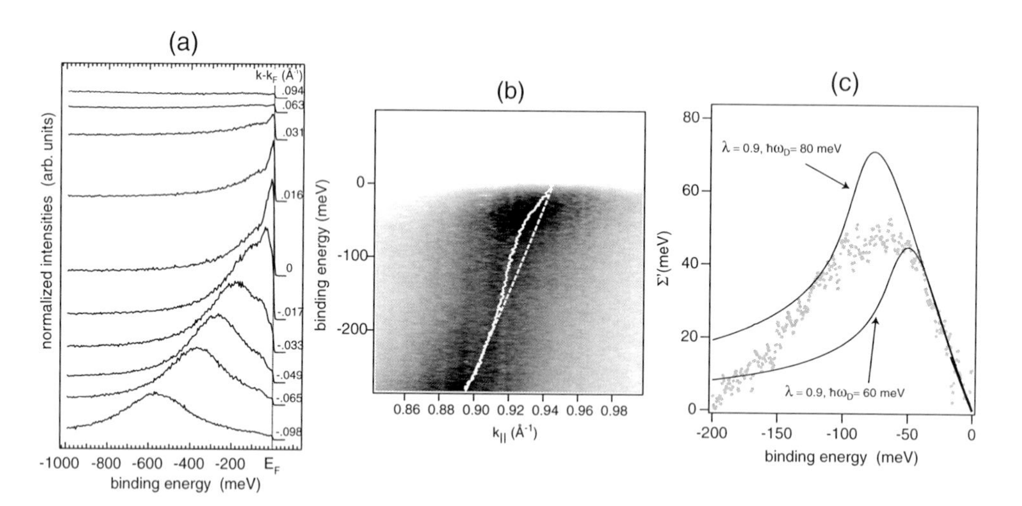

Figure 7.6 (a) High-resolution photoemission spectra of the Be(0001) surface state near k_F at $T = 12$ K. Reprinted with permission from [50]. Copyright 1999 by the American Physical Society. (b) Photoemission intensity of the same state near E_F at $T = 70$ K. The dashed line shows the bare dispersion $\varepsilon(k)$ and the solid line tracks the MDC maxima, giving the renormalized dispersion $E(k)$. (c) Resulting Σ'. The different lines are models to extract λ as described in the text. From E.D.L. Rienks *et. al.* to be published.

of this equation ($\Sigma'(\omega, T) = 0$ and $\Sigma''(\omega, T) =$ const.) are violated. This complicated line shape is a direct confirmation of an old prediction [51]. In this case, a simple analysis of temperature-dependent data to extract information about e–ph would be clearly aggravated by the complicated line shape. Instead, Figure 7.6b and c show the determination of Σ' and how information about λ is extracted from more recent data. The renormalized dispersion $E(k)$ is represented by the solid line that tracks the maxima of the MDCs, according to (7.16). The bare dispersion $\varepsilon(k)$ is found from two conditions: (1) it has to cross E_F at the same k_F unless there is a significant distortion of the band by a finite energy resolution and (2) it must coincide with $E(k)$ for high binding energies as $\Sigma'(\omega)$ approaches zero. In the present case, $\varepsilon(k)$ is described by a second-order polynomial. The resulting self-energy $\Sigma'(E)$ can now be determined using (7.18). Alternatively, $\Sigma'(\varepsilon)$ could be determined from a fit to the width of the state, as in Figure 7.5, but the position of a peak is generally more stable in noisy data than its width. The resulting $\Sigma'(E)$ is given in Figure 7.6c. From this it is possible to extract λ in several ways. The simplest is to use (7.15) and to extract λ from the slope of Σ' close to E_F. This is illustrated by the short bold line near E_F that corresponds to a $\lambda = 0.9$. It is crucial to keep in mind the conditions for this approach to be valid: the temperature must be very low compared to Θ_D (fulfilled for Be) and the energy range used must be very small because only the slope *at the Fermi energy* is of interest.

Alternatively, the entire Σ' can be fitted with a model self-energy, for example, using a Debye model (7.21) to calculate α^2F and then (7.14) to calculate Σ'. In the Debye model, this calculation contains two parameters λ and ω_D as well as the sample temperature. Two such calculations are shown in Figure 7.6c for $\lambda = 0.9$ and $\hbar\omega_D = 80$ meV as well as for $\lambda = 0.9$ and $\hbar\omega_D = 60$ meV. Roughly spoken, and at low temperature, λ gives the slope of the curve at E_F and ω_D determines the maximum of the curve. In the present case, it is evident that the Debye model is too simple to account for the detailed shape of Σ'. No set of parameters can be found to result in a satisfactory overall fit.

Recently, a different approach to e–ph data analysis has been proposed, which potentially solves several of the problems mentioned above. The experimentally determined self-energy Σ' or Σ'' is not analyzed using a model for α^2F combined with (7.14) or (7.10). Rather, $\alpha^2F(\omega)$ is directly obtained from Σ' using an integral inversion of (7.14) based on a maximum entropy approach [52, 53]. This method directly yields $\alpha^2F(\omega)$, that is, the most fundamental property for the description of e–ph, and it has the potential to provide an interesting fine structure in this function. For its reliable application, very high-quality data are needed.

7.4
Some Examples

In this section, we present some results for the e–ph coupling at surfaces. We focus on simple and noble metals and discuss only cases in which both experimental data and *ab initio* calculations are available. An overview of both experimental and theoretical results is given in Table 7.1. Comparing the calculated and experimentally

Table 7.1 Electron–phonon coupling parameter λ and linewidth (in meV) for surface electronic states.

	E ((eV))	λ(calc.)	λ(expt.)	$\Gamma_{e-ph}(T=0\ K)$	$\Gamma_{Expt.}$
Al(001)					
$\overline{\Gamma}$	−2.8 eV [54]	0.51 [6]		35 [6]	267 [5]
		0.23 [19]		18 [19]	
				26 [55]	
$\overline{X}$	−4.55 eV [56]	0.78 [6]		50 [6]	
Be(0001)					
$\overline{\Gamma}$	−2.78 eV [57]	0.38 [58]	1.15 [38]		281 [59]
			0.87 [35]		
Mg(0001)					
$\overline{\Gamma}$	−1.63 eV [60]	0.28 [61]	0.27 [41]	19 [61]	133 [41]
$\overline{M}$	−0.95 eV [60]	0.38 [61]		20 [61]	
Ag(111)					
$\overline{\Gamma}$	−0.04 eV [62]	0.12 [19]	0.12 [14]	3.7 [19]	6 [8]
				7.2 [8]	6 ± 0.5 [12]
Cu(111)					
$\overline{\Gamma}$	−0.4 eV [63]	0.16 [19]	0.14 ± 0.02 [44]	7.3 [19]	24 [8]
			0.11 [14]	21.7 [8]	23 ± 1[12]
				5.67 [55]	
Au(111)					
$\overline{\Gamma}$	−0.5 eV [64]	0.11 [19]	0.34 ± 0.01 [42]	3.6 [19]	18 [8]
				18.9 [8]	21 ± 1 [12]

determined mass enhancement parameters λ shows a very satisfactory agreement for this class of materials. Details of experiments and calculations are discussed in the following section.

7.4.1
The (111) Surface of the Noble Metals

The (111) surfaces of the noble metals Ag, Cu, and Au all support a similar Shokley-type surface state in the bulk L-gap of the metal. This surface state has long been an important model system for studying electronic structure and line shapes by ARPES (see Ref. [12] both for recent high-resolution data and for a historic overview of the field). The surface state is well localized within a few layers of the surface and has a small binding energy such that the e–ph interaction for these states should be strongly influenced by surface phonon modes. The electron–phonon coupling turns out to be very similar for all three surfaces.

Results of a theoretical investigation of the phonon-mediated decay of surface states on Ag(111), Cu(111), and Au(111) were presented in Refs [14, 19]. The electronic states were defined using model potentials [21, 65] that reproduce the correct surface projected bandgap at the $\overline{\Gamma}$-point and the surface-state energies for the systems. The

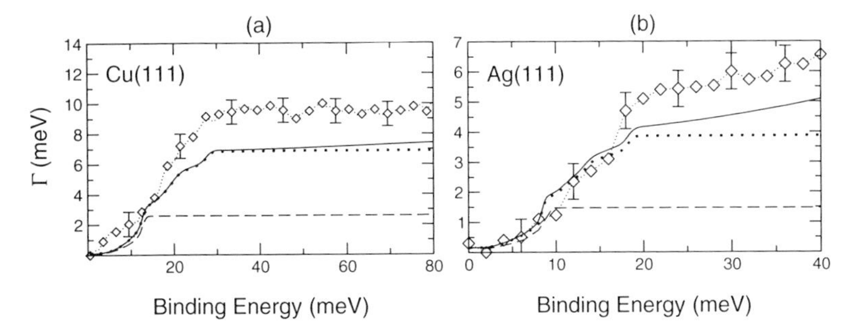

Figure 7.7 (a) Lifetime broadening of the Cu(111) surface hole state as a function of binding energy, $\Gamma_{e-e} + \Gamma_{e-ph}$ (solid line), Γ_{e-ph} (dotted line), Rayleigh mode contribution to Γ_{e-ph} (dashed line), and photoemission data (diamonds). (b) The same as in (a) for Ag(111). Reprinted with permission from [14]. Copyright 2002 by the American Physical Society.

phonon modes were obtained from a single force constant model, where the force constant was fitted to reproduce the elastic constants and the maximum bulk phonon energy.

Figure 7.7 shows the result of such a calculation for Cu(111) and Ag(111) and the comparison to experimental data, obtained from the linewidth of the state near E_F [14]. The overall agreement between calculation and experiment is very good. As expected because of the small penetration and as shown in Figure 7.3, the Rayleigh mode gives a very significant contribution to $\alpha^2 F$ and hence to Γ_{e-ph} for these surfaces. It is the dominant mechanism for hole decay at very small energies for which e–e scattering is insignificant. Note also that Γ_{e-ph}, or equivalently Σ'', which results from this calculation shows considerable fine structure.

Another important result from the calculations is that the coupling strength λ is relatively independent of the binding energy of the hole. Again, this is in good agreement with experimental data that do not point toward any strong binding energy dependence of λ [14, 42, 44].

It should be mentioned that an initial experimental study of the e–ph on Au(111) gave a value of $\lambda = 0.33$ in rather poor agreement with the calculated $\lambda = 0.11$ [19]. Later on, this discrepancy was ascribed to the thermal excitation of defects at elevated temperature and the experimental linewidth could be reconciled with $\lambda = 0.11$. This effect is discussed in more detail in connection with e–ph on Al(001) below.

7.4.2 Be(0001)

The Be(0001) surface is a nearly ideal system to test the electron–phonon coupling of surfaces because the $\overline{\Gamma}$ surface state resides in a wide gap and contributes significantly to the total density of states (DOS) at the Fermi level. In fact, the bulk DOS of Be is not free electron-like due to the strongly covalent bonding character in this metal but the surface DOS is, justifying the view of the surface state as two-dimensional free electron gas that is decoupled from the bulk [66].

Several experimental studies and one *ab initio* calculation of the e–ph coupling on Be(0001) have been published [35, 38, 49, 50, 53, 58]. The experimental values of λ spread over an unsatisfactory large range between 0.7 and 1.18, even though some care has to be exercised here because the lowest reported value by Tang *et al.* might be caused by oxygen contamination of the sample [46], and not all values have been measured at E_F or at the same direction of $\vec{k}_F$. The theoretical value of λ at E_F was found to be 0.9 in good agreement with the available data. In any event, the coupling is much stronger than in bulk Be for which $\lambda = 0.21$–0.23 (theory, [67, 68]).

7.4.3
Mg(0001) and Al(001)

The $\bar{\Gamma}$ surface states of Mg(0001) and Al(001) have a rather different character from those of the noble metal (111) surfaces and of Be(0001). Both reside in a narrow projected bandgap and penetrate deeply into the bulk. Thus, one would expect a certain similarity to actual bulk states, both in their electronic character and in their e–ph interaction.

An *ab initio* study of the electron–phonon coupling and its contribution to the lifetime broadening of $\bar{\Gamma}$ surface state on Al(001) was reported in Ref. [6]. As expected, the largest contribution to the electron–phonon coupling comes from the scattering of excited electrons with bulk phonon modes. In general, the surface phonons contribute less than 30% to the e–ph coupling. This fact was also proved by model potential calculations [19] where it was shown that the interband scattering in the $\bar{\Gamma}$ surface state makes the most important contribution to the Eliashberg function.

Another important finding was that the low- and middle-energy phonons are more involved in the scattering processes of electrons than the high-energy phonon modes unlike the case of bulk Al, as well as other simple metals Be and Mg, where the lower energy part of the phonon spectrum is strongly suppressed by e–ph matrix elements [27, 61, 68–70].

The calculated $\lambda(\bar{\Gamma}) = 0.51 \pm 0.01$ [6] is somewhat higher than the e–ph coupling parameter averaged over momenta both at the Fermi level of bulk Al, $\lambda\,(E_F) = 0.43$, and at the Fermi energy of the Al(001) surface, $\lambda\,(E_F) = 0.45$.

Experimental results for this surface state at $\bar{\Gamma}$ were presented in Ref. [5]. The temperature-dependent linewidth of the state had been measured over a wide temperature range but the data could not be accounted for using (7.10) or (7.12) plus a temperature-independent offset. Indeed, a fit to the high-temperature part of the data, in the range where (7.12) should be applicable, resulted in $\lambda = 0.84$, in very poor agreement with theory. This problem could be resolved by taking into account the possibility of e–df scattering from thermally excited defects. While the e–df scattering strength is still assumed to be independent of temperature, the number of defects is not and there is an exponentially increasing probability of thermally excited defects at elevated temperatures. With this assumption, a satisfactory fit to the data could be obtained that was consistent with the theoretical value for λ.

The $\bar{\Gamma}$ surface state on Mg(0001) is similar in character to the one on Al(001) in that it penetrates very deeply into the bulk. The temperature-dependent linewidth of the state has been analyzed along the same lines as discussed above and the results have been interpreted using a three-dimensional Debye model for the $\alpha^2 F$. A problem in this interpretation was the unknown surface Debye temperature Θ_D: a good fit to the data could be achieved for a wide range of λ values depending the choice of Θ_D. This is not surprising: both λ and Θ_D appear in the model $\alpha^2 F$ (7.21) and a change in one value can almost entirely be compensated by a corresponding change in the other. The problem was resolved by defining an effective Θ_D based on experimental data on the surface vibrations combined with a calculated probability density function of the surface state. This resulted in a value of $\lambda = 0.27$.

A detailed *ab initio* study of the electron–phonon interaction and phonon-mediated contribution to the linewidth of surface electronic states on Mg(0001) was reported in Ref. [61]. The results are very similar to those obtained for Al(001): there is a strong interaction of electrons with bulk phonon modes because the surface electronic states in both cases lie very close to bulk electronic bands. λ was found to have a value of 0.28, in excellent agreement with the experimental data.

7.5 Conclusions

A brief conclusion from these examples is that the e–ph coupling strength for these simple systems is now reasonably well understood. *Ab initio* calculations are of such a high quality that they compare well with the available experimental data. The *ab initio* study not only allows to evaluate the Eliashberg function, the e–ph self-energy, the e–ph coupling strength parameter λ, and the phonon-mediated contribution Γ_{e-ph} to the linewidth Γ, but also allows to discriminate between Γ_{e-ph} and thermally induced defect contribution Γ_{e-df} to the temperature dependence of the surface-state linewidth. From the experimental side, extracting information on the e–ph interaction from the data is not straightforward and the choice of approach depends on the properties of the system (coupling strength, Debye temperature). The biggest challenge for the experiment is to provide data of sufficient quality to extract fine structure in the self-energy that can then be related to individual phonon modes contributing to the e–ph. Ideally, it would be possible to compare the fine structure in measured and calculated Eliashberg functions. For these purposes, the use of such spectroscopies as photoemission and two-photon time- and energy-resolved photoemission as well as scanning tunneling spectroscopy may be very beneficial.

References

1 Grimvall, G. (1981) *The Electron–Phonon Interaction in Metals*, North-Holland, Amsterdam.

2 Anderson, P.W. and Schrieffer, J.R.R. (1991) A dialouge on the theory of high Tc. *Phys. Today*, **44** (6), 54–61.

3 Damascelli, A., Hussain, Z., and Shen, Z.-X. (2003) Angle-resolved photoemission studies of the cuprate supercondcuctors. *Rev. Mod. Phys.*, **75**, 473.

4 Chulkov, E.V., Borisov, A.G., Gauyacq, J.P., Sanchez-Portal, D., Silkin, V.M.,

Zhukov, V.P., and Echenique, P.M. (2006) Electronic excitations in metals and at metal surfaces. *Chem. Rev.*, **106**, 4160–4206.

5 Fuglsang Jensen, M., Kim, T.K., Bengio, S., Sklyadneva, I.Y., Leonardo, A., Eremeev, S.V., Chulkov, E.V., and Hofmann, P. (2007) Thermally induced defects and the lifetime of electronic surface states. *Phys. Rev. B*, **75** (15), 153404.

6 Sklyadneva, I.Y., Chulkov, E.V., and Echenique, P.M. (2008) Electron–phonon interaction on an Al(001) surface. *J. Phys. Condens. Matter*, **20**, 165203

7 Echenique, P.M., Berndt, R., Chulkov, E.V., Fauster, T., Goldmann, A., and Höfer, U. (2004) Decay of electronic excitations at metal surfaces. *Surf. Sci. Rep.*, **52**, 219.

8 Kliewer, J., Berndt, R., Chulkov, E.V., Silkin, V.M., Echenique, P.M., and Crampin, S. (2000) Dimensionality effects in the lifetime of surface states. *Science*, **288**, 1399–1402.

9 Kroger, J., Limot, L., Jensen, H., Berndt, R., Crampin, S., and Pehlke, E. (2005) Surface state electron dynamics of clean and adsorbate-covered metal surfaces studied with the scanning tunnelling microscope. *Prog. Surf. Sci.*, **80**, 26–48.

10 Hofer, U., Shumay, I.L., Reuss, Ch., Thomann, U., Wallauer, W., and Fauster, T. (1997) Time-resolved coherent photoelectron spectroscopy of quantized electronic states on metal surfaces. *Science*, **277** (5331), 1480–1482.

11 Weinelt, M. (2002) Time-resolved two-photon photoemission from metal surfaces. *J. Phys. Condens. Matter*, **14**, R1099–R1141.

12 Reinert, F., Nicolay, G., Schmidt, S., Ehm, D., and Hüfner, S. (2001) Direct measurements of the *L*-gap surface states on the (111) face of noble metals by photoelectron spectroscopy. *Phys. Rev. B*, **63**, 115415.

13 Chulkov, E.V., Kliewer, J., Berndt, R., Silkin, V.M., Hellsing, B., Crampin, S., and Echenique, P.M. (2003) Hole dynamics in a quantum-well state at Na/Cu(111). *Phys. Rev. B*, **68** (19), 195422.

14 Eiguren, A., Hellsing, B., Reinert, F., Nicolay, G., Chulkov, E.V., Silkin, V.M., Hüfner, S., and Echenique, P.M. (2002) Role of bulk and surface phonons in the decay of metal surface states. *Phys. Rev. Lett.*, **88**, 066805.

15 Hellsing, B., Carlsson, J., Wallden, L., and Lindgren, S.A. (2000) Phonon-induced decay of a quantum-well hole: one monolayer Na on Cu(111). *Phys. Rev. B*, **61**, 2343–2348.

16 Hellsing, B., Eiguren, A., and Chulkov, E.V. (2002) Electron–phonon coupling at metal surfaces. *J. Phys.: Condens. Matter*, **14**, 5959–5977.

17 Sklyadneva, I.Y., Leonardo, A., Echenique, P.M., Eremeev, S.V., and Chulkov, E.V. (2006) Electron–phonon contribution to the phonon and excited electron (hole) linewidths in bulk Pd. *J. Phys. Condens. Matter*, **18**, 7923–7935.

18 Allen, P.B. and Cohen, M.L. (1969) Pseudopotential calculation of mass enhancement and superconducting transition temperature of simple metals. *Phys. Rev.*, **187**, 525.

19 Eiguren, A., Hellsing, B., Chulkov, E.V., and Echenique, P.M. (2003) Phonon-mediated decay of metal surface states. *Phys. Rev. B*, **67**, 235423.

20 Chulkov, E.V., Silkin, V.M., and Echenique, P.M. (1997) Image potential states on lithium, copper and silver surfaces. *Surf. Sci.*, **391** (1–3), L1217–L1223 11.

21 Chulkov, E.V., Silkin, V.M., and Echenique, P.M. (1999) Image potential states on metal surfaces: binding energies and wave functions. *Surf. Sci.*, **437**, 330.

22 Black, J.E., Shanes, F.C., and Wallis, R.F. (1983) Surface vibrations on face-centered cubic metal-surfaces: the (111) surfaces. *Surf. Sci.*, **133**, 199–215.

23 Borisova, S.D., Rusina, G.G., Eremeev, S.V., Benedek, G., Echenique, P.M., Sklyadneva, I.Y., and Chulkov, E.V. (2006) Vibrations in submonolayer structures of Na on Cu(111). *Phys. Rev. B*, **74**, 165412.

24 Ashcroft, N.W. (1966) Electron–ion pseudopotentials in metals. *Phys. Lett.*, **23**, 48.

25 Baroni, S., Giannozzi, P., and Tesla, A. (1987) Green's-function approach to linear response in solids. *Phys. Rev. Lett.*, **58**, 1861–1864.

26 Zein, N.E. (1984) *Sov. Phys. Solid State*, **26**, 3028.

27 Liu, A.Y. and Quong, A.A. (1996) Linear-response calculation of electron–phonon coupling parameters. *Phys. Rev. B*, **53**, R7575–R7579.

28 Savrasov, S.Y., Savrasov, D.Y., and Andersen, O.K. (1994) Linear-response calculations of electron–phonon interactions. *Phys. Rev. Lett.*, **71**, 372–375.

29 Eberhardt, W. and Plummer, E.W. (1980) Angle-resolved photoemsission determination of the band structure and multielectron excitations in Ni. *Phys. Rev. B*, **21**, 3245–3255.

30 Hüfner, S. (2003) *Photoelectron Spectroscopy*, 3rd edn, Springer, Berlin.

31 Kevan, S.D. (ed.) (1992) Angle-resolved photoemission, in *Studies in Surface Chemistry and Catalysis*, vol. **74**, Elsevier, Amsterdam.

32 Kirkegaard, C., Kim, T.K., and Hofmann, P. (2005) Self-energy determination and electron–phonon coupling on Bi(110). *N. J. Phys.*, **7**, 99.

33 Matzdorf, R. (1998) Investigation of line shapes and line intensities by high-resolution UV-photoemission spectroscopy: some case studies on noble-metal surfaces. *Surf. Sci. Rep.*, **30**, 153–206.

34 Hansen, E.D., Miller, T., and Chiang, T.-C. (1998) Observation of photoemission line widths narrower than the inverse lifetime. *Phys. Rev. Lett.*, **80**, 1766–1769.

35 LaShell, S., Jensen, E., and Balasubramanian, T. (2000) Nonquasiparticle structure in the photoemission spectra from the Be(0001) surface and determination of the electron self energy. *Phys. Rev. B*, **61**, 2371–2374.

36 Smith, N.V., Thiry, P., and Petroff, Y. (1993) Photoemission linewidth and quasiparticle lifetimes. *Phys. Rev. B*, **47**, 15476–15481.

37 Balasubramanian, T., Glans, P.-A., and Johansson, L.I. (2000) Electron–phonon mass-enhancement parameter and the Fermi-line eccentricity at the Be($10\bar{1}0$) surface from angle-resolved photoemission. *Phys. Rev. B*, **61**, 12709–12712.

38 Balasubramanian, T., Jensen, E., Wu, X.L., and Hulbert, S.L. (1998) Large value of the electron–phonon coupling parameter ($\lambda = 1.15$) and the possibility of surface superconductivity at the Be(0001) surface. *Phys. Rev. B*, **57**, R6866–R6869.

39 Gayone, J.E., Hoffmann, S.V., Li, Z., and Hofmann, P. (2003) Strong energy dependence of the electron–phonon coupling on Bi(100). *Phys. Rev. Lett.*, **91**, 127601.

40 Hofmann, P., Cai, Y.Q., Grütter, C., and Bilgram, J.H. (1998) Electron–lattice interactions on α-Ga(010). *Phys. Rev. Lett.*, **81**, 1670–1673.

41 Kim, T.K., Sorensen, T.S., Wolfring, E., Li, H., Chulkov, E.V., and Hofmann, P. (2005) Electron–phonon coupling on the Mg(0001) surface. *Phys. Rev. B*, **72** (7), 075422.

42 LaShell, S., McDougall, B.A., and Jensen, E. (2006) Electron–phonon mass enhancement parameter of the [overline gamma] surface state on Au(111) measured by photoemission spectroscopy. *Phys. Rev. B*, **74** (3), 033410.

43 Matzdorf, R., Meister, G., and Goldmann, A. (1996) Phonon contributions to photohole linewidth observed for surface states on copper. *Phys. Rev. B*, **54**, 14807.

44 McDougall, B.A., Balasubramanian, T., and Jensen, E. (1995) Phonon contribution to quasiparticle lifetimes in Cu measured by angle-resolved photoemission. *Phys. Rev. B*, **51**, R13891.

45 Paggel, J.J., Miller, T., and Chiang, T.-C. (1999) Temperature dependent complex band structure and electron–phonon coupling in Ag. *Phys. Rev. Lett.*, **83**, 1415–1418.

46 Tang, S.J., Ismail, Sprunger, P.T., Plummer, E.W. (2002) Electron–phonon coupling and temperature-dependent shift of surface state on Be($10\bar{1}0$). *Phys. Rev. B*, **65**, 235428.

47 Valla, T., Fedorov, A.V., Johnson, P.D., and Hulbert, S.L. (1999) Many-body effects in angle-resolved photoemission: quasiparticle energy and lifetime of a Mo(110) surface state. *Phys. Rev. Lett.*, **83**, 2085–2088.

48 Kordyuk, A.A., Borisenko, S.V., Koitzsch, A., Fink, J., Knupfer, M., and Berger, H. (2005) Bare electron dispersion from experiment: self-consistent self-energy analysis of photoemission data. *Phys. Rev. B*, **71** (21), 214513.

49 Hengsberger, M., Frésard, R., Purdie, D., Segovia, P., and Baer, Y. (1999) Electron–phonon coupling in photoemission spectra. *Phys. Rev. B*, **60** (15), 10796–10802.

50 Hengsberger, M., Purdie, D., Segovia, P., Garnier, M., and Baer, Y. (1999) Photoemission study of a strongly coupled electron–phonon system. *Phys. Rev. Lett.*, **83**, 592–595.

51 Engelsberg, S. and Schrieffer, J.R. (1963) Coupled electron–phonon systems. *Phys. Rev.*, **131**, 993–1008.

52 Shi, Junren., Tang, S.-J., Biao Wu, P.T., Sprunger, W.L., Yang, V., Brouet, X.J., Zhou, Z., Hussain, Z.-X., Shen, Z.Z., and Plummer, E.W. (2004) Direct extraction of the Eliashberg function for electron–phonon coupling: a case study of Be($10\bar{1}0$). *Phys. Rev. Lett.*, **92**, 186401.

53 Tang, S.-J., Shi, Junren., Biao Wu, P.T., Sprunger, W.L., Yang, V., Brouet, X.J., Zhou, Z., Hussain, Z.X., Shen, Z.Z., and Plummer, E.W. (2004) A spectroscopic view of electron–phonon coupling at metal surfaces. *Phys. Status Solidi (b)*, **241** (10), 2345–2352.

54 Plummer, E.W. (1985) Deficiencies in the single particle picture of valence band photoemission. *Surf. Sci.*, **152/153**, 162–179.

55 Nojima, A., Yamashita, K., and Hellsing, B. (2008) Model Eliashberg function for surface states. *Appl. Surf. Sci.*, **254**, 7938–7941.

56 Levinson, H.J., Greuter, F., and Plummer, E.W. (1983) Experimental band structure of aluminium. *Phys. Rev. B*, **27**, 727–747.

57 Bartynski, R.A., Jensen, E., Gustafsson, T., and Plummer, E.W. (1985) Angle-resolved photoemission investigation of the electronic structure of Be: surface states. *Phys. Rev. B*, **32**, 1921–1926.

58 Eiguren, A., de Gironcoli, S., Chulkov, E.V., Echenique, P.M., and Tosatti, E. (2003) Electron–phonon interaction at the Be(0001) surface. *Phys. Rev. Lett.*, **91**, 166803.

59 Silkin, V.M., Balasubramanian, T., Chulkov, E.V., Rubio, A., and Echenique, P.M. (2001) Surface-state hole decay mechanisms: the Be(0001) surface. *Phys. Rev. B*, **64** (8), 085334.

60 Schiller, F., Heber, M., Servedio, V.D.P., and Laubschat, C. (2004) Electronic structure of Mg: from monolayers to bulk. *Phys. Rev. B*, **70**, 125106.

61 Leonardo, A., Sklyadneva, I.Y., Silkin, V.M., Echenique, P.M., and Chulkov, E.V. (2007) Ab *initio* calculation of the phonon-induced contribution to the electron-state linewidth on the Mg(0001) surface versus bulk Mg. *Phys. Rev. B*, **76**, 035404.

62 Hulbert, S.L., Johnson, P.D., Stoffel, N.G., and Smith, N.V. (1985) Unoccupied bulk and surface-states on Ag(111) studied by inverse photoemission. *Phys. Rev. B*, **32**, 3451–3455.

63 Gartland, P.O. and Slagsvold, B.J. (1975) Transitions conserving parallel momentum in photoemission from the (111) face of copper. *Phys. Rev. B*, **12**, 4047–4058.

64 Heimann, P., Neddermeyer, H., and Roloff, H.F. (1977) Ultraviolet photoemission from intrinsic surface-states of noble-metals. *J. Phys. C*, **10** (1), L17–L22.

65 Chulkov, E.V., Sarria, I., Silkin, V.M., Pitarke, J.M., and Echenique, P.M. (1998) Lifetime of image-potential states on copper surfaces. *Phys. Rev. Lett.*, **80**, 4947.

66 Plummer, E.W. and Hannon, J.B. (1994) The surfaces of beryllium. *Prog. Surf. Sci.*, **46**, 149.

67 McMillan, W.L. (1968) Transition temperature of strong-coupled superconductors. *Phys. Rev.*, **167**, 331–334.

68 Sklyadneva, I.Y., Chulkov, E.V., Schöne, W.-D., Silkin, V.M., Keyling, R., and Echenique, P.M. (2005) Role of electron–phonon interactions versus electron–electron interactions in the broadening mechanism of the electron and hole linewidths in bulk Be. *Phys. Rev. B*, **71** (17), 174302.

69 Bauer, R., Schmid, A., Pavone, P., and Strauch, D. (1998) Electron–phonon coupling in the metallic elements Al, Au, Na, and Nb: a first-principles study. *Phys. Rev. B*, **57**, 11276–11282.

70 Savrasov, S.Y. and Savrasov, D.Y. (1996) Electron–phonon interactions and related physical properties of metals from linear-response theory. *Phys. Rev. B*, **54**, 16487–16501.

Part Two
Collective Excitations

Dynamics at Solid State Surface and Interfaces Vol.1: Current Developments
Edited by Uwe Bovensiepen, Hrvoje Petek, and Martin Wolf

ISBN: 978-3-527-40937-2

8
Low-Energy Collective Electronic Excitations at Metal Surfaces

Vyacheslav M. Silkin, Evgueni V. Chulkov, and Pedro M. Echenique

8.1
Introduction

How solids interact with environment depends in a significant, and often crucial, way on the properties of their surfaces. Sometimes, the effect of truncation of a crystal is limited by little modification of its bulk properties, but quite frequently, the surface introduces new phenomena that cannot be deduced from knowledge of the properties of extended systems only. It is known for a long time that the presence of surface can introduce a new kind of one-particle electronic states in addition to the bulk ones. The examples of such states are intrinsic surface states that are often classified as Tamm [1] and Shockley [2] surface states. These states with their wave functions localized at the surface are of great interest in surface science due to their relevance in many phenomena and have been subject of intensive theoretical and experimental studies for years [3]. A well-known example is an s–p surface state at the (111) surface of noble metals. This state has a partly occupied free electron-like parabolic dispersion and is frequently considered a prototype of a quasi two-dimensional (2D) electron gas with 2D Fermi energy equal to the surface state binding energy at the center of the 2D Brillouin zone (2DBZ). However, more precisely, this 2D electron gas is spatially localized together with the three-dimensional (3D) bulk electrons and, actually, constitutes only a small fraction of the total charge density in the vicinity of surface atomic layers. Nevertheless, due to their 2D character and more slow decay into the vacuum in comparison with the bulk states, the surface states play an important role in many phenomena that take place at metal surfaces. Despite the very detailed knowledge about the properties of the surface states being available, they still can throw up a surprise. Thus, recently it was demonstrated that they can dramatically affect the low-energy dynamical screening properties of the surfaces, which support such kind of localized states, providing a mechanism for the existence of a novel phenomenon – collective charge density oscillations involving "out-of-phase" fluctuations between the surface and the bulk electronic subsystems with peculiar linear energy dispersion at small 2D momenta. The properties of this mode constitute the subject of the present chapter.

Dynamics at Solid State Surface and Interfaces Vol.1: Current Developments
Edited by Uwe Bovensiepen, Hrvoje Petek, and Martin Wolf

ISBN: 978-3-527-40937-2

8.2
Analytical and Numerical Calculations

8.2.1
Some Analytical Results

In the description of basic properties of electron systems, a great success was achieved considering them as a free electron gas with a charge density corresponding to the average valence electronic density. In numerous studies of such electronic systems, the importance of their geometry was demonstrated. Their shape, in particular, leads to qualitatively different collective charge density excitations, which are important in many fields of science. Below some examples of such modes in some highly symmetrical geometries are given. Let us start with a 3D infinitely expanded electron system presented in Figure 8.1a. At the beginning of the 1950s, it was demonstrated that its valence electrons can experience collective oscillations [4], called plasmons, for which the frequency is given by

$$\omega_{\rm p} = \left(\frac{4\pi N e^2}{m^*}\right)^2, \tag{8.1}$$

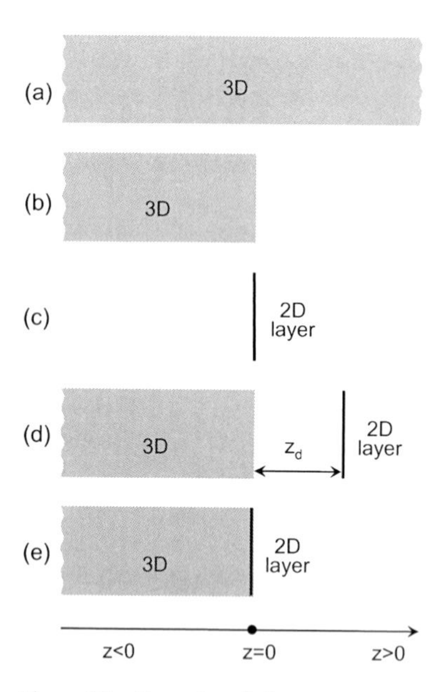

Figure 8.1 Examples of electron gas geometries for which collective charge density excitations are discussed: (a) an infinite homogeneous 3D electron gas, (b) a bounded semi-infinite electron gas, (c) an in-plane infinite 2D electron gas, (d) a 2D electron gas sheet placed in front of a 3D electron gas at a distance z_d, and (e) the same as in (d) for $z_d = 0$.

where N is the electron density, e and m^* are, respectively, electron charge and effective mass. Several years later, it was predicted [5] that the boundary effect (Figure 8.1b) for the 3D electron gas produces another mode at lower energy, $\omega_{sp} = \omega_p/\sqrt{2}$. Later on, the quantum of this mode was called surface plasmon [6]. After its experimental detection [7, 8], this surface mode has been a subject of intense theoretical and experimental investigations in the course of which a deep understanding of its properties was obtained [9–16]. A strong continuous interest in the surface plasmon phenomenon can be explained by its key role in a great variety of fundamental properties of solids. Nevertheless, because the typical energy of bulk and surface plasmons in metals is in the range of several electron volts (in metals, with valence density corresponding[1)] to $2 \leq r_s \leq 6$, plasmons have energies in the range $3\ \text{eV} \leq \omega_p \leq 20\ \text{eV}$), they do not influence directly the phenomena characterized by small excitation energy like that occurs near the Fermi level.

By contrast, the dynamical screening in a 2D electron gas is fundamentally different from that of a 3D electron system. Thus, in a 2D electron gas (Figure 8.1c), the 2D plasmon mode energy shows a $\sqrt{q_{||}}$ dispersion [17] at small 2D momenta[2)] $q_{||}$ and there is no energy threshold for its excitation in the long-wavelength limit. This kind of plasmons were intensively investigated in dilute electronic systems [17, 18] and recently were observed in metallic monolayers adsorbed on a silicon surface [19, 20]. A model in which the 2D electron gas sheet placed in front of the 3D electron system at some distance z_d, as shown in Figure 8.1d, was also considered [21, 22]. While describing the 3D substrate by a Drude dielectric function

$$\varepsilon_{\text{Drude}}^{3D} = 1 - \frac{\omega_p^2}{\omega(\omega + i\eta)}, \tag{8.2}$$

one finds that both the electron–hole and the collective excitations within the 2D sheet are described by an effective 2D dielectric function $\varepsilon_{\text{eff}}^{2D}$, which in a long-wavelength limit has two zeros [23]. One zero corresponds to a high-frequency collective oscillation at energy $\omega^2 = \omega_{sp}^2 + \omega_{2D}^2$, in which the 2D and 3D electrons oscillate in phase and the second zero corresponds to a low-energy oscillation in which the 2D and 3D electrons oscillate out of phase. The energy ω_{AP} of this low-frequency mode (called an acoustic mode due to the *linear* dependence of its energy on the momentum as seen below) at small momenta $q_{||}$ is found to be of the form [23]

$$\omega_{AP} = \alpha \upsilon_F^{2D} q_{||}, \tag{8.3}$$

where υ_F^{2D} is the Fermi velocity of the 2D electron gas and

$$\alpha = \sqrt{1 + \frac{[I(z_d)]^2}{\pi[\pi + 2I(z_d)]}} \tag{8.4}$$

1) The valence electron density N is usually characterized by the density parameter $r_s = (3/4\pi N)^{1/3}$.

2) In this chapter, we use the next notation for the 3D and 2D vectors $\mathbf{r} \equiv \{\mathbf{r}_{||}, z\}$.

with

$$I(z_d) = \lim_{q_{\|} \to 0} W(z_d, z_d, q_{\|}, \omega \to 0), \tag{8.5}$$

where $W(z, z', q_{\|}, \omega)$ is the dynamical screened Coulomb interaction [23]. If the 3D substrate is characterized by the Drude dielectric function of Eq. (8.2), $W(z_d, z_d, q_{\|}, \alpha v_F^{2D} q_{\|})$ has a form

$$W\left(z_d, z_d, q_{\|}, \alpha v_F^{2D} q_{\|}\right) = \begin{cases} 0, & \text{for} \quad z_d \leq 0, \\ 4\pi z_d, & \text{for} \quad z_d > 0. \end{cases} \tag{8.6}$$

Hence, when the 2D electron gas is spatially separated from the 3D substrate ($z_d > 0$), the use of $I(z_d) = 4\pi z_d$ in Eq. (8.4) yields at large separation distances z_d

$$\alpha = \sqrt{2 z_d}, \tag{8.7}$$

a result obtained in a study of charge carrier crystallization in low-density inversion layers [21]. By contrast, if the 2D sheet is located inside the 3D substrate ($z_d \leq 0$), within this model $I(z_d) = 0$, and the effective 2D dielectric function $\varepsilon_{\text{eff}}^{2D}$ has no second zero at low energies. Hence, if the 3D dielectric response is described within a local picture these low-energy collective oscillations of the 2D electron gas immersed in the 3D system would be completely screened by the surrounding 3D electron gas. Therefore, for a long time it was believed that plasmons with acoustic-like dispersion could only exist in the case of spatially separated plasmas [22]. Only recently, it was demonstrated that indeed acoustic-like dispersing collective oscillations can appear [23] as a result of the nonlocal character of the 3D dynamical screening (which is a more realistic description of dielectric properties of an electron system in comparison to the Drude-like dielectric function (8.2)). Nevertheless, we would like to stress here that these analytical considerations can give insight only into the nature of possible collective excitations in a given electronic system, whereas their predictions for real systems should be taken with some caution. In realistic systems, numerous interband transitions can occur. This can have significant, and often, crucial effect on the electronic excitation spectra. This problem acquires special importance when one is interested in the investigation of low-energy collective electronic excitations. In this case, the decay into incoherent electron–hole pairs may prevent the existence of such collective excitations. Therefore, *ab initio* calculation methods have been developed in the last two decades in order to incorporate the band structure of real solids, and their ability to describe the properties of collective electronic excitations in real bulk materials and their surfaces have been demonstrated. Some details of such kind of calculations applied to metal surfaces will be given in the next section.

8.2.2
Self-Consistent Dielectric Response

In linear response theory, the central quantity is the density response function $\chi(\mathbf{r}, \mathbf{r}', \omega)$ of an interacting electron system defined as [24]

$$n_{\text{ind}}(\mathbf{r},\omega) = \int d\mathbf{r}'\chi(\mathbf{r},\mathbf{r}',\omega)\phi^{\text{ext}}(\mathbf{r}',\omega), \tag{8.8}$$

where $n_{\text{ind}}(\mathbf{r},\omega)$ is the electron density induced in the system by an external perturbation $\phi^{\text{ext}}(\mathbf{r},\omega)$. Within the time-dependent density functional theory [25], χ can be obtained from its noninteracting counterpart χ^0 via a Dyson-like equation

$$\chi(\mathbf{r},\mathbf{r}',\omega) = \chi^0(\mathbf{r},\mathbf{r}',\omega) + \int\int d\mathbf{r}_1 d\mathbf{r}_2 \chi^0(\mathbf{r},\mathbf{r}_1,\omega)\left[V(\mathbf{r}_1,\mathbf{r}_2) + K^{\text{xc}}(\mathbf{r}_1,\mathbf{r}_2)\right]\chi(\mathbf{r}_2,\mathbf{r}',\omega). \tag{8.9}$$

Here, the kernel, in general, contains the bare Coulomb interaction V and the exchange correlation term, K^{xc}. In the following, we shall adapt these formulas to the surface calculations considering the description of electronic system by the so-called repeated slabs geometry. Hence, one can use a 3D Fourier expansion for all the quantities. Then, the expression for $\chi^0_{\mathbf{GG}'}(\mathbf{q}_{||},\omega)$, that is, a spatial Fourier transform of the density response function $\chi^0(\mathbf{r},\mathbf{r}',\omega)$ of noninteracting electrons, adapted to the $\mathbf{k}_{||}$ summation over the 2DBZ, has a form [26, 27]

$$\begin{aligned}\chi^0_{\mathbf{GG}'}(\mathbf{q}_{||},\omega) &= \frac{2}{\Omega}\sum_{\mathbf{k}_{||}}^{\text{2DBZ}}\sum_{n,n'}\frac{f_{n\mathbf{k}_{||}} - f_{n'\mathbf{k}_{||}+\mathbf{q}_{||}}}{E_{n\mathbf{k}_{||}} - E_{n'\mathbf{k}_{||}+\mathbf{q}_{||}} + \omega + i\eta}\\ &\times\left\langle\psi_{n\mathbf{k}_{||}}\left|e^{-i(\mathbf{q}_{||}\mathbf{r}_{||}+\mathbf{G}\mathbf{r})}\right|\psi_{n'\mathbf{k}_{||}+\mathbf{q}_{||}}\right\rangle\left\langle\psi_{n'\mathbf{k}_{||}+\mathbf{q}_{||}}\left|e^{i(\mathbf{q}_{||}\mathbf{r}_{||}+\mathbf{G}'\mathbf{r})}\right|\psi_{n\mathbf{k}_{||}}\right\rangle.\end{aligned} \tag{8.10}$$

Here, Ω is a normalization volume, the factor 2 accounts for the spin, $f_{n\mathbf{k}_{||}}$ is the Fermi factor, and $E_{n\mathbf{k}_{||}}$ ($\psi_{n\mathbf{k}_{||}}$) represent the one-particle eigenenergies (eigenfunctions) of the ground state. The summation over $\mathbf{k}_{||}$ runs over the whole 2DBZ and summation over n, n' includes both the occupied and the unoccupied bands. The numerical evaluation of $\chi^0_{\mathbf{GG}'}(\mathbf{q}_{||},\omega)$ matrices is the most time-consuming part of the total calculation and for surface evaluations requires rather long-time computations. An efficient scheme for the reduction of computing consists of, instead of direct use of Eq. (8.10), the evaluation of the spectral function $S^0_{\mathbf{GG}'}(\mathbf{q}_{||},\omega)$ defined on a uniform grid of energies ω_i as [28]

$$\begin{aligned}S^0_{\mathbf{GG}'}(\mathbf{q}_{||},\omega) &= \frac{2}{\Omega}\sum_{\mathbf{k}_{||}}^{\text{2DBZ}}\sum_{n,n'}(f_{n\mathbf{k}_{||}} - f_{n'\mathbf{k}_{||}+\mathbf{q}_{||}})\,\delta\left(E_{n\mathbf{k}_{||}} - E_{n'\mathbf{k}_{||}+\mathbf{q}_{||}} + \omega\right)\\ &\times\left\langle\psi_{n\mathbf{k}_{||}}\left|e^{-i(\mathbf{q}_{||}\mathbf{r}_{||}+\mathbf{G}\mathbf{r})}\right|\psi_{n'\mathbf{k}_{||}+\mathbf{q}_{||}}\right\rangle\left\langle\psi_{n'\mathbf{k}_{||}+\mathbf{q}_{||}}\left|e^{i(\mathbf{q}_{||}\mathbf{r}_{||}+\mathbf{G}'\mathbf{r})}\right|\psi_{n\mathbf{k}_{||}}\right\rangle.\end{aligned} \tag{8.11}$$

Subsequently, the imaginary part of $\chi^0_{\mathbf{GG}'}(\mathbf{q}_{||},\omega)$ is calculated as

$$S^0_{\mathbf{GG}'}(\mathbf{q}_{||},\omega) = \frac{1}{\pi}\,\text{sgn}\,(\omega)\,\text{Im}\left[\chi^0_{\mathbf{GG}'}(\mathbf{q}_{||},\omega)\right] \tag{8.12}$$

and the real part of $\chi^0_{\mathbf{GG}'}(\mathbf{q}_{||},\omega)$ is obtained from the imaginary one through the Hilbert transform.

The information about the surface collective excitations can be obtained from the imaginary part of the surface response function, $g\left(\mathbf{q}_{||}, \omega\right)$, defined as [29]

$$g\left(\mathbf{q}_{||}, \omega\right) = \int \mathrm{d}\mathbf{r} n_{\mathrm{ind}}(\mathbf{r}, \omega) \mathrm{e}^{q_{||} z}, \tag{8.13}$$

where $n_{\mathrm{ind}}(\mathbf{r}, \omega)$ is the density induced at the crystal surface by an external potential of the form

$$\phi^{\mathrm{ext}}(\mathbf{r}, \omega) = -\frac{2\pi}{q_{||}} \mathrm{e}^{q_{||} z} \mathrm{e}^{\mathrm{i}\mathbf{q}_{||}\mathbf{r}_{||}} \mathrm{e}^{-i\omega t}. \tag{8.14}$$

Further details of calculations of $g\left(\mathbf{q}_{||}, \omega\right)$ at the *ab initio* level can be found elsewhere [30–32].

If one assumes translational invariance in the plane parallel to the surface, a set of single-particle states with wave functions $\psi_{n\mathbf{k}_{||}}(\mathbf{r})$ and energies $E_{n\mathbf{k}_{||}}$ can be expressed in the following manner:

$$\psi_{n\mathbf{k}_{||}}(\mathbf{r}) = \frac{1}{\sqrt{S}} \mathrm{e}^{\mathrm{i}\mathbf{k}_{||}\mathbf{r}_{||}} \phi_n(z) \tag{8.15}$$

and

$$E_{n\mathbf{k}_{||}} = E_n + \frac{k_{||}^2}{2m_n^*}, \tag{8.16}$$

where S is a normalization area. In Eqs. (8.15) and (8.16), $\phi_n(z)$ and E_n are the eigenfunctions and eigenvalues, respectively, of a one-dimensional (1D) Schrödinger equation of the form

$$\left[-\frac{1}{2}\frac{\mathrm{d}^2}{\mathrm{d}z^2} + V_{\mathrm{MP}}(z)\right] \phi_n(z) = E_n \phi(z), \tag{8.17}$$

where V_{MP} is a 1D model potential describing interaction of electrons with an ionic system. In principle, it can be constructed in many ways. Thus, very frequently, the so-called "jellium" model is used. Nonetheless, this model has an important drawback as it does not allow the appearance of localized electronic states at a metal surface, which actually arise at many real surfaces [3], because of the absence of an energy gap for bulk electrons in the calculated energy spectrum. One possibility to overcome this problem, keeping advantage of a 1D-potential approach at the same time, consists in construction of a model potential where relevant information about the surface electronic structure of a material is incorporated. One kind of such model potentials was proposed some time ago [33]. This potential reproduces the key features of the surface band structure, which, for example, in the case of the (111) surface of the noble metals, include the existence of a bandgap at the center of 2DBZ and presence of the s–p_z surface and image potential states in it [33].

In order to solve Eq. (8.17) one can consider a thick slab with a given number of atomic layers, although it can be done for a semi infinite crystal as well. Hence, the 1D wave function $\phi_n(z)$ can be expanded in a Fourier series. Once one has an accurate description of $\phi_n(z)$ and E_n, the evaluation of the 2D Fourier transform $\chi^0(z, z', q_{\|}, \omega)$ of noninteracting electrons moving in the potential V_{MP} can be performed as

$$\chi^0\left(z, z', q_{\|}, \omega\right) = \frac{2}{S} \sum_{n,n'} \phi_n(z) \phi^*_{n'}(z) \phi_{n'}(z') \phi^*_n(z') \sum_{\mathbf{k}_{\|}} \frac{f_{n\mathbf{k}_{\|}} - f_{n'\mathbf{k}_{\|}+\mathbf{q}_{\|}}}{E_{n\mathbf{k}_{\|}} - E_{n'\mathbf{k}_{\|}+\mathbf{q}_{\|}} + \omega + i\eta}. \tag{8.18}$$

Here, the sum over n and n' includes both the occupied and the unoccupied states, $f_{n\mathbf{k}_{\|}}$ are the Fermi factors, and η is a positive infinitesimal. The explicit expressions for χ^0 in the reciprocal space can be found for the case of $m^*_n = 1$ in Ref. [34] and for the case of $m^*_n \neq 1$ in Ref. [35].

8.3 Results of Numerical Calculations

8.3.1 Monolayers

The first calculation of the *ab initio* surface response function, as briefly described in Section 8.2.2, was done considering some free-standing metallic monolayers [30]. In particular, it was demonstrated in the case of a hexagonal Be monolayer that the presence of two types of carriers in parabolic σ- and π-like bands crossing the Fermi level leads to, besides conventional 2D intraband and interband plasmons [36], an additional low-energy mode. Figure 8.2 presents the calculated surface loss function, Im g, for this system at some 2D momenta $\mathbf{q}_{\|}$, where the peaks AP corresponding to this mode can be observed. Note that we expect of this mode to exist at all small momenta, but the finite broadening procedure for the representation of the δ-function in Eq. (8.11) does not allow to trace it down to vanishing $q_{\|}$. The energy of this mode has an acoustic-like dispersion with momentum $\mathbf{q}_{\|}$ and its slope is determined by the Fermi velocity of electrons in the σ band. Indeed, as seen in Figure 8.3, two lowest energy bands in the Be monolayer along the $\bar{\Gamma}\bar{\mathrm{K}}$ symmetry direction have Fermi vectors, $k^{\sigma}_{\mathrm{F}} = 0.81\,\mathrm{a.u.}^{-1}$ and $k^{\pi}_{\mathrm{F}} = 0.27\,\mathrm{a.u.}^{-1}$ and their parabolic-like dispersions can be fitted with the effective masses $m^*_{\sigma} = 1.48$ and $m^*_{\pi} = 1.04$. As a result, the Fermi velocity of the carriers in the slower π band $\upsilon^{\pi}_{\mathrm{F}}$, (and subsequently, the group velocity of the acoustic plasmon, υ^{AP}, according to Eq. (8.3)), is $\upsilon^{\mathrm{AP}} \approx \upsilon^{\pi}_{\mathrm{F}} = 0.26\,\mathrm{a.u.}$ Note that essentially the same results regarding the acoustic plasmon for isolated Be monolayer have been obtained very recently in the *ab initio* calculations performed by Yuan and Gao [32].

At the same time, in the case of a Li monolayer only conventional 2D plasmons were obtained [30] because only one energy band is crossing the Fermi level in this

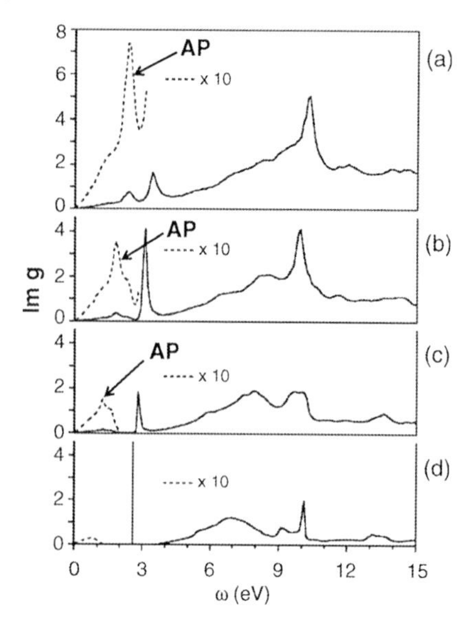

Figure 8.2 Surface loss function of the free-standing beryllium monolayer for several values of 2D momenta $q_{||}$ in the $\overline{\Gamma}\overline{\mathrm{K}}$ symmetry direction of 2DBZ: (a) $q_{||} = 0.216\,\mathrm{a.u.}^{-1}$; (b) $q_{||} = 0.162\,\mathrm{a.u.}^{-1}$; (c) $q_{||} = 0.108\,\mathrm{a.u.}^{-1}$; and (d) $q_{||} = 0.054\,\mathrm{a.u.}^{-1}$. The peaks shown by arrows correspond to the acoustic plasmon as a result of the presence of two types of energy bands at the Fermi level. Reprinted with permission from [30]. Copyright 2003 by the American Physical Society.

case. Actually, very similar situation is realized in a silver monolayer [19] and a graphene sheet [37] where only a 2D plasmon was experimentally observed. In the case of boron monolayers, the acoustic plasmon could not be obtained [30] although up to three energy bands could cross the Fermi level. This may be explained by the fact that all these bands have similar Fermi velocities.

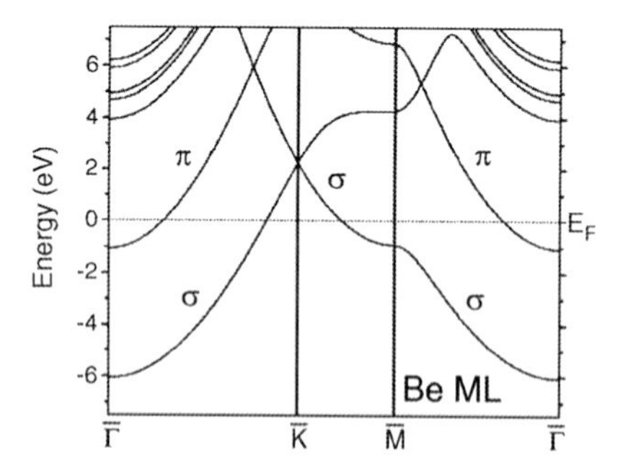

Figure 8.3 Band structure of the free-standing hexagonal Be monolayer. The bands in vicinity to the Fermi level, E_F, are denoted by symbols according to the plane reflection symmetry. Reprinted with permission from [30]. Copyright 2003 by the American Physical Society.

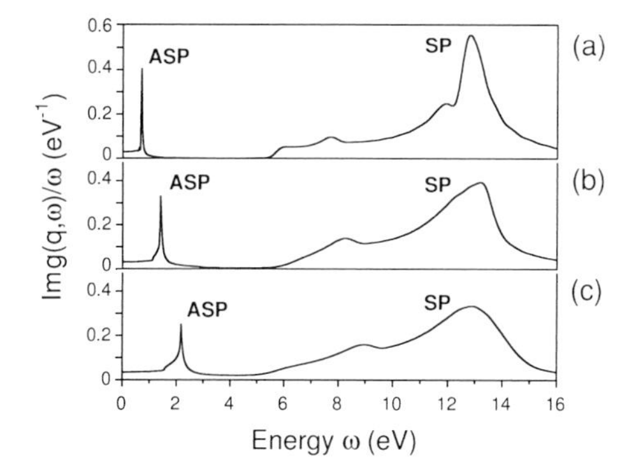

Figure 8.4 Normalized surface loss function, $\mathrm{Im}\,[g\,(q_{||},\omega)]/\omega$, of Be(0001) as a function of energy ω for three values of the 2D momentum $q_{||}$: (a) $q_{||} = 0.05\,\mathrm{a.u.}^{-1}$, (b) $q_{||} = 0.10\,\mathrm{a.u.}^{-1}$, and (c) $q_{||} = 0.15\,\mathrm{a.u.}^{-1}$. Peaks corresponding to the acoustic surface plasmon (surface plasmon) are denoted by ASP (SP). Reprinted from [38] . Copyright 2004 EDP Sciences.

8.3.2
Metal Surfaces

8.3.2.1 1D Calculations

The first self-consistent calculation of the electronic excitations at a metal surface taking into account its realistic electronic structure was performed for Be(0001) surface [38]. This surface is characterized by a strong s–p_z surface state with quasiparabolic dispersion localized in a wide bulk energy gap at the 2DBZ center around the Fermi level [39, 40]. This information was used for construction of the 1D model potential [33] for this surface. Figure 8.4 shows the imaginary part of the surface response function, Im g, of Be(0001) obtained with the surface band structure, which was calculated with this 1D model potential. As seen in the figure, the excitation spectrum is clearly dominated by two distinct features. In the upper energy part, one can see a rather broad peak around $\omega = 13$ eV, which corresponds to the conventional surface plasmon. This mode would also be present in the absence of a surface state, that is, if the 1D model potential V_{MP} in Eq. (8.17) is replaced by the jellium Kohn–Sham potential. On the low-energy side of the spectra, one can see a well-defined peak that disperses quasi-linearly with momentum. The quasi-linear energy dispersion of this lower acoustic surface plasmon mode [38] is demonstrated in Figure 8.5.

Here, one can see how the ASP dispersion follows closely the upper border of continuum for electron–hole pair excitations in the surface state band. At the small momenta, this dispersion presents a form of Eq. (8.3). With increasing momentum, the quadratic term [17] starts to contribute and, respectively, modify this initially linear dispersion.

Subsequent work [35] demonstrated that existence of an acoustic surface plasmon is not a special property of the Be(0001) surface only. For example, Figure 8.6 presents from Ref. [35] the calculated surface energy loss function for the (111) surfaces of

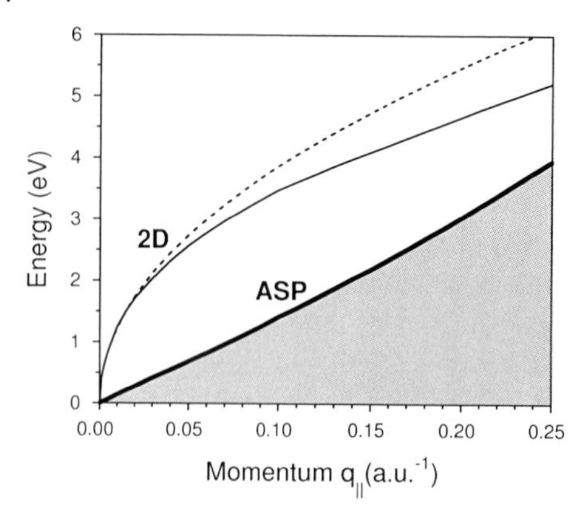

Figure 8.5 Dispersion of the acoustic surface plasmon ASP energy versus 2D momentum $q_{||}$ (thick solid line). Thin solid line shows dispersion of the conventional 2D plasmon in a 2D electron system with the charge density corresponding to the Be(0001) surface state. Dashed line corresponds to the 2D plasmon dispersion $\omega_{\mathrm{p}}^{\mathrm{2D}} = \upsilon_{\mathrm{F}}^{\mathrm{2D}}\sqrt{q_{||}}$ with $\upsilon_{\mathrm{F}}^{\mathrm{2D}} = 0.45$ a.u. The gray region corresponds to the 2D electron–hole pairs continuum with the upper edge $\omega_{\mathrm{up}}^{\mathrm{2D}} = \upsilon_{\mathrm{F}}^{\mathrm{2D}}q_{||} + q_{||}^2/2m_{\mathrm{2D}}^*$. Note that the electron–hole pairs inside the 3D electron system can be excited at any energy ω and momentum $q_{||}$. Reprinted from [38] . Copyright 2004 EDP Sciences.

noble metals Cu, Ag, and Au at $q_{||} = 0.01$ a.u.$^{-1}$. In this figure, one can see a sharp peak in Im g corresponding to acoustic surface plasmon for all the three surfaces. The calculations confirmed that this mode has also the quasi-linear dispersion of the form of Eq. (8.3) with the coefficient α being very close to unity at small $q_{||}$ [35]. In Figure 8.7, one can see that the surface loss function, Im $g(q_{||}, \omega)$, for Cu(111), Ag(111), and Au(111) surfaces (where the realistic surface band structure is included into the model) is strongly inhomogeneous in contrast to the featureless behavior expected from the jellium model calculations [15]. Thus, one observes that in this

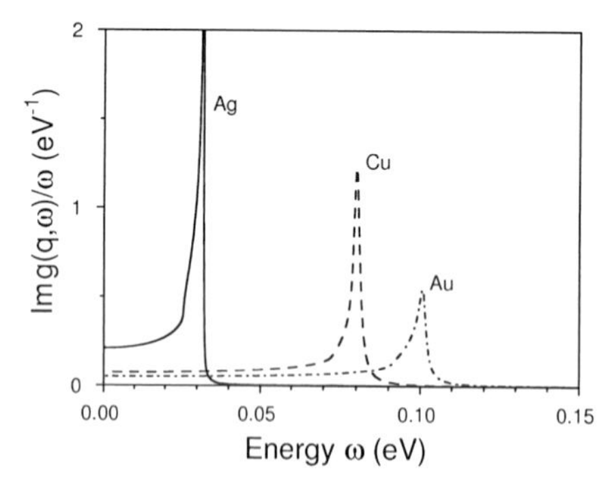

Figure 8.6 Calculated normalized surface loss functions, Im $[g(q_{||}, \omega)]/\omega$, for the (111) surfaces of Cu, Ag, and Au at $q_{||} = 0.01$ a.u.$^{-1}$ as a function of energy ω. Reprinted from [41]. Copyright 2008 IOP Publishing Ltd.

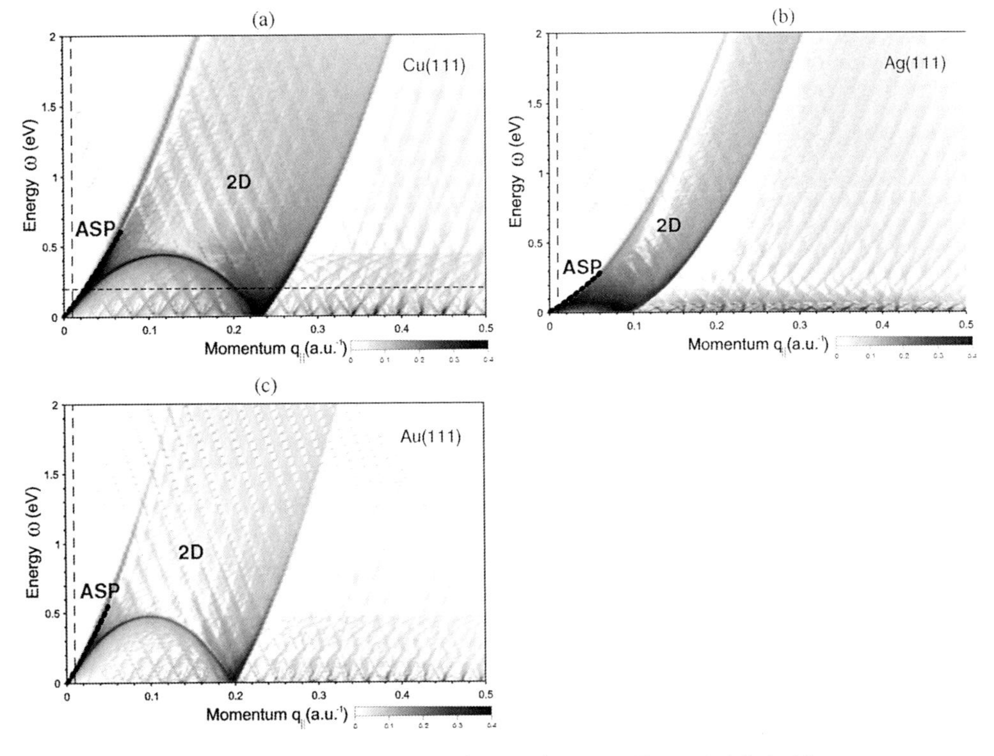

Figure 8.7 2D plot of the calculated normalized surface loss function, $\mathrm{Im}\,[g(q_{||}, \omega)]/\omega$, for (a) Cu(111), (b) Ag(111), and (c) Au(111) surfaces as a function of momentum $q_{||}$ and energy ω. The phase-space region where the intraband electron–hole excitations within the surface state are possible is denoted by "2D." Intraband bulk excitations can occur everywhere. The thick solid line denoted by "ASP" shows the acoustic surface plasmon dispersion. The vertical dashed line shows the cut along which $\mathrm{Im}\,[g(q_{||}, \omega)]/\omega$ is displayed in Figure 8.6. The horizontal dotted line in the case of Cu(111) shows the cut along which the dynamical induced density is calculated in Figure 8.11. Visible faint features in the figure are due to the finite thickness slab effects. Reprinted from [41]. Copyright 2008 IOP Publishing Ltd.

"energy–momentum" phase space the surface loss function is dominated by the regions denoted by "2D" that correspond to intraband electron–hole pair excitations in the surface state. Note that in Figure 8.7 the "2D" regions in the momentum–energy phase space are different for all three surfaces. This is explained by the fact that the positions of the left and right borders of this region are determined by the equations $\omega_{\mathrm{L}}^{2\mathrm{D}} = \upsilon_{\mathrm{F}}^{2D} \cdot q_{||} + q_{||}^2/2m_{2\mathrm{D}}^*$ and $\omega_{\mathrm{R}}^{2\mathrm{D}} = -\upsilon_{\mathrm{F}}^{2\mathrm{D}} \cdot q_{||} + q_{||}^2/2m_{2\mathrm{D}}^*$, respectively. The different values for surface state Fermi velocities $\upsilon_{\mathrm{F}}^{2\mathrm{D}}$ and effective masses $m_{2\mathrm{D}}^*$ on each surface are translated to the material specific "2D" regions. In addition, in Figure 8.7, a strong dispersing peak structure highlighted by the solid lines "ASP" at small $q_{||}$ and ω in Im g corresponding to the acoustic surface plasmon can be clearly resolved.

According to Eq. (8.3), the initial slope of the acoustic surface plasmon dispersion is determined by the Fermi velocity of the surface state, υ_F^{2D}. The examples of numerical self-consistent calculations reported above confirm this analytical result. At the same time, it is well known that the surface state properties (including its Fermi velocity) can be modified effectively, for example, by changing the sample temperature or by adsorption of various species, such as adatoms, adlayers, molecules, and so on. This suggests that the properties of an acoustic surface plasmon, and in particular its dispersion, might correspondingly be modified. A well-known system for such kind of observation might be the deposition of alkali atoms on noble metal surfaces. With small amount of deposited atoms, the main effect is a shift down of the surface state with the corresponding increase in the surface state Fermi velocity. In turn, this should increase the slope of the acoustic surface plasmon dispersion in a similar fashion as it was observed for 2D plasmon in an Ag monolayer adsorbed on the Si(111) surface [19]. Upon an increase in the alkali atom coverage, an electronic band induced by adsorbates, that is, a quantum well state, starts to be occupied [42]. As the wave function of the quantum well state is mainly localized at the alkali adsorbate layer, it must dominate the dielectric properties of the whole system in the "low-ω–low-$q_{||}$" domain. This effect was shown recently in the case of the Na/Cu(111) system in Refs [43–45], where the evolution of the surface loss function and screened dynamical Coulomb interaction in Na/Cu(111) upon the change in the Na coverage was studied.

8.3.3 3D Calculations

As was pointed out above, the simplified calculations of collective electronic excitations in metallic systems can give important theoretical insight into the nature of such modes. Nevertheless, as the main source for decay of such excitations – the realistic band structure [31] – is lacking in such considerations, the predictive power of such theories is rather poor. Even in the 1D self-consistent calculations reported above the band structure effects are taken into account only in the direction perpendicular to the surface. Therefore, to be maximally predictive the calculations should explicitly include the information about the real surface electronic structure, as can be done from *ab initio* band structure calculations. Such calculations for the Be(0001) surface were performed very recently [46]. The calculated dispersion of the acoustic surface plasmon is in excellent agreement with electron energy loss experimental data, as one can see in Figure 8.8. At the same time, this dispersion deviates from that (solid line in Figure 8.8) obtained with the 1D model potential (see Section 8.3.2.1). This fact is explained by details of the surface state dispersion. On one hand, as it is already mentioned above, the energy dispersion of acoustic surface plasmons follows very closely in the "energy–momentum" phase space the upper edge of a region for electron–hole excitations within the surface state band. On the other hand, this region is determined by the surface state dispersion. In particular, the upper edge of the continuum is determined by the steepest part of the surface state dispersion, that is, by an unoccupied part of the surface state dispersion. As seen in Figure 8.9, the *ab initio* dispersion of the SS surface state deviates from a quasi-free

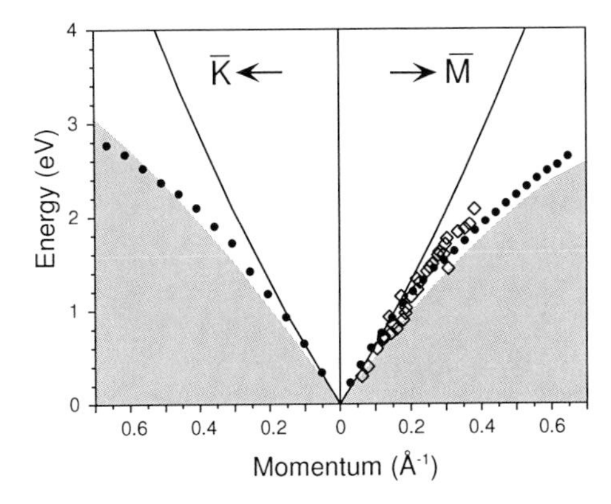

Figure 8.8 *Ab initio* acoustic surface plasmon dispersion (dots) along two symmetry directions. The gray areas show regions where intraband electron–hole excitations within the surface state SS of Figure 8.9 are possible. The experimental data of Ref. [46] are shown by open diamonds. Thin solid line shows acoustic surface plasmon dispersion calculated using the 1D model potential [33] for Be(0001) and effective mass $m^*_{2D} = 1.2$ for the surface state. Reprinted from [47]. Copyright Wiley-VCH GmbH & Co. KGaA. Reproduced with permission.

electron-like parabolic behavior. *Ab initio* calculations of Ref. [47] predicted that indeed a small anisotropy of the acoustic plasmon dispersion at the Be(0001) surface should exist due to anisotropy in its surface electronic structure observed in Figure 8.9. Hence, in the case of Be(0001) surface there exist both *ab initio* and experimental confirmations of existence of an acoustic surface plasmon at a real metal surface. Moreover, an excellent agreement between *ab initio* calculations and experimental data regarding its energy dispersion confirms the feasibility of this theoretical approach in the evaluation of low-energy excitation spectra for real

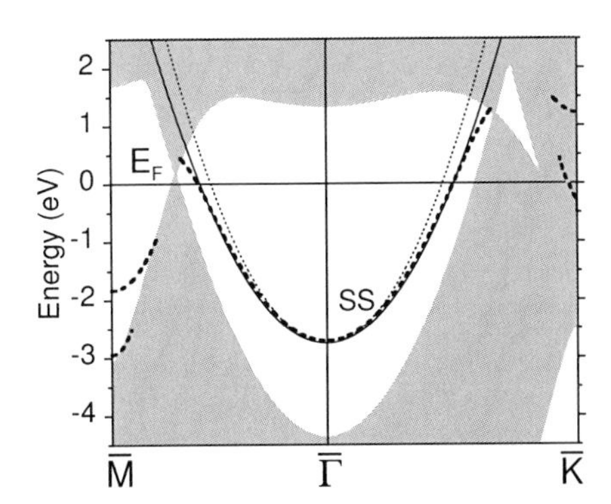

Figure 8.9 *Ab initio* band structure of the Be(0001) surface along two symmetry directions of the 2DBZ. The gray regions show projected bulk electronic states. The dispersion of the surface states is shown by dashed lines. The surface state relevant for the present discussion is denoted as SS. Thin solid line shows parabolic fit of the SS dispersion with the experimental value [39] for its effective mass $m^*_{2D} = 1.2$. The SS surface state dispersion with $m^*_{2D} = 1$ (used in the calculations of Ref. [38]) is shown with thin dotted line. Reprinted from [47]. Copyright Wiley-VCH GmbH & Co. KGaA. Reproduced with permission.

surfaces. The Be(0001) surface with a high value of ϑ_F^{2D} is favorable for experimental observation of the acoustic surface plasmon [46]. On other surfaces, this mode is expected to exist at lower energies, thus making its experimental detection more difficult [48]. Special care in this case should be taken in selecting the energies of incident electrons and the scattering geometry.

8.3.4
Dynamical Charge Density Oscillations

Excitation of the acoustic surface plasmon corresponds to creation of charge density oscillations at a metal surface. In order to visualize how this charge density is distributed in space, $\delta n_{\text{ind}}(z, q_{||}, \omega)$ was calculated in Ref. [38] in response to an external perturbing potential of Eq. (8.14). The calculations were performed for the Be(0001) surface with the use of the 1D model potential approach. The evaluated imaginary part of the induced charge density is shown in Figure 8.10. In the figure one can see that the induced charge density corresponding to the acoustic surface plasmon is localized in the vicinity of the surface and its amplitude follows nicely the amplitude of the surface state charge density.

There is another possibility to see how the charge density corresponding to the acoustic surface plasmon behaves in real space and time. In Ref. [49], the charge density oscillations caused by a point charge oscillating in time with some frequency ω_0 and fixed at some position z_0 were calculated taking as an example the Cu(111)

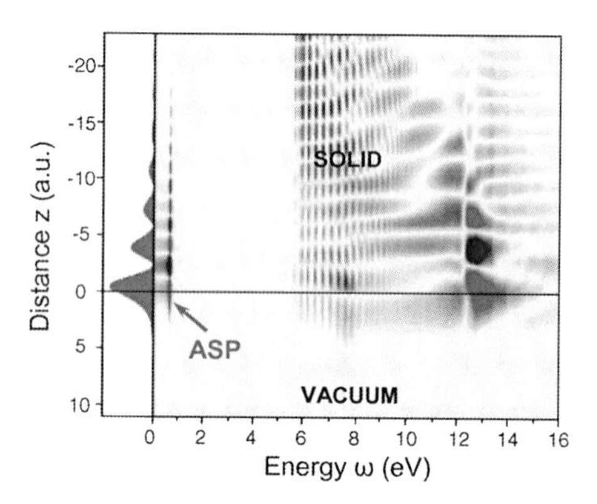

Figure 8.10 2D plot of the imaginary part of the induced charge density, $\text{Im}\left[n_{\text{ind}}(z, q_{||}, \omega)\right]$ at $q_{||} = 0.05$ a.u.$^{-1}$ versus energy ω and coordinate z, at the Be(0001) surface, as obtained from Eq. (8.8) in response to an external potential of Eq. (8.14). The solid (vacuum) is located at negative (positive) distances z. Positive (negative) values of $\text{Im}\left[n_{\text{ind}}(z, q_{||}, \omega)\right]$ are represented by red (blue) color. More intense color corresponds to larger values. The surface state charge density is shown on the left-hand side of the figure. The charge density oscillations corresponding to the acoustic surface plasmon at this momentum are marked by arrow. Oscillations at energies above approximately 5.8 eV are due to interband transitions. The strong oscillating features around energy of 13 eV correspond to the conventional surface plasmon that produce a peak "SP" in the surface loss function of Figure 8.4. Reprinted from [38]. Copyright 2004 EDP Sciences. (Please find a color version of this figure on the color plates.)

surface. In this case, one is interested in the evaluation of

$$\begin{aligned} n_{\text{ind}}(\varrho, z, t) &= \text{Re}\left[n_{\text{ind}}(\varrho, z, \omega_0)\right] \cos(\omega_0 t) + \text{Im}\left[n_{\text{ind}}(\varrho, z, \omega_0)\right] \sin(\omega_0 t) \\ &= \left|n_{\text{ind}}(\varrho, z, \omega_0)\right| \cos(\omega_0 t - \theta(\varrho, z, \omega_0)). \end{aligned} \tag{8.19}$$

Here, ϱ is a distance from the point charge along the surface and θ is a phase of delay. Note that in a static case, $(\omega_0 = 0)$, $\text{Im}\left[n_{\text{ind}}(\varrho, z, \omega_0 = 0)\right] = 0$, and subsequently $\theta = 0$. Nevertheless, in the dynamical case (when $\omega_0 \neq 0$) the induced charge density $n_{\text{ind}}(\mathbf{r}, \omega)$ is, in general, a complex function. The use of the 1D model potential allows to obtain the following expression for n_{ind} in this case:

$$n_{\text{ind}}(z, q_{||}, \omega) = \int dz' \chi(z, z', q_{||}, \omega) \phi^{\text{ext}}(z', q_{||}, \omega), \tag{8.20}$$

where $\phi^{\text{ext}}(z, q_{||}, \omega)$ is the 2D Fourier transform of an external Coulomb potential. Finally, one obtains

$$n_{\text{ind}}(\varrho, z, \omega) = \frac{1}{2\pi} \int_0^\infty dq_{||} q_{||} J_0(q_{||}\varrho) n_{\text{ind}}(z, q_{||}, \omega), \tag{8.21}$$

where J_0 is the Bessel function of the zero order. Figure 8.11 presents the real and imaginary parts of $n_{\text{ind}}(\varrho, z, \omega_0)$ calculated for Cu(111) at $\omega_0 = 0.44$ eV. In the figure,

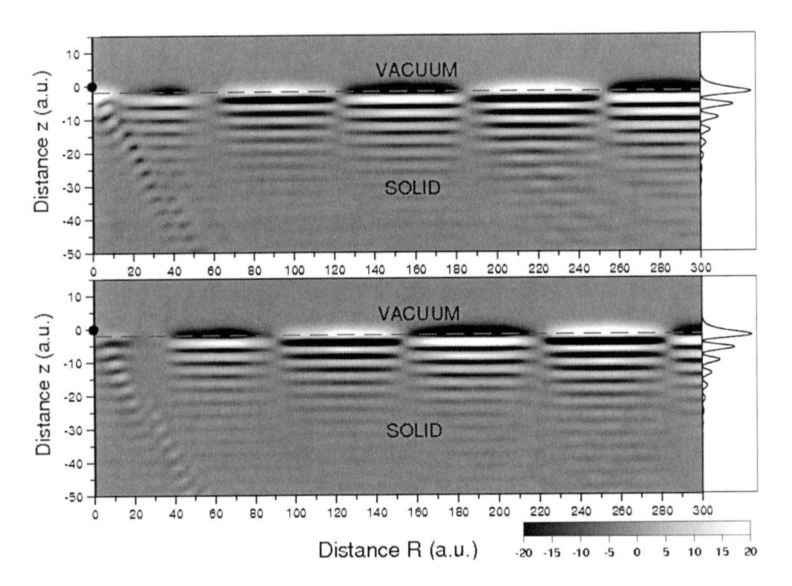

Figure 8.11 2D plot of real (*top*) and imaginary (*bottom*) parts of the charge density $n_{\text{ind}}(\varrho, z, \omega_0)$ (multiplied by R^2, where $R = \sqrt{\varrho^2 + z^2}$ is distance from the impurity) induced be the time-varying external point charge with a frequency $\omega_0 = 0.44$ eV versus distance ϱ and distance z. In each panel, the position of the point charge at $R = 0$ and $z = 0$ is shown by a filled circle. Horizontal dashed line shows the crystal border. Surface-state charge density averaged in the parallel plane is shown on the right-hand side. Reprinted from [49]. Copyright 2005, with permission from Elsevier.

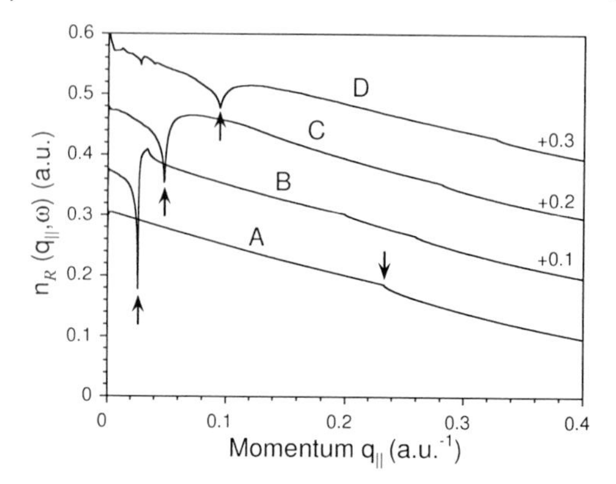

Figure 8.12 Real part of the Fourier transform of induced charge density, Re $[n_{\text{ind}}(z=0, q_{||}, \omega_0)]$, for some frequencies ω_0: (a) $\omega_0 = 0$; (b) $\omega_0 = 0.22$ eV; (c) $\omega_0 = 0.44$ eV; and (d) $\omega_0 = 1.0$ eV. The features corresponding to the acoustic surface plasmon are highlighted by vertical arrows. Note that in the static case ($\omega_0 = 0$) the only weak feature (shown by inverted vertical arrow) is presented at $q_{||} = 2k_F^{2D} = 0.218$ a.u.$^{-1}$ giving origin to the conventional static Friedel charge density oscillations. Reprinted from [49]. Copyright 2005, with permission from Elsevier.

one can see how the charge density along the surface is dominated by density oscillations with wavelength $\lambda = 2\pi/q_{||}^{ASP}$ (where $q_{||}^{ASP} \approx \omega_0/\vartheta_F^{2D}$) corresponding to the acoustic surface plasmon. Note that the amplitude of these dominant oscillations decays into the solid in concert with the surface-state amplitude. A close inspection of Figure 8.11 also reveals other oscillations with smaller amplitude due to the features in $n_{\text{ind}}(z, q_{||}, \omega_0)$, which can be seen in Im $g(q_{||}, \omega)$ of Figure 8.7 along the line $\omega = \omega_0 = 0.44$ eV. These features in $n_{\text{ind}}(z, q_{||}, \omega_0)$ can be detected in Figure 8.12, where one can observe the strongest singularities related to the acoustic surface plasmon. At small ω_0's, these singularities due to the acoustic surface plasmon are much more prominent compared to the other ones and lead to the characteristic $\propto \cos(q_{||}^{ASP}\varrho)/\varrho$-like charge oscillations along the surface, that is, with a spatial decay inversely proportional to the distance ϱ from the impurity only, in contrast to a $\propto 1/\varrho^2$decay law for the conventional static Friedel oscillations [50].

8.3.5
Acoustic Surface Plasmon Interaction with Light

At long wavelength (i.e., when $q_{||} \rightarrow 0$), the dispersion of the acoustic surface plasmon takes the form of Eq. (8.3) with coefficient α close to unity. As the surface-state Fermi velocities are typically about three orders of magnitude lower than the light velocity, the acoustic surface plasmon cannot be excited with light on the atomically flat surfaces. However, some kind of imperfections on the surface might be able to provide the necessary momentum to allow this process. For this purpose, one can consider a periodic grating with some period L as shown at the bottom of Figure 8.13. Hence, if light hits such a surface, a periodic structure can provide the incident free electromagnetic wave with some momentum determined

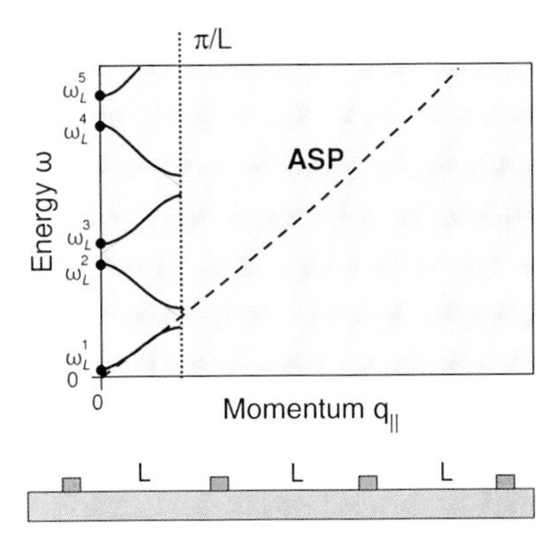

Figure 8.13 Interaction of light with acoustic surface plasmon with "ASP" dispersion. A periodic grating structure on the surface with period L can provide a free electromagnetic radiation impinging the surface with additional momentum $2\pi/L$ at frequencies ω_L^i.

by the grating period L. From the dispersion relation of Eq. (8.3), one can easily determine the necessary size of the grating period in order to match the required frequencies ω_L^i for $q_{||} = 0$ (see Figure 8.13), at which light can interact with the acoustic surface plasmon. In an analogy with the electronic or photonic band structure, the magnitude of the splitting at $q_{||} = 0$ and $q_{||} = \pi/L$ is determined by the strength of a disturbing potential related to grating. It would be interesting to investigate the magnitude of the potential that could be introduced, for example, by adsorbed atoms or molecules. The extreme situation of rather large perturbing potential can be obtained when a localized object such as an atomic cluster is deposited on the surface. This example is illustrated in Figure 8.14. Note that in this case our discussion refers to the surface of this object. Due to localized geometry,

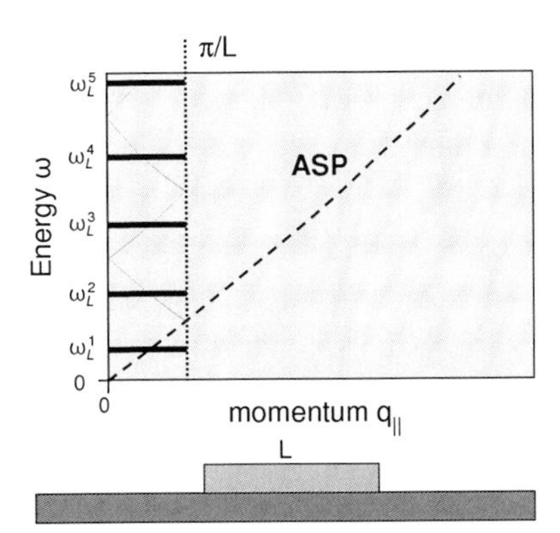

Figure 8.14 Interaction of light with acoustic surface plasmon with "ASP" dispersion. A finite metallic object of size L can produce a set of localized charge density oscillations interacting with light, whose frequencies ω_L^i are determined by the ASP dispersion and the size L.

instead of continuous acoustic surface plasmon dispersion, one can obtain a set of localized energy levels with separation between them determined by the shape and size of the adsorbed object. Indeed, this scenario was proposed for explanation of results of recent experiments [51] on the infrared reflection from the metallic nanoparticles, although the energies of the observed signals are rather difficult to contrast with the surface-state properties obtained for the respective clean extended surfaces. At the same time, the surface states on the faces of nanoparticles (if they still can exist) can be dramatically affected by the confinement with corresponding changes in the excitation spectra. It would be of great interest to investigate this issue in more details.

8.4
Concluding Remarks

In the last few years, self-consistent calculations of the dynamical electronic response properties on a variety of metal surfaces have revealed unexpected features in the low-energy domain (i.e., at energies small in comparison with the energy of conventional surface plasmons). Thus, it was found that the surface electronic states, which actually constitute only a small fraction of the total electronic charge at metal surfaces, can produce strong effect in the surface response due to their two-dimensionality. Their impact can be as strong enough to generate a novel collective surface mode, called acoustic surface plasmon, which corresponds to "out-of-phase" resonant charge density oscillations in surface and bulk electronic subsystems. This phenomenon has been predicted theoretically to exist at real metal surfaces and recently it was experimentally confirmed in electron energy loss measurements. Self-consistent calculations performed for several metal surfaces show that such acoustic surface plasmon mode must be a rather general phenomenon. Hence, in addition to the widely known and frequently referred electron–hole pair's excitations at metal surfaces, a strong collective excitation with low energies can participate in dynamical surface processes. It would be of great interest to determine a possible role of this novel mode in various phenomena at surfaces and find a way to use it in a controlled manner.

References

1 Tamm, I.E. (1932) *Phys. Z. Sowjet.*, **1**, 733–746.
2 Shockley, W. (1939) *Phys. Rev.*, **56**, 317–323.
3 Inglesfield, J.E. (1982) *Rep. Prog. Phys.*, **45**, 223–284.
4 Pines, D. and Bohm, D. (1952) *Phys. Rev.*, **85**, 338–353.
5 Ritchie, R.H. (1957) *Phys. Rev.*, **106**, 874–881.
6 Stern, E.A. and Ferrel, R.A. (1960) *Phys. Rev.*, **120**, 130–136.
7 Powell, C.J. and Swan, J.B. (1959) *Phys. Rev.*, **115**, 869–875.
8 Powell, C.J. and Swan, J.B. (1959) *Phys. Rev.*, **116**, 81–83.
9 Ritchie, R.H. (1973) *Surf. Sci.*, **34**, 1–19.
10 Raether, H. (1980) *Excitation of Plasmons and Interband Transitions by Electrons*,

vol. 88, Springer Tracks in Modern Physics, Springer, New York.
11 Feibelman, P.J. (1982) *Prog. Surf. Sci.*, **12**, 287–407.
12 Plummer, W., Tsuei, K.-D., and Kim, B.-O. (1995) *Nucl. Instrum. Methods Phys. Res. B*, **96**, 448–459.
13 Raether, H. (1988) *Surface Plasmons on Smooth and Rough Surfaces and on Gratings*, vol. 111, Springer Tracks in Modern Physics, Springer, New York.
14 Rocca, M. (1995) *Surf. Sci. Rep.*, **22**, 1–71.
15 Liebsch, A. (1997) *Electronic Excitations at Metal Surfaces*, Plenum Press, New York.
16 Pitarke, J.M., Silkin, V.M., Chulkov, E.V., and Echenique, P.M. (2007) *Rep. Prog. Phys.*, **70**, 1–87.
17 Stern, F. (1967) *Phys. Rev. Lett.*, **18**, 546–548.
18 Ando, T., Fowler, A.B., and Stern, F. (1982) *Rev. Mod. Phys.*, **54**, 437–672.
19 Nagao, T., Hildebrandt, T., Henzler, M., and Hasegawa, S. (2001) *Phys. Rev. Lett.*, **86**, 5747–5750.
20 Rugeramigabo, E.P., Nagao, T., and Pfnur, H. (2008) *Phys. Rev. B*, **78**, 155402.
21 Chaplik, A.V. (1972) *Zh. Eksp. Teor. Fiz.*, **62**, 746; (1972) *Soviet Phys. JETP*, **35**, 395.
22 Das Sharma, S. and Madhukar, A. (1981) *Phys. Rev. B*, **23**, 805–815.
23 Pitarke, J.M., Nazarov, V.U., Silkin, V.M., Chulkov, E.V., Zaremba, E., and Echenique, P.M. (2004) *Phys. Rev. B*, **70**, 205403.
24 Fetter, A.L. and Walecka, J.D. (1971) *Quantum Theory of Many-Particle Systems*, McGraw-Hill, New York.
25 Runge, E. and Gross, E.K.U. (1984) *Phys. Rev. Lett.*, **52**, 997–1000.
26 Alder, S.L. (1962) *Phys. Rev.*, **126**, 413–420.
27 Wiser, N. (1963) *Phys. Rev.*, **129**, 62–69.
28 Miyke, T. and Aryasetiawan, F. (2000) *Phys. Rev. B*, **61**, 7172–7175.
29 Persson, B.N.J. and Zaremba, E. (1985) *Phys. Rev. B*, **31**, 1863–1872.
30 Bergara, A., Silkin, V.M., Chulkov, E.V., and Echenique, P.M. (2003) *Phys. Rev. B*, **67**, 245402.
31 Silkin, V.M., Chulkov, E.V., and Echenique, P.M. (2004) *Phys. Rev. Lett.*, **93**, 176801.
32 Yuan, Z. and Gao, S. (2009) *Comput. Phys. Commun.*, **180**, 466–473.
33 Chulkov, E.V., Silkin, V.M., and Echenique, P.M. (1999) *Surf. Sci.*, **437**, 330–352.
34 Eguiluz, A.G. (1983) *Phys. Rev. Lett.*, **51**, 1907–1910.
35 Silkin, V.M., Pitarke, J.M., Chulkov, E.V., and Echenique, P.M. (2005) *Phys. Rev. B*, **72**, 115435.
36 Jain, J.K. and Das Sarma, S. (1987) *Phys. Rev. B*, **36**, 5949–5952.
37 Liu, Y., Willis, R.F., Emtsev, K.V., and Seyller, T. (2008) *Phys. Rev. B*, **78**, 201403(R).
38 Silkin, V.M., García-Lekue, A., Pitarke, J.M., Chulkov, E.V., Zaremba, E., and Echenique, P.M. (2004) *Europhys. Lett.*, **66**, 260–264.
39 Bartynski, R.A., Jensen, E., Gustafsson, T., and Plummer, E.W. (1985) *Phys. Rev. B*, **32**, 1921–1926.
40 Chulkov, E.V., Silkn, V.M., and Shirykalov, E.N. (1987) *Surf. Sci.*, **188**, 287–300.
41 Silkin, V.M., Alducin, M., Juaristi, J.I., Chulkov, E.V., and Echenique, P.M. (2008) *J. Phys.: Condens. Matter*, **20**, 304209.
42 see, for example, Algdal, J., Balasubramanian, I., Breitholtz, M., Chis, V., Hellsing, B., Lindgren, S.Å., and Walldén, L. (2008) *Phys. Rev. B*, **78**, 085102.
43 Silkin, V.M., Quijada, M., Vergniory, M.G., Alducin, M., Borisov, A.G., Díez Muiño, R., Juaristi, J.I., Sánchez-Portal, D., Chulkov, E.V., and Echenique, P.M. (2007) *Nucl. Instrum. Methods Phys. Res. B*, **258**, 72–78.
44 Silkin, V.M., Quijada, M., Díez Muiño, R., Chulkov, E.V., and Echenique, P.M. (2007) *Surf. Sci.*, **601**, 4546–4552.
45 Silkin, V.M., Balassis, A., Leonardo, A., Chulkov, E.V., and Echenique, P.M. (2008) *Appl. Phys. A*, **92**, 453–461.
46 Diaconescu, B., Pohl, K., Vattuone, L., Savio, L., Hofmann, Ph., Silkin, V.M., Pitarke, J.M., Chulkov, E.V., Echenique, P.M., Farías, D., and Rocca, M. (2007) *Nature (London)*, **448**, 57–59.

47 Silkin, V.M., Pitarke, J.M., Chulkov, E.V., Diaconescu, B., Pohl, K., Vattuone, L., Savio, L., Hofmann, Ph., Farías, D., Rocca, M., and Echenique, P.M. (2008) *Phys. Status Solidi (a)*, **205**, 1307–1311.

48 Politano, A., Chiarello, G., Formoso, V., Agostino, R.G., and Colavita, E. (2006) *Phys. Rev. B*, **74**, 081401(R).

49 Silkin, V.M., Nechaev, I.A., Chulkov, E.V., and Echenique, P.M. (2005) *Surf. Sci.*, **588**, L239–L245.

50 Lau, K.W. and Kohn, W. (1978) *Surf. Sci.*, **75**, 69–85.

51 Traverse, A., Girardeau, T., Prieto, C., Meneses, D.D.S., and Zanghi, D. (2008) *Europhys. Lett.*, **81**, 47001.

9
Low-Dimensional Plasmons in Atom Sheets and Atom Chains

Tadaaki Nagao

9.1
Introduction

When the size of the object shrinks beyond micrometer scale and it reaches down to nanometer or subnanometer scale, novel effects that originate from its smallness come into view. First, the most common and important effect is the size effect, or the confinement effect. This effect sets in when the object size (thickness, width, etc.) becomes comparable or smaller than the characteristic lengths of material, such as Fermi wavelength λ_F, effective Bohr radius a_B^*, effective Wigner–Seitz radius r_s^*, or screening length λ_D. For example, wavelength shift in photoluminescence and phonon frequency shifts are often observed in semiconductor quantum dots. Also, metal-to-insulator transition or even the change in the superconductivity transition temperature is reported for nanoparticles and nanofilms with sizes close to λ_F [1, 2].

Second, when the object size goes down to the atomic scale, salient change in the structural properties appear such as reduction in melting temperature, magic atom number in cohesive energy, structural phase transitions, and so on. The multiply twinned particles (MTPs) [3] and shell structures in nanowires [4], two-dimensional metallic alloy phases on crystal surfaces [5], and nanofilm allotropes [6] showed distinct structural properties that are very different from the bulky three-dimensional (3D) objects. As a result, the electronic and magnetic structures of these nanophases can be completely different from the ordinary bulk ones and thus such new phases are regarded to constitute novel classes of materials.

As shown above, by shrinking down the size of the objects, confinement and dimensionality as well as "novel nanophase" effect become operative and we can utilize these effects as new tuning parameters for tailoring the material properties in addition to the conventional chemical synthesis for bulk materials. This is the biggest advantage of utilizing the nanoscale objects, and by further combining and integrating these nanoscale building blocks, we can explore novel functional properties in a rather extensive manner.

Dynamics at Solid State Surface and Interfaces Vol. 1: Current Developments
Edited by Uwe Bovensiepen, Hrvoje Petek, and Martin Wolf

ISBN: 978-3-527-40937-2

Among many functional properties, excitonic and plasmonic properties show very strong size and shape effects in nanometer scale, and in this sense, these properties have attracted large interest in nanotechnology. Because λ_F and λ_D in most metals are in nanometer and subnanometer scale, plasmons in metals show the maximum tunability in nano/atomic scale by changing the shape and the dimensionality of the objects. This fits nicely to the strategy of "plasmonics": tailoring the optical properties at the nanoscale for the next-generation nanophotonics/optics devices. In this context, not only from the scientific point of view but also from the technical point of view, it is of great importance to clarify whether plasmons are confined and can propagate in atomic scale tiny objects such as in atom chains and so on. Such investigation will be the first step toward the realization of the ultimately small atomic scale optical device components.

9.2 Difference between the Surface Plasmons and the Atomic Scale Plasmons

The light with micrometer-scale vacuum wavelengths can be confined and propagated in nanometer scale at the metal–air or metal–dielectric interfaces in the form of surface plasmon polariton. Thanks to this remarkable characteristic, surface plasmon now enjoys a renewed interest in the field of nanophotonics and chemo/biosensing. The concept of the surface plasmon is safely adopted when the size of a metallic object is large enough compared to λ_F and λ_D, and the charge density oscillation of the plasmon decays into the material in a more or less three-dimensional fashion [7]. However, when the metal object is downsized and the plasmon is confined to nanoscale small object, it is expected that the plasmon come to possess very different properties than in ordinary 3D media. This is not only because of the confined motion of the oscillating charge but also because of the largely modified electromagnetic field distribution in the surrounding media (air, water, and dielectrics). Such change means that the restoring force acting on the induced surface charge of the nanoobject becomes very different from that of the larger object and thus leads to a very different plasma frequency: in most cases, the frequency redshifts dramatically. Figure 9.1 shows the charge density oscillation in materials with different dimensionality. The upper two cases show the 3D-type plasmons in bulk and semi-infinite media. The lower three cases show the plasmons confined in low-dimensional media. For a nanofilm, two surface-localized modes exist at the top and the bottom, but in an ideal two-dimensional medium, such as in an atom sheet, there is no difference between the top and the bottom surfaces and the charge oscillation takes place in a sheer 2D manner. In a one-dimensional system, such as in an atom chain, the electron motion is restricted in the chain and the charge oscillation takes place only in one-dimensional manner.

Since the Fermi wavelength λ_F of a typical metal is in the subnanometer scale, shape and dimensionality effects are most clearly seen in the ultimately small systems like in atom sheets and atom chains. Taking advantage of the modern technique of molecular beam epitaxy, we can fabricate wide variety of such systems at

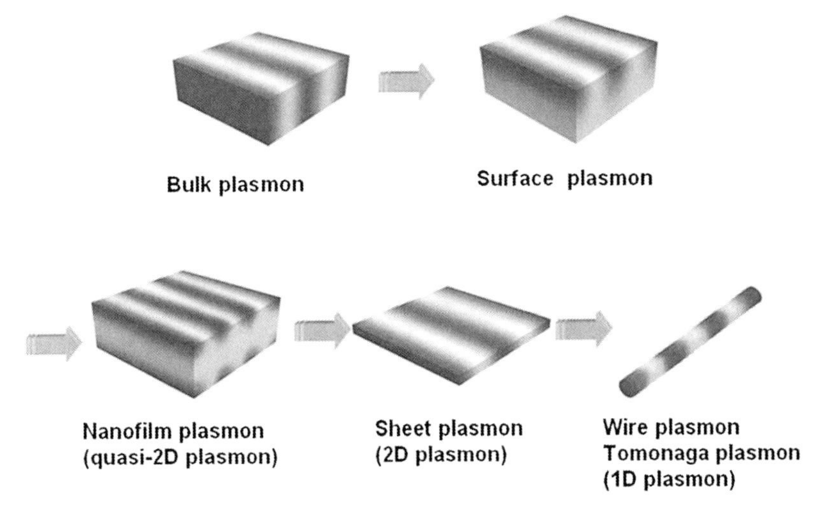

Figure 9.1 Schematic illustration of the charge density oscillation of plasmon in metallic media with different shapes and dimensionalities.

the surfaces of semiconductors. When the bandgap energy of the substrate is large enough, the excited plasmons in the metallic objects are free from hybridizing with the substrate interband transition and so on, and the excitation with purely collective nature is observable. Such electronically clean and simple systems serve as good touchstones of the theory of low-dimensional solid-state plasma in metallic density regime [8–10].

9.3 Measurement of Atomic Scale Low-Dimensional Metallic Objects

In the following section, we report on the direct measurement of plasmon dispersion relation for ultrathin metallic objects such as metallic nanofilms (quasi-2D) and metallic monoatomic layers (2D) fabricated on single-crystal silicon surfaces. Spectroscopy and diffraction technique with slow electron beam are the most suitable techniques to investigate the electronic excitation and dielectric properties of such ultimately thin system. Their excellent atomic level sensitivity gives access to information about structures and electronic or vibronic properties of ultimately small systems such as adsorbed gases or monolayer of metal sheets supported on crystal surfaces. Among these techniques, high-resolution electron energy loss spectroscopy (HREELS) showed its power and versatility in studying localized vibrational modes, phonon dispersion, and dielectric response not only at the surfaces but also at the interfaces several nanometers deep below the surface [11]. Electromagnetic waves of the plasmons are transmitted far from the metallic objects and strongly interact with slow electrons by the so-called dipole scattering mechanism. Also, the energy and the momentum ranges of EELS are orders of magnitudes

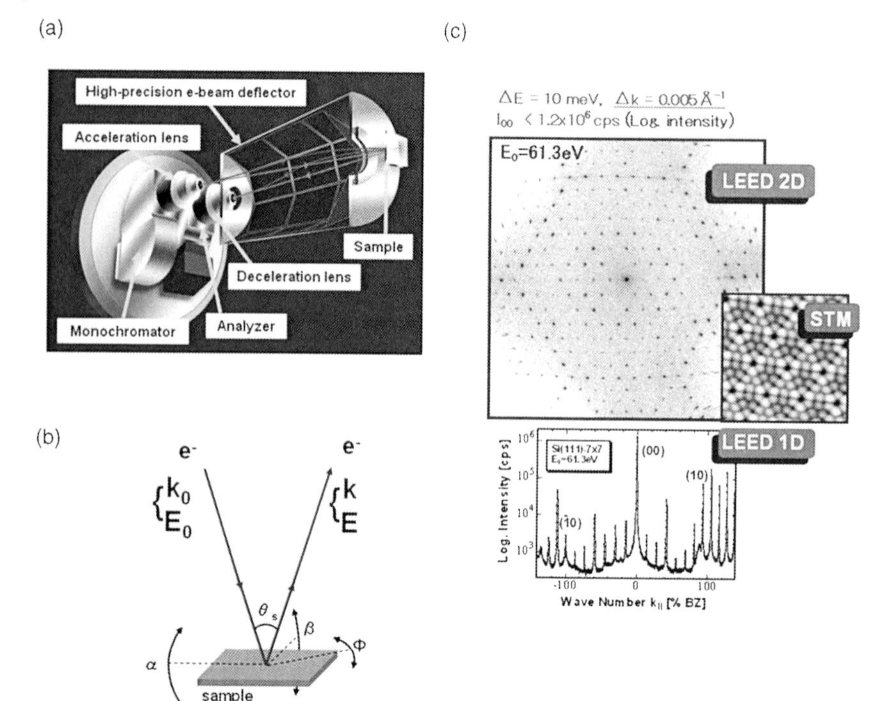

Figure 9.2 (a) An electron spectrometer equipped with an electron monochromator, analyzer, and a high-precision electrostatic beam deflector. (b) Geometry of the incident and outgoing electron beam from the sample is shown. Electron beams are freely tilted with respect to the sample while keeping the angle θ_s constant. This is identical to the situation where the sample is freely tilted by angles α, β, and ϕ with respect to the fixed incident and outgoing electron beams. (c) An example of the low-energy electron diffraction pattern taken with the elastically reflected beam from the clean Si(111)-7 × 7 surface. Incident energy was 61.3 eV. Inset in the middle right shows the real space image taken by scanning tunneling microscopy (10 nm × 10 nm). Bottom is the diffraction profile of the same sample.

larger compared to conventional photon probes. These advantages make this technique ideal for determining the plasmon dispersion in ultimately small objects like atom chains.

Figure 9.2 shows the schematic illustration of the electron spectrometer used in the present EELS experiments. Our custom-built spectrometer consists of a double-pass monochromator, a single-pass analyzer, and an electrostatic high-precision electron beam deflection unit. Electron beams are transferred between these modules through acceleration and deceleration focusing lenses [12, 13]. The spectrometer is installed in a permalloy ultrahigh vacuum chamber and the sample is mounted on a sample manipulator equipped with a custom-built flow-type cryostat with temperature control ranging from 9 to 1500 K. The typical momentum resolution was 0.005 Å^{-1} (0.06° angular resolution) with energy resolution of 10–25 meV. The

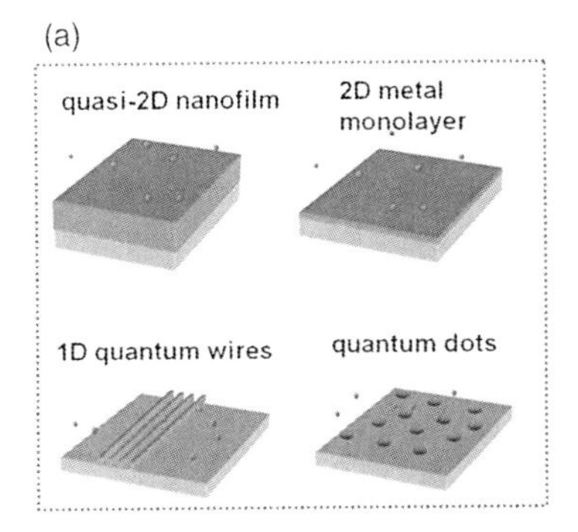

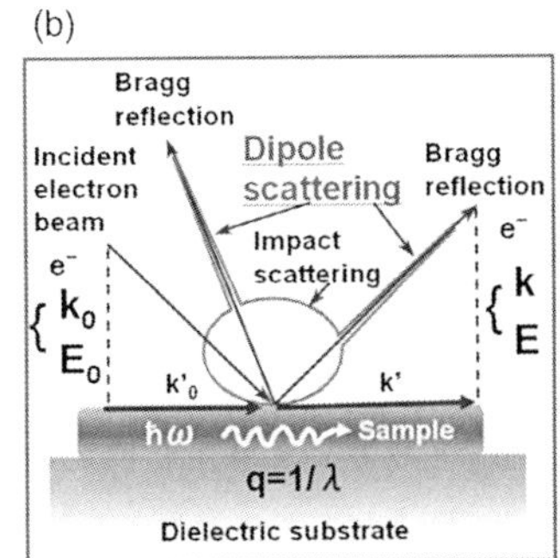

Figure 9.3 (a) Atomic scale low-dimensional structures prepared by molecular beam epitaxy. (b) Geometry of the EELS measurement in reflection mode. Slow electrons strongly interact with plasmons by the so-called dipole scattering mechanism. The plasmon losses take place near the direction of Bragg beams.

incident electron energy can be varied from 1 to 250 eV. The typical electron count rate was 0.1–1 million counts per second (cps) at the Bragg spots. The experiments were usually performed in ultrahigh vacuum (6×10^{-11} Torr base pressure).

Figure 9.3 shows the schematics of the epitaxially grown atomic-scale plasmonic structures on semiconductor surfaces. The crystallographic orientation and the quality of the grown sample are characterized *in situ* by the spectrometer shown in Figure 9.2 in the ultrahigh vacuum chamber. The slow electrons incident on the samples interact strongly with the electromagnetic wave from the plasmons via dipole scattering mechanism and are inelastically reflected back in the directions near the Bragg beams. The momentum and the excitation energy of the plasmons are determined by the energy and the momentum conservation rules:

$$\Delta E = E_0 - E = \hbar\omega, \tag{9.1}$$

$$\Delta k = k_0' - k' + G = q, \tag{9.2}$$

where G is the lattice vector of the sample.

The plasmons confined in atomic scale low-dimensional materials are expected to disperse steeply from far- to near-infrared region in a narrow momentum window. So high momentum and energy resolutions are essential for clearly resolving the loss peaks from these plasmons [8–10, 14–16].

9.4 Plasmons Confined in Ag Nanolayers

To elucidate the plasmon confinement in ultrathin media with atomic scale thickness, we have measured the thickness dependence of the plasmon dispersion curves of Ag nanolayer directly supported on the clean Si(111) surface. Ag is known to be

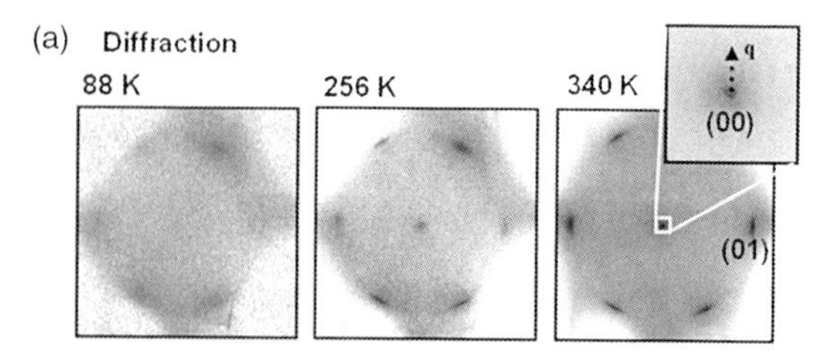

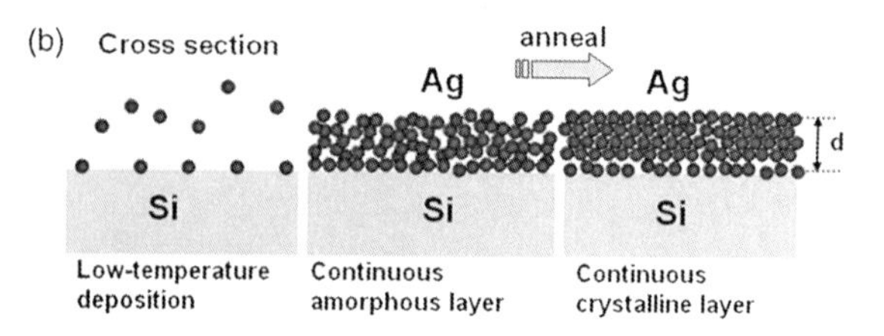

Figure 9.4 (a) Electron diffraction pattern taken with the elastic beam reflected from the 15 ML Ag nanofilm prepared on the single-crystal Si(111) surface. The film was prepared by low-temperature deposition and subsequent annealing. (b) Schematic illustration of the low-temperature deposition (shown for a few ML Ag nanofilm).

a paradigm system for the studies of surface plasmons by EELS owing to its very high intensity and sharp linewidth of the loss peak. Here the atomic structure of the system is nearly the same as in bulk crystal, so the nanoscale confinement effect is the main issue for this system.

Figure 9.4a shows the electron diffraction pattern taken from a 15 monolayer thick Ag film (hereafter referred to as 15 ML Ag nanofilm) prepared on the clean Si(111)-7 $\times$ 7 surface as shown in Figure 9.2c. The pattern was taken through the energy-filtered elastically reflected beam. The deposition of Ag nanofilm was carried out at low temperature (88 K) and then annealed up to 340 K and cooled down to room temperature (300 K). The crystallinity of this nanofilm becomes very high after annealing, which is evidenced by the sharpness of the (00) Bragg spot at the center. The lattice constant derived from the pattern (2.9 Å) indicates that this film is a Ag(111) film with the crystal structure similar to that of the bulk Ag crystal. Lateral grain size is about tens of nanometers as judged by the half-width of the (00) spot. Presence of the arc-shape spots indicates the slight rotational disorder in the azimuthal direction of the ultrathin films with respect to the Si(111) crystal substrate. After checking the crystallinity of the fabricated Ag films, we zoomed into the region near the (00) spot, as shown in the inset, and then measured the EELS spectra at various $\mathbf{q}$ points along the direction indicated as a dotted arrow in this figure.

Figure 9.5 shows the EELS spectra taken as a function of momentum transfer $\mathbf{q}$ (identical to $\mathbf{k}'{-}\mathbf{k}'_0$). The momentum transfer value is indicated with respect to the size of the primitive translational vector $G = 1.88\,\text{Å}^{-1}$ (= 100 BZ%) of the 2D

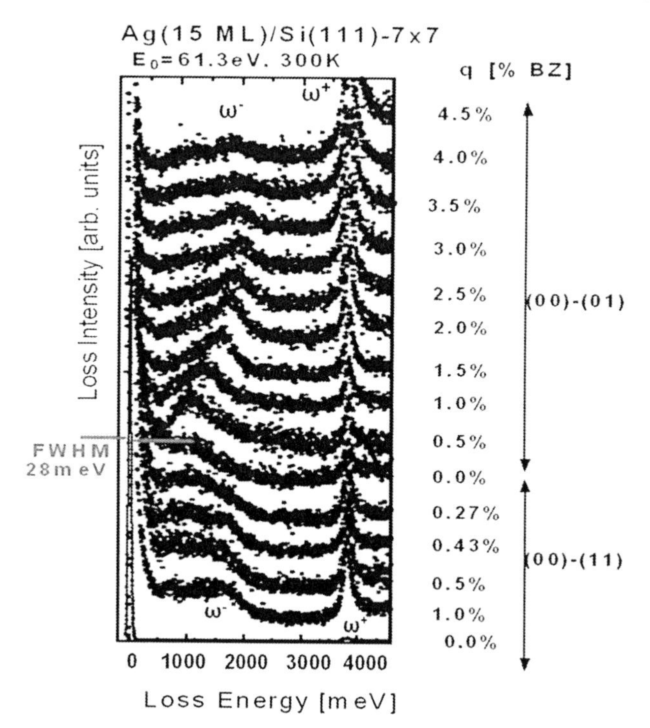

Figure 9.5 Momentum-resolved EELS spectra measured from a 15-layers-thick (3.4 nm thick) Ag nanofilm. Energy resolution was 28 meV. The momentum transfer value is indicated with respect to the 1.88 Å^{-1}, which is the primitive translational vector in the reciprocal space of the Si(111) surface. Momentum was scanned to (00)-(01) and (00)-(11) directions.

Brillouin zone of the Si(111) surface. Near the (00) Bragg spot ($\mathbf{q} \approx 0$), we observed two loss peaks that rapidly decay in intensity and broaden in peak width within a narrow momentum window of $\Delta q = 0.1$ Å^{-1}. The energy dispersion curves of these two modes can be explained in terms of a quasi-2D plasmon in the Ag nanofilm. The lower branch ω^- is steeply dispersing up to 2.0 eV and the higher branch ω^+ was nearly nondispersive and stays around 3.8 eV. The upper nondispersive ω^+ mode can be assigned to a plasmon mode similar to the Ag surface plasmon since its energy (3.8 eV) is very close to that of the semi-infinite surface plasmon of Ag. Also, its high intensity, asymmetric line shape, as well as its narrow linewidth (200 meV wide) are very analogous to those of Ag surface plasmon. The lower ω^- mode can be assigned as "an interface plasmon" since its loss energy approaches 2.0 eV, which is the energy of the interfacial plasmon mode at the semi-infinite Ag/Si system.

Thickness dependence of plasmon energy of these two modes is shown in Figure 9.6. Momentum value was fixed at $q = 0.04$ Å^{-1}. As the thickness increases, the upper mode ω^+ undergoes slight redshift. On the other hand, the lower ω^- mode

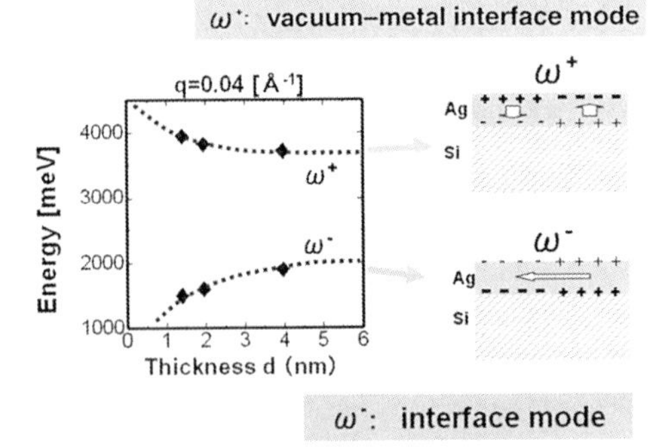

Figure 9.6 *Left*: Thickness dependence of the two plasmon branches taken at a fixed momentum value $q = 0.04\,\text{Å}^{-1}$. *Right*: Charge distribution and the polarization of the two modes.

undergoes slight blueshift. These behaviors are consistent with the theoretically expected feature of interfering two surface modes at the two sides of the film [17, 18]. The two modes are expected to approach asymptotically to different energies (2.0 and 3.7 eV) since the nanofilm is asymmetric, that is, the dielectric media in contact with the two surfaces are different (vacuum and Si). Such behavior is actually seen in the present experiment. According to a two-layer model [11, 14], plasmon dispersion is expressed as follows:

$$\exp(-2|\mathrm{q}|\sqrt{}\mathrm{d}) = (\varepsilon_{\mathrm{Ag}}+1)(\varepsilon_{\mathrm{Si}}+\varepsilon_{\mathrm{Ag}})/(1-\varepsilon_{\mathrm{Ag}})(\varepsilon_{\mathrm{Si}}-\varepsilon_{\mathrm{Ag}}) \tag{9.3}$$

Here $\varepsilon_{\mathrm{Ag}}$ and $\varepsilon_{\mathrm{Si}}$ are the dielectric constants of Ag and Si and d is the thickness of the nanofilm. As a preliminary evaluation, the plasmon dispersion and the charge distribution in the film were calculated for each mode by using bulk Si and Ag dielectric constants. The upper mode ω^+ has larger amount of the oscillating charge at Ag–vacuum interface and is mostly polarized in the film perpendicular direction. On the other hand, the lower mode ω^- has large charge oscillation at the Si–Ag interface and its polarization is mainly in the film parallel direction. Since the restoring force acting on the induced charge will be larger for the case of the former mode than for the latter mode, frequency of the ω^+ mode is higher than ω^- mode. When the film thickness becomes thinner and thinner, the top and the bottom surfaces will finally become identical and thus only the ω^--like mode with lateral polarization is expected to survive.

In Figure 9.7, the dispersion relation of the plasmon in the film is shown together with the simulated result derived by Eq. (9.3). The overall agreement is rather good in spite of the fact that the theory does not include any quantum effects such as band quantization in the thickness direction and so on. This means that the plasmonic properties of the film-based material with thicknesses even below 2 nm can be predicted rather precisely by just using bulk dielectric constants and macroscopic theory. Such information is surprising, but at the same time rather important for predicting the performance of the plasmonic devices with nanometer size.

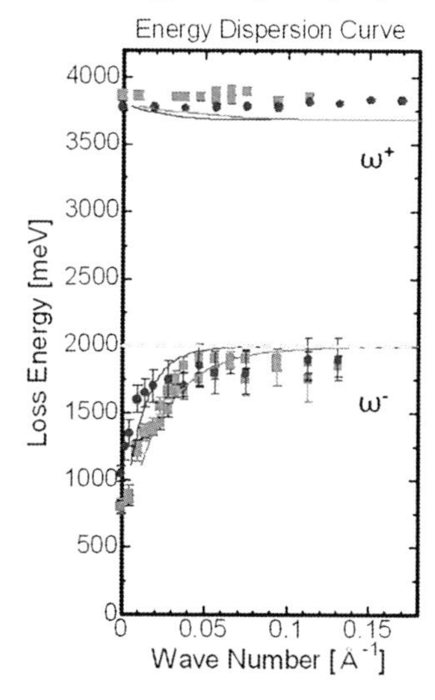

Figure 9.7 Plasmon dispersion relation of the Ag nanofilm is shown. Thickness dependence of the two plasmon branches was taken.

9.5
Plasmon in a Two-Dimensional Monoatomic Ag Layer

In contrast to the Ag nanolayer with finite thickness, ultimately thin 2D media should support ω^--like mode since there is no polarization in the thickness direction. Two-dimensional plasmon, or sheet plasmon, corresponds to this case since its charge density distribution is restricted in the 2D space and thus shows very different electrodynamical properties compared with those of the bulk plasmons, surface plasmons, and quasi-2D-type plasmons in metal nanofilms. For example, bulk and surface contributions are always inseparable in the properties of the surface plasmons. On the contrary, in the cases of 2D plasmons, properties of the outermost atomic layer can be reflected selectively and separately from the bulk ones if the coupling to the substrate excitation is weak. We can take advantage of this characteristic for utilizing the 2D plasmon as a sensitive noncontact probe of the electrodynamics in atomically thin region, such as metallic monolayer or surface-state bands supported on less conductive dielectric media. Such 2D plasmon mode was found in a monoatomic layer of Ag fabricated on the Si surface [8, 14].

A single-crystalline monolayer of Ag(111) film is difficult to obtain on semiconductor substrates. Therefore, to fabricate a single-crystalline 2D metallic monolayer, we prepared a "2D nanophase" composed of Ag and Si by depositing precisely 1 ML of Ag onto the heated Si(111)-7 $\times$ 7 surface (around 800 K) in ultrahigh vacuum ($<1 \times 10^{-9}$ Torr). This new "2D nanophase" or "2D alloy phase" shows $\sqrt{3}$ times periodicity with respect to that of the Si(111) substrate, different from that of the Ag(111) film. The crystallinity of this monolayer is very high as was evaluated by electron diffraction (grain size larger than 600 Å, corresponds to 0.01 $Å^{-1}$ in reciprocal space). This guarantees that the plasmon can propagate freely without being scattered by grain boundaries up to a very long wavelength. The energy resolution of the spectra was 20–30 meV, which was mostly limited by the broadening due to multiplasmon and intraband single-particle excitations in the metallic Ag layer as well as in the subsurface space charge layer.

Figure 9.8a (top) shows electron diffraction patterns measured by using elastic and (bottom) inelastic beams of the spectrometer. The data are displayed in logarithmic intensity. The former (Figure 9.8a, top) shows sharp diffraction pattern with $\sqrt{3}$ times periodicity with respect to the Si(111) surface. The outermost layer is a Ag–Si two-dimensional alloy formed at 800 K, and the epitaxial relationship of the overlayer perfectly matches to that of the silicon substrate. The latter (Figure 9.8a, bottom) shows an inelastic 2D scan obtained from this monolayer at a loss energy of 400 meV taken from the wave number region shown by a white square from the elastic one

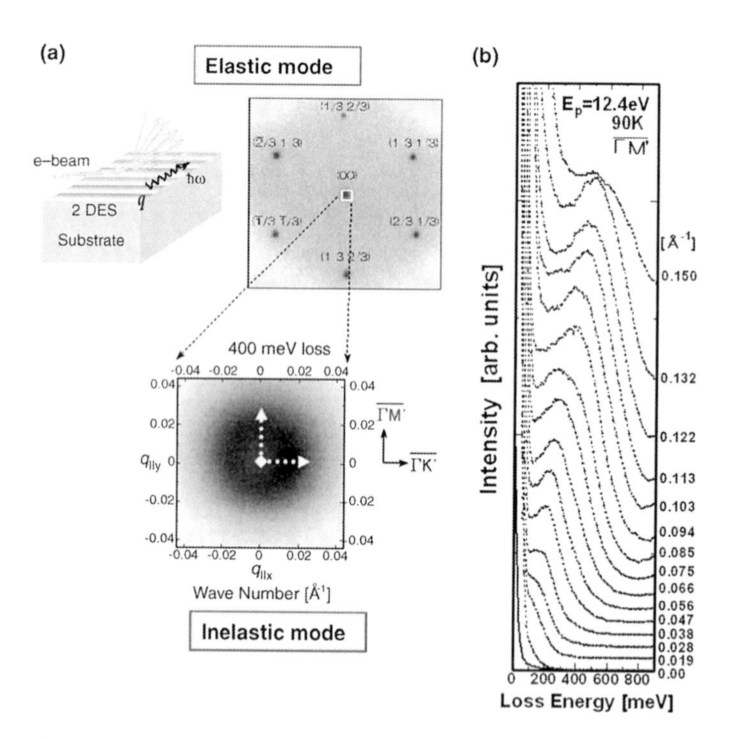

Figure 9.8 (a) (upper left) Schematic illustration of the excitation of charge density wave oscillation in Ag monolayer on Si. (upper right) Electron diffraction pattern taken from the Ag monolayer on the Si(111) surface. (bottom) "Inelastic 2D scan" of the beam scattered from the same sample with an energy loss of 400 meV. Electron primary energy was 50.3 and 12.3 eV for the two patterns, respectively. (b) Momentum-resolved EELS spectra from the 2D plasmon of a monolayer.

(Figure 9.8a, top). The intensity distribution exhibits ring-like nearly isotropic feature with respect to the azimuthal orientation, which indicates the 2D isotropic nature of the excitation.

Figure 9.8b shows typical momentum-resolved spectra obtained by scanning the $\mathbf{q}$ along the $\Gamma M'$ symmetry direction. The excitation energy goes down to 0 at small $\mathbf{q}$ values. Nearly identical slopes are also observed in other directions such as in $\Gamma K'$ symmetry direction that evidences the 2D nature of the excitation. Quickly increasing linewidth as a function of $\mathbf{q}$ is clearly perceived; such feature is not likely to be explained by single-particle excitation. All these features are consistent with the 2D-type plasmonic excitation and evidence the confinement of the electromagnetic wave in atomically thin media [8, 14, 19].

The plasmon energy dispersion is plotted as a function of momentum transfer $\mathbf{q}$ in Figure 9.9. In a simplest approximation, based on a local response theory,

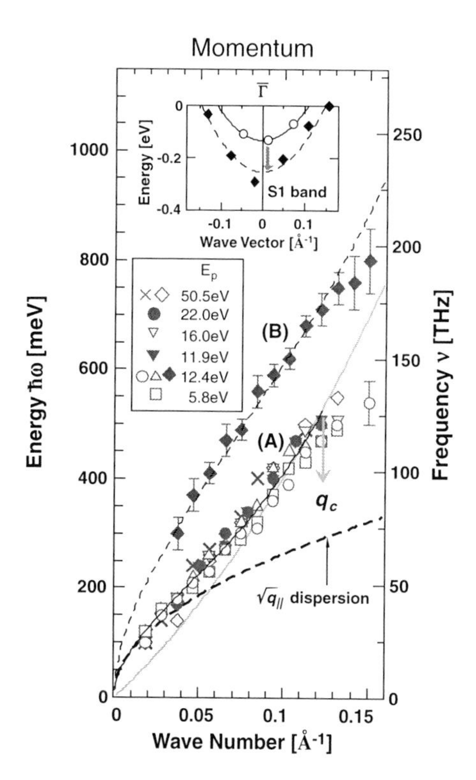

Figure 9.9 Plasmon dispersion curve determined by the momentum-resolved EELS spectra. Experimental curve (A) shows the data before the deposition of donor adatoms. The bold solid curve at curve (A) is the best fit to the nonanalytic full RPA dispersion by Stern [21]. Bold dashed curve below the curve (A) shows the $\sqrt{|\mathbf{q}|}$-type dispersion by Eq. (9.4). The thin solid curve just beneath the curve (A) is the upper edge of the single-particle excitation continua. Curve (B) is the data taken after 0.15 ML of donor Ag adatoms are deposited. Inset shows the "band sinkage" of this 2DES upon additional Ag deposition. (Please find a color version of this figure on the color plates.)

longitudinal plasmon in an infinitely thin film on a semi-infinite dielectric medium is known to be expressed as follows:

$$\omega_{2D}(\mathbf{q}) = [4\pi N_{2D} e^2 m^{*-1} (1+\varepsilon_{Si})^{-1} \mathbf{q}]^{1/2}. \tag{9.4}$$

Here, $\omega_{2D}(\mathbf{q})$ is the 2D plasma frequency, N_{2D} is the areal density of electrons in the two-dimensional electron system (2DES), ε_{Si} is the dielectric constant of the Si substrate that is nearly dispersionless within the frequency range of interest ($\varepsilon_{Si} \approx 11.5$) [20], m^* is the electron's effective mass, and e is the elementary electric charge.

The dashed curve shows the $\sqrt{|\mathbf{q}|}$-type dispersion by Eq. (9.4) (for a semiconductor substrate). Experimental dispersion seems to be rather far from the $\sqrt{|\mathbf{q}|}$-type dispersion of Eq. (9.4). The discrepancy between the $\sqrt{|\mathbf{q}|}$ dispersion and the experiment becomes larger as the momentum increases. This is because the local response theory is valid only at small wave number (momentum) region $|\mathbf{q}| \ll |\mathbf{k}_F|$, where $\mathbf{k}_F$ is the Fermi wave number of the 2DES. Since our measurement is capable of determining directly the plasmon dispersion at large $\mathbf{q}$ values, such discrepancy is clearly detected in the experiments. This is a big advantage compared to the measurement using optical probes.

Within the framework of "nonlocal" response theory using random phase approximation (RPA), energy dispersion of a longitudinal plasma in a 2D nearly free electron (NFE) system was calculated by Stern [21]. Based on this theory, we analyzed the plasmon in a 2DES on a semi-infinite dielectric substrate. Just for convenience, we show an approximation up to the second-order term in $\mathbf{q}$:

$$\omega_{2D}(\mathbf{q}) = [4\pi N_{2D} e^2 m^{*-1} (1+\varepsilon_{Si})^{-1} |\mathbf{q}| + 6 N_{2D} \hbar^2 \pi (2m^*)^{-2} |\mathbf{q}|^2 + O(|\mathbf{q}|^3)]^{1/2}. \tag{9.5}$$

The first term is identical to the $\sqrt{|\mathbf{q}|}$ dispersion from the local response theory (see Eq. (9.4)). It is interesting to note that if the dielectric constant of the substrate becomes very large, that is, $(1+\varepsilon_{Si}) \to \infty$, such as for the case of metal substrate, the first term simply goes to zero and it results in the $|\mathbf{q}|$ dispersion. This is exactly the case of the acoustic surface plasmon where the 2DES is dynamically screened by the substrate 3D charge, as described in Chapter 8 [22]. The second and higher order terms reflect the nonlocal effects and quantum statistics of the 2DES; for example, the second term is rewritten as $(3/4)v_F^2 \mathbf{q}^2$ with the Fermi velocity v_F of the 2DES and exhibits the degeneracy of the electron system. Also, this term is free from substrate polarization (i.e., no ε_{Si} contained in this term) and mainly reflects the character of the 2DES itself. It should be noted that fitting to the approximated dispersion, including only up to the second-order term in Eq. (9.2), is not yet accurate enough compared with the fit to the nonanalytic full RPA dispersion. This is because the $\mathbf{q}$ range in our experiment is wide enough, ranging up to around $\mathbf{k}_F$, and the effect of higher order term becomes very significant.

The black solid curve at curve A in Figure 9.9 is the best fit to the nonanalytic full RPA dispersion. The overall fit is excellent. The electron density and the electron effective mass are determined to be $N_{2D} = 1.9 \times 10^{13}/\text{cm}^2$ and $m^* = 0.30\, m_e$, respectively (m_e is vacuum electron mass). Compared with the $\sqrt{|\mathbf{q}|}$-type dispersion calculated using the same N_{2D} and m^* values (shown by the dashed curve), far better agreement between the full RPA dispersion and the experiment is perceived, especially at larger $\mathbf{q}$ region. This means that *nonlocal effects and quantum statistics play essential roles in atomic scale low-dimensional plasmons at larger* $\mathbf{q}$ *region*. With the theoretical approach here, by fitting over a wide $\mathbf{q}$ range, the "nonlocal" full RPA dispersion yields the values of N_{2D} and m^* simultaneously as fitting parameters. Meanwhile, in the "local" $\sqrt{|\mathbf{q}|}$ case, only their ratio N_{2D}/m^* is obtained.

It is surprising that we can develop the above discussions without encountering any difficulty due to the presence of the background ion lattice that is regarded as the main cause of the disagreements between the NFE theories and the experiments for volume plasmons, even for the simplest alkali metals [7, 23]. The reasons why our 2DES is free from such complexity are as follows: (1) The plasmon energy is small enough compared with any possible excitation energy for the core and the valence band electrons. (2) Ion potential is partly screened by the 2DES with high effective density. (3) Because of the long Fermi wavelength ($\lambda_F = 42$–58 Å) of our 2DES, the wavelengths of the constituent electronic wave functions in the ground state and of the plasmon modes are substantially longer than the ion spacing, thus rendering the excited state of the 2DES insensitive to the short-period lattice. Because of these advantages, all that must be considered is the mean static polarizability contribution at the substrate–vacuum interface, which is well represented by $(\varepsilon_{Si} + 1)/2$. This "clean and ideal" situation makes this 2DES a nearly perfect experimental realization of an ideal 2D Fermi gas with a metallic density.

It should be noted that the effective density parameter, or the effective Wigner–Zeitz radius $r_s^* = (\pi N_{2D})^{-1/2} a_B^{*-1}$, of this system is as small as 1.2 (Bohr radius a_B^* is defined by $a_B^* = \{(\varepsilon_{Si} + 1)/2\}(m_e/m^*)\, a_B$ with the Bohr radius $a_B = 0.529$ Å). This value is rather small compared to most of the free electron metals like bulk aluminum, sodium, potassium, and so on (in case of aluminum, it is still 2.1). This fact means that the electron–electron coupling in this 2D monolayer is considerably smaller than any bulk 3D materials.

We can examine the effect of "carrier doping" on this atomic scale 2DES by depositing randomly adsorbed Ag adatoms onto the "pristine" $\sqrt{3} \times \sqrt{3}$ Ag monolayer. As seen from curves (A) and (B) in Figure 9.9, dramatic blueshift in the plasmon dispersion is observed. This clearly demonstrates that adsorbed atoms donate the electrons to this 2D monolayer and thus lead to the "swelling" of the 2D electron pocket. After the theoretical analysis, effective electron mass of the 2D band was found to be nearly the same; this means that the "swelling" of the electron pocket is due to the "sinkage" of the two-dimensional metallic band, as displayed in the inset in Figure 9.9 [8].

According to the result of our calculation, the plasmon can exist up to the momentum $|\mathbf{q}_c| = 0.125$ Å^{-1} as shown in Figure 9.9 without decaying into electron–hole pairs by strong wave–particle interaction (Landau damping) [8]. In the energy loss spectra, the loss feature also disappears approximately above at larger $\mathbf{q}$ region

$|\mathbf{q}| = 0.15\,\text{Å}^{-1}$, accompanied with a significant linewidth broadening as can be seen in Figure 9.8b. It should also be noted that at $\mathbf{q} < \mathbf{q}_c$, we can see monotonic linewidth broadening that cannot be explained by Landau damping. The most possible explanation will be the decay via two (and higher multiple) electron–hole pairs, induced by Coulomb collision or electron–electron scattering [24–26]. Presence of this type of damping was strongly suggested for volume plasmons in free electron metals, where the electron density r_s values are more or less of the same order as the present system [23, 25]. Thus, the observed linewidth dispersion strongly suggests that the electronic correlation effect plays a more important role and manifest itself more clearly in the aspect of plasmon damping than in the energy dispersion of our 2D plasmon.

9.6
Plasmons in Atomic Scale Quantum Wires

What would happen if the dimensionality of the object is further reduced to one dimension? If we talk about the materials in metallic density regime, such object should be ultimately thin object like atom chain or atomic scale quantum wire. It is very interesting to see whether in such ultimately small system, plasmon can ever exist and propagate in it. If it really exists, it must assume the form of a low-energy sound wave-like excitation that Tomonaga has introduced in his early theory of one-dimensional electron system [27]. However, at small $\mathbf{q}$, its dispersion must be different either from the $\sqrt{|\mathbf{q}|}$-type dispersion of the 2D plasmons on semiconductor substrates [8, 21] or the $|\mathbf{q}|$-type dispersion of the acoustic surface plasmon on metallic substrates (Figure 9.10) [22, 28].

To realize such situation, 1DES should be free from intermixing with the electronic states of the surrounding 3D media. One way to realize such situation is to self-assemble the metallic atom chains on the semiconductor/insulator single-crystal surfaces. For example, high-index surfaces of silicon are now widely exploited and rich variety of 1D metallic phases are obtained experimentally.

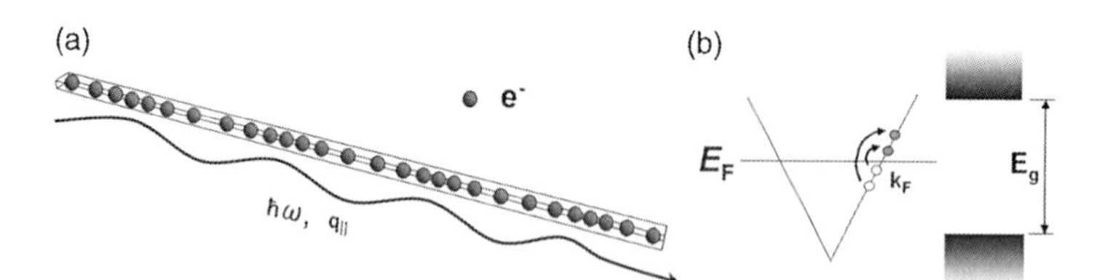

Figure 9.10 (a) Schematic illustration of the electron confinement in one-dimensional space. Plasmon in such systems is also expected to show dispersion relation starting from zero energy at zero momentum (Tomonaga plasmon). (b) Schematic band diagram of the 1D electron system similar to that of Tomonaga's model 1D electronic band near the Fermi level E_F is free from coupling to the electronic states of the surrounding media (E_g: energy gap of the surrounding media).

9.6.1 Plasmons in Self-Assembled Au Atom Chains

One of the most prototypical atomic scale 1D systems is the self-assembled Au atom chains on a silicon surface (the Si(557)–Au system) (Figure 9.11). The high-index Si(557) surface serves as an ideal anisotropic template with high crystallinity to sustain a perfectly ordered atom chains that can be viewed as the ultimate 1DES. The electronic structure near E_F is known to exhibit two adjacent parabolic electron pockets with a sheer 1D character with large Fermi wave number $k_F = 0.41\,\text{Å}^{-1}$, which is close to the half-filling point in the Brillouin zone [29, 30]. This 1D band is free from coupling to the electronic states of the Si substrate. Such a prototype 1D electronic band offers a diversity of intriguing topics such as the spin–charge separation, Peierls instability, and Rashba-type spin–orbit (SO) coupling effect of the 1D band [31–33]. The exchange correlation (XC) hole radius of the electrons in this 1DES is $r_{xc} \approx 1/k_F = 2.4\,\text{Å}$, and the electrons are densely packed into atom chains with an average distance of 7.7 Å. This situation makes this system ideal for investigating dimensionality effect and 1D confinement effect on the dynamic electronic correlation in metallic high-density regime.

The surface of Si(557) wafer was cleaned by resistive heating using DC current fed through the wafer. 0.2 ML of Au was evaporated, while the substrate Si(557) wafer was heated at 800 K in ultrahigh vacuum. Prior to the EELS measurement, diffraction

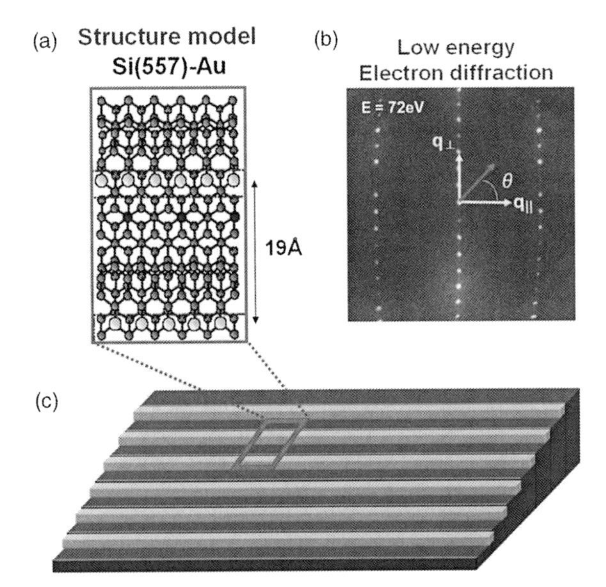

Figure 9.11 (a) Au atom chain structure on the Si(557) substrate. Large spheres represent Au atoms and the smaller ones Si atoms. (b) An example of high-resolution LEED pattern taken from this sample. (c) Schematic illustration of the atom wire array on the high-index Si(557) face. $\mathbf{q}_{||}$ denotes chain-parallel and $\mathbf{q}_\perp$ denotes momentum transfer in chain-perpendicular direction.

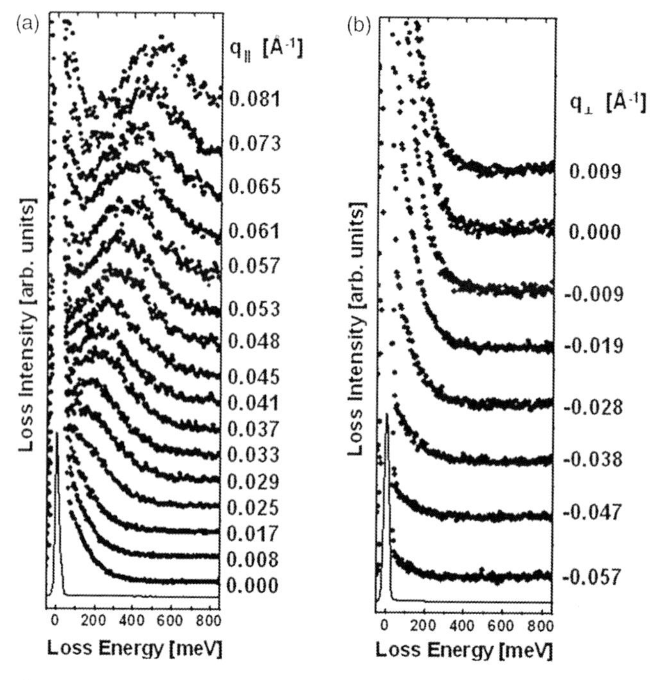

Figure 9.12 Momentum-resolved EELS spectra taken (a) in the chain-parallel direction (the $[1\bar{1}0]$ direction, or $\mathbf{q}_{||}$ direction) and (b) chain-perpendicular direction (the $[\bar{1}\bar{1}2]$ direction, or $\mathbf{q}_{\perp}$ direction).

pattern was observed to check the crystallinity and the crystallographic orientation of the atom chains (Figure 9.11b). The sharp Bragg spots ($< 0.01\,\text{Å}^{-1}$ in full width at half maximum) and their spacing indicate a highly ordered atom chains grown along the $[1\bar{1}0]$ direction ($\mathbf{q}_{||}$ direction).

Figure 9.12 shows two sets of the momentum-resolved EEL spectra taken along the wire-parallel ($\mathbf{q}_{||}$) and the wire-perpendicular ($\mathbf{q}_{\perp}$) directions. In the wire-parallel direction (Figure 9.12a), a single loss peak is clearly observed to disperse rapidly from mid- to near-infrared region as a function of $q_{||}$ $(= |\mathbf{q}_{||}|)$ accompanied by monotonic linewidth broadening. On the other hand, in the $\mathbf{q}_{\perp}$ direction (Figure 9.12b), no prominent feature was observed. This means that the observed excitation freely propagates only along the wires, but not across the wires.

Figure 9.13 shows a series of EELS spectra taken by varying the probing depth of the electron beam with the change in its incident energy E_{p}. As clearly seen in Figure 9.13, there is no significant probing depth dependence of the loss peak, and thus the excitation is assigned to the one confined in the atom chains formed on top of the crystal surface.

Figure 9.14 shows an energy dispersion curve determined from the energy positions of the peaks in the EELS spectra in comparison with the theoretical curve described below. As is clearly seen, the peak position approaches zero as $q_{||}$ tends

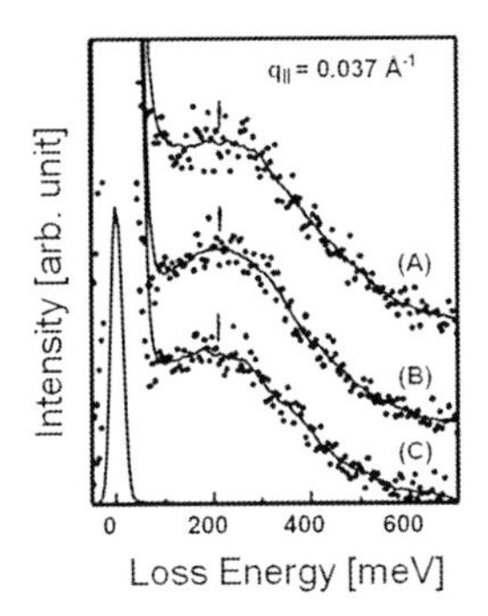

Figure 9.13 Incident energy dependence (probing depth dependence) of the EELS spectra measured at $q_{||} = 0.037\,\text{Å}^{-1}$. The electron incident energy was varied from (A) 26.0 eV to (B) 35.0 eV to (C) 45 eV. No significant probing depth dependence was detected.

toward zero, which is the expected sound wave-like feature first discussed by Tomonaga [27]. Below $q_{||} = 0.02\,\text{Å}^{-1}$ and $\hbar\,\omega = 150$ meV, each loss peak merges with the elastic peak and features a high Drude tail reflecting the highly conductive nature of the chains.

The 1D propagation of the excitation can be examined more quantitatively by changing the azimuthal direction θ (Figure 9.11b) of the in-plane momentum $\mathbf{q}$ from $\theta = 0°$ (along the chains) to $\theta = 90°$ (perpendicular to the chains). In Figure 9.14, data

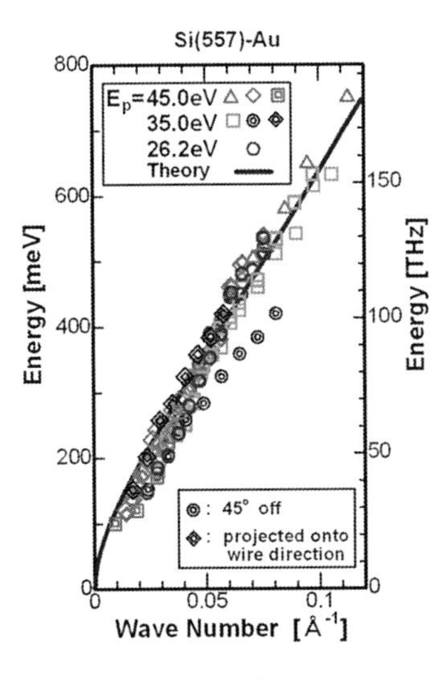

Figure 9.14 Energy dispersion curve determined from the peak positions in the momentum-resolved EELS spectra. Bold curve is the theoretical curve, including spin–orbit coupling (SOC) and exchange correlation effects. The precision in energy is roughly represented by the size of each data point. Below $q_{||} = 0.02\,\text{Å}^{-1}$, the precision is ± 30 meV due to the high tail from the quasielastic peak. Double circles and double diamonds are the data points taken at $\theta = 45°$ and plotted against $q = |\mathbf{q}|$ and against $q \cos(45°)$, respectively (θ is defined in Figure 9.11b, upper right). (Please find a color version of this figure on the color plates.)

points taken at $\theta = 45°$ are plotted against $q = |\mathbf{q}|$ (double circles). After replotting the data against the component of $\mathbf{q}$ projected onto the chain direction, $q\cos(45°)$, they fall again onto the same dispersion curve as for $\theta = 0°$ (shown by double diamonds). Such behavior is consistent with the expected behavior of a 1D plasmon, which should follow the relation $\omega(q, \theta) = \omega(q\cos\theta, 0)$. By taking into account all these behaviors, the observed excitation is thus unambiguously assigned to the 1D plasmon in the metallic atom chain array.

At small momentum $\mathbf{q}_{||}$, the plasmon dispersion in an isolated wire is expressed approximately as

$$\omega = \omega_0 q_{||} a \left[\ln\left(\frac{2\sqrt{2}}{q_{||}a}\right)\right]^{1/2} + O(q_{||}^2), \quad (q_{||}a \ll 1), \tag{9.6}$$

$$\omega_0 = \left(\frac{4N_{1D}e^2}{(1+\varepsilon_{Si})m^* a^2}\right)^{1/2}. \tag{9.7}$$

Here $2a$ and ε_{Si} denote the wire width and the dielectric constant of the substrate Si ($\varepsilon_{Si} = 11.5$), respectively, and N_{1D} and m^* signify the linear electron density and the electron effective mass, respectively. This relation is different from that for a 2D plasmon where the dispersion is proportional to the $\sqrt{|\mathbf{q}|}$ for small momentum values [8, 21]. However, the real system here is not an isolated wire and the crosstalk with the neighboring wires (with 1.9 nm separation) will be operative. Especially at small $\mathbf{q}_{||}$ where the electromagnetic wave from oscillating 1D charge transmits far away and thus the ensemble of the wires should give rise to the plasmonic mode similar to the 2D plasmon. On the other hand, as $q_{||}$ becomes larger, the effect of higher order terms becomes significant and electronic correlation effects, such as dynamic exchange correlation effects, become more operative. Such an effect, which is normally inaccessible using conventional optical probes, can be examined in detail by the present EELS in a wide q range and with a high q resolution.

To examine the XC effect quantitatively, we analyzed the data with a theory beyond the RPA theory. The analysis includes a self-consistent local-field correction (LFC) developed by Singwi *et al.* [34] and applied to the one-dimensional system by Friesen and Bergersen [35]. The XC effects can be described by the XC holes that make the electrons steer clear of each other within the distance of the XC hole radius $r_{XC} \sim k_F^{-1} = 2.4$ Å. This avoidance effect is considered to be more serious in 1D than in higher dimensions where the electrons have a greater choice of detours. The XC effects become more significant as $q_{||}$ becomes more comparable to k_F, that is, r_{XC}^{-1}. In our analysis, the electron effective mass m^* and the wire width $2a$ are the fitting parameters. We used $\varepsilon_{Si} = 11.5$ and $k_F = 0.41$ Å^{-1} taken from the literature [20, 30].

The best theoretical fit to the experimental data is shown as the bold solid curve in Figure 9.14. We can obtain markedly good agreement over the entire range of the observed $q_{||}$ up to 0.11 Å^{-1}. The parameters m^* and $2a$ are adjusted to $m^* = 0.60\, m_e$ and $2a = 4$ Å. Here, m_e stands for the free electron mass. The wire width $2a = 4$ Å is reasonable considering that the 1D electronic bands are associated with the

Au-related atomic chains structure (about 3.3 Å wide) that is shown in Figure 9.11. The effective mass $m^* = 0.60\, m_e$ is also in reasonable agreement with the value of $m^* = 0.45\, m_e$ determined by photoemission experiments [30, 32]. In photoemission, m^* reflects the band curvature near the band minimum, while in EELS, plasmons are composed of electronic transitions near the Fermi level E_F and reflect the band dispersion near E_F, so the slight difference in m^* value comes from this difference. While the theoretical dispersion curve is very sensitive to change in m^*, it is less sensitive to change in $2a$, since $2a$ (4 Å) is much shorter than the plasmon wavelength $2\pi/q_{||}$.

In our analysis, we assumed only one spin component in each 1D band, taking into account the proposal by Portal *et al.* [33]. They proposed that "the two parallel and adjacent 1D bands are Rashba SO-split bands," owing to its strong Au 6p character and the structural asymmetry of the chain structure. If we consider each of the 1D band as twofold spin degenerate, the adjusted effective mass becomes $m^* = 0.85\, m_e$, far from the value reported by photoemission experiments. Also, the degree of fit is degraded [10]. From these facts, our plasmon is likely to be an intraband plasmon in one of the SO split bands with an electron density of $N_{1D} = k_F/\pi = 0.13\,\text{Å}^{-1} = 1.3 \times 10^7\,\text{cm}^{-1}$.

We further investigated another Au-induced atom chain structure on the flat Si(111) surface. It exhibited very similar 1D plasmonic excitation, but, in contrast to the case of the Si(557) substrate, the observed plasmon was assigned to that in the metallic 1D band without SO splitting [16].

Although the qualitative features of the observed plasmon are close to that of an ideal free electron plasma, our quantitative analysis has revealed that the dynamic XC effects are, in fact, significant. The effective density parameter (electron density) of this system is as small as $r_s^* = 1/(2N_{1D}a_B^*) = 0.52$ for $m^* = 0.45\, m_e$, and $r_s^* = 0.70$ for $m^* = 0.60\, m_e$. This indicates a very high average velocity of electrons and, seemingly, the electron–electron coupling is negligible. However, in spite of this naive expectation, with an increase in $q_{||}$, the XC effects start to redshift the plasmon frequency noticeably. For example, at $q_{||} = 1/4\, k_F$, this redshift is as large as one tenth of the plasmon frequency, and the m^* value adjusted without incorporating the XC effect becomes more unreasonably large ($m^* > 0.85\, m_e$). When we compare with the 2D case, the redshift is smaller, which means that the *dynamic XC effects become more conspicuous in 1D than in 2D*, owing to stronger electron confinement in 1D.

9.6.2
Plasmons in In-Induced Atom Chains

In chains on the Si(111) surface (the Si(111)-4 × 1-In system) is also known as an atomic scale high-density 1D electron system and has been studied by many experimental techniques [36, 37]. Its band structure consists of three-electron pockets m_1, m_2, and m_3, the center of which is located at the same position in the **k**-axis (Figure 9.16). Its atomic structure model is shown in Figure 9.15c. The periodicity of the chain array is four times of the Si(111)-1 × 1 unit cell as can be seen in the diffraction pattern in Figure 9.15b. Compared to the Au chains on the Si(557)–Au

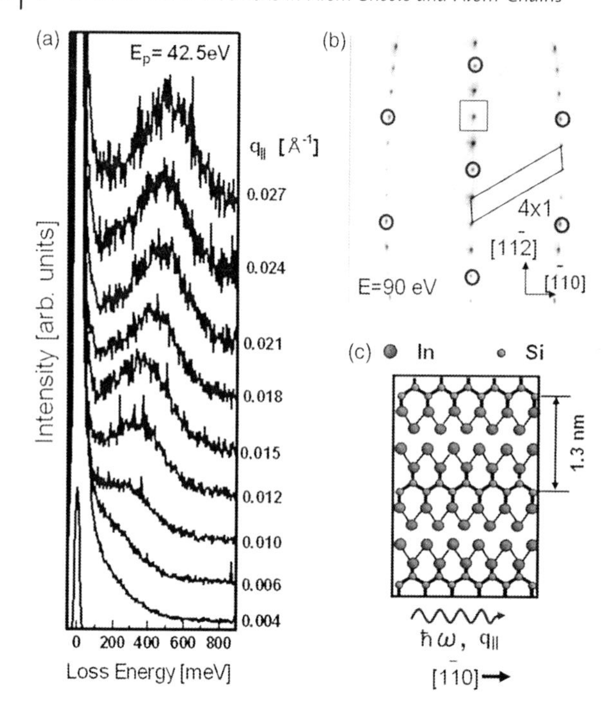

Figure 9.15 (a) Momentum-resolved EELS spectra in the In chain direction taken as a function of $\mathbf{q}_{||}$ with electron primary energy of $E_0 = 42.5$ eV are shown. (b) Diffraction pattern from an array of In atom chains on the Si(111) taken with electron primary energy of $E_0 = 90$ eV. 4×1 unit cell is indicated by a parallelogram. EELS is taken near the superreflection spot indicated by a square. (c) Atomic structure model of the In atom chains on the Si(111)-4×1-In surface. The separation between the chain is 1.3 nm (the width of the chain is about 0.9 nm

system (0.3 nm Au chain width, 1.6 nm gap between the chains), the Si(111)-4×1–In system has larger chain width (1.0 nm) and the gap between the In chains is narrower (0.3 nm).

Momentum-resolved EELS spectra taken from the In chains are shown in Figure 9.15a [13, 15]. In $\mathbf{q}_{\perp}$ direction, no loss feature was observed like the Au chains on the Si(557). Qualitative behavior of the $\mathbf{q}_{||}$ dependencies of the loss peak position and the linewidth is basically the same as that of Si(557)–Au, indicating that the observed loss is also attributed to the 1D plasmon. However, the rising behavior (slope) and the energy of the plasmon dispersion are very different. For example, when we compare the plasmon frequency near $\mathbf{q}_{||} \sim 0.03$ Å^{-1}, the frequency of the In chain is higher more than a factor of 2. This difference can be attributed to the difference in the electronic band structure of these two 1D systems. For example, both 1D systems have large electron pocket that cross at the Fermi level at $k_F \times 0.4$ Å^{-1} (k_F is the Fermi wave number). However, the electron pocket of Si(557)–Au is Rashba SO split, which means the electrons with only one type of spin in the 1D band can contribute to the plasmonic excitation. On the other hand, that of In chains on the

Si(111) is spin degenerate, which means the electron density in this band is two times larger than the Si(557)–Au case. Moreover, the Si(111)-4 × 1-In system possesses two additional electron pockets with smaller k_F values (0.28 and 0.078 Å^{-1}). This indicates that the 1D electron density is larger by more than a factor of 3 compared to that of Si(557)–Au. In addition, the entire band shape is different and the effective electron mass of the largest band of Si(111)-4 × 1-In is only about 60% of that of Si(557)–Au. All these differences give rise to the steeper slope and the higher plasmon frequencies of the Si(111)-4 × 1-In system. Also, larger width of the In chains and the smaller gaps between them can lift the plasmon frequency by increasing the Coulomb interaction between the In strips.

The same theoretical scheme as was used in the Si(557)–Au system was also used to analyze the plasmon dispersion curve. The observed plasmon dispersion curve was theoretically reproduced quantitatively by assuming that its oscillating charge mainly originates from the m_3 band. The other plasmons from m_1 and m_2 bands were not observed in the EELS spectra most possibly because the plasmon intensity is rather small reflecting the small electron density of these m_1 and m_2 bands. Also, these weaker plasmons are presumably dynamically screened by the most intense plasmon from m_3 and thus the observation of these two weaker modes become rather difficult. The schematic illustration of the 1D electron pocket is shown in Figure 9.16.

This system is also well known as a prototypical 1D system that undergoes the metal-to-insulator transition due to the Peierls instability. Momentum-resolved EELS measurement of this system was carefully performed as a function of temperature [15]. The metal-to-insulator transition was clearly detected and studied in detail from the standpoint of plasmon. The plasmon loss disappeared completely in the entire momentum and energy region when the sample was cooled down below the transition temperature. However, it reappeared clearly above the transition temperature. This fact indicates that the magnitude of the opened energy gap below the transition temperature is of the order of 500 meV or even larger since their gap size should be larger at least than the highest plasmon energy. Such wide energy gap is somewhat too large compared to the standard Peierls scenario and suggest some novel mechanism.

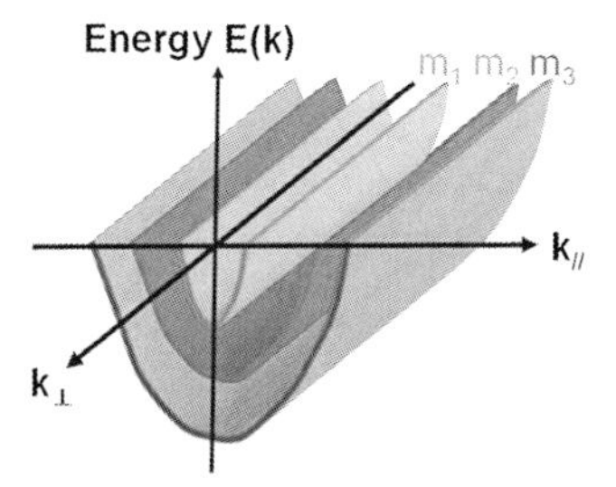

Figure 9.16 Schematic band diagram of the electron pockets of the In chain systems near the Fermi level E_F. The observed plasmon is attributed to the excitation in the largest m_3 band.

9.7
Conclusions

Compared to most of the conventional optical probes, the biggest advantages of the low-energy EELS are the atomic level sensitivity and the wide accessible range both in energy and in momentum space. By further improving this technique through realizing the energy resolution in the meV range and the momentum resolution in the 1/1000 $Å^{-1}$ range, characteristic plasmon dispersion curves of atomic scale low-dimensional metallic objects were successfully investigated. From an atomically flat nanofilms of Ag, two plasmon branches – nondispersing ω^+ and steeply dispersing ω^- – were observed. These modes are the split plasmon modes due to the interference between the top and the bottom surfaces. On the other hand, from an atomically flat monoatomic Ag layer, two-dimensional plasmon of a 2D free electron gas was observed. This mode is the sheer two-dimensional mode that the charge density oscillation is strictly confined in the atomic scale thin metal layer. Finally, one-dimensional plasmons confined in ultimately thin atomic scale quantum wires were experimentally detected and clarified to show the expected low-energy feature that Tomonaga had mentioned in 1950s. The measured data were successfully utilized as touchstones for the nonlocal response theory with local field correction, and further clarified the significance of the dimensionality effect in the exchange correlation effect on atomic scale plasmons.

References

1 Rosetti, R. *et al.* (1983) *J. Chem. Phys.*, **79**, 1086.
2 Guo, Y. *et al.* (2004) *Science*, **306**, 1915.
3 Ino, S. and Ogawa, S. (1967) *J. Physical Soc. Jpn.*, **22**, 1365.
4 Kondo, Y. and Takayanagi, K. (2000) *Science*, **289**, 606.
5 Barth, J.V. *et al.* (1993) *Surf. Sci.*, **292**, L769.
6 Nagao, T. *et al.* (2004) *Phys. Rev. Lett.*, **93**, 105501.
7 Raether, H. (1980) *Excitation of Plasmons and Interband Transitions by Electrons*, Springer-Verlag, Berlin.
8 Nagao, T. *et al.* (2001) *Phys. Rev. Lett.*, **86**, 5747.
9 Nagao, T. *et al.* (2006) *Phys. Rev. Lett.*, **97**, 116802.
10 Nagao, T. *et al.* (2007) *J. Physical Soc. Jpn.*, **76**, 114714.
11 Ibach, H. and Mills, D.L. (1982) *Electron Energy Loss Spectroscopy and Surface Vibrations*, Academic Press.
12 Nagao, T. and Hasegawa, S. (2000) *Surf. Interface Anal.*, **30**, 488.
13 Nagao, T. (2008) *Surf. Interface Anal.*, **40**, 1764.
14 Nagao, T. (2004) *Oyo Buturi*, **73**, 1312.
15 Liu, C., Inaoka, T., Yaginuma, S., Nakayama, T., Aono, M., and Nagao, T. (2008) *Phys. Rev. B*, **77**, 205415.
16 Liu, C., Inaoka, T., Yaginuma, S., Nakayama, T., Aono, M., and Nagao, T. (2008) *Nanotechnology*, **19**, 355204.
17 Ritchie, R.H. (1957) *Phys. Rev.*, **106**, 874.
18 Ishida, H. (1987) *Solid State Commun.*, **63**, 701.
19 Inaoka, T. *et al.* (2002) *Phys. Rev. B*, **66**, 245320.
20 Philipp, H.R. and Ehrenreich, H. (1963) *Phys. Rev.*, **129**, 1550.
21 Stern, F. (1967) *Phys. Rev. Lett.*, **18**, 546.

22 Silkin, V.M. *et al.* (2004) *Europhys. Lett.*, **66**, 260; Diaconescu, B. et al. (2007) *Nature*, **448**, 57.
23 vom Felde, A., Sprosser-Prou, J., and Fink, J. (1989) *Phys. Rev. B*, **40**, 10181.
24 Ichimaru, S. (1986) *Plasma Physics: An Introduction to Statistical Physics of Charged Particles*, Chapter 7, Benjamin/Cummings.
25 Sturm, K. and Oliveira, L.E. (1981) *Phys. Rev. B*, **24**, 3054.
26 Totsuji, H. (1976) *J. Physical Soc. Jpn.*, **40**, 857.
27 Tomonaga, S. (1950) *Prog. Theor. Phys.*, **5**, 544.
28 Silkin, V.M. *et al.* (2005) *Phys. Rev. B*, **72**, 15435.
29 Robinson, I.K., Bennett, P.A., and Himpsel, F.J. (2002) *Phys. Rev. Lett.*, **88**, 096104.
30 Losio, R. *et al.* (2001) *Phys. Rev. Lett.*, **86**, 4632.
31 Segovia, P. *et al.* (1999) *Nature*, **402**, 504.
32 Ahn, J.R. (2003) *Phys. Rev. Lett.*, **91**, 196403.
33 Sánchez-Portal, D., Riikonen, S., and Martin, R.M. (2004) *Phys. Rev. Lett.*, **93**, 146803.
34 Singwi, K.S., Tosi, M., Land, R.H., and Sjölander, A. (1968) *Phys. Rev.*, **176**, 589.
35 Friesen, W.I. and Bergersen, B. (1980) *J. Phys. C*, **13**, 6627.
36 Bunk, O. *et al.* (1999) *Phys. Rev. B*, **59**, 12228.
37 Yeom, H.W. *et al.* (1999) *Phys. Rev. Lett.*, **82**, 4898.

10
Excitation and Time-Evolution of Coherent Optical Phonons

Muneaki Hase, Oleg V. Misochko, and Kunie Ishioka

Coherent optical phonons are the lattice vibrations of the optical branches excited in-phase over a macroscopic spatial region by ultrashort laser pulses. Their generation and relaxation dynamics have been investigated by optical pump–probe techniques in a broad range of solid materials. When probed with a linear optical process such as reflection, the probing depth is from nanometers to micrometers determined by the absorption coefficient and the laser wavelength. By employing a nonlinear optical process such as second harmonic generation, however, it is possible to monitor coherent phonons at surfaces and interfaces exclusively. When the photoexcitation creates electron–hole plasma, coherent phonons can also serve as an ultrafast probe of nonequilibrium electron–phonon coupling. Using an optical pulse train with a designed repetition rate, coherent phonons can be excited to large amplitudes, which is expected to be a promising strategy toward nonthermal phase transitions in solids. New femtosecond beam sources such as X-ray and THz pulses offer complementary techniques to probe the ultrafast phononic and electronic dynamics under extremely nonequilibrium conditions.

10.1
Coherent Phonons in Group V Semimetals

Bismuth (Bi) and antimony (Sb) have been model systems for optical studies on coherent phonons because of their relatively low frequencies and large amplitudes in the optical reflectivity. They both have an A7 crystalline structure and sustain two Raman-active optical phonon modes shown in Figure 10.1e. The coherent phonons of A_{1g} and E_g symmetries have been detected as periodic modulations of the reflectivity (Figure 10.1a) in the order of $\Delta R/R = 10^{-6}$ to 10^{-2} depending on the excitation density. Recently, the coherent A_{1g} phonons of Bi have gained a renewed interest as the benchmark for femtosecond X-ray diffraction experiments.

Dynamics at Solid State Surface and Interfaces Vol.1: Current Developments
Edited by Uwe Bovensiepen, Hrvoje Petek, and Martin Wolf

ISBN: 978-3-527-40937-2

10.1.1
Generation and Relaxation of Different Symmetry Phonons

Early reflectivity measurements on Bi and Sb by Cheng *et al.* [1, 2] detected only the fully symmetric (A_{1g}) mode but not the other Raman-active (E_g) mode. The initial phase of the coherent A_{1g} phonons was very close to a cosine function of time (maximum amplitude at $t = 0$). The observations led the authors to propose a non-Raman generation mechanism, that is, displacive excitation of coherent phonons (DECP) [3]. A sudden shift of the vibrational potential kick-starts the coherent oscillation on the electronic excited state in DECP. The absence of the E_g mode was attributed to an exclusive coupling between the A_{1g} mode and the photoexcited electrons via Peierls distortion. A later theoretical study by Zijlstra *et al.* [4] confirmed the DECP generation of the A_{1g} phonons by calculating the potential energy surface in the electronic excited states. The minimum of the potential surface was shifted significantly along the trigonal axis (the direction of the A_{1g} motion) by electronic excitation, while it was

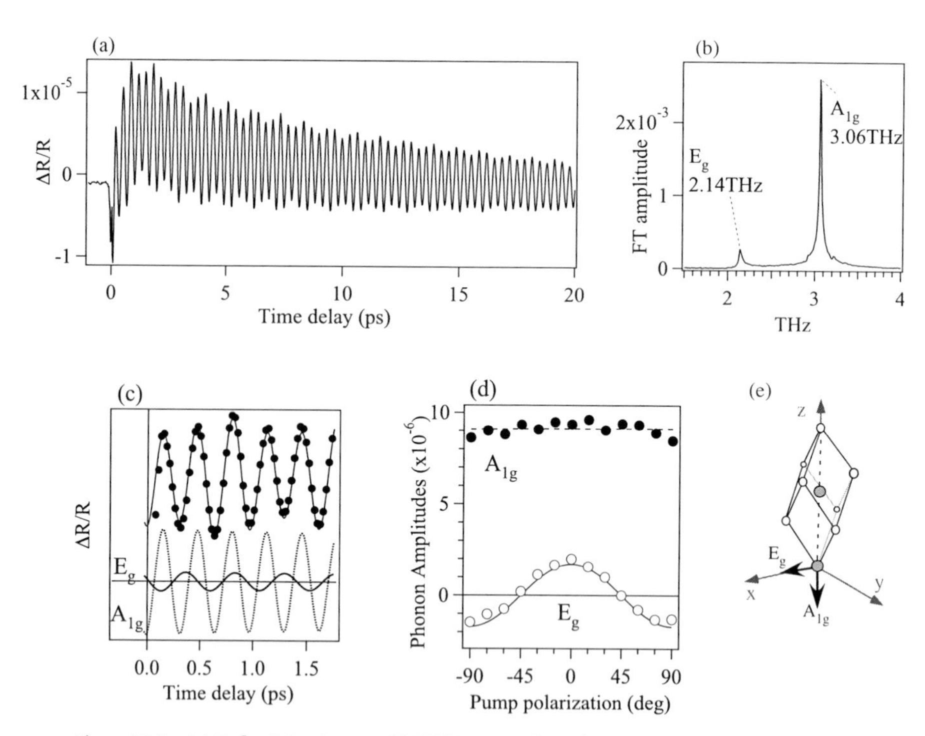

Figure 10.1 (a) Reflectivity change of Bi(0001) surface at 8 K. (b) Fourier transform (FT) of its oscillatory part. (c) Decomposition of the oscillatory part (filled circles) into the coherent A_{1g} (dotted curve) and E_g (bold solid curve) modes. (d) Pump polarization dependence of the amplitudes of the coherent A_{1g} and E_g phonons. (e) Crystalline structure of Bi and Sb indicating the directions of the A_{1g} and E_g modes.

little affected perpendicular to the trigonal axis (the E_g direction; see Figure 10.1e).

Coherent E_g phonons of Sb were observed later in the transient reflectivity by Merlin and coworkers [5]. The authors attributed the generation of both coherent A_{1g} and E_g phonons to impulsive stimulated Raman scattering (ISRS) based on the similar relative intensity of the A_{1g} and E_g modes in the time-resolved and Raman measurements. They proposed that the initial phase of the coherent oscillation should become cosine-like under resonant photoexcitation [6]. However, the significant difference in the initial phases (by 66°) between the A_{1g} and E_g modes was not fully understood.

Coherent E_g phonons of Bi were observed first in the anisotropic reflectivity change $\Delta R_x - \Delta R_y$ of a polycrystalline film at room temperature [7] and later in the reflectivity change ΔR of the (0001) surface of a single crystal below 200 K [8] (Figure 10.1a and b). As shown in Figure 10.1c, the coherent E_g phonon was a sine function of time (zero amplitude at $t = 0$). Its amplitude exhibited a $\cos 2\theta$ pump polarization dependence (Figure 10.1d), in spite of isotropic optical absorption within the x–y plane. The results, together with the similar observations for Sb [9], confirmed that the coherent E_g phonons of the group V semimetals are generated exclusively by ISRS [8, 9].

Recently, Boschetto *et al.* [10] proposed that DECP excitation of the coherent A_{1g} phonons of Bi was retarded from the Raman excitation by a few tens of femtoseconds, corresponding to the finite time required for photoexcited electrons to thermalize. This is unlikely, however, at least under moderate photoexcitation in which the thermalization takes longer than a phonon period. The "initial negative dip" in the reflectivity, which they attributed to the Raman-generated coherent phonons, is more likely to be the coherent electronic coupling of the pump and probe fields, commonly referred to as the coherent artifact [11].

Under moderate ($\ll 1\,\mathrm{mJ/cm^2}$) photoexcitation, time evolution of coherent A_{1g} and E_g phonons of Bi and Sb was respectively described by a damped harmonic oscillation:

$$\Delta R_{cp}(t) = \Delta R_{cp}(t=0)\exp(-\Gamma t)\sin(\Omega_0 t - \varphi). \tag{10.1}$$

The decay rate Γ and the frequency Ω_0 were independent of photoexcitation density, while the amplitude $\Delta R_{cp}(t=0)$ increased linearly [8, 9]. Both Γ and Ω_0 depended on temperature, in good agreement with those obtained from Raman scattering measurements (Figure 10.2a and b). The temperature dependence of Γ was perfectly reproduced by assuming the decay of an A_{1g} phonon into two isoenergetic acoustic phonons (Figure 10.2c):

$$\Gamma(T) = \Gamma_0[1 + 2n(\Omega_0/2)] = \Gamma_0\left[1 + \frac{2}{\exp(\hbar\Omega_0/2k_B T) - 1}\right]. \tag{10.2}$$

Defects in crystals can be an additional source for the dephasing of coherent phonons. This was demonstrated by introducing point defects into Bi by means of

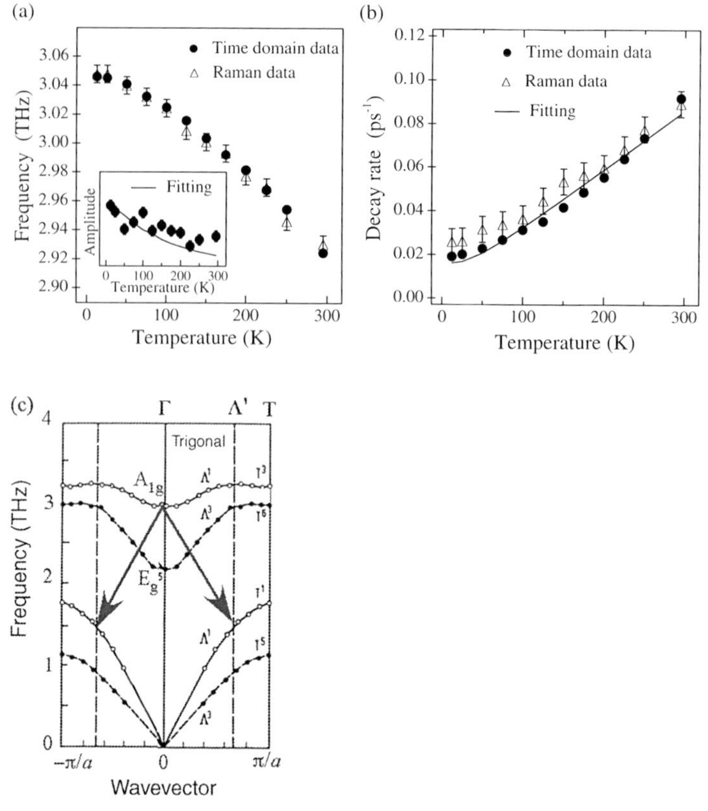

Figure 10.2 Temperature dependence of the frequency (a) and decay rate (b) of the A_{1g} phonon of Bi observed in transient reflectivity and in Raman scattering. Solid curve in (b) represents calculation after Eq. (10.2) (from Ref. [13]). (c) Schematic illustration showing the decay path of an A_{1g} phonon into two isoenergetic acoustic phonons with opposite wave vectors.

self-ion implantation [12]. The results (Figure 10.3) showed that the coherent A_{1g} phonons dephase faster with increasing ion fluence due to additional scattering by point defects.

10.1.2
Coherent Phonons in Extreme Nonequilibrium Conditions

Under intense (1–10 mJ/cm^2) photoexcitation, where the photoexcited carrier density exceeds the intrinsic (thermally excited) carrier density, the coherent A_{1g} and E_g phonons dephased faster, and their frequencies redshifted, with increasing excitation density [14–16]. In this intense excitation regime, the amplitude of the coherent phonons in the reflectivity ΔR_{cp} becomes a sublinear function of excitation density. The coherent oscillation is better described as a *chirped* damped function

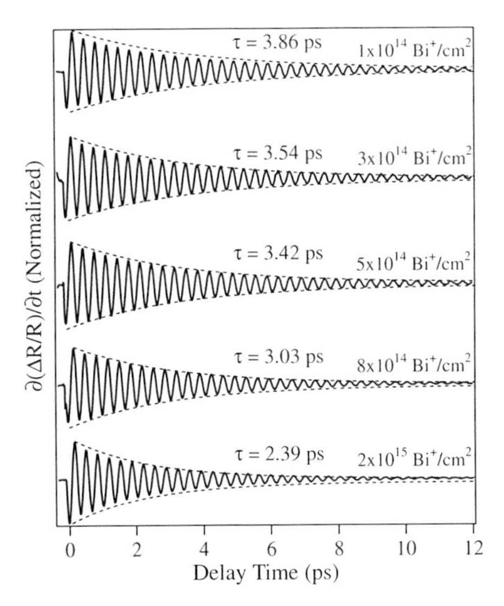

Figure 10.3 Oscillatory part of the reflectivity change of a polycrystalline Bi film after implantation of 500 keV Bi^+ ions at different fluences. Reprinted with permission from [12]. Copyright 2000, American Institute of Physics.

$$\Delta R_{cp}(t) = \Delta R_{cp}(t=0)\exp(-\Gamma t)\sin((\Omega_0 + \alpha t)t - \varphi), \quad (10.3)$$

with the frequency varying at the *phonon chirp* α with time delay. This is demonstrated in Figure 10.4 for the A_{1g} phonons of Bi under photoexcitation at 7.6 mJ/cm^2. The frequency of the A_{1g} phonons of Bi was redshifted to 2.45 THz at $t \simeq 0$, and then recovered to its equilibrium value of 2.92 THz in picosecond timescale [14]. At this excitation density, the amplitude of the coherent A_{1g} phonon reached as large as

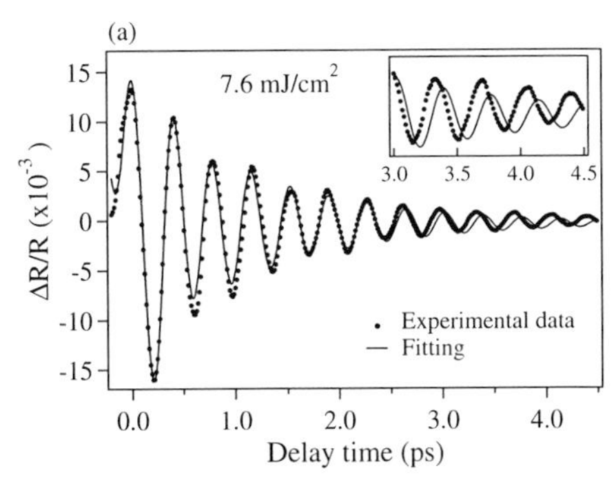

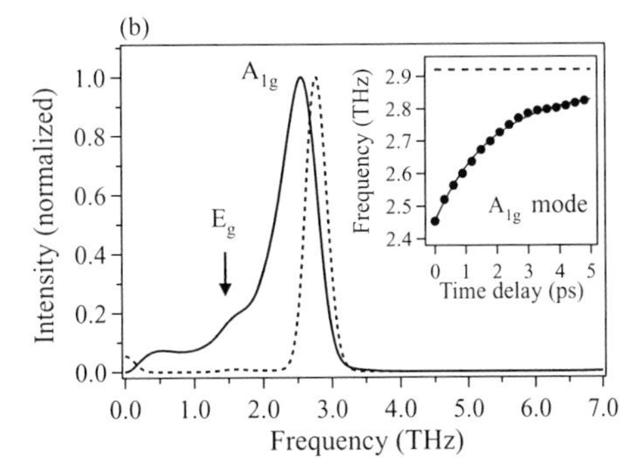

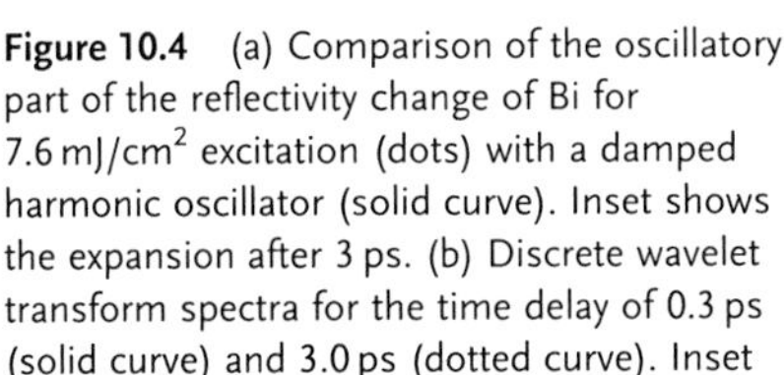

Figure 10.4 (a) Comparison of the oscillatory part of the reflectivity change of Bi for 7.6 mJ/cm^2 excitation (dots) with a damped harmonic oscillator (solid curve). Inset shows the expansion after 3 ps. (b) Discrete wavelet transform spectra for the time delay of 0.3 ps (solid curve) and 3.0 ps (dotted curve). Inset shows the A_{1g} frequency as a function of the time. The dashed line in the inset indicates the A_{1g} frequency obtained under weak excitation ($\leq$ 0.1 mJ/cm^2). Reprinted with permission from [14]. Copyright 2002 by the American Physical Society.

$\Delta R_{cp}/R \sim 10^{-2}$, and the excited carrier density was estimated to be 2% of the valence electrons in Bi.

There are two possible explanations for the transient redshift of the coherent A_{1g} phonons: lattice anharmonicity and electronic softening. On one hand, Hase *et al.* [14] found a correlation between the transient frequency and the amplitude, and concluded that the anharmonicity of the vibrational potential is responsible for the redshift. On the other hand, Murray *et al.* performed double-pulse experiments [17] combined with density functional theory (DFT) calculations [18], and revealed a significant softening of the potential due to electronic excitation. More recently, Misochko *et al.* [16] demonstrated that the redshift and the asymmetric broadening can be understood in terms of a Fano interference between two paths: one directly from the discrete (phonon) state and another mediated by a (electronic) continuum. The interference leads to an asymmetric line shape in the frequency domain given by

$$f(\varepsilon) = \frac{(\varepsilon + q)^2}{1 + \varepsilon^2}, \tag{10.4}$$

with q being the asymmetry parameter and ε the dimensionless energy. The parameter q is a function of the phonon self-energy, whose real and imaginary parts correspond to the carrier density-dependent phonon frequency and broadening, respectively. Its temperature and pump power dependences strongly suggested that the continuum includes both the lattice and electronic degrees of freedom. The importance of the phonon anharmonicity under intense photoexcitation was also demonstrated by the observation of the two-phonon combination modes (e.g., $A_{1g} - E_g$ and $A_{1g} + E_g$) in the experiment [19] and simulation [4], though the electronic softening undoubtedly plays a substantial role in the transient frequency shift.

Under even more intense photoexcitation ($>10\,\mathrm{mJ/cm^2}$) at low temperature, the coherent A_{1g} and E_g phonons of Bi and Sb exhibited a collapse–revival in their amplitudes (Figure 10.5) [19, 20]. This phenomenon has a clear threshold in the pump density, which is common to the two phonon modes but depends on temperature and the crystal (Bi or Sb). At first glance, the amplitude collapse–revival appears to be analogous to the fractional revival in nuclear wave packets in molecules. However, the pump power dependence may be an indication of a polarization, not quantum, beating between different spatial components of the coherent response within the laser spot [21].

10.1.3
Time-Resolved X-Ray Detection

The relatively low frequency of the A_{1g} mode made Bi feasible as the first target of the time-resolved X-ray diffraction (TRXRD) detection of coherent optical phonons. Sokolowski-Tinten and coworkers observed an abrupt drop of diffracted intensity at photoexcitation, followed by a periodic modulation, at A_{1g} frequency (Figure 10.6) [22, 23]. Since the diffraction efficiency in this direction is a linear function of atomic displacement, the observation confirmed the DECP nature of

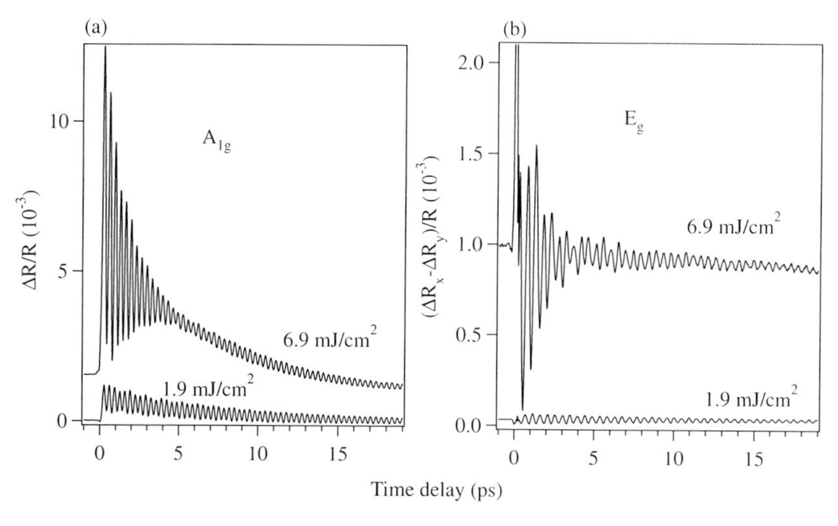

Figure 10.5 Amplitude collapse–revival of (a) the A_{1g} and (b) E_g phonons observed in the transient reflectivity change $\Delta R/R$ and its anisotropic component $(\Delta R_p - \Delta R_s)/R$, respectively, from Bi (0001) excited at 6.9 mJ/cm^2 at 8 K. Reflectivity traces for less intense excitation are also shown. Traces are offset for clarity.

the generation of the coherent A_{1g} phonons in Bi. The shift in the equilibrium position was estimated to be 2% of the interatomic separation when 1.8% of valence electrons were photoexcited [23].

The A_{1g} phonon frequency measured by X-ray diffraction was strongly redshifted (inset of Figure 10.6) to a value much lower than the optically observed one under a similar excitation density [14, 20]. The discrepancy may arise from the different

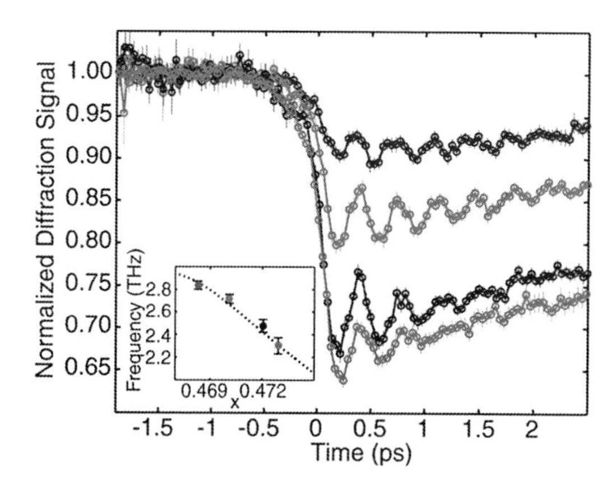

Figure 10.6 Bi (111) X-ray diffraction efficiency as a function of time delay between the optical pump pulse and X-ray probe pulse. The excitation fluence ranges from 0.7 to 2.3 mJ/cm^2. The inset displays the A_{1g} frequency as a function of the normalized atomic equilibrium position along the trigonal axis. Adapted from [23]. Reprinted with permission from AAAS.

sample thicknessi TRXRD experiments used a 50 nm thick film of Bi, because otherwise the X-ray would probe much deeper than the photoexcited layer.

Coherent phonons of Bi were also used as a standard for subpicosecond synchronization between the laser pulse and X-ray pulse from the accelerator [24]. TRXRD experiments with grazing incidence revealed the depth dependence of the X-ray signal from Bi [25].

10.1.4 Optical Coherent Control

By making use of a classical or quantum mechanical interference, one can control the temporal evolution of nuclear wave packets in molecules and crystals. An appropriately timed sequence of femtosecond light pulses can selectively excite single vibrational mode. The main objective of such coherent control is to prepare an extremely nonequilibrium vibrational state in molecules and crystals, and to drive them into nonthermal chemical reactions and phase transitions.

Bi has been one of the prime targets for the optical control experiments. Hase *et al.* [7] demonstrated the enhancement and cancellation of the amplitude of the coherent A_{1g} phonons of Bi by changing the separation Δt between the two pump pulses. Maximum enhancement and complete cancellation was achieved for $\Delta t = nT$ and $(n+1/2)T$, respectively, with an integer n and the phonon period T (Figure 10.7). When the system sustains more than one Raman-active phonon

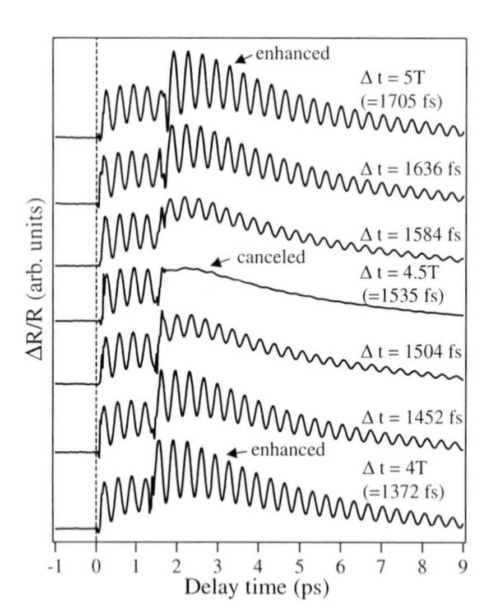

Figure 10.7 Double-pump and probe reflectivity signal of Bi, showing the enhancement and cancellation of the coherent A_{1g} phonons. The horizontal axis gives the time delay of the probe pulse with respect to the first pump pulse. Δt is the time delay between the two pump pulses in units of the A_{1g} phonon period $T = 341$ fs. Reprinted with permission from [7]. Copyright 1996, American Institute of Physics.

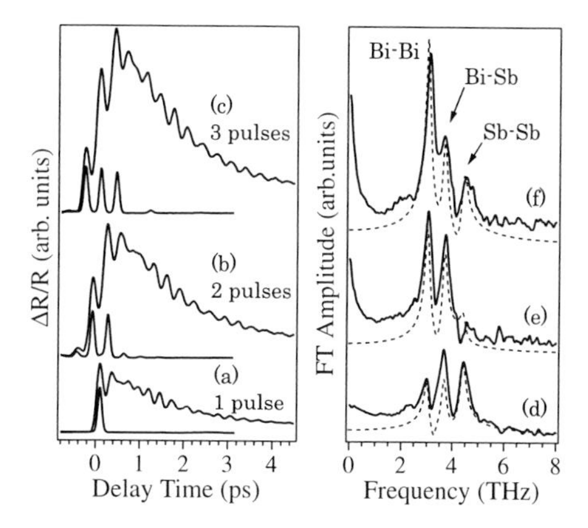

Figure 10.8 Selective enhancement of the coherent A_{1g} phonons of Bi–Bi bonds in $Bi_{0.31}Sb_{0.69}$ mixed crystal. (a–c) The reflectivity change excited with a single, double, and triple pump pulses, respectively. Autocorrelations of the pump pulses are also shown. (d–f) The corresponding FT spectra. Reprinted from [26]. Copyright 1998 by the Japan Society of Applied Physics.

mode, pulse trains can selectively enhance one of the modes and weaken the others by appropriately choosing Δt. This idea was effectively demonstrated for Bi–Bi, Sb–Sb, and Bi–Sb local vibrations in a Bi–Sb mixed crystal (Figure 10.8) [26].

Coherent control experiments were also performed on Bi with intense photoexcitation near the Lindemann stability limit [15, 17, 27, 28]. DeCamp *et al.* [15] demonstrated that when irradiated with mJ/cm^2 pulses, the amplitude of the coherent A_{1g} phonons in Bi was canceled by the second pump pulse at $\Delta t = 1.5T$, but was hardly enhanced by that at $\Delta t = T$. The results clearly indicated that the pump fluence was already in the "nonlinear regime," in which the phonon amplitude is a sublinear function of the fluence. Wu and Xu [28] performed a reflectivity measurements with 400 nm double-pump and 800 nm probe pulses, and observed a much larger A_{1g} phonon amplitude than that under a similar excitation density with 800 nm. From the difference in the temperature increase between the enhancement and the cancellation, they estimated that 4.4% of the pump pulse energy is coupled to coherent A_{1g} phonons. They also obtained an experimental relation between the reflectivity change and temperature, which would translate their long-time lattice temperature rise to be <75 K.

10.2
Ultrafast Electron–Phonon Coupling in Graphitic Materials

Graphite possesses highly anisotropic, layered crystal structure, which translates to a quasi-2D electronic structure with electronic bands dispersing linearly near the Fermi level (E_F) and forming point-like Fermi surfaces. Visible light excites vertical transitions from the valence (π) to the conduction (π^*) bands near the K point of the

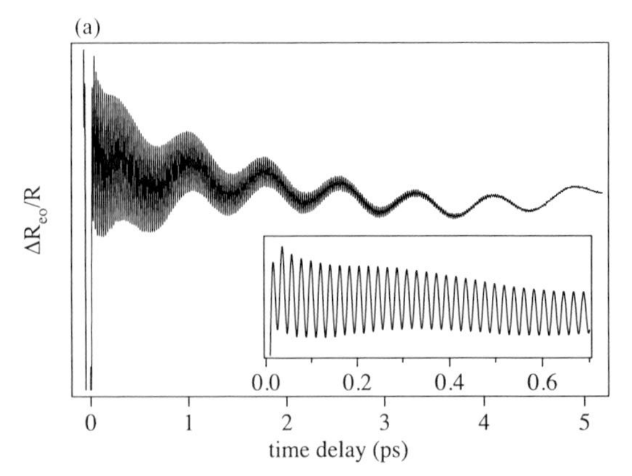

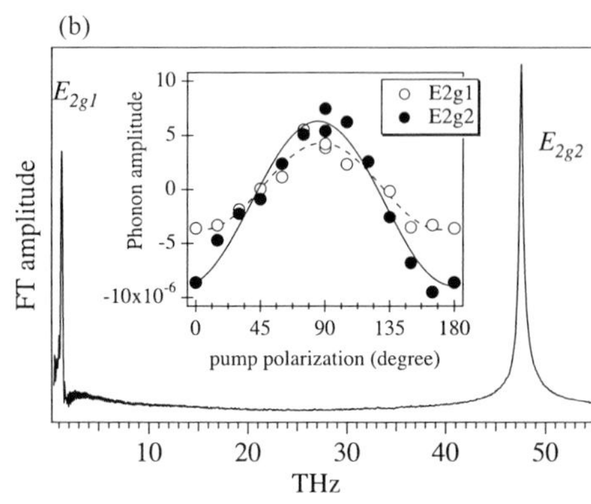

Figure 10.9 (a) Anisotropic reflectivity change of graphite, featuring periodic modulations due to the in-plane stretching (E_{2g2}, 21 fs period) and the interlayer shear (E_{2g1}, 770 fs period) modes. (b) FT spectrum of its oscillatory part. Inset in (b) shows the amplitudes of the two coherent phonons as a function of pump polarization.

Brillouin zone. Anisotropic reflectivity change of graphite (Figure 10.9a) was modulated at two disparate periods of 21 and 770 fs, assigned to the in-plane carbon stretching (Raman *G* peak or E_{2g2}) [29] and the interlayer shear (E_{2g1}) [30, 31], respectively. The amplitudes of both phonons exhibited a cos2θ dependence on the pump polarization angle θ (inset of Figure 10.9b), confirming their ISRS generation. The coherent in-plane (E_{2g2}) phonon of graphite has a frequency of 47 THz. With increasing excitation density, the frequency of the E_{2g2} phonon blueshifted, and the dephasing rate decreased (Figure 10.10a). This was opposite to the phonon softening, as one would expect from thermal expansion by laser heating [32],

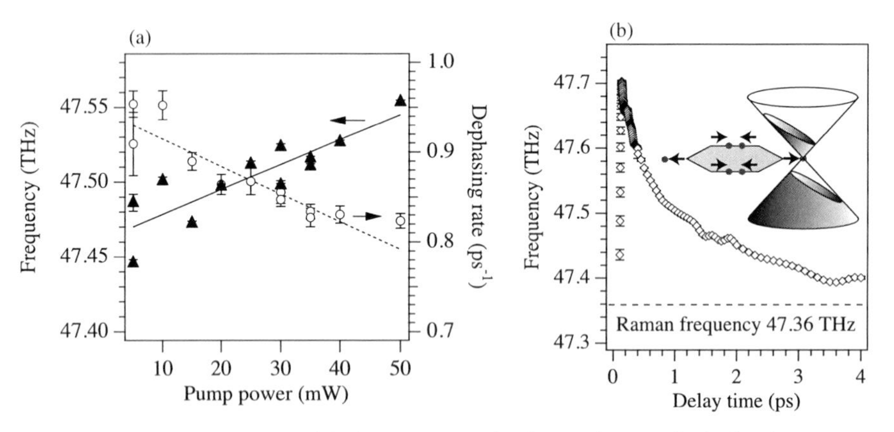

Figure 10.10 (a) Frequency and dephasing rate of the E_{2g2} mode obtained from fitting to the transient reflectivity. (b) Time evolution of the E_{2g2} phonon frequency obtained from time-windowed FT. Inset in (b) shows schematically the band structure of graphite in the presence of an E_{2g2} lattice distortion in a nonadiabatic approximation.

Fano interference [16], electronic weakening of crystal bonding [23, 33], or lattice anharmonicity [14]. A time-windowed FT analysis (Figure 10.10b) revealed that the phonon frequency blueshifted promptly at the photoexcitation. The recovery time of the frequency to its near-equilibrium value (47.36 THz) corresponded to that of carrier–lattice equilibration, suggests that the frequency shift is dominated by the excited electron dynamics. A DFT calculation demonstrated that nonadiabaticity plays an essential role in the phonon stiffening because of the low dimensional band structure of graphite [29]. For unexcited graphite, the E_{2g2} phonon at the Γ point is softened due to electronic screening (a Kohn anomaly). For photoexcited graphite, the softening is weakened because the excited electrons near the E_F cannot follow the fast in-plane nuclear motion (breakdown of Born–Oppenheimer approximation).

With a structure of rolled up graphene sheets, carbon nanotubes (CNTs) have much in common in their electronic and phonon properties with those of graphite, for example, the linearly dispersing band structure near E_F and the Raman G peak at $\sim$1600 cm^{-1}. Yet, extensive experimental and theoretical studies have found clear evidence of quantum size effects in CNTs. CNTs sustain the Raman-active radial breathing modes (RBMs), whose frequency gives a good measure of their diameter. The tangential modes, which contribute to the Raman G peak, can split into different symmetries depending on the chirality. Transient transmittance of single-walled carbon nanotubes (SWNTs) was modulated at two periods of $\sim$140 and 21 fs, corresponding to the RBM and G mode, respectively, as shown in Figure 10.11.

The most striking feature of the coherent G mode was the periodic modulation in its frequency (Figure 10.11c) [34]. Since the modulation period coincided with that of the RBM, a straightforward interpretation was given in terms of the anharmonic coupling between the two vibrational modes on the excited excitonic state. The FT spectrum of the transient transmission signal gave almost equally spaced sidebands around the relatively sharp G peak at 1588 cm^{-1}, while the Raman spectrum of the

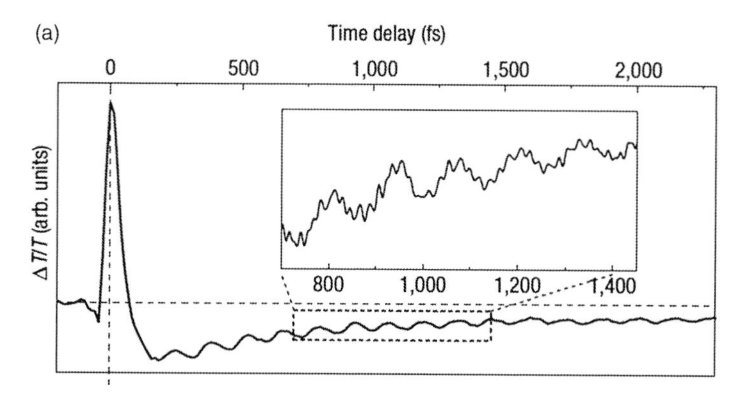

Figure 10.11 (a) Transient transmittance change $\Delta T/T$ of SWNTs excited and probed at 2.1 eV. (b) FT spectrum obtained from the time domain trace in (a). The inset represents enlargement around the G mode. (c) G mode frequency as a function of pump–probe time delay. Adapted by permission from Macmillan Publishers Ltd.: Nature Physics [34], Copyright 2006.

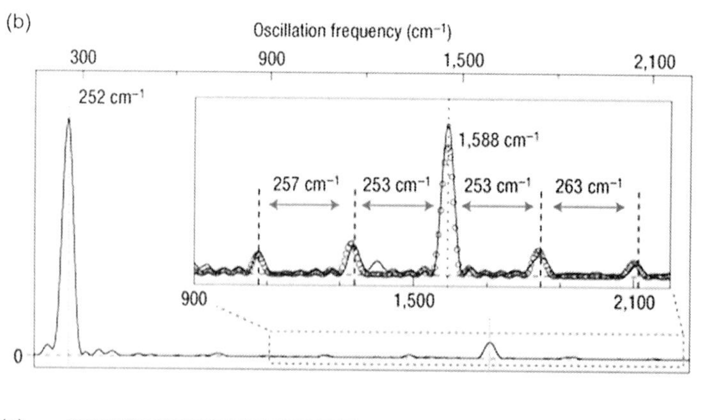

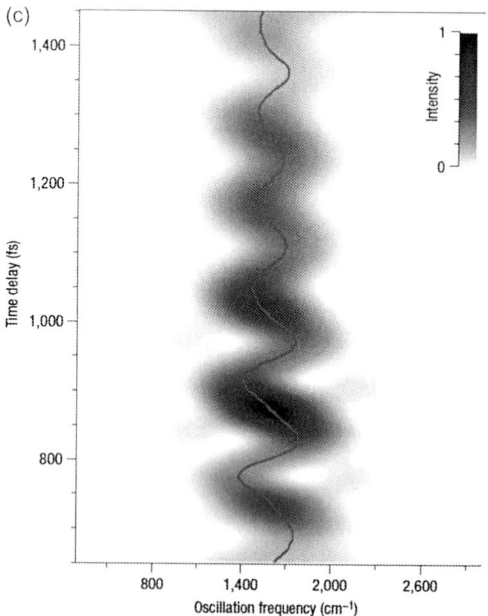

Figure 10.11 (*Continued*).

same sample consisted of much broader G and D bands. Quantum chemical modeling showed that a corrugation of the SWNT surface due to an exciton self-trapping leads to a coupling between longitudinal and radial modes.

The frequency of the coherent RBMs exhibited a clear excitation wavelength dependence (Figure 10.12) [35]. The different frequencies were attributed to SWNTs with different diameters coming to the excitonic resonance. The FT spectra of the coherent RBMs in Figure 10.12 had noticeable differences from the resonant Raman spectra, such as the different intensities and better frequency resolution.

Chirality-selective excitation of coherent RBM was demonstrated by using multiple pulse excitation technique (Figure 10.13) [36]. Here pulse trains at a THz

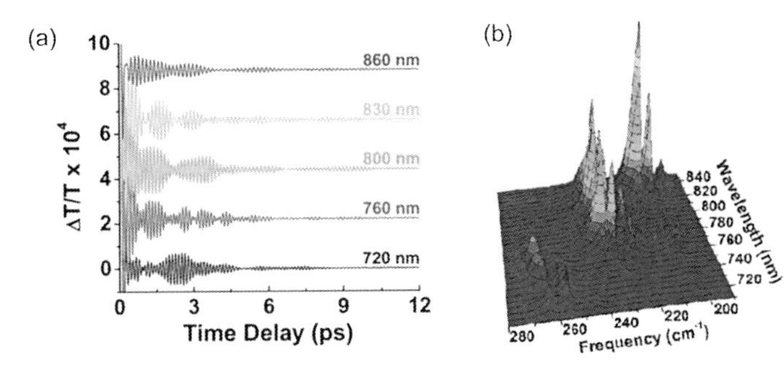

Figure 10.12 (a) Oscillatory part of the transient transmittance of SWNTs in a suspension excited and measured with 50 fs pulses at different photon energies. (b) A 3D plot of its fast FT obtained over a photon energy range of 1.746–1.459 eV (wavelength of 710–850 nm). Reprinted with permission from [35]. Copyright 2006 American Chemical Society. (Please find a color version of this figure on the color plates.)

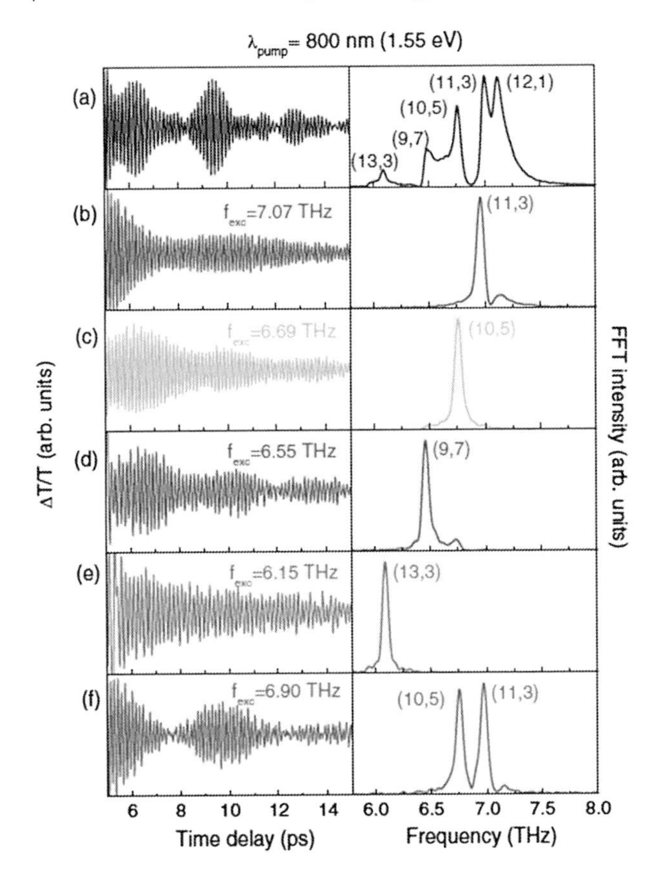

Figure 10.13 *Left*: Transmission modulations due to coherent RBM vibrations in an ensemble SWNTs in solution. *Right*: Their FT spectra together with the chirality of the assigned peaks. (a) Excited by a single pump pulse. (b–f) Excited by pulse trains at repetition rate between 7.07 and 6.15 THz. Reprinted with permission from [36]. Copyright 2009 by the American Physical Society.

repetition rate were created using a spatial light modulator (SLM) [37]. The success in the selective excitation is expected to serve as the first step toward surface functionalization via a selective cap opening, as has been predicted by a molecular dynamics simulation [38]. Such selective surface functionalization would improve the performance of CNT-based electronic devices [39] and chemical sensors [40], in which the quality of the surfaces and interfaces matters in the atomic level.

10.3
Quasiparticle Dynamics in Silicon

Coherent optical phonons of Si revealed the ultrafast formation of renormalized quasiparticles [41]. The anisotropic transient reflectivity of n-doped Si(001) featured the coherent optical phonons with a frequency of 15.3 THz (Figure 10.14a). Rotation of the sample by 45° from $\Gamma_{25'}$ to Γ_{12} configuration led to disappearance of the coherent oscillation (Figure 10.14b and d), which suggests a phase shift (π) of

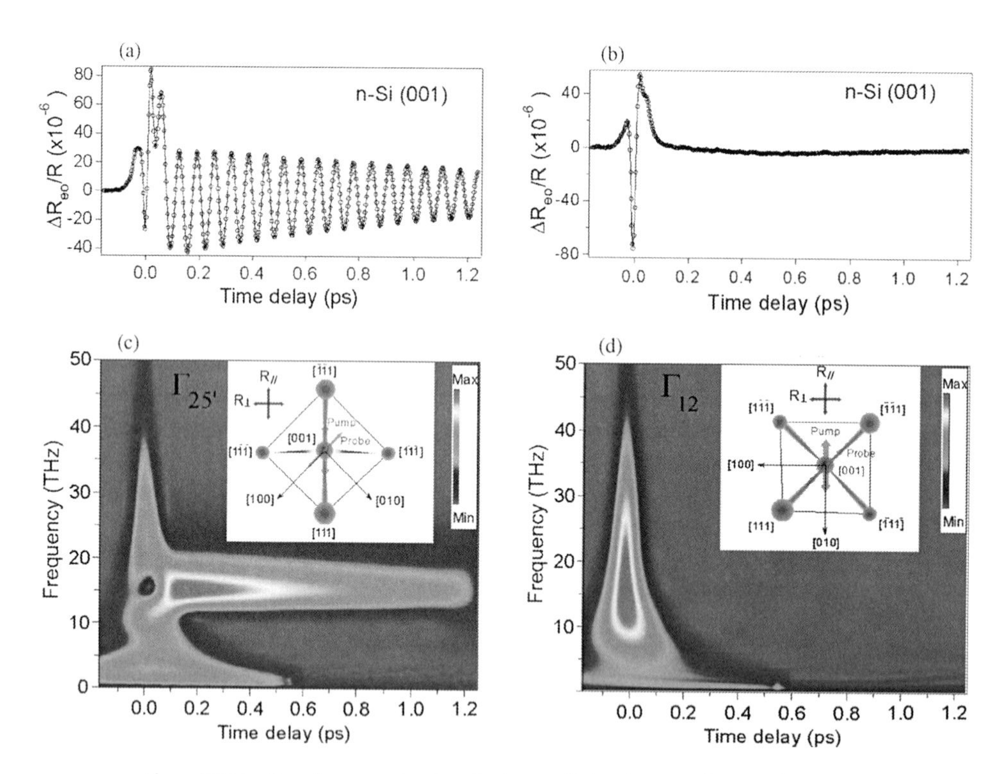

Figure 10.14 Transient anisotropic reflectivity change for Si(001) in $\Gamma_{25'}$ (a) and Γ_{12} (b) geometry. (c and d) Continuous wavelet transform of (a) and (b), respectively. Insets define the polarization of the pump and probe light relative to the crystalline axes. Adapted by permission from Macmillan Publishers Ltd.: Nature [34], Copyright 2006. (Please find a color version of this figure on the color plates.)

the coherent phonons in the $R_{\|}$ and $R_{\perp}$ components, resulting in the cancellation of the oscillation with Γ_{12} configuration (Figure 10.14d inset) [41, 42]. Similar polarization dependence was also observed for the coherent optical phonons of Ge [42] and diamond [43]. In time-dependent spectral amplitude (chronogram) in Figure 10.14c, the optical phonon was seen as the horizontal band at 15.3 THz for $t > 0$. The chronogram also revealed the broadband response at $t \simeq 0$ due to the coherent electronic coupling of the pump and probe fields via the third-order susceptibility during coherent Stokes and anti-Stokes Raman scattering.

The most intriguing aspect of the chronogram was the antiresonance at 15.3 THz slightly after $t = 0$. It revealed interference effects leading to the coherent phonon generation and subsequent "dressing" by electron–hole pairs photoexcited near the Γ point and along the Λ direction. Many-body time-dependent approach showed that the deformation potential scattering is the origin of the destructive interference [44]. These results clearly demonstrated the possibility of observing the quantum mechanical manifestations of carrier–phonon interactions in the real time [45], which until now could only be deduced from transport measurements and spectral line shape analysis.

The initial phase of the coherent optical phonons was obtained to be $\varphi = -113°$ from fitting to Eq. (10.1), when the excitation photon energy (3.05 eV) was close to the E_0' critical point [41]. The initial phase was shifted to $\varphi = -178°$ when excited far below the direct bandgap with 1.55 eV photon [46]. The results indicated that the coherent oscillation is a sine function of time for nonresonant excitation, but approaches a cosine function for near-resonant excitation. Based on these observations, Riffe and Sabbah [46] proposed a model in which the initial phase of the coherent phonons depends on the electronic relaxation rate γ:

$$\varphi_{\mathrm{RSFL}} = \tan^{-1}\left[\frac{4\pi^2\nu^2\varepsilon_1' + (\gamma-\Gamma)2\varepsilon_2}{2\pi\nu(2\varepsilon_2-\gamma\varepsilon_1')}\right] + \frac{\pi}{2}, \tag{10.5}$$

where $\varepsilon = \varepsilon_1 + i\varepsilon_2$ is the dielectric function at the laser frequency ω_0 and $\varepsilon_1' = d\varepsilon_1/d\omega$. Since it is in general difficult to determine γ from an experiment or a theory, the RSFL model may be used to estimate γ from the experimentally observed φ. In the case of Si, $\gamma = 32\,\mathrm{ps}^{-1}$, corresponding to the electronic relaxation time of 32 ± 5 fs, was obtained for excitation at 1.55 eV photon [46].

10.4
Coherent Optical Phonons in Metals

Raman-active phonons in Zn and Cd were studied by means of pump–probe reflectivity measurements [47]. The oscillation in the reflectivity change (Figure 10.15) was attributed to the coherent E_{2g} phonons. The amplitude of the coherent phonons of

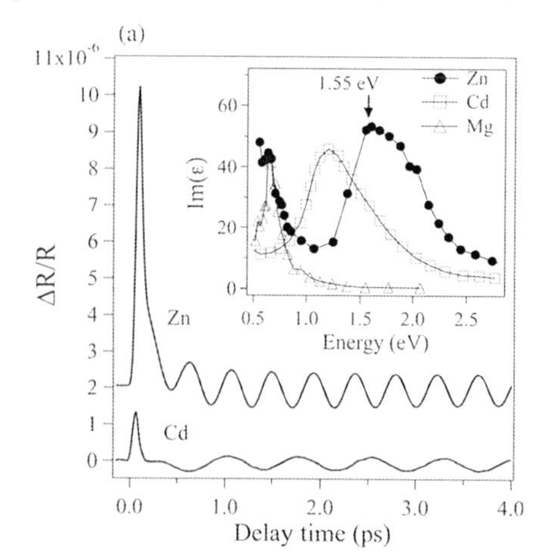

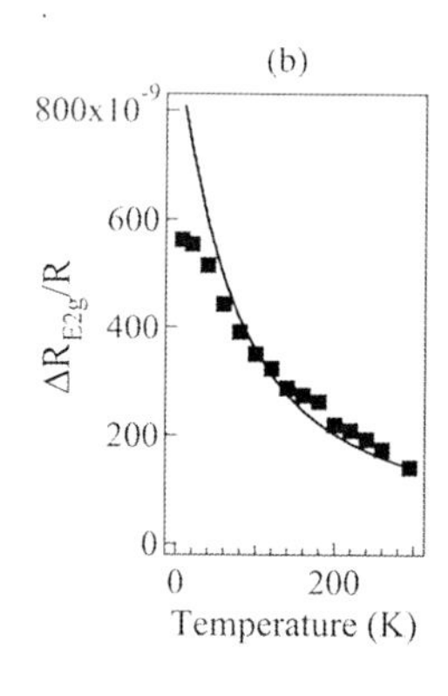

Figure 10.15 (a) Transient reflectivity change of Zn and Cd at 7 K. Inset shows the imaginary part of the dielectric function of Zn, Cd, and Mg. (b) Amplitude of the coherent E_{2g} phonon of Zn as a function of temperature. Solid curve in (b) represents the fit to Eq. (10.6). Reprinted with permission from [47]. Copyright 2005 by the American Physical Society.

Zn decreased with rising temperature, in accordance with that of the nonoscillatory reflectivity change due to the photoexcited electrons. The latter temperature dependence was explained in terms of that of the photoinduced quasiparticle density n_{p}:

$$n_{\mathrm{p}} = n(T_{\mathrm{e}}') - n(T_{\mathrm{e}}) \propto T_{\mathrm{e}}' - T_{\mathrm{e}} = \sqrt{T_{\mathrm{e}}^2 + 2U_{\mathrm{l}}/\gamma} - T_{\mathrm{e}}, \tag{10.6}$$

where $n(T_{\mathrm{e}})$ and $n(T_{\mathrm{e}}')$ are the quasiparticle densities at the electronic temperatures before photoexcitation (T_{e}) and after photoexcitation (T_{e}'), γ is the Sommerfeld constant, and U_{l} is the absorbed energy. With U_{l}/γ as a fitting parameter, the temperature dependence of the coherent phonon amplitude was well reproduced by that of n_{p} (Figure 10.15). The result indicated the resonant ISRS generation of coherent phonons in Zn.

Under intense (mJ/cm^2) photoexcitation, the coherent E_{g} phonons of Zn exhibited transient frequency shift and broadening (Figure 10.16), which was similar to that of Bi and was also explained in terms of the Fano-type resonance [48]. The Fano parameter $1/q$ depended critically on the pump fluence, which indicated that in addition to the carrier density-dependent phonon self-energy effect, the ratio between the phononic and electronic Raman scattering cross sections, both of which contribute to $1/q$, significantly varied. The latter effect presumably suggested pronounced change in the electronic structure, such as band structure collapse, or the solid-to-liquid structural phase transition.

In contrast, the coherent transverse optical (TO) phonon in polycrystalline Zr film showed a transient frequency shift in spite of very weak photoexcitation ($\leq 2\mu$ J/cm^2) [49].

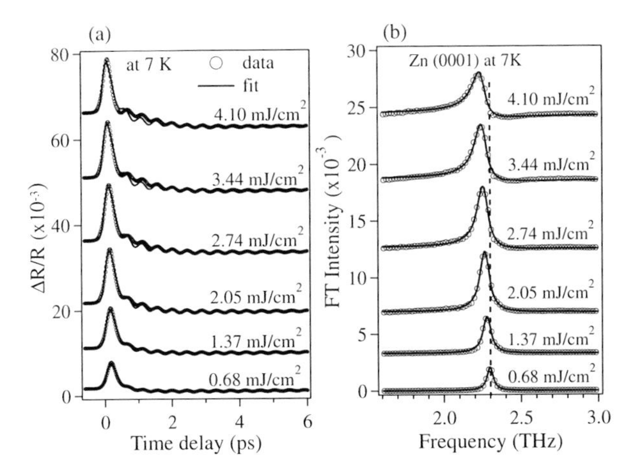

Figure 10.16 (a) Transient reflectivity changes of single-crystal Zn at 7 K for different pump fluences and (b) their FT spectra. In both panels, open circles are the experimental data. Solid curves in (a) and (b) represent fits to Eq. (10.3) with an exponential baseline and to Eq. (10.4), respectively. Reprinted with permission from [47]. Copyright 2005 by the American Physical Society.

10.5 Coherent Phonon-Polaritons in Ferroelectrics

Ferroelectric crystals undergo a phase transition with increasing temperature from the polarized ferroelectric phase to the nonpolarized paraelectric phase. In perovskite ferroelectrics such as $LiTaO_3$ and $LiNbO_3$, coherent optical phonons couple with photons to form coherent phonon-polaritons [50]. These are mixed light–polarization states that arise from the coupling of the far-infrared light to phonon resonances. At frequencies far away from the phonon resonance, the polaritons are strongly light-like and their group velocity is close to the velocity of light. At frequencies close to the resonance, the polarization have strong phonon character and the polariton dispersion is very flat, which implies that the polaritons will propagate very slowly. Temperature-driven phase transitions were monitored by observing coherent phonon-polaritons in these perovskite ferroelectrics using four-wave mixing [50] and ultrafast polarization spectroscopy [51]. The former technique enables the wave vector-selected generation of phonon-polaritons by varying the angle between the two pump beams.

Spatiotemporal control of coherent phonon-polaritons have been demonstrated using femtosecond pulse shaping technique [52]. Since the coherent phonon-polaritons propagate in the ferroelectric samples with light-like speeds, multiple excitation beams arriving at different sample locations at different times can be utilized to design the coherent matter waves in the far field (Figure 10.17).

Because of the reduced symmetry of the ferroelectric phase, $LiTaO_3$ has a large optical nonlinearity, and consequently femtosecond visible pulses can generate THz radiation by means of ISRS. Though the resulting time-dependent electrical

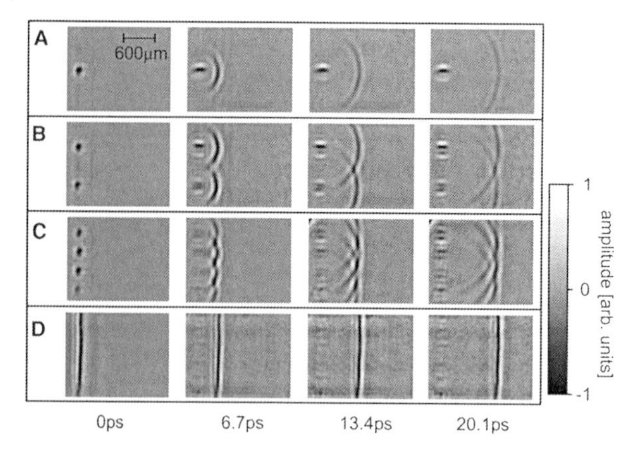

Figure 10.17 Spatiotemporal coherent control of phonon-polaritons in $LiTaO_3$. Responses to impulsive excitation with (A) one, (B) two, (C) four, and (D) nine excitation regions are shown. The time delay between successive frames is 6.7 ps. Phonon-polariton responses moving from left to right are shown. Adapted from [52], Reprinted with permission from AAAS.

polarization has been measured with electro-optic sampling and so on, the corresponding lattice motion remained undetected by optical measurements. A recent TRXRD experiment on $LiTaO_3$ [53] demonstrated the coherent modulation of the X-ray intensity at 1.5 THz, which was assigned as the phonon-polariton mode of the A_1 symmetry (Figure 10.18).

Structural phase transitions are often characterized by a "soft" optical phonon, whose frequency decreases substantially as the transition temperature is approached from above or below. In GeTe, which undergoes the phase transition at 657 K, the transverse optical (TO) phonon of the A_{1g} symmetry is such a soft phonon. Using optical pulse trains, the coherent A_{1g} phonons were excited to a large amplitude

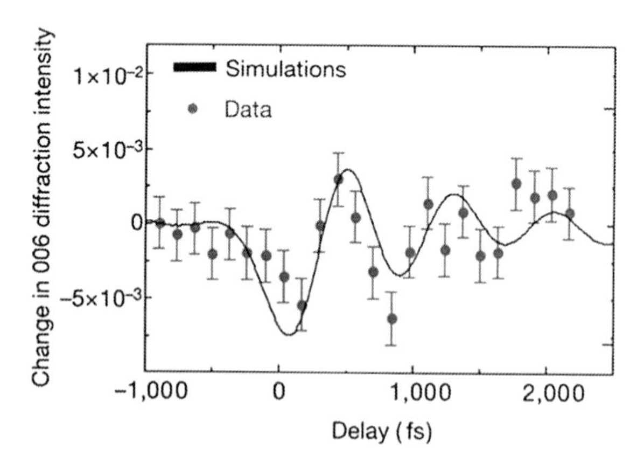

Figure 10.18 Time-dependent change in the 006 X-ray diffraction intensity of $LiTaO_3$. Error bars represent one standard deviation of the shot noise limited distribution of detected photons. Adapted by permission from Macmillan Publishers Ltd.: Nature [53], Copyright 2006.

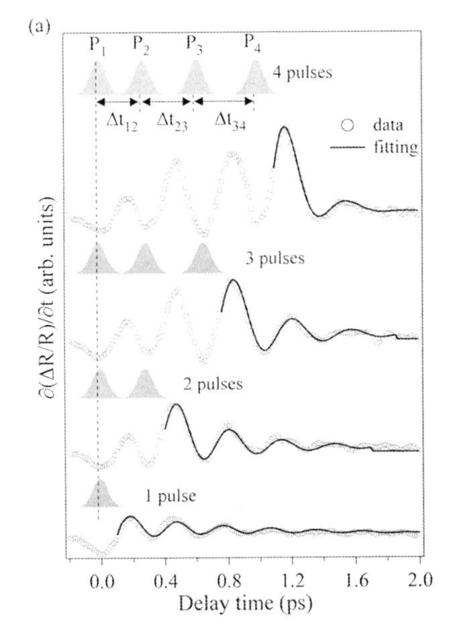

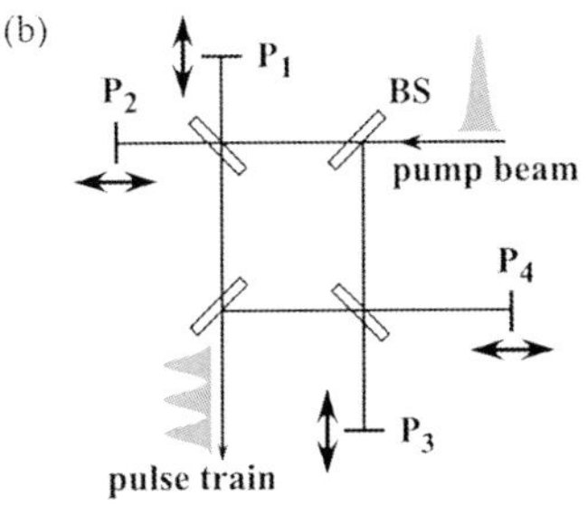

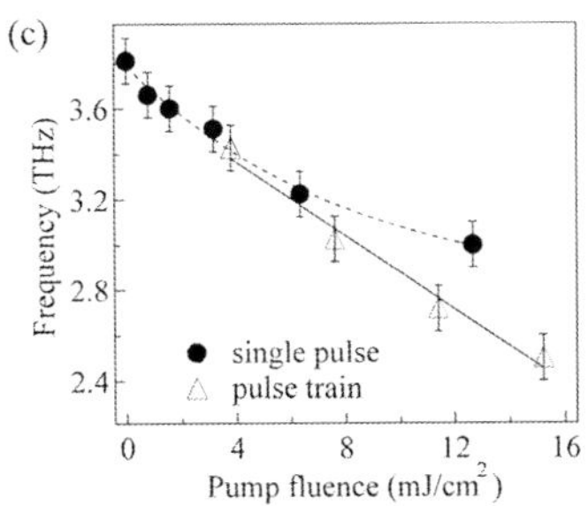

Figure 10.19 (a) Repetitive excitation of the A_{1g} mode of GeTe using a THz-rate pulse train. The intervals between the pump pulses, $\Delta t_{12}, \Delta t_{23}$, and Δt_{34}, are set to 290, 320, and 345 fs, respectively. (b) Optical layout of a twin Michelson interferometer for generation of the pulse train. (c) The pump fluence dependence of the frequency of the A_{1g} mode for excitation with a single pulse and the pulse train. Reprinted with permission from [54]. Copyright 2003, American Institute of Physics.

(Figure 10.19) [54]. Under single pump pulse excitation, the frequency of the soft mode redshifted and saturated at high pump fluences. In contrast, under a sequence of pump pulses, a large redshift was observed without saturation. In the latter situation, the coherent phonon was strongly damped, indicating that the lattice is driven to a precursor state of the phase transition.

10.6
Current Developments in Other Materials

Coherent optical phonons have also been investigated extensively in other solid materials using optical and nonoptical detection techniques.

Coherent phonons in III–V semiconductors and their heterostructures have several different generation mechanisms as well as optical detection processes, on which Ref. [55] has given a recent comprehensive review. It is characteristic of the longitudinal optical (LO) phonons in polar semiconductors to couple with plasmons and form coherent LO phonon–plasmon coupled (LOPC) modes. Ultrafast transition of an optical phonon resonance to a coupled phonon–plasmon system has been detected using time-resolved THz transmission spectroscopy [56].

Strongly correlated systems offer one of the most intriguing targets in the current solid state physics. Temperature-driven phase transitions were monitored by observing coherent phonons in Mott insulators V_2O_3 [57] and VO_2 [58], and in a low-dimensional spin system α'-NaV_2O_5 [59]. Photoinduced phase transition (PIPT) in the strongly correlated systems has attracted much attention as a potential approach to novel electronic and magnetic phases. Real-time observation of coherent phonons in these systems has been revealing the ultrafast dynamics of PIPT. For example, photodoping of holes into VO_2 induced an insulator-to-metal transition, which exhibited a delay of several tens of femtoseconds associated with the phonon connecting the two crystallographic phases [60]. Optical measurements on magnetoresistive manganite $Pr_{0.7}Ca_{0.3}MnO_3$ using sub-10 fs pulses revealed photoinduced insulator–metal phase transition, which was accompanied by an oscillation suggestive of coherent orbital wave [61]. In one-dimensional extended Peierls–Hubbard system $[Pd(chxn)_2Br]Br_2$, photoinduced formation of Mott–Hubbard state was observed, followed by a coherent oscillation of the bridging Br [62]. Organic conductors, including TTF-CA [63], K-TCNQ [64], and $(EDO\text{-}TTF)_2PF_6$ [65] are specifically under investigation in association with the photoinduced charge–state transitions. A recent time- and angle-resolved photoemission spectroscopy (trARPES) study on $TbTe_3$ [66] revealed a periodic modulation of the binding energy at different frequencies for different electronic states: 2.3 and 3.6 THz for the Te conduction band and Te band, respectively. The former oscillation was attributed to the charge density wave (CDW) coupled to long-wavelength optical phonons, and the latter to the Te phonon. The results demonstrated that a large-amplitude coherent-phonon excitation can drive melting of the CDWs.

Development of ultrashort intense laser pulses has enabled the observation of small-amplitude, high-frequency phonons in wide gap materials. Typical examples include coherent optical phonons in diamond [43], GaN [67], and TiO_2 [68]. Coherent optical phonons can couple with localized excitations such as excitons and defect centers. For example, strong exciton–phonon coupling was demonstrated for lead phthalocyanine [69] and CuI [70] as an intense enhancement of the coherent phonon amplitude at the excitonic resonances. In alkali halides [71], nuclear wave packets localized near F centers were observed as periodic modulations of the luminescence spectra.

10.7
Concluding Remarks

In this chapter, we have reviewed recent developments in the experimental researches on coherent optical phonons in selected solid materials. Group V semimetals Bi and Sb offer model systems, in which the generation mechanism of the coherent phonons depends critically on the symmetry of the phonon mode. The relaxation of the coherent phonons in these semimetals is dominated by lattice anharmonicity under weak photoexcitation and by electron–phonon coupling under intense excitation. Sub-10 fs optical pulses have made possible the real-time observation of coherent phonons of graphitic materials with frequencies up to 50 THz and of their ultrafast interactions with nonequilibrium electrons and other phonon modes. Transforming the transient reflectivity into frequency–time space reveals the coherent phonon generation and subsequent dressing of the phonon by electron–hole plasma in Si. Pulse shaping techniques have stimulated pioneering studies on optical control of coherent phonons, which aims at the mode-selective excitation of large-amplitude coherent phonons toward the goal of nonthermal phase transition. New detection schemes such as TRXRD and trARPES have contributed to attract a renewed, broad interest on ultrafast dynamics of coherent optical phonons.

References

1 Cheng, T.K., Brorson, S.D., Kazeroonian, A.S., Moodera, J.S., Dresselhaus, G., Dresselhaus, M.S., and Ippen, E.P. (1990) Impulsive excitation of coherent phonons observed in reflection in bismuth and antimony. *Appl. Phys. Lett.*, **57**, 1004–1006.

2 Cheng, T.K., Vidal, J., Zeiger, H.J., Dresselhaus, G., Dresselhaus, M.S., and Ippen, E.P. (1991) Mechanism for displacive excitation of coherent phonons in Sb, Bi, Te and Ti_2O_3. *Appl. Phys. Lett.*, **59**, 1923–1925.

3 Zeiger, H.J., Vidal, J., Cheng, T.K., Ippen, E.P., Dresselhaus, G., and Dresselhaus, M.S. (1992) Theory for displacive excitation of coherent phonons. *Phys. Rev. B*, **45** (2), 768–778.

4 Zijlstra, E.S., Tatarinova, L.L., and Garcia, M.E. (2006) Laser-induced phonon–phonon interactions in bismuth. *Phys. Rev. B*, **74** (22), 220301.

5 Garrett, G.A., Albrecht, T.F., Whitaker, J.F., and Merlin, R. (1996) Coherent THz phonons driven by light pulses and the Sb problem: What is the mechanism? *Phys. Rev. Lett.*, **77** (17), 3661–3664.

6 Stevens, T.E., Kuhl, J., and Merlin, R. (2002) Coherent phonon generation and the two stimulated Raman tensors. *Phys. Rev. B*, **65**, 144304.

7 Hase, M., Mizoguchi, K., Harima, H., Nakashima, S., Tani, M., Sakai, K., and Hangyo, M. (1996) Optical control of coherent optical phonons in bismuth films. *Appl. Phys. Lett.*, **69** (17), 2474–2476.

8 Ishioka, K., Kitajima, M., and Misochko, O.V. (2006) Temperature dependence of coherent A_{1g} and E_g phonons of bismuth. *J. Appl. Phys.*, **100**, 093501.

9 Ishioka, K., Kitajima, M., and Misochko, O.V. (2008) Coherent A_{1g} and E_g phonons of antimony. *J. Appl. Phys.*, **103**, 123505.

10 Boschetto, D., Gamaly, E.G., Rode, A.V., Luther-Davies, B., Glijer, D., Garl, T., Albert, O., Rousse, A., and Etchepare, J. (2008) Small atomic displacements recorded in bismuth by the optical reflectivity of femtosecond laser-pulse excitations. *Phys. Rev. Lett.*, **100** (2), 027404.

11 Misochko, O.V., Kitajima, M., and Ishioka, K. (2009) Comment on "small atomic displacements recorded in bismuth by the optical reflectivity of femtosecond laser-pulse excitations". *Phys. Rev. Lett.*, **102** (2), 029701.

12 Hase, M., Ishioka, K., Kitajima, M., Ushida, K., and Hishita, S. (2000) Dephasing of coherent phonons by lattice defects in bismuth films. *Appl. Phys. Lett.*, **76** (10), 1258–1260.

13 Hase, M., Mizoguchi, K., Harima, H., Nakashima, S., and Sakai, K. (1998) Dynamics of coherent phonons in bismuth generated by ultrashort laser pulses. *Phys. Rev. B*, **58** (9), 5448–5452.

14 Hase, M., Kitajima, M., Nakashima, S., and Mizoguchi, K. (2002) Dynamics of coherent anharmonic phonons in bismuth using high density photoexcitation. *Phys. Rev. Lett.*, **88** (6), 067401.

15 DeCamp, M.F., Reis, D.A., Bucksbaum, P.H., and Merlin, R. (2001) Dynamics and

coherent control of high-amplitude optical phonons in bismuth. *Phys. Rev. B*, **64**, 092301.

16 Misochko, O.V., Ishioka, K., Hase, M., and Kitajima, M. (2007) Fano interference for large-amplitude coherent phonons in bismuth. *J. Phys. Condens. Matter*, **19** (15), 156227.

17 Murray, É.D., Fritz, D.M., Wahlstrand, J.K., Fahy, S., and Reis, D.A. (2005) Effect of lattice anharmonicity on high-amplitude phonon dynamics in photoexcited bismuth. *Phys. Rev. B*, **72**, 060301(R).

18 Murray, É.D., Fahy, S., Prendergast, D., Ogitsu, T., Fritz, D.M., and Reis, D.A. (2007) Phonon dispersion relations and softening in photoexcited bismuth from first principles. *Phys. Rev. B*, **75** (18), 184301.

19 Misochko, O.V., Ishioka, K., Hase, M., and Kitajima, M. (2006) Fully symmetric and doubly degenerate coherent phonons in semimetals at low temperature and high excitation: similarities and differences. *J. Phys. Condens. Matter*, **18**, 10571.

20 Misochko, O.V., Hase, M., Ishioka, K., and Kitajima, M. (2004) Observation of an amplitude collapse and revival of chirped coherent phonons in bismuth. *Phys. Rev. Lett.*, **92** (19), 197401.

21 Diakhate, M.S., Zijlstra, E.S., and Garcia, M.E. (2009) Quantum dynamics study of the amplitude collapse and revival of coherent A_{1g} phonons in bismuth: a classical phenomenon? *Appl. Phys. A*, **96**, 5–10.

22 Sokolowski-Tinten, K., Blome, C., Blums, J., Cavalleri, A., Dietrich, C., Tarasevitch, A., Uschmann, I., Förster, E., Kammler, M., Horn von Hoegen, M., and von der Linde, D. (2003) Femtosecond X-ray measurement of coherent lattice vibrations near the Lindemann stability limit. *Nature*, **422** (6929), 287–289.

23 Fritz, D.M., Reis, D.A., Adams, B., Akre, R.A., Arthur, J., Blome, C., Bucksbaum, P.H., Cavalieri, A.L., Engemann, S., Fahy, S., Falcone, R.W., Fuoss, P.H., Gaffney, K.J., George, M.J., Hajdu, J., Hertlein, M.P., Hillyard, P.B., Horn-von Hoegen, M., Kammler, M., Kaspar, J., Kienberger, R., Krejcik, P., Lee, S.H., Lindenberg, A.M., McFarland, B., Meyer, D., Montagne, T., Murray, É.D., Nelson, A.J., Nicoul, M., Pahl, R., Rudati, J., Schlarb, H., Siddons, D.P., Sokolowski-Tinten, K., Tschentscher, Th., von der Linde, D., and Hastings, J.B. (2007) Ultrafast bond softening in bismuth: mapping a solid's interatomic potential with X-rays. *Science*, **315**, 633–636.

24 Beaud, P., Johnson, S.L., Streun, A., Abela, R., Abramsohn, D., Grolimund, D., Krasniqi, F., Schmidt, T., Schlott, V., and Ingold, G. (2007) Spatiotemporal stability of a femtosecond hard-X-ray undulator source studied by control of coherent optical phonons. *Phys. Rev. Lett.*, **99**, 174801.

25 Johnson, S.L., Beaud, P., Milne, C.J., Krasniqi, F.S., Zijlstra, E.S., Garcia, M.E., Kaiser, M., Grolimund, D., Abela, R., and Ingold, G. (2008) Nanoscale depth-resolved coherent femtosecond motion in laser-excited bismuth. *Phys. Rev. Lett.*, **100** (15), 155501.

26 Hase, M., Itano, T., Mizoguchi, K., and Nakashima, S. (1998) Selective enhancement of coherent optical phonons using THz-rate pulse train. *Jpn. J. Appl. Phys.*, **37** (3A), L281–L283.

27 Misochko, O.V., Lu, R., Hase, M., and Kitajima, M. (2007) Coherent control of the lattice dynamics of bismuth near the Lindemann stability limit. *JETP*, **104** (2), 245–253.

28 Wu, A.Q. and Xu, X. (2007) Coupling of ultrafast laser energy to coherent phonons in bismuth. *Appl. Phys. Lett.*, **90** (25), 251111.

29 Ishioka, K., Hase, M., Kitajima, M., Wirtz, L., Rubio, A., and Petek, H. (2008) Ultrafast electron–phonon decoupling in graphite. *Phys. Rev. B*, **77** (12), 121402.

30 Mishina, T., Nitta, K., and Masumoto, Y. (2000) Coherent lattice vibration of interlayer shearing mode of graphite. *Phys. Rev. B*, **62** (41), 2908–2911.

31 Ishioka, K., Hase, M., Kitajima, M., and Ushida, K. (2001) Ultrafast carrier and phonon dynamics in ion-irradiated graphite. *Appl. Phys. Lett*, **78**, 3965–3967.

32 Tan, P.-H., Deng, Y.-M., and Zhao, Q. (1998) Temperature-dependent Raman spectra and anomalous Raman

phenomenon of highly oriented pyrolytic graphite. *Phys. Rev. B*, **58** (9), 5435–5439.

33 Hunsche, S., Wienecke, K., Dekorsy, T., and Kurz, H. (1995) Impulsive softening of coherent phonons in tellurium. *Phys. Rev. Lett.*, **75** (9), 1815–1818.

34 Gambetta, A., Manzoni, C., Menna, E., Meneghetti, M., Cerullo, G., Lanzani, G., Tretiak, S., Piryatinski, A., Saxena, A., Martin, R.L., and Bishop, A.R. (2006) Real-time observation of nonlinear coherent phonon dynamics in single-wall carbon nanotubes. *Nat. Phys.*, **2**, 515–520.

35 Lim, Y.-S., Yee, K.-J., Kim, J.-H., Haroz, E.H., Shaver, J., Kono, J., Doorn, S.K., Hauge, R.H., and Smalley, R.E. (2006) Coherent lattice vibrations in single-walled carbon nanotubes. *Nano Lett.*, **6** (12), 2696–2700.

36 Kim, J.-H., Han, K.-J., Kim, N.-J., Yee, K.-J., Lim, Y.-S., Sanders, G.D., Stanton, C.J., Booshehri, L.G., Hařoz, E.H., and Kono, J. (2009) Chirality-selective excitation of coherent phonons in carbon nanotubes by femtosecond optical pulses. *Phys. Rev. Lett.*, **102** (3), 037402.

37 Weiner, A.M. and Leaird, D.E. (1990) Generation of terahertz-rate trains of femtosecond pulses by phase-only filtering. *Opt. Lett.*, **15** (1), 51–53.

38 Dumitrica, T., Garcia, M.E., Jeschke, H.O., and Yakobson, B.I. (2006) Breathing coherent phonons and caps fragmentation in carbon nanotubes following ultrafast laser pulses. *Phys. Rev. B*, **74** (19), 193406.

39 Anantram, M.P. and Léonard, F. (2006) Physics of carbon nanotube electronic devices. *Rep. Prog. Phys.*, **69** (3), 507–561.

40 Modi, A., Koratkar, N., Lass, E., Wei, B.Q., and Ajayan, P.M. (2003) Miniaturized gas ionization sensors using carbon nanotubes. *Nature*, **424** (6945), 171–174.

41 Hase, M., Kitajima, M., Constantinescu, A.M., and Petek, H. (2003) The birth of a quasiparticle in silicon observed in time–frequency space. *Nature*, **426**, 51–54.

42 Pfeifer, T., Kütt, W., Kurz, H., and Scholz, R. (1992) Generation and detection of coherent optical phonons in germanium. *Phys. Rev. Lett.*, **69** (22), 3248–3251.

43 Ishioka, K., Hase, M., Kitajima, M., and Petek, H. (2006) Coherent optical phonons in diamond. *Appl. Phys. Lett.*, **89**, 231916.

44 Lee, J.D., Inoue, J., and Hase, M. (2006) Ultrafast Fano resonance between optical phonons and electron–hole pairs at the onset of quasiparticle generation in a semiconductor. *Phys. Rev. Lett.*, **97** (15), 157405.

45 Huber, R., Tauser, T., Brodschelm, A., Bichler, M., Abstreiter, G., and Leitenstorfer, A. (2001) How many-particle interactions develop after ultrafast excitation of an electron–hole plasma. *Nature*, **414** (6861), 286.

46 Riffe, D.M. and Sabbah, A.J. (2007) Coherent excitation of the optic phonon in Si: transiently stimulated Raman scattering with a finite-lifetime electronic excitation. *Phys. Rev. B*, **76** (8), 085207.

47 Hase, M., Ishioka, K., Demsar, J., Ushida, K., and Kitajima, M. (2005) Ultrafast dynamics of coherent optical phonons and nonequilibrium electrons in transition metals. *Phys. Rev. B*, **71** (18), 184301.

48 Hase, M., Demsar, J., and Kitajima, M. (2006) Photoinduced Fano resonance of coherent phonons in zinc. *Phys. Rev. B*, **74** (21), 212301.

49 Kruglyak, V.V., Hicken, R.J., Srivastava, G.P., Ali, M., Hickey, B.J., Pym, A.T.G., and Tanner, B.K. (2007) Optical excitation of a coherent transverse optical phonon in a polycrystalline Zr metal film. *Phys. Rev. B*, **76** (1), 012301.

50 Bakker, H., Hunsche, S., and Kurz, H. (1998) Coherent phonon polaritons as probes of anharmonic phonons in ferroelectrics. *Rev. Mod. Phys.*, **70** (2), 523–536.

51 Kohmoto, T., Tada, K., Moriyasu, T., and Fukuda, Y. (2006) Observation of coherent phonons in strontium titanate: structural phase transition and ultrafast dynamics of the soft modes. *Phys. Rev. B*, **74** (6), 064303.

52 Feurer, T., Vaughan, J.C., and Nelson, K.A. (2003) Spatiotemporal coherent control of lattice vibrational waves. *Science*, **299** (5605), 374–377.

53 Cavalleri, A., Wall, S., Simpson, C., Statz, E., Ward, D.W., Nelson, K.A., Rini, M., and Schoenlein, R.W. (2006) Tracking the

motion of charges in a terahertz light field by femtosecond X-ray diffraction. *Nature*, **442** (7103), 664–666.

54 Hase, M., Kitajima, M., Nakashima, S., and Mizoguchi, K. (2003) Forcibly driven coherent soft phonons in GeTe with intense THz-rate pump fields. *Appl. Phys. Lett.*, **83**, 4921–4923.

55 Först, M. and Dekorsy, T. (2007) Coherent phonons in bulk and low-dimensional semiconductors, in *Coherent Vibrational Dynamics* (eds S. De Silvestri, G. Cerullo, and G. Lanzani), Practical Spectroscopy, CRC, Boca Raton, pp. 129–172.

56 Huber, R., Kübler, C., Tübel, S., Leitenstorfer, A., Vu, Q.T., Huang, H., Köhler, F., and Amann, M.-C. (2005) Femtosecond formation of coupled phonon–plasmon modes in InP: ultrabroadband THz experiment and quantum kinetic theory. *Phys. Rev. Lett.*, **94** (02), 027401.

57 Misochko, O.V., Tani, M., Sakai, K., Kisoda, K., Nakashima, S., Andreev, V.N., and Chudnovsky, F.A. (1998) Optical study of the Mott transition in V_2O_3: comparison of time-and frequency-domain results. *Phys. Rev. B*, **58** (19), 12789–12794.

58 Kim, H.T., Lee, Y.W., Kim, B.J., Chae, B.G., Yun, S.J., Kang, K.Y., Han, K.J., Yee, K.J., and Lim, Y.S. (2006) Monoclinic and correlated metal phase in VO_2 as evidence of the Mott transition: coherent phonon analysis. *Phys. Rev. Lett.*, **97** (26), 266401.

59 Kamioka, H., Saito, S., Isobe, M., Ueda, Y., and Suemoto, T. (2002) Coherent magnetic oscillation in the spin ladder system α'-NaV_2O_5. *Phys. Rev. Lett.*, **88** (12), 127201.

60 Cavalleri, A., Dekorsy, Th., Chong, H.H.W., Kieffer, J.C., and Schoenlein, F R.W. (2004) Evidence for a structurally-driven insulator-to-metal transition in VO_2: a view from the ultrafast timescale. *Phys. Rev. B*, **70** (16), 161102.

61 Polli, D., Rini, M., Wall, S., Schoenlein, R.W., Tomioka, Y., Tokura, Y., Cerullo, G., and Cavalleri, A. (2007) Coherent orbital waves in the photo-induced insulator–metal dynamics of a magnetoresistive manganite. *Nat. Mater.*, **6**, 643–647.

62 Matsuzaki, H., Yamashita, M., and Okamoto, H. (2006) Ultrafast photoconversion from charge density wave state to Mott–Hubbard state in one-dimensional extended Peierls–Hubbard system of Br-bridged Pd compound. *J. Physical Soc. Japan*, **75** (12), 123701.

63 Iwai, S., Tanaka, S., Fujinuma, K., Kishida, H., Okamoto, H., and Tokura, Y. (2002) Ultrafast optical switching from an ionic to a neutral state in tetrathiafulvalene-*p*-chloranil (TTF-CA) observed in femtosecond reflection spectroscopy. *Phys. Rev. Lett.*, **88** (5), 057402.

64 Okamoto, H., Ikegami, K., Wakabayashi, T., Ishige, Y., Togo, J., Kishida, H., and Matsuzaki, H. (2006) Ultrafast photoinduced melting of a spin-Peierls phase in an organic charge-transfer compound, K-tetracyanoquinodimethane. *Phys. Rev. Lett.*, **96** (3), 037405.

65 Chollet, M., Guerin, L., Uchida, N., Fukaya, S., Shimoda, H., Ishikawa, T., Matsuda, K., Hasegawa, T., Ota, A., Yamochi, H., Saito, G., Tazaki, R., Adachi, S., and Koshihara, S. (2005) Gigantic photoresponse in 1/4-filled-band organic salt $(EDO\text{-}TTF)_2PF_6$. *Science*, **307** (5706), 86–89.

66 Schmitt, F., Kirchmann, P.S., Bovensiepen, U., Moore, R.G., Rettig, L., Krenz, M., Chu, J.H., Ru, N., Perfetti, L., Lu, D.H., Wolf, M., Fisher, I.R., and Shen, Z.X. (2008) Transient electronic structure and melting of a charge density wave in $TbTe_3$. *Science*, **321** (5896), 1649–1652.

67 Yee, K.J., Lee, K.G., Oh, E., and Kim, D.S. (2002) Coherent optical phonon oscillations in bulk GaN excited by far below the band gap photons. *Phys. Rev. Lett.*, **88** (10), 105501.

68 Fujiyoshi, S., Ishibashi, T., and Onishi, H. (2005) Fourth-order Raman spectroscopy of wide-band gap materials. *J. Phys. Chem. B*, **109** (18), 8557–8561.

69 Mizoguchi, K., Fujita, S., and Nakayama, M. (2004) Resonance effects on coherent phonon generation in lead phthalocyanine crystalline films. *Appl. Phys. A*, **78** (4), 461–464.

70 Kojima, O., Mizoguchi, K., and Nakayama, M. (2005) Intense coherent longitudinal optical phonons in CuI thin films under exciton-excitation conditions. *J. Lumin.*, **112** (1–4), 80–83.

71 Koyama, T., Takahashi, Y., Nakajima, M., and Suemoto, T. (2007) Nuclear wave-packet dynamics on nearly degenerate two adiabatic potential energy surfaces in the excited state of KI F centers. *Phys. Rev. B*, **76**, 115122.

11
Photoinduced Coherent Nuclear Motion at Surfaces: Alkali Overlayers on Metals

Yoshiyasu Matsumoto and Kazuya Watanabe

11.1
Introduction

Electron–phonon coupling is ubiquitous and is at the heart of many phenomena in physics and chemistry of condensed matter. Adsorbates on metal surfaces are no exception. Electron–phonon coupling plays a central role in various nuclear dynamics at metal surfaces: vibrational relaxation, diffusion, chemical reaction, and so on. Because of the coupling, transient changes in the distribution of electron density at metal surfaces induce adsorbate nuclear motions. Coupling also plays an important role in the reverse process: adsorbate nuclear motions are damped via electron–hole pair excitations in addition to phonon emissions into the bulk.

One of the simplest processes in the study of the electron–phonon coupling at surfaces is adsorbate vibrational relaxation. Information of vibrational relaxation has been deduced from the analysis of a vibrational spectral line observed with frequency domain methods: high-resolution electron energy loss spectroscopy (HREELS), infrared reflection absorption spectroscopy (IRAS), and helium atom scattering (HAS).

Thanks to the advent of ultrafast laser technology, time domain spectroscopy is now also a powerful means to investigate vibrational dynamics at metal surfaces. When electrons near a metal surface is excited by a femtosecond laser pump pulse with a duration extremely shorter than that of a vibrational mode of adsorbates, this ultrafast electronic excitation leads to in-phase collective vibrational motions of adsorbates, coherent surface phonons, through the electron–phonon coupling. Then, the evolution of coherent vibrations is monitored by a probe pulse as a function of pump–probe delay. Because transient electron temperatures achieved by the pump pulse excitation can be extremely higher than thermal desorption temperatures of adsorbates, the time domain pump–probe spectroscopy allows to explore the electron–phonon coupling under high electronic excitation conditions. Thus, the photoinduced coherent nuclear motions provide an interesting opportunity for studying the electron–phonon coupling and its effects on vibrational dynamics of

Dynamics at Solid State Surface and Interfaces Vol.1: Current Developments
Edited by Uwe Bovensiepen, Hrvoje Petek, and Martin Wolf

ISBN: 978-3-527-40937-2

adsorbed systems under highly nonequilibrium conditions that cannot be accessed by stationary thermal excitations.

Coherent bulk phonons, created by the impulsive excitation of a solid, and their dephasing dynamics have been investigated by measuring transient changes in reflectivity of the solid surface [1]. However, this method is not sensitive enough to detect coherent surface phonons because reflection signals come from the entire probe pulse penetration depth in the solid. Thus, it is necessary to employ a surface-sensitive technique for studying vibrational dynamics at surfaces. One of the surface-sensitive methods is second harmonic (SH) generation. Because SH generation from the bulk with centrosymmetry is forbidden within the dipole approximation, SH signals are contributed only from the surface, where the centrosymmetry is inherently broken. SH intensity is modulated by coherent nuclear motions at the surface because the surface electronic structure, and hence the second-order nonlinear susceptibility responsible for SH generation, is affected by the nuclear motions [2]. Consequently, time-resolved second harmonic generation (TRSHG) spectroscopy is very versatile to monitor the coherent nuclear motions at surfaces [3, 4].

Overlayers of alkali metal atoms on metal surfaces are typical model systems of metals on metals. Extensive experimental and theoretical studies in the past allow us to access detailed information on the adsorption geometries [5], the vibrational [6, 7] and electronic structures [8], and the electronic excitation [9] of alkali metal overlayers on metal surfaces. Moreover, it has been known that alkali overlayers enhance the conversion efficiency of SH generation by a few orders of magnitude in comparison with clean metal surfaces [10]. Therefore, TRSHG is best suited for studying surface vibrational dynamics at alkali metal overlayers on metal surfaces. In this section, we describe the recent progress in understanding the dynamics of coherent surface phonons at alkali metal-covered metal surfaces [11–16]. Throughout this section, we use two definitions of alkali atom coverage: θ and θ'. Coverages denoted by θ are defined as the ratio of the number of alkali atoms to that of metal surface atoms per unit area; coverages denoted by θ' are defined as $\theta' = \theta/\theta_{sat}$, where θ_{sat} is the saturation coverage. If it is useful to represent coverage in both ways, the value of θ' is given in parentheses.

11.2
Impulsive Excitation

When a force $F(t)$ with duration shorter than the oscillation period of a surface phonon mode is exerted to a surface covered with alkali adatoms, the adatoms and surface atoms start oscillating coherently. It is convenient to apply the density matrix method [17] to a statistical ensemble of vibrating adatoms that dissipates the energy into a bath system. With this method the population of a vibrational state $|\psi_n\rangle$ is represented by a diagonal term $\hat{\varrho}_{nn} = \langle\psi_n|\hat{\varrho}|\psi_n\rangle$ and the coherence between the states $|\psi_n\rangle$ and $|\psi_m\rangle (n \neq m)$ is represented by an off-diagonal term $\hat{\varrho}_{nm}$, where $\hat{\varrho}$ is a density matrix operator. The expectation value of nuclear displacement along a normal coordinate, $Q(t) = \langle \hat{Q}(t) \rangle$, can be expressed by the created coherence:

$$Q(t) = Tr\{\hat{Q}\hat{\varrho}(t)\} = \sum_{n,m} \hat{\varrho}_{nm}(t)\hat{Q}_{mn}. \tag{11.1}$$

Here, $\hat{Q}_{mn}$ is the vibrational matrix element between the mth and nth vibrational states, $\langle \psi_m | \hat{Q} | \psi_n \rangle$.

The equation of motion for the displacement of coherent vibration given by Eq. (11.1) can be described phenomenologically as

$$\frac{d^2 Q}{dt^2} + 2\beta \frac{dQ}{dt} + \Omega_0^2 Q = \frac{F(t)}{\mu}, \tag{11.2}$$

where $\Omega_0/2\pi$ is the frequency of an undamped oscillator, μ is the effective mass of the coherent oscillator, and β is the damping rate of the oscillator. Damping of the coherent vibration is caused by population decay and pure dephasing as a result of interactions with a bath system, that is, the substrate on which atoms are adsorbed. The formal solution of Eq. (11.2) can be written as

$$Q(t) = \frac{1}{\mu\Omega_1} \int_{-\infty}^{t} dt' F(t')\, e^{-\beta(t-t')} \sin[\Omega_1(t-t')], \tag{11.3}$$

where $\Omega_1 = \sqrt{\Omega_0^2 - \beta^2}$.

When the force is turned off, the system shows free induction decay with a damping rate β:

$$Q(t) \propto \sin(\Omega_1 t - \phi)\, e^{-\beta t}, \tag{11.4}$$

where ϕ is the initial phase. In general, the initial phase is described as [18]

$$\tan\phi = \frac{\mathrm{Im}[i\tilde{F}(-\Omega_1 - i\beta)]}{\mathrm{Re}[i\tilde{F}(-\Omega_1 - i\beta)]}. \tag{11.5}$$

Here, $\tilde{F}(\Omega)$ is the Fourier transform of $F(t)$. If the force is impulsive, that is, $F(t) \propto \delta(t)$, the initial phase is zero: the nuclear oscillation is described as $Q(t) \propto \exp(-\beta t)\sin\Omega_1 t$. If the force has a finite duration, the initial phase deviates from zero, depending on the temporal profile of $F(t)$. Therefore, the initial phase provides useful information about $F(t)$. We discuss this point in more detail in Section 11.5.4.

One of the important issues in the study of coherent surface phonons at metal surfaces is to clarify the origin of the force responsible for the coherent excitation. When the bandwidth of a pump laser pulse is larger than a phonon frequency of adsorbate, photons in the pump pulse can couple to adsorbate vibration via virtual electronic excitations: impulsive stimulated Raman scattering. According to Eq. (11.5), the initial phase of coherent phonons is zero. In opaque systems such as metals, electron–hole pairs are created at and close to the surface by absorption of incident photons. Transient changes in the electron density distribution around adsorbates induced by the real electronic transitions also provide a force to displace the adsorbates. If a duration of force can be shorter than the oscillating period of a vibrational mode of adsorbates, the transient electronic excitations also initiate coherent surface phonons. In contrast to the stimulated Raman scattering, the initial phase is not zero because of a finite duration of the force. Because absorption cross sections are usually much larger than Raman scattering cross sections on metal surfaces, real electronic excitations with ultrashort laser pulses play a dominant role in creating coherent phonons at metal surfaces; hence, it is important to specify which electronic transitions really induce adsorbate nuclear motions.

11.3 Alkali Metal Overlayers

11.3.1 Adsorbate Structures

Geometric structures of alkali metal overlayers on metal surfaces have been reviewed in detail by Diehl and McGrath [5]. Here, we briefly summarize them, particularly those on fcc(111) surfaces, for the sake of discussions in the following sections. Surface structures of alkali adatoms are determined by a balance between adsorbate–substrate and interadsorbate interactions. If the former interaction is stronger than the latter, alkali atoms are registered on specific adsorption sites or even intermix with substrate atoms. For example, Li often leads to massive substrate reconstructions particularly on surfaces with open structures. In contrast, heavier alkali atoms usually adsorb on metal surfaces to form overlayers without massive reconstruction. On fcc(111) surfaces, light alkali atoms tend to adsorb at hollow sites, whereas heavier atoms adsorb on atop sites. At low coverages, alkali atoms adsorb with a well-defined nearest neighbor distance, but do not have long-range orders, if the temperature is low enough. As coverage increases, alkali adsorbates form almost exclusively hexagonal or quasihexagonal structures. The hexagonal structures are commonly observed on fcc(111) metal surfaces and even on surfaces with square or rectangular symmetry such as fcc(001). In most alkali metal overlayers, the distance between nearest alkali atoms at the saturation coverage is smaller than that of the bulk alkali metal. The compression is due to strong intralayer metallic bonding.

More specifically, we describe the geometric structures of alkali-covered metal surfaces of interest in this section: Na, K, and Cs either on Pt(111) or on Cu(111) surfaces. For sodium overlayers on Cu(111), Na atoms locally form a $p(3 \times 3)$ hexagonal phase at $\theta \sim 0.11$ (0.25). As coverage increases, a hexagonal $(3/2 \times 3/2)$ structure is formed at the saturation coverage via a $p(2 \times 2)$ structure. Figure 11.1 shows the $(3/2 \times 3/2)$ structure: it has a (3×3) symmetry with four atoms per unit cell. There are two adsorption sites in this structure: threefold hollow and distorted hollow sites.

Potassium adatoms also form hexagonal structures on Cu(111), but the interadsorbate interaction plays a more dominant role in determining the overlayer structures in comparison with Na on Cu(111). Although a hexagonal $p(2 \times 2)$

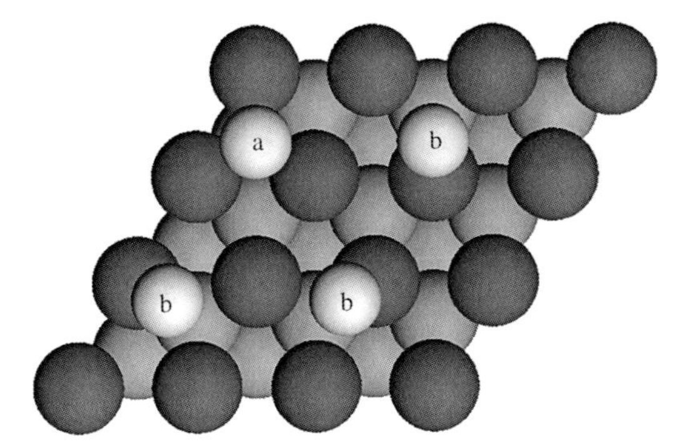

Figure 11.1 Adsorption structure of Na on Cu (111) at the saturation coverage, $(3/2 \times 3/2)$-Na. Na atoms are depicted with white; Cu atoms in the top layer of Cu substrate with dark gray; Cu atoms below the top layer with gray. The unit cell has four Na atoms: Atom a is Na on the threefold hollow site; atoms b are on the distorted hollow sites.

commensurate structure is formed at $\theta = 0.25$ (0.63), potassium overlayers are incommensurate with the substrate lattice in other coverages. The K–K spacing of nearest neighbors decreases as coverage increases, while the overlayers at low temperatures keep the orientationally ordered hexagonal structure [19]. Thus, no specific adsorption sites exist except for atop sites for the $p(2 \times 2)$ overlayer. At the saturation coverage, the K–K spacing is 4.0 Å: 15% reduction from the bulk potassium. Cesium adsorption on Cu(111) is similar to potassium. It forms the most stable structure of $p(2 \times 2)$ and saturates the first layer at $\theta \sim 0.28$, corresponding to a nearest neighbor Cs–Cs spacing of 4.86 Å. The compression at the saturation coverage is 11% in comparison with bulk cesium [5].

There is a subtle difference in the overlayer structures of potassium between Pt (111) and Cu(111). Potassium forms commensurate structures on Pt(111): (2×2), $(\sqrt{3} \times \sqrt{3})R30°$, and so on; the hexagonal structure is not compressed continuously with coverage. This indicates that the adsorbate–substrate interaction is stronger on Pt(111) than Cu(111). This stronger interaction with the substrate is responsible for couplings between the K–Pt stretching mode with substrate phonons as described later. The saturation coverage of potassium on Pt(111) is $\theta = 0.44$, where the K–K spacing is 4.16 Å, a 13% reduction of the K–K spacing from bulk potassium. Cesium adsorption on Pt(111) also shows the superstructures as potassium. It saturates at $\theta = 0.41$. This corresponds to a Cs–Cs spacing of 4.36 Å, that is, a compression of 20% relative to the bulk.

11.3.2
Surface Phonon Bands

Adsorption of alkali metal atoms introduces new surface phonon modes: an out-of-plane stretching mode (S-mode) and in-plane frustrated translational modes (T-modes) of alkali atoms. The S-modes of various alkali adsorption systems have been observed mostly by HREELS, whereas the T-modes by HAS. In particular, HAS

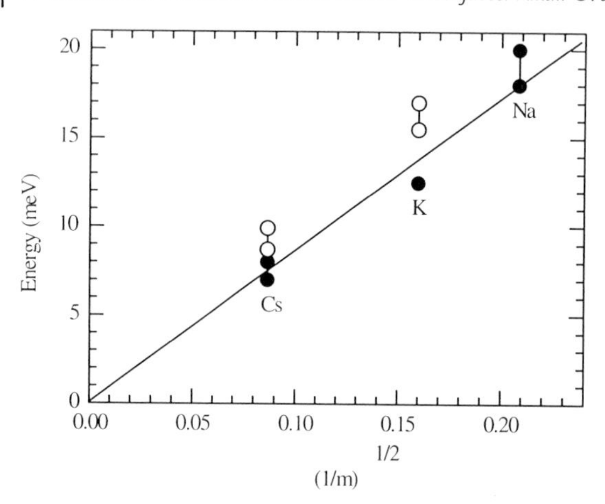

Figure 11.2 Vibrational energies of the stretching modes of alkali atoms on Cu(111) and Pt(111) surfaces as a function of inverse square root of the mass of alkali atom. Open circles are for Pt(111) and filled circles are for Cu(111). The solid line is a linear fit of vibrational energies for Cu(111). Because vibrational energy generally depends on alkali coverage, the energy range is depicted by two markers for each system where the coverage dependence is reported.

provides the dispersion curves of alkali-induced surface phonons with respect to the wave vector parallel to the surface. The energies of vibrational modes of various alkali adsorption systems measured experimentally are compiled by Finberg *et al.* [6].

In Figure 11.2, we summarize the S-mode vibrational energies of alkali adsorption systems relevant to this section: Na, K, and Cs on Cu(111) and Pt(111) surfaces. The vibrational energies of S-modes are nearly proportional to $1/\sqrt{m}$, where m is the mass of alkali atom. This indicates that the potential energy curves along the surface normal are almost the same for the series of alkali atoms on these surfaces, and the stretching vibration is roughly described by the displacement of alkali atom.

The vibrational structures of alkali adsorbates have been most extensively studied on copper surfaces. On these surfaces, the vibrational energies of S-modes do not depend strongly on surface orientation and alkali coverage. Furthermore, HAS measurements of Cs/Cu(001) [20] and lattice dynamical calculations with interatomic interaction potentials for Na/Cu(111) [21] revealed that the S-modes of Na and Cs have little dispersion. Thus, the vertical-polarized alkali oscillators couple weakly to each other on copper surfaces.

The vibrational energy of the S-mode of K on Pt(111) depends strongly on coverage: it shifts from 16.7 meV at $\theta = 0.02$ to 21.7 meV at $\theta = 0.16$ and decreases to 19.2 meV at $\theta = 0.33$ [22]. The increase in vibrational energy at $0.02 \leq \theta \leq 0.16$ is accounted for by the dipole–dipole coupling. As described in Section 11.3.3, the alkali overlayer becomes metallic at $\theta \sim 0.16$ (0.36). Thus, the redshift above $\theta = 0.16$ is related to the change in the electronic structure of the alkali overlayer.

Although the S-mode is roughly described by the displacement of alkali atom itself, this does not necessarily mean that the S-mode is completely isolated from substrate phonons. The coupling of S-mode with substrate phonons has been discussed in terms of lattice dynamical calculations with interatomic interaction potentials: Na/Cu(111) [21] and K/Pt(111) [7]. We show that TRSHG spectroscopy provides direct evidence of the couplings between the S-mode and substrate phonon modes in Section 11.5.3.

The T-modes of alkali atoms have been measured on copper surfaces with HAS. In contrast to the S-modes, the T-modes show larger coverage dependence. For example, the vibrational energy of the T-modes of Na/Cu(111) increases by a factor of 3 in the range from $\theta = 0.11$ to 0.44 [21]. In addition, the radial Cs–Cs force constant in the Cs overlayer was 25% larger than that of bulk Cs [20]. These facts are consistent with compression in alkali overlayers; the lateral interatomic interaction increases with coverage. Because alkali adsorption does not induce large displacement of surface atoms particularly on Cu(111), the corrugation of potential along the surface plane is small. Thus, the anharmonic coupling between S- and T-modes will be small.

11.3.3 Electronic Structures

The electronic structure of alkali-covered metal surfaces has been one of the central issues in surface science. Here, we briefly summarize general features. More detailed information on this subject is available in excellent reviews [23, 24].

The electronic structure and, hence, the bonding of alkali atoms on metal surfaces strongly depend on alkali coverage. Figure 11.3 schematically illustrates

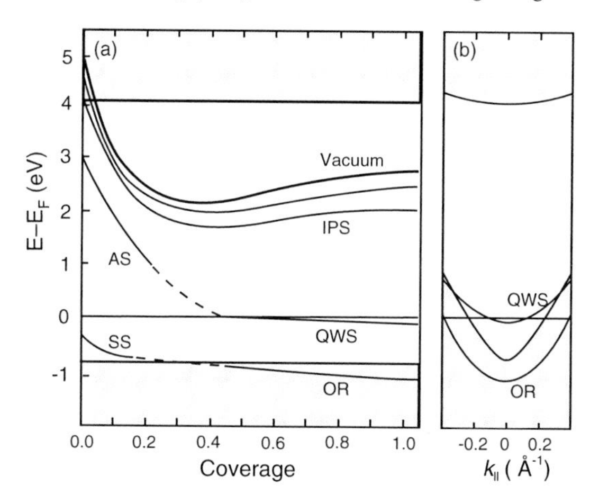

Figure 11.3 (a) Electronic states on an alkali-covered Cu(111) surface as a function of coverage. IPS: image potential state; AS: antibonding state; SS: Shockely surface state; QWS: quantum-well state; and OR: overlayer resonance. (b) The surface-projected band structures at the saturation coverage as a function of electron momentum parallel to the surface. The shaded regions show the band structure of bulk copper projected onto the surface Brillouin zone. The energy structures are adapted from the experimental and theoretical works on Na [25, 26] and K [27]. Because this figure is to provide the electronic structure qualitatively, the energies plotted here are not exact.

a typical electronic structure of alkali adsorption systems as a function of coverage based on the experimental and theoretical works for Na [25, 26] and K [27] on Cu surfaces.

Because of the s, p-inverted L-projected bandgap characterizing the clean Cu (111) surface, a two-dimensional potential well is formed between the substrate and the vacuum barriers at this surface. Thus, the Shockley surface state (SS) located in the L-bandgap has an electron density mostly localized in the potential well. In addition, there are a series of image potential states (IPSs) close to the vacuum level.

The most striking feature is a remarkable coverage dependence of the work function. At low coverages, $\theta' \leq 0.1$, the work function decreases sharply with increasing coverage. As coverage increases further, the work function reaches a minimum at $\theta' \sim 0.4$ and increases toward that of the bulk alkali metal. The remarkable changes in work function can be understood in terms of the classical model of Gurney [28]. In this model, an alkali atom donates an outermost occupied s electron to the metal to form a positive ion at the low coverages; this results in a surface dipole layer that is responsible for the large drop in work function. As coverage increases, the lateral interaction of alkali atoms becomes stronger; alkali atoms are depolarized, and the alkali overlayer becomes metallic. IPSs follow the work function changes because they are pinned to the vacuum.

Because the electronic configuration at low coverages ($\theta' \leq 0.1$) considerably differs from that at high coverages ($\theta' \geq 0.5$), we separately describe it for each coverage range. At the low coverage limit, an alkali-induced electronic state, absent on the clean surface, appears at an energy between the IPS and the Fermi level E_F. This state correlates to the electronic levels of an isolated alkali atom in vacuum: the occupied outermost s and unoccupied p_z levels. As the alkali atom approaches the metal surface, the s and p_z orbitals are hybridized as a result of interaction with the metal to form the nonstationary state [24]. Because the wave function of this state has an antibonding character with respect to the alkali–substrate bond, this state is denoted as the antibonding state (AS). AS is located at $\sim$ 3 eV above E_F irrespective of the alkali-atom period [29]. AS steeply shifts toward E_F with increasing coverage. SS also shifts downward with increasing coverage and it crosses the lower energy edge of the L-bandgap.

At high coverages, the alkali overlayers are metallic because the orbitals of neighboring alkali atoms extensively overlap with each other. Two bands characteristic to the overlayer are an overlayer resonance (OR) located below the L-bandgap and a quantum-well state (QWS) at around E_F. These bands correlate with those of a free-standing alkali monolayer in the vacuum: the s-like lowest and the p_z-like second lowest bands [30]. When the monolayer is brought closer to the metal surface, they are stabilized by interaction with the metal, while these bands maintain the integrity. The s-like band correlates with OR; the p_z-like band correlates with QWS. Because QWS is located in the L-bandgap, its wave function is localized at the surface. In contrast, the wave function of OR extends more into the substrate because it is located below the lower edge of L-bandgap.

11.4
Time-Resolved SHG Spectroscopy

11.4.1
Enhancement of SHG Intensity by Alkali Adsorption

When an optical field $\mathbf{E}(t)$ increases, the macroscopic polarization induced in a medium $\mathbf{P}(t)$ shows nonlinearity with respect to the field. The nonlinear polarization responsible for second harmonic generation is given by

$$\mathbf{P}^{(2)}(t) = \chi^{(2)} : \mathbf{E}(t) \cdot \mathbf{E}(t), \tag{11.6}$$

where $\chi^{(2)}$ is the second-order nonlinear susceptibility. In a medium with centrosymmetry, $\chi^{(2)}$ of the medium is zero within the electric dipole approximation. In contrast, $\chi^{(2)}$ at a surface, where centrosymmetry is generally broken, is not zero; dipole-allowed SH signals are generated from the surface. Thus, SH generation spectroscopy provides an inherent surface sensitivity if bulk materials have centrosymmetry.

Figure 11.4 shows how the SH intensity of 800 nm ($\hbar\omega = 1.55$ eV) photons depends on the coverage of alkali atoms on Cu(111) and Pt(111) surfaces. The SH intensity is enhanced by a few orders of magnitude compared to the clean surface when alkali atoms are adsorbed. Although the fine details are different, there are some common features in the coverage dependence of SH intensity among the

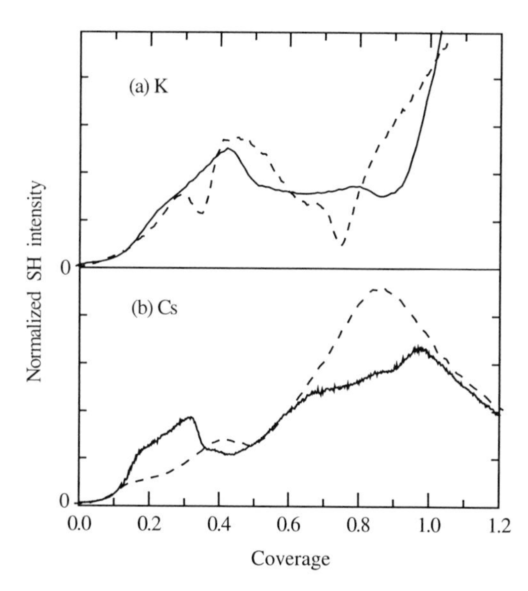

Figure 11.4 Variations in second harmonic intensity as a function of coverage θ' of (a) potassium and (b) cesium on Pt(111) (solid curve) and Cu(111) (dashed curve) surfaces. The surfaces are irradiated with photons at $\hbar\omega = 1.55$ eV. Sample temperatures range from 90 to 110 K during the irradiation. SH intensities are normalized at $\theta' = 0.6$.

alkali-covered surfaces. Potassium adsorbates show a peak at $\theta' \sim 0.4$ and a strong enhancement at $\theta' \sim 1.0$; cesium adsorbates show basically two peaks at $\theta' \sim 0.3$–0.4 and 0.8–1.0. Sodium adsorbates on Cu(111) (not shown) also show two "resonance" peaks below $\theta' = 1$. These trends indicate that the SH enhancement mostly depends on alkali atom, not on metal substrate.

There are two major origins of the SH enhancement associated with alkali adsorption: interband transitions between surface electronic states and multipole plasmon excitation [9, 31]. The interband transitions play an important role in low coverages ($\theta' < 0.5$), whereas the multipole plasmon excitation at high coverages ($\theta' \sim 1$). At the low coverages, AS and IPSs are steeply stabilized with increasing coverage. Thus, the interband transitions from SS to AS or from SS to IPS become resonant to $2\hbar\omega(= 3.10\,\text{eV})$ below $\theta' = 0.4$. Therefore, the "resonant" peaks observed at $\theta' \leq 0.5$ in Figure 11.4 are likely due to resonant or near-resonant transitions in which the surface localized bands, SS, AS, and IPSs, are involved.

As coverage increases over $\theta' \sim 0.4$, the alkali overlayer is depolarized and the metallic QWS and OR bands are formed. Because the electron density at the alkali overlayer increases as a result of neutralization, the polarizability of the layer becomes much larger than that of the clean metal surface [32]. Thus, the transitions from the occupied-OR band to the s, p band of the substrate contribute greatly to the SH intensity. Furthermore, because QWS is partly filled at $\theta' \sim 0.5$, the intraband transitions of QWS and the interband transitions from QWS to the substrate bands also contribute to the SH intensity. However, the SH intensity decreases in $\theta' =$ 0.4–0.8 for K and $\theta' = 0.3$–0.5 for Cs. This could be due to the destructive interference in SH signals between the transitions from the OR and QWS bands. The destructive interference takes place if the phases of $\chi^{(2)}$ terms relevant to the transitions from the two bands are different.

At $\theta' \sim 1.0$, the contribution to SH intensity from multipole plasmon excitation may be larger than that from the interband transitions because the photon energy $2\hbar\omega(= 3.1\,\text{eV})$ is close to plasmon resonance energies: $\hbar\omega_p = 3.8$ and 3.5 eV for K and Cs, respectively, where ω_p is the bulk plasma frequency. Irradiation of a metal surface with an oscillating electromagnetic field induces a dynamic screening field [33]. If the optical frequency ω is close to $0.8\omega_p$, electronic transitions at the surface excite resonantly a damped collective mode along the surface normal. This coupling between the optical field and multipole plasmons at the surface results in the local field enhancement of the nonlinear response such as SH generation. Liebsch calculated the frequency dependence of SH dipole moments of alkali overlayers at $\theta' = 1$ by using the time-dependent density functional method [31]. According to the calculations, the resonant peak of the imaginary part of the dipole moment is located at 2.0, 1.4, and 1.3 eV for Na, K, and Cs on Al, respectively; the widths of resonances are 0.3–0.5 eV. Because these resonances are inherent to alkali overlayers, the resonance energies do not depend on substrate. The photon energy of $\hbar\omega = 1.55$ eV is close to the resonance of the potassium overlayer at $\theta' = 1$. This is consistent with the significant increase of SH intensity at the saturation coverage of K on Cu(111) and Pt(111) as shown in Figure 11.4.

11.4.2 The Principle of TRSHG Spectroscopy

In TRSHG spectroscopy, the SH intensity of a probe pulse is measured as a function of pump–probe delay time t. Transient changes in the SH intensity $\Delta I_{SH}(t)$ are defined as $\Delta I_{SH}(t) = [I_{SH}(t) - I_{SH}^{0}(t)]/I_{SH}^{0}(t)$, where $I_{SH}(t)$ and $I_{SH}^{0}(t)$ are the SH intensities at a delay time t with and without a pump pulse, respectively. From now on, we denote $\Delta I_{SH}(t)$ as TRSHG signals.

There are two origins contributing to $\Delta I_{SH}(t)$: pump pulse-induced electron and nuclear dynamics. First, because electron populations in the alkali-induced bands are modulated by electronic excitation by the pump pulse, transient changes in electron populations δn directly contribute to TRSHG signals. Because the electronic response to the pump pulse is much faster than that of nuclear response induced by the electronic excitation, the TRSHG signals originating in the electron dynamics appear close to $t \sim 0$ ps.

Second, a response due to coherent nuclear dynamics follows the electronic response in TRSHG signals. As described in Section 11.3.3, when a metallic alkali monolayer is brought from the vacuum to a metal surface, the electronic bands of the free-standing alkali monolayer shift their binding energies as a result of interactions with the metal substrate. Thus, the binding energies of the alkali-induced bands depend on the displacement of the overlayer along the surface normal δQ. Consequently, the oscillation of δQ due to coherent excitation of the S-mode alters the binding energies and populations of the alkali-induced bands; in this way, the coherent vibration of alkali atoms indirectly contributes to TRSHG signals. Because the lateral displacements of the alkali overlayer are not expected to shift the binding energies of alkali-induced bands as large as the vertical ones, the lateral motions of alkali adsorbates contribute little to TRSHG signals. Therefore, the TRSHG spectroscopy on alkali-covered metal surfaces should be more sensitive to the S-mode than the T-modes.

Keeping these origins in mind, we assume that the second-order susceptibility under ultrashort pulse excitation can be expanded in terms of the electron population in the kth alkali-induced band δn_k and the nuclear displacement of the ith phonon mode δQ_i up to the first order as

$$\chi^{(2)} = \chi^{(2)}\Big|_0 + \sum_k \frac{\partial \chi^{(2)}}{\partial n_k}\Big|_0 \delta n_k + \sum_i \frac{\partial \chi^{(2)}}{\partial Q_i}\Big|_0 \delta Q_i. \tag{11.7}$$

Here, the first term is a constant independent of the effects induced by the electronic excitation, and the second and third terms are due to the variations in $\chi^{(2)}$ by electron population changes in alkali-induced bands and by nuclear displacements, respectively.

Because the SH intensity is proportional to the square of $|\chi^{(2)}|$, that is, $I_{SH} \propto |\chi^{(2)}|^2$, $\Delta I_{SH}(t)$ is mainly determined by the following first-order terms with respect to δQ_i and δn_k:

$$\Delta I_{\mathrm{SH}}(t) \propto \sum_i \chi^{(2)}|_0 \cdot \left.\frac{\partial \chi^{(2)}}{\partial Q_i}\right|_0 \delta Q_i + \sum_k \chi^{(2)}|_0 \cdot \left.\frac{\partial \chi^{(2)}}{\partial n_k}\right|_0 \delta n_k, \tag{11.8}$$

where all the terms higher than the second order are neglected. Equation (11.8) indicates that the TRSHG signals caused by coherent nuclear motion are proportional to its displacement.

11.5
Electronic and Nuclear Responses in TRSHG Signals

11.5.1
Representative TRSHG Traces

Figure 11.5 shows representative TRSHG traces obtained for Na, K, and Cs/Cu (111) [15, 16]. As described in Eq. (11.8), TRSHG traces from alkali-covered surfaces are contributed by both population changes in surface and substrate electronic bands and coherent nuclear vibrations at the surface. The TRSHG traces in Figure 11.5 clearly show these responses of the adsorption systems. The electron population changes give rise to a sharp spike at $t \sim 0$ ps and nonoscillatory decaying components extending to $t \sim 1$–2 ps; the coherent nuclear dynamics emerge as damped oscillatory components following the fast electronic response at $t \sim 0$ ps. Note that the oscillation frequencies of damped oscillatory components decrease with increasing

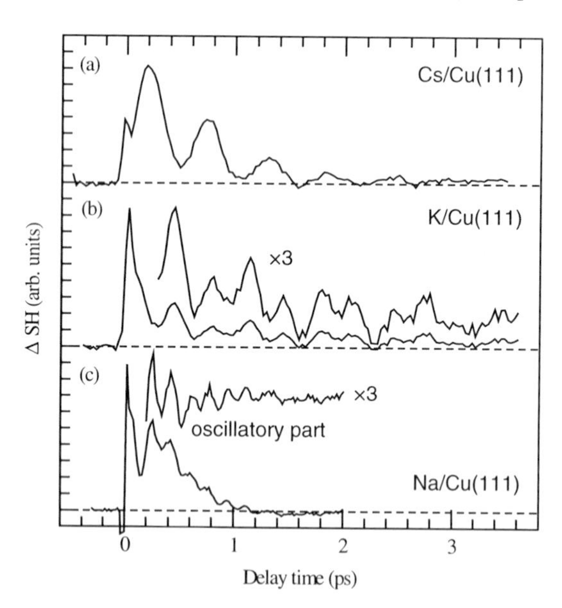

Figure 11.5 Typical TRSHG traces for (a) Cs/Cu(111) at $\theta = 0.25$, (b) K/Cu(111) at $\theta = 0.35$, and (c) Na/Cu(111) at $\theta = 0.44$. Excitation photon energies are 3.10 eV, 2.20 eV, and 2.14 eV, respectively.

alkali mass. We discuss both the electronic and the nuclear responses deduced from TRSHG traces in more detail in the following sections.

11.5.2
Electronic Response

TRSHG signals at $t \sim 0$ ps, reflecting the electronic response to the pump pulse, show peculiar changes in sign for both Cu(111) and Pt(111) surfaces as alkali coverage increases. A typical example is shown in Figure 11.6 observed for K/Cu (111) [16]. The TRSHG signals at $t \sim 0$ ps change the sign from negative to positive in a very narrow coverage range: $\theta = 0.20$ (0.5)–0.25 (0.63). Below $\theta = 0.20$ (0.5), the TRSHG trace at $t \sim 0$ ps shows an almost instantaneous response to the pump pulse: the negatively going profile has a decay time constant of $\tau = 20$ fs, which is almost identical to the duration of the pump pulse. At the coverages between $\theta = 0.20$ (0.5) and 0.25 (0.63), the TRSHG traces show a differential-like profile, which can be reproduced by the linear combination of two components: a negative component

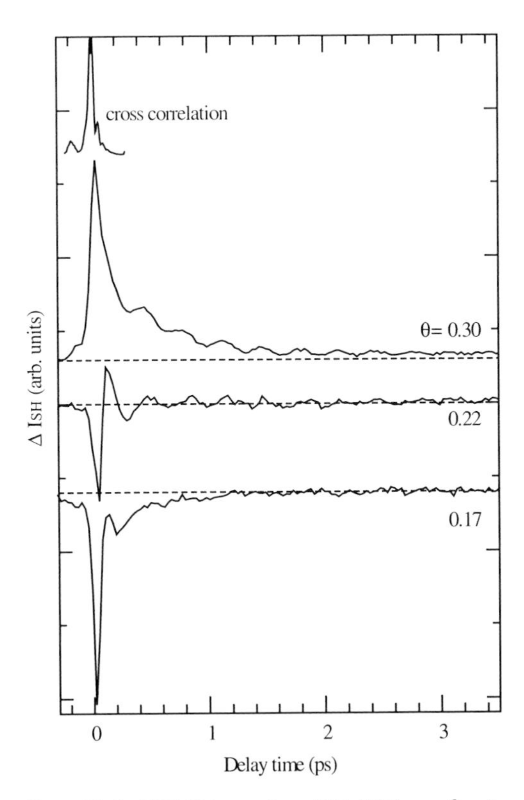

Figure 11.6 TRSHG traces from K/Cu(111) as a function of K coverage. The traces are vertically shifted arbitrarily and dashed lines indicate zeros for each curve. The top trace represents cross correlation of pump and probe pulses.

with $\tau = 20$ fs and a positive one with $\tau = 40$ fs. Above $\theta = 0.25$ (0.63), only the positive component appears.

These changes in the electronic response in the TRSHG traces are qualitatively explained in terms of the coverage dependence of alkali-induced electronic states. As described in Section 11.3.3, the energies of alkali-induced states change with coverage. The OR band is occupied at $\theta < 0.2$ (0.5), but the QWS band is unoccupied. Thus, electrons in the OR band mainly contribute to the SH intensity of the probe pulse. Because the pump pulse excites electrons in the OR band to the unoccupied s, p -bulk band, the electron population of the OR band is depleted. This makes the SH intensity of the probe pulse smaller than that without the pump pulse; hence, the depletion of electron population of the OR band results in a negative peak in TRSHG traces. Because the pump photon energy ($\hbar\omega = 2.2$ eV) is in near resonance with one- and two-photon transitions from OR to AS and from OR to IPS, respectively (see Figure 11.3), the near-resonant transitions may promote the depletion of electron population of the OR band. A lifetime of $\tau < 20$ fs is not unreasonable for holes in the OR band because holes in the surface resonance can be filled rapidly by bulk electrons.

When coverage increases to $\theta \sim 0.25$ (0.63), the QWS band shifts down toward E_F [27]; thus, the electronic transitions from the bulk d-bands to the QWS band are energetically possible. Furthermore, transitions from the d-band to the unoccupied s, p -bulk band just above E_F may be followed by a rapid resonant transfer of hot electrons to the QWS band. This indirect transition also increases the population of the QWS band transiently. The extra electrons in the QWS band increase the SH intensity of the probe pulse, giving rise to a positive peak in TRSHG traces.

Note that the oscillatory component due to the coherent K–Cu stretching mode is clearly visible at $\theta \geq 0.22$ (0.55) but not at $\theta = 0.17$ (0.43). The distinct difference is associated with the change in the electronic structure: the QWS band becomes partly filled at $\theta \sim 0.25$ (0.63). We discuss this point in more detail in Section 11.6.

11.5.3
Nuclear Response

The oscillating components in TRSHG traces following the electronic response at $t \sim 0$ ps originate in the nuclear response due to coherently excited nuclear vibrations. To deduce the properties of excited coherent phonon modes, we analyze the TRSHG traces with linear prediction singular value decomposition [34] by using the following function composed of damped cosinusoidals and exponential decay components:

$$\Delta I_{SH}(t) = \sum_i A_i \exp\left(-t/\tau_i^{(p)}\right)\cos(\Omega_i t + \phi_i) + \sum_j B_j \exp\left(-t/\tau_j^{(e)}\right). \qquad (11.9)$$

Here, the ith damped oscillatory component is characterized with the phonon frequency Ω_i, the initial phase ϕ_i, and the dephasing time $\tau_i^{(p)}$; the jth exponential decay component is characterized with the electron decay time $\tau_i^{(e)}$.

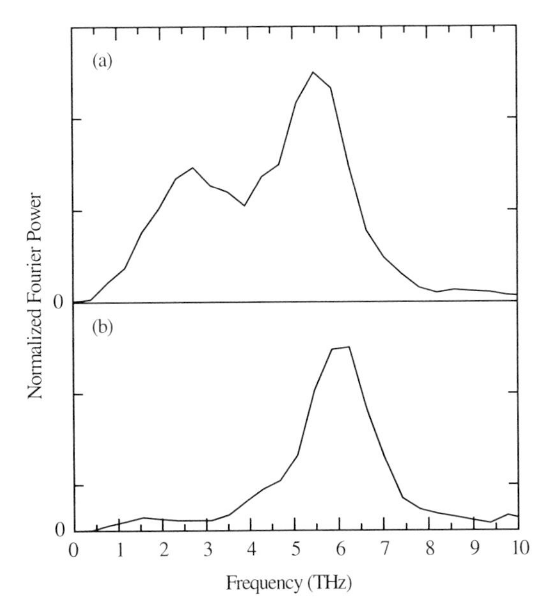

Figure 11.7 Fourier power spectra of oscillatory components observed from Na/Cu(111). (a) Na coverage $\theta = 0.44$ and (b) $\theta = 0.14$.

In addition to the fitting procedure, we can obtain phonon spectra by Fourier transformation (FT) of oscillatory parts in TRSHG traces. Figure 11.7 shows the Fourier power spectra of Na/Cu(111) [15]. At $\theta = 0.14$ (0.32), the FT spectrum shows a single peak at 6 THz (20 meV). The frequency is consistent with that observed in HREELS measurements [35]; this was attributed to the S-mode of Na. When the surface is saturated by Na, where the overlayer has a $(3/2 \times 3/2)$ structure, two prominent peaks at 2.7 THz (9.0 meV) and 5.5 THz (18 meV) appear in the spectrum.

Borisova *et al.* [21] calculated the dispersion curves and the local density (LDOS) distributions of phonon modes for Na/Cu(111) with various commensurate periodic structures of Na overlayers. All the Na-covered Cu(111) surfaces have a common surface resonance with pronounced LDOS at 22 meV (5.5 THz) originating in the phonon mode of Na along the surface normal z, that is, the S-mode. While the vibrational energy of the S-mode is almost independent of Na coverage, the LDOS distribution of this vibrational resonance becomes broader as coverage increases. This indicates that the coupling between the S-mode and substrate phonon modes increases with coverage. At the saturation coverage, the LDOS of z-polarized phonons of adatoms spread in a broad energy range. In addition to the S-mode resonance at 22 meV (5.5 THz), another S-mode resonance at $\bar{\Gamma}$ appears at 26.3 meV (6.5 THz) corresponding to the shear vertical displacements of the Na adatoms at the distorted hollow sites (see Figure 11.1). Moreover, several phonon bands with z polarization appear in the range of 9–16 meV (2.3–4.0 THz). These extra bands, which are missing at low coverages, originate from a strong mixing between the S-mode and surface phonon modes of Cu. Consequently, the theoretical LDOS distribution spreads around 3 and 6 THz with substantial widths. These LDOS features are consistent with the observed FT spectrum (Figure 11.7a) composed of two broad peaks at 2.7 and 5.5 THz. Therefore, the observed phonon bands are assignable to the coupled modes between the S-mode and surface phonon modes of Cu.

K- and Cs-covered Cu(111) surfaces show a peak due to the S-mode at 3.1 THz (10 meV) and 1.8 THz (6.0 meV), respectively. Because no extra bands appear as in the low coverage of Na on Cu(111), the S-modes of K and Cs on Cu(111) couple weakly with surface phonon modes of Cu.

In contrast, multiple peaks appear in all the FT spectra of K/Pt(111) as shown in Figure 11.8 [14]. The fine structure in the vibrational spectra depends on coverage in a complex manner. At $\theta \leq 0.31$ (0.70), the prominent peaks are at 4.5 THz (19 meV) and 5.2 THz (22 meV). At $\theta \geq 0.34$ (0.77), the major peak is at 4.7 THz (19 meV) and a new peak appears at 3.8 THz (16 meV). In addition, there is another weak peak in the range of 2–3 THz in almost all the spectra. The fine structures observed indicate that the frequency resolution of TRSHG is much higher than the earlier HREELS measurements that showed only one broad peak at 5 THz (21 meV) [22]. More importantly, these fine structures indicate that the S-mode character is distributed in various surface phonon modes on Pt(111). The Cs overlayers on Pt(111) also show multiple peaks in the FT spectra around 2.5 THz. These findings indicate that K and Cs interact more strongly with Pt than with Cu surfaces, resulting in mixing of the S-mode with surface phonon modes. The distributions of S-character among substrate phonon modes were also revealed by calculations with the embedded atom

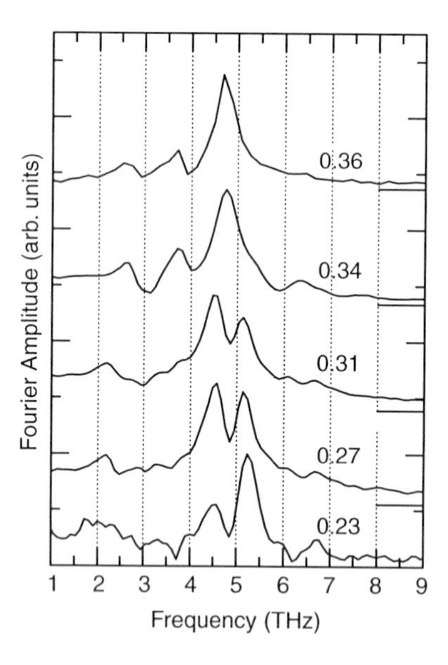

Figure 11.8 Fourier amplitude spectra of K/Pt (111) as a function of coverage. Potassium coverages θ are denoted in the figure. The spectra are normalized at their peak intensities [14]. The multiple peaks imply that the K–Pt stretching mode is coupled with surface phonon modes. Reprinted in part with permission from [14]. Copyright 2006 by the American Physical Society.

method [7]. The stronger adsorbate–substrate interaction on Pt(111) also manifests itself in the coverage dependence of overlayer structure: as coverage increases, the hexagonal structure is continuously compressed on Cu(111), but commensurate structures are kept in certain coverage ranges on Pt(111).

11.5.4 Initial Phase

As described in Section 11.2, the initial phase of a coherent phonon mode is determined by the time profile of force exerted on ions [18]. Thus, it provides valuable information on the force responsible for creation of coherent phonons.

The stimulated Raman scattering with virtual electronic excitations is responsible for coherent excitation in transparent materials. If an excitation pulse has a broad spectral width, stimulated Raman scattering occurs as a result of mixing between the frequency components ω_i and ω_j that satisfy the vibrational resonant condition of $\omega_i - \omega_j = \Omega_1$; in other words, the duration of the pump pulse has to be less than the vibrational period for the impulsive stimulated Raman scattering (ISRS).

For opaque materials, real excitations also take place. Changes in the population of electrons in an empty band or holes in an occupied band or both modify the deformation potential of ions in solids and at surfaces. If a pump pulse produces electrons (holes) in an empty (occupied) band with a sufficiently rapid rising edge, the ions cannot follow the change in the deformation potential and start to oscillate around new equilibrium positions. This excitation mechanism is called as displacive excitation of coherent phonons (DECP) [36]. Merlin and coworkers developed the transient stimulated Raman scattering (TSRS) model [37], including both the ISRS and the DECP processes in equal footing. In this model, the Fourier transform of the driving force $F(t)$ is approximately written as

$$\tilde{F}(\omega) \propto \left(\varepsilon_1' + \frac{2\varepsilon_2}{-i\omega}\right)\tilde{I}(\omega). \tag{11.10}$$

Here, $\varepsilon (= \varepsilon_1 + i\varepsilon_2)$ is the dielectric function of the material, $\varepsilon_1' = d\varepsilon_1/d\omega$, and $\tilde{I}(\omega)$ is the Fourier transform of a pump pulse $I(t)$. The initial phase of the created coherent phonon mode with a renormalized frequency Ω_1 and a damping rate β is given by

$$\phi_{\mathrm{TSRS}} = \arctan\,(\bar{\varepsilon} - \bar{\beta}), \tag{11.11}$$

where $\bar{\varepsilon} = \Omega_1 {\varepsilon'}_1/2\varepsilon_2$ and $\bar{\beta} = \beta/\Omega_1$.

In the original TSRS model, the decay of coupled charge density, that is, the density of excited charges responsible for displacement of ions, is neglected. Riffe and Sabbah [18] modified the model to include the effect of a finite lifetime of coupled charge density. According to this extended model, that is, Raman scattering finite lifetime (RSFL), the initial phase is given by

$$\phi_{\mathrm{RSFL}} = \arctan\left(\frac{\bar{\varepsilon} + \bar{\gamma} - \bar{\beta}}{1 - \bar{\varepsilon}\bar{\gamma}}\right). \tag{11.12}$$

Here, $\bar{\gamma} = \gamma/\Omega_1$ and γ are the damping rate of coupled charge density. If the force due to the stimulated Raman scattering with virtual electronic transitions is negligibly small, that is, $\bar{\varepsilon} \sim 0$, the initial phase of the RSFL model converges to the one predicted by the DECP model:

$$\phi_{\rm DECP} = \arctan\,(\bar{\gamma} - \bar{\beta}). \tag{11.13}$$

Let us consider which model is most appropriate for coherent excitation of alkali overlayers. We take Cs overlayers as an example. Because the wavelength dependence of the dielectric constant of a Cs monolayer is not available, we are forced to estimate the parameter $\bar{\varepsilon}$ on the basis of the data of bulk Cs: $\bar{\varepsilon} \sim 0.03$ at $\hbar\omega = 1.55$ eV and ~ 0.01 at $\hbar\omega = 3.1$ eV [38]. These values are smaller than the typical value of $\bar{\beta} = 0.97$ by a couple of orders of magnitude, that is, $\bar{\beta}, \bar{\gamma} \gg \bar{\varepsilon}$. This relation may not hold because $d\varepsilon_1/d\omega$ can be substantially large if pump photon energy is in resonance with a transition between alkali-induced surface states. However, as we note in Section 11.6, there is no resonance enhancement of coherent phonon excitation at $\hbar\omega \sim 2$ eV; that is, near resonance with the transitions of QWS $\rightarrow$ IPS. Thus, the contribution of the Raman scattering to the initial phase is negligibly small. Consequently, the DECP model is likely appropriate to apply to the coherent phonon excitation at alkali overlayers at $\hbar\omega \sim 2$ eV.

Because the decays of coupled charge density are supposed to be faster than those of coherent phonons, the contribution of γ to the initial phase would be larger than that of β. Under these circumstances, we note two limiting cases for the initial phase with regard to γ. In the case of $\gamma \gg \Omega_1$, that is, $\bar{\gamma} \gg 1$, the initial phase approaches $\pi/2$: the coherent phonon motions oscillate in a sin-like manner. This is the impulsive limit. On the other hand, in the case of $\gamma \ll \Omega_1$, that is, $\bar{\gamma} \ll 1$, the initial phase approaches 0, that is, cos-like. This is the displacive limit. Figure 11.9 shows how initial phases depend on $\bar{\gamma}$ and $\bar{\beta}$ in the DECP model. The initial phases measured are also plotted as a function of $\bar{\beta}$ of the various alkali adsorption systems at the pump photon energies $\hbar\omega = 2.0$–2.2 eV. This plot shows that $0.17 \leq \bar{\gamma} \leq 0.57$ for the alkali adsorption systems studied, indicating that they are closer to the displacive limit. Moreover, the obtained values of $\bar{\gamma}$ imply that the decay time of coupled charge density ranges from 0.06 to 0.41 ps. This is consistent with the finding that the TRSHG traces for both on Cu(111) and Pt(111) surfaces contain an exponentially damped nonoscillating component with $\tau^{(e)} \sim 0.4$ ps.

11.6 Excitation Mechanism

Real electronic excitations by the pump pulse create hot electrons and holes in substrate and surface bands. But, not all electronic excitations are effective in coupling with motions of alkali atoms. An important issue is to determine which electronic excitation is most effective in coupling and, thus, the creation of coherent phonons. There are two extreme cases in photoinduced nuclear dynamics at surfaces,

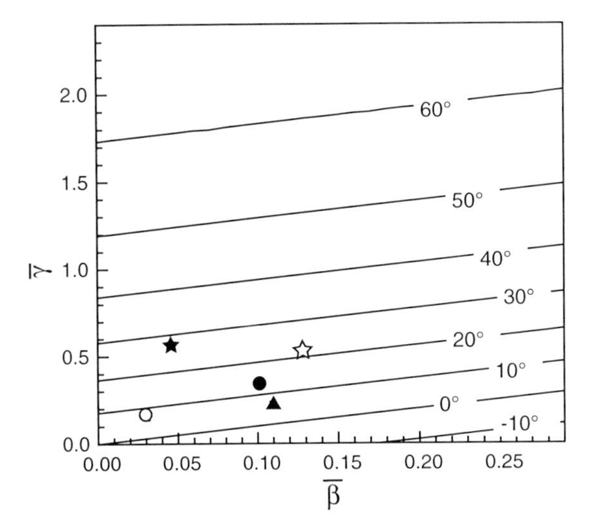

Figure 11.9 Contours of initial phase in the DECP model as a function of $\bar{\beta}$ and $\bar{\gamma}$. Observed initial phases are also plotted as a function of $\bar{\beta}$ for alkali adsorption systems: K/Pt(111) (solid star), Cs/Pt(111) (open star), Na/Cu(111) (solid circle), K/Cu(111) (open circle), and Cs/Cu(111) (solid triangle). The data were obtained in the case of excitation photon energy of ~ 2 eV.

as has been frequently discussed in the studies of surface photochemistry: adsorbate-localized excitation versus substrate-mediated excitation [39]. For the alkali adsorption systems in the coverage range from $\theta' = 0.5$ to 1.0, transitions of OR $\rightarrow$ QWS, OR $\rightarrow$ IPS, and QWS $\rightarrow$ IPS are candidates for the adsorbate-localized excitation. Intra- and interband excitations of s, p and d-bands of bulk are involved in the substrate-mediated excitation: electrons or holes created by the electronic excitation of bulk bands transiently transfer to the alkali-induced electronic state, resulting in modulation of the electron density near alkali adatoms. In addition, another possible excitation mechanism specific to alkali overlayers is the multipole plasmon excitation described in Section 11.4.1. This excitation produces a longitudinally oscillating electron density at an alkali-covered surface, which may also initiate the coherent motion of alkali adsorbates.

Because each excitation mechanism cited above has its own photon energy dependence quite different from each other, a useful way to clarify the electronic transitions responsible for the creation of coherent surface phonons is to measure TRSHG traces as a function of pump photon energy. Plotting the initial amplitude against the pump phonon energy gives an action spectrum. Figure 11.10 shows an action spectrum obtained for Na and K-covered Cu(111) surfaces.

If the adsorbate-localized excitation is responsible, resonant structures in the action spectrum are expected at the resonant energies of transitions: 2.12 eV (Na, QWS $\rightarrow$ $n = 1$ IPS), 2.19 eV (K, QWS $\rightarrow$ $n = 2$ IPS), 2.32 eV (K, QWS $\rightarrow$ $n = 3$ IPS), and 2.53 eV (Na, QWS $\rightarrow$ $n = 2$ IPS). In addition, the amplitudes at 2.12 and 2.19 eV should be much larger than those at 2.32 and 2.53 eV because the wave

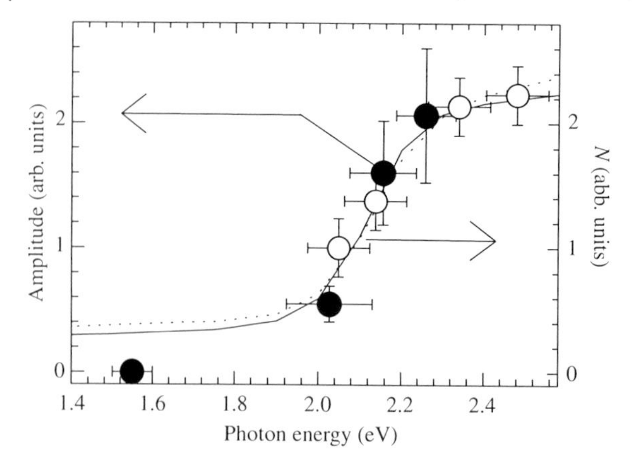

Figure 11.10 Excitation photon energy dependence of initial amplitudes of coherent surface phonons on Na/Cu(111) (open circles) [15] and K/Cu(111) (filled circles) [16]. Photocarrier number densities estimated by Eq. (11.14) are plotted for $d = 10$ nm (dashed curve) and $d = 1000$ nm (solid curve). Note that the action spectra and the carrier densities are scaled at 2.34 eV. Reprinted in part with permission from [15] and [16]. Copyright 2007 and 2009 by the American Physical Society.

function of QWS has a better spatial overlap with IPS with lower n than that with higher n in each adsorption system. However, the observed action spectrum is inconsistent with these expectations. Thus, it is unlikely that the electronic transitions between the surface states play a major role in the coherent phonon excitation for the adsorption systems of Na and K on Cu(111).

Multipole plasmon excitation has a resonance for the one-monolayer alkali overlayer at around $0.8\omega_p$: 4.7, 3.0, 2.9 eV for Na, K, and Cs, respectively. According to calculations by using the time-dependent local density approximation [40], the surface excitation spectra of Na overlayers on Al show a broad peak at $\theta' \leq 1$. Because the calculated spectra show a slow rising edge in the photon energy range of 2.0–2.5 eV, the action spectrum observed might be explained by this excitation mechanism. However, potassium overlayers should have a steeper rising edge in this photon energy range because the resonance is located at the energy lower than that of Na overlayer by 0.7 eV. This is not consistent with the observations: the action spectra of Na and K overlayers reveal a common photon energy dependence. Two-photon excitation of multipole plasmon could also be responsible for creating the coherent phonons because 2ω spans from 3.0 to 5.0 eV, covering all the multipole plasmon resonances of the alkali overlayers. In contrast to one-phonon excitation, the excitation spectra of two-photon processes have a much narrower resonance feature at around $0.4\omega_p$: the widths are in the range of 0.3 to 0.5 eV. Thus, if this excitation is responsible for coherent phonon excitation, the action spectra would have a narrow peak at 2.0 eV for Na and 1.3 eV for K. These expectations again contradict the observed results as follows: first, there is no resonance enhancement at $\hbar\omega = 2.0$ eV for Na overlayers. Second, two-photon excitation of multipole plasmon at 1.5–2.5 eV should not takes place effectively at potassium-covered Cu(111) because the photon energy exceeds the plasmon resonance at 1.3 eV. However, the observed action spectrum clearly reveals that the coherent phonons are excited in this energy range and the amplitude increases with photon energy. Therefore, multipole plasmon excitation either by one-photon or two-photon process is not responsible for coherent phonon excitation.

Instead, the action spectrum in Figure 11.10 rather resembles the absorption curve of bulk copper. We estimated the density of photoexcited carriers N that are generated in the Cu substrate and propagate to the surface by using the following equation [41]:

$$N = IA[1-\exp(-2\pi k\delta/\lambda)], \tag{11.14}$$

where I is the incident photon number; A is the absorbance of the substrate, δ is the hot carrier mean free path, λ is the wavelength of pump pulses, and k is the extinction coefficient of bulk copper [38]. Photon energy dependences of N for $\delta = 10$ and 1000 nm are plotted in Figure 11.10, where they are normalized at 2.34 eV. The action spectrum corresponds well to the carrier density curve estimated from substrate absorption. Therefore, the substrate electronic excitation is likely responsible for the coherent phonon excitation of alkalis on copper.

Because the emergence of oscillatory modulations in TRSHG traces is concomitant with the stabilization of the QWS band to E_F, it is likely that the population change in the QWS band is responsible for a driving force on alkali atoms along the alkali atom–Cu coordinate. One possible excitation path for generating the impulsive force is hole creation in the QWS. Electron–hole pairs are formed in the substrate by the d-band $\rightarrow$ s, p -band transition, followed by subsequent Auger recombination of an electron in the QWS band with the d-band hole, resulting in creation of holes in the QWS. An appreciable coupling between the hole creation at the alkali-derived QWS and surface (or interface) phonon excitation has been suggested [42]. Alternatively, hot electrons created by the d-band $\rightarrow$ s, p -band transition could be injected into the QWS above E_F, resulting in abrupt fluctuations of electron density at alkali atoms. In either case, the rapid change in charge density at the surface, particularly in the QWS band, are likely the origin for the driving force along the alkali atom–Cu stretching coordinate.

It is interesting to compare the excitation mechanism described here with that for photoinduced "frustrated" desorption of Cs on Cu(111) at low coverages [43, 44]. Nuclear motions of Cs adsorbates in an electronic excited state were monitored by using interferometric time-resolved two-phonon photoemission. At the low coverages, a Cs adatom rests on the ionic potential energy surface (PES) of the ground electronic state. When an ultrafast pump pulse resonantly excites from SS to AS, Cs is excited to a repulsive PES; thus, it starts to move normal to the surface. However, the lifetime of the excited state is so short that the Cs atom is quenched back to the deep ionic PES; thus, it cannot desorb, that is, desorption is frustrated. This comparison leads us to conclude that the alkali-induced electronically excited states play a central role in generating photoinduced coherent nuclear motions at alkali-covered surfaces: AS for "frustrated" desorption in low coverage, on the one hand, and QWS for coherent phonons of the S-mode in high coverage, on the other.

11.7
Summary and Outlook

After introducing the basic properties of a typical alkali metal overlayer on metals, namely, adsorption geometry, phonon bands, and electronic structures, we describe the principle of time-resolved SHG spectroscopy and its application to the alkali metal adsorption systems. Electronic excitation of the adsorption systems by ultrafast laser pulses induces coherent surface phonons. The large enhancement in SH intensity at the alkali-covered metal surfaces allows to precisely monitor the time evolution of the coherent phonons by measuring intensity modulations in SH of probe pulses. Electronic excitations of both adsorbate-induced and substrate bands by using ultrafast light pulses can induce abrupt fluctuations of charge density around alkali adatoms, resulting in coherent nuclear motions of adsorbates via electron–phonon couplings. For Na and K/Cu(111), excitation photon energy dependence of coherent surface phonons clearly indicates that substrate electronic excitation at $h\nu > 2$ eV is primarily responsible for creation of the coherent motions of adsorbates. The analysis of initial phases of coherent phonons indicates that the coherent phonons observed are created in the DECP limit.

Alkali metal overlayers introduce metallic bands such as overlayer resonances and quantum-well states having large electron density at the alkali overlayers. Thus, direct excitation of these adsorbate-induced states should also initiate coherent surface phonons. Because the substrate absorption of Cu(111) significantly increases above $h\nu > 2$ eV, the contributions of transitions between surface-localized states might be obscured by those of substrate excitation. It would be worth investigating the excitation photon energy dependence of coherent phonons below $h\nu = 2$ eV more carefully.

TRSHG is a versatile tool to investigate coherent phonon dynamics at surfaces. This method is ideally suited for alkali overlayers because SH intensity is enhanced tremendously by adsorption of alkali atoms. Surface plasmons are largely responsible for the enhancements. However, applications of TRSHG to other adsorption systems have been very limited, mainly because they lack such large enhancement in SH intensity. This obstacle can be removed if SH intensity is enhanced by tuning the photon energy of probe pulses, $h\nu$ or $2h\nu$, to an electronic resonance of adsorption systems. In addition, using much shorter pump pulses definitely extends the applicability of TRSHG to coherent surface phonons at higher frequencies.

References

1 Dekorsy, T., Cho, G.C., and Kurz, H. (2000) Chapter 4, in *Coherent Phonons in Condensed Media*, vol. 76, Topics in Applied Physics, Springer Verlag, Berlin, pp. 169–209.

2 Chang, Y.-M., Xu, L., and Tom, H.W.K. (1997) Observation of coherent surface optical phonon oscillations by time-resolved surface second-harmonic generation. *Phys. Rev. Lett.*, **78**, 4649–4652.

3 Chang, Y.-M., Xu, L., and Tom, H.W.K. (2000) Coherent phonon spectroscopy of GaAs surfaces using time-resolved second-harmonic generation. *Chem. Phys.*, **251**, 283–308.

4 Matsumoto, Y. and Watanabe, K. (2006) Coherent vibrations of adsorbates induced by femtosecond laser excitation. *Chem. Rev.*, **106**, 4234–4260.

5 Diehl, R.D. and McGrath, R. (1996) Structural studies of alkali metal adsorption and coadsorption on metal surfaces. *Surf. Sci. Rep.*, **23**, 43–171.

6 Finberg, S.E., Lakin, J.V., and Diehl, R.D. (2002) He-atom scattering study of the vibrational modes of alkalis on Al(1 1 1). *Surf. Sci.*, **496**, 10–20.

7 Rusina, G.G., Eremeev, S.V., Echenique, P.M., Benedek, G., Borisova, S.D., and Chulkov, E.V. (2008) Vibrations of alkali metal overlayers on metal surfaces. *J. Phys. Condens. Matter*, **20** (22), 224007.

8 Lindgren, S.-Å. and Walldén, L. (2000) Some properties of metal overlayers on metal substrates, Chapter 13, in *Handbook of Surface Science*, vol. 2, Elsevier, Amsterdam, pp. 899–951.

9 Liebsch, A. (1997) *Electronic Excitations at Surfaces, Physics of Solids and Liquids*, Plenum Press, New York.

10 Tom, H.W.K., Mate, C.M., Crowell, J.E., Shen, Y.R., and Somorjai, G.A. (1986) Studies of alkali adsorption on Rh(111) using optical second-harmonic generation. *Surf. Sci.*, **172**, 466.

11 Watanabe, K., Takagi, N., and Matsumoto, Y. (2004) Direct time-domain observation of ultrafast dephasing in adsorbate–substrate vibration under the influence of a hot electron bath: Cs adatoms on Pt(111). *Phys. Rev. Lett.*, **92** (5), 057401.

12 Watanabe, K., Takagi, N., and Matsumoto, Y. (2005) Femtosecond wavepacket dynamics of Cs adsorbates on Pt(111): coverage and temperature dependences. *Phys. Rev. B*, **71**, 085414.

13 Watanabe, K., Takagi, N., and Matsumoto, Y. (2005) Mode selective excitation of coherent surface phonons on alkali-covered metal surfaces. *Phys. Chem. Chem. Phys.*, **7**, 2697–2700.

14 Fuyuki, M., Watanabe, K., and Matsumoto, Y. (2006) Coherent surface phonon dynamics at K-covered Pt(111) surfaces investigated by time-resolved second harmonic generation. *Phys. Rev. B*, **74** (19), 195412.

15 Fuyuki, M., Watanabe, K., Ino, D., Petek, H., and Matsumoto, Y. (2007) Electron–phonon coupling at an atomically defined interface: Na quantum well on Cu(111). *Phys. Rev. B*, **76** (11), 115427.

16 Watanabe, K., Inoue, K.-I., Nakai, I.F., Fuyuki, M., and Matsumoto, Y. (2009) Ultrafast electron and lattice dynamics at potassium-covered Cu(111) surfaces. *Phys. Rev. B*, **80** (7), 075404.

17 Blum, K. (1981) *Density Matrix Theory and Applications*, Plenum Press, New York.

18 Riffe, D.M. and Sabbah, A.J. (2007) Coherent excitation of the optic phonon in Si: transiently stimulated Raman scattering with a finite-lifetime electronic excitation. *Phys. Rev. B*, **76** (8), 085207–085212.

19 Fan, W.C. and Ignatiev, A. (1988) Growth of an orientationally ordered incommensurate potassium overlayer and its order–disorder transition on the Cu (111) surface. *Phys. Rev. B*, **37** (10), 5274.

20 Witte, G. and Toennies, J.P. (2000) Phonons in a quasi-two-dimensional solid: cesium monolayer on Cu(001). *Phys. Rev. B*, **62**, R7771.

21 Borisova, S.D., Rusina, G.G., Eremeev, S.V., Benedek, G., Echenique, P.M., Sklyadneva, I.Y., and Chulkov, E.V. (2006) Vibrations in submonolayer structures of Na on Cu(111). *Phys. Rev. B*, **74** (16), 165412.

22 Klünker, C., Steimer, C., Hannon, J.B., Giesen, M., and Ibach, H. (1999) Vibrations of potassium at Pt(111) and formation of KOH studied by electron energy-loss spectroscopy. *Surf. Sci.*, **420**, 25–32.

23 Bonzel, H.P. (1989) *Physics and Chemistry of Alkali Metal Adsorption*, Elsevier, Amsterdam.

24 Gauyacq, J.P., Borisov, A.G., and Bauer, M. (2007) Excited states in the alkali/noble metal surface systems: a model system for the study of charge transfer dynamics at surfaces. *Prog. Surf. Sci.*, **82** (4–6), 244.

25 Fischer, N., Schuppler, S., Fauster, T., and Steinmann, W. (1994) Coverage-dependent electronic structure of Na on Cu(111). *Surf. Sci.*, **314**, 89–96.

26 Carlsson, J.M. and Hellsing, B. (2000) First-principles investigation of the

quantum-well system Na on Cu(111). *Phys. Rev. B*, **61**, 13973.
27 Schiller, F., Corso, M., Urdanpilleta, M., Ohta, T., Bostwick, A., McChesney, J.L., Rotenberg, E., and Ortega, J.E. (2008) Quantum well and resonance-band split off in a K monolayer on Cu(111). *Phys. Rev. B*, **77** (15), 153410.
28 Gurney, R.W. (1935) Theory of electrical double layers in adsorbed films. *Phys. Rev.*, **47**, 479–482.
29 Zhao, J., Pontius, N., Winkelmann, A., Sametoglu, V., Kubo, A., Borisov, A.G., Sanchez-Portal, D., Silkin, V.M., Chulkov, E.V., Echenique, P.M., and Petek, H. (2008) Electronic potential of a chemisorption interface. *Phys. Rev. B*, **78** (8), 085419.
30 Wimmer, E. (1983) All-electron local density functional study of metallic monolayers: I. Alkali metals. *J. Phys. F: Met. Phys.*, **13**, 2313–2321.
31 Liebsch, A. (1989) Second-harmonic generation from alkali-metal overlayers. *Phys. Rev. B*, **40** (5), 3421–3424.
32 Persson, B.N.J. and Dubois, L.H. (1989) Work function, optical absorption, and second-harmonic generation from alkali-metal atoms adsorbed on metal surfaces. *Phys. Rev. B*, **39** (12), 8220–8235.
33 Feibelman, P.J. (1982) Surface electromagnetic fields. *Prog. Surf. Sci.*, **12** (4), 287–407.
34 Barkhuijsen, H., de Beer, R., Boveé, W.M.M.J., and Van Ormondt, D. (1985) Retrieval of frequencies, amplitudes, damping factors, and phases from time-domain signals using a linear least-squares procedure. *J. Magn. Reson.*, **61**, 465.
35 Lindgren, S.-Å., Svensson, C., and Walldén, L. (1993) Some vibrational properties of clean and water vapor exposed Cu(111)/Na, K. *J. Electron Spectrosc. Relat. Phenom.*, **64–65**, 483–490.
36 Zeiger, H.J., Vidal, J., Cheng, T.K., Ippen, E.P., Dresselhaus, G., and Dresselhaus, M.S. (1992) Theory for displacive excitation of coherent phonons. *Phys. Rev. B*, **45**, 768.
37 Stevens, T.E., Kuhl, J., and Merlin, R. (2002) Coherent phonon generation and the two stimulated Raman tensors. *Phys. Rev., B*, **65**, 144304.
38 Palik, E.D. (ed.) (1998) *Handbook of Optical Constants of Solid I–III*, Acadamic Press.
39 Ho, W. (1995) Chapter 24, in *Surface Photochemistry*, World Scientific, pp. 1047–1140.
40 Ishida, H. and Liebsch, A. (1992) Electronic excitations in thin alkali-metal layers adsorbed on metal surfaces. *Phys. Rev. B*, **45**, 6171.
41 Ying, Z.C. and Ho, W. (1990) Thermoinduced and photoinduced reactions of NO on Si(111)7×7. II. Photoreaction mechanisms. *J. Chem. Phys.*, **93**, 9089–9095.
42 Chulkov, E.V., Kliewer, J., Berndt, R., Silkin, V.M., Hellsing, B., Crampin, S., and Echenique, P.M. (2003) Hole dynamics in a quantum-well state at Na/Cu(111). *Phys. Rev. B*, **68** (19), 195422.
43 Petek, H., Weida, M.J., Nagano, H., and Ogawa, S. (2000) Real-time observation of adsorbate atom motion above a metal surface. *Science*, **288**, 1402–1404.
44 Petek, H., Nagano, H., Weida, M.J., and Ogawa, S. (2001) Surface femtochemistry: frustrated desorption of alkali atoms from noble metals. *J. Phys. Chem. B*, **105**, 6767–6779.

12
Coherent Excitations at Ferromagnetic Gd(0001) and Tb(0001) Surfaces

Alexey Melnikov and Uwe Bovensiepen

12.1
Introduction

It is shown that ferromagnetic lanthanide surfaces like Gd(0001) and Tb(0001) present good model systems to investigate excitation, interaction, and relaxation of electronic, phononic, and magnetic subsystems in the time domain. In particular, analysis of an optically excited coherent phonon (CP)–magnon mode by pump–probe experiments gives detailed insight into the intriguing interaction between lattice and magnetization dynamics on femtosecond timescales.

The lanthanide series of elements (Ce–Lu) is well known for its consecutive filling of the 4f shell, which is localized at the ion core. Due to the poorly screened 4f orbitals, an effective increase of nuclear charge occurs with increasing electron count, which leads to the reduction in the ionic radius with increasing electron count. The orbital and the spin magnetic moments per ion are also related to the number of the 4f electrons by Hund's rules. Accordingly, the spin magnetic moment is largest for the half-filled configuration $4f^7$ of gadolinium (Gd), $S=7/2$. The orbital magnetic moment vanishes ($L=0$) for this configuration. In terbium (Tb) with $4f^8$, the spin and orbital magnetic moments are given by $S=3$ and $L=3$, respectively.[1)]

Both elements crystallize in the hexagonal close-packed structure with the lattice constant $a_{\mathrm{Gd}} = 3.63$ Å and $a_{\mathrm{Tb}} = 3.60$ Å. Their structural and vibrational properties are accordingly similar [1]. The lanthanide elements exhibit rich and interesting magnetic ordering phenomena [2], which are closely related to the indirect (or RKKY) exchange interaction. This type of exchange couples the magnetic moments that are localized at the ion cores and, hence, have no direct wave function overlap. This indirect exchange acts through polarization of the delocalized conduction band electrons. As depicted in Figure 12.1b, for the simplified situation of a spin impurity embedded in free electron gas, the spin polarization (SP) surrounding the localized spin exhibits a damped oscillation as a function of distance r from the ion core. Since

1) The ionic configuration in Tb is actually $4f^9 6s^2$ while it is $4f^7 5d^1 6s^2$ for Gd. In a solid, which is of interest here, the number of 4f electrons increases as an integer count along the lanthanide series [1].

Dynamics at Solid State Surface and Interfaces Vol.1: Current Developments
Edited by Uwe Bovensiepen, Hrvoje Petek, and Martin Wolf

ISBN: 978-3-527-40937-2

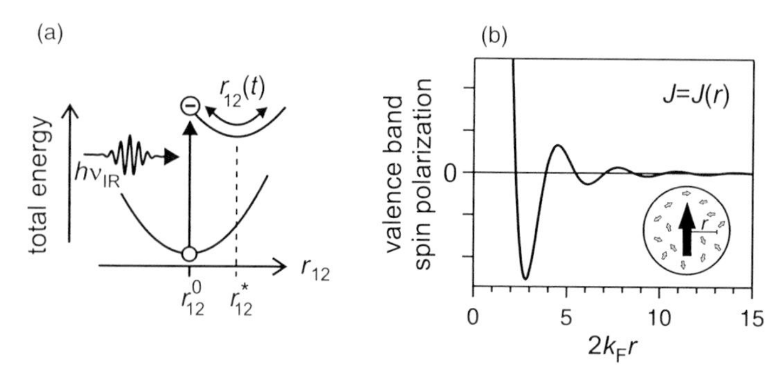

Figure 12.1 (a) Displacive optical excitation of coherent optical phonons for resonant absorption. A femtosecond laser pulse populates unoccupied electronic states, which resides in a displaced potential surface with respect to the ground state. Not shown is the decay of the excited potential surface to the ground states, which proceeds with relaxation of the excess energy. (b) Spin polarization in a quasifree electron gas in the vicinity of a spin impurity. The damped oscillatory behavior with distance leads to the indirect (or RKKY) exchange interaction, which is responsible for the magnetic order in lanthanide materials.

this spin polarization mediates the exchange coupling among neighboring atomic magnetic moments, the exchange constant J strongly depends on r.

In this textbook, the excitation of coherent optical phonons has already been discussed in the article by Hase *et al.* Briefly, absorption of femtosecond laser pulses in the infrared spectral region generates a nonequilibrium population of carriers in an excited-state potential surface, which has a potential minimum shifted with respect to the ground state, as illustrated in Figure 12.1a. The related vibration of the interatomic distance can modify – if appropriately chosen – the exchange coupling J since J depends on the interatomic distance (Figure 12.1). In such an excited state, the exchange coupling constant J should become time dependent, as the respective coherent variation in the lattice constant r might affect the instantaneous J and modify the magnetic state of the system.

In this chapter, we discuss such a coherent coupling between magnetic and phononic excitations for the lanthanide surfaces Gd(0001) and Tb(0001). We investigate the transient magnetic and phonon dynamics by optical second harmonic generation (SHG) and photoelectron (PE) spectroscopy in femtosecond time-resolved experiments.

12.2
Relaxation of the Optically Excited State

Before we go into the details of the SHG experiment, it is helpful to investigate the optically excited state of Gd(0001) and its relaxation with femtosecond time-resolved photoelectron spectroscopy. Figure 12.2c sketches the employed experimental technique. An intense infrared femtosecond laser pulse (800 nm, absorbed fluence

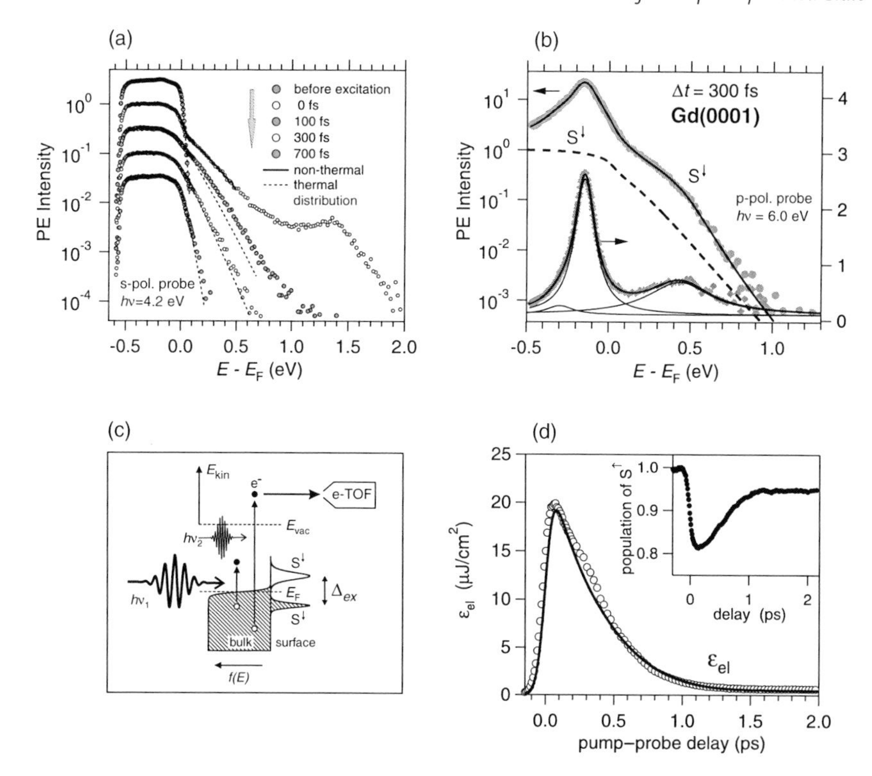

Figure 12.2 Time-resolved photoelectron spectroscopy of Gd(0001)/W(110) illustrated in panel (c). The photoelectron intensity is depicted by symbols as a function of energy in the vicinity of the Fermi level in panel (a) for bulk-sensitive photoemission using s-polarized pulses and in panel (b) for a surface- and bulk-sensitive photoemission using p-polarized pulses at the given pump–probe delays. The dotted line in panel (a) describes Fermi–Dirac distribution functions and the solid lines represent nonthermalized distributions [3]. In panel (b), the spectral function of the $5d_{z^2}$ surface state is retrieved by normalizing the spectrum to a nonthermalized distribution function (dashed line). Panel (d) shows the excess energy as a function of delay for a surface-sensitive measurement. The solid line is a simulation by a two-temperature model. The inset gives the population of the majority surface state component ((b–d) are reproduced with permission from Refs [6, 10]). Copyright 2005 IOP Publishing Ltd and American Institute of Physics.

$\approx 1\,\mathrm{mJ/cm^2}$) excites electron–hole pairs. The evolution of the nonequilibrium distribution function and transient variations in the spectral function of the surface are probed by a time-delayed second laser pulse at a photon energy larger than the work function $E_{\mathrm{vac}} - E_{\mathrm{F}}$. In Figure 12.2a, photoelectron spectra in the vicinity of E_{F} are shown on a logarithmic scale for different pump–probe delay times before and after optical excitation. At maximum temporal overlap of pump and probe pulses (0 fs), the spectrum exhibits a clear deviation from a Fermi–Dirac distribution function (FDDF), as highlighted by the kink at E_{F} [3]. After 100 fs, this deviation becomes considerably weaker (compared to the dashed line) and after 300 fs it can barely be recognized. Hence, a description by a thermalized distribution function and a two-temperature model [4, 5], which considers the excess energy by an instantaneous increase of the electron temperature T_{el} and subsequent cooling by electron–phonon scattering, is expected to hold after about 100 fs [6]. Before that, under conditions without a well-established FDDF, the excess energy ε_{el} cannot be described by the electronic temperature. However, ε_{el} can be described independent of the distribution function by integrating over the number of excited electrons $N(E)$ multiplied by their single-particle energy E [3, 6].

The ferromagnetic Gd(0001) surface features an exchange split $5d_{z^2}$ surface state, which is localized not only along the surface normal but also in the lateral directions [7]. At low temperature, the majority surface component $S^{\uparrow}$ is occupied and the minority one $S^{\downarrow}$ unoccupied (Figure 12.2c) [8, 9]. Employing a p-polarized UV laser pulse at $\hbar\omega = 6$ eV, the majority component of the surface state $S^{\uparrow}$ gives a pronounced line in the photoelectron spectrum (Figure 12.2b). Pumping by a preceding IR laser pulse populates unoccupied states and the unoccupied component $S^{\downarrow}$ can be distinguished as a shoulder on an exponential background. Normalization of the spectrum to the respective distribution function yields the spectral function of the surface, which is shown in Figure 12.2b. Note that it is essential for this analysis to record the spectrum with a dynamic range of more than 10^4, which is facilitated here by electron counting. On this basis, the pump-induced variation of the line intensity, the linewidth, and the binding energy have been analyzed for the surface spectrum.

Supposing a constant photoemission matrix element, the intensity of the $S^{\uparrow}$ PE line represents the population of that state. As depicted in the inset of Figure 12.2d, this quantity is reduced within the first few 100 fs after excitation at an absorbed fluence of 1 mJ/cm^2 by almost 20%, which indicates a strong optical excitation of the surface [10]. The population partially recovers within the first picoseconds, during which the crystal lattice is heated by electron–phonon scattering [6]. The population does not fully recover after 1 ps because the lattice temperature increases by 100 K. The transient linewidth and the exchange splitting have also been analyzed to obtain detailed information about the respective scattering processes. For details on these topics, the reader can refer to the publications [10, 11].

Before we proceed to the coherent lattice dynamics induced by the optical excitation, we discuss the excess energy. The excess energy ε_{el} in the probed sample volume in the vicinity of the surface can be described by

$$\varepsilon_{el}(t) = 2\int_{E_F=0}^{1.5\text{ eV}} N(E,t)\cdot|E|\mathrm{d}E. \tag{12.1}$$

The factor 2 accounts for a symmetric distribution of electrons and holes [3]. In Figure 12.2d, the experimentally obtained result is shown. The energy density at 100 fs is 20 μJ/cm^2, which is two orders of magnitude larger than in thermal equilibrium with $T_0 = 100$ K. The maximum converts into $E_{el}/k_B = 1900$ K and ε_{el} relaxes very close to the initial value within 1 ps. The time-dependent $\varepsilon_{el}(t)$ can be successfully described by a two temperature model, in which ε_{el} is defined by γT_{el}^2 with γT_{el} being the electronic part of the heat capacity [6, 10]. This means that we describe the transient energy by a heat bath model and an energy flow from the hot electron system to the lattice as derived for equilibrium processes. However, we avoid here the ambiguity related to thermalized distribution functions. As shown in Figure 12.2d, this description reproduces $\varepsilon_{el}(t)$ reasonably well.

12.3 Coupled Lattice and Spin Excitations

The above results represent a rather generic behavior of femtosecond electron dynamics in metals after intense optical excitation [12–14]. The localized surface state of lanthanide surfaces adds a coherent contribution to the response of metals to optical excitation, which is presented in the following.

12.3.1 Transient Binding Energy Variations of the Surface State

We now focus on the analysis of the binding energy E_B of the $S^{\uparrow}$ surface state component after optical excitation. In Figure 12.3a, the pump-induced change of E_B is depicted as a function of pump–probe delay. Two contributions of this response can be distinguished at a first glance. (i) A continuous shift of E_B toward E_F by 37 meV is explained by lattice heating of the surface after excitation and the respective expansion of the crystal. The resulting more shallow potential well lowers the binding

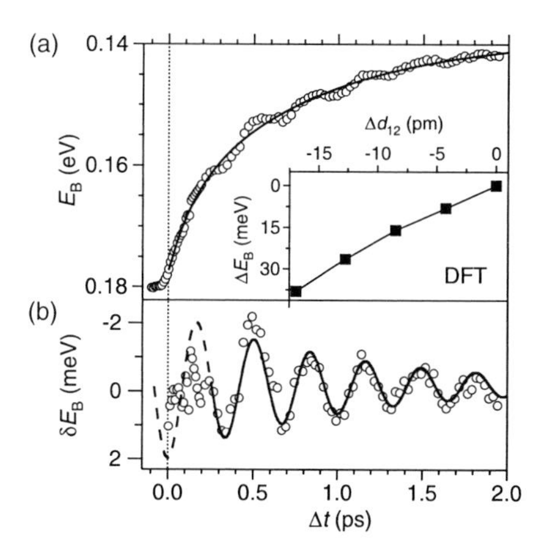

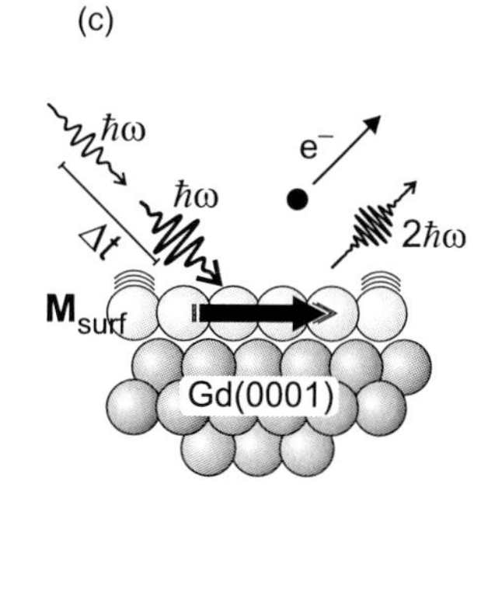

Figure 12.3 (a) Binding energy of the occupied surface state component as a function of delay after optical excitation. The solid line represents the continuous shift due to heating of the surface. On top of this a binding energy oscillation is observed, which is shown in (b) after subtracting the continuous shift. The inset shows the variation in binding energy upon surface contraction in the electronic ground state, as calculated by density functional theory (reproduced with permission from Ref. [11]. Copyright 2007 by the American Physical Society.). (c) This shows an artist's view of the surface excitation and its detection.

energy as observed here. (ii) On top of this thermal expansion, we observe an oscillatory response with a frequency of 3.0(1) THz and an initial amplitude of 2 meV, which is damped with a characteristic time of 1.0(1) ps. The latter is depicted in Figure 12.3b after the continuous shift has been subtracted. The oscillation is attributed to a phonon at the surface, which is derived from the longitudinal optical Γ_{3^+} mode of bulk Gd [15]. This mode leads to a vibration of the basal (0001) surface plane with respect to the underlying bulk, as illustrated in Figure 12.3c. As discussed above, the optical excitation couples electronic surface and bulk states. Hence, an optically induced transient displacement and successive excitation of a coherent phonon represents a plausible scenario. However, as will be shown below, it is essential to consider the ferromagnetic order of Gd(0001), which dominates the efficiency of the coherent phonon excitation mechanism in this system.

The ability of time-resolved photoemission to determine the oscillation amplitude in the binding energy variation quantitatively facilitates an estimation of the real space amplitude. With this aim, the variation in E_B as a function of lattice contraction has been calculated (see inset in Figure 12.3a) using density functional theory [11]. We cannot describe the nonequilibrium state explicitly. Furthermore, anharmonic effects such as thermal expansion might not be reproduced. However, small harmonic variations of interlayer spacing can be expected to be estimated with this approach rather well. For the range of the oscillatory binding energy variation, we consider a linear approximation and conclude that the 2 meV oscillation in E_B represents an oscillation in real space of 1 pm or 0.3% of the interlayer spacing d_{12} between the surface plane and the adjacent layer.

12.3.2
Nonlinear Optics as a Simultaneous Probe of Lattice Vibrations and Magnetic Excitations

The fact that these dynamics occur for a spin-polarized electronic state raises the question whether these processes are actually spin dependent. The question has been answered by optical second harmonic (SH) generation experiments, which enable for simultaneous detection of lattice vibrations and magnetization dynamics. Optical SHG is the nonlinear process of frequency doubling arising from an anharmonicity of the electronic potential in the studied medium. Due to higher order of the corresponding susceptibility tensor with respect to that for the linear optical response, SH has higher sensitivity to the symmetry properties of the medium. In particular, the SHG is forbidden in the bulk of centrosymmetric media but allowed at the interfaces, which provide its unique surface and interface sensitivity. That is why SHG can be successfully used for the studies of adsorption and structural modifications of a surface. Owing to the same reason, it is extremely sensitive, for example, to electric charge gradients and mechanical strains induced in the vicinity of interfaces of centrosymmetric media. More information about nonlinear optical processes can be found in Ref. [16]. In this section, we give a review of experimental studies performed at Gd(0001) and Tb(0001) surfaces by time-resolved SHG.

12.3.2.1 Magneto-Induced SHG: Experimental Scheme and Data Analysis

The electric field at SH consists of two contributions [17]: (i) The electronic (or crystallographic) part that monitors the variation in the band structure and the electron distribution function. It also reflects the lattice dynamics since the adiabatic approximation is valid on the timescales of interest. (ii) The magnetic part that is in first-order approximation proportional to the local magnetization M or spin polarization in the place of localization of nonlinear sources, which is the surface here. In the following, we will refer to these two contributions as the even and the odd components as they behave symmetric and antisymmetric with respect to the magnetization reversal. In combination, they form the total electric field at the double frequency:

$$\vec{E}(2\omega) = \vec{E}_{\text{even}}(2\omega) + \vec{E}_{\text{odd}}(2\omega) = \vec{\beta} + \vec{\alpha}M. \tag{12.2}$$

Here $\vec{\beta} \equiv \vec{E}_{\text{even}}(2\omega)$ and $\vec{\alpha} \equiv \vec{E}_{\text{odd}}(2\omega)/M$, where $M \propto \vec{M}$ is the projection of magnetization to the easy axis of the magnetization in the case of strong magnetic anisotropy or to the direction defined by the experimental geometry in the case of weak anisotropy. For example, in the case of *p*-in, *P*-out SHG (i.e., the polarization of both fundamental and SH waves is in the plane of incidence) at cubic, hexagonal, or isotropic surfaces, this direction is perpendicular to the optical plane of incidence. We assume $\vec{\alpha}, \vec{\beta} = \text{const}|_M$, which means that $\vec{E}_{\text{even}}$ is independent of M and $\vec{E}_{\text{odd}}$ is proportional to M. According to our knowledge, an experimental observation of higher order terms in the expansion of $\vec{E}(2\omega)$ over M was not reported up to now.

In the case of $\left|\vec{E}_{\text{even}}\right| \gg \left|\vec{E}_{\text{odd}}\right|$, which was demonstrated experimentally [18], the SH intensity can be approximated by

$$I(2\omega) \propto \left|\vec{E}_{\text{even}}(2\omega) + \vec{E}_{\text{odd}}(2\omega)\right|^2 \approx \left|\vec{\beta}\right|^2 + 2\vec{\alpha}\vec{\beta}M = kM + c, \tag{12.3}$$

with $c \equiv |\vec{\beta}|^2$ and $k \equiv 2\vec{\alpha}\vec{\beta}$. Equation (12.3) describes the so-called nonlinear magneto-optical Kerr effect, which in the present case monitors the surface magnetization. Hysteresis loops measured at Gd(0001) films have a rectangular shape indicating a single-domain state [19]. This opens the way to investigate the dynamics of the saturated magnetization, when the reversal of the external magnetic field $\vec{H}$ leads to the change of the sign of the cross-term in (12.3). Then, the SH intensity is given by

$$I^{\uparrow\downarrow}(2\omega) = E^2_{\text{even}}(2\omega) + E^2_{\text{odd}}(2\omega) \pm 2E_{\text{even}}(2\omega)E_{\text{odd}}(2\omega)\cos\varphi, \tag{12.4}$$

where φ is the relative phase between even and odd SH fields and the arrows indicate the direction of $\vec{H}$, and thereby of the magnetization.

The commonly used scheme for measurements of pump-induced variations of the SHG signal is based on the modulation technique and contains a chopper in the pump beam working at the frequency that is much lower than the repetition rate of the laser but much higher than frequencies of variations of all other experimental parameters. The SH intensity is measured for opposite directions of $\vec{H}$ for time delay t between pump and probe pulses by a detector synchronized with the chopper. Thus, the signals measured with opened and closed chopper give the intensities in the presence ($I^{\uparrow\downarrow}(t)$) and in the absence ($I_0^{\uparrow\downarrow}(t)$) of excitation. The latter quantity does not depend on the pump–probe delay t itself, but may vary with real time due to the fluctuations of the laser output, alignment, sample conditions, and so on. To describe the pump-induced variations, we combine these measured intensities into the quantity

$$D^{\pm}(t) = \frac{I^{\uparrow}(t) \pm I^{\downarrow}(t)}{I_0^{\uparrow}(t) \pm I_0^{\downarrow}(t)}. \tag{12.5}$$

To separate E_{even} and E_{odd}, the phase φ (see Eq. (12.4)) is a decisive quantity. It was measured for $\hbar\omega = 1.52$ eV in a single-beam interferometry scheme described in Ref. [19] to be smaller than 15°. Therefore, small pump-induced variations of φ will affect Eq. (12.4) insignificantly, as was also the case for Ni reported in Ref. [20]. Using conditions $E_{\text{odd}}^2 \ll E_{\text{even}}^2$ and $\cos\varphi \approx 1$, we can describe the pump-induced variations of the SH field components monitoring the electron/lattice and magnetization dynamics, respectively, by the quantities

$$\Delta_{\text{even}}(t) = \sqrt{D^{+}(t)} - 1 \approx \frac{E_{\text{even}}(t) - E_{\text{even}}^0}{E_{\text{even}}^0} \tag{12.6}$$

and

$$\Delta_{\text{odd}}(t) = \frac{D^{-}(t)}{\sqrt{D^{+}(t)}} - 1 \approx \frac{E_{\text{odd}}(t) - E_{\text{odd}}^0}{E_{\text{odd}}^0} \approx \frac{M(t) - M_0}{M_0}, \tag{12.7}$$

where E_{even}^0, E_{odd}^0, and M_0 denote the components of the SH optical field and the magnetization before optical excitation.

The surface sensitivity of SHG arises from the prohibition on SH generation in bulk crystals with inversion symmetry within the dipole approximation. Moreover, at lanthanide surfaces, SHG can be resonantly enhanced by the surface state [19] as illustrated in Figure 12.4 for the case of electronic transitions in the vicinity of the $\bar{\Gamma}$ point at Gd(0001) driven by 1.5 eV laser pulses. Here SH is generated via two channels, involving electronic transitions within the majority and minority spin subsystems. In the first case, the occupied surface state component of majority character $S^{\uparrow}$ serves as an initial state and a real unoccupied intermediate state of the SHG process is represented by the majority valence subband. In the second case, the unoccupied component of minority character $S^{\downarrow}$ serves as a real unoccupied intermediate state, while the initial state is provided by occupied minority valence subband. At slightly larger electronic momenta parallel to the surface resonant conditions for these processes are fulfilled even better [7]. Oxidation of Gd(0001) surface shifts the surface state to higher binding energy, so that the minority component also occurs below the Fermi level. This decreases the SH intensity by an order of magnitude and changes the sign of the *odd* component [19], which strongly supports the above considerations.

The experimental geometry is shown in Figure 12.5a, inset, where the saturating magnetic field is applied perpendicular to the optical plane of incidence. The p-polarized laser pulses of about 35 fs duration, 800 nm wavelength, and 40 nJ/pulse energy from a cavity dumped Ti:sapphire oscillator operated at 1.52 MHz repetition

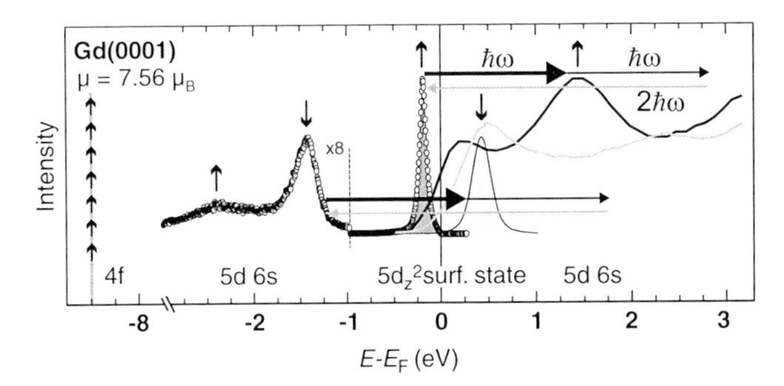

Figure 12.4 Majority (up arrow) and minority (down arrow) valence electron states of Gd(0001) measured by photoemission (circles) [21], spin-resolved inverse photoemission (black (up) and gray (down) curves) [8], and scanning tunneling spectroscopy (gray area above E_F) [22]. The exchange split surface state (filled) occurs within an orientational bandgap around the Fermi level. Two main pathways (within majority and minority subbands) for optical excitation and SHG detection with 1.52 eV photons are indicated (reproduced from [23] Copyright 2008 IOP Publishing Ltd.).

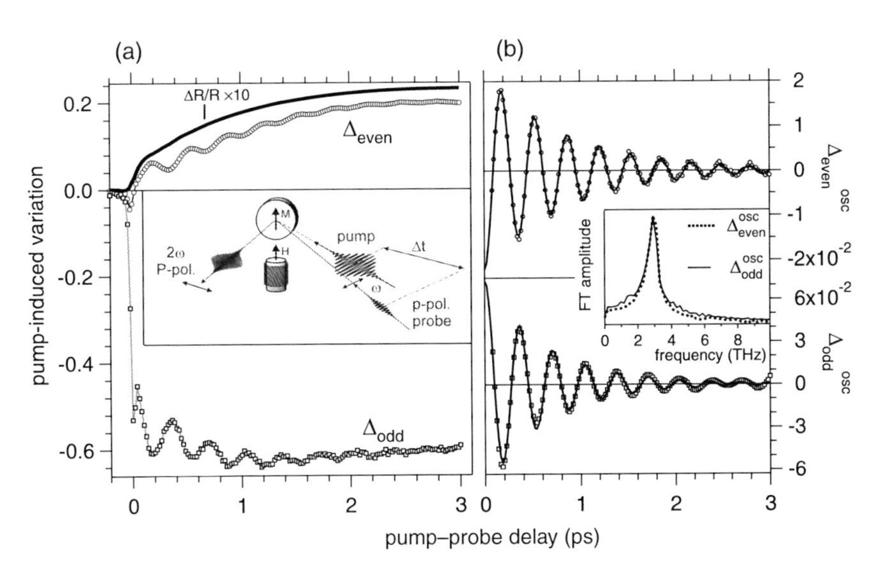

Figure 12.5 (a) Time dependence of pump-induced change in the even (upper) and odd (lower) SH components at $T = 90$ K. The transient linear reflectivity $\Delta R/R$ is also displayed by the solid line. The inset shows the experimental scheme of the transversal magneto-optical geometry. (b) Coherent part of the even and odd SH fields depicted in (a) after smoothing and subtracting the incoherent part. Solid lines represent fits to Eq. (12.10). Inset (b) shows the Fourier transform of both the oscillations (see Ref. [21]). (Figure is reproduced from [24] with kind permission of Springer Science & Business Media.)

rate are split at a power ratio of 4:1 into pump and probe pulses. The pump beam is chopped with a frequency of 700 Hz and the intensity of p-polarized SH generated by the probe pulse is recorded by a two-channel photon counter. At the same time, the relative pump-induced variation of the linear reflectivity

$$\Delta R(t)/R = (I_\omega(t) - I_\omega^0(t))/I_\omega^0(t) \tag{12.8}$$

is obtained from the intensity of reflected probe detected by a photodiode and a lock-in amplifier.

12.3.2.2 Magneto-Induced SHG: Coherent Surface Dynamics

Typical results for the pump-induced variations of the even and odd SH components are displayed in Figure 12.5a together with the transient linear reflectivity. The absorbed pump fluence is about 1 mJ/cm^2. In both SH components within the first 3 ps *oscillatory* Δ^{coh} and *nonoscillatory* Δ^{inc}, contributions due to *coherent* and *incoherent* processes, respectively, are clearly discernible:

$$\Delta_i(t) = \Delta_i^{coh}(t) + \Delta_i^{inc}(t), \quad i = \text{even or odd}. \tag{12.9}$$

The *linear* reflectivity increases upon excitation and saturates around 2 ps, reflecting the elevated electron temperature T_e by electron–electron scattering and subsequent equilibration with the lattice by electron–phonon interaction, as discussed in Section 12.2 [6, 11, 21]. The T_e buildup is reflected in the fast rise of $\Delta R/R$ within the first 100 fs. The following reduction of $\Delta R/R$ due to decreasing T_e is overcompensated by the rise due to increasing lattice temperature T_l on a 1 ps timescale of electron–lattice thermalization. Apparently, the reflectivity is more sensitive to T_l than to T_e. These two timescales manifest themselves also in $\Delta_{even}^{inc}(t)$. The initial jump is hardly visible in Figure 12.5a but is more pronounced at shorter fundamental wavelengths [18]. Since $\Delta_{even}^{inc}(t)$ evolves similarly in time to $\Delta R/R$, we attribute this contribution to electron–phonon scattering processes in the surface layer. The behavior of $\Delta_{odd}(t)$ differs significantly from that of $\Delta_{even}(t)$ in showing an ultrafast (i.e., within 50 fs resolution of the experiment) drop of E_{odd} by about 50%. Based on Eq. (12.2), we associate it with a 50% reduction of M at the surface, which we discuss in detail in Section 12.3.2.5. After that $\Delta_{odd}^{inc}(t)$ keeps a nearly constant level of 40% of the equilibrium value over about 50 ps and recovers on a 500 ps timescale [10]. Further analysis of the incoherent contributions can be found in Refs [6, 18, 21].

In the following, we focus on the periodic variations in the *even* and *odd* SH components, which are both well resolved in Figure 12.5a. The fact that an *oscillatory* component is significant in the SH response, but weak in $\Delta R/R$, indicates the surface nature of the *coherent* process. We note that $\Delta R/R$ also shows a small oscillatory component that enables a detailed comparison of coherent surface and bulk dynamics [25]. Here we restrict ourselves to the surface phenomena and describe the coherent contributions by a damped harmonic function with a linear frequency chirp:

$$\Delta^{coh}(t) = Ae^{-t/\tau}\cos(\Omega t + \xi t^2 + \varphi). \tag{12.10}$$

To fit the coherent contribution to Eq. (12.10), a nontrivial mathematical procedure described in Ref. [23] is used.

Oscillatory components of the even and odd transient SH components are displayed in Figure 12.5b (symbols) together with the fit to Eq. (12.10) (solid curves). The inset displays Fourier-transformed spectra showing a peak around 3 THz broadened due to the decay and frequency chirp. The fit to Eq. (12.10) gives slightly different decay times for the even and odd components, $\tau_{\text{even}} = 0.86 \pm 0.02$ ps and $\tau_{\text{odd}} = 0.64 \pm 0.04$ ps, which coincide with the timescale of electron–lattice thermalization, frequency $\Omega/2\pi = 2.80 \pm 0.05$ THz for both components, and a chirp $\xi/\pi = 0.17 \pm 0.01$ THz/ps. A more detailed analysis [25] reveals that while the frequency increases linearly with time for $t < 2$ ps, after 2 ps it becomes a constant of 3.15 ± 0.08 THz, which is in good agreement with the theoretical frequency of the Γ_{3+} phonon mode near the Γ point [15]. This mode is the vibration of the whole (0001) plane with respect to the adjacent layers. Therefore, we attribute $\Delta^{\text{coh}}_{\text{even}}$ to the *optical excitation* of a *coherent phonon* at the Gd(0001) surface, which is derived from the bulk Γ_{3+} mode. Since the phase in Eq. (12.10) is found to be $\varphi \approx 0$ for $\Delta^{\text{coh}}_{\text{odd}}$ and $\varphi \approx \pi$ for $\Delta^{\text{coh}}_{\text{even}}$, we consider a *displacive* generation mechanism [26] as reported for the fully symmetric modes in absorbing media (see article by Ishioka, Hase, Misochko and Figure 12.1).

This surface CP is derived from a longitudinal phonon mode in the bulk with the wave vector $q \to 0$ directed along the normal to the surface (c-axis) representing the motion of basal atomic planes with respect to adjacent layers. Owing to the weak gradient of the pump intensity along the surface (the laser spot size is about 10^5 times larger than the lattice period), the translation symmetry in the surface plane is conserved, which is satisfied by the geometry of considered vibration. This mode can be hardly excited in the bulk due to the translation symmetry between adjacent planes: even and odd planes are almost indistinguishable since the light penetration depth of 10–20 nm is much larger than the interlayer distance. This is apparently the reason for a small oscillatory component in $\Delta R/R$ [25]. At the surface, this symmetry is broken and the excitation becomes possible. In other words, the interlayer vibrational potential along the surface normal is weakly affected by the electronic excitation in the bulk, but it is significantly affected at the surface due to the structural relaxation and the break of inversion symmetry making this potential essentially anharmonic. Considering the origin of SHG sensitivity to the vibration of surface atomic plane with respect to the neighboring one, it is worth to look to Figures 12.3–12.5. As shown by time-resolved photoemission, the coherent phonon leads to the oscillation of the $S^{\uparrow}$ binding energy with the amplitude 2 meV that gives an estimate for the respective vibration amplitude of about 1 pm (Figure 12.3). Obviously (Figure 12.4), such periodic energy shift will periodically modify the resonant condition for the SHG in the majority channel due to the change of the gap between initial and intermediate states. This leads to the observed periodic variation of the SHG yield (Figure 12.5). Moreover, the variation of $\Delta^{\text{inc}}_{\text{even}}$ on a 2 ps timescale is similar to that of the nonoscillatory component of $S^{\uparrow}$ binding energy, which supports the given explanation.

To create a physical picture describing the behavior of Δ_{odd}, we recall that the strength of exchange interaction between neighboring 4f spins depends on the interionic distance (Figure 12.1). Since the Γ_{3+} mode does not affect the ion configuration within the basal plane, we can consider the energy of exchange interaction of spin $\vec{S}_i$ (averaged over the layer) in the ith plane with $\vec{S}_j$ in the adjacent jth layer ($j = i \pm 1$)

$$U_{ij}^S = f(d_{ij})\vec{S}_i\vec{S}_j, \tag{12.11}$$

where the prefactor f depends sensitively on the interlayer spacing d_{ij} [27]. For a small variation of d_{ij}, the coupling between the neighboring planes is approximated by $f(d_{ij}) \approx f(d_{ij}^0) + f'(d_{ij}^0)\Delta d_{ij}$ (here and later $f'(x) \equiv \partial f(x)/\partial x$), which decreases with increasing d_{ij} and leads to spin disorder. In the bulk, $\vec{S}_i$ in the first-order approximation is not affected by the Γ_{3+} mode since for this mode $\Delta d_{ii+1} = -\Delta d_{ii-1}$ and therefore $\Delta U_{ii+1}^S = -\Delta U_{ii-1}^S$. In contrast, at the surface where such antisymmetry is broken, the interlayer motion leads to a variation in the spin polarization, which is observed as the oscillation in the *odd* SH component. According to these considerations, the maximum of Δ_{odd}^{coh} at $t = 0$ (Figure 12.5b) corresponds to the minimum of d_{12}, that is, a contraction of the surface interlayer spacing with respect to the equilibrium value d_{12}^0. This contraction leads to about 2 meV downward shift of $S^\uparrow$ energy increasing the energy gap between $S^\uparrow$ and unoccupied bulk majority valence subband (Figure 12.4), which decreases the SHG yield leading to the minimum of $\Delta_{even}^{coh}(t = 0)$ (Figure 12.5b). Note that the surface atomic plane cannot be displaced instantaneously. In terms of Figure 12.1a, we are considering the *contraction* with respect to the (transient) *quasiequilibrium* position r_{12}^* defined by the minimum of the *excited-state* potential, but not to the *equilibrium* position r_{12}^0 of the *ground state*. Since in Gd, the frequency of the CP mode matches nearly the frequency of the optical magnon of the similar wave vector [28], we can consider a resonant phonon–magnon coupling and attribute the observed oscillations $\Delta_{even}^{coh}(t)$ and $\Delta_{odd}^{coh}(t)$ to laser-induced excitation of a coherent resonantly coupled optical phonon–magnon mode at the Gd(0001) surface.

12.3.2.3 **The Mechanism of Coherent Phonon Excitation**

The nature and mechanism of this excitation was discussed since the first observation [21]. The surface nature of the excitation discussed above from basic symmetry consideration was shown experimentally by an Y overlayer experiment: 3 ML of Y quench the oscillation. This observation shows the crucial role of the exchange splitting of the surface state, since the only difference of the electronic structure of Y(0001) from that of Gd(0001) is the absence of the exchange splitting due to the absence of 4f electrons, making Y paramagnetic. The spin polarization of the exchange split surface state thus plays a crucial role.

We now look to the temperature dependence of the coherent dynamics at Gd(0001) and Tb(0001) [23, 29]. Typical $\Delta_{even}^{Gd}(t)$ traces at different temperatures are presented in Figure 12.6a. Amplitudes A_{even}^{Gd}, A_{even}^{Tb}, and A_{odd}^{Gd} are obtained by fitting the oscillatory components $\Delta_{even}^{coh,Gd}(t)$, $\Delta_{even}^{coh,Tb}(t)$, and $\Delta_{odd}^{coh,Gd}(t)$ to Eq. (12.10),

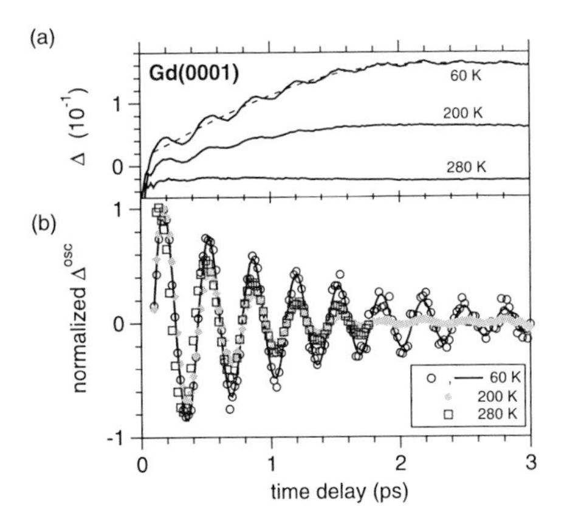

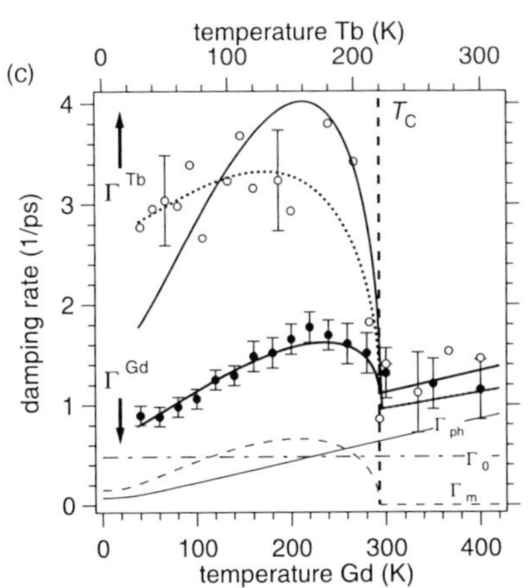

Figure 12.6 (a) Typical $\Delta_{\mathrm{even}}(t)$ traces of Gd(0001) at various temperatures. (b) Oscillatory components $\Delta^{\mathrm{coh}}_{\mathrm{even}}(t)$ normalized to their initial amplitude. (c) Temperature dependence of the decay rates $\Gamma = 1/\tau$ obtained from the fit of $\Delta^{\mathrm{coh}}_{\mathrm{even}}(t)$ to Eq. (12.10) for Gd (solid dots, to the bottom axis) and Tb (open circles, to the top axis). Solid curves in (c) present the fit to Eq. (12.16), which considers a constant contribution Γ_0 and two temperature-dependent contributions for phonon- (second term in Eq. (12.16)) and magnon-mediated damping (third term in Eq. (12.16)). For Gd, these contributions are displayed at the bottom and marked by Γ_0, Γ_{ph}, and Γ_{m}, respectively. For Tb, an additional magnetic scattering channel (fourth term in Eq. (12.16)) is required to describe the data at low temperatures, as shown by dotted curve. Reprinted with permission from [29]. Copyright 2008 by the American Physical Society.

respectively. They are shown in Figure 12.7a as a function of the reduced temperature T/T_{C} with the Curie temperature $T^{\mathrm{Gd}}_{\mathrm{C}} = 293$ K and $T^{\mathrm{Tb}}_{\mathrm{C}} = 221$ K. Here we exclude $\Delta^{\mathrm{Tb}}_{\mathrm{odd}}$ since it was not accessible owing to the strong magnetic anisotropy of Tb(0001) films on W(110), which does not allow to switch M by an external magnetic field. All the amplitudes in Figure 12.7 show similar temperature dependence: they reduce by an order of magnitude while the temperature rises from 40 K to T_{C} and then remain nearly constant above T_{C}. One of the reasons to perform such studies also for Tb(0001) is to check the importance of the *resonant* character of phonon–magnon coupling (see discussion above) for the CP *excitation*. In Tb, the phonon spectrum is widely identical to that in Gd, but the magnon frequency is 30% lower, which makes the phonon–magnon coupling nonresonant for the considered mode. However, Figure 12.7 clearly shows that $A^{\mathrm{Gd}}_{\mathrm{even}}(T/T_{\mathrm{C}}) \approx A^{\mathrm{Tb}}_{\mathrm{even}}(T/T_{\mathrm{C}})$. Since both were obtained under similar experimental conditions, we can conclude that the CP excitation is not linked to the *resonant* character of phonon–magnon coupling. However, the pronounced temperature dependence in Figure 12.7 resembles the temperature-dependent magnetization and suggests a considerable influence of the *magnetization* on the CP excitation.

The mechanism of this excitation can be sketched as follows. The pump laser pulse excites 5d electrons in the minority and majority valence subbands, as indicated by bold arrows in Figure 12.4. While in the majority channel the laser field excites

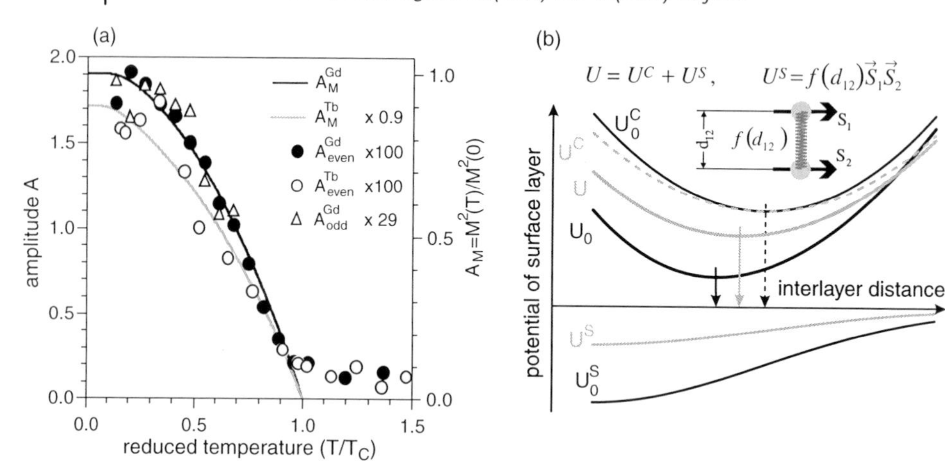

Figure 12.7 (a) Initial amplitudes of $\Delta_{even}^{coh}(t)$ and $\Delta_{odd}^{coh}(t)$ of Gd(0001) and $\Delta_{even}^{coh}(t)$ of Tb(0001) obtained from the fit to Eq. (12.10) (symbols, to the left axis) as a function of the reduced temperature (note the identical scaling coefficients for A_{even}^{Gd} and A_{even}^{Tb}); squared magnetization for the 4f electrons given by the Brillouin function with $J_{Gd} = 7/2$ and $J_{Tb} = 6/2$ (solid curves, to the right axis). (b) Schematic illustration for the interlayer oscillation potential modified by laser-induced spin disorder, explaining the spin-induced CP excitation. The ground-state potential U_0 is the superposition of "nonmagnetic" (or charge) and spin-dependent contributions U_0^C and U_0^S, respectively. After optical excitation, the spin-dependent part U^S has a smaller absolute value than U_0^S, but the same dependence on the interlayer distance. The charge component U^C might be slightly broadened with respect to U_0^C leading to lower frequency. Finally, the minimum of the excited-state potential $U = U^C + U^S$ is shifted with respect to that of U_0, which launches the CP. Reprinted from [23]. Copyright 2008 IOP Publishing Ltd.

electrons from the occupied $S^{\uparrow}$ component of the *localized* surface state into the unoccupied bulk band with *delocalized* character, minority electrons are excited from the occupied bulk band into the $S^{\downarrow}$ surface state component. As a result, the spin of the valence electrons is transferred from the surface layer into the underlying bulk, and the surface-state spin polarization is reduced considerably. However, owing to the strong *intra*atomic exchange interaction between 5d and 4f electrons [30], the 5d spin polarization will be partially recovered by the cost of 4f magnetic moment in the surface layer. This process of angular momentum transfer has a characteristic timescale of 10 fs [31]. Finally, the laser excitation leads to a considerable reduction of the total (i.e., 4f and 5d) spin $\vec{S}_1$ via 4f to 5d spin transfer in the surface layer and, subsequently, surface-to-bulk transport of 5d spin polarization. The reduction is proportional to the initial effective spin:

$$\Delta\vec{S}_1 = \gamma\vec{S}_1, \qquad \vec{S}_1^* \equiv \vec{S}_1 - \Delta\vec{S}_1 = (1-\gamma)\vec{S}_1. \tag{12.12}$$

Here the factor γ increases with increasing pump fluence. From the initial drop in Δ_{odd} shown in Figure 12.5a, γ is estimated to be ≈ 0.5. Now we consider the potential of the interlayer vibration in the form

$$U_{12}(d_{12}) = U_{12}^{C}(d_{12}) + U_{12}^{S}(d_{12}), \tag{12.13}$$

where U^C depends only on the electronic density (charge-dependent term) and can be approximated by a parabola $U_{12}^C(d_{12}) = k(d_{12}-d_{12}^0)^2/2$ in the vicinity of its minimum at d_{12}^0, which is indicated by a dashed arrow in Figure 12.7b. The spin-dependent contribution U^S defined by Eq. (12.11) is calculated after Ref. [27] and schematically shown in Figure 12.7b. In the vicinity of d_{12}^0, it can be approximated by $U_{12}^S(d_{12}) \approx \alpha(d_{12}-d_{12}^0)\vec{S}_1\vec{S}_2 + f(d_{12}^0)\vec{S}_1\vec{S}_2$, where $\alpha = f'(d_{12}^0)$. The minimum of U_{12} defines the equilibrium interlayer distance of the nonperturbed system $d'_0 = d_0 - \alpha\vec{S}_1\vec{S}_2/k$ (black arrow in Figure 12.7b). The excitation shifts the minimum to $d_0'^{*} = d_0 - \alpha\vec{S}_1^{*}\vec{S}_2/k$ (gray arrow in Figure 12.7b) with $\vec{S}_1^{*}$ defined by Eq. (12.12). Since the spin $\Delta\vec{S}_1$ is redistributed within the large number of bulk atomic planes, we neglect $\Delta\vec{S}_2$ here. The shift

$$\Delta d'_0 = d_0'^{*} - d'_0 = \alpha\left(\vec{S}_1 - \vec{S}_1^{*}\right)\vec{S}_2/k = \alpha\gamma\vec{S}_1\vec{S}_2/k \tag{12.14}$$

gives the amplitude of the displacively excited coherent interlayer vibration around a new (quasi-)equilibrium position. In addition, the electronic excitation may lead to the softening of the potential $U^C(d_{12})$ with respect to $U_0^C(d_{12})$ (gray dashed curve in Figure 12.7b), which would reduce the CP frequency. Taking into account that for a small lattice displacement $\Delta_{\text{even}} \propto \Delta d_{12}$ (see Figures 12.3 and 12.4) and in Eq. (12.14)$S_{1,2} \propto M$, we obtain the temperature dependence of CP amplitude:

$$A(T) \propto M^2(T). \tag{12.15}$$

To calculate $M(T)$, we use the Brillouin function with $S_{\text{Gd}} = 7/2$ and $S_{\text{Tb}} = 6/2$ representing the magnetic moment of 4f electrons. The calculated squared magnetizations for Gd and Tb are plotted to the right axis of Figure 12.7a. They demonstrate an excellent agreement with the temperature dependence of the experimental amplitude at $T < T_C$. The small nonzero amplitude at $T > T_C$ can be attributed to the nonmagnetic displacive excitation mechanism (Figure 12.1) [21]. It is based on the different efficiency of pump-induced electronic transitions involving the surface state among majority and minority d-bands (thick arrows in Figure 12.4) and leading to a redistribution of the charge density between the surface and the bulk, which might also shift the vibration potential. Hence, we conclude that at lanthanide surfaces, two excitation mechanisms contribute to CP generation: a nonmagnetic displacive excitation and a spin-dependent mechanism. The latter is active only when the system is ferromagnetic, that is, at $T < T_C$.

To make a quantitative estimation, *ab initio* calculations were performed for the case of Gd in Ref. [23]. The experimentally observed reduction of the surface spin polarization by 50% ($\gamma = 0.5$), cannot be reproduced. Nevertheless, the metastable state of the surface characterized by antiferromagnetic ordering with respect to the bulk ($\gamma = 2$) can be realized in the model *ab initio* calculations, which give the CP amplitude in the real space $\delta = 4.5$ pm [23]. By linear interpolation, we can estimate $\delta \approx 1.1$ pm for 50% reduction of spin polarization, which is in excellent agreement with the results presented in Section 12.3.1.

12.3.2.4 **Damping of the Coherent Phonons**

Temperature dependence of CPs gives us more insight. Figure 12.6a shows that the CP relaxation time τ (cf. Eq. (12.10)) at Gd(0001) depends on the temperature T. This is emphasized in Figure 12.6b presenting the coherent part $\Delta_{\text{even}}^{\text{coh}}(t)$ normalized to the initial amplitude. One can see in Figure 12.6c that for both Gd and Tb, the damping rate $\Gamma(T) = 1/\tau(T)$ first increases with increasing T, but drops abruptly right below T_C. The observation indicates a significant contribution of the magnetically ordered spin subsystem to the CP damping at $T < T_C$. Above T_C, the damping in Tb is identical to that in Gd, which can be understood by considering their very similar valence electronic and lattice structures. To understand why at low temperatures the damping in Tb is about three times larger, we recall that in the 4f shell of Gd the orbital momentum $L = 0$, while in the case of Tb $L = 3$. This means that in Tb, the large spin–orbit interaction occurs, leading to stronger spin–lattice coupling, in contrast to weak indirect coupling through 5d electrons for Gd. Such difference in the spin–lattice coupling leads to about two orders of magnitude larger static magnetoelastic constants in the bulk Tb than those in the bulk Gd [1]. In transient dynamics, however, the difference appears to be moderated.

To elucidate the nature of CP damping further, we develop a phenomenological model [23, 29]. We can classify the channels of the CP damping into two: (i) spin-independent ones and (ii) those associated with the magnetic order. We assume the efficiency of the latter damping to be proportional to the spontaneous magnetization $M(T)$. We consider the temperature-dependent damping in the form of

$$\Gamma(T) = \Gamma_0 + \Gamma_p F_B(T) + \Gamma_m F_B(T) M(T) + \Gamma_s M(T), \tag{12.16}$$

$$F_B(T) = 1 + 2\exp(\hbar\Omega/2kT) - 1. \tag{12.17}$$

The spin-independent contribution to CP damping originates from interaction with defects, electrons, and thermal (incoherent) phonons [32]. Since the dynamics is driven by light, optically excited electrons and incoherent phonons excited by electron–phonon scattering contribute in particular to metallic systems where low-energy excitations are allowed around E_F [33, 34]. We approximate the contribution of defects and electrons to $\Gamma(T)$ by a temperature-independent Γ_0 in Eq. (12.16), because the excitation density is kept constant with T. Following Refs [33, 35], the thermal phonon contribution is attributed to anharmonic decay of an optical CP with a frequency Ω and a momentum $\boldsymbol{q}_0 = 0$ into two acoustic phonon modes with $\Omega_1 = \Omega_2 = \Omega/2$ and $\boldsymbol{q}_1 = -\boldsymbol{q}_2$. It is described by the second term in Eq. (12.16), where the factor (12.17) accounts for a temperature-dependent population of acoustic modes according to Bose statistics. These two contributions lead to a monotonic increase in Γ with T [33, 35].

Analogous to spin-independent damping, we introduce a temperature-dependent and temperature-independent scattering rate in the spin-dependent contribution. The spin dependence is represented by the factor $M(T)$ defining the density of spins available for the excitation, which is calculated by the Brillouin function. Since this factor itself depends on the temperature, both spin-dependent contributions to CP damping are finally temperature dependent. The first contribution is associated with anharmonic CP decay with creation of two acoustic magnons. It is represented by the third term in Eq. (12.16), which is the spin-dependent analogue of the second term. This analogy with the phonon–phonon decay is valid due to high density of acoustic magnon modes (with opposite wave vectors due to the inversion symmetry) at frequencies about $\Omega/2$ (see Figure 4c and d in Ref. [23]). The factor Γ_m reflects therefore the efficiency of spin–lattice coupling provided by the interaction of *collective* excitations of the lattice and spin subsystems, which are (optical) phonons and (acoustic) magnons, respectively.

These three contributions give a good fit in the case of Gd (Figure 12.6c) with $\Gamma_0 = 0.48 \pm 0.05\ \text{ps}^{-1}$, $\Gamma_p = 0.075 \pm 0.009\ \text{ps}^{-1}$, and $\Gamma_m = 0.16 \pm 0.02\ \text{ps}^{-1}$ for $M(0) = 1$, $\Omega/2\pi = 3$ THz. To fit the results for Tb we fix the values of Γ_0 and Γ_p to those for Gd because of the very similar valence electronic and lattice structures. The fit (upper solid curve in Figure 12.6c) describes the pronounced variation at T_C giving $\Gamma_m = 1.03 \pm 0.02\ \text{ps}^{-1}$, but at $T < 150$ K the weak variation of $\Gamma(T)$ is overestimated. Owing to that, we add the fourth term to Eq. (12.16), which is the spin-dependent analogue of the first one. If the first term describes CP scattering through electrons and defects, the fourth one describes the CP scattering through *local* perturbations in the orientation of 4f magnetic moments. The factor Γ_s reflects therefore the efficiency of spin–lattice coupling provided by the interaction of *collective* excitations of the lattice (phonons) with *localized* 4f magnetic moments. Since in Tb 4f moments are coupled to static lattice deformations much stronger than in Gd [1], we suppose that this "local" fourth term in Eq. (12.16) is essential in the case of Tb but not of Gd. The combination of two spin-dependent contributions gives an excellent fit (dotted curve in Figure 12.6c) with $\Gamma_m = 0.38 \pm 0.05\ \text{ps}^{-1}$ and $\Gamma_s = 1.8 \pm 0.1\ \text{ps}^{-1}$. However, its origin requires further investigations.

12.3.2.5 **Future Developments**

The developed model description of coherent dynamics optically excited at lanthanide surfaces is self-consistent and brings into agreement the known experimental and theoretical results. However, to exclude all possible ambiguities, additional experimental proofs would be beneficial. Basically, there are two aspects lacking further investigations. First, our conclusions on the surface-state spin dynamics are based on SHG. Here an independent experimental proof such as time- and spin-resolved photoemission would be useful. Second, since the loss of the surface spin polarization observed by SHG regards the polarization of valence electrons and in the model consideration we operate with the total magnetic moment, we need to *postulate* an identical dynamics of 5d and 4f spins, basing on the strong f–d coupling. Thus, to proof this scenario, we need to perform experiments directly addressing the 4f shell of the surface layer. Developing this line of studies, we have already investigated the 4f spin dynamics in the *bulk* Gd by magnetic linear dichroism in time-resolved 4f core-level photoemission [36] and X-ray magnetic circular dichroism. The comparison to the spin dynamics in bulk Gd valence band investigated by the magneto-optical Kerr effect, which will be discussed in future work, shows that these two spin subsystems behave very similar as a function of time delay after the optical excitation, which supports our current understanding.

12.4 Conclusion

The combination of such complementary surface-sensitive techniques as time-resolved PE and magneto-induced SHG is powerful for detailed studies of surface electron, spin, and lattice dynamics. The exchange split surface state provides an optical excitation of a CP mode at lanthanide surfaces coupled to a coherent response of surface spin polarization and its sensitive detection by both techniques. This facilitates a detailed insight into ultrafast processes leading to excitation and damping of this particular mode.

Two different mechanisms of CP excitation at Gd(0001) and Tb(0001) surfaces are identified and compared quantitatively. (i) A temperature-independent process enables CP generation above T_C. It is attributed to the redistribution of optically excited charge carriers among surface and bulk states. In a simplified picture, it resembles electronic excitation from a ground-state potential to an excited-state one with the potential minimum shifted along the vibration coordinate. (ii) The second process clearly dominates below T_C and increases the efficiency of CP excitation with decreasing temperature. Since it resembles the temperature-dependent magnetization, we conclude that it is based on the ferromagnetic order. This conclusion is confirmed by *ab initio* calculations showing that inversion of spin polarization of the surface atomic plane results in an increase in the equilibrium interlayer distance at the surface. This effect can be viewed as a "dynamic form of surface magnetostriction" if we consider the response of the lattice to the change of surface magnetization while the bulk magnetization remains unchanged.

The magnetic state also opens additional damping processes that are absent in a *para*- or diamagnetic material, which we identify as phonon–magnon scattering. At Gd(0001) with its vanishing orbital momentum of the 4f shell, the temperature-dependent damping of CP is well described by a temperature-dependent anharmonic decay, including phonon–phonon scattering and phonon–magnon term vanishing at T_C and a temperature-independent contribution attributed to the scattering at defects and electron–hole pair excitations. At Tb(0001) surfaces, where the large orbital momentum of the 4f shell enhances the spin–lattice interaction compared to Gd, the damping of the coherent phonon is several times larger below T_C than found for the Gd(0001) surface. This confirms the decisive role of spin-dependent relaxation channels.

Overall, our systematic and complementary data on Gd and Tb demonstrate the coupling of lattice motion to the dynamics of the spin subsystem. We expect that our study triggers further theoretical work because, as we have shown here, a rather simple parametrization of optically induced effects in the magnetization leads to a reasonable description of the lattice rearrangement. Thus, an improved theoretical treatment of the optically excited state might be a promising pathway to arrive at a comprehensive description of the coupled electron, lattice, and spin dynamics in solids.

Acknowledgments

The authors would like to thank all coworkers who contributed to the presented results over the recent years. In particular, they are grateful to M. Lisowski and P.A. Loukakos who carried out the TRPE experiments with their remarkable skills and dedication, I. Radu and A. Povolotskiy who made various essential contributions to the nonlinear optical experiments, and T.O. Wehling and A.I. Lichtenstein who performed *ab initio* calculations of spin-dependent CP excitation. The numerous fruitful discussions with them and with K. Starke, E. Matthias, and M. Wolf laid the foundation of the comprehensive study presented here. Financial support of the Deutsche Forschungsgeminschaft through SPP 1133 and SFB 668 is gratefully acknowledged.

References

1 Coqblin, B. (1977) *The Electronic Structure of Rare-Earth Metals and Alloys*, Academic Press, London.
2 Chikazumi, S. (1964) *Physics of Magnetism*, John Wiley & Sons, Inc., New York.
3 Lisowski, M., Loukakos, P.A., Bovensiepen, U., Stähler, J., Gahl, C., and Wolf, M. (2004) *Appl. Phys. A*, **78**, 165.
4 Anisimov, S.L., Kapeliovich, B.L., and Perel'man, T.L. (1974) *Sov. Phys. JETP*, **39**, 375.
5 Allen, P.B. (1987) *Phys. Rev. Lett.*, **59**, 1460.
6 Bovensiepen, U. (2007) *J. Phys. Condens. Matter*, **19**, 083201.
7 Kurz, Ph., Bihlmayer, G., and Blügel, S. (2002) *J. Phys. Condens. Matter*, **14**, 6353.
8 Donath, M., Gubanka, B., and Passek, F. (1996) *Phys. Rev. Lett.*, **77**, 5138.
9 Maiti, K., Malagoli, M.C., Dallmeyer, A., and Carbone, C. (2002) *Phys. Rev. Lett.*, **88**, 167205.
10 Lisowski, M., Loukakos, P.A., Melnikov, A., Radu, I., Ungureanu, L., Wolf, M., and Bovensiepen, U. (2005) *Phys. Rev. Lett.*, **95**, 137402.
11 Loukakos, P.A., Lisowski, M., Bihlmayer, G., Blügel, S., Wolf, M., and Bovensiepen, U. (2007) *Phys. Rev. Lett.*, **98**, 097401.
12 Brorson, S.D., Kazeroonian, A., Moodera, J.S., Face, D.W., Cheng, T.K., Ippen, E.P., Dresselhaus, M.S., and Dresselhaus, G. (1990) *Phys. Rev. Lett.*, **64** (18), 2172.
13 Hohlfeld, J., Wellershoff, S.-S., Güdde, J., Conrad, U., Jähnke, V., and Matthias, E. (2000) *Chem. Phys.*, **251**, 237.
14 Del Fatti, N., Voisin, C., Achermann, M., Tzortzakis, S., Christofilos, D., and Vallée, F. (2000) *Phys. Rev. B*, **61** (24), 16956.
15 Rao, R.R. and Menon, C.S. (1974) *J. Phys. Chem. Solids*, **35**, 425.
16 Shen, Y.R. (1984) *The Principles of Nonlinear Optics*, John Wiley & Sons, Inc., New York.
17 Pan, R.-P., Wei, H.D., and Shen, Y.R. (1989) *Phys. Rev. B*, **39**, 1229.
18 Melnikov, A., Radu, I., Bovensiepen, U., Starke, K., Matthias, E., and Wolf, M. (2005) *J. Opt. Soc. Am. B*, **22**, 204.
19 Melnikov, A., Krupin, O., Bovensiepen, U., Starke, K., Wolf, M., and Matthias, E. (2002) *Appl. Phys. B*, **74**, 723.
20 Melnikov, A., Güdde, J., and Matthias, E. (2002) *Appl. Phys. B*, **74**, 735.
21 Melnikov, A., Radu, I., Bovensiepen, U., Krupin, O., Starke, K., Matthias, E., and Wolf, M. (2003) *Phys. Rev. Lett.*, **91**, 227403.
22 Rehbein, A., Wegner, D., Kaindl, G., and Bauer, A. (2003) *Phys. Rev. B*, **67**, 033403.
23 Melnikov, A., Radu, I., Povolotskiy, A., Wehling, T., Lichtenstein, A., and Bovensiepen, U. (2008) *J. Phys. D*, **41**, 164004.
24 Bovensiepen, U. (2006) *Appl. Phys. A*, **82**, 395.

25 Bovensiepen, U., Melnikov, A., Radu, I., Krupin, O., Starke, K., Wolf, M., and Matthias, E. (2004) *Phys. Rev. B*, **69**, 235417.

26 Zeiger, H.J. *et al.* (1992) *Phys. Rev. B*, **45**, 768.

27 Turek, I., Kudrnovski, J., Bihlmayer, G., and Blügel, S. (2003) *J. Phys. Condens. Matter*, **15**, 2771.

28 Koehler, W.C., Child, H.R., Niklow, R.M., Smith, H.G., Moon, R.M., and Cable, J.W. (1974) *Phys. Rev. Lett.*, **24**, 16.

29 Melnikov, A., Povolotskiy, A., and Bovensiepen, U. (2008) *Phys. Rev. Lett.*, **100**, 247401.

30 Ahuja, R., Auluck, S., Johansson, B., and Brooks, M.S.S. (1994) *Phys. Rev. B*, **50**, 5147.

31 Calculated by Lichtenstein, A.I., Katsnelson, M.I., Wehling, T.O. (private communications).

32 Matsumoto, Y. and Watanabe, K. (2005) *Chem. Rev.*, **106**, 4234.

33 Hase, M., Ishioka, K., Demsar, J., Ishida, K., and Kitajima, M. (2005) *Phys. Rev. B*, **71**, 184301.

34 Watanabe, K., Takagi, N., and Matsumoto, Y. (2005) *Phys. Rev. B*, **71**, 085414.

35 Hase, M., Mizoguchi, K., Harima, H., Nakashima, S., and Sakai, K. (1998) *Phys. Rev. B*, **58**, 5448.

36 Melnikov, A., Prima-Garcia, H., Lisowski, M., Giessel, T., Weber, R., Schmidt, R., Gahl, C., Bulgakova, N.M., Bovensiepen, U., and Weinelt, M. (2008) *Phys. Rev. Lett.*, **100**, 107202.

Part Three
Heterogeneous Electron Transfer

Dynamics at Solid State Surface and Interfaces Vol. 1: Current Developments
Edited by Uwe Bovensiepen, Hrvoje Petek, and Martin Wolf

ISBN: 978-3-527-40937-2

13
Studies on Auger Neutralization of He^+ Ions in Front of Metal Surfaces

Stephan Wethekam and Helmut Winter

13.1
Introduction

Charge transfer phenomena play an important role during the interactions of atoms and ions in front of solid surfaces. As a consequence, a large body of work has been performed over the last decades on this topic that is of interest both in fundamental research and in a variety of applications [1–5]. As typical examples, we mention particle detection, ion beam analysis, plasma wall interactions, catalytic processes, or spectroscopic tools in surface science. The transfer of electrons for atomic particles in front of the surface of a solid proceeds via tunneling mechanisms and is initiated as soon as the potential barrier is sufficiently lowered. In front of a metal surface, two different mechanisms play a dominant role: (1) resonant electron transfer where the energy of the tunneling electrons is conserved and (2) the Auger capture process, an inelastic tunneling mechanism where energy is conserved by the uptake of energy by a second electron or electronic excitations of the surface (e.g., plasmon).

Here, we will focus on the neutralization of He^+ ions in front of metal surfaces that proceeds owing to the large binding energy of He atoms via the Auger capture process as shown in Figure 13.1. The suppression of resonant electron transfer processes for the ground and excited states of He atoms makes the neutralization of He^+ ions a model system for studies on Auger neutralization (AN) in front of metal surfaces [6]. From conservation of energy, it is straightforward to derive the energy transfer to a second electron. For neutralization of an atom of binding energy E_a in front of a metal with work function W and Fermi energy E_F, the energy of Auger electrons with respect to vacuum is

$$E_e = E_a - 2W - 2E_F + E_1 + E_2, \tag{13.1}$$

where the reference of the initial electron energies E_1, E_2, and E_F is the bottom of the conduction band. For Auger transitions with two Fermi electrons involved ($E_1 = E_2 = E_F$), one finds $E_e = E_a - 2W$ and thus $E_a > 2W$ as the condition for the Auger emission of electrons.

Dynamics at Solid State Surface and Interfaces Vol.1: Current Developments
Edited by Uwe Bovensiepen, Hrvoje Petek, and Martin Wolf

ISBN: 978-3-527-40937-2

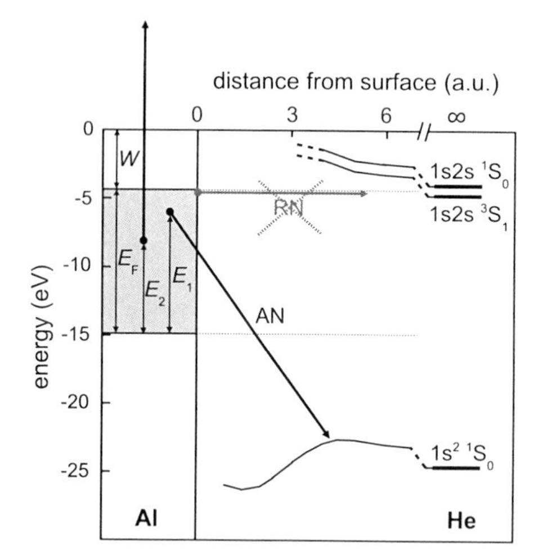

Figure 13.1 Sketch of energy diagram for interaction of He atom with aluminum surface and of Auger neutralization(RN: resonant neutralization). Note energy level shift when the atom approaches the surface.

For the neutralization of He$^+$ ions ($E_a = 24.6$ eV), this condition is generally fulfilled so that the Auger process was studied, initiated by the pioneering work of Hagstrum [7, 8], in detail via the spectroscopy of emitted electrons and analyzed in terms of a microscopic modeling of the interaction mechanisms. In this description of the charge transfer dynamics, transition rates for AN play a pivotal role where the calculation of AN rates for ions in front of a metal surface is even for a "simple" atomic system such as He atoms a challenging task. Since such rates enter into electron spectra in a fairly indirect manner, recent progress in studies on AN was achieved by scattering experiments from metal surfaces under a grazing angle of incidence. In these studies, the observation of tiny fractions of ions, which have survived the complete collision process in front of the surface, as well as effects of dielectric response phenomena on projectile trajectories provided new insights into the AN process. It turns out that besides providing an accurate information on the AN transition rates, the level shifts of atoms in front of the surface play a key role in understanding the neutralization of noble gas atoms in front of metal surfaces.

13.2 Concept of Method

The method applied here for the study of the charge transfer dynamics for He$^+$ ions in front of a metal surface is based on grazing ion–surface collisions [9] where the projectiles are scattered from the surface under glancing angles of incidence of typically 1°. In this regime of scattering, the projectiles are steered in terms of small angle collisions by atoms of the topmost surface layer, so-called "surface channeling," giving rise to well-defined trajectories for an ensemble of incident atomic particles [9, 10]. The effective scattering potential for projectiles results to a major extent from the superposition of individual pair potentials between projectile and surface atoms, where for channeling conditions this potential can be well approximated by a continuum approach [11]. Then, for a random azimuthal orientation of a crystal surface, this continuum potential depends on the distance from the surface z only so that the motion of projectiles during the scattering event can be separated in (1) a "slow" one along the surface normal with energy $E_z = E_0 \sin^2\Phi_{in}$ and (2) a "fast" one parallel with respect to the surface plane with energy $E_{||} = E_0 \cos^2 \Phi_{in} \approx E_0$ with E_0 being the kinetic energy of the incident projectile and Φ_{in} the grazing angle of incidence. As a representative example, one has for a 2 keV beam of He atoms/ions and $\Phi_{in} = 1°$ a normal energy $E_z = 0.6$ eV corresponding to a velocity for the approach to the surface $v_z = 2.45 \cdot 10^{-3}$ a.u. = 1 Å/19 fs. This intrinsic high time resolution will play an important role for the further discussion.

Typical AN rates Γ close to a metal surface amount to $\Gamma \approx 10^{-2}$ a.u. [6] that correspond to lifetimes for inner shell holes of the interacting ion in the fs domain. A variety of studies outlined within this book demonstrate that such a high time resolution can be achieved routinely in laser-driven pump–probe experiments [12], and for atoms in the gas phase the Auger decay has been resolved recently [13]. In the upper panel of Figure 13.2, we present a simple sketch of the concept to study the dynamics of AN in front of a metal surface that contains the essential ingredients of the pump–probe technique. In the first step, an inner shell hole in the adsorbed atom is generated where for He atoms a photon in the EUV is needed to reach vacuum energies. After a time delay Δt in the fs domain, the filling of the hole is probed by a second pulse via the emission of another electron.

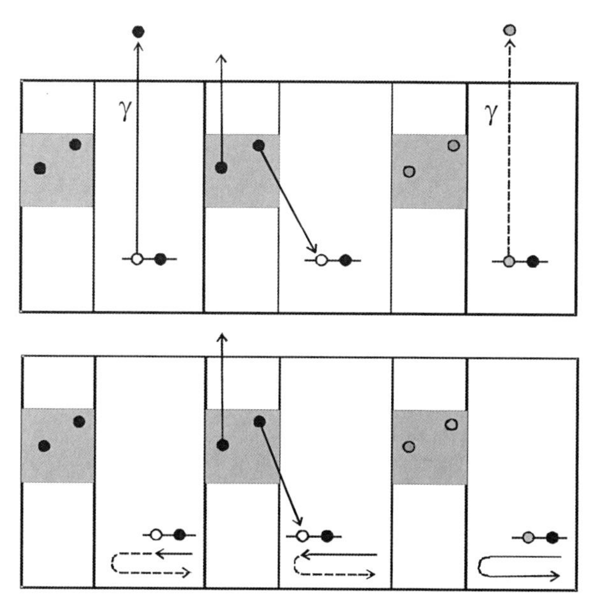

Figure 13.2 Schemes for investigating Auger neutralization dynamics by means of pump–probe (upper panels) and ion scattering (lower panels) experiments.

In the lower panel of Figure 13.2, we show the concept of similar studies using ion scattering where the high time resolution results from the fast motion of the projectiles. Here, the atoms/ions probe a much wider range of distances from the surface compared to the specific adsorption sites in the other scheme. During the approach of the incident ion toward the surface, the 1s hole of the He^+ ions is filled with enhanced probability (rate). For sufficiently small energies/velocities, the projectiles reach a distance of closest approach z_{min} and escape thereafter from the close vicinity of the surface. The key issue for the method described here is the feature that for suitably chosen scattering conditions the AN rates and collision times are such that small fractions of incident He^+ ions survive the complete scattering process. Then, these fractions can be used to study the interaction dynamics in detail. We will demonstrate that important new aspects with respect to the neutralization of He^+ ions in front of metal surfaces can be revealed by application of ion scattering.

13.2.1
Trajectories and Time Regime

In the study of charge transfer in grazing ion surface collisions, the trajectories play an essential role since these determine the regime of distances from the surfaces and the interaction times with the solid. In the Thomas–Fermi model of electronic screening for interatomic pair potentials, one obtains the averaged planar continuum potential based on the approximation proposed by Moliére [14] $U(z) = 2\,\pi\, n_s Z_1 Z_2\, a\, \Sigma_i a_i/b_i \exp(-b_i z/a)$, where n_s is the area density of atoms in the topmost layer, Z_1 and Z_2 are the atomic numbers of projectile and surface atoms, $a_i = (0.35, 0.55, 0.1)$, $b_i = (0.3, 1.2, 6)$, and the screening length $a = 0.885\,(Z_1^{1/2} + Z_2^{1/2})^{-2/3}$ [11].

For grazing scattering, typical distances of closest approach amount to more than 1 a.u. The "critical energy" for planar channeling is $E_{crit} = 2\pi\, a\, n_s Z_1 Z_2$ [11]. This energy relates to a critical angle $\Phi_{crit} = (E_{crit}/E_0)^{1/2}$ that amounts for keV ions scattered from metal surfaces to grazing angles Φ_{in} of several degrees. This means that all the experiments discussed here are performed in the regime of planar channeling, that is, the scattering process proceeds in well-defined trajectories for an ensemble of projectiles in front of the topmost layer of surface atoms. For the detailed structure of trajectories, additional contributions may have a considerable influence on their exact shape. In particular, the effects of embedding the projectiles into the electron gas in the selvedge of a metal surface [15, 16], thermal vibrations of lattice atoms, and dielectric response phenomena [17] (predominantly the image potential) can play an important role).

Basic features of the Auger neutralization of ions impinging on metal surfaces were already pointed out by Hagstrum in his pioneering studies starting around 1950 [7]. It turns out that the AN-transition rates are sufficiently high so that a dominant fraction of incident ions gets neutralized before the projectiles have reached the surface or, for grazing scattering, the distance of closest approach to the solid. In passing we note that an interesting detail of minor interest in the early work turns out to provide essential information on the interaction scenario: (small) fractions of ions that survived the complete scattering event [18].

On the basis of Hagstrum's work, an estimate of the time regime for the AN process can be performed straightforward. For sufficiently large distances from the surface, the approach of ions can be assumed to proceed with constant velocity v_z so that for an AN rate approximated by the exponential decay $\Gamma_A(z) = \Gamma_A(0)\exp(-z/z_c)$ the solution of the rate equation for the ion fraction $dP^+/dt = -\Gamma_A(z)\, P^+$ is $P^+(z) = P^+(z(t = -\infty))\exp[-\exp(z_s - z)/z_c)]$. Then, the remaining fraction of ions shows the strongest change with distance at $z_s = z_c \log(\Gamma_A(0) z_c / v_z)$, where distance z and time t are related via v_z. For hyperthermal scattering or grazing collisions of keV ions, v_z amounts to some 10^{-3} a.u. so that for typical parameters for AN $((\Gamma_A(0) = 0.1$ a.u., $z_c = 1.5$ a.u.) z_s is about 5 a.u., that is, well in front of the surface.

In Figure 13.3, we show for the scattering of He^+ ions with an energy for the normal motion of $E_z = 4$ eV the evolution of P^+, z, and the AN rate as function of interaction time ($t = 0$ is adjusted to apex of trajectory = distance of closest approach to surface). The plot reveals that electron transfer from the solid to the ion proceeds in about 20 fs. The time resolution intrinsic to ion scattering is substantially higher and allows one to apply this technique for a detailed study of the AN process for He^+ ions.

The ion fractions after grazing surface collisions are generally small; however, for many ion–metal systems, they are sufficiently high for a reliable detection of P^+. An interesting aspect of (planar) surface channeling is the presence of defined projectile trajectories that can be varied in a controlled manner by changing the angle of incidence Φ_{in} as shown in Figure 13.4. For larger angles of incidence, the projectiles reach distances closer to the surface, but the path length and the interaction time in the vicinity of the surface is shorter than for scattering under smaller angles. In this manner, contributions of different distances from the surface on charge transfer can be probed where the outcome of the scattering process is given by

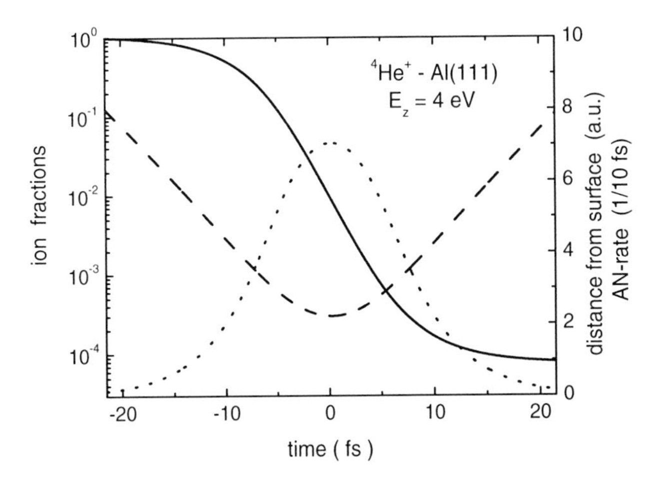

Figure 13.3 Fraction of surviving ions (solid curve), distance from surface (corresponds to trajectory, dashed curve), and AN rate (dotted curve) as function of time (apex of trajectory defines $t = 0$) for scattering of $^4He^+$ ions from Al(111) surface with $E_z = 4$ eV.

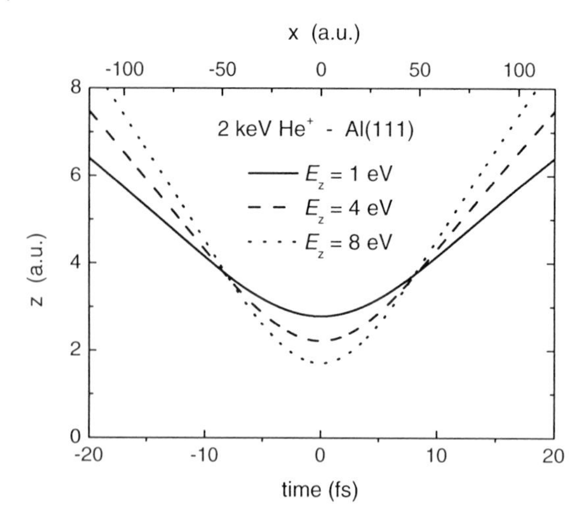

Figure 13.4 Classical projectile trajectories for grazing scattering of 2 keV He$^+$ ions from Al(111) surface under different normal energies.

$$P^+ = \exp\left(-\int_{\text{traj}} \Gamma_{\text{AN}}(z)\text{d}t\right) = \exp\left(-\int_{\text{traj}} \Gamma_{\text{AN}}(z)\frac{\text{d}z}{v_z(z)}\right). \quad (13.2)$$

This can be used in applying concepts of tomography to the measurements of ion survival. By this type of technique, information on the AN rate Γ_{AN} over the interval of distances can be obtained where the dominant part of charge transfer takes place. In this respect, ion scattering provides more general information than methods that probe the interaction dynamics of adsorbed atoms at fixed equilibrium positions only.

The trajectories are affected by the charge state of the projectiles. This can be understood simply by the role of the image charge of ionized species in front of a metal surface that gives rise to an attractive image potential. Then, ions are slightly accelerated toward the surface plane prior to their neutralization so that angular deflections with respect to the exit angle for incident neutral and charged projectiles can be found [9, 19]. In general, however, the resulting potentials for ions and atoms in front of metal surfaces are an intricate problem. This holds, in particular, for the region close to the surface where the classical approximation for the image potential $-1/4(z - z_{\text{im}})$ fails (z_{im} is distance of image reference plane from surface). On a more general scope, it is straightforward with the help of the sketch in Figure 13.5 to relate the difference in potential energies for neutral He^0 atoms and He$^+$ ions $\Delta E_z = V_{\text{He}^0}(z) - V_{\text{He}^+}(z)$ to the level shift of the 1s atomic ground state $\Delta E_{1s}(z)$ [6]. This level shift can be derived from the exit angles for scattering of neutral and charged projectiles $\Phi^{\text{o}}_{\text{out}}$ and Φ^{+}_{out}, respectively, from $\Delta E_z = E_0\,(\sin^2\Phi^{+}_{\text{out}} - \sin^2\Phi^{\text{o}}_{\text{out}})$.

After discussing the specific features of the experimental method used in our work, we will now outline detailed studies of the charge transfer dynamics of He$^+$ ions in front of metal surfaces. It will be evident that the level shift for the ground state of the

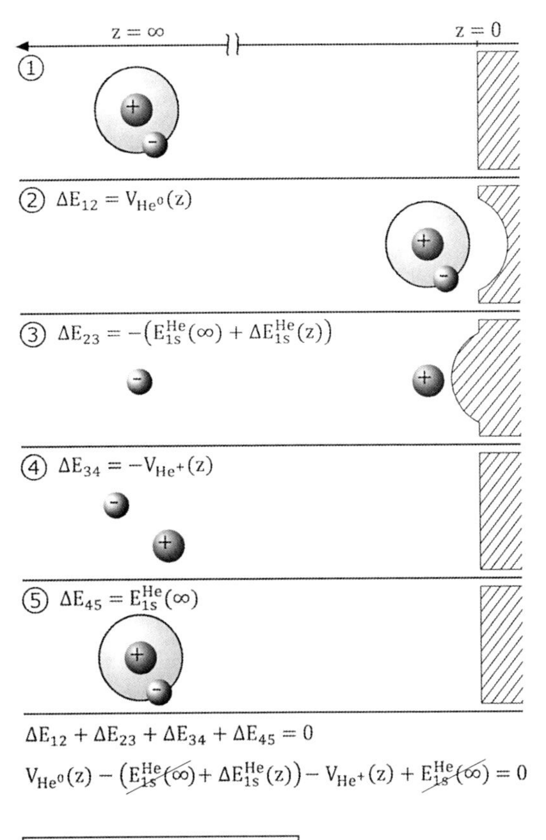

$$\boxed{\Delta E_{1s}^{He}(z) = V_{He^0}(z) - V_{He^+}(z)}$$

Figure 13.5 Sketch of cycle for illustration of relation between interaction potentials for He^0 atoms and He^+ ions and energy shift of atomic ground state. He^0 in ground state moved in front of surface (distance z) and ionized. Core (He^+) removed from surface and atom recombined, so that initial situation recovered. Energy balance yields $\Delta E_{1s}(z) = V_{He}{}^0(z) - V_{He}{}^+(z)$.

He atom plays an important role in the microscopic understanding of the neutralization of He^+ ions in front of metal surfaces.

13.3
Studies on Auger Neutralization of He^+ Ions

The method applied here for studying the charge transfer dynamics for He^+ ions in front of a metal surface is based on an ion scattering method where two specific features of importance under a grazing angle of incidence are essential: (1) the presence of fractions of surviving ions in the reflected beam and (2) the shift of the angular distributions after scattering of neutral and ionized projectiles. The ion

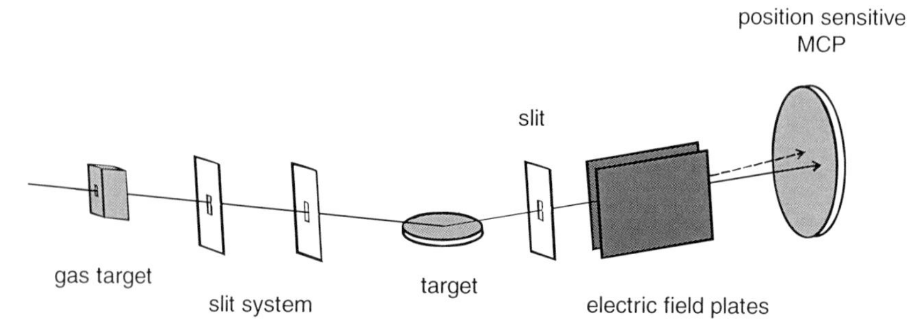

Figure 13.6 Experimental setup for studying AN of He$^+$ ions during grazing scattering from metal surface.

fractions are closely related to the AN rates, whereas the angular shifts provide information on the level shift of the atomic ground state.

In Figure 13.6, we show a sketch of the experimental setup for studying grazing ion surface scattering. Typical angles of incidence Φ_{in} amount to some degrees so that a fairly high angular resolution is needed. This is achieved by collimation of the projectile beam by sets of vertical and horizontal slits to a divergence in the 0.01° domain. The slit systems are also used as components of two differential pumping stages for operating in the beam line a gas cell for charge transfer in order to produce fast neutral projectiles and to maintain a base pressure in the lower 10^{-11} mbar regime within the UHV chamber. The metal surfaces, for example, Al(111) and Ag (111), are prepared by cycles of grazing sputtering with 25 keV Ar^+ ions and subsequent annealing. In addition to ion scattering, the surfaces are characterized by standard surface analytical tools such as Auger electron spectroscopy, LEED, or measurements of the target work function. The preferential emission of electrons for scattering of the projectile beam along low indexed directions in the surface plane (axial surface channeling) provides a simple online method to adjust the setting of the target azimuth [9, 20] to high index axial channels (random orientation). The charge fractions of the beam reflected from the target surface are measured by means of a pair of electric field plates in order to disperse the beam with respect to charge. The detection of particles is achieved with a position-sensitive microchannel plate detector that allows one to simultaneously record complete angular distributions.[1)]

In Figure 13.7, we show polar angular distributions for scattered particles obtained in our studies. In the upper panel, we have plotted for scattering of 1 keV He^0 atoms (full circles) and He$^+$ ions (open circles) the intensity of backscattered atoms (beam is almost completely neutralized, see discussion below) as a function of the angle of exit for $\Phi_{in} = 1.6°$ ($E_z = 0.8$ eV). The angular distribution for incident ions reveals a pronounced shift compared to the distribution of neutral atoms toward larger angles. This can be basically attributed to the effect of the image force acting on ions on their incident path that accelerates the projectiles toward the surface so that the angles of incidence and thus also of exit are enhanced compared to neutral species where this

1) Roentdek Handels-GmbH, Kelkheim-Ruppertshain, Germany.

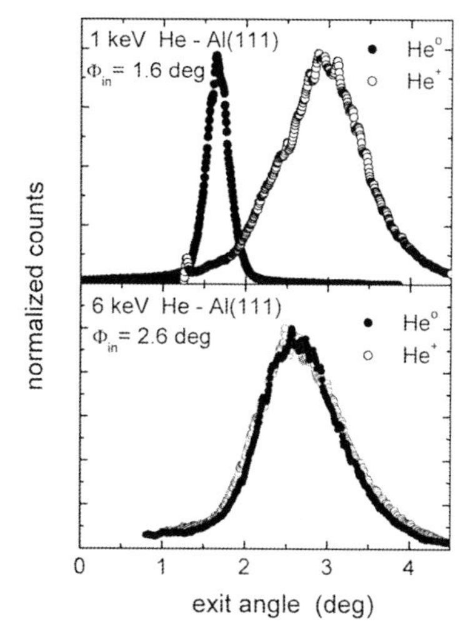

Figure 13.7 Polar angular distributions for scattering of 1 keV (upper panel) and 6.5 keV (lower panel) He atoms (full circles) and He^+ ions (open circles) from Al(111) under $\Phi_{in} = 1.6°$ (upper panel) and 2.6° (lower panel). Peaks are normalized to same heights.

force is absent [19]. From the angular splitting, we derive an energy gain $\Delta E_z = 1.8$ eV that corresponds for the classical image shift $\Delta E_{im} = 1/4(z - z_{im})$ (z_{im} is the position of image charge reference plane, located for Al(111) about 3 a.u. in front of the topmost surface layer) to a predominant neutralization of the incident ions at a distance of about $z_s \approx 3.5$ a.u. as already considered in the early studies by Hagstrum [7]. Thus, one might conclude that AN takes place at a distance of about 6–7 a.u. in front of the surface plane.

For the angular distributions shown in the lower panel of Figure 13.7, the normal energy amounts to $E_z = 13.4$ eV. Within the resolution of the experiment, we observe here no shift between the two distributions, that is, closer to the surface ΔE_z, and thus the 1s level shift is substantially reduced.

Hecht *et al.* [21] have analyzed the neutralization of He^+ ions for grazing scattering from Al(111) through image charge effects. From computer simulations of the angular distributions, an AN rate was derived plotted as function of distance from the surface by the solid line in the upper panel of Figure 13.8. For further discussion, it is important to note that for simulations the gain for the normal energy ΔE_z was deduced using the expression for the classical image potential. The resulting AN rates shown in Figure 13.8 led for He^+ ions with initial kinetic energies E_z in the eV domain to distances of predominant neutralization as estimated from the energy gain. The comparison with theoretical AN rates [22, 23] reveals that these rates were about three orders of magnitude smaller than the rates derived from the experimental normal energy gain. This discrepancy was initially attributed to deficiencies in the theory owing to the complexity of the problem as, for example, the realistic modeling of the potential barrier between surface and atom/ion and the (anisotropic) screening of the Coulomb interaction.

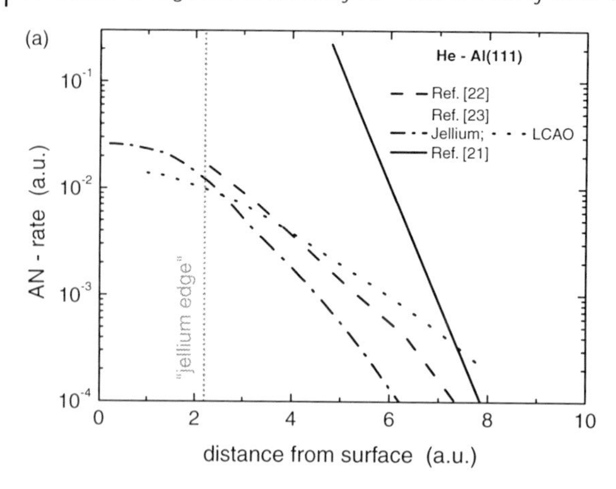

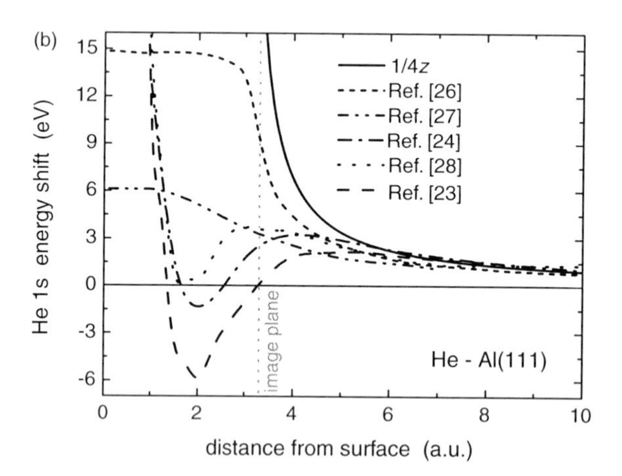

Figure 13.8 (a) AN rate as function of distance from surface derived from scattering experiments based on classical image potential (solid curve) and from calculations (broken curves). (b) Energy shift of 1s He ground state. Classical image shift 1/4z (solid curve) and calculations (broken curves).

On the basis of the theoretical work by More *et al.* [24], van Someren *et al.* [25] pointed out that the level shift of the He ground state plays a pivotal role and proposed an alternative scenario to describe the measured energy gain. In the lower panel of Figure 13.8, we show level shifts for the He atoms in front of Al(111) as derived by various authors in recent years [23, 24, 26–28]. The key issue is a clear-cut deviation from the classical image shift $1/4z$ for distances $z \leq 6$ a.u. In fact, in two calculations the level shift changes sign at distances below about 3 a.u. caused by chemical interactions of the projectiles with the conduction band. Then, the experimental energy gain can also be interpreted by the neutralization of ions at a closer distance from the surface and clearly smaller AN rates are necessary to be in line with the experiment. With these assumptions, the discrepancy between theoretical and experimental transition rates could be resolved.

It turns out that it is possible to differentiate between the two alternative scenarios via the fractions of incident ions that survive the complete scattering event with the surface. In the former analysis of AN, the resulting rates close to the surface were so large that fractions of surviving ions are about 100 orders of magnitude smaller compared to the incident ion fractions. However, for the alternative scenario of clearly reduced rates, surviving fractions up to about 1 per mill could be found in the reflected beams. The experimental findings of small but defined fractions of surviving ions showed clear evidence for the alternative scenario and demonstrated the relevance of the atomic level shift for the detailed understanding of the AN process for noble gas atoms in front of a metal surface [18, 29].

In Figure 13.9, we display a two-dimensional intensity pattern recorded with a position-sensitive microchannel plate detector after scattering of 2 keV $^3He^+$ ions from a Ni(110) surface under $\Phi_{in} = 1.7°$. The reflected beam has passed a narrow slit and is then dispersed with respect to charge by a pair of electric field plates. The intense dark bar in the middle of the plot stems from events for neutral atoms, whereas events for singly charged ions are represented within the faint bar on the left-hand side. Since the charge fractions generally differ by several orders of magnitude, sufficiently large count rates and a high dynamic range in counting the particles are important prerequisites. Furthermore, one has to check for different efficiencies in the detection of atoms and ions with the microchannel plate.

In Figure 13.10, we show ion fractions obtained in this manner for scattering of both He^0 atoms (open circles) and He^+ ions (full circles) from an Al(111) surface under $\Phi_{in} = 1.25°$ (upper panel) and 2.05° (lower panel) as function of projectile energy. For energies $E > 5$ keV, the ion fractions are within the accuracy of the measurements independent of the charge of the incident projectiles, whereas for smaller energies a

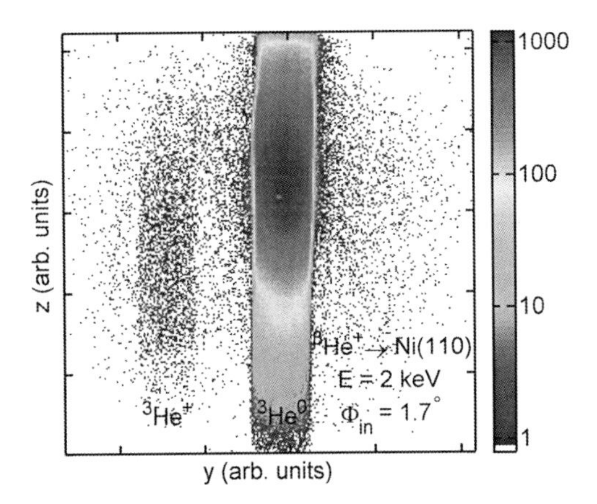

Figure 13.9 Intensity pattern (color scale) as recorded with position-sensitive microchannel plate detector for scattering of 2 keV $^3He^+$ ions from Ni(110) surface under $\Phi_{in} = 1.7°$. Reflected beam is dispersed normal to scattering plane by means of pair of electric field plates. *Intense bar in middle*: signal from neutral atoms; *faint bar on left-hand side*: signal from singly charged ions that have survived complete scattering event. (Please find a color version of this figure on the color plates.)

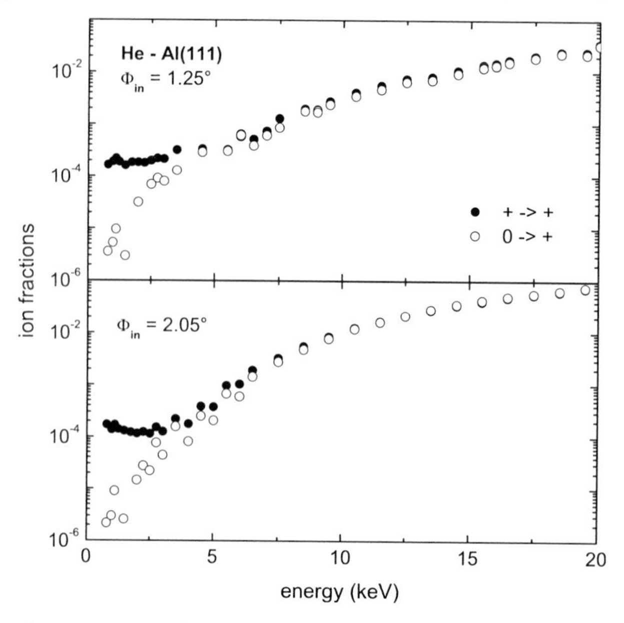

Figure 13.10 Ion fractions as function of projectile energy for scattering of He^0 atoms (open circles) and He^+ ions (full circles) from Al(111) surface under $\Phi_{in} = 1.25°$ (upper panel) and $\Phi_{in} = 2.05°$ (lower panel).

memory effect to the incident charge is present. This latter finding can be understood only if fractions of about 10^{-4} of He^+ ions survive the collision with the surface. At higher projectile energies, contributions from reionization via electron promotion and kinematic effects dominate the ion fractions and a separation with respect to surviving ions is not possible then. Therefore, studies on the AN dynamics based on ion survival can be performed only at projectile energies below some keV.

In Figure 13.11, we have plotted the surviving ion fractions for scattering of He^+ ions from Al(111) as function of the normal energy E_z for data sets recorded at projectile energies between 0.8 and 3 keV (full circles). The scan of E_z is achieved under planar surface channeling conditions by the controlled variation of the incidence angle according to $E_z = E_0 \sin^2\Phi_{in}$. The curves with open symbols represent ion fractions calculated from the integration over complete trajectories (Eq. (13.2)) by using different theoretical AN rates $\Gamma(z)$. In these calculations, $V_{He}{}^0(z)$ was obtained from Ref. [30] and the 1s level shift from Ref. [23]. The AN rates were calculated using the LCAO method (up triangles) and the jellium model (open squares) [23].

The calculation of the theoretical AN rates results in ion fractions P^+ that are about one and two orders of magnitude larger than the experiment, that is, the rates are generally too low or the interaction potentials are slightly too repulsive in order to lead to a sufficiently high neutralization of the incident ion beam. In this respect, however, we have to stress that the fractions of surviving ions are extremely sensitive to the size of the AN rates. This high sensitivity is demonstrated by the fact that the rates calculated with the LCAO method have to be enhanced by only a

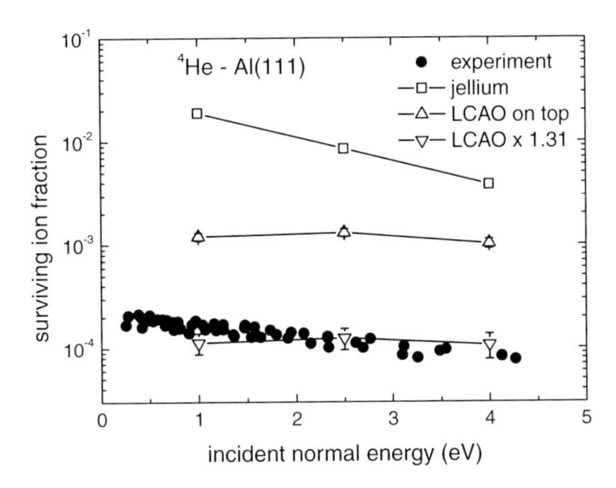

Figure 13.11 Ion fractions as function of normal energy for scattering of $^4He^+$ ions from Al(111). *Full circles*: experimental data; *curves with open symbols*: simulations based on theoretical AN rates obtained with LCAO method for on-top positions and jellium model [23]. *Down triangles*: simulations with LCAO AN rates [23] multiplied by factor of 1.31. For details see Ref. [31].

factor of about 1.3 in order to bring the theoretical surviving ion fractions to the experimental ones.

Owing to the high sensitivity of the ion fractions on the AN rates, it is beyond the scope of the accuracy of the available theoretical methods for calculations of the Auger neutralization process to achieve a close quantitative agreement with the experiments.

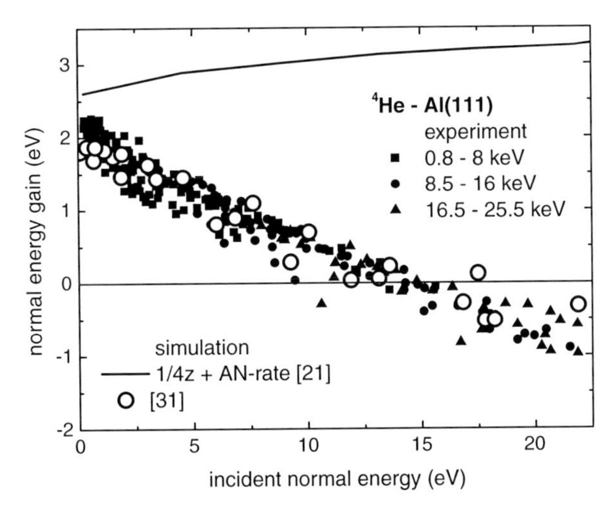

Figure 13.12 Normal energy gain as function of normal energy for incident projectiles derived from shifts of polar angular distributions for He^0 atoms and He^+ ions. *Full symbols*: experimental data; *open circles*: simulations based on theoretical AN rates and 1s level shifts from Ref. [23, 31]. *Solid curve*: 1/4z level shift and AN rate from [21]. For details see Ref. 31.

In Figure 13.12, measured normal energy gains, derived from shifts of angular distributions as shown in Figure 13.7, as function of incident normal energy are shown (full symbols). We compare with simulations based on the classical image charge potential and AN rates from Ref. [21] (solid curve) as well as the HE 1s level shift and LCAO AN rate from Ref. [23] (open circles). As for the ion fractions, the refined model based on Ref. [23] is in good accord with experiments whereas results for the classical level shift are in sheer contrast with experiment. For further details we refer the reader to Ref. [31].

13.3.1
Studies on Auger Neutralization Making use of Isotope Effect

Ion scattering experiments under channeling conditions provide an interesting feature to study the neutralization dynamics of He^+ by using the two stable isotopes, ^3He and ^4He. Since the total interaction potential is the same for isotopes of a given sort of atom, one expects for different isotopes at the same kinetic energy to follow the same trajectories. The interesting aspect of studies on the AN dynamics of He^+ ions is that the velocities, and thus the timescales for the interaction with the surface, scale with the square root of the mass of the isotopes. For two He^+ isotopes of same energy, the timescale differs by a factor $(M(^4He)/M(^3He))^{1/2} = (4\,amu/3\,amu)^{1/2} = 1.15$, that is, for $^3He^+$ ions the interaction times with the surface are shorter than for $^4He^+$. Therefore, it is straightforward to envision that the fractions of surviving ions are larger for scattering of $^3He^+$ ions than for $^4He^+$ ions. This is indeed observed in the experiments as shown in Figure 13.13 for the ion fractions after scattering of $^3He^+$ (full symbols) and $^4He^+$ (open symbols) from an Al(100) surface where the grazing angle of incidence Φ_{in} was adjusted to normal energies E_z ranging from 0.5 to 8 eV

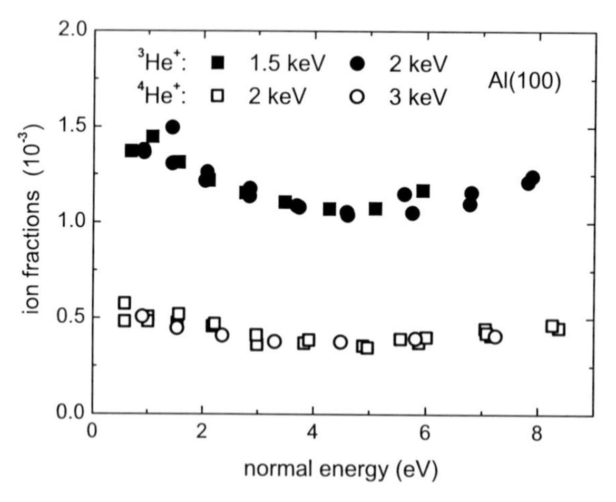

Figure 13.13 Ion fractions as function of normal energy for scattering of $^3He^+$ (full symbols) and $^4He^+$ ions (open symbols) from Al(100) at projectile energies of 1.5 keV (full squares), 2 keV (full circles, open squares), and 3 keV (open circles). Reprinted with permission from [30]. Copyright 2006 by the American Physical Society.

for projectile energies of 1.5, 2, and 3 keV. The ion fractions for the lighter isotope are about a factor of three larger than for the heavier isotope [32].

In order to analyze the data, we replace in Eq. (13.2)v_z by E_z^{kin} so that

$$\frac{\log P^+}{\sqrt{M}} = \left(- \int_{\text{traj}} \Gamma_{\text{AN}}(z) \frac{\mathrm{d}z}{\sqrt{2E_z^{\text{kin}}(z)}} \right). \tag{13.3}$$

Then, for charge transfer dominated by a single process with a defined transition rate, the right side of Eq. (13.3) is the same for different isotopes with the same E_z, that is, for the same settings of projectile energy and angle of incidence. This leads to the important simple relation $\log P^+ \sim M^{1/2}$, and we have plotted the data for the He$^+$ ions from Figure 13.13 in Figure 13.14 as $\log P^+/M^{1/2}$ versus normal energy E_z. Within the scatter of data, we find for this scaling a remarkable agreement between the two data sets that can serve as an indication that electron transfer is indeed dominated by a single process, that is, AN.

In order to obtain further evidence for this interpretation, we also performed experiments with He^{2+} ions where the neutralization process is clearly more intricate. Here, charge transfer proceeds via the intermediate of a singly charged ion where the excited and doubly excited states of the ion and atom might also play a role [33]. While skipping a detailed discussion of this topic, we state that the electron transfer dynamics is fairly complex that can hardly be approximated by a single dominant process as for He$^+$ ions. As a consequence, it is not astonishing that the scaling of data according to Eq. (13.3) fails as shown in the upper part of Figure 13.14. In this respect, we note that the detailed interaction scenario of He^{2+} ions with metal surfaces is still a matter of discussion and not cleared up in detail [34].

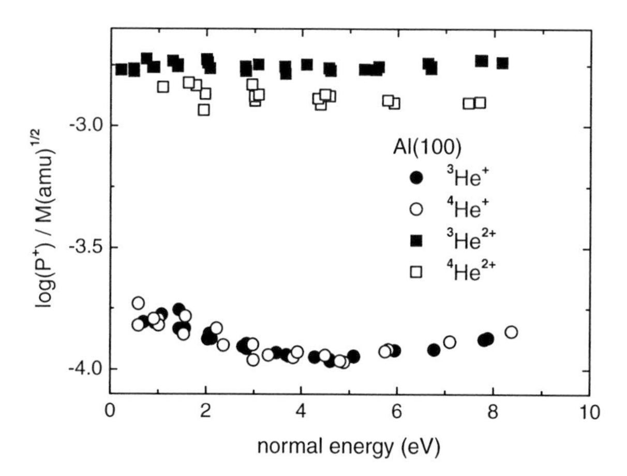

Figure 13.14 Log(P^+)/M as function of normal energy E_z for scattering of ^{3}He$^+$ (full circles). ^{4}He$^+$ (open circles), ^{3}He^{2+} (full squares), and ^{4}He^{2+} (open squares) from Al(100). Reprinted with permission from [30]. Copyright 2006 by the American Physical Society.

13.3.2
Face Dependence of Auger Neutralization

For charge transfer in front of metal surfaces, the spill out of conduction electrons to vacuum plays an important role. The electron densities are given by the corresponding electronic wave functions that show an exponential decay with the distance from the surface. A simple important reference to the electron wave function is the so-called "jellium edge" where the density of itinerant electrons is reduced to about one half of the bulk value and that is positioned at half the interatomic spacing between the first layers of a crystal surface. Therefore, the electron densities in the selvedge of a metal surface differ for various faces of a crystal of a given element. This also affects the AN rates in front of surfaces with different Miller indices. Furthermore, since the planar scattering potential for surface channeling of the fast projectiles depends on the areal density of atoms of the target surface, one expects a face dependence of the ion survival.

In experiments with He$^+$ ions scattered from Ag(111) and Ag(110) surfaces, Bandurin *et al.* [35] observed differences for the fractions of survived ions of about one order of magnitude. Similar results were also reported for different faces of a Cu crystal [36]. In Figure 13.15, we show ion fractions from our recent work for scattering from Al(111), Al(110), and Al(100) surfaces where similar effects were found [31]. For the low surface Miller index faces of crystals with fcc structure and lattice constant a, the positions of the jellium edge are $z_j = a/4 = 1.91$ a.u., $(a/4)(2)^{-1/2} = 1.35$ a.u., $(a/4)\,2(3)^{-1/2} = 2.21$ a.u., and the densities of surface atoms $n_s = 2/a^2$, $(2)^{1/2}/a^2$, $4/(3)^{1/2}a^2$ for the (100), (110), and (111) faces, respectively, that is, $d_j\,(111) > d_j(100) > d_j(110)$ and $n_s(111) > n_s(100) > n_s(110)$.

The fractions of surviving ions for the different faces in Figure 13.15 follow primarily the shift of the reference for the AN rate, that is, the positions of the jellium

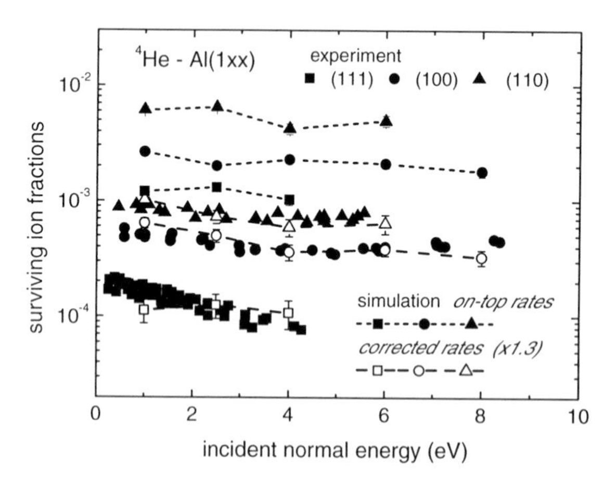

Figure 13.15 Ion fractions as function of normal energy E_z for scattering of He$^+$ ions from Al(111) (full squares), Al(100) (full circles), and Al(110) (full triangles). *Full symbols with dotted curves*: simulations with AN rates from Ref. [23]; *open symbols*: simulations with AN rates from Ref. [23] corrected by factors of 1.27–1.37.

edge. Thus, the AN rates dominate the outcome of ion survival compared to the scattering potential that affects the trajectory, in particular, the distance of closest approach to the topmost surface layer, and would result in the inverse face dependence of surviving ion fractions.

13.3.3
Effect of Magnetization of Target Surface on Auger Neutralization

In the work on the neutralization of He^+ ions discussed so far, the electronic spin was not explicitly considered. Recently, it was suggested to make use of the neutralization of He^{2+} ions in order to obtain information on the (short range) magnetic ordering of ferromagnetic surfaces [37]. It turns out, however, that the neutralization scenario of He^{2+} in front of metal surface is too complex in order to obtain clear-cut information on magnetic properties. The first studies on this topic were further affected by uncontrolled adsorption phenomena of the target that modify the work function of the surface and affect charge transfer [34].

For the neutralization dynamics of He^+ ions, the interaction mechanisms could be cleared up in the studies outlined above. In the resulting conceptually much simpler scenario, it is therefore straightforward to take into account the effect of the polarization of the electronic spin within domains of a ferromagnetic surface. This effect can be simply illustrated by the presence of highly spin-polarized domains of lateral extension exceeding the interaction length of the ions. Since the incident ions have a 1s hole, for the capture of an electron and the formation of the $1s^2$ singlet ground term, an electron of opposite spin orientation is needed. Then, for a highly spin-polarized target only one half of the electrons will contribute to AN and the fractions of surviving ions will increase to 50%. In Figure 13.16, we show calculations

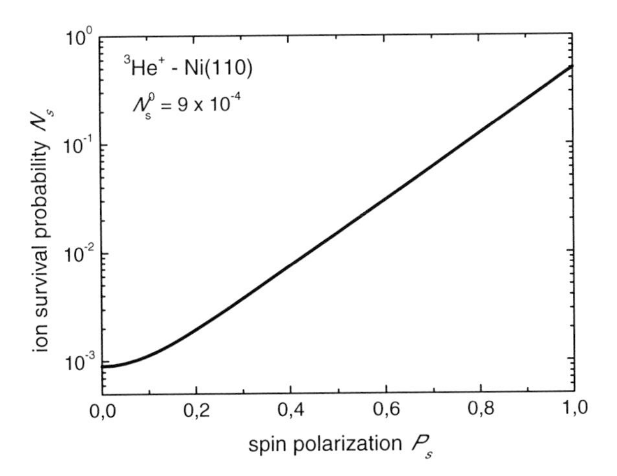

Figure 13.16 Calculated ion survival probability as function of spin polarization P_S for capture of electrons by He^+ ions into ground term of neutral He atom. Reprinted from [36]. Copyright 2007, with permission from Elsevier.

on ion survival based on a simple spin blocking model, where the ion fraction for the unpolarized target is taken from the experiment of scattering of He^+ from a Ni(110) target heated above the Curie temperature.

In an experiment with a Ni(110) surface, we tested this method [38]. The degree of spin polarization in the surface region of Ni(110) is a controversial topic, as different authors using different techniques report spin polarizations from a few percent to 100%. The idea of the experiment on the grazing scattering of He^+ ions from Ni(110) is to raise the target temperature above the Curie temperature (627 K) in order to destroy the magnetic order of the target. From the dependence shown in Figure 13.16, one would expect a reduction of the ion survival with increasing temperature and, in particular, for a target with a large spin polarization substantial ion fractions. In the experiments, we observed ion fractions in the 10^{-4} domain that can be understood by a fairly low spin polarization up to a few percent only. The slight increase in the ion fractions with temperature cannot be described by effects of the magnetic order. The analysis of the scattering process in terms of classical trajectory computer simulations revealed that thermal vibrations of atoms of the crystal lattice lead to a slight enhancement of the distances of closest approach for projectiles to the surface resulting in an increase in ion survival. The solid curve in Figure 13.17 represents calculations taking into account AN and the modification of trajectories owing to interactions with vibrating lattice atoms. Therefore, we conclude that the spin polarization of electrons captured in the AN process has to be small. For demonstration of the method, a ferromagnetic target with a higher spin polarization for the conduction electrons would be of interest. From Figure 13.16, we estimate that a spin polarization above about 20% could be detected via the ion fractions.

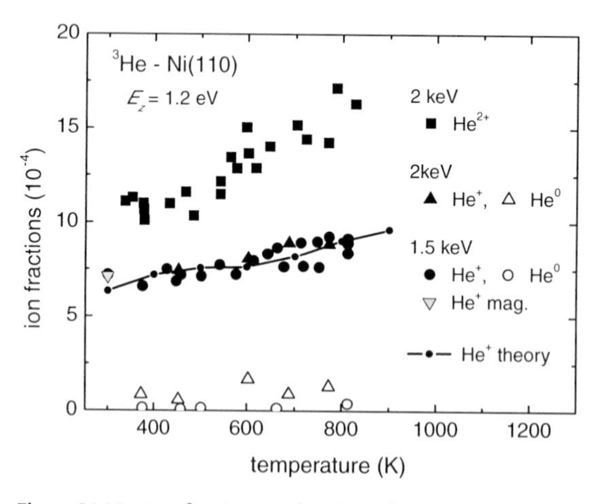

Figure 13.17 Ion fractions as function of target temperature for scattering of He^+ (full triangles, full circles), He^0 (open triangles, open circles), and He^{2+} (full squares) from Ni(110) with $E_z = 1.2$ eV. Reprinted from [36]. Copyright 2007, with permission from Elsevier.

13.4
Summary and Conclusions

In summarizing our discussions on the neutralization dynamics during the collision of He^+ ions with a metal surface, we state that considerable progress in the microscopic understanding on this topic has been achieved. A key feature in revealing the interaction mechanism is the survival of ions during the complete scattering event from the surface that can be favorably detected in the collision geometry under glancing angle of incidence. In this regime of planar channeling, the projectiles are steered in terms of small-angle scattering by atoms of the topmost surface layer giving rise to well-defined trajectories. Then, the charge transfer between ion and metal can be modeled in detail with a high intrinsic time resolution in the fs domain. Based on the capability of grazing surface scattering to derive information on the atomic level shift and its important role in the charge transfer process, we nowadays have arrived at a microscopic understanding of Auger neutralization in front of metal surfaces.

Acknowledgments

The contributions by K. Maass, M. Busch, G. Lindenberg, A. Schüller, and D. Blauth are gratefully acknowledged. We thank R.C. Monreal and D. Valdés for a fruitful collaboration on this topic. The studies were supported by the Deutsche Forschungsgemeinschaft (DFG) under contract Wi1336.

References

1 Los, J. and Geerlings, J.J.C. (1990) *Phys. Rep.*, **190**, 133.
2 Rabalais, J.W. (ed.) (1994) *Low Energy Ion–Surface Interactions*, John Wiley & Sons, Inc., New York.
3 Brako, R. and Newns, D.M. (1989) *Rep. Prog. Phys.*, **52**, 655.
4 Gauyacq, J.P., Borisov, A.G., and Bauer, M. (2007) *Prog. Surf. Sci.*, **82**, 244.
5 Brongersma, H.H., Draxler, M., de Ridder, M., and Bauer, P. (2007) *Surf. Sci. Rep.*, **62**, 63.
6 Monreal, R.C. and Flores, F. (2004) *Adv. Quantum Chem.*, **45**, 175.
7 Hagstrum, H.D. (1954) *Phys. Rev.*, **96**, 336.
8 Hagstrum, H.D. (1966) *Phys. Rev.*, **150**, 495.
9 Winter, H. (2002) *Phys. Rep.*, **367**, 387.
10 Niehus, H., Heiland, W., and Taglauer, E. (1993) *Surf. Sci. Rep.*, **17**, 213.
11 Gemmell, D.S. (1974) *Rev. Mod. Phys.*, **46**, 129.
12 Zacharias, H. (ed.) (2007) *Prog. Surf. Sci.*, **82**, 161.
13 Drescher, M., Hentschel, M., Kienberger, R., Uiberacker, M., Yakovlev, V., Scrinzi, A., Westerwalbesloh, T., Kleineberg, U., Heinzmann, U., and Krausz, F. (2002) *Nature*, **419**, 803.
14 Moliére, G. (1947) *Z. Naturforschung*, **A2**, 133.
15 Puska, M. and Nieminen, R. (1991) *Phys. Rev. B*, **43**, 12221.
16 Schüller, A., Adamov, G., Wethekam, S., Maass, K., Mertens, A., and Winter, H. (2005) *Phys. Rev. A*, **69**, 050901.
17 Echenique, P.M., Flores, F., and Ritchie, R.H. (1990) *Solid State Physics*, vol. **43**, Academic Press, New York, p. 229.
18 Wethekam, S., Mertens, A., and Winter, H. (2003) *Phys. Rev. Lett.*, **90**, 037602.

19 Winter, H. (1996) *J. Phys. Condens. Matter*, 8, 10149.
20 Pfandzelter, R., Bernhard, T., and Winter, H. (2003) *Phys. Rev. Lett.*, **90**, 036102.
21 Hecht, T., Winter, H., and Borisov, A.G. (1998) *Surf. Sci.*, **406**, L607.
22 M.A. Cazalilla, N. Lorente, R. Díez Muiño, J.-P. Gauyacq, D. Teillet-Billy, and P.M. Echenique, (1998) *Phys. Rev. B*, **58**, 13991.
23 Valdés, D., Goldberg, E.C., Blanco, J.M., and Monreal, R.C. (2005) *Phys. Rev. B*, **71**, 245417.
24 More, W., Merino, J., Monreal, R., Pou, P., and Flores, F. (1998) *Phys. Rev. B*, **58**, 7385.
25 van Someren, B., Zeijlmans van Emmichoven, P.A., and Niehaus, A. (2000) *Phys. Rev. A*, **61**, 022902.
26 M. Kato, R.Williams, and M. Hono (1988), Nucl. Instrum. *Methods Phys. Res. B*, **33**, 462.
27 Merino, J., Lorente, N., More, W., Flores, F., and Yu Gusev, M. (1997) *Nucl. Instrum. Methods Phys. Res. B*, **125**, 250.
28 Wang, N.P., Garcia, E.A., Monreal, R., Flores, F., Goldberg, E.C., Brongersma, H.H., and Bauer, P. (2001) *Phys. Rev. A*, **64**, 012901.
29 Monreal, R.C., Guillemot, L., and Esaulov, V.A. (2003) *J. Phys. Condens. Matter*, **15**, 1165.
30 O'Connor, D.J. and Biersack, J.P. (1986) *Nucl. Instrum. Methods Phys. Res. B*, **15**, 14.
31 Wethekam, S., Valdés, D., Monreal, R.C., and Winter, H. (2008) *Phys. Rev. B*, **78**, 075423.
32 Wethekam, S. and Winter, H. (2006) *Phys. Rev. Lett.*, **96**, 207601.
33 Zeijlmans van Emmichoven, P.A., Wouters, P.A.A.F., and Niehaus, A. (1988) *Surf. Sci.*, **195**, 115.
34 Wethekam, S., Busch, M., and Winter, H. (2009) *Surf. Sci.*, **603**, 209.
35 Bandurin, Yu., Esaulov, V.A., Guillemot, L., and Monreal, R.C. (2004) *Phys. Rev. Lett.*, **92**, 017601.
36 Primetzhofer, D., Markin, S.N., Juaristi, J.I., Taglauer, E., and Bauer, P. (2008) *Phys. Rev. Lett.*, **100**, 213201.
37 Unipan, M., Robin, A., Morgenstern, R., and Hoekstra, R. (2006) *Phys. Rev. Lett.*, **96**, 177601.
38 Wethekam, S., Busch, M., Monreal, R.C., and Winter, H. (2009) *Nucl. Instrum. Methods Phys. Res. B*, **267**, 571.

14
Electron Transfer Investigated by X-Ray Spectroscopy

Wilfried Wurth and Alexander Föhlisch

Electron transfer at surfaces and interfaces plays an important role in many different processes such as catalytic, biophysical, electrochemical, and photoinduced reactions, molecular electronics, and solar energy conversion. Transfer can occur either resonantly or inelastically, involving more than one electron. Transfer processes occur typically on femtosecond and subfemtosecond timescales for systems where the available density of states is high such as metals. To study the fundamental aspects of the dynamics of electron transfer, it is convenient to create an electron wave packet by photoexcitation and follow its evolution by suitable probe techniques. X-rays are particularly well suited in this respect because they allow site- and element-specific localized excitations in complex systems, lead to direct excitation to a specific state, and can probe the subsequent evolution of this particular state.

In this chapter, we will focus on two very different approaches using soft X-rays to investigate electron transfer, namely, high-resolution core hole spectroscopy at third-generation synchrotron sources, the so-called core hole clock method [1–3], and stroboscopic pump–probe experiments using ultrashort X-ray pulses from free electron laser sources. The examples that are chosen to illustrate the potential of X-ray spectroscopy to study electron transfer are taken exclusively from our own recent work. Hence, this is by no means meant to be an extensive and representative review of the field.

14.1
Core Hole Clock Spectroscopy

The term core hole clock spectroscopy reflects the fact that in this type of spectroscopy, the intrinsic lifetime of a core hole state is used as an internal reference against which competing relaxation processes are timed. Basically, the decay of a resonantly excited core hole state is studied spectroscopically (excellent reviews with many references can be found in the literature [4–7]). We will discuss here only the nonradiative decay channels, which for core hole states created by the absorption of soft X-rays represent the majority of decay channels.

Dynamics at Solid State Surface and Interfaces Vol.1: Current Developments
Edited by Uwe Bovensiepen, Hrvoje Petek, and Martin Wolf

ISBN: 978-3-527-40937-2

14.1.1 Basics

Resonant X-ray absorption close to an ionization threshold of a core state leads to a state with a core hole and an extra electron in a formerly empty valence state (see Figure 14.1b). At a surface or an interface, either such a core-excited state may decay via autoionization (Figure 14.1c) or the excited electron has the possibility to delocalize (Figure 14.1a) before the core hole decay takes place.

The two decay scenarios will lead to distinctly different sets of final states. In the decay spectra, the autoionization channels will have higher kinetic energy due to the screening of the two holes in the final state by the initially excited electron (the so-called spectator electron). This autoionization process can be viewed as an Auger–Raman process where the incoming photon leads to the emission of an outgoing electron and leaves the system in an electronically excited state. In contrast, the core hole decay that follows the electron delocalization or transfer is a normal Auger decay. Under the assumption that core hole decay and electron delocalization (transfer) are processes with a constant rate, that is, follow an exponential decay law, it can be easily shown that the intensity ratio of the spectral features for the two decay scenarios is given by the ratio of the respective transition rates R:

$$\frac{I_{\text{Raman}}}{I_{\text{Auger}}} = \frac{R_{\text{Raman}}}{R_{\text{Auger}}} = \frac{\tau_{\text{CT}}}{\tau_{\Gamma}}.$$

The ratio of the transition rates is of course inversely proportional to the respective decay times. A key observation here is that the relevant time for the Auger–Raman part is the natural core hole lifetime τ_{Γ}. Hence, if we experimentally determine the respective intensity ratios, we obtain a value for the charge transfer time τ_{CT} in units of τ_{Γ}. Since the core hole lifetime is to a very good approximation a quantity that depends only weakly on the surrounding [8] of a specific atom, its value can be taken from atomic data (e.g., Ref. [9]). Therefore, using this spectroscopic approach, we can quantitatively determine electron transfer times [1, 7, 10].

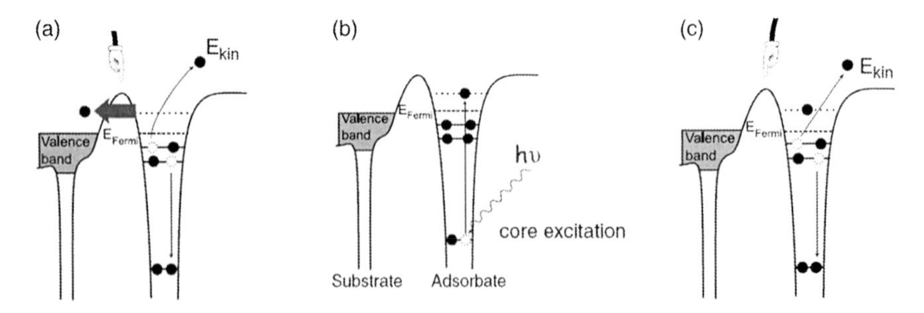

Figure 14.1 The cartoons depict resonant excitation of an adsorbate core electron into a bound unoccupied resonance (b) and the possible nonresonant decay channels before (c) and after (a) transfer of the resonantly excited electron from the adsorbate to the substrate.

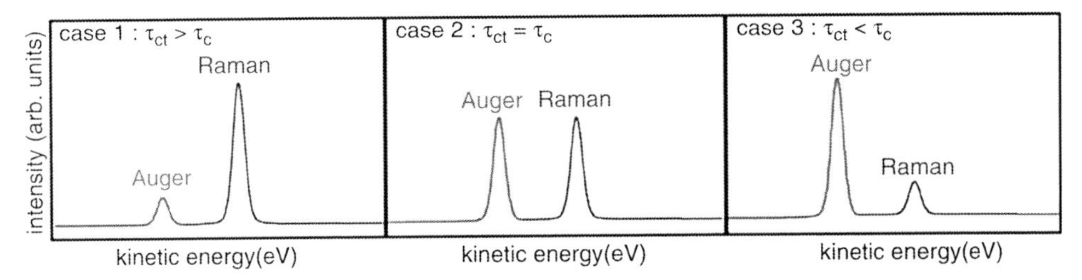

Figure 14.2 The cartoons show qualitatively the relative intensities for Raman and Auger channels for different relations of the charge transfer time τ_{CT} to the core hole lifetime τ_Γ. (Please find a color version of this figure on the color plates.)

In Figure 14.2, the expected relative intensities for three different cases are depicted schematically. Unfortunately, the spectroscopic signatures of Raman versus Auger decay channels are not so easy to disentangle in reality. Usually nonradiative decay leads to many different final states depending on the configuration of the final two-hole one-electron (Raman) or two-hole (Auger) states, respectively. Even if the corresponding Raman and Auger final states are shifted in energy as mentioned above, different configurations may overlap energetically (see Figure 14.3 as an example).

This makes a precise determination of intensity ratios from single decay spectra rather difficult.

However, apart from screening of the hole states through the excited electron, there is another important difference between the Raman and the normal Auger final states. If the excitation energy is varied, the kinetic energy of the spectral features resulting from the Raman decay channels (where the excited electron is localized at

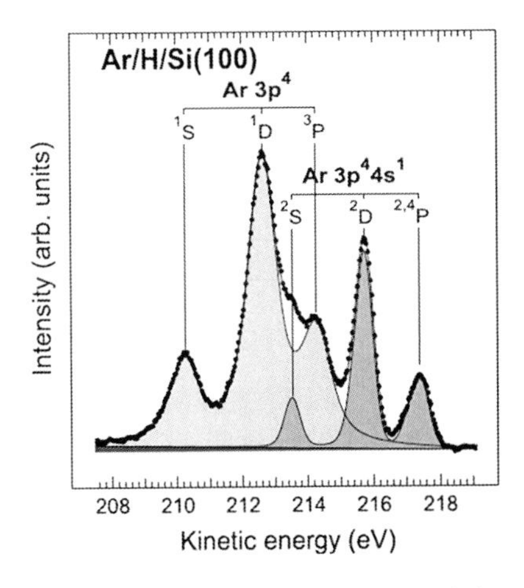

Figure 14.3 The figure shows the core hole decay spectrum after Ar $2p_{3/2} \rightarrow 4s$ resonant excitation for Ar adsorbed on H/Si(100). Included is the deconvolution into the different decay channels (Raman ($3p^4 4s^1$) versus Auger ($3p^4$) obtained from a complete analysis of the photon energy dependence. Reprinted from [11]. Copyright 2009 IOP Publishing Ltd.

the core hole site) changes linearly with the exciting photon energy, while the kinetic energy of the spectral features resulting from the normal Auger decay (where the excited electron is already delocalized) remains constant [7, 12]. This behavior is shown schematically in Figure 14.4.

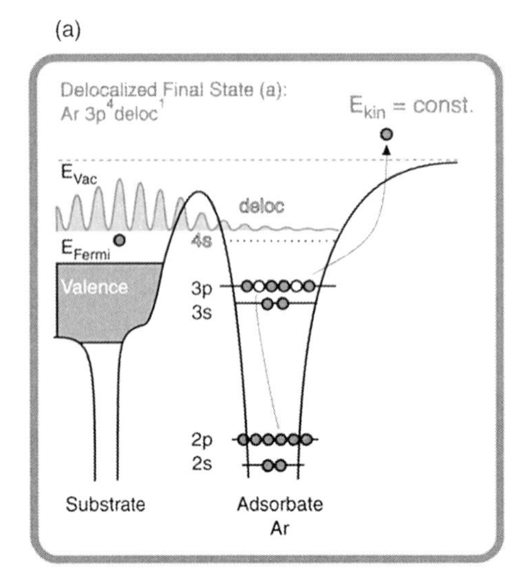

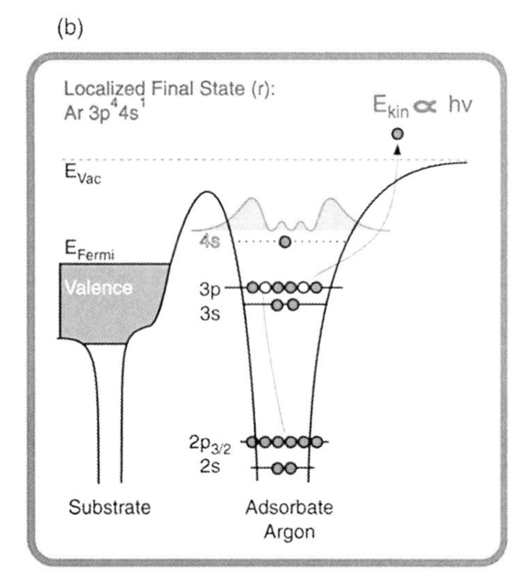

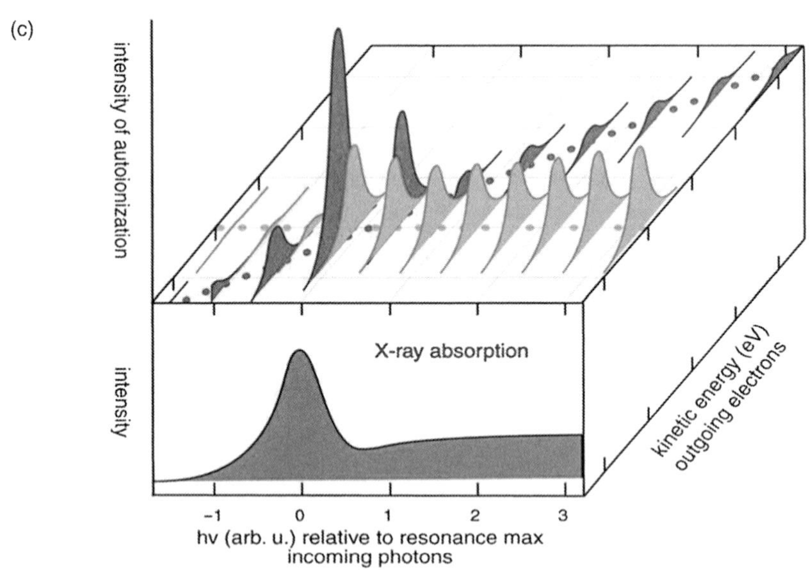

Figure 14.4 The figure (c) shows schematically the photon energy dependence of the kinetic energy of the outgoing electrons when the photon energy is tuned across the absorption resonance for a Raman channel where the excited electron is localized at the site of the excitation ((b), green) and for an Auger channel where the excited electron is delocalized ((a), red). (Please find a color version of this figure on the color plates.)

Using subnatural linewidth resonant excitation as it is nowadays possible with high-resolution beamlines at third-generation synchrotron sources and taking decay spectra while tuning the photon energy across the resonance, one is in a position to accurately determine the different components of the decay spectra and evaluate their intensity ratio. Thus, it is possible to determine electron transfer times quantitatively from high-resolution core hole spectroscopy. Since the accuracy is limited by the evaluation of the intensity ratios, the range of electron transfer times that are accessible spans roughly from a tenth of the core hole lifetime to about ten times this value. This is, however, not a serious drawback since core hole lifetimes are typically in the femtosecond or subfemtosecond regime.

In the following sections, a few examples will be given demonstrating the use of core hole clock spectroscopy to investigate some aspects of electron transfer between adsorbates and metal substrates.

14.1.2
Electron Stabilization

We will first consider the case of a simple weakly coupled, that is, physisorbed, adsorbate on a metal surface. A prototype system for systematic studies of electron transfer from an adsorbate into a metal substrate in the past has been Argon [2, 3, 7, 12–17]. Resonantly exciting an Ar $2p_{3/2}$ core electron into the empty Ar 4s level creates a core-excited state (Ar $2p^5 4s^1$), which for Ar adsorbed on a metal surface typically is energetically located a few eV below the ionization threshold. Note that the excited Ar atom in the $Z + 1$ approximation often used for core-excited states would correspond electronically to a K atom hypothetically adsorbed at the equilibrium position of Ar in front of the surface. However, it is clear that the $Z + 1$ model for an adsorbate is a crude approximation if one wants to estimate the time for delocalization of an excited electron because besides the aforementioned differences in adsorption geometry, differences in coverage and single electron potential are neglected. Depending on the energetic position of the core-excited state with respect to the Fermi level E_F of the substrate and the respective substrate bands, the core-excited electron will rapidly delocalize into the substrate. Figure 14.5 shows the energetic position of the Ar resonance with respect to the surface-projected bulk band structure and the surface states for the case of the Cu(111) and Cu(100) surfaces.

Note that the Ar resonance is a flat band showing no dispersion in $k_{||}$ due to the localization of the core-excited states. It is clear that the resonance position in both cases coincides with empty states in the bulk bands. However, for the Cu(111) surface, a much larger $k_{||}$ is needed to enter the bulk bands as in the case of the Cu (100) surface.

As a consequence, the wave packet of the excited electron will evolve differently in time for the two surfaces. This can be nicely seen in Figure 14.6 where a snapshot of the wave packet evolution is depicted. While in the Cu(111) case the wave packet has only slightly progressed into the substrate with a large component parallel to the surface, in the Cu(100) surface delocalization into the bulk is much faster. The electron transfer times determined by core hole clock spectroscopy for the two

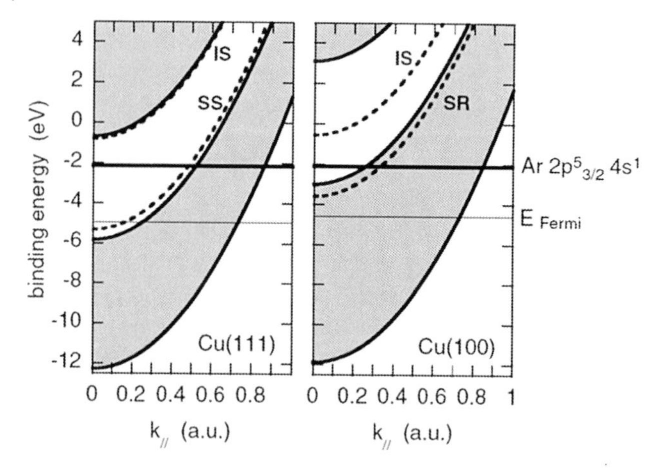

Figure 14.5 Energetic position of the Ar $2p_{3/2}^{-1}4s^{+1}$ resonance with respect to the surface projected density of states of Cu(111) and Cu(100), respectively. Reprinted from [17]. Copyright 2006, with permission from Elsevier.

surface orientations and calculated results for the two surfaces and a structureless jellium model, respectively, are given in Table 14.1.

In particular, comparison of the experimental and theoretical values for the two surfaces with the value obtained when Cu is modeled using structureless jellium

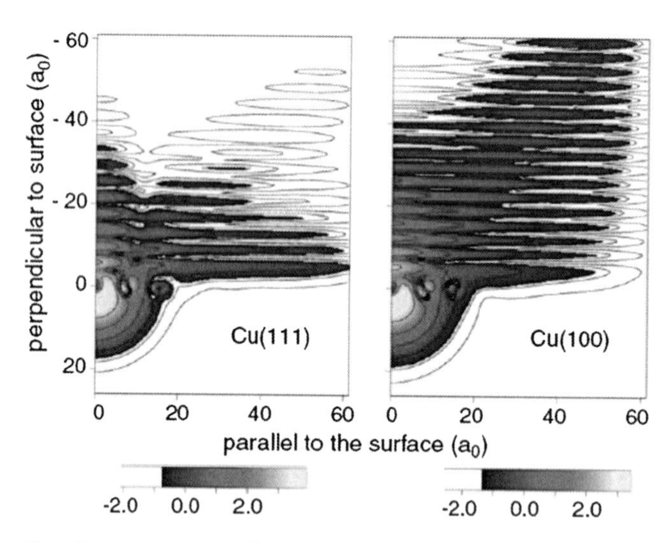

Figure 14.6 Contour plot of a cut of the resonant 4s wave packet in the Ar $2p_{3/2}^{-1}4s^{+1}$/Cu systems. The plot presents the logarithm of the electron density in cylindrical coordinates parallel (ϱ) and perpendicular (z) to the surface. The z-coordinate is positive in vacuum. The adsorbate is centered at the origin. The ϱ-axis is in a plane that contains one of the nearest Ar neighbors. The coordinates are in atomic units $a_0 = 0.529$ Å. The electron density decreases when going from yellow to dark red. White corresponds to very small densities. The contour lines are separated by a factor e. Reprinted from [17]. Copyright 2006, with permission from Elsevier. (Please find a color version of this figure on the color plates.)

Table 14.1 Electron transfer time.

	Cu(111)	Cu(100)	Jellium
Theory	12 fs	6 fs	1.2 fs
Experiment	5.6 ± 0.1 fs	3.5 ± 0.1 fs	—

From Ref. [17].

shows the strong stabilization of the excited electron on the adsorbate (an effect first discussed for Cs on Cu [18]) due to the fact that evolution of the wave packet along the surface normal is not possible because of the respective gaps in the substrate electronic structure. While theory and experiment agree on the increase of the electron transfer time by nearly a factor of 2 from Cu(100) to Cu(111), the absolute values of the electron transfer time observed experimentally are smaller than the calculated values. This discrepancy is still to be understood in more detail.

To conclude this section, from the results presented here it is obvious that in a coupled adsorbate/substrate system, resonant single electron transfer, which is the dominating contribution here – transfer processes involving inelastic e–e scattering are found to be of minor importance theoretically – is strongly influenced by band structure effects.

14.1.3
Attosecond Charge Transfer and the Effect of Orbital Polarization

In case of strong coupling, that is, chemisorption, of an adsorbate to a substrate, there is strong hybridization of the electronic states of the adsorbate with the substrate system. Hence, delocalization of an electronic wave packet created locally at the position of the adsorbate by resonant core excitation will be extremely fast. To study the timescale for such a process, we have made use of the fact that the natural lifetime of core holes that can decay nonradiatively via Coster–Kronig or even super Coster–Kronig processes is one or two orders of magnitude smaller than for normal decay processes. Here the initial and final state vacancies are in the same electronic shell (same principal quantum number n); the probability for these transitions is higher and the corresponding core hole lifetimes shorter than in the case of decay processes involving different values of n. Hence, the range of accessible electron transfer times that depends on the core hole lifetime (see above) will shift from the femtosecond to the attosecond regime.

As a test case, we have investigated the survival probability of an excited S $3p_z$ electron after S 2s to $3p_z$ core excitation for the case of S atoms adsorbed on a Ru(001) surface where the z-axis denotes the axis perpendicular to the surface [19]. The S 2s core hole will decay via Coster–Kronig processes leading to $2p^{-1}(3s, 3p)^{-1}$deloc or $2p^{-1}(3s, 3p)^{-1}3p_z^{+1}$ final states depending on the delocalization probability of the excited $3p_z$ electron on the timescale of the core hole decay that is here given by the 2s hole lifetime $\tau_\Gamma = 0.5$ fs.

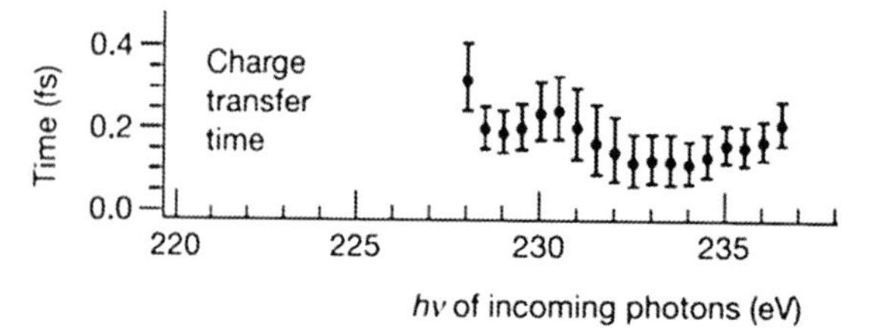

Figure 14.7 Electron transfer times obtained after S 2s → 3p excitation for chemisorbed S on Ru(001).

Figure 14.7 shows the electron transfer times as a function of excitation energy determined from the intensity ratio of the $2p^{-1}3s^{-1}$deloc and the $2p^{-1}3s^{-1}3p_z^{+1}$ final states multiplied by the 2s hole lifetime of 0.5 fs. At the position of the resonance maximum that is given by the first data point, we obtain $\tau_{CT} = 0.32 \pm 0.09$ fs. For excitation above resonance closer to the vacuum level, the transfer times are even shorter.

In Figure 14.8, the computational results of first-principles calculations of the charge transfer dynamics for the $c(4 \times 2)$S/Ru(001) system are shown. The results for two different adsorption sites (fcc and hcp sites) are depicted as a function of the orientation of the electric field vector of the exciting radiation with respect to the coordinate frame given in the figure. Note that the orientation of the electric field vector in the coordinate frame determines the relative contribution of the S 3p orbitals parallel and perpendicular to the surface, respectively, that is, the observed dependence shows the influence of orbital polarization on charge transfer dynamics.

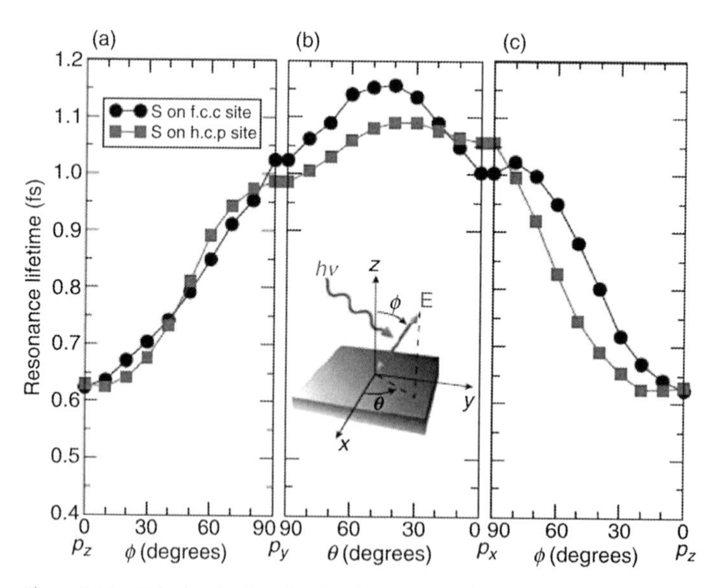

Figure 14.8 Calculated values for the electron transfer time from the 3p state of S on Ru(001) for two different S adsorption sites (fcc versus hcp) and different orientations of the 3p orbitals. Adapted by permission from Macmillan Publishers Ltd.: Nature [19], Copyright 2005. (Please find a color version of this figure on the color plates.)

It is clear from the calculations that the influence of the specific adsorption site is very weak. The value obtained for the situation where the *E* vector is parallel to the surface normal, that is, when the S 3s electron is excited into the $3p_z$ (the experimental situation), is 0.65 ± 0.15 fs. The theoretical time constant is also well below 1 fs and the agreement with experiment is very satisfying if one takes into account that in the calculations the core vacancy is not treated explicitly. Furthermore, theory predicts that for excitation into the S 3p orbitals parallel to the surface ($3p_{x,y}$), the survival probability of the excited electron should be significantly enhanced due to the smaller overlap with the substrate.

In a second experiment [20], we have checked this theoretical prediction by measuring and evaluating the decay spectra obtained for two different polarization directions of the incoming soft X-ray radiation with respect to the substrate surface. From the resulting data points in Figure 14.9, one can see that excitation into the 3p

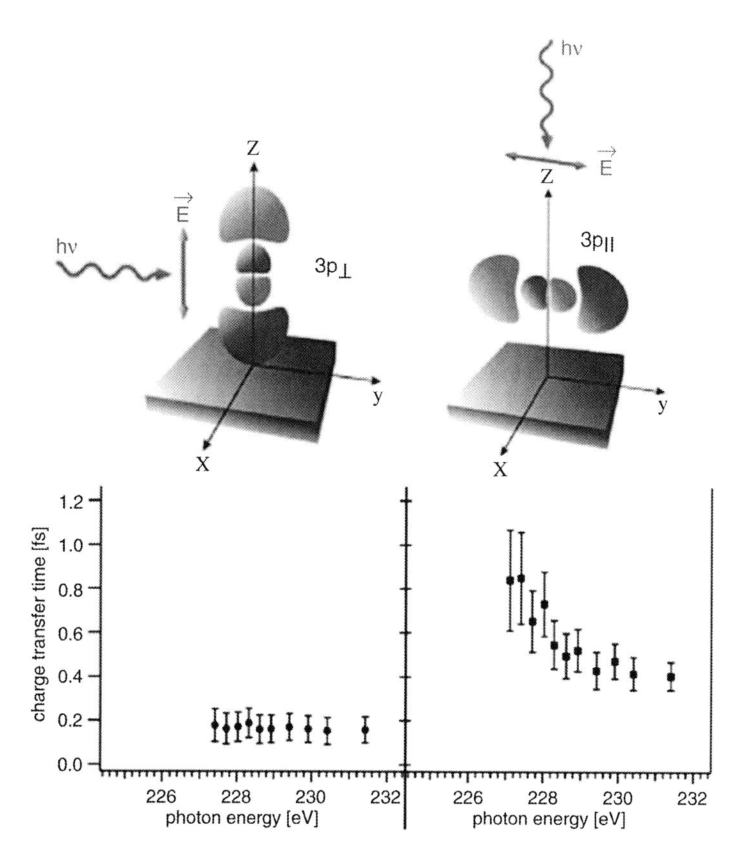

Figure 14.9 The figure shows the electron transfer time obtained for electrons excited to the 3p orbitals of S adsorbed on Ru(001). On the left side the polarization of the exciting photons is chosen such that the 3p orbital perpendicular to the surface is populated, while on the right side the 3p orbitals oriented parallel to the surface are selected in the excitation process. (Please find a color version of this figure on the color plates.)

orbital perpendicular to the surface (left side) leads to very different delocalization of the excited electron as compared to excitation into the orbitals parallel to the surface (right side). As predicted, CT is much faster for the 3p orbital perpendicular to the surface than for the in-plane orbitals, due to the larger overlap of the wave functions and therefore stronger coupling between adsorbate and substrate states. The observed decrease of CT with increasing detuning away from the resonance maximum may be related to the change of the band structure of Ru(001) 2 eV above the Fermi energy from being dominated by states with a strong contribution from the localized Ru 4d orbitals below to display more dispersive and delocalized states with main sp character.

To conclude this section, we would like to emphasize that by choosing the appropriate decay processes, one can even apply core hole clock spectroscopy down to the attosecond regime. Resonant excitation with variable polarization allows to study subtle effects related to the overlap of specific orbitals of an adsorbate with the substrate.

14.1.4
Clocks with Different Timing

An important question in the context of the use of core hole clock spectroscopy to measure relaxation times in complex systems is the influence of the core hole created in the excitation process on the relaxation process. A core-excited state $c^{-1}\,\mathrm{unocc}^{+1}$ on an adsorbate certainly has a different energetic position with respect to, for example, the Fermi level of the substrate compared to a valence-excited state $v^{-1}\mathrm{unocc}^{+1}$ that involves the same unoccupied level. As discussed above, to zeroth order, this can be included by using the $Z + 1$ approximation locally at the site of the core hole, that is, by representing the core hole by a positive point charge. However, the difference in Coulomb interaction U between the core hole and the excited electron versus the valence hole and the excited electron that determines the relative level shift can be strongly reduced as a result of the interatomic many-electron screening response of the system. Apart from the shift in energy of the resonance with respect to the other relevant levels that certainly has an influence on electron transfer time, the wave function of the localized excited state and hence the coupling to other states will also be influenced by the presence of a core hole. Hence, one has to conclude that a theoretical description of results obtained with core hole clock spectroscopy and a comparison with other techniques such as time-resolved two-photon photoemission have to take these effects explicitly into account. Quantitatively similar results will only be obtained if the core hole effects are of minor importance due to very efficient interatomic screening. Experimentally, a comparison between core hole clock spectroscopy and two-photon photoemission is also difficult because one has to find suitable systems where both techniques can be applied. Recently we have tried such a comparison for molecular C_6F_6 layers on Cu (111) [21]. In this study, we find similar trends in charge transfer time as a function of coverage from submonolayer to multilayers with both techniques. While the absolute values differ by roughly a factor of 2 for the mono- and bilayer cases, the values agree

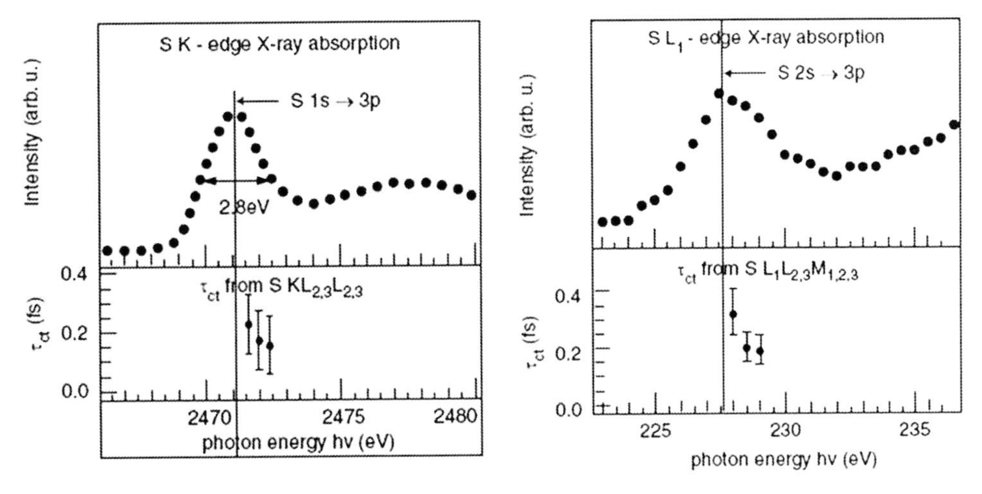

Figure 14.10 Comparison of the charge transfer times τ_{CT} extracted from independent core hole clock spectroscopic investigations after S 1s → 3p and S 2s → 3pexcitation for S on Ru(001).

for thick molecular layers. We can qualitatively explain these differences when we consider the experimentally determined position of the respective excited state bands with respect to substrate band structure.

In two other experimental studies [22, 23], we have also investigated core hole effects, in particular the influence of different core holes, on the measured electron transfer times. A nice system to perform such a study is again S on Ru(001). Here we can realize two excited states, namely, S 1s → 3p and S 2s → 3p, where only the core levels involved are different, the 1s core hole being of course more localized than the 2s core hole.

Figure 14.10 shows the results obtained for the electron transfer times from the core hole clock studies using the two different core holes. Within the error bars, the results are identical despite the fact that the lifetime of the S 1s core hole is about a factor of 2 longer than the S 2s core hole due to the Coster–Kronig transitions involved in the latter case and the different localization of the core holes.

Therefore, we can conclude that the electron transfer times obtained from core hole clock spectroscopy do not depend on the specific core level used. This enables the use of hard X-rays and deep-lying core levels to study bulk properties and buried interfaces. It also allows to some extent to adjust the core hole lifetime with respect to the transfer time of interest if different core hole states can be prepared.

14.2 Time-Resolved Soft X-Ray Spectroscopy

In this section, we will focus on the newly emerging field of time-resolved soft X-ray spectroscopy. With new light sources such as high harmonic generation (HHG) laser sources and the accelerator-based free electron laser sources, powerful XUV and soft

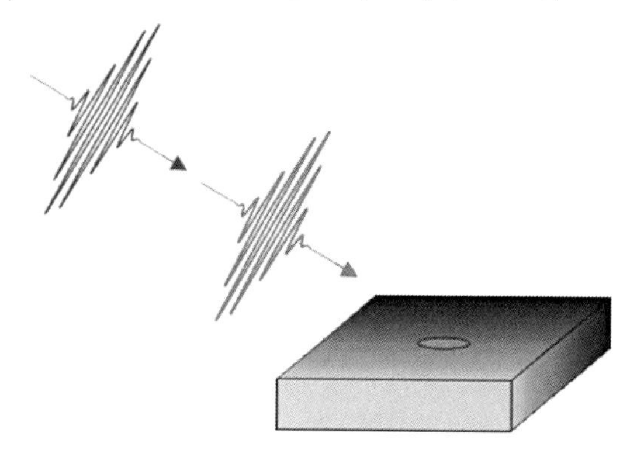

Figure 14.11 A two-color pump–probe experiment is schematically shown, where two pulses with different photon energies and a fixed delay in time impinge on a surface. In our experiment, the first optical pulse photoexcites the system, while the second soft X-ray pulse probes the system using core-level photoelectron spectroscopy. (Please find a color version of this figure on the color plates.)

X-ray sources with ultrashort pulses are available that enable stroboscopic pump–probe experiments with one or more XUV or soft X-ray pulses, as schematically depicted in Figure 14.11. This opens a very exciting new field because it brings extreme time resolution to soft X-ray spectroscopy and its capability to probe the electronic (and magnetic) structure of complex systems in very fine detail. Here, we will only discuss first steps toward time-resolved core-level photoemission at free electron laser sources based on our own experiments at FLASH, the free electron laser at DESY in Hamburg [24]. Exciting results from HHG sources, which in the moment deliver the shortest pulses down to hundred attoseconds but with less photons per pulse and limited in photon energy so far, will be covered elsewhere in this book.

The pulses currently available at FLASH have been characterized according to the pulse duration and the temporal coherence. In Figure 14.12, we show the longitudinal (temporal) coherence and the average pulse length of the FLASH pulses. It is to be noted that the number of photons per pulse at FLASH [24] is on the order of 10^{12}, and the photon energy can currently be tuned from less than 20–200 eV in the first harmonic and from 60–600 eV in the third harmonic, making this a very powerful source that ultimately will be able to deliver up to 10^3 pulses per second. Evaluation of the autocorrelation traces shows that the temporal coherence is on the order of 10 fs [25, 26] (note that these results for the electric field autocorrelation agree very well with the measured frequency spectrum as expected [26]), while the average pulse width is around 35 fs in agreement with other recent experiments [27, 28]. For two-color pump–probe experiments, an optical laser (800 nm, pulse width 120 fs) can be used that is synchronized to the free electron laser. The jitter between the two sources depends on the operation conditions of the accelerator. While it is typically around 400–500 fs, it has been demonstrated that it can be stabilized to below 35 fs rms using a fiber laser-based feedback system. Since the number of photons/pulse is so high, quite often the single shot experiments are feasible. In this case, the jitter between the

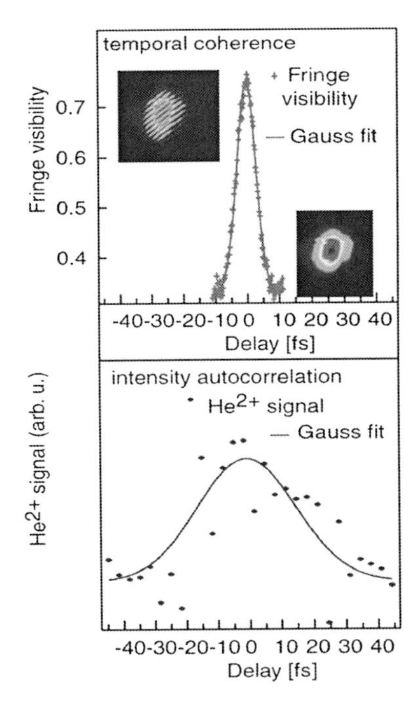

Figure 14.12 The figure shows a measurement of the temporal coherence (top) and the pulse length (bottom) of pulses from the free electron laser FLASH performed with our XUV split and delay unit based on a Mach–Zehnder interferometer [26]. The temporal coherence was obtained from the visibility of interference fringes recorded at 32 nm, while the intensity autocorrelation was obtained monitoring two-photon double ionization of He atoms at 47 eV. (Please find a color version of this figure on the color plates.)

two pulses can be measured for each individual shot using, for example, electro-optical sampling to determine the electron bunch arrival time in the FEL with respect to the optical pulse [29]. The time resolution for these experiments is then close to the limit given by the cross-correlation between the optical pulse and the FEL pulse [29–31]. With the parameters mentioned above, FLASH is an extremely powerful source for time-resolved soft X-ray spectroscopy. In the first period of user operation at FLASH, we started to investigate the possibilities for time-resolved core-level photoemission spectroscopy and time-resolved X-ray emission spectroscopy. In collaboration with other groups, we have been able to perform first time-resolved core-level photoemission experiments on optically pumped surfaces [32, 33] and solids [34]. Furthermore, we have been able to demonstrate the possibility to follow electronic structure changes in a semiconductor after optical excitation of electron–hole pairs using X-ray emission spectroscopy [35].

In Sections 14.2.1 and 14.2.2, we will discuss two examples for time-resolved core-level spectroscopy that clearly show that femtosecond chemistry at surfaces using time-resolved ESCA (electron spectroscopy for chemical analysis) is within reach.

14.2.1
Laser-Assisted Photoemission

Prior to performing the first two-color pump–probe photoemission experiments, we have studied the possible space charge effects on the measured line shape of core levels due to the large number of electrons emitted as a result of the extremely brilliant XUV or soft X-ray pulse. In this study, we have monitored the line shape of the spin–orbit split W4f core levels excited with 118.5 eV photons from the third harmonic of FLASH as a function of the number of photons/pulse [33]. From the experimental results, it is obvious that we can observe strong space effects above a certain number of incoming photons that can also be modeled theoretically [33, 36]. However, it is also clear that photoemission experiments can be performed with a free electron laser source such as FLASH in a regime of fluences where space charge effects can be avoided completely.

Using these conditions, we have investigated effects related to the simultaneous presence of the field of the optical laser in the W4f photoemission process induced by the FLASH pulses. Laser-assisted photoemission at surfaces, that is, the appearance of sidebands in the XUV photoemission lines due to the interaction of the outgoing photoelectron with a strong optical laser field [37], has recently attracted a lot of interest. Using high harmonic generation XUV sources, it has been clearly demonstrated that this effect can be observed for clean metal surfaces [38–40] as well as for adsorbates on metal surfaces [41] and that it can be used for time-resolved experiments at surfaces [37, 41]. In our experiments, we observe sideband formation when the pulses of the optical laser and the pulses from the free electron laser overlap in time. Figure 14.13 shows the W4f photoemission lines obtained for a scan where the delay between the optical laser pulses and the FLASH pulses was varied.

Although the optical laser in our case is relatively weak (5×10^{10} W/cm^2), an analysis of the ratio of the area under the main W4f$_{7/2}$ peak (shaded region A in Figure 14.13b) and the area where the first sideband is expected (shaded region B in Figure 14.13b) as a function of the delay between the two pulses shows a clear dip in the ratio A/B indicative of sideband formation when the two pulses overlap (Figure 14.13c, note that $t = 0$ was independently measured using the cross-correlation technique developed by us [30]). The width of the dip is around 400 fs in perfect agreement with the expectations based on the known pulse width and the respective jitter of the two sources. From the discussion above, it is clear that the main source of the width in time of the overlap between the two sources is the jitter between the two sources. However, as also mentioned above, this can in principle be significantly reduced using single pulse detection and sorting according to the measured delay or by temporal stabilization of the free electron laser.

Hence, this study clearly shows the feasibility of time-resolved two-color pump–probe photoemission experiments at FLASH. This opens the way for first experiments on time-resolved core-level photoemission to monitor carrier dynamics in semiconductors as discussed in the next section.

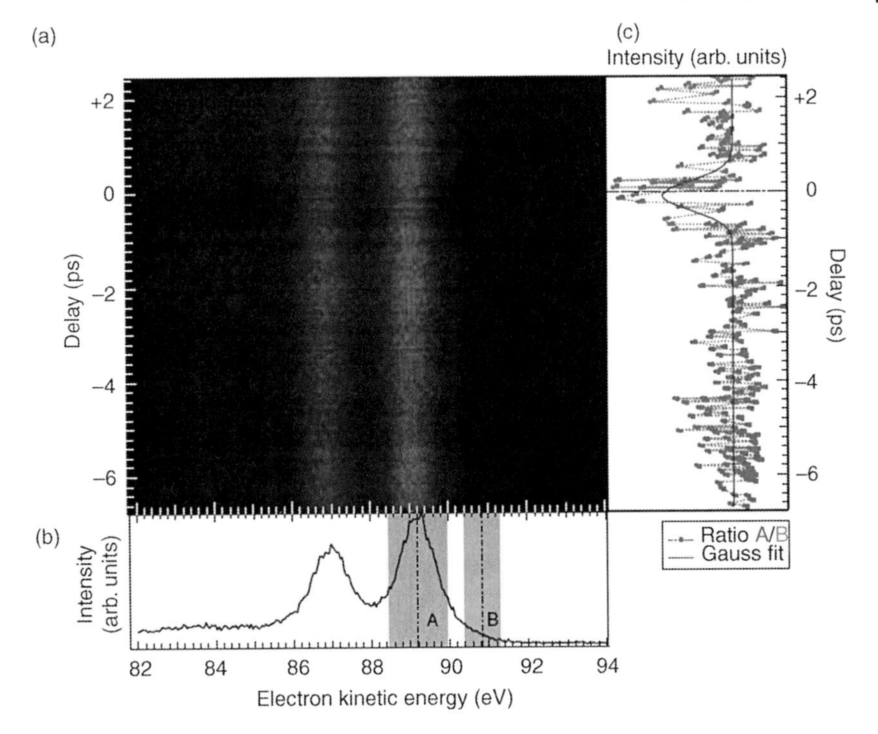

Figure 14.13 The figure shows a density plot of the spin–orbit split W4f photoemission lines as a function of the delay between an optical (800 nm$\hat{=}$1.55 eV) laser pulse and a XUV pulse (118.5 eV) from FLASH impinging on a W(110) surface. At the bottom, a horizontal cut showing a W4f photoemission spectrum is given where two regions of interest are highlighted. Region A includes the unperturbed W4f$_{7/2}$ line, while Region B is shifted to higher energies by the photon energy of the optical laser. To the right, the intensity ratio A/B is shown as a function of delay between the two pulses. Around zero delay, a deep can be seen with a width of about 400 fs indicative of the appearance of sidebands when both pulses are present. Reprinted from [33]. Copyright 2009 IOP Publishing Ltd. (Please find a color version of this figure on the color plates.)

14.2.2 Surface Carrier Dynamics

In an attempt to perform the first time-resolved femtosecond ESCA experiment at surfaces with FLASH, we have investigated the surface carrier dynamics in germanium [32]. At the surface of a doped semiconductor, typically band bending is observed that is due to the change in electrostatic potential at the surface induced by the existence of excess carriers resulting from unsaturated bonds at the surface. Photodoping the semiconductor, that is, creating electron–hole pairs by optical excitation of valence band electrons to the conduction band, will transiently reduce the band bending due to screening of excess surface charges. The screening response will be dominated by the transport properties of the carriers [42]. This photoinduced shift may partially compensate the intrinsic band bending at the surface, as shown schematically in Figure 14.14.

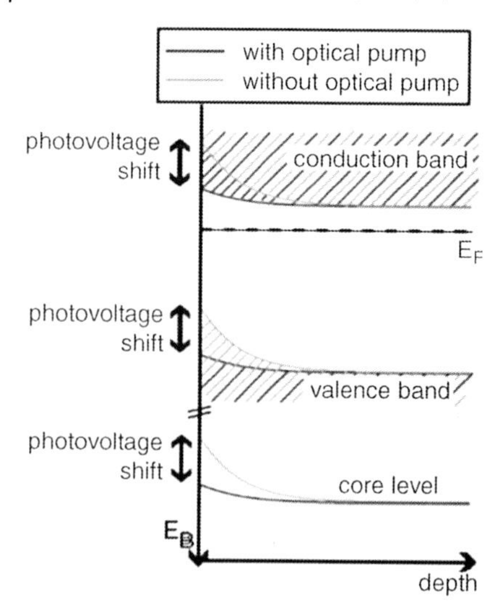

Figure 14.14 The figure shows schematically the transient change in surface potential, that is, the so-called surface photovoltage, induced by the creation of electron–hole pairs by an optical laser pulse. Note that as a result all electronic levels, including core levels, are shifted with respect to the Fermi level. (Please find a color version of this figure on the color plates.)

Since all electronic levels, including the core levels, follow the changes in the electrostatic potential, core-level photoemission spectroscopy where a semiconductor core level (in our case Ge 3d) is measured with respect to the Fermi level can monitor transient changes in the electrostatic potential induced by photodoping [43–48].

In a first attempt to perform such an experiment at a free electron laser [32], we have measured the transient changes in the energetic position of the Ge 3d core level as function of the delay between an optical pulse (1.55 eV, 120 fs, 6 mJ/cm^2) creating mobile carriers within the surface depletion zone of n-doped Ge and a XUV pulse from FLASH (121.6 eV) photoexciting the Ge 3d core level. Figure 14.15 shows the result of this experiment. We can clearly observe a small change in the core-level binding energy with respect to the Fermi level that reflects the ultrafast lifting of the band bending at the surface due to the screening of the surface charges by the carriers created by the optical pulse. The timescale of the observed surface carrier dynamics is 2ps. Detailed modeling of the experimental results taking into account carrier mobilities is currently underway.

14.3 Summary

In this chapter, we have shown how X-ray spectroscopy can be used to obtain rather detailed information on electron transfer processes. Using high-resolution spectroscopy and monitoring the decay of resonant core-excited states, that is, performing

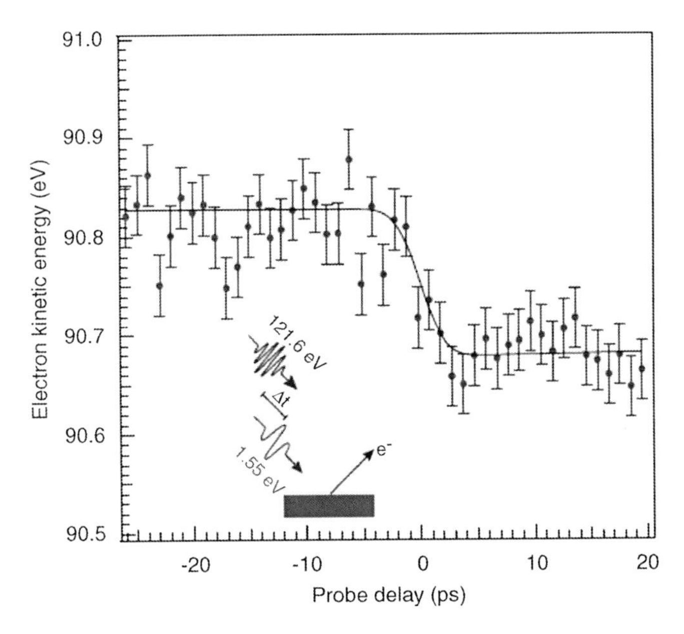

Figure 14.15 Femtosecond time-resolved ESCA of the Ge 3d photoemission lines of an n-doped Ge crystal using XUV pulses (photon energy 121.6 eV) from the free electron laser FLASH at DESY in Hamburg. The optical pump pulse (1.55 eV) creates mobile charge carriers within the surface depletion layer. The difference in electron and hole mobility leads to an ultrafast lifting of the band bending at the surface, which is probed through the binding energy shift of the photoemitted Ge 3d electrons relative to the Fermi level. (Please find a color version of this figure on the color plates.)

core hole clock spectroscopy, one can quantitatively determine electron transfer times even in the attosecond regime.

While this approach relies on the use of the core hole lifetime as an internal reference and therefore has some limitations, the time-resolved X-ray spectroscopy using new sources such as free electron lasers delivering intense ultrashort XUV and soft X-ray pulses offers great potential for the study of electron transfer processes in complex systems. Here the full potential of established spectroscopic techniques monitoring the electronic structure such as X-ray absorption spectroscopy, photoemission, and X-ray emission spectroscopy can be used for the first time in a time-dependent fashion on a femtosecond timescale to study ultrafast changes induced in the electronic states.

Acknowledgments

We would like to acknowledge the excellent work of our diploma students, Ph.D. students, and Post-Docs Martin Beye, Martin Deppe, Franz Hennies, Mitsuru Nagasono, Annette Pietzsch, William Schlotter, Florian Sorgenfrei, Edlira Suljoti, and Sethuraman Vijayalakshmi. Very fruitful collaborations with the groups of Dietrich Menzel and Peter Feulner at the Technical University Munich, Martin Wolf at the Free University Berlin, Wolfgang Drube at HASYLAB/DESY in Hamburg, and Markus Drescher at the University Hamburg are gratefully acknowledged. We would like to thank Daniel Sanchez-Portal and Pedro Echenique (San Sebastian) as well as Jean-Pierre Gauyacq and Andrei Borisov (Paris) for their excellent theoretical support and for many enlightening discussions. Financial support for this work has been given by the Deutsche Forschungsgemeinschaft DFG and the German Ministry for Education and Research BMBF.

References

1 Bjorneholm, O., Nilsson, A., Sandell, A., Hernnas, B., and Martensson, N. (1992) Determination of time scales for charge-transfer screening in physisorbed molecules. *Phys. Rev. Lett.*, **68** (12), 1892–1895.

2 Bjorneholm, O., Sandell, A., Nilsson, A., Martensson, N., and Andersen, J.N. (1992) Autoionization of adsorbates. *Phys. Scr.*, **T41**, 217–225.

3 Wurth, W., Feulner, P., and Menzel, D. (1992) Resonant excitation and decay of adsorbate core holes. *Phys. Scr.*, **T41**, 213–216.

4 Brühwiler, P.A., Karis, O., and Martensson, N. (2002) Charge-transfer dynamics studied using resonant core spectroscopies. *Rev. Mod. Phys.*, **74** (3), 703–740.

5 Menzel, D. (2008) Ultrafast charge transfer at surfaces accessed by core electron spectroscopies. *Chem. Soc. Rev.*, **37** (10), 2212–2223.

6 Wang, L., Chen, W., and Wee, A.T.S. (2008) Charge transfer across the molecule/metal interface using the core hole clock technique. *Surf. Sci. Rep.*, **63** (11), 465–486.

7 Wurth, W. and Menzel, D. (2000) Ultrafast electron dynamics at surfaces probed by resonant Auger spectroscopy. *Chem. Phys.*, **251** (1–3), 141–149.

8 Coville, M. and Thomas, T.D. (1991) Molecular effects on inner-shell lifetimes: possible test of the one-center model of Auger decay. *Phys. Rev. A*, **43** (11), 6053–6056.

9 Campbell, J.L. and Papp, T. (2001) Widths of the atomic K-N7 levels. *At. Data Nucl. Data Tables*, **77** (1), 1–56.

10 Ohno, M. (1994) Deexcitation processes in adsorbates. *Phys. Rev. B*, **50** (4), 2566–2575.

11 Lizzit, S., Zampieri, G., Kostov, K.L., Tyuliev, G., Larciprete, R., Petaccia, L., Naydenov, B., and Menzel, D. (2009) Charge transfer from core-excited argon adsorbed on clean and hydrogenated Si (100): ultrashort timescales and energetic structure. *New J. Phys.*, **11**, 053005.

12 Karis, O., Nilsson, A., Weinelt, M., Wiell, T., Puglia, C., Wassdahl, N., Martensson, N., Samant, M., and Stöhr, J. (1996) One-step and two-step description of deexcitation processes in weakly interacting systems. *Phys. Rev. Lett.*, **76** (8), 1380–1383.

13 Föhlisch, A., Menzel, D., Feulner, P., Ecker, M., Weimar, R., Kostov, K.L., Tyuliev, G., Lizzit, S., Larciprete, R., Hennies, F., and Wurth, W. (2003) Energy dependence of resonant charge transfer from adsorbates to metal substrates. *Chem. Phys.*, **289** (1), 107–115.

14 Keller, C., Stichler, M., Comelli, G., Esch, F., Lizzit, S., Menzel, D., and Wurth, W. (1998) Femtosecond dynamics of adsorbate charge-transfer processes as probed by high-resolution core-level spectroscopy. *Phys. Rev. B*, **57** (19), 11951–11954.

15 Keller, C., Stichler, M., Comelli, G., Esch, F., Lizzit, S., Menzel, D., and Wurth, W. (1998) Resonant Auger processes in adsorbates. *J. Electron Spectros. Relat. Phenomena*, **93** (1–3), 135–141.

16 Sanchez-Portal, D., Menzel, D., and Echenique, P.M. (2007) First-principles calculation of charge transfer at surfaces: the case of core-excited Ar(2p(3/2)(−1)4s) on Ru(0001). *Phys. Rev. B*, **76** (23), 235406.

17 Vijayalakshmi, S., Föhlisch, A., Hennies, F., Pietzsch, A., Nagasono, M., Wurth, W., Borisov, A.G., and Gauyacq, J.P. (2006) Surface projected electronic band structure and adsorbate charge transfer dynamics: Ar adsorbed on Cu(111) and Cu (100). *Chem. Phys. Lett.*, **427** (1–3), 91–95.

18 Borisov, A.G., Gauyacq, J.P., Kazansky, A.K., Chulkov, E.V., Silkin, V.M., and Echenique, P.M. (2001) Long-lived excited states at surfaces: Cs/Cu(111) and Cs/Cu (100) systems. *Phys. Rev. Lett.*, **86**, 488.

19 Föhlisch, A., Feulner, P., Hennies, F., Fink, A., Menzel, D., Sanchez-Portal, D., Echenique, P.M., and Wurth, W. (2005) Direct observation of electron dynamics in the attosecond domain. *Nature*, **436** (7049), 373–376.

20 Deppe, M., Föhlisch, A., Hennies, F., Sanchez-Portal, D., Echenique, P.M., and Wurth, W. (2007) Ultrafast charge transfer and atomic orbital polarization. *J. Chem. Phys.*, **127** (17), 174708.

21 Föhlisch, A., Vijayalakshmi, S., Pietzsch, A., Nagasono, M., Wurth, W., Kirchmann, P.S., Loukakos, P.A., Bovensiepen, U., Wolf, M., Tchaplyguine, M., and Hennies, F. Charge transfer dynamics in molecular solids and adsorbates driven by local and non-local excitations, in press.

22 Föhlisch, A., Vijayalakshmi, S., Hennies, F., Wurth, W., Medicherla, V.R.R., and Drube, W. (2007) Verification of the core-hole-clock method using two different time references: attosecond charge transfer in c(4×2)S/Ru(0001). *Chem. Phys. Lett.*, **434** (4–6), 214–217.

23 Keller, C., Stichler, M., Fink, A., Feulner, P., Menzel, D., Föhlisch, A., Hennies, F., and Wurth, W. (2004) Electronic transfer processes studied at different time scales by selective resonant core hole excitation of adsorbed molecules. *Appl. Phys. Mater. Sci. Process.*, **78** (2), 125–129.

24 Ackermann, W. *et al.* (2007) Operation of a free-electron laser from the extreme ultraviolet to the water window. *Nat. Photonics*, **1** (6), 336–342.

25 Mitzner, R., Siemer, B., Neeb, M., Noll, T., Siewert, F., Roling, S., Rutkowski, M., Sorokin, A.A., Richter, M., Juranic, P., Tiedtke, K., Feldhaus, J., Eberhardt, W., and Zacharias, H. (2008) Spatio-temporal coherence of free electron laser pulses in the soft X-ray regime. *Opt. Express*, **16** (24), 19909–19919.

26 Schlotter, W.F., Sorgenfrei, F., Beeck, T., Beye, M., Gieschen, S., Meyer, H., Nagasono, M., Föhlisch, A., and Wurth, W., Longitudinal coherence measurements of an XUV free electron laser. *Opt. Lett.*, **35**, 372, (2010).

27 Frühling, U., Wieland, M., Gensch, M., Gebert, T., Schütte, B., Krikunova, M., Kalms, R., Budzyn, F., Grimm, O., Rossbach, J., Ploenjes, E., and Drescher, M. (2009) Single-shot terahertz-field-driven X-ray streak camera. *Nat. Photonics*, **3** (9), 523–528.

28 Mitzner, R., Sorokin, A.A., Siemer, B., Roling, S., Rutkowski, M., Zacharias, H., Neeb, M., Noll, T., Siewert, F., Eberhardt, W., and Richter, M. (2009) Direct autocorrelation of soft-X-ray free-electron-laser pulses by time-resolved two-photon double ionization of He. *Phys. Rev. A*, **80**, 025402.

29 Azima, A., Düsterer, S., Radcliffe, P., Redlin, H., Stojanovic, N., Li, W., Schlarb, H., Feldhaus, J., Cubaynes, D., Meyer, M., Dardis, J., Hayden, P., Hough, P., Richardson, V., Kennedy, E.T., and Costello, J.T. (2009) Time-resolved pump–probe experiments beyond the jitter limitations at FLASH. *Appl. Phys. Lett.*, **94** (14), 144102.

30 Gahl, C., Azima, A., Beye, M., Deppe, M., Döbrich, K., Hasslinger, U., Hennies, F., Melnikov, A., Nagasono, M., Pietzsch, A., Wolf, M., Wurth, W., and Föhlisch, A. (2008) A femtosecond X-ray/optical cross-correlator. *Nat. Photonics*, **2** (3), 165–169.

31 Maltezopoulos, T., Cunovic, S., Wieland, M., Beye, M., Azima, A., Redlin, H., Krikunova, M., Kalms, R., Frühling, U., Budzyn, F., Wurth, W., Föhlisch, A., and Drescher, M. (2008) Single-shot timing measurement of extreme-ultraviolet

free-electron laser pulses. *New J. Phys.*, **10**, 033026.

32 Bostedt, C., Chapman, H.N., Costello, J.T., Crespo Lopez-Urrutia, J.R., Düsterer, S., Epp, S.W., Feldhaus, J., Föhlisch, A., Meyer, M., Möller, T., Moshammer, R., Richter, M., Sokolowski-Tinten, K., Sorokin, A., Tiedtke, K., Ullrich, J., and Wurth, W. (2009) Experiments at FLASH. *Nucl. Instrum. Methods Phys. Res. A*, **601** (1–2), 108–122.

33 Pietzsch, A., Föhlisch, A., Beye, M., Deppe, M., Hennies, F., Nagasono, M., Suljoti, E., Wurth, W., Gahl, C., Doebrich, K., and Melnikov, A. (2008) Towards time resolved core level photoelectron spectroscopy with femtosecond X-ray free-electron lasers. *New J. Phys.*, **10**, 033004.

34 Hellmann, S., Beye, M., Sohrt, C., Rohwer, T., Sorgenfrei, F., Redlin, H., Kalläane, M., Marczynski-Bühlow, M., Bauer, M., Föhlisch, A., Kipp, L., Wurth, W., and Rossnagel, K. Ultrafast melting of a charge-density wave probed on the atomic scale, in press.

35 Beye, M., Sorgenfrei, F., Schlotter, W.F., Wurth, W., and Föhlisch, A. Silicon melts in two steps: femtosecond snapshots of the valence electrons, in press.

36 Hellmann, S., Rossnagel, K., Marczynski-Buehlow, M., and Kipp, L. (2009) Vacuum space–charge effects in solid-state photoemission. *Phys. Rev. B*, **79** (3), 035402.

37 Baggesen, J.C. and Madsen, L.B. (2008) Theory for time-resolved measurements of laser-induced electron emission from metal surfaces. *Phys. Rev. A*, **78** (3), 032903.

38 Miaja-Avila, L., Lei, C., Aeschlimann, M., Gland, J.L., Murnane, M.M., Kapteyn, H.C., and Saathoff, G. (2006) Laser-assisted photoelectric effect from surfaces. *Phys. Rev. Lett.*, **97** (11), 113604.

39 Miaja-Avila, L., Yin, J., Backus, S., Saathoff, G., Aeschlimann, M., Murnane, M.M., and Kapteyn, H.C. (2009) Ultrafast studies of electronic processes at surfaces using the laser-assisted photoelectric effect with long-wavelength dressing light. *Phys. Rev. A*, **79** (3), 030901.

40 Saathoff, G., Miaja-Avila, L., Aeschlimann, M., Murnane, M.M., and Kapteyn, H.C. (2008) Laser-assisted photoemission from surfaces. *Phys. Rev. A*, **77** (2), 022903.

41 Miaja-Avila, L., Saathoff, G., Mathias, S., Yin, J., La-o vorakiat, C., Bauer, M., Aeschlimann, M., Murnane, M.M., and Kapteyn, H.C. (2008) Direct measurement of core-level relaxation dynamics on a surface–adsorbate system. *Phys. Rev. Lett.*, **101** (4), 046101.

42 Dekorsy, T., Pfeifer, T., Kütt, W., and Kurz, H. (1993) Subpicosecond carrier transport in GaAs surface–space–charge fields. *Phys. Rev. B*, **47**, 3842.

43 Bröcker, D., Giessel, T., and Widdra, W. (2004) Charge carrier dynamics at the SiO_2/Si(100) surface: a time-resolved photoemission study with combined laser and synchrotron radiation. *Chem. Phys.*, **299** (2–3), 247–251.

44 Pietzsch, A., Föhlisch, A., Hennies, F., Vijayalakshmi, S., and Wurth, W. (2007) Interface photovoltage dynamics at the buried BaF_2/Si interface: Time resolved laser-pump/Synchroton-probe photoemission, *Appl. Phys. A*, **88**, 587.

45 Siffalovic, P., Drescher, M., and Heinzmann, U. (2002) Femtosecond time-resolved core-level photoelectron spectroscopy tracking surface photovoltage transients on p-GaAs. *Europhys. Lett.*, **60** (6), 924–930.

46 Tokudomi, S., Azuma, J., Takahashi, K., and Kamada, M. (2007) Ultrafast decay of surface photo-voltage effect on n-type GaAs(100) surface. *J. Physical Soc. Japan*, **76** (10), 104710.

47 Tokudomi, S., Azuma, J., Takahashi, K., and Kamada, M. (2008) Ultrafast time dependence of surface photo-voltage effect on p-type GaAs(100) surface. *J. Physical Soc. Japan*, **77** (1), 014711.

48 Widdra, W., Bröcker, D., Giessel, T., Hertel, I.V., Krüger, W., Liero, A., Noack, F., Petrov, V., Pop, D., Schmidt, P.M., Weber, R., Will, I., and Winter, B. (2003) Time-resolved core level photoemission: surface photovoltage dynamics of the SiO_2/Si(100) interface. *Surf. Sci.*, **543** (1–3), 87–94.

15
Exciton Formation and Decay at Surfaces and Interfaces

Matthias Muntwiler and Xiaoyang Zhu

15.1
Introduction

Excitons are the primary optical excitations of a wide range of low dielectric materials, including semiconductors, insulators, and molecular solids. In contrast to free, independent charge carriers in the conduction or valence band of a semiconductor, excitons consist of a correlated pair of an electron in the conduction band and a hole in the valence band. Correlation arises due to their mutual Coulomb interaction, which lowers the total energy of the quasiparticle with respect to two independent particles. In molecules, correlation is particularly strong because electron wave functions are intrinsically confined to a small spatial region. These two limiting cases are described by models known as the Mott–Wannier (MW) exciton [1] and the Frenkel exciton (FE) [2], respectively. Although the physics of excitons in inorganic semiconductors and isolated molecules has been developed since the 1930s, recent progress in the development of applications of organic optoelectronic devices has renewed scientific interest in excitons and their formation, decay, and energy transfer dynamics in organic semiconductors, and, particularly, at interfaces of organic semiconductors.

In this chapter, we will discuss recent experimental results of time-resolved two-photon photoelectron spectroscopy (TR-2PPE) of excitons at the interface of two model organic semiconductors. We start with an introduction about classification of excitons and photophysics of organic semiconductors. In Section 15.2, we will review a few, mostly phenomenological, theoretical concepts that have been used to model excitons in organic semiconductors. In Section 15.3, the technique of time-resolved two-photon photoelectron spectroscopy, particularly its application to excitons, will be discussed. The results from two model systems, thin films of C_{60} and of pentacene, will then be presented and discussed in Sections 15.4 and 15.5, respectively.

Dynamics at Solid State Surface and Interfaces Vol.1: Current Developments
Edited by Uwe Bovensiepen, Hrvoje Petek, and Martin Wolf

ISBN: 978-3-527-40937-2

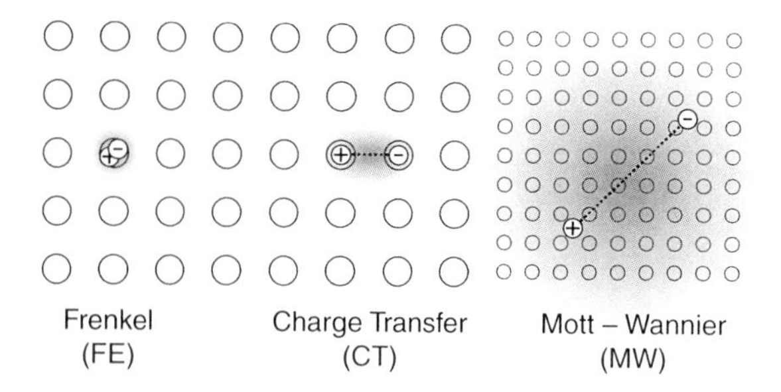

Figure 15.1 Three different exciton flavors are characterized by the separation distance between the electron and the hole with respect to lattice dimensions. While in a Frenkel exciton both hole and electron are confined to the same site, they are delocalized over many lattice sites in a Mott–Wannier exciton. Charge transfer excitons represent an intermediate case where the electron and the hole are localized on different sites.

15.1.1 Exciton Flavors

Excitons are divided into two main categories according to the extent of their wave function with respect to lattice dimensions as illustrated in Figure 15.1. In a *Frenkel exciton*, both the electron and the hole are confined to the same lattice site. This is typically the case for molecules and molecular solids with weak intermolecular (van der Waals) interactions. In these materials, the electronic structure of individual molecules is similar to the gas phase and the molecular wave functions overlap only weakly. The electronic structure of van der Waals molecular crystals can be described in a tight binding model based on molecular energy levels.

In the limit of delocalized excitons, the wave function of a *Mott–Wannier exciton* is spread over many lattice sites. With increasing electron–hole separation, the states occupied by the two charges approach the free carrier states. This situation is prevalent in inorganic semiconductors. The large extent of the wave function allows the treatment of Mott–Wannier excitons in a dielectric continuum approximation (Section 15.2.1).

Between the two extremes is a *charge transfer (CT) exciton* with the electron and the hole localized on two different lattice or basis sites and bound by their mutual Coulomb interaction. This situation occurs in similar, weakly interacting molecular systems as the Frenkel excitons. The name is derived from the creation process where the electron is transferred from one site to an adjacent site. The distance between the electron and the hole can in principle be larger than that between nearest neighbors because dielectric screening in organic materials is typically weak ($\varepsilon = 2, \ldots, 5$), and the Coulomb interaction is long ranged.

15.1.2 Photophysics of Organic Semiconductors

Organic materials are being widely investigated for potential applications in molecular electronics and optoelectronics. Their appeal draws from large-scale, inexpensive manufacturing techniques, abundant raw materials, and novel device features such as mechanical flexibility. There are two main morphologies of the materials: (1) Molecular crystals consist of molecular building blocks that are bound through van der Waals, electrostatic, or hydrogen bond interaction [3]. (2) In conducting polymers, the building blocks are bound by covalent bonds to form long linear chains [4]. For optoelectronic applications, there are several device structures: organic light-emitting diodes (OLED), organic photovoltaic (OPV) cells, organic photodetectors, and organic charge injection lasers. Here we focus on OPV. Experiments have demonstrated working photovoltaic cells with both kinds of materials [5, 6]. For studies of fundamental processes, however, crystalline materials are preferred because of reduced structural complexity. For the two semiconductors discussed in this chapter, the molecular and crystalline structures including growth conditions are well known, which, on the theoretical side, has prompted improved calculations of their electronic structure. Nevertheless, first-principles calculations of excited states, even more so at interfaces, still remain a formidable task.

Figure 15.2 shows a schematic of the fundamental processes involved in organic photovoltaics. The photovoltaic cell is made of two materials with different energy gaps between the lowest unoccupied (LUMO) and the highest occupied molecular orbital (HOMO). The two phases are named (electron) donor and (electron) acceptor according to the relative alignment of their LUMO energy levels. In the first step, photon absorption leads to the formation of a tightly bound Frenkel exciton. Ideally, the exciton diffuses toward the donor–acceptor interface. The potential step at the interface constitutes the energetic driving force for dissociating the Frenkel exciton into charge carriers. Following dissociation the carriers are collected at the electrodes. However, all these processes are subject to losses. For instance, the absorption cross section may be low, the exciton may decay before it reaches the interface, and charge carriers may experience transport losses

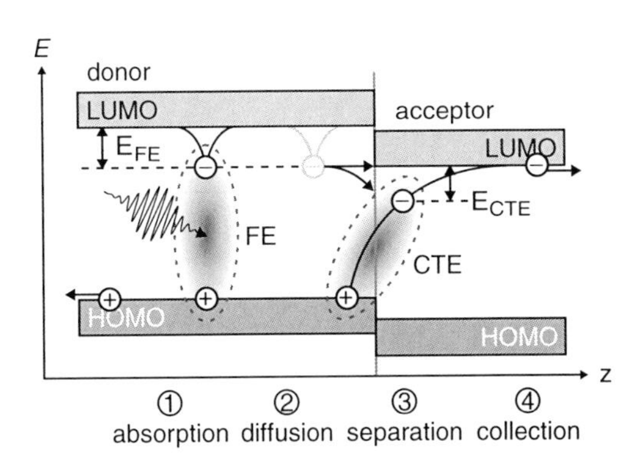

Figure 15.2 Energy diagram of a donor–acceptor interface in organic photovoltaics.(1) A Frenkel exciton is created; (2) the exciton reaches the donor/acceptor interface; (3) the electron transfers into the acceptor, dissociating the exciton or transforming into a charge transfer exciton (CTE); and (4) free carriers are collected.

or recombination. At the interface, energy is lost due to the potential drop. In addition, rather than dissociating into free carriers, the Frenkel exciton may transform into a bound charge transfer exciton (CTE) [7]. The ultimate efficiency of an organic photovoltaic cell depends critically on the energetics and dynamics of each of these processes [8, 9].

As an experimental technique, TR-2PPE is ideally suited for measuring the energetics and dynamics of exciton dissociation because it combines optical absorption with spectroscopy of electron levels and femtosecond time resolution. 2PPE is a pump–probe scheme where, in reference to Figure 15.2, the first photon would create an exciton and the second photon would probe the electron by photoemission in one of the states 1–4, depending on the delay between the two photons. Most importantly, the exciton energy levels can be compared to single-electron (or transport) levels by hot electron injection from a metallic substrate. The particularities of the technique concerning excitons will be discussed in detail in Section 15.3. A disadvantage of any photoemission technique with regard to the study of buried interfaces is the small probing depth of a few atomic layers. This aspect needs to be compensated for by making a careful choice and comparative study of slightly different model systems, such as the ones discussed later in this chapter.

15.2
Exciton Models

Since by nature they are a many-body excited state, excitons are intrinsically complicated quasiparticles that are difficult to model using first principles. Although a number of first-principles approaches exist [10–12], we will concentrate on simpler, parametrized models. These models still provide a basic understanding of the phenomena and can be used to qualitatively explain the energy levels observed in the experimental data presented in later sections. Most importantly, the models (including most DFT calculations available to date) do not describe the dynamic aspects of excitons. And since the calculations deal with the ground state and lowest excited states, exciton dissociation or the photoemission processes that couple excitons to single-particle states are not described. At interfaces, where electron transfer constitutes the dominant decay channel over recombination, one might start with a single-electron description such as in the dielectric models described below, and treat the exciton decay as an electron-transfer, taking into account the long-range Coulomb attraction.

We start this section with the dielectric models, based on a pair of opposite point charges embedded in a homogeneous dielectric medium. For a discussion of excitons in molecular crystals, we will review the Frenkel [2] and Merrifield [13] models. For simplicity, coupling to vibrational modes [14], self-trapping, and polaron formation [15, 16] will not be discussed in detail here, although the magnitude of these effects is expected to be of the order of up to a few tens of meV on a timescale of picoseconds, and could thus well be resolved by TR-2PPE. Self-trapping and polaron formation are discussed in Chapter 4.

15.2.1 Dielectric Models

In the simplest representation, an exciton can be understood as a positive and a negative point charge that are bound by their mutual Coulomb interaction in a dielectric medium. This corresponds to a hydrogen atom problem with the charge reduced by $1/\varepsilon$, where ε is the relative dielectric constant, and reduced mass μ. The Coulomb potential experienced by the electron is then

$$V_{\text{e-h}} = -\frac{e^2}{4\pi\varepsilon_0}\frac{1}{\varepsilon r}, \tag{15.1}$$

where r is the distance between the electron and the hole. The resulting binding energies are then defined by the exciton Rydberg energy

$$R_{\text{ex}} = \frac{\mu}{m_{\text{e}}}\frac{1}{\varepsilon^2}R_\infty, \tag{15.2}$$

and the exciton radius

$$a_{\text{ex}} = \frac{m_{\text{e}}}{\mu}\varepsilon a_0, \tag{15.3}$$

where R_∞ and a_0 are the Rydberg energy and Bohr radius of the hydrogen atom with infinite nuclear mass, respectively, and m_{e} is the free electron mass. Figure 15.3 shows these two quantities as a function of the dielectric constant ε, assuming free electron masses for the electron and the hole. It can be seen from the energy curve that the binding energies span the range from meV to eV for the range of dielectric constants of inorganic and organic semiconductors. Correspondingly, the exciton size ranges from a single to hundreds of lattice sites. Since the dielectric constant is a macroscopic property, one should expect the model to fail at very short distances. By contrast, in inorganic semiconductors where the wave function spans many unit

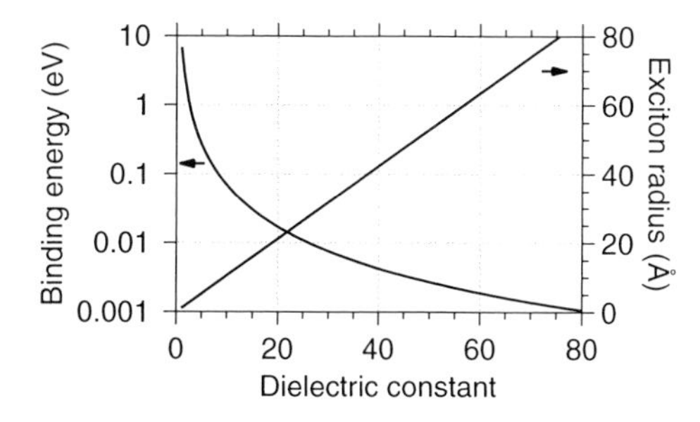

Figure 15.3 Binding energy and size of an electron–hole pair in a homogeneous dielectric, assuming free electron masses for the electron and the hole. The curve shows the binding energy versus dielectric constant, the line indicates the maximum of the radial probability distribution of the lowest wave function. Typical organic semiconductors are in the range $\varepsilon = 2, \dots, 5$, while inorganic semiconductors are in the range $\varepsilon = 10, \dots, 50$.

cells, the microscopic structure of the solid can be neglected and the dielectric constant is well defined. In addition, one can still consider using the dielectric function at optical frequencies rather than the static constant for smaller excitons to account for different dynamics of the polarizability [15]. Thus, the model is often used to describe the binding energies of the delocalized Mott–Wannier exciton [1]. The model has also proven useful for smaller excitons, for instance, to explain a general trend of Frenkel exciton binding energy in small molecules versus molecular size [17] or to explain the energy levels and wave functions of charge transfer excitons at organic semiconductor surface, as will be discussed in Section 15.5.

15.2.1.1 Mott–Wannier Exciton

The Mott–Wannier exciton combines the dielectric model with solutions of the periodic potential of the crystal lattice. This is achieved by creating a wave packet out of Bloch functions of the conduction band and the valence band, see, for example, [15]. In the effective mass approximation, eigenvalues can be obtained by solving a hydrogen-like Schrödinger equation for the envelope function F_n of the wave packet:

$$\left(-\frac{\hbar^2\nabla_r^2}{2\mu}-\frac{e^2}{\varepsilon|\vec{r}|}\right)F_n(\vec{r}) = \left(E_n - E_g - \frac{\hbar^2\vec{k}^2}{2(m_e^* + m_h^*)}\right)F_n(\vec{r}), \tag{15.4}$$

which is based on the bandgap E_g and the effective masses of the hole m_h^* and the electron m_e^*. The resulting excitation energies E_n form a Rydberg-like series ($n = 1, 2, \ldots$) of eigenstates that are referenced to the parabolic conduction band according to

$$E_n(\vec{k}) = E_g + \frac{\hbar^2\vec{k}^2}{2(m_e^* + m_h^*)} - \frac{R_{ex}}{n^2}, \tag{15.5}$$

where the last term corresponds to the exciton binding energy obtained from the dielectric model in Eq. (15.2). For a dielectric constant of $\varepsilon = 10$ that is typical of inorganic semiconductors, R_{ex} is of the order of 10 meV (Figure 15.3), and thus, Mott–Wannier excitons are very weakly bound. A high-resolution technique is required in order to resolve them from regular band-to-band transitions. It also becomes clear that Mott–Wannier excitons show dispersion with crystal momentum, which derives from the curvature of the related conduction and valence bands.

15.2.1.2 Charge Transfer Exciton at Dielectric Interfaces

We approximate a charge transfer exciton at the interface of a homogeneous dielectric in the same formalism as introduced above. This model will be used to explain the observation of charge transfer excitons at the pentacene surface in Section 15.5. We start with the interface of a homogeneous dielectric slab with dielectric constant ε, and vacuum, as illustrated in Figure 15.4. We approximate the hole by a point charge that is fixed in space (e.g., localized at a molecular site), neglecting the microscopic structure of the molecule and the crystal and assuming infinite mass of the hole. The electron and the hole induce polarization charge at the interface due to the

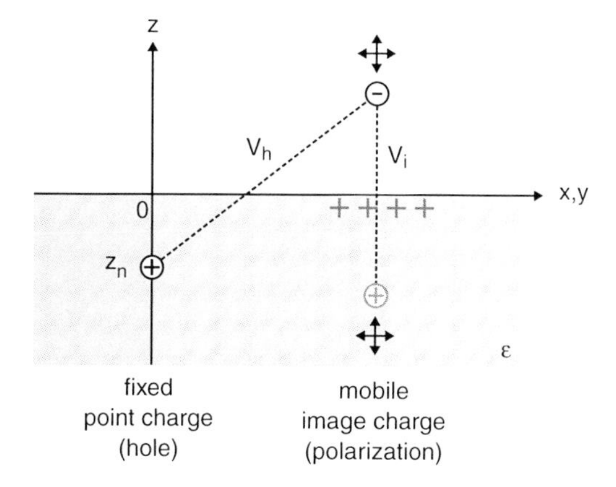

Figure 15.4 Charge transfer exciton at the surface of a dielectric medium. The electron induces an attractive surface polarization that gives rise to the image potential V_i. In addition, the electron is also attracted by the screened Coulomb potential V_h of the hole. The position of the hole is fixed in space at $\vec{r} = (x, y, z) = (0, 0, z_h)$ (localized on a molecule), whereas the electron is allowed to move.

discontinuity of the dielectric constant. We constrain the electron to the positive half-space by introducing an infinite potential barrier at the interface.

At the surface of the dielectric, the electron is attracted by two electrostatic potentials. First, the image potential formed by the surface polarization is [18]

$$V_i(z) = -\frac{e^2}{4\pi\varepsilon_0}\frac{\beta}{4z}, \quad \beta = \frac{\varepsilon - 1}{\varepsilon + 1}, \tag{15.6}$$

where the z-axis corresponds to the surface normal and is referenced to the image plane at $z = 0$. Assuming a hard wall at the surface, the Schrödinger equation yields a series of *image potential states* (IPS) that are bound in front of the surface but can move freely parallel to the surface. Their energies are

$$E_n = E_{\text{vac}} - \frac{1\,\text{Ry}\,\beta^2}{16\;n^2}, \quad n = 1, 2, 3, \ldots \tag{15.7}$$

with respect to the vacuum level E_{vac}. Notice the factor $\beta^2 \leq 1$ that accounts for the reduced polarizability of the surface [18]. For an ideal metal, $\beta = 1$, in which case the binding energy of the lowest state can be as high as 0.85 eV.

The second contribution originates from the positive charge of the hole, screened by the dielectric:

$$V_h(\varrho, z) = -\frac{e^2}{4\pi\varepsilon_0}\frac{\gamma}{\sqrt{\varrho^2 + (z - z_h)^2}}, \quad \gamma = \frac{2}{\varepsilon + 1}, \tag{15.8}$$

where the hole is fixed at $(x, y, z) = (0, 0, z_h)$, and $\varrho = \sqrt{x^2 + y^2}$ is the horizontal distance from the hole. The resulting potential

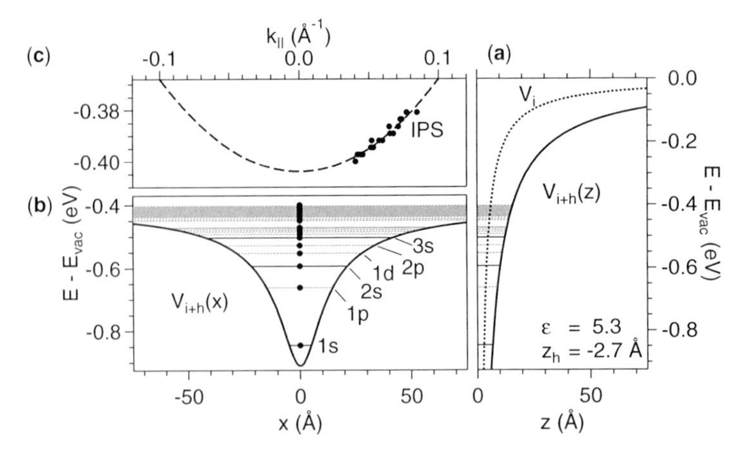

Figure 15.5 Potential and energy levels of the charge transfer exciton at the surface of a dielectric medium with $\varepsilon = 5.3$ and $z_h = -2.7$ Å. (a) Profiles of the image potential V_i (Eq. (15.6), dotted line), and the CTE potential V_{i+h} (Eq. (15.9), solid line) as a function of z above the surface ($x = y = 0$). (b) Profile of the CTE potential V_{i+h} as a function of x parallel to the surface for $y = 0$ and $z = 6.2$ Å. Along this line, $V_{i+h}(x)$ converges toward the lowest eigenvalue of the pure IPS at $E_1 = -0.404$ eV. Large dots and horizontal lines indicate energy levels of the localized eigenstates as explained in the text. (c) Eigenvalues of the delocalized CTE states versus wave vector $\vec{k}_{\|}$ parallel to the surface. The dashed line is a parabolic fit corresponding to free electron dispersion. The energy scales are referenced to the vacuum level.

$$V_{i+h} = V_i + V_h \tag{15.9}$$

has cylindrical symmetry. Profiles of the potential along x and z are shown in Figure 15.5.

We solve the time-independent Schrödinger equation of an electron in the model potential V_{i+h} numerically using the finite element method. On the basis of the potential shown in Figure 15.5, we expect two regimes for bound solutions: The lowest states must be localized in all three spatial directions. Above an energy level corresponding to the lowest ($n = 1$) image potential state, the electron must be able to escape the attraction of the hole and thus have a delocalized wave function. This is indeed seen in the calculated results: The large dots in Figure 15.5 represent calculated eigenvalues from a full three-dimensional calculation, and Figure 15.6 shows a few select wave functions. Above the image potential level given by Eq. (15.7), the wave functions are delocalized across the whole domain, and their dispersion relation $E(|\vec{k}_{\|}|)$, where $\vec{k}_{\|}$ is extracted from a Fourier transform of the wave function, follows a free electron parabola. The series of states below the image potential state are localized in all spatial directions. These states represent charge transfer excitons. Similar to the hydrogen atom problem, a whole series of states exists that converges toward the continuum, in this case the two-dimensional continuum of image potential states. We can thus identify the exciton binding energy as the energy eigenvalue with respect to the image state energy. According to the symmetry of the wave functions, discrete principal j and angular momentum l_z quantum numbers

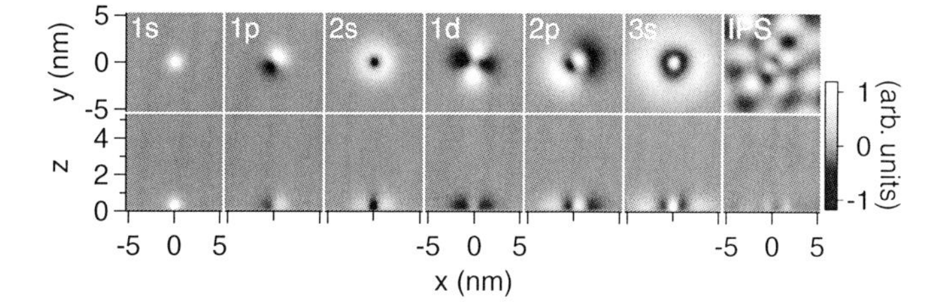

Figure 15.6 Gray scale representations of calculated electron wave functions $\psi_{j,l_z}(x, y, z)$ of selected charge transfer exciton states at the surface of a dielectric medium with $\varepsilon = 5.3$ and $z_h = -2.68$ Å. *Top row*: Projections onto the surface plane. The images are the integrals $\int \psi(x, y, z)\, dz$ of the three-dimensional wave functions along the *z*-axis. *Bottom row*: Cross sections $\psi(x, 0, z)$. Note that the *z*-axis is expanded by a factor 2. The wave functions are not normalized. The calculated IPS wave function (*far right*) contains rotationally equivalent Fourier components with $|\vec{k}_{||}| = 0.085$ Å^{-1}. The finite element calculations induce slight artifacts: The azimuthal orientation of the patterns (and so the azimuthal distribution of Fourier components in the IPS pattern) is arbitrary. The 2*p* and 3*s* patterns show slight deviations from the nominal symmetry. These deviations correspond to a numerical precision of a few meV.

can be assigned to the individual states. Given the cylindrical symmetry of the potential, l_z is the only angular momentum quantum number and is always defined with respect to the *z*-axis. Unlike the hydrogen atom problem, where the form of the Coulomb potential restricts the range of angular momentum eigenvalues, solutions with angular momentum higher than $j-1$ exist for the potential $V_{\mathrm{i+h}}$.

The calculations shown in Figures 15.5 and 15.6 were done using a dielectric constant of 5.3 as calculated by Tsiper and Soos for pentacene along the long molecular axis [19]. The position of the hole z_h is then treated as a fit parameter. Measured data of the 1s CT exciton (cf. Section 15.5) is well represented by $z_\mathrm{h} \approx -2.7$ Å, which is of the order of the size of a phenyl ring of the molecule. Calculated exciton binding energies are then $E_{\mathrm{B,CT}} = 0.44$ eV (1s), 0.25 eV (1p), and 0.19 eV (2s) for the lowest three states [20]. These binding energies are of the same order of magnitude as the general trend obtained for the dielectric model of Eq. (15.2) (see Figure 15.7). Compared to thermal excitation energies of the order of $k_\mathrm{B}T$, where k_B is the Boltzmann constant, the CT excitons are very stable against spontaneous dissociation.

The presented model can be modified to represent a real donor–acceptor junction if the vacuum is replaced by a second dielectric material [21]. If we take the dielectric constant to be $\varepsilon = 3.15$ in both materials, it equals the $1/\gamma$ screening factor of the dielectric vacuum model (15.8), and the potential becomes

$$V_h(\varrho, z) = -\frac{e^2}{4\pi\varepsilon_0}\frac{1}{\varepsilon}\frac{1}{\sqrt{\varrho^2 + (z - z_h)^2}}. \tag{15.10}$$

The position of the hole is still fixed, and the electron wave function is constrained to the opposite side of the interface. Qualitatively, the results are similar. Because the image potential is absent in this case, however, the binding energies reduce to

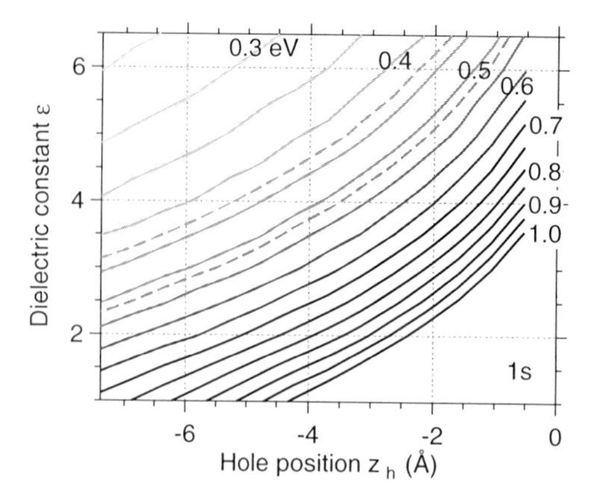

Figure 15.7 Binding energy of the 1s CT exciton at the surface of a homogeneous dielectric medium as a function of the dielectric constant ε and the *z*-position of the hole. The contour interval is 0.05 eV. These energies were calculated in a simplified, two-dimensional geometry of cylindrical symmetry. Dashed contours indicate experimental energies of monolayer (0.43 eV) and multilayer (0.52 eV) pentacene films (cf. Section 15.5).

$E_{B,CT} = 0.20$ eV (1s), 0.12 eV (1p), and 0.10 eV (2s), and the wave functions extend much further into the acceptor. The results are summarized in Table 15.1.

15.2.2
Merrifield Model of Frenkel and Charge Transfer Excitons

The dielectric continuum approach to model small excitons is expected to break down if the microscopic structure of the medium cannot be neglected. This is particularly the case in organic molecular crystals where the electronic structure of individual molecules is largely unperturbed and determines most of the optical properties. We present here a parameterized description of Frenkel and charge transfer excitons in

Table 15.1 Numerical results from calculations of the dielectric CT exciton model. The *surface* model consists of a dielectric with ε = 5.3 in one half-space and vacuum in the other. The *D/A interface* consists of two dielectrics with ε = 3.15 and an impenetrable potential wall between them. The hole is set at $z = -2.68$ Å. $\langle z \rangle$ is the expectation value of the distance of the electron wave function from the interface.

State	Surface		D/A Interface	
	$E_{B,CT}$ (meV)	$\langle z \rangle$ (Å)	$E_{B,CT}$ (meV)	$\langle z \rangle$ (Å)
1*s*	441	3.8	200	10.1
1*p*	252	4.2	118	14.4
2*s*	188	4.4	103	21.4
1*d*	148	4.4	—	—
3*s*	88	4.6	64	36.1

molecular semiconductors due to Merrifield [13], which has been successfully developed to reproduce optical and electroabsorption spectra of weakly bound crystals of π-conjugated molecules [14], polyacenes [22, 23], and C_{60} [24]. We restrict ourselves to the most basic features and conclusions, for details the reader may consult the referenced literature.

15.2.2.1 **Frenkel Exciton**

We start with the simple case of two van der Waals bound molecules with negligible orbital overlap. The total Hamiltonian consists of three terms

$$H_{\text{dimer}} = H_{\text{mol},1} + H_{\text{mol},2} + V_{1,2}, \tag{15.11}$$

where $H_{\text{mol},n}$ represents the Hamiltonian of molecule n, containing the kinetic energies and Coulomb interactions between all nuclei and electrons. The operator $V_{1,2}$ contains Coulomb interactions between pairs of charges one of which belongs to molecule 1 and the other one to molecule 2. Thus, even though the molecules have no significant orbital overlap, this last term establishes an electrostatic interaction between them.

We now extend the model to a linear chain of molecules and assume that each molecule can be in either the ground state or its lowest excited state, and there is at most one excited molecule in the system, so that the system is completely described by the state vector $|n\rangle$, $n = 0, 1, 2, \ldots$, where $n \geq 1$ refers to the excited molecule, and $n = 0$ denotes the ground state. We restrict interactions to nearest neighbors only (a zeroth order approximation as electrostatic interaction is long ranged), and write the Hamiltonian in second quantization:

$$H_{\text{NN}}^{\text{FE}} = E_{\text{FE}} \sum_n a_n^\dagger a_n + J \sum_n \left(a_n^\dagger a_{n+1} + a_{n+1}^\dagger a_n\right). \tag{15.12}$$

The creation operator $a_n^\dagger|0\rangle = |n\rangle$ creates an exciton at site n, and the annihilation operator $a_n|n\rangle = |0\rangle$ destroys one, according to Figure 15.8a. Because $\langle k|a_l^\dagger a_m|n\rangle =$

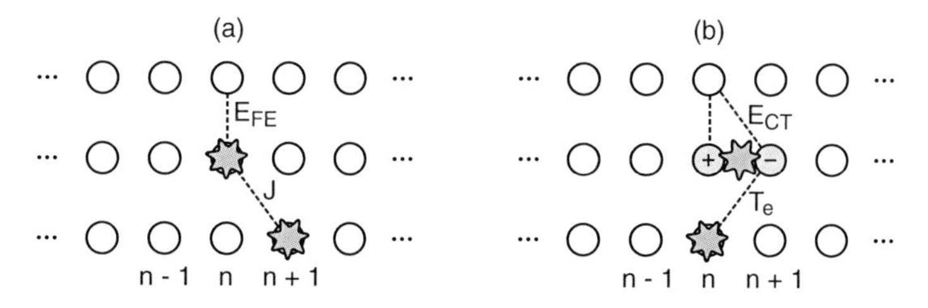

Figure 15.8 Exciton coupling in the Frenkel and Merrifield Hamiltonians of a linear chain. (a) Creation and nearest neighbor coupling of a Frenkel exciton. The first row represents the ground state. In the transition to the second row, a Frenkel exciton is created at site n, and hops to site $n+1$ in the third row with coupling strength J. (b) Creation of a charge transfer exciton, and nearest neighbor charge transfer (F-CT) coupling. The CT-exciton (second row) can be created from the ground state (first row) or from a Frenkel exciton (third row) by electron transfer or hole transfer (not shown).

$\delta_{kl}\delta_{mn}$, the first term yields E_{FE}, that is, the amount of exciton energy stored in the system. The second term describes the coupling, or exciton transfer, between neighboring sites corresponding to $V_{1,2}$ in Eq. (15.11). In the dipole approximation, the interaction strength is

$$J_{\mathrm{dipole}} = \frac{(\vec{\mu}_1 \cdot \vec{\mu}_2) r_{12}^2 - 3(\vec{\mu}_1 \cdot \vec{r}_{12})(\vec{\mu}_2 \cdot \vec{r}_{12})}{r_{12}^5}, \qquad (15.13)$$

where $\vec{\mu}_i$ denotes the dipole moment of molecule i. Given this interaction, the Hamiltonian has a structure that is analogous to the tight binding model of electron interactions, and is solved analogously. First, using the translational symmetry, we can Fourier-transform the Hamiltonian into a sum over momentum k, where

$$H_{\mathrm{NN}}^{\mathrm{FE}}(k) = (E_{\mathrm{FE}} + 2J\cos k) a_k^{\dagger} a_k. \qquad (15.14)$$

The k-dependence shows the formation of an exciton band, that is, a momentum-dependent excitation energy, with a bandwidth of $4J$ for the linear chain,

$$E(k) = E_{\mathrm{FE}} + 2J\cos k, \quad k \in [0, \pi]. \qquad (15.15)$$

Such a Frenkel exciton band is shown in Figure 15.9a (curved line) for $J = 0.1$ eV. Note that Frenkel exciton band formation occurs even though the electronic basis functions are localized on molecular units, in contrast to Mott–Wannier excitons where the electronic basis functions are delocalized. The coupling strength J is typically of the order of 1–100 meV. With respect to dynamics, this translates into a hopping time of the order of a 100 fs according to the Heisenberg uncertainty principle.

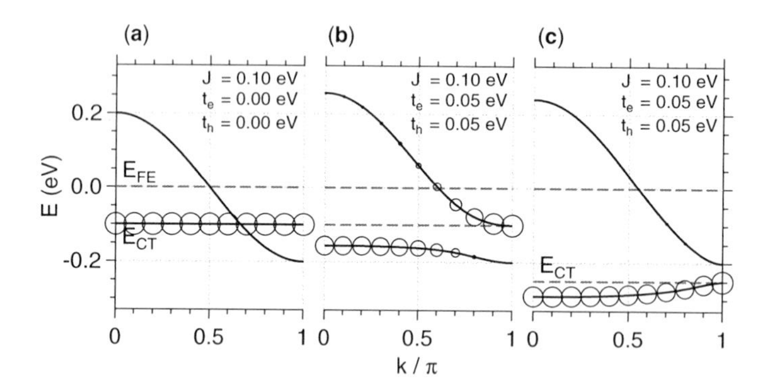

Figure 15.9 Exciton band structure (solid lines) based on the linear, nearest neighbor Merrifield model of FE-CT mixing (Eq. (15.20)) for three different sets of parameters J, t_e, and t_h. (a) Noninteracting case: FE and CT excitons are independent particles. FE excitons form a dispersive band. (b) FE-CT mixing: Both bands show dispersion but do not cross. Each band has partial, k-dependent FE and CT character. (c) FE-CT mixing at larger energetic separation. The individual bands retain their FE or CT character, but both show dispersion. The energy scale is referenced to the Frenkel exciton energy E_{FE}. The exciton energies E_{FE} and E_{CT} are indicated by broken lines. The marker size indicates the charge transfer character, where the largest size corresponds to 100% CT. k spans the positive half of the Brillouin zone.

Given the momentum conservation law, the exciton momentum is connected to experimental parameters. In particular, if the excitation is induced by absorption of a photon, $k \approx 0$ and the excitation energy has its highest energy value. The exciton band can be observed only indirectly in optical spectroscopy, for example, in a time-resolved experiment where the exciton relaxes to a lower band state before it is probed. In order to directly probe the exciton band structure, higher momentum needs to be transferred. Such an approach has been demonstrated using electron energy loss spectroscopy [25].

15.2.2.2 Charge Transfer Exciton

The Frenkel Hamiltonian (15.14) is often not sufficient to explain the exciton band structure observed in absorption and energy loss spectra [14, 23, 25]. To improve the model, charge transfer excitons need to be taken into account. The few available first-principles calculations also suggest that excitons in polyacenes have a significant charge transfer character [26]. The Merrifield model [13] extends the Frenkel Hamiltonian to include the creation of a charge transfer exciton on adjacent sites from the ground state with coupling constant E_{CT}, and from a Frenkel exciton with the electron (hole) coupling constant $t_{\mathrm{e}}(t_{\mathrm{h}})$ according to Figure 15.8b.

$$H_{\mathrm{NN}} = \sum_{k} \left(H_{\mathrm{NN}}^{\mathrm{FE}}(k) + H_{\mathrm{NN}}^{\mathrm{CT}}(k) + H_{\mathrm{NN}}^{\mathrm{FE-CT}}(k)\right) \tag{15.16}$$

The first term describes again the isolated Frenkel exciton of (15.14). A charge transfer exciton of momentum k and electron transfer to the neighbor on the positive side is created from the ground state by the operator $c^{\dagger}_{k,+1}$. The corresponding term of the Hamiltonian is given by

$$H_{\mathrm{NN}}^{\mathrm{CT}}(k) = E_{\mathrm{CT}}\left(c^{\dagger}_{k,+1}c_{k,+1} + c^{\dagger}_{k,-1}c_{k,-1}\right). \tag{15.17}$$

The last term describes the coupling of a Frenkel exciton to a charge transfer exciton by electron (e) or (h) hole transfer:

$$H_{\mathrm{NN}}^{\mathrm{FE-CT}}(k) = a^{\dagger}_{k}\left[(t_{\mathrm{e}} + t_{\mathrm{h}}\mathrm{e}^{-ik})c_{k,+1} + (t_{\mathrm{e}} + t_{\mathrm{h}}\mathrm{e}^{ik})c_{k,-1}\right] + \mathrm{h.c.} \tag{15.18}$$

In this term, the Frenkel and CT exciton terms mix. For instance, a Frenkel exciton of momentum k is created from a charge transfer exciton $|k, +1\rangle$ according to

$$a^{\dagger}_{k}c_{k,+1}|k, +1\rangle = |k, 0\rangle, \tag{15.19}$$

and the Hamiltonian is no longer diagonal. After a basis transformation to symmetric and antisymmetric states with respect to the direction of charge transfer [14], the FE–CT mixing part of the Hamiltonian can be brought to the matrix form

$$H_{\mathrm{NN}+}(k) = \begin{pmatrix} E_{\mathrm{FE}} + 2J\cos k & \sqrt{2}t_k \\ \sqrt{2}t_k & E_{\mathrm{CT}} \end{pmatrix}, \tag{15.20}$$

where the mixing terms are given by

$$t_k = \sqrt{(t_e + t_h)^2 \cos^2\tfrac{k}{2} + (t_e - t_h)^2 \sin^2\tfrac{k}{2}}. \tag{15.21}$$

The new basis states are now linear combinations of Frenkel and CT excitons, that is, they have Frenkel and CT character at the same time. The eigenvalues of the Hamiltonian (15.20) are displayed in Figure 15.9 for three situations with $J = 0.1$ eV. In panel (a) the charge transfer coupling constants t_e and t_h are zero, and the Frenkel and charge transfer exciton bands do not mix. Only the Frenkel band shows dispersion. In panel (b) charge transfer is enabled. The bands do not cross any more, both bands disperse, and charge transfer character is distributed over the two bands. Panel (c) shows a situation where the FE and CT excitons are further separated but are still weakly mixing.

15.2.2.3 Finite Size Effect

Analogous to electron wave functions in the tight binding model, the exciton bandwidth depends on the size of the system. Based on (15.14), for a linear chain of N molecules the excitation energy of the exciton state is

$$E_j(N) = E_{FE} + 2J\cos\frac{\pi j}{N+1} \quad j = 1, 2, \ldots, N. \tag{15.22}$$

In an isolated molecule $N = 1$, and the excitation energy equals E_{FE}. For the infinite chain, the bandwidth approaches $4J$, and the lowest excitation energy is $E_{FE} - 2J$. Thus, being larger for shorter chains the excitation energy shows a clear finite size effect. An example of the finite size effect in molecular semiconductors was observed in quantum well structures of PTCDA/NTCDA [27].

15.3 Photoelectron Spectroscopy of Excitons

Traditionally, excitons in organic semiconductors have been studied by all-optical absorption and luminescence spectroscopy. After pulsed laser sources were developed, these techniques have been extended to probe exciton dynamics, that is, the decay of excitons by recombination, intersystem crossing from singlet to triplet states, and vibrational dynamics. The charge transfer character of excitons can be probed by applying a modulated electric field in electroabsorption spectroscopy [28]. Since optical techniques can probe only the exciton band structure close to the center of the Brillouin zone, electron energy loss spectroscopy has been applied to resolve momentum information [25]. In order to study exciton dynamics at interfaces or surfaces, a surface-sensitive technique such as photoelectron spectroscopy is necessary. As is discussed elsewhere in this book, TR-2PPE is a method of choice for studying the dynamics of unoccupied electronic states between the Fermi and the vacuum levels of metal and semiconductor surfaces. Since it involves an optical pump step, the creation of excitons is an inherent feature of the technique. Early photoemission experiments of excited states of bulk organic semiconductors have used ultraviolet photoelectron spectroscopy (UPS) [29] or two-photon photoelectron yield as a probe [30]. Since then, development of femtosecond lasers, advances in the understanding of 2PPE and dynamical processes, and the improvements in sample preparation have enabled detailed energy- and time-resolved spectroscopy of excitons. In this section, we lay out the framework to determine exciton binding energies and dynamics from 2PPE spectra.

15.3.1 Energy Levels

For a discussion of the energy levels involved in probing excitons with 2PPE, we start by comparing excitonic and single-particle, or free carrier, states. In molecular solids where the molecules are weakly bound by van der Waals forces, the electronic system can be described in a tight binding picture based on molecular orbitals, that is, the highest occupied molecular orbitals HOMO, HOMO-1, ..., and the lowest unoccupied molecular orbitals LUMO, LUMO + 1, ... Their bandwidth is typically narrower than the energetic spacing between the levels. Although we restrict the discussion to the HOMO and LUMO levels in this section, it is possible to expand it to the next few lower and higher states as long as the wave function overlap, and thus the bandwidth, is small. Besides band formation, another difference to isolated molecules is that the molecular energy levels shift toward the chemical potential, that is, the ionization potential (IP) decreases, and the electron affinity (EA) increases, due to polarizability of the surrounding medium [31, 32]. This is because the energy levels represent ionic states. As Figure 15.10 shows, the HOMO level is defined as the energy difference between a neutral ground state (M^0) and the positive ion (M^+) that misses an electron in the HOMO. Correspondingly, the LUMO level is defined by adding an electron to form a negative ion (M^-). Polarization induced by the ionic states is a dynamic process that includes a fast and a slow component: The electronic system responds on the timescale of 10^{-16} s and causes a shift of up to 1 eV. This response is faster than the photoemission process, and thus energy levels measured with photoemission include the shift. The nuclear lattice responds on a much slower timescale of 100–1000 fs, and a small shift of typically a few tens of meV. This process leads to polaron formation and self-trapping of the charge. With femtosecond time-resolved 2PPE, the time-dependent energy shift of the self-trapping process can be observed in the time domain [33].

The lowest exciton state is created by promoting an electron from the HOMO (creating a hole) to the LUMO, while the molecule remains neutral. The energy needed for this process, E_{opt}, is lower than the difference $IP-EA$ between the HOMO and LUMO levels,

$$IP-EA = E_B + E_{opt}, \tag{15.23}$$

where E_B is the *exciton binding energy* (cf. Figure 15.10). With respect to organic semiconductors, E_{opt} is often called the *optical gap*, whereas $E_t = IP-EA$ is called the

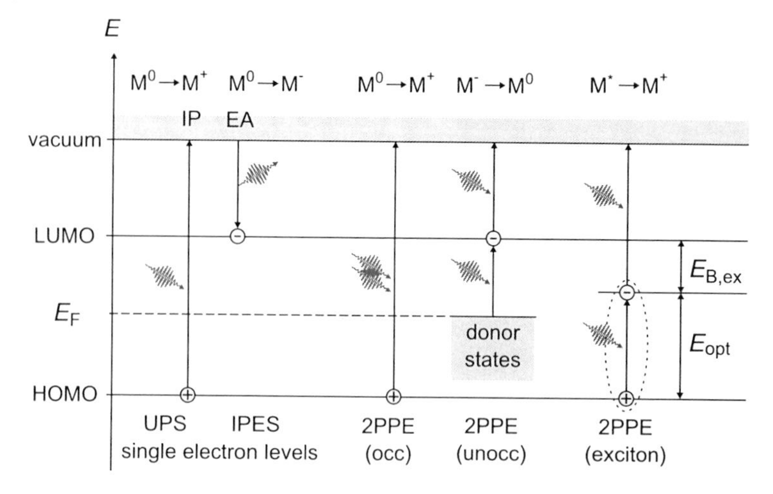

Figure 15.10 Energy level diagram of two-photon photoelectron spectroscopy of excitons. An exciton is a correlated electron–hole state that requires lower excitation energy than the creation of an electron in the LUMO and a hole in the HOMO of two infinitely separated molecules. IP: ionization potential; EA: electron affinity; UPS: ultraviolet photoelectron spectroscopy; IPES: inverse photoelectron spectroscopy; 2PPE: two-photon photoelectron spectroscopy; $E_{B,ex}$: exciton binding energy; and E_F: Fermi energy. The header indicates the molecular configurations of the initial and final states that determine the probed energy. M^0: ground state; $M^{\pm}$: ion; and $M^{☆}$: exciton. Reprinted from [38]. Copyright 2009, with permission from Elsevier.

transport gap. This distinction reflects the fact that the ionic HOMO and LUMO levels constitute charge carriers, whereas an exciton does not transport charge. The exciton binding energy can be broken up into three contributions according to Knupfer [17],

$$E_{\mathrm{B}} = E_{\mathrm{B}}^{\mathrm{intra}} + U - W, \tag{15.24}$$

by comparing the bound exciton state with two uncorrelated charge carriers, an electron and a hole, placed on two molecules that are an infinite distance apart. Starting from the exciton state, in the first step, the bound electron–hole pair is promoted to an unbound electron–hole pair on the same molecule. This step compensates the intramolecular binding energy $E_{\mathrm{B}}^{\mathrm{intra}}$ that arises due to correlation of the electrons in the molecule. In the second step, the charges are separated onto two molecules, which means that charge is added to each of the target molecules. This costs the on-site Coulomb repulsion energy U also known as the Hubbard U [34]. In the third step, in the HOMO and LUMO bands of the intermolecular band structure the charges can gain energy by delocalization of the amount of the bandwidth W. In many cases, $E_{\mathrm{B}}^{\mathrm{intra}}$ and W are small and of similar value so that the two contributions cancel [17], and the exciton binding energy can be approximated as

$$E_{\mathrm{B}} \approx U. \tag{15.25}$$

For molecular crystals, the U term is the proper account of the exciton binding energy, rather than the Coulomb energy we used in the heuristic approach at the beginning of the chapter. Nevertheless, for π-conjugated molecular systems the Coulomb repulsion can be roughly understood as the field energy of a capacitor $U \approx e^2/C$, where the capacity depends roughly on the size of the molecule [17]. Because polarization of the surrounding medium causes screening of the charge on a molecule, the polarization energy is factored into U. For C_{60}, for example, U decreases from 3 eV in the gas phase to 1.4–1.6 eV in the bulk, and to 1.0 eV in a monolayer adsorbed on a metallic gold surface [35–37].

Figure 15.10 also shows how charge carrier and exciton levels are probed in 2PPE. Occupied states such as the HOMO are probed by coherent absorption of two photons. Probing the LUMO first requires electron injection from a metallic substrate to create a transient M^- state that is not perturbed by the photohole. In order to probe an exciton, the pump photon energy needs to be tuned to its excitation energy E_{opt}. While the unoccupied charge carrier states can be measured only in ultrathin films on metallic substrates, exciton formation occurs independent of film thickness and is also observable in single-crystal samples. One particularity of measuring exciton levels with 2PPE is the following: In principle, the exciton binding energy with respect to the charge carrier states is shared by the hole and the electron, as the whole system should be regarded as a quasiparticle. However, since 2PPE ultimately detects the electron, the total energy cost to break up the exciton reduces the kinetic energy of the photoelectron. Therefore, it is most convenient to attribute the whole exciton binding energy to the electron as shown in Figure 15.10 [38].

15.4 Frenkel Excitons in C_{60}

Solid C_{60} is a prototypical system of an organic semiconductor consisting of weakly bound molecules. The structural and electronic properties of C_{60} are well known from studies of the gas, bulk, and thin film phases, in pristine and doped forms. Excitation energies of excitons in the gas and bulk phase, and their relation to ionic states, are known from a variety of experimental and theoretical techniques [39]. From these studies, the on-site Coulomb interaction U is known for many systems. As we have seen in Section 15.3.1, U is approximately related to the exciton binding energy. In this section, we first demonstrate how excitons are measured with 2PPE and then discuss the dynamics of excitons at the C_{60}–metal interface.

15.4.1 Structural Overview

We choose Au(111) as a substrate to grow C_{60} thin films because the films grow epitaxially in a structure that is very close to bulk C_{60}. In addition, chemical interaction of C_{60} with the gold surface and static charge transfer are smaller than those on most other metal surfaces. Due to the proximity of the metal surface, however, the on-site Coulomb interaction U is reduced from the bulk value of 1.6 eV to about 1.0 eV in the first monolayer [37, 40]. The structure of the system is shown

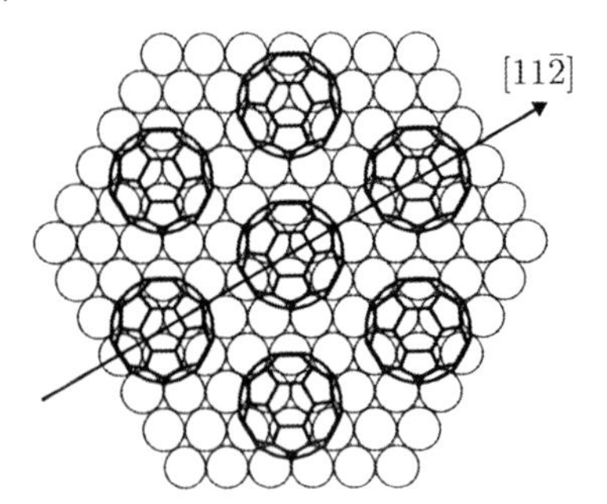

Figure 15.11 Structure of monolayer C_{60} on Au (111) (*top view*). The lattice constant of the $(2\sqrt{3} \times 2\sqrt{3})R30^{\circ}$ superlattice of C_{60} sites is 0.4% smaller than the bulk lattice constant of C_{60}. Multilayer C_{60} grows layer-by-layer in the bulk-like fcc structure. The $[11\bar{2}]$ direction is indicated. Reprinted with permission from [43]. Copyright 2006 by the American Physical Society.

schematically in Figure 15.11. Monolayer and multilayer thin films of C_{60} grow epitaxially in a $(2\sqrt{3} \times 2\sqrt{3})R30^{\circ}$ superstructure. This structure has a very small lattice mismatch of 0.4% with respect to bulk C_{60} [37].

15.4.2 Energy Levels

2PPE spectra of C_{60}/Au(111) at different film thicknesses and of the clean Au(111) surface are shown in Figure 15.12 [41]. There is a clear distinction between monolayer ($\theta = 1$ ML) and multilayer ($\theta > 1$ ML) spectra, and no major change occurs above 5 ML. On the basis of the finite mean free path of photoelectrons, we assign the visible features as originating mainly from the top few C_{60} layers. Thus, at high coverage, the multilayer on Au(111) approximates the situation of the few topmost C_{60} layers at the surface of a bulk crystal. In contrast, the monolayer is strongly influenced by the metal substrate.

The origin of the various peaks is analyzed by variation of the photon energy, Figure 15.13, which allows one to distinguish occupied states from unoccupied states, and final-state resonances [41]. Among the unoccupied state peaks, variation of the film thickness allows to distinguish excitons from states that are populated by charge injection from the metal substrate, as the latter process is not observed in the spectrum above a certain layer thickness.

A summary of the energy levels is given in Figure 15.14 [41]. We start our discussion with the HOMO and LUMO + x levels that correspond to hole and electron states measurable in UPS and IPES. At multilayer coverage, a peak from two-photon excitation of the occupied h_u HOMO is resolved at 2.0 eV below the Fermi level. This binding energy agrees well with UPS measurements of the same system [37]. The peak is not resolved in the 1 ML spectrum because its spectral region is dominated by the LUMO + 1 peak. The LUMO + 1 and LUMO + 2 are transiently populated by charge transfer from the substrate at 1 ML coverage. At higher coverage, the probability of charge transfer into the topmost layer, and thus

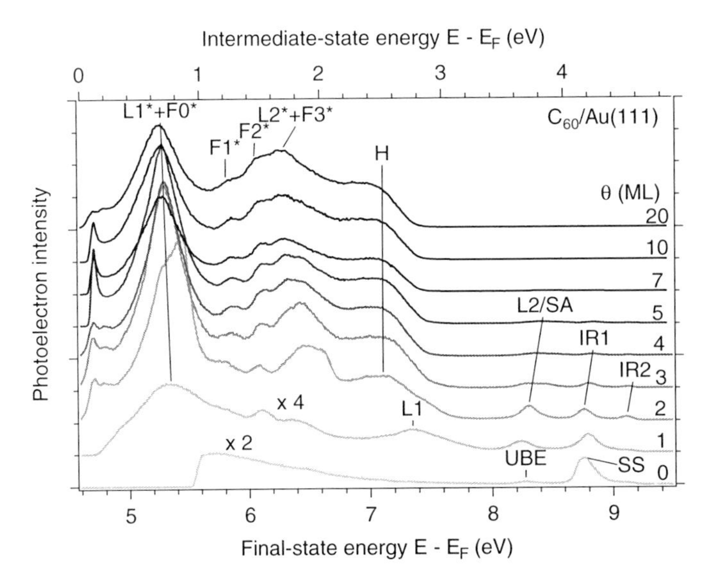

Figure 15.12 Monochromatic 2PPE spectra of C_{60}/Au(111) at various coverages measured at a photon energy of $h\nu = 4.593$ eV. The 0 ML spectrum is taken on the bare Au(111) surface. SS: occupied Shockley surface state; UBE: upper band edge; H*x*: HOMO $- x$, $(x+1)$th highest occupied molecular orbital; L*x*: LUMO $+ x$, $(x+1)$th lowest unoccupied molecular orbital; SA: superatom molecular orbital (SAMO); IR*n*: image resonance; and F*n*: final state resonance. Asterisks denote excitons. The energy scales are referenced to the Fermi level of the substrate. The intermediate-state scale applies to unoccupied states (LUMO $+ x$ and LUMO $+ x^*$). Reprinted with permission from [41]. Copyright 2005 by the American Physical Society.

peak intensity, is reduced. In an otherwise completely flat spectral region, the LUMO + 2 peak is still resolved on the first few monolayers. This peak can have multiple origins. First, the LUMO + 2 and LUMO + 3 may mix because they lie close in terms of energy. Second, a charge outside the C_{60} cage induces polarization of the extended π system of the molecule that can be treated as a positive image charge at the center of the molecule. This spherical image potential establishes hydrogen-like superatomic orbitals (SAMO) that are located in the same energetic region starting at 3.9 eV above the Fermi level [42]. In the thin film, this state has a delocalized wave function that shows dispersion versus the electron wave vector and thickness-dependent quantum confinement [43].

The LUMO is not visible at all in the 2PPE spectrum. We infer its position from the HOMO–LUMO separation of 3.5 eV measured by UPS and IPES [35], and the position of the HOMO observed in 2PPE. Thus, the LUMO is located at 1.51 eV, in the middle of the projected bandgap of the substrate, which might be a reason why it is not observed in 2PPE. On the basis of UPS and IPES data [35, 37], we add further levels that are not observed in 2PPE to the diagram. We can see that, in the monolayer, the separation between HOMO and LUMO + 1 is 0.23 eV lower (4.71 instead of 4.48 eV) than for bulk C_{60} due to screening of the charge by the metallic substrate. It appears that the decrease affects mainly the HOMO level but not the LUMO + 1

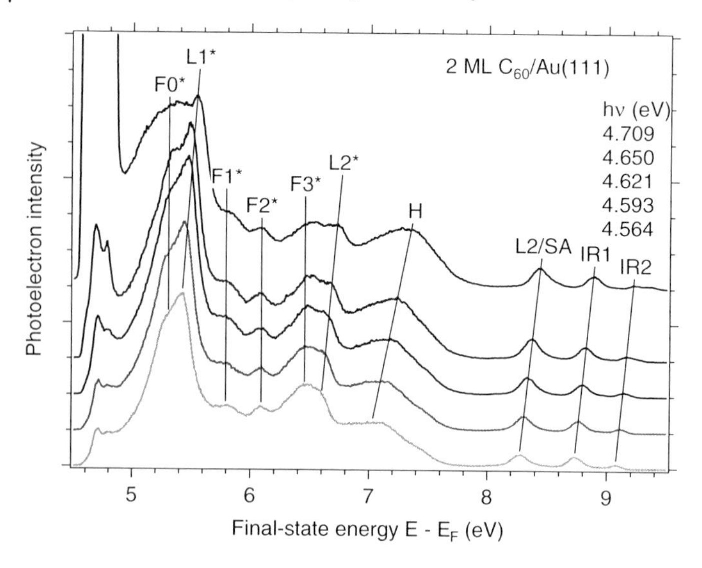

Figure 15.13 Monochromatic 2PPE spectra of 2 ML C_{60}/Au(111) at various photon energies. The spectra are vertically offset by an amount proportional to the photon energy. Lines indicate the energy shift of the peaks versus photon energy. The HOMO peak shifts with $2h\nu$, unoccupied LUMO + x^* excitons shift with $\approx h\nu$, and final state resonances do not shift. See the caption of Figure 15.12 for an explanation of the acronyms. Reprinted with permission from [41]. Copyright 2005 by the American Physical Society.

level. This indicates that the molecular orbitals hybridize with the gold surface so that the LUMO + 1 level becomes pinned.

Excitons arise from direct optical excitation of C_{60} molecules and are observed regardless of film thickness. Here, we are mainly interested in the LUMO + 1* and LUMO + 2* excitons. The photon energy used for the displayed 2PPE spectra is set at a strong resonance in the optical absorption spectrum at 4.6 eV [44, 45]. Since this resonance is observed both in the solution and in the solid, the excitations are mainly intramolecular, that is, Frenkel excitons. As explained in the previous section, in contrast to optical transitions in absorption spectra, 2PPE can relate the exciton transitions to single-electron states. First, we note that the energy levels of the transient LUMO + 1* and LUMO + 2* states are about 2 eV below the independent electron states LUMO + 1 and LUMO + 2. This exciton binding energy is of the same order as the on-site Coulomb energy $U \approx 1.6$ eV in solid C_{60}, and leads us to assign the excitons to the respective LUMO + x levels. The reason why the energy difference is larger than 1.6 eV may be that the observed state is not the directly populated state but a relaxed state that has additional binding energy. Assuming a direct transition to the observed peaks would require occupied states at 2.8 and 3.9 eV below E_F. However, even considering spectral broadening, there are no occupied states in that spectral region. Assuming that the HOMO-1 and the HOMO must be the ground states for the LUMO + 1* and LUMO + 2* excitons, respectively, the transient states would be located approximately 0.4 eV above the observed exciton

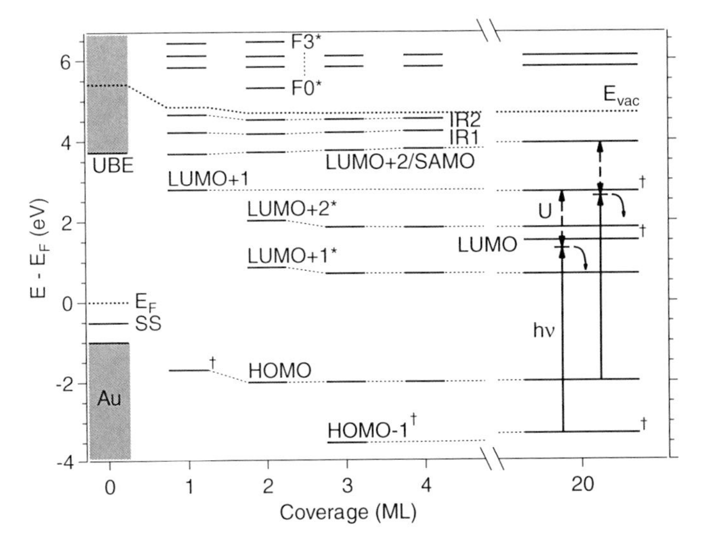

Figure 15.14 Energy levels (bold horizontal lines) of the observed spectral features of C_{60} on Au(111) as a function of film thickness. Levels marked with † are not observed in 2PPE but taken from literature values [35, 37]. Vertical arrows indicate the proposed excitation process of the LUMO + 1* and LUMO + 2* excitons via a rapidly decaying intermediate state. The 0 ML region depicts the projected band structure of the gold substrate along the [111] direction that features a gap between −1 and 3.72 eV. See the caption of Figure 15.12 for an explanation of the acronyms. Reprinted with permission from [41]. Copyright 2005 by the American Physical Society.

states, and their binding energies would be closer to 1.6 eV. As to the mechanism that populates the observed, relaxed states, we can only speculate. One possibility would be an electronic intraband or interband transition in the electron or exciton band structure. Such a transition could involve a crossover to a charge transfer exciton, as described by the Merrifield model in Section 15.2.2 [24]. Another possibility would be the formation of an exciton-polaron due to coupling to phonon modes. Such interaction, however, typically occurs on a timescale of picoseconds [39], which is not observed experimentally (see Section 15.4.3).

Evidence of band formation is also found in the downward energy shift of the exciton levels as a function of increasing film thickness, as the exciton transitions from quantum confinement to a fully developed band. Apart from the monolayer–bilayer transition, no such trend is observed in the single-particle peaks of the HOMO, the LUMO + 1, and the LUMO + 2 levels. Band formation occurs both in the electronic structure, according to, for example, the tight binding model, and in the exciton structure, as we have seen in Section 15.2.2.3. However, exciton bandwidths are small, of the order of 50 meV [24]. A shift of that order is seen for $\theta \geq 3$ ML. The large shift of 150 meV between two and three monolayers must also include effects of the electronic band structure, that is, of a finite-size effect in the single-particle reference levels. Another reason might be that, just as the first C_{60} layer, the bilayer system is too strongly perturbed by structural interaction with the metal substrate.

As a final remark we note that several other exciton peaks, labeled F0*–F3*, appear in the spectrum. They are populated by two-photon excitation with a total excitation energy of 9.2 eV, according to their photon energy dependence. The electron energy lies above the vacuum level, and they spontaneously ionize. Since not much is known about such high-lying exciton states, it is not possible to make an assessment.

15.4.3
Exciton Dynamics

We discuss the dynamics of the two Frenkel excitons LUMO + 1* and LUMO + 2* that are well resolved in the 2PPE spectra, whereas the LUMO* exciton is not observed because it energetically lies too deep to be probed. So far, the dynamics of Frenkel excitons in solid C_{60} have been studied by optical absorption and luminescence spectra [39]. The lowest singlet exciton is optically pumped and decays on a timescale of 1 ns by recombination (luminescence) or by crossing into a triplet state that has a very long lifetime (≈ μs) [46]. Higher singlet excitons decay rapidly (subpicoseconds) into the lowest singlet exciton, although the exact timescale is not known. At the C_{60}–metal interface, we expect mainly the decay channel into lower excitons, and a decay channel due to resonant charge transfer to the metal substrate.

Furthermore, since lattice relaxation occurs on the 100 fs timescale, optical excitation occurs in a *vertical transition*, which initially leaves the exciton in a higher vibronic state. The exciton may subsequently relax into a lower vibronic state due to lattice deformation. This process is called *self-trapping*. Polaron self-trapping is observed in 2PPE as a time-dependent shift of the intermediate state energy [47].

In order to discuss the dynamic aspects, we first have a look at the 2PPE spectra of a thick film as a function of pump–probe delay in Figure 15.15. For this figure, the incoherent background spectrum taken at a large delay time $t = 2000$ fs is subtracted. The amplitude of the HOMO peak follows the decay of the autocorrelation signal. The LUMO + 2* exciton decays quickly with a lifetime of < 50 fs that is not resolved. However, the LUMO + 1* exciton peak has a clear lifetime of several hundred femtoseconds and will be discussed in more detail.

We first notice that there is no significant energy shift in the spectra of Figure 15.15 as a function of delay time. Such a shift would be expected as a result of self-trapping. The shift could be faster than 50 fs or smaller than ≈ 10 meV, which would be experimentally unresolvable.

Figure 15.16 shows cross-correlation curves measured on the LUMO + 1* peak at various coverage. Lifetimes τ are extracted by fitting the raw data with the convolution of the temporal pulse profile AC and a symmetric exponential function $e^{-|t|/\tau}$. The measured lifetimes depend both on coverage and on photon energy.

15.4.3.1 Coverage Dependence: Distance-Dependent Quenching

The lifetime of the LUMO + 1* exciton exhibits a remarkable exponential dependence on film thickness, displayed in Figure 15.17. We separate the decay rate $1/\tau$ into an asymptotic component τ_{in} and a coverage-dependent component τ_q due to quenching at the metal surface according to

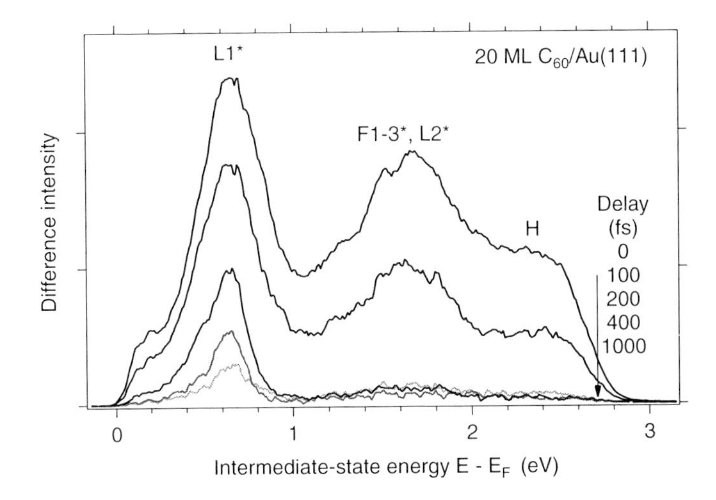

Figure 15.15 Monochromatic 2PPE spectra ($h\nu = 4.593$ eV) of 20 ML C_{60}/Au(111) at various pump–probe delays. A background spectrum at $t = 2000$ fs has been subtracted from each spectrum. Only the LUMO + 1* exciton shows remaining intensity at delays larger than the pulse length of the laser (100 fs). See the caption of Figure 15.12 for an explanation of the acronyms. Reprinted with permission from [41]. Copyright 2005 by the American Physical Society.

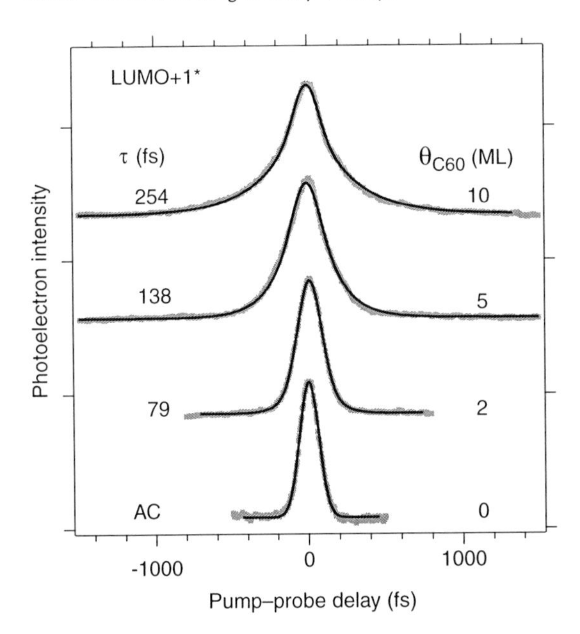

Figure 15.16 2PPE cross-correlation curves (photoelectron intensity versus pump–probe delay at fixed kinetic energy) of the LUMO + 1* exciton on C_{60}/Au(111) at various coverage, measured at $h\nu = 4.593$ eV pump and probe. Dots represent raw data, solid lines are fits as described in the text. Intensities are normalized and offset for clarity. The lowest curve is the autocorrelation (AC) function, measured on the Au(111) Shockley surface state. Reprinted with permission from [41]. Copyright 2005 by the American Physical Society.

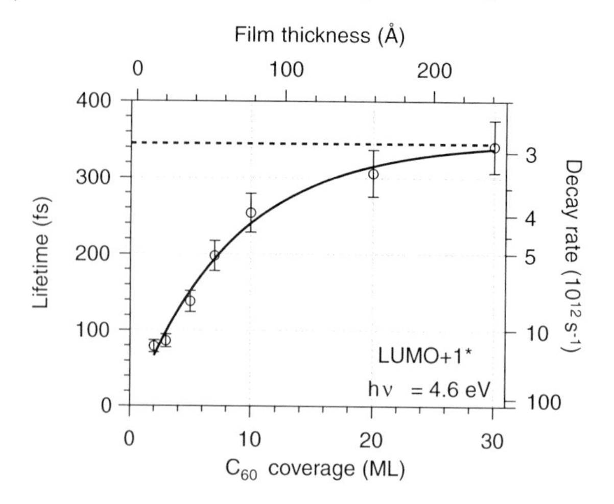

Figure 15.17 Lifetime (τ) and decay rate (1/τ) of the LUMO + 1* exciton as a function of C_{60} film thickness. The solid line is an exponential fit of Eq. (15.27), where $\beta = (0.023 \pm 0.005)$ Å^{-1} and $\tau_0^{-1} = 1.2 \times 10^{13}$ s^{-1}. The dashed line indicates the asymptotic value $\tau_{in} = (345 \pm 10)$ fs.

$$\frac{1}{\tau(d)} = \frac{1}{\tau_{in}} + \frac{1}{\tau_q(d)}. \tag{15.26}$$

We interpret the intrinsic component $\tau_{in} = (345 \pm 10)$ fs as caused by decay into a lower exciton, for example, the HOMO/LUMO + 1, or HOMO/LUMO exciton because this decay channel is considered the most efficient. Unfortunately, the build-up of a LUMO* population is not observable in the 2PPE spectrum because its energy nominally lies 0.1–0.3 eV below the Fermi level [35, 48], which is too low to be probed reliably.

Since 2PPE is most sensitive to the top C_{60} layer, the coverage-dependent part is interpreted as a dependence on the distance d of the top layer from the metal substrate. The exponential dependence looks like a tunneling behavior where the electron tunnels through the C_{60} film into the metal. Given

$$\tau_q^{-1}(d) = \tau_0^{-1} e^{-\beta d}, \tag{15.27}$$

the parameters of the exponential component are $\beta = (0.023 \pm 0.005)$ Å^{-1} and $\tau_0^{-1} = 1.2 \times 10^{13}$ s^{-1} [41]. The β value, or strength of the distance dependence, obtained here is much lower than for typical molecular tunneling barriers, where it lies in the range from 1.3 Å^{-1} for saturated alkyl chains down to 0.2 Å^{-1} for π-bonded systems [49]. This means that electron transfer from the top layer to the metal interface must be mediated by a more efficient transport mechanism, such as an exciton band.

τ_0^{-1} is the quenching rate at zero distance and can be converted into the exciton–metal coupling strength, or spectral width, using the uncertainty principle $\Gamma = \hbar/\tau_0 \approx 8$ meV. This value is almost an order of magnitude smaller than the 60 meV measured at the structurally similar C_{60}/Cu(111) interface and is in line with smaller static electron transfer, that is, weaker chemical interaction of C_{60} with the metal surface.

15.5
Charge Transfer Excitons at the Surface of Pentacene

Charge transfer excitons have originally been observed in charge transfer salts where two molecules form donor–acceptor pairs with a static driving force for charge separation [50]. Later, it was found that excitons in single-compound solids may also have a charge transfer character (cf. Section 15.2.2) [14, 51]. Evidence of their occurrence at the interface of two different semiconductors has recently been obtained in applications of organic photovoltaics [7, 52]. In this section, we present a direct experimental observation of charge transfer excitons at the surface of a crystalline pentacene thin film [20]. This section refers to the dielectric model and numerical results presented in Section 15.2.1.2.

15.5.1
System Overview

Thin films of pentacene are widely studied with many methods, their structure and morphology are thus well known. A structural summary of a crystalline monolayer of pentacene on a Bi(111) surface is presented in Figure 15.18 [53–55]. Due to a point-on-line matching of the *b*-axis with the Bi(111) substrate, the crystal structure of the thin film is the same as the low-density bulk polymorph of pentacene [56]. The molecules stand up and arrange in the typical herringbone pattern. The films are grown by vapor deposition of pentacene molecules from a temperature-stabilized Knudsen cell. With the substrate at room temperature, pentacene molecules have significant mobility, and thus the films grow roughly in a layer-by-layer fashion although the next layer starts to grow before completion of the first one. The crystalline domains are up to 100 nm in diameter. Due to the sixfold symmetry of the Bi(111) surface and the chirality of the Pc/Bi interface, 12 equivalent domains may be observed.

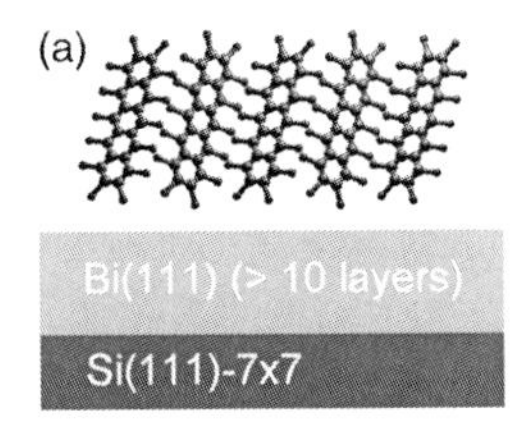

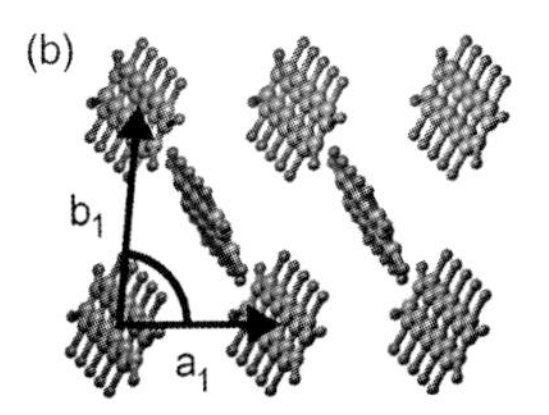

Figure 15.18 Structure of monolayer pentacene on Bi(111): (a) side view and (b) top view. On Bi(111), pentacene adopts the low-density bulk crystal structure with lattice constants $a_1 = 6.06$ Å, $b_1 = 7.90$ Å, and oblique angle of $\gamma = 85.8°$. The Bi(111) surface is prepared by vapor deposition of bismuth on a clean 7 × 7-reconstructed Si(111) surface. Reprinted with permission from [54]. Copyright 2005, American Institute of Physics.

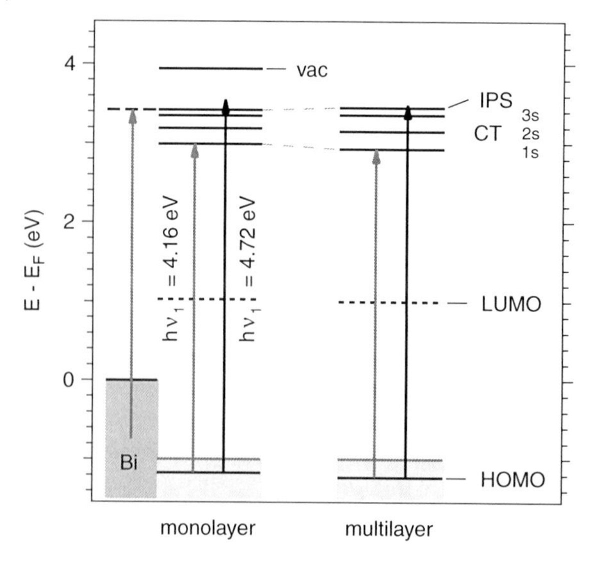

Figure 15.19 Experimental energy levels of the image potential state and the *s*-symmetry charge transfer excitons for monolayer and multilayer pentacene on Bi(111). Vertical arrows indicate optical transitions (pump) at $h\nu_1 = 4.16$ and 4.72 eV observed in 2PPE. The HOMO level is calculated from the HOMO → CT 1*s* transition. In UPS measurements, the HOMO spans the shaded spectral region [58]. The LUMO levels were not measured but calculated from the HOMO position and a transport gap of 2.2 eV. Reprinted from [38]. Copyright 2009, with permission from Elsevier.

15.5.2
Energy Levels

Charge transfer excitons on pentacene/Bi(111) are probed with two-photon photoemission. A summary of the measured energy levels and explanation of the excitation process is shown in Figure 15.19. 2PPE spectra at low pump photon energy (4.16 eV) are displayed in Figure 15.20. On the clean Bi(111) surface, the dominant features are image potential states [57] that disappear upon completion of the first monolayer of pentacene. In parallel, the work function decreases from 4.27 eV for the clean Bi(111) surface to 3.94 eV. Due to the lower work function, an image potential state on top of pentacene (Pc IPS) appears at lower energy. According to the energy level diagram, this peak lies too high to be pumped from an occupied state of pentacene. Instead, it must be pumped from the bismuth substrate. At higher film thickness, the wave function overlap with the substrate is reduced and the peak disappears.

A prominent peak due to the 1*s* charge transfer exciton is visible at a binding energy of 1.0 eV with respect to the vacuum level. According to Eq. (15.7), the image potential cannot produce such a high binding energy. Rather, as we have seen in Section 15.2.1.2, the electron is bound by both the image potential and the Coulomb potential of the photohole. With respect to the $n = 1$ IPS, the exciton binding energy amounts to 0.43 eV, which is in agreement with calculations (Table 15.1). Considering the photon energy of 4.16 eV and the energetic alignment of the HOMO at

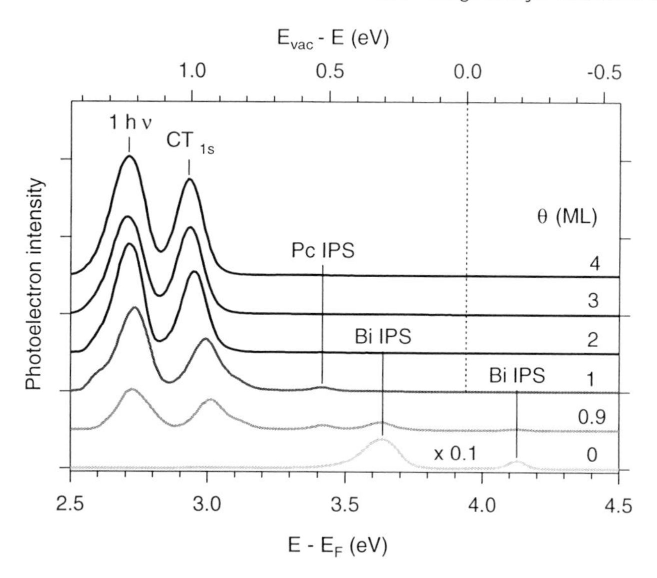

Figure 15.20 2PPE spectra from pentacene/Bi (111) surfaces ($h\nu_1 = 4.16$ eV; $h\nu_2 = 1.39$ eV) in the film thickness range from 0 to 4 ML. CT: charge transfer exciton; Pc IPS: image potential state on the pentacene surface; Bi IPS: image potential states ($n = 1, 2$) on clean Bi(111). $E_{vac} - E$ is the binding energy referenced to the vacuum level. For clarity, the curves are vertically offset, and the clean bismuth spectrum is scaled down by a factor of 10. Reprinted from [38]. Copyright 2009, with permission from Elsevier.

1.2 eV below the Fermi level of bismuth [58], the initial state of the peak must be the HOMO, and thus, the photohole is created in the molecule. The momentum dependence of the IPS and CTE states is clearly different: While the IPS has a nearly free electron dispersion with effective mass $m^* \approx 1.5m_e$, the CTE peak does not shift versus momentum [20]. This can be attributed to the localized nature of the state or to the destruction of the quasiparticle in the photoemission process. To establish the surface-related nature of the CTE state, and to distinguish it from intramolecular states, an overlayer of another dielectric material is deposited on top of pentacene. *n*-Nonane (C_9H_{20}) has a large bandgap of >8 eV and expels electron wave functions that would occupy the space in front of the substrate. We find that both the CTE and the IPS peaks are quenched upon adsorption of one monolayer of nonane. This clearly demonstrates that the CTE wave function is concentrated in the vacuum in front of the surface [20].

At increased pentacene film thickness, the 2PPE spectra show a shift of about 0.1 eV to larger binding energy. The multilayer regime corresponds most closely to the bulk regime where only occupied pentacene states are available as initial states. That is why the pentacene IPS disappears, and the CTE remains visible. In the monolayer regime, however, the net binding energy of the CT exciton is lower because the charges are closer to the highly polarizable bismuth surface (the dielectric constant is $\varepsilon \geq 100$ [57]). Analogous to the discussion of the HOMO level shift in monolayer C_{60} (Section 15.4.2), the positive hole induces a negative image charge in bismuth that screens the charge of the hole. The true CTE binding energy, corresponding to the dielectric slab, is thus observed in the multilayer regime.

The numerical calculations in Section 15.2.1.2 predict a whole series of states. States above the 1*s* require higher photon energy to be pumped from the HOMO of pentacene. Indeed, these states are observed if the photon energy is increased, as shown in Figure 15.21. To reduce single-photon photoemission from bismuth and associated space charge effects on the spectrum, these spectra are measured on 7 ML samples. The peak positions correspond roughly to the calculated 2s and 3s states. Higher states seem to be present in the broad spectral features at $h\nu \geq 4.6$ eV but are not resolved. At the highest photon energy, the pentacene IPS reappears; in this case, it is directly pumped from the HOMO of pentacene. At this energy, the electron has sufficient energy to escape the potential well formed by the photohole.

It is notable that states with *p* and higher angular momentum are not observed in the spectrum. The 1*p* state of Figure 15.5 would be located between the 1s and 2s peaks but is clearly absent. A 1d peak is not required for the decomposition of the spectrum at $h\nu_1 = 4.47$ eV, and thus not resolved, either. The reason could be dipole selection rules in the pump or probe steps of the 2PPE process. Pumping into *p* ($l_z = 1$) states via dipole transition is forbidden because the pentacene HOMO has π symmetry, and the molecular plane is a nodal plane. For *d* states, the pump step is allowed but the probe step is suppressed because it corresponds to photoemission from an oriented d_{xy} orbital at normal emission [59].

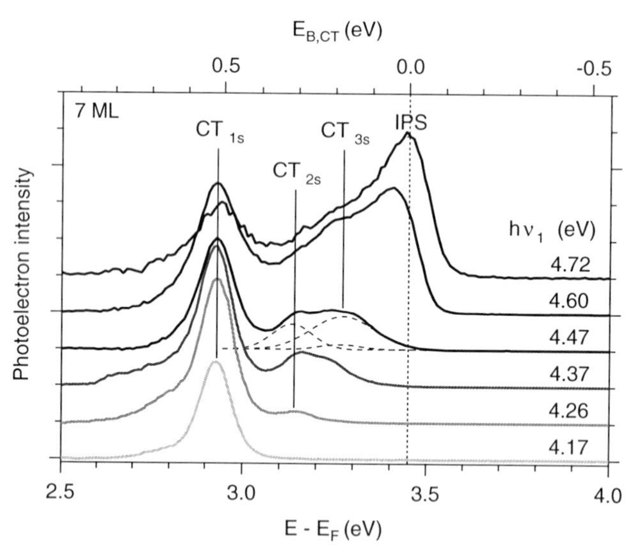

Figure 15.21 2PPE spectra of 7 ML pentacene on Bi(111) at various photon energies, showing the 1*s*–3*s* CT exciton peaks. The exciton binding energy scale ($E_{B,CT}$) is referenced to the IPS on multilayer pentacene; on this scale, the CTE binding energies are 0.52 (1s), 0.32 (2s), and 0.18 eV (3s). A background spectrum containing the single-photon peak has been subtracted. Dashed lines indicate the decomposition of the $h\nu_1 = 4.47$ eV spectrum into three peaks of which the two most intense are assigned to the 2s and 3s exciton states. Reprinted from [38]. Copyright 2009, with permission from Elsevier.

15.5.3
Dynamics

2PPE cross-correlation curves of the CTE and IPS peaks of 1 ML pentacene on Bi(111) are shown in Figure 15.22 for three different photon energies. The curves are fit to a rate equation model using a Gaussian pump pulse profile. The fits yield both the lifetime τ of the transient state and the time offset t_0 of the pump pulse. The lifetimes of the CT excitons are of the order of 40–50 fs, and thus quite a bit shorter than those of the IPS at 130 fs. Since the CT excitons are bound to the photohole, recombination is probably the major decay channel. The IPS decays into unoccupied pentacene states or back into the substrate. This decay channel that requires charge transfer through the pentacene layer is slower than recombination of a charge transfer exciton.

The correlation curves exhibit several peculiar features: (1) The lifetimes of the CT excitons do not vary monotonously from the 1*s* to the 3*s* state; (2) the individual curves are shifted with respect to each other; and (3), the time offset of the CTE curves

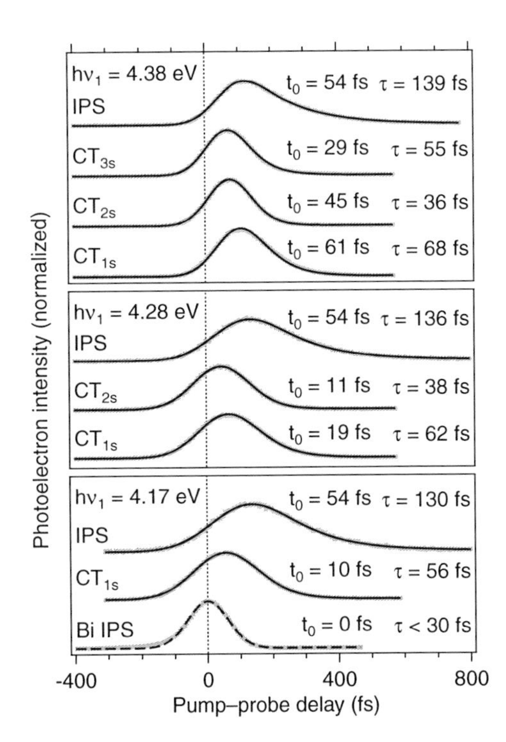

Figure 15.22 Cross-correlation curves of the CT excitons and the IPS on 1 ML pentacene on Bi(111) measured at three different photon energies. $h\nu_1$ is the pump and $h\nu_2 = h\nu_1/3$ the probe energy. Transient-state lifetimes τ are determined by fitting a rate equation model using a Gaussian pump pulse profile. t_0 is the temporal offset of the Gaussian with respect to zero delay $t = 0$. $t = 0$ is calibrated on the Bi IPS that was measured at 0.9 ML pentacene coverage. Gray dots represent raw data, black lines are fits. The curves are normalized and vertically offset for clarity.

changes with photon energy. These observations can be explained as due to relaxation from higher CTE states into lower ones. According to the energy level diagram in Figure 15.19, at $h\nu_1 = 4.17$ eV, the 1*s* CTE state is pumped directly from the HOMO. At this photon energy, we see the intrinsic lifetime of the 1*s* CTE that decays with a time constant of 56 fs. At $h\nu_1 = 4.28$ eV, the 2*s* CTE is pumped directly at time $t_0 \approx 10$ fs. Within 40 fs, it decays into the 1*s* state. Due to this indirect population process, t_0 of the 1*s* state appears to be populated at a later time. At $h\nu_1 = 4.38$ eV, analogously, an unresolved state above the 3*s* is pumped directly. It decays to the 3*s* and the lower states whose time offsets are shifted even further. From these measurements, we conclude that hot excitons (> 1*s*) decay within a few tens of femtoseconds into lower states, and the lowest one decays within 56 fs by recombination with the hole. In addition, the CT excitons may decay into intramolecular Frenkel excitons that are not observed in the 2PPE spectrum. The apparent longer lifetime of the 1s CTE at higher photon energy also indicates that its population is fed by the decay of higher states.

In Figure 15.22, we have used the IPS on pentacene as a reference clock, assuming that its time offset does not depend significantly on electron energy. However, this reference clock runs late by 54 fs with respect to the short-lived $n = 1$ Bi IPS on domains of bare bismuth. In the common picture of an IPS on a metal surface, the IPS wave function decays exponentially into the substrate. This exponential tail has a small but finite spatial overlap with the substrate, which allows the IPS to be instantaneously populated in a direct optical transition [60] (see also Chapter 3). This is obviously not the case in the pentacene/Bi system where the IPS is populated by charge transfer from the bismuth substrate through the pentacene layer. While the details of this process are not yet fully understood, we speculate that the time delay occurs due to an indirect process either via an interface state at the bismuth pentacene interface or via a molecular state of pentacene. It is known from charge transfer theory (see Chapter 4) that the traversal time through a metal–molecule–metal interface can increase significantly with respect to the 1 fs of pure tunneling if the initial and final states are in resonance with a molecular state [61].

15.6 Conclusions

In this chapter, we have studied the energetics and dynamics of Frenkel and charge transfer excitons at the interfaces of two organic semiconductors. In thin films of C_{60} on Au(111), we observe two Frenkel excitons associated with the LUMO + 1 and LUMO + 2 levels. We see indirect evidence of exciton band formation in the variation of their energy, and their decay rates as a function of film thickness. The excitons decay on a timescale of 10–100 fs into lower excitons, and by resonant electron transfer to the metal substrate. At the surface of a pentacene thin film, a series of charge transfer excitons are observed where the electron is bound by the photohole but located in front of the surface. These charge transfer excitons are explained in a simple model of a homogeneous dielectric slab, assuming that the photohole is localized in the dielectric, and the electron resides in front of the surface. The model predicts that similar CT excitons exist generally at interfaces of organic semiconductors.

The existence of CT exciton states at organic interfaces has implications on photovoltaic applications. In Figure 15.2, we saw that at an organic heterojunction a Frenkel exciton may relax into a CT exciton rather than dissociate. Due to large binding energy and fast recombination of the CT exciton states, charge separation is suppressed. Based on thermodynamic arguments, the probability of charge separation is proportional to $\exp(-E_B/k_B T)$ and thus very small for the lowest CTE states. Efficient charge separation at an organic D/A interface therefore must avoid the formation of charge transfer excitons with large binding energy. The probability for charge separation is much higher when the intermediates are hot CT excitons with binding energies lower than $k_B T$. On the basis of these conclusions, we identify the following key factors that govern hot CTE formation and free carrier generation [21]:

- A potential drop (due to the difference between transport levels in the donor and acceptor) at the interface is necessary to overcome the binding energy of the Frenkel exciton in the donor phase and to enable population of the lowest free carrier states. If the potential drop is larger, excess energy may transform into kinetic energy (or group velocity in the case of a crystalline material) that transports the carriers away from the interface.
- Crystallinity of the organic semiconductor can lead to delocalized free carrier states. Delocalized states have high mobility so that the carriers can be quickly removed from the interface.
- The electronic coupling between the exciton in the donor and the CT excitons across the interface is increased if the two exciton states are energetically in resonance and have significant spatial overlap between their wave functions. By tuning the energy level alignment at the interface to the binding energy of the Frenkel exciton, the coupling to hot CT excitons can be increased. To maximize spatial overlap, the symmetry of the individual wave functions needs to be taken into account.

As a final remark, we have also demonstrated the ability of 2PPE to measure energy levels, population and decay dynamics of Frenkel and charge transfer excitons in organic semiconductors. In particular, it is possible to relate properties of exciton and single-particle states within the same experimental technique. While relatively simple dielectric models are sufficient to qualitatively explain observed energy levels, they naturally fail to explain the many-body effects inherent to excitons. For instance, since the exciton is a correlated electron–hole pair, properties of the hole, such as band formation, decay dynamics, and polaron formation, may have an observable effect on the photoelectron, in a similar way as the hole lifetime and electron–phonon interactions modify the spectral function of one-photon photoemission. In this respect, the full potential as well as limitations of the 2PPE technique, as applied to excitons, remain to be explored.

References

1 Wannier, G.H. (1937) The structure of electronic excitation levels in insulating crystals. *Phys. Rev.*, **52** (3), 191–197.

2 Frenkel, J. (1931) On the transformation of light into heat in solids. I. *Phys. Rev.*, **37** (1), 17–44.

3 Silinsh, E.A. and Čápek, V. (1994) *Organic Molecular Crystals*, American Institute of Physics.

4 Yu, G., Gao, J., Hummelen, J.C., Wudl, F., and Heeger, A.J. (1995) Polymer photovoltaic cells: enhanced efficiencies via a network of internal donor–acceptor heterojunctions. *Science*, **270** (5243), 1789–1791.

5 Arkhipov, V.I. and Bässler, H. (2004) Exciton dissociation and charge photogeneration in pristine and doped conjugated polymers. *Phys. Status Solidi (a)*, **201** (6), 1152–1187.

6 Yoo, S., Domercq, B., and Kippelen, B. (2004) Efficient thin-film organic solar cells based on pentacene/C_{60} heterojunctions. *Appl. Phys. Lett.*, **85** (22), 5427–5429.

7 Morteani, A.C., Sreearunothai, P., Herz, L.M., Friend, R.H., and Silva, C. (2004) Exciton regeneration at polymeric semiconductor heterojunctions. *Phys. Rev. Lett.*, **92** (24), 247402.

8 Scharber, M.C., Mühlbacher, D., Koppe, M., Denk, P., Waldauf, C., Heeger, A.J., and Brabec, C.J. (2006) Design rules for donors in bulk-heterojunction solar cells: towards 10% energy-conversion efficiency. *Adv. Mater.*, **18** (6), 789–794.

9 Rand, B.P., Burk, D.P., and Forrest, S.R. (2007) Offset energies at organic semiconductor heterojunctions and their influence on the open-circuit voltage of thin-film solar cells. *Phys. Rev. B*, **75** (11), 115327.

10 Combescot, M. and Pogosov, W. (2008) Microscopic derivation of Frenkel excitons in second quantization. *Phys. Rev. B*, **77** (8), 085206.

11 Hummer, K., Puschnig, P., Sagmeister, S., and Ambrosch-Draxl, C. (2006) *Ab-initio* study on the exciton binding energies in organic semiconductors. *Mod. Phys. Lett. B*, **20** (6), 261–280.

12 Shen, Z. and Forrest, S.R. (1997) Quantum size effects of charge-transfer excitons in nonpolar molecular organic thin films. *Phys. Rev. B*, **55** (16), 10578–10592.

13 Merrifield, R.E. (1961) Ionized states in a one-dimensional molecular crystal. *J. Chem. Phys.*, **34** (5), 1835–1839.

14 Hoffmann, M. (2003) Mixing of Frenkel and charge-transfer excitons and their quantum confinement in thin films, Chapter 5, in *Electronic Excitations in Organic Multilayers and Organic Based Heterostructures* (eds V. M. Agranovich and G. F. Bassani), 1st edn, vol. 31, Thin Films and Nanostructures, Elsevier, pp. 221–292.

15 Song, K.S. and Williams, R.T. (1993) *Self-Trapped Excitons*, vol. 105, Springer Series in Solid-State Sciences, Springer, Berlin.

16 O'Brien, M.C.M. (1996) Vibronic energies in C_{60}^{n-} and the Jahn–Teller effect. *Phys. Rev. B*, **53** (7), 3775–3789.

17 Knupfer, M. (2003) Exciton binding energies in organic semiconductors. *Appl. Phys. A*, **77** (5), 623–626.

18 Cole, M.W. (1971) Electronic surface states of a dielectric film on a metal substrate. *Phys. Rev. B*, **3** (12), 4418.

19 Tsiper, E.V. and Soos, Z.G. (2003) Electronic polarization in pentacene crystals and thin films. *Phys. Rev. B*, **68** (8), 085301.

20 Muntwiler, M., Yang, Q., Tisdale, W.A., and Zhu, X.-Y. (2008) Coulomb barrier for charge separation at an organic semiconductor interface. *Phys. Rev. Lett.*, **101** (19), 196403.

21 Zhu, X.-Y., Yang, Q., and Muntwiler, M. (2009) Charge transfer excitons at organic semiconductor surfaces and interfaces. *Acc. Chem. Res.*, **42** (11), 1779–1787.

22 Petelenz, P., Slawik, M., Yokoi, K., and Zgierski, M.Z. (1996) Theoretical calculation of the electroabsorption spectra of polyacene crystals. *J. Chem. Phys.*, **105** (11), 4427–4440.

23 Petelenz, P. and Mazur, G. (1999) Band gap and binding energies of charge-transfer excitons in organic molecular

crystals. *Chem. Phys. Lett.*, **301** (3–4), 223–227.

24 Pac, B., Petelenz, P., Eilmes, A., and Munn, R.W. (1998) Charge-transfer exciton band structure in the fullerene crystal-model calculations. *J. Chem. Phys.*, **109** (18), 7923–7931.

25 Knupfer, M. and Berger, H. (2006) Dispersion of electron–hole excitations in pentacene along (1 0 0). *Chem. Phys.*, **325** (1), 92–98, Electronic Processes in Organic Solids.

26 Hummer, K., Puschnig, P., and Ambrosch-Draxl, C. (2004) Lowest optical excitations in molecular crystals: bound excitons versus free electron–hole pairs in anthracene. *Phys. Rev. Lett.*, **92** (14), 147402.

27 So, F.F. and Forrest, S.R. (1991) Evidence for exciton confinement in crystalline organic multiple quantum wells. *Phys. Rev. Lett.*, **66** (20), 2649–2652.

28 Sebastian, L., Weiser, G., and Bässler, H. (1981) Charge transfer transitions in solid tetracene and pentacene studied by electroabsorption. *Chem. Phys.*, **61** (1–2), 125–135.

29 Salaneck, W.R., Gibson, H.W., Plummer, E.W., and Tonner, B.H. (1982) Ultraviolet photoelectron spectroscopy of optically excited states in *trans*-polyacetylene. *Phys. Rev. Lett.*, **49** (11), 801–804.

30 Katoh, R. and Kotani, M. (1990) External photoemission by singlet-exciton photoionization in an anthracene single crystal. *Chem. Phys. Lett.*, **174** (6), 537–540.

31 Hill, I.G., Kahn, A., Soos, Z.G., and Pascal, R.A., Jr. (2000) Charge-separation energy in films of π-conjugated organic molecules. *Chem. Phys. Lett.*, **327** (3–4), 181–188.

32 Neaton, J.B., Hybertsen, M.S., and Louie, S.G. (2006) Renormalization of molecular electronic levels at metal–molecule interfaces. *Phys. Rev. Lett.*, **97** (21), 216405.

33 Ge, N.-H., Wong, C.M., Lingle, R.L., Jr., McNeill, J.D., Gaffney, K.J., and Harris, C.B. (1998) Femtosecond dynamics of electron localization at interfaces. *Science*, **279**, 202.

34 Mott, N.F. (1990) *Metal-Insulator Transitions*, 2nd edn, Taylor & Francis, London.

35 Lof, R.W., van Veenendaal, M.A., Koopmans, B., Jonkman, H.T., and Sawatzky, G.A. (1992) Band gap, excitons, and Coulomb interaction in solid C_{60}. *Phys. Rev. Lett.*, **68** (26), 3924–3927.

36 Rudolf, P., Golden, M.S., and Brühwiler, P.A. (1999) Studies of fullerenes by the excitation, emission, and scattering of electrons. *J. Electron Spectrosc. Relat. Phenom.*, **100**, 409.

37 Tzeng, C.-T., Lo, W.-S., Yuh, J.-Y., Chu, R.-Y., and Tsuei, K.-D. (2000) Photoemission, near-edge X-ray-absorption spectroscopy, and low-energy electron-diffraction study of C_{60} on Au(111) surfaces. *Phys. Rev. B*, **61**, 2263.

38 Muntwiler, M., Yang, Q., and Zhu, X.-Y. (2009) Exciton dynamics at interfaces of organic semiconductors. *J. Electron Spectrosc. Relat. Phenom.*, **174**, 116–124.

39 Dresselhaus, M.S., Dresselhaus, G., and Eklund, P.C. (1996) *Science of Fullerenes and Carbon Nanotubes*, Academic Press, San Diego.

40 Lu, X., Grobis, M., Khoo, K.H., Louie, S.G., and Crommie, M.F. (2004) Charge transfer and screening in individual C_{60} molecules on metal substrates: a scanning tunneling spectroscopy and theoretical study. *Phys. Rev. B*, **70** (11), 115418.

41 Dutton, G., Quinn, D.P., Lindstrom, C.D., and Zhu, X.-Y. (2005) Exciton dynamics at molecule–metal interfaces: C_{60}/Au(111). *Phys. Rev. B*, **72** (4), 045441.

42 Zhao, J., Feng, M., Yang, J., and Petek, H. (2009) The superatom states of fullerenes and their hybridization into the nearly free electron bands of fullerites. *ACS Nano*, **3** (4), 853–864.

43 Zhu, X.-Y., Dutton, G., Quinn, D.P., Lindstrom, C.D., Schultz, N.E., and Truhlar, D.G. (2006) Molecular quantum well at the C_{60}/Au(111) interface. *Phys. Rev. B*, **74** (24), 241401.

44 Krätschmer, W., Lamb, L.D., Fostiropoulos, K., and Huffman, D.R. (1990) Solid C_{60}: a new form of carbon. *Nature*, **347**, 354–358.

45 Ren, S.L., Wang, Y., Rao, A.M., McRae, E., Holden, J.M., Hager, T., Wang, K., Lee, W.-T., Ni, H.F., Selegue, J., and Eklund, P.C. (1991) Ellipsometric determination of the optical constants of

C_{60} (Buckminsterfullerene) films. *Appl. Phys. Lett.*, **59** (21), 2678–2680.

46 Kuhnke, K., Becker, R., Epple, M., and Kern, K. (1997) C_{60} exciton quenching near metal surfaces. *Phys. Rev. Lett.*, **79** (17), 3246–3249.

47 Muntwiler, M. and Zhu, X.-Y. (2007) Formation of two-dimensional polarons that are absent in three-dimensional crystals. *Phys. Rev. Lett.*, **98** (24), 246801.

48 Matus, M., Kuzmany, H., and Sohmen, E. (1992) Self-trapped polaron exciton in neutral fullerene C_{60}. *Phys. Rev. Lett.*, **68** (18), 2822–2825.

49 Adams, D.M., Brus, L., Chidsey, C.E.D., Creager, S., Creutz, C., Kagan, C.R., Kamat, P.V., Lieberman, M., Lindsay, S., Marcus, R.A., Metzger, R.M., Michel-Beyerle, M.E., Miller, J.R., Newton, M.D., Rolison, D.R., Sankey, O., Schanze, K.S., Yardley, J., and Zhu, X. (2003) Charge transfer on the nanoscale: current status. *J. Phys. Chem. B*, **107** (28), 6668–6697.

50 Jerome, D. and Schulz, H.J. (1982) Organic conductors and superconductors. *Adv. Phys.*, **31** (4), 299.

51 Hummer, K. and Ambrosch-Draxl, C. (2005) Oligoacene exciton binding energies: their dependence on molecular size. *Phys. Rev. B*, **71** (8), 081202.

52 Hallermann, M., Haneder, S., and Da Como, E. (2008) Charge-transfer states in conjugated polymer/fullerene blends: below-gap weakly bound excitons for polymer photovoltaics. *Appl. Phys. Lett.*, **93** (5), 053307.

53 Al-Mahboob, A., Sadowski, J.T., Nishihara, T., Fujikawa, Y., Xue, Q.K., Nakajima, K., and Sakurai, T. (2007) Epitaxial structures of self-organized, standing-up pentacene thin films studied by LEEM and STM. *Surf. Sci.*, **601** (5), 1304–1310.

54 Sadowski, J.T., Nagao, T., Yaginuma, S., Fujikawa, Y., Al-Mahboob, A., Nakajima, K., Sakurai, T., Thayer, G.E., and Tromp, R.M. (2005) Thin bismuth film as a template for pentacene growth. *Appl. Phys. Lett.*, **86** (7), 073109.

55 Thayer, G.E., Sadowski, J.T., Meyer zu Heringdorf, F., Sakurai, T., and Tromp, R.M. (2005) Role of surface electronic structure in thin film molecular ordering. *Phys. Rev. Lett.*, **95** (25), 256106.

56 Campbell, R.B., Robertson, J.M., and Trotter, J. (1961) The crystal and molecular structure of pentacene. *Acta Cryst.*, **14**, 705–711.

57 Muntwiler, M. and Zhu, X.-Y. (2008) Image potential states on the metallic (111) surface of bismuth. *New J. Phys.*, **10**, 113018.

58 Kakuta, H., Hirahara, T., Matsuda, I., Nagao, T., Hasegawa, S., Ueno, N., and Sakamoto, K. (2007) Electronic structures of the highest occupied molecular orbital bands of a pentacene ultrathin film. *Phys. Rev. Lett.*, **98** (24), 247601.

59 Goldberg, S.M., Fadley, C.S., and Kono, S. (1981) Photoionization cross-sections for atomic orbitals with random and fixed spatial orientation. *J. Electron Spectrosc. Relat. Phenom.*, **21** (4), 285–363.

60 Shumay, I.L., Höfer, U., Reuß, C., Thomann, U., Wallauer, W., and Fauster, Th. (1998) Lifetimes of image-potential states on Cu(100) and Ag(100) measured by femtosecond time-resolved two-photon photoemission. *Phys. Rev. B*, **58** (20), 13974–13981.

61 Nitzan, A. (2001) Electron transmission through molecules and molecular interfaces. *Annu. Rev. Phys. Chem.*, **52** (1), 681–750.

16
Electron Dynamics at Polar Molecule–Metal Interfaces: Competition between Localization, Solvation, and Transfer

Julia Stähler, Uwe Bovensiepen, and Martin Wolf

16.1
Introduction

Electron-induced processes and reactions in aqueous systems are of key relevance in diverse fields, ranging from electron transfer and light harvesting in biological systems, photochemistry in the atmosphere, electrochemistry and corrosion, radiation chemistry, and nuclear waste remediation to medical diagnosis and therapy [1–3]. These processes usually involve the creation, rearrangement, and transfer of electrons or charged particles in a polar environment. Therefore, solvation and localization of these charges by the reorientation of the surrounding molecular dipoles are key processes that govern the dynamics and reaction rates in aqueous systems or other polar solvents. The solvation reorganization energy, which is associated with the polarization of the solvent environment, determines the energetics and stabilization of charged particles in the solvent. Its magnitude is on the order of several eV [4] and provides – facilitated by thermal fluctuations of the solvent environment – the driving force for electron transfer in solution. Electron transfer can therefore take place spontaneously at room temperature, as described by Marcus theory [5]. On the other hand, if an excess charge or dipole is suddenly created in a polar solvent (e.g., by photoexcitation), its environment will respond on ultrafast timescales, which are characteristic of the molecular rearrangements of the solvent [6]. In bulk liquid water, for example, the formation dynamics of hydrated electrons involve several intermediate transient species, which can be measured by femtosecond time-resolved IR spectroscopy [7].

Solvation processes also play an important role in the context of electron transfer across interfaces between different materials. Such processes are of both fundamental interest and technological significance: The solvation dynamics in a quasi two-dimensional solvent at an interface is expected to differ from isotropic systems due to the reduced dimensionality and the resulting differences in solvent structure and motions [8]. In addition, the electronic coupling at polar molecule–metal interfaces opens a new decay channel for the excited-state population, leading to

Dynamics at Solid State Surface and Interfaces Vol.1: Current Developments
Edited by Uwe Bovensiepen, Hrvoje Petek, and Martin Wolf

ISBN: 978-3-527-40937-2

a competition between localization (and solvation) of the excess electrons in the molecular adlayer and charge transfer back to the metal substrate. This interplay makes molecule–metal interfaces model systems for the study of fundamental processes such as charge injection, localization, and transport that play central roles in various optoelectronic and molecular devices, which exhibit coupling between a conducting electrode and the molecular system [9, 10]. Furthermore, interfacial electron transfer is also identified as the initial step in surface photochemistry and femtochemistry, where chemical reactions are initiated by transient negative ion formation via photoinduced transfer of hot electrons from the metal substrate to the adsorbate [11]. Such transient negative ion states play a central role in dissociative electron attachment (DEA) of molecules into neutral and negatively charged fragments. In the upper atmosphere, electron-sensitive molecules (like fluorochlorocarbons) adsorbed on the surface of water droplets or ice particles can interact with solar photons and other energetic particles and undergo dissociative electron attachment to generate chemical species detrimental to environment [2, 12]. Solvated electrons at the surface of water/ice particles may thereby serve as an important intermediate in these reactions.

Polar adsorbate layers on single-crystal metal surfaces provide a model system to study the various processes discussed above under well-defined conditions. Here, the substrate acts as a source of the excess electrons that are injected into the polar adlayer after photoexcitation. The remaining hole is thereby completely screened in the metal substrate – in contrast to studies in solution or condensed molecular systems, where excitonic processes prevail. Harris and coworkers have made pioneering contributions to this field and were first to study the dynamics of small polaron formation [13, 14] and electron solvation processes at various molecule–metal interfaces [15]. They used time- and angle-resolved two-photon photoelectron spectroscopy to study dynamic localization and energetic stabilization of electrons on monolayers and thin multilayers of both polar and nonpolar adsorbates on single-crystal Ag(111) surfaces. An overview of this work can be found in several review articles [8, 14] and in chapter 4 in this book. Related studies were also made on wide bandgap semiconductor (TiO_2) surfaces by Petek and coworkers [16].

This chapter focuses on the electron dynamics at various ice– and ammonia–metal interfaces. Four elementary steps are identified: (1) electron injection and (2) localization in the adsorbate layer (island), (3) solvation by molecular rearrangement, and (4) electron back transfer to the metal substrate. These processes are common features of the electron dynamics at *all* polar molecule–metal interfaces investigated in this work and, moreover, are highly interconnected. The excited electron population is, on the one hand, localized and solvated by reorganization of the solvent molecules, but on the other hand is continuously reduced by back transfer of electrons to the metal substrate where the electrons relax by electron–hole pair formation (comparable to an Auger process). As will be shown below, we observe a *competition* between charge transfer and solvation, which is strongly influenced by the screening capability of the adsorbate and the electronic coupling strength of the solvated electron state with the continuum of unoccupied states in the metal substrate. By time-resolving the transient balance between electron solvation and transfer, this competition is utilized to study the impact of the substrate, the adsorbate, and its molecular structure on interfacial electron dynamics.

Starting with the electron dynamics in amorphous D_2O films on the Cu(111) and Ru(001) surfaces in Section 16.2.1, the four elementary steps of electron injection, localization, solvation, and transfer are introduced. By comparison of the two substrates, the competing channels for electron relaxation, solvation, and transfer are illustrated. Furthermore, it is shown that electron transfer is initially, in the first 300 fs, dominated by the substrate's electronic properties, but at later times it is determined by tunneling through a potential barrier at the interface that results from the screening of the excess charge by the solvent molecules.

Section 16.2.2 demonstrates that a very similar crossover occurs on comparable timescales in the case of amorphous NH_3 layers on Cu(111). Still, the different screening properties of ammonia and the change of solvation site allow a more concerted analysis of the data toward the characteristics of the solvated electron state: Initially, in the *substrate-dominated* regime, the wave function of the excess electron extends through the NH_3 layer and has significant overlap with the metal. Due to electron solvation, however, a *transient* barrier forms at the interface that increasingly screens the electrons from the metal. From there on, electron transfer is mediated by tunneling through that barrier and therefore termed *barrier-determined*. The analogy of the femto- and picosecond dynamics in amorphous NH_3 and D_2O on Cu(111) suggests that the transition between the two regimes of electron transfer, between strong and weaker coupling, is a characteristic of heterogeneous charge transfer at amorphous polar molecule–metal interfaces.

Crystallization of the solvent leads to dramatic changes of the electron dynamics as discussed in Section 16.3.1. Excited electrons are observed *minutes* after optical excitation at the crystalline D_2O/Ru(001) interface; this lifetime is up to 16 orders of magnitude longer than those of electrons in amorphous ice. This effect is promoted by very efficient screening of the excess charge in preexisting, deep traps at the ice–vacuum interface, which is a clear contrast to the case of amorphous ice where electron localization occurs directly beneath the conduction band (CB) bottom. The competition is dominated by electron solvation, and thus charge transfer back to the substrate becomes extremely small. Furthermore, the extraordinarily long lifetime of excess electrons at the surface of crystalline ice makes them a precursor for (low-energy) electron-driven chemical reactions. In this context, their dissociative attachment to coadsorbed $CFCl_3$ is discussed in Section 16.3.2.

16.2
Competing Channels of Electron Relaxation in Amorphous Layers

In the following, we present a time-resolved two-photon photoelectron (2PPE) spectroscopy study of ultrathin films of *amorphous* D_2O and NH_3 ice on Cu(111) and Ru(001). The sample preparation has been described in detail elsewhere [17, 18]. In 2PPE spectroscopy (cf. Figure 16.1c, top), metal electrons are photoexcited from below the Fermi level E_F by a femtosecond laser pulse $h\nu_1$ (*pump*) to unoccupied,

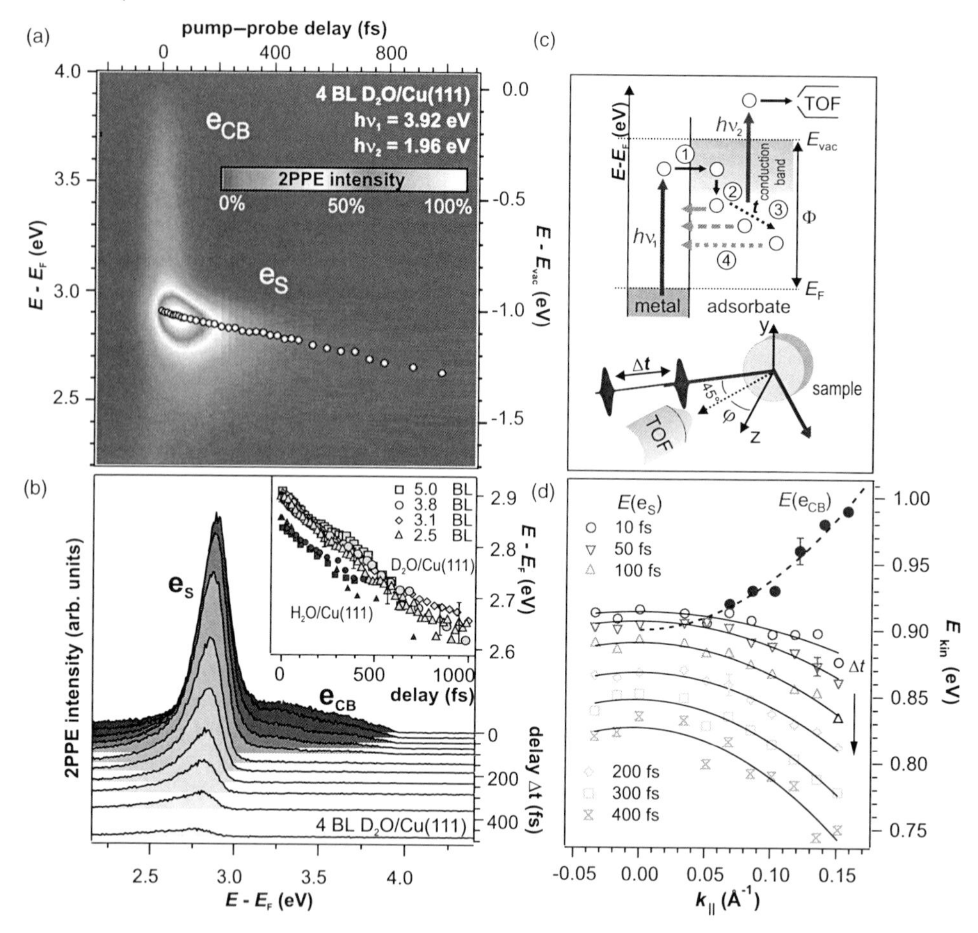

Figure 16.1 (a) Time-resolved 2PPE intensity of amorphous multilayers D_2O/Cu(111) in false color representation. The energy shift of the solvated electrons is indicated by white markers (modified from Ref. [19]). (b) 2PPE spectra of the same data set (modified from Ref. [39]). *Inset*: Energy shift of e_S for various coverages of H_2O and D_2O on Cu(111). (c) Time- and angle-resolved 2PPE scheme. (d) Dispersion of conduction band and solvated electrons at a series of time delays. (a), (d) adapted with permission from [19]. Copyright 2003 American Chemical Society. (b) adapted with permission from [30]. Copyright 2002 by the American Physical Society. (Please find a color version of this figure on the color plates.)

bound states below the vacuum level E_{vac}. The subsequent nonequilibrium dynamics of the electron population are monitored by a second, time-delayed laser pulse $h\nu_2$ (*probe*): It excites the electrons to the vacuum where their kinetic energy and momentum parallel to the surface are detected using an electron time-of-flight (TOF) spectrometer. By variation of the pump–probe time delay, the photoinduced nonequilibrium femtosecond dynamics can be monitored in real time. Spectra are plotted as a function of intermediate-state energy referring to the Fermi level $E - E_F = E_{kin} + \Phi - h\nu_2$ (with work function Φ).

16.2.1
Amorphous Ice on Metal Surfaces

Figures 16.1a and b depicts time-resolved 2PPE data of amorphous multilayers of D_2O on Cu(111) as a function of intermediate-state energy and femtosecond pump–probe delay. The data show a distinct peak, termed e_S, at $E - E_F = 2.9$ eV, on top of an energetically broad continuum e_{CB}. It will be shown below that the short-lived e_{CB} results from delocalized conduction band electrons in the adsorbate layer and that the long-lived feature e_S originates from solvated electrons that are localized in the adlayer. Note that the peak maximum of e_S shifts toward the Fermi level with increasing time delay as indicated by the white circles in Figure 16.1a; this is a result of energy transfer to molecular degrees of freedom (electron solvation).

Angular resolution of the 2PPE spectra provides important information on the momentum distribution of the photoelectrons and can be achieved by rotation of the sample as sketched in Figure 16.1c (bottom). The parallel momentum of a photoelectron is connected to the emission angle φ by

$$k_{||}(\varphi, E_{kin}) = \sin\varphi \cdot \sqrt{\frac{2m_e}{\hbar} \cdot E_{kin}}. \tag{16.1}$$

The distribution $\Delta k_{||}$ of an electronic state in momentum space facilitates an estimate of its (lateral) real space dimension Δx by a lower bound due to Heisenberg's uncertainty principle. Using time- *and* angle-resolved 2PPE, the temporal evolution of charge localization can be analyzed and used to understand the origin of the longer lifetime of e_S compared to e_{CB} as shown in the following. Figure 16.1d depicts the dispersion of e_{CB} (solid markers) and e_S (open symbols) for different pump–probe delays. e_{CB} has a positive dispersion with the free electron mass $m_{eff} = 1.0(2) \times m_e$, that is, the electrons in this band are delocalized parallel to the surface. The peak e_S, on the contrary, results from a localized state. Actually, it appears to have an effective negative dispersion, which becomes stronger with increasing time delay. This curvature results from the finite bandwidth $\Delta k_{||}$ of the interfacial electrons in momentum space and the experimental constraint that spectra are taken at a specific angle and not a specific $k_{||}$. As discussed in detail in Ref. [19], the bandwidth $\Delta k_{||}$ broadens as solvation proceeds. Following Heisenberg's uncertainty principle, the bandwidth gives a lower limit for the spatial extend of the solvated electron. The resulting lateral diameter of the wave function of 10–20 Å is attributed to an increased spatial constriction of the excess electron at larger binding energies.

With the above observations, we can now fully characterize the four elementary steps of (1) charge injection from the metal into the D_2O adlayer, (2) localization in the ice, (3) electron solvation due to molecular reorientation, and (4) electron back transfer to the substrate (see Figure 16.1c for illustration of 1–4):

1) **Electron injection**: Due to the low absorption cross section of the ultrathin layers and because of the large bandgap of ice (8.2 eV, [20]), photoexcitation within the adsorbate can be ruled out at the photon energies (<4 eV) and fluences (<2 μJ/cm^2) used, as it would require a three-photon process. On the other hand,

the electronic band structure of Cu(111) is well known [21]: Both, e_S and e_{CB}, are located in the Cu(111) bandgap and can therefore not be attributed to the substrate. Hence, they must be adsorbate-induced. The only remaining pathway for e_S and e_{CB} excitation is thus excitation of metal electrons from below the Fermi level followed by charge injection into the adsorbate layer (see step 1 in Figure 16.1c). This is driven by electronic coupling between the metal states and the continuum e_{CB}. As e_{CB} is delocalized parallel to the interface, we assign it to the conduction band of the ice. The coupling strength between e_{CB} and the substrate states is responsible for the rapid decay of the e_{CB} population back to the metal, which occurs within the laser pulse duration and is estimated to be faster than 10 fs [19].

2) **Electron localization**: The feature e_S is observed at the bottom of the conduction band and exhibits a different dispersion in as short as 10 fs, compatible with a localized state (open circles in Figure 16.1d). It is concluded that in addition to electron transfer, part of the CB electron population localizes at favorable sites in the D_2O layer (step 2 in Figure 16.1c) leading to the spectral signature e_S. Plausible candidates for such localization sites are preexisting bond angle fluctuations or defects that cause minima in the interfacial potential [22]. After the ultrafast population of these sites, the degree of localization of the electron distribution e_S changes with time delay due to the continuous reorientation of the surrounding polar water molecules.
3) **Electron solvation**: In parallel with localization, a transient energy shift of e_S toward the Fermi level is observed (step 3 in Figure 16.1c). This relaxation also results from the reorganization of the D_2O molecules that accommodate the excess charge. Having observed both characteristic properties, energetic stabilization and electron localization, of electron solvation in a polar solvent, we assign the peak e_S to solvated electrons in the amorphous D_2O layer. The inset of Figure 16.1b depicts the peak shift of the e_S feature for different coverages of amorphous D_2O and H_2O on the Cu(111) substrate. Apparently, the dynamics are very similar for the two isotopes. On the first sight, this observation is astounding, as the energetic stabilization is driven by molecular motion, which should be influenced by changes of the hydrogen mass. However, the absence of an isotope effect is in agreement with related measurements in the liquid phase [23]. The isotope-determined librational modes of ice play a role on shorter timescales ($<$40 fs) only [24], while the dynamics observed here result from frustrated translational motion of the entire molecules[1].
4) **Electron back transfer**: Being in an excited state in front of a metal surface, the excess electrons continuously decay back to the Cu(111) substrate (step 4 in Figure 16.1c), as seen in the intensity decay of e_S as a function of pump–probe time delay in Figure 16.1a and b. This back transfer depends strongly on the coupling between the solvated electrons and the accepting metal states. Although a detailed discussion will be provided in the following, we would like to point out

1) However, the energies of the H_2O are 50 meV lower than for D_2O. This could result from the 50 meV smaller band gap of (bulk) H_2O compared to D_2O (cf. Ref. [20]).

that the continuous change of the excess electron's environment induced by solvation reduces the transfer probability of the excess charge during the time-resolved measurement.

Summarizing the above, excess electrons are injected into the amorphous ice layers where they localize and solvate due to molecular rearrangement. Simultaneously, the population decays back to the substrate. This back transfer of (excited) excess electrons in the D_2O layer to the substrate is driven by the high density of unoccupied states in the metal at low energies. At the same time, the remaining population of e_S is on its pathway to equilibrium with the solvent network. The concurrence of these two processes and their opposite consequences (depopulation of e_S and stabilization of the e_S population, respectively) results in a *competition* between electron transfer and solvation.

Charge transfer phenomena that go along with molecular reorientations are often discussed on the basis of Marcus theory [5]. In this picture, the solvent reorganization is described by harmonic potentials as a function of the global nuclear (reaction) coordinate q, and charge transfer is mediated by thermal activation leading to solvent reorganization. Also, the electronic coupling of the initial and the final states influences the electron transfer rate, which is particularly important in the case of *heterogeneous* charge transfer to a continuum of unoccupied metal states. As discussed in detail in Ref. [25], the observed electron transfer rate is a result of (i) thermally activated motion within the solvent network that facilitates the charge transfer and (ii) the electronic coupling term between the initial and the respective final states in the metal. The former, however, is negligible if the temperature of the system is too low (frozen lattice) or if the charge transfer occurs on faster timescales than molecular reorientation does. In such a scenario, which is the case for the amorphous D_2O and NH_3 morphologies discussed in this chapter [25, 26], the electron transfer is determined by the electronic coupling only. As mentioned above, the electronic coupling strength changes due to electron solvation, and can be described by tunneling through potential barriers at the interface. The competition between electron transfer and solvation is reflected by a *transient* potential as discussed below. To get to the bottom of this phenomenon, the different contributions to electron transfer and solvation are disentangled in the following by systematic variation of the substrate, adsorbate, and structure of the adsorbate layer.

Starting with a comparison of the electron dynamics at two different ice–metal interfaces, we show that the contributions of charge transfer and solvation to the electron dynamics can be distinguished and that these processes are highly interconnected. This is accomplished by choosing a substrate that exhibits a different surface electronic band structure than Cu(111): Using Ru(001), a metal surface with a narrower projected bandgap than that of Cu(111), the electronic coupling between the solvated electrons and the accepting metal states is increased. The population decay traces of solvated electrons in amorphous D_2O on Cu(111) (diamonds) and Ru(001) (circles) are displayed in Figure 16.2a. Comparison of the two data sets unveils that the electron decay differs significantly for early delays (<300 fs) when the electrons decay four times faster to Ru(001) ($\tau = 34(5)$ fs) than to Cu(111)

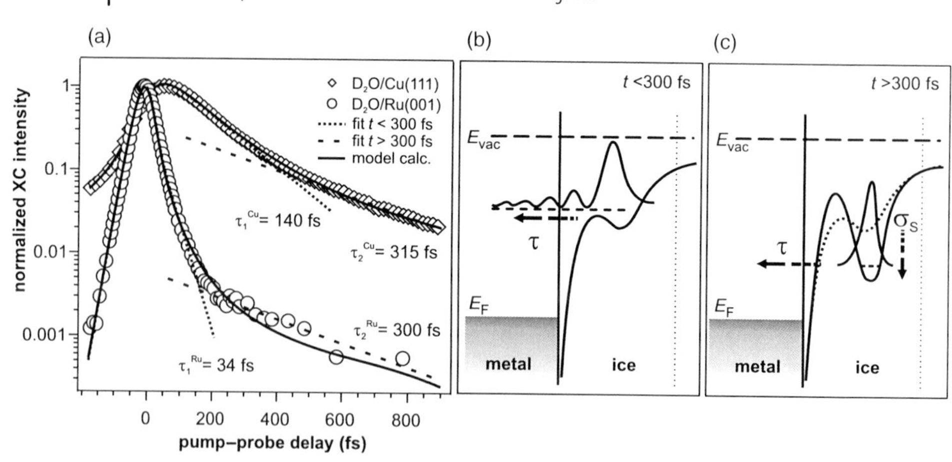

Figure 16.2 (a) Time-dependent 2PPE intensity of e_S for amorphous D_2O on Cu(111) and Ru (001). Two regimes of electron transfer are observed: In the early stages of solvation ($t < 300$ fs), the transfer is substrate-dominated (b) and later (c) barrier-determined (modified from Ref. [17]). τ_i are determined by least square fits of exponential decays convoluted with laser pulse envelopes. Adapted with permission from [17]. Copyright 2006 American Chemical Society.

($\tau = 140(15)$ fs). At later delays, however, the electron back transfer slows down for both substrates and becomes comparable with 315(20) fs and 300(20) fs for the copper and the ruthenium substrate, respectively. Apparently, the influence of the electronic structure of the metal on the transfer dynamics is reduced upon electron solvation. The origin of these two regimes of electron transfer, *substrate-dominated* for $t < 300$ fs and *barrier-determined* for $t > 300$ fs, will be discussed in the following.

At early delays ($t < 300$ fs), the electron dynamics for Cu(111) and Ru(001) substrates differ. This difference is solely induced by the substrate change and then diminished at later delays when the population decay becomes similar for both metal templates. Initially, in the *substrate-dominated* regime, as illustrated in Figure 16.2b, the excited electron still exhibits considerable wave function overlap with the substrate. Therefore, details of the substrate electronic band structure determine the rate of electron back transfer. The narrower orientational bandgap of Ru(001) requires a smaller momentum change upon back transfer, and thus the electron population in the adlayer decays faster than on Cu(111). However, at the same time, the remaining electron population of e_S is further localized and stabilized, which leads to a reduced electronic coupling of the electrons to the metal (cf. Figure 16.2c). Now, electron solvation considerably influences the charge transfer by screening of the excess charge with the surrounding solvation shell. In this *barrier-determined* charge transfer regime (Figure 16.2c), the population decay slows down and becomes comparable for the substrates studied, as differences of the substrate's electronic properties are increasingly screened from the excess electrons by the transient potential barrier. This concept of a dynamically evolving potential is illustrated by the sketch in Figure 16.2c. Solvation leads not only to a binding energy increase of the electron but also gives rise to a growing tunneling barrier that separates the charge from the metal substrate.

The interplay between the heterogeneous charge transfer and electron solvation can be summarized as follows: (i) The electron transfer is driven by the electronic coupling between excess charge and metal states. (ii) Thus, it is influenced by the electron's degree of localization. (iii) The localization is in turn determined by the conformation of the solvation shell and therefore changed upon solvation. The *competition* between charge transfer and solvation is best illustrated when considering the determining factors of the two processes: On the one hand, the transfer reduces the electron population and therefore hinders the electrons from further solvation, and on the other hand, the solvation leads to an enhanced screening of the charge and therefore to a decreased probability of back transfer.

In conclusion, electron transfer and solvation compete with each other, because transfer is determined by the transient potential that is varied upon molecular rearrangement. The further the electron solvation proceeds, the weaker the electron population decay, as illustrated by the gray arrows in Figure 16.1c (step 4)[2]. However, the coupling of the excited electrons to the substrate also influences the transfer dynamics. As both molecular coordination and electron–metal distance influence the coupling to the substrate, it is important to know where the electron is located, in the bulk or on the surface of the ice layers.

Using scanning tunneling microscopy, Morgenstern and Mehlhorn have shown that amorphous D_2O grows in closed *layers* on top of Cu(111) for coverages larger than 3 BL (bilayers) [27], but in nonwetting *islands* below 2 BL D_2O/Cu(111) [28]. Here, xenon overlayers on top of D_2O/Cu(111) are used to probe whether the electrons are localized in the bulk of the adsorbed D_2O or at the ice–vacuum interface: After preparation of the desired morphology of D_2O, we adsorb Xe on top of the sample (cf. Figure 16.3a) to probe the resulting changes in the 2PPE spectra. Depending on whether the electrons are localized in the bulk of the ice or at the ice–vacuum interface, we anticipate different effects on the photoelectron spectra: In the case of bulk solvation, we do not expect major impact of the Xe atoms on the solvated electrons, because they are surrounded and screened by D_2O molecules. For electrons localized at the ice–vacuum interface on the other hand, the rare gas atoms will directly interact with the excess charge and alter the 2PPE spectra due to spatial confinement and/or polarization effects. Figure 16.3b depicts 2PPE spectra of closed D_2O layers ($\theta > 3$ BL) before (solid curve) and after Xe adsorption (dotted), showing that the signature e_S of the solvated electrons remains unaltered [29]. The work function increase of 90 meV, indicated by the shift of the low-energy edge of the spectrum, results from the change of the surface dipole moment induced by the rare gas adsorption. This originates from the different electron affinity of Xe and D_2O.

2) As discussed in detail in Ref. [17], this effect can lead to additional energy shifts of the inhomogeneously broadened e_S peak. Electrons at higher energies (i.e., less solvated) experience higher charge transfer probabilities, so the maximum of e_S is additionally shifted toward lower energies.

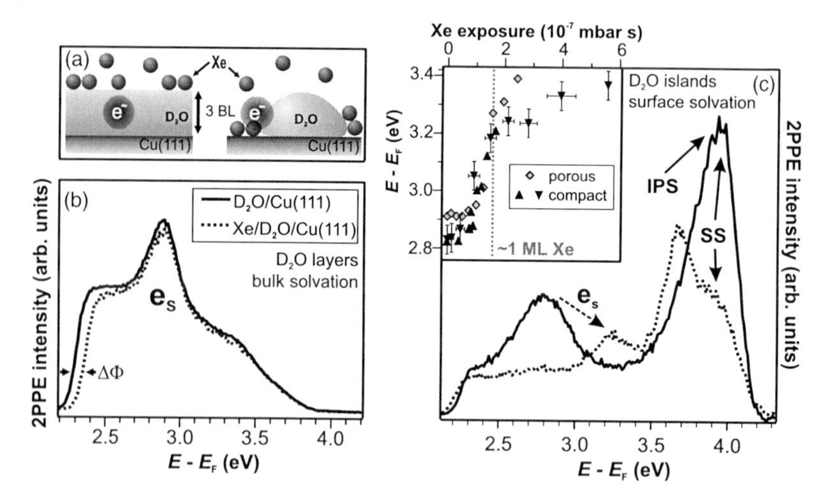

Figure 16.3 (a) Xe titration of solvated electrons for amorphous D_2O layers and islands on Cu (111). (b) Closed layers: e_S is not affected, because the electrons are localized and screened in the bulk of the ice. (c) Islands: the e_S feature shifts by 400 meV to higher energies upon Xe adsorption, as the excess electrons are localized at the ice–vacuum interface where it interacts with the rare gas atoms. Adapted with permission from [29] and [30]. [29] Reproduced by permission of the PCCP Owner Societies. [30] Copyright 2007 by the American Physical Society.

The spectra of D_2O islands on Cu(111) are shown in Figure 16.3c [30][3]. In contrast to the closed layers, the e_S peak changes significantly when Xe is adsorbed on top of the system: Its intensity decreases, and it continuously shifts toward E_{vac}. The e_S peak position as a function of xenon exposure is depicted in the inset of Figure 16.3c. The presence of the rare gas atoms thus obviously influences the environment of the solvated electrons. We therefore conclude that for the D_2O layers, where the e_S peak was unaffected by the rare gas adsorption, electrons are localized in the bulk and, on the other hand, at the ice–vacuum interface in the case of D_2O islands.

The origin of the two different species of solvated electrons, located at the surface and in the bulk of D_2O on Cu(111), can intuitively be understood by consideration of two factors that determine the attraction of an electron solvation site: (i) The density of unsaturated hydrogen bonds is higher at ice–*vacuum* interfaces than in the bulk. (ii) The image potential of the metal, although modified by the adlayer, drives the electron toward the ice–*metal* interface. The latter argument leads to bulk solvation in the case of closed D_2O layers, while we find solvation at the ice–vacuum interface for D_2O islands. However, electron localization at the surface of the D_2O islands does not necessarily mean that the electrons reside *on top* of the islands, but rather includes the possibility of electron solvation at the island edges as illustrated in Figure 16.3a (right). This scenario is supported by both factors: (i) unsaturated H-bonds at the

3) As the bare Cu(111) surface is partially exposed, the image potential state and the surface state (SS) appear in the spectra. For details, see Ref. [30].

ice–vacuum interface and (ii) proximity to the metal driven by the image potential. A detailed discussion of this matter can be found in Ref. [29].

We also observe that the electron dynamics do not change for closed D_2O layers when the layer thickness is increased further than 3 BL, which is consistent with the picture of electron solvation in the bulk of the ice layers: The electron–metal distance as well as the local surrounding of the excess charge remain unaltered if additional D_2O is adsorbed on top of the multilayer. This is different for NH_3/Cu(111) where excess electrons always localize at the ammonia–vacuum interface as shown in Section 16.2.2. This attribute is used in the following to vary the distance of the solvated electrons to the metal substrate and in this way utilized to tune the electronic coupling of the excited electrons to the substrate states. As discussed in the following, systematic variation of the layer thickness therefore enables quantitative characterization of the tunneling barrier and thus quantification of the competition between electron solvation and transfer.

16.2.2
Amorphous NH_3 on Cu(111)

The tunneling probability of an electron through a potential barrier is described by

$$\Gamma(d) \propto \exp[-\beta \cdot d], \tag{16.2}$$

where d is the barrier thickness and the inverse range parameter that reflects the shape and height of this barrier. Assuming that d can be varied systematically and supposing that the resulting transfer probabilities are measured, it would be possible to extract the β and thereby gain information about the tunneling potential. Although not possible in the case of wetting amorphous ice layers where electrons localize in the bulk of the adlayer (cf. Section 16.2.3), this type of experiment is feasible when turning to amorphous NH_3 layers on Cu(111). Here, the electrons always localize at the NH_3–vacuum interface, as demonstrated by Xe overlayer experiments similar to the ones shown in Figure 16.3 [31].

Very similar to ice–metal interfaces, the dynamics of excess electrons in amorphous NH_3 on Cu(111) are characterized by the four elementary steps described earlier: electron injection, localization, stabilization, and back transfer. Here, we will solely focus on the population dynamics of the excess electrons; a detailed presentation of the electron dynamics at NH_3/Cu(111) can be found in Ref. [31]. Figure 16.4a depicts an exemplary population decay trace for a 12 Å thick layer of NH_3 on Cu(111). It was fitted with a simple rate equation model (solid curve, see flowchart) that is discussed below. To demonstrate the qualitative similarities of these transfer dynamics with D_2O/Cu(111), the time-dependent background of hot electrons in the Cu(111) substrate (dotted curve) is subtracted from the NH_3/Cu(111) data: The resulting population decay (dashed curve) of excess electrons in amorphous NH_3/Cu(111) is – similar to D_2O – characterized by a fast component ($t < 300$ fs) and a slower part ($t > 300$ fs), which are labeled e_P and e_S, respectively. The data are analyzed by an empirical model basing on rate equations as illustrated by the

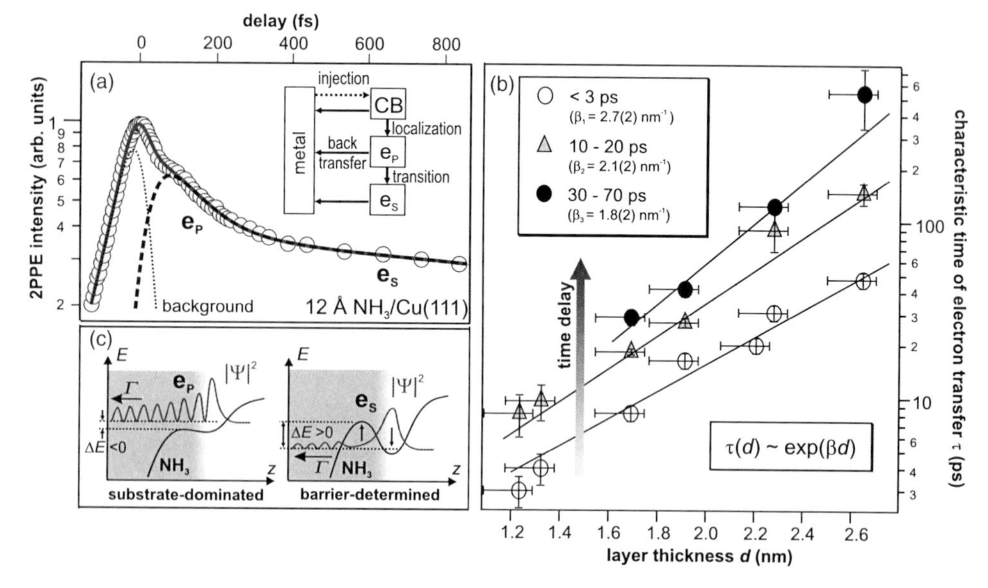

Figure 16.4 (a) The population decay at $NH_3/Cu(111)$ interfaces is characterized by two species of solvated electrons, a precursor state e_P with femtosecond decay and the solvated state e_S with dynamics on picosecond timescales. (b) The residence time of solvated electrons depends exponentially on the ammonia layer thickness. (c) e_P is a scattering state and strongly coupled to the electronic structure of the substrate. In the barrier-determined transfer regime, e_S is separated from the substrate by a transient potential barrier. Reprinted with permission from [31]. Copyright 2008 American Chemical Society.

flowchart in Figure 16.4a[4]. Briefly, the electrons are injected (dotted arrow) into the adsorbate layer via the conduction band and localize on fs timescales (36(6) fs) at favorable sites e_P in the adlayer. The species e_P decays with 140(20) fs, similar to the initial decay time of solvated electrons in $D_2O/Cu(111)$ (140(15) fs, see Figure 16.2a). This observation strongly suggests that the initial back transfer ($t < 300$ fs) is also substrate-dominated for the $NH_3/Cu(111)$ interface. Then, as indicated by the flowchart in Figure 16.4a, the transition from e_P to e_S occurs. The characteristic time of this process is 180 fs. In contrast to the solvated electron dynamics in amorphous ice layers, however, the transfer of electrons in e_S back to the substrate ($t > 300$ fs) depends here on the thickness of the NH_3 film, because the electrons are localized at the ammonia–vacuum interface. The characteristic time of the e_S population decay (determined using the rate equation model) for various NH_3 layer thicknesses is plotted in Figure 16.4b (gray circles). It depends exponentially on the layer thickness, allowing the extraction of the inverse range parameter β. This finding shows that charge transfer is mediated by tunneling through an interfacial barrier that becomes wider with increasing layer thickness in agreement with Eq. (16.2).

4) The following time constants are the average values of coverages between 12.5 and 32 Å, because only the back transfer of e_S depends on the layer thickness, as discussed subsequently.

Consistent with our earlier discussion for amorphous ice, the back transfer of electrons in e_S at the NH_3–vacuum interface is thus barrier-determined.

The other two datasets in Figure 16.4b result from the electron population decay at two other intervals of pump–probe delays as indicated in the legend. Analysis according to Eq. (16.2) yields the β-values for the corresponding stages of solvation. Two observations should be mentioned: (i) For any specific NH_3 layer thickness, the transfer is slowed down with increasing pump–probe delay (cf. arrow in Figure 16.4b), which is indicated by τ_d (○) < τ_d (▲) < τ_d (●). This continuous deceleration of electron back transfer shows that the ongoing solvation process decreases the electronic coupling of the excess charge. The molecular rearrangement leads to an enhanced screening of the electron and thus to a reduction of the transfer probability. (ii) The corresponding inverse range parameters β_i become smaller for the three different data sets. Since the β_i are characterized by the height and shape of the tunneling barrier and because they change with time delay, it can be concluded that the interfacial potential is time-dependent, being varied by electron solvation: $\beta = \beta(t)$. In other words, the transient barrier can be monitored by the time evolution of $\beta(t)$, that is, by the electron population decay of e_S.

As described in detail in Ref. [31], it is possible to extract information about the tunneling barrier from these experimentally determined $\beta(t)$. First, a key conclusion is that the potential barrier actually *rises* with ongoing solvation due to enhanced screening by molecular reorientation. Approximating the tunneling barrier by a parabola, this evolution could be quantified: Every 10 meV binding energy gain of the solvated electrons result in additional 26(10) meV increase of the tunneling barrier[5]. Second, the physical reason for the transition from the substrate-dominated to the barrier-determined regime of charge transfer is identified: The transition from e_P to e_S occurs exactly when the energy of the solvated electron state coincides with the potential barrier maximum ($E_{barrier} - E_{electron} = \Delta E = 0$). This is illustrated in Figure 16.4c. Excited electrons in e_P exhibit a wave function that extends through the ammonia layer (Figure 16.4c, left). The electron density is highest at a potential minimum at the NH_3–vacuum interface, but not yet separated from the metal by a potential barrier. Therefore, e_P is identified to be a scattering state. At this point, the electrons still exhibit a considerable amount of wave function overlap with the substrate, which therefore dominates the back transfer dynamics. Thus, electron transfer is, similar to the ice–metal interfaces, substrate-dominated. With ongoing solvation, the reorientation of polar molecules results in an enhanced screening and stronger localization of the solvated electron, which is manifested by the buildup of an interfacial potential barrier (Figure 16.4c, right). The solvated electron wave function is now separated from the metal by the transient barrier and its energy below maximum of the tunneling barrier ($\Delta E > 0$). In this barrier-determined transfer regime, the electron transfer rate results from the tunneling probability through this barrier, which evolves with progressing solvation.

5) This result is based on the assumption that the parabola width equals the NH_3 layer thickness at $E = E_F$; for details, see Ref. [31].

It should be noted that the very similar characteristics of the electron transfer dynamics in amorphous D_2O and NH_3 on metal surfaces suggest that the crossover from substrate-dominated to barrier-determined transfer is a general pathway of heterogeneous electron transfer at molecule–metal interfaces. This crossover is a result of the competition between charge transfer and solvation, as the transfer always depends on the coupling of the excited electron to the substrate states, which is modified by the transient potential landscape resulting from electron solvation/localization. Such a generalized view on interfacial electron transfer would exhibit two *limiting cases* for (i) electron dynamics dominated by charge transfer due to strong electronic coupling to the substrate combined with weak screening by solvation and (ii) vice versa. In the former case, the electron population decay would be much more rapid than any polarization or even molecular reorientation of the solvent, and the lifetime of the excess electron would be accordingly short. In the latter case of very efficient screening of the excited electron, that is, solvation and strong localization dominating the competition, the electron transfer would vanish, giving rise to infinite[6)] electron lifetimes. This limiting case will be subject of Section 16.3 where we show dramatic suppression of electron transfer at crystalline ice structures leading to extraordinarily long lifetimes.

16.3
Ultrafast Trapping and Ultraslow Stabilization of Electrons in Crystalline Solvents

All investigations of the electron dynamics in *amorphous* NH_3 and D_2O ice structures at metal surfaces suggest a general scenario that is governed by the competition between the localization and subsequent solvation of the excess charge and the electron transfer back to the metal substrate. The resulting dynamics observed for various amorphous structures and different adsorbates occur on timescales of femtoseconds up to several 10–100 ps. This raises the question if pronounced structural changes, like the transition from amorphous to crystalline ice, would lead to similar observations, or if entirely new processes would occur. In particular, it is *a priori* not even clear whether electrons can be solvated at all in the crystalline phase, as it is well known that ions can be solvated in *liquid water* but not in *crystalline ice* [32]. On the other hand, molecular dynamics studies suggest that the surface of crystalline ice is disordered, which might provide trapping sites for excess electrons at the surface [33, 34]. These structural differences should have a profound influence on the dynamics of excess electrons at the crystalline ice/metal interface, which has been studied for D_2O on Ru(001).

16.3.1
Crystalline Ice on Ru(001)

Crystalline D_2O on Ru(001) is prepared by adsorption at 150 K followed by sample annealing to 162 K and concomitant observation of the isothermal desorption yield.

6) If thermal fluctuations of the solvent are excluded.

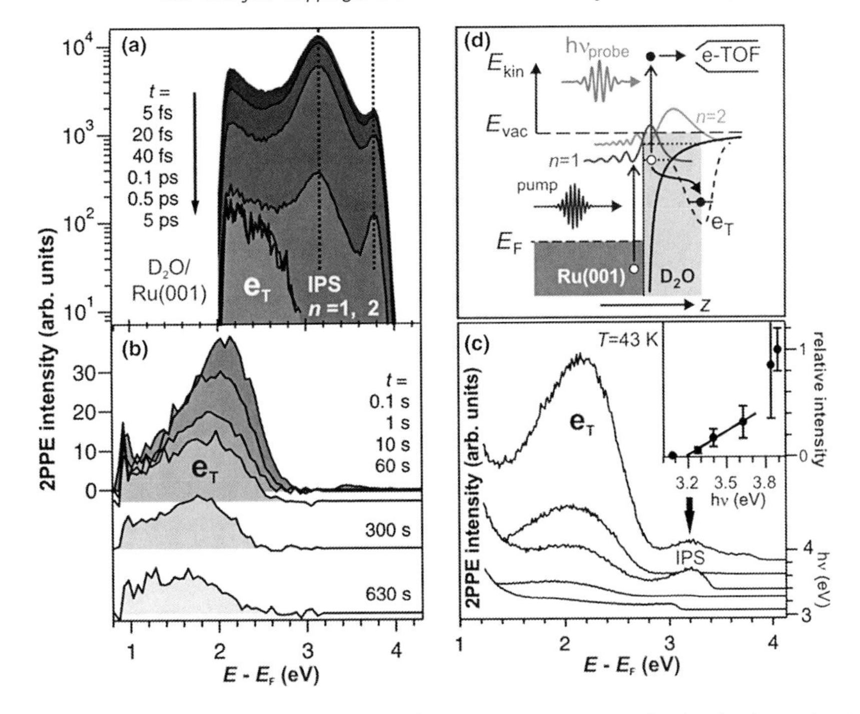

Figure 16.5 Time-resolved 2PPE spectra of D_2O/Ru(001) at 43 K on timescales of (a) femtosecond to picosecond and (b) seconds. The peak e_T at energies below the $n = 1$ IPS persists for minutes. (c) Variation of the pump photon energy unveils injection of excess electrons via the delocalized IPS. Intensities are normalized to the photon density in the laser pulses. (d) Fundamental processes of electron trapping. Localization in deep traps after injection via $n = 1$ IPS. Reprinted with permission from [36]. Copyright 2009 American Chemical Society.

At a nominal coverage of 2–4 BL, crystallites form on top of the wetting first layer [35]. Figure 16.5a depicts 2PPE spectra of 2 BL D_2O on the Ru(001) surface for different pump–probe time delays up to 5 ps. Besides the hot electron distribution at the low-energy cutoff ($E - E_F = 2$ eV), the data exhibit contributions from the first two image potential states (IPS) ($n = 1, 2$) and a peak, termed e_T, ~2 eV above the Fermi level that has a significantly lower intensity (note the logarithmic axis). Due to the time-dependent background of hot electrons and the contribution from the IPS, this signal is only visible after 0.5 ps, because it is as small as about 1% of the IPS intensity at $t = 0$ fs, but then remains unchanged up to 5 ps. Closer investigation of the dynamics of the feature e_T furthermore unveils that it persists even much longer: Figure 16.5b presents spectra of e_T at time delays on the timescale of *seconds*[7] [36] that show an excited electronic state that persists for minutes. Compared to the $n = 1$ IPS at the D_2O/Ru(001) interface, which is delocalized parallel to the surface and decays with a time constant <10 fs, these dynamics are slowed down by 17 orders of magnitude

7) The spectra are recorded making use of the formation of a photostationary state between excitation and depopulation of e_T. For details, see Ref. [36].

in time. Concomitantly, the peak shifts down toward the Fermi level, even on minute timescales.

In the following, we characterize the properties of the exceptionally long-lived electrons e_T with respect to their excitation, localization, relaxation, and stabilization. As the time-dependent background in Figure 16.5a hampers the analysis of the excitation dynamics, we elucidate the excitation pathway by the variation of the pump (and probe) photon energy. Note that the extraordinarily long lifetime of the trapped electrons exceeds the laser pulse separation of our system (5 μs at 200 kHz) by far. The trapped electrons can thus be excited with one laser pulse and probed by a second, subsequent one. These processes of population and depopulation of e_T lead to the formation of a photostationary state that is characterized by the equilibrium of photoexcitation and probing of the trapped electron population.

Figure 16.5c depicts 2PPE spectra of e_T for various excitation photon energies as indicated by the right axis. The spectra are normalized to the number of photons per pulse and analyzed according to the intensity of the e_T feature. This analysis is presented in the inset of Figure 16.5c and reveals that the intensity of e_T decreases with decreasing photon energy and vanishes for $h\nu \leq 3.2$ eV. This threshold coincides with the energy of the IPS (3.15 eV) with respect to the Fermi level E_F. Hence, we conclude that the e_T feature is excited through the delocalized image potential state.

On the first sight, the initial electron dynamics appear to be similar for crystalline and amorphous ice on metal surfaces: The charge injection into the adsorbate occurs via a state that is delocalized parallel to the surface. This is followed by localization of the excess charge and a peak shift toward the Fermi level. The crucial difference is, however, the extraordinarily long residence time in the case of crystalline D_2O. To get to the bottom of this phenomenon, it is important to consider the following observations: (i) The trapped electron state is populated through the $n = 1$ image potential state that exhibits a lifetime below 10 fs, indicating that electron localization occurs on the same timescale. (ii) Figure 16.5a shows e_T at 2.3 eV after 500 fs, which documents that 850 meV (energy difference to the IPS at 3.15 eV) have been dissipated since photoexcitation. (iii) The intensity of e_T remains nearly constant between 0.5 and 5 ps, showing that the major difference in molecular screening of crystalline and amorphous D_2O ice must have occurred for $t < 500$ fs (compare Figure 16.1). Subsuming these three characteristics, we conclude the following scenario for electron localization, which is illustrated by Figure 16.5d. In the case of crystalline D_2O, electrons localize after charge injection through the $n = 1$ image potential state on ultrafast timescales (i) in deep traps that rapidly screen the excess charge from the metal (ii). The high potential barrier (cf. Figure 16.5d) efficiently suppresses the population decay to the substrate (iii), enabling the observation of further relaxation even on minute timescales.

The above finding of ultrafast localization in a deep trap in the case of crystalline ice on Ru(001) has considerable relevance for understanding the ultralong persistence of e_T. The rapidness of the electron localization combined with the large energy loss indicates that electron trapping occurs at preexisting sites. Furthermore, we have shown by Xe titration experiments, as introduced before, that these traps must

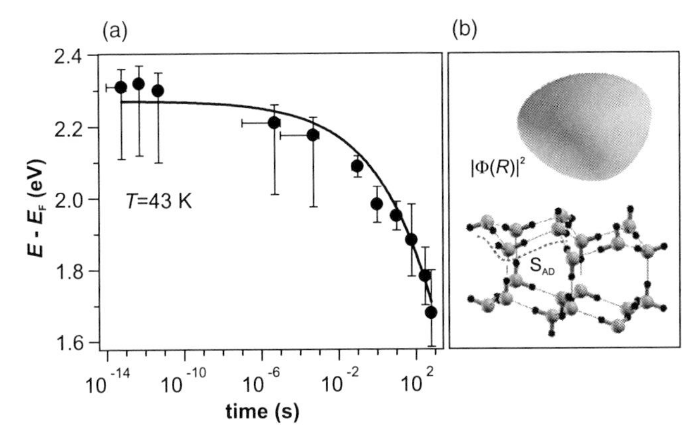

Figure 16.6 (a) Shift of the e_T peak maximum over 17 orders of magnitude in time. (b) Calculated orientational defect structure S_{AD} at the ice surface and corresponding density of the trapped electron at a contour line of 70%. Reprinted with permission from [36]. Copyright 2009 American Chemical Society.

be located at the ice–vacuum interface (see Ref. [36]). *Ab initio* density functional theory (DFT) calculations of Bockstedte and coworkers [36] suggest orientational defects at the ice surface as trapping sites: Figure 16.6b displays exemplarily the electron density localized above such an orientational defect (S_{AD}), which arises from reorientation of surface molecules and the reestablished hydrogen-bonded network in the surrounding. The energetic stabilization is mediated by sequential population of such conformational substates and proceeds over 17 orders of magnitude in time (see Figure 16.6a). The screening of the excess charge in front of the surface by such molecular rearrangements leads to substantial drop of the electron density $|\Phi(R)|^2$ within the ice layer and hence an almost vanishing tunneling rate to the unoccupied states of the metal substrate [36]. This strong localization of the electrons in a deep trap in front of the surface explains their extraordinary long lifetime in front of the metal surface. Such deep traps are either not formed for amorphous ice or are rapidly depopulated due to competing relaxation channels in amorphous ice layers, where such long-lived electrons have not been observed.

It is noteworthy that – even though the lifetime of excess electrons in crystalline D_2O exceeds the one of amorphous ice by far – the electron dynamics can be described by the general scenario for electron dynamics developed in Section 16.2: The competition between electron transfer and electron solvation/localization, which gave rise to the transition from the substrate-dominated to the barrier-determined regime of charge transfer for amorphous solvents, is also valid for the long-lived electrons at the vacuum interface of crystalline D_2O/Ru(001). The different dynamics result from the strongly differing initial conditions of electron trapping; while the evolution of the interfacial barrier can be monitored in the case of amorphous solvents, electron localization occurs in preexisting, deep traps at crystalline ice surfaces that rapidly screen the excess charge once it is trapped. This process goes along with ultrafast energy dissipation of 850 meV, which may be mediated by electron–hole pair formation in the metal. As soon as it is localized in the preexisting trap, it is very efficiently screened and the rate of back transfer is strongly reduced by the interfacial barrier. It can hence be concluded that the electron population decay of trapped electrons at the crystalline ice–vacuum interface represents one of the limiting cases of the general scenario of interfacial electron transfer suggested earlier in this work: Extremely efficient screening of the excess charge wins the competition with electron transfer leading to extraordinarily long electron lifetimes.

16.3.2
Reactivity of Trapped Electrons on Ice

Localized electrons in front of an ice surface, which exhibit long lifetimes in their excited state, can interact with reactive species impinging from the gas phase. It is well known that electrons with excess energy above the vacuum level can induce chemical reactions, for example, via dissociative electron attachment [37]. The long-lived, trapped electrons on the ice surface allow elucidating the reactivity of excited electrons below the vacuum energy at the ice vacuum interface. One potential reaction in this context is the dissociation of chlorofluorocarbons in stratospheric clouds that plays a key role in ozone layer depletion.

As a test of the reactivity of the trapped electrons on crystalline ice surfaces, we therefore chose the DEA of $CFCl_3$. The inset in Figure 16.7 illustrates the principle of the experiment: Trapped electrons at a crystalline ice–vacuum interface are prepared by illumination of the sample with UV photons. At the same time, $CFCl_3$ molecules impinge the sample. The resulting 2PPE spectra for increasing $CFCl_3$ exposure

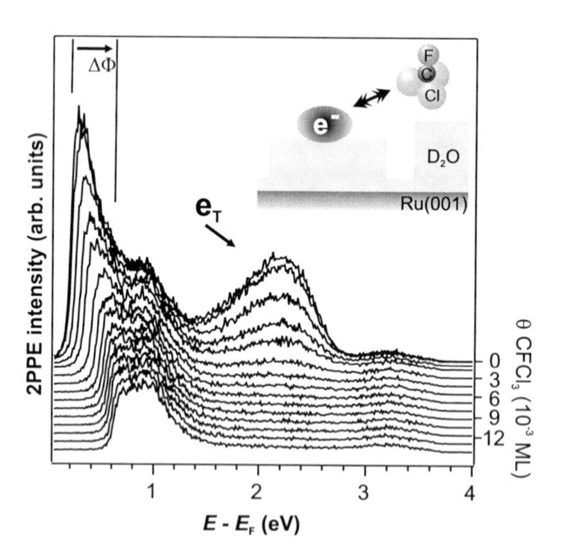

Figure 16.7 Photoelectron spectra of crystalline ice on Ru(001) as a function of $CFCl_3$ exposure Θ (right axis). The population of the trapped electrons e_T is quenched after adsorption of $\Theta = 4 \times 10^{-3}$ ML, which is accompanied by a work function increase $\Delta\Theta = 250$ meV. Adapted from [38]. Reproduced by permission of the PCCP Owner Societies.

(right axis) are shown in Figure 16.7. Two consequences of the $CFCl_3$ exposure are observed: (i) The signal of the trapped electrons e_T is quenched, demonstrating the interaction of the electrons with the reactant. (ii) The system's work function is significantly increased (by 250 meV) when adsorbing only a small amount of $CFCL_3$ (0.004 ML). This suggests a corresponding buildup of negative charge at the ice–vacuum interface (presumably due to formation of Cl^- ions) [38]. While the details of underlying chemical reactions need further investigations, the magnitude of the work function change at such low reactant density demonstrates the high cross section of this process. From a general point of view, these experiments suggest further studies of the role of excited electrons in surface chemical reactions and the influence of the electron lifetimes and energetics on their reactivity.

16.4 Conclusion

In the present chapter, we have portrayed the fundamental aspects of the electron dynamics at a variety of polar molecule–metal interfaces. In this context, one of the main challenges was the derivation *and* separation of the four elementary processes, electron injection, localization, solvation, and transfer, and the determination of their respective receptiveness to systematic variation of the experimental parameters. The investigated interfaces comprise ultrathin amorphous and crystalline D_2O and NH_3 on Cu(111) and Ru(001). All these systems have in common that charge injection proceeds by excitation of metal electrons and charge injection via electronic states of the adlayer, which are delocalized parallel to the surface. The electron localization occurs at preexisting potential minima on ultrafast timescales and is promoted by molecular reorganization due to the presence of the excess charge. This process is accompanied by an energetic stabilization of the electrons and a continuous back transfer of the excited-state population to the unoccupied states of the metallic substrate. The above comparability of the investigated interfaces allows the following conclusions:

1) Electron solvation and transfer back to the substrate *compete*: While the electron transfer continuously reduces the excited-state population, the solvation is characterized by the transient evolution of a potential barrier that increasingly suppresses the population decay to the substrate.
2) The competition between solvation and transfer leads to the observation of two regimes of charge transfer: Electron transfer prevails in the competition directly after the creation of the excess electron and the population decay is *substrate-dominated*. Charge separation from the metal substrate occurs when solvation becomes stronger in the competition. Then, the tunneling barrier forms at the interface and electron transfer is *barrier-determined*.
3) The initial trapping of the excess charge determines the balance between transfer and solvation and therefore influences the subsequent electron dynamics: The extremely long lifetime of excess electrons at the crystalline ice–vacuum inter-

face would not be possible without the ultrafast and efficient localization of electrons in deep traps right after excitation.

The preceding three key results present a comprehensive view of the electron dynamics at polar molecule–metal interfaces. They illustrate the equilibration pathway of an excess electron in a solvent environment that is subject to the balance of two competing relaxation channels that depend crucially on the initial conditions as, for instance, the structure of the adlayer. Such studies of electron transfer processes in model systems as discussed in this work provide general insights into fundamental aspects of charge transfer and carrier dynamics also in molecular systems relevant to applications, for example, in organic electronics. Besides the ample relevance for electronic processes in materials, however, the detailed understanding of the specific processes also enables the direct application to low-energy electron-driven chemical reactions. The observed reactivity of the highly localized electrons at a molecule–vacuum interface might also be of relevance in atmospheric chemistry. Understanding and controlling such processes by modification of electron lifetimes, localization, and density is a challenging goal and will be subject of future investigations.

Acknowledgments

This study was performed in collaboration with M. Bockstedte, M. Bertin, C. Gahl, T. Klamroth, D.O. Kusmierek, M. Meyer, A. Rubio, and P. Saalfrank. We would like to thank K. Morgenstern, H. Petek, X.-Y. Zhu, and A. Nitzan for inspiring discussions. The work was funded by the Deutsche Forschungsgemeinschaft through SPP 1093 and Sfb 450 as well as the German–Israeli Foundation (GIF).

References

1 Nitzan, A. (2006) *Chemical Dynamics in Condensed Phases*, Oxford University Press.
2 Madey, T.E., Johnson, R.E., and Orlando, T.M. (2002) *Surf. Sci.*, **500**, 838.
3 Garrett, B.C. *et al.* (2005) *Chem. Rev.*, **105**, 355.
4 Miller, R.J.D., McLendon, G.L., Nozik, A.J., Schmickler, W., and Willig, F. (1995) *Surface Electron-Transfer Processes*, Wiley-VCH Verlag GmbH, New York.
5 Marcus, R.A. (1964) *Annu. Rev. Phys. Chem.*, **15**, 155.
6 Fleming, G.R. and Cho, M.H. (1996) *Annu. Rev. Phys. Chem.*, **47**, 109.
7 Laenen, R., Roth, T., and Lauberau, A. (2000) *Phys. Rev. Lett.*, **85**, 50.
8 Szymanski, P., Garrett-Roe, S., and Harris, C.B. (2005) *Prog. Surf. Sci.*, **78**, 1.
9 Nitzan, A. and Ratner, M.A. (2003) *Science*, **300**, 1384.
10 Zhu, X.-Y. (2004) *J. Phys. Chem. B*, **108**, 8778.
11 Frischkorn, C. and Wolf, M. (2006) *Chem. Rev.*, **106**, 4207.
12 Perry, C.C., Faradzhev, N.S., Madey, T.E., and Fairbrother, D.H. (2007) *J. Chem. Phys.*, **126**, 204701.
13 Ge, N.-H., Wong, C.M., Lingle, R.L., Jr., McNeill, J.D., Gaffney, K.J., and Harris, C.B. (1998) *Science*, **279**, 202.
14 Ge, N.-H., Wong, C.M., and Harris, C.B. (2000) *Acc. Chem. Res.*, **33**, 111.

15 Miller, A.D., Bezel, I., Gaffney, K.J., Garrett-Roe, S., Liu, S.H., Szymanski, P., and Harris, C.B. (2002) *Science*, **297**, 1163.

16 Zhao, J., Li, B., Onda, K., Feng, M., and Petek, H. (2006) *Chem. Rev.*, **106**, 4402.

17 Stähler, J., Gahl, C., Bovensiepen, U., and Wolf, M. (2006) *J. Phys. Chem. B*, **110**, 9637.

18 Stähler, J. (2007) Freezing hot electrons: electron transfer and solvation dynamics at D_2O– and NH_3–metal interfaces. Dissertation. Freie Universität Berlin. http://www.diss.fu-berlin.de/diss/receive/FUDISS_thesis_000000002622.

19 Bovensiepen, U., Gahl, C., and Wolf, M. (2003) *J. Phys. Chem. B*, **107**, 8706.

20 Shibaguchi, T., Onuki, H., and Onaka, R. (1977) *J. Physical Soc. Japan*, **42**, 152;E_{gap} taken at a threshold of 50%.

21 Smith, N.V. (1985) *Phys. Rev. B Condens. Matter*, **32**, 3549.

22 Nordlund, D., Ogasawara, H., Bluhm, H., Takahashi, O., Odelius, M., Nagasono, M., Pettersson, L.G.M., and Nilsson, A. (2007) *Phys. Rev. Lett.*, **99**, 217406.

23 Kimura, Y., Alfano, J.C., Walhout, P.K., and Barbara, P.F. (1994) *J. Phys. Chem.*, **98**, 3450.

24 Emde, M.F., Baltuška, A., Kummrow, A., Pshenichnikov, M.S., and Wiersma, D.A. (1998) *Phys. Rev. Lett.*, **80** (21), 4645.

25 Stähler, J., Meyer, M., Zhu, X.-Y., Bovensiepen, U., and Wolf, M. (2007) *New J. Phys.*, **9**, 394.

26 Stähler, J., Bovensiepen, U., Meyer, M., and Wolf, M., in preparation.

27 Gahl, C., Bovensiepen, U., Frischkorn, C., Morgenstern, K., Rieder, K.-H., and Wolf, M. (2003) *Surf. Sci.*, **532**, 108.

28 Mehlhorn, M., Gawronski, H., and Morgenstern, K. (2008) *Phys. Rev. Lett.*, **101**, 196101.

29 Meyer, M., Stähler, J., Kusmierek, D.O., Wolf, M., and Bovensiepen, U. (2008) *Phys. Chem. Chem. Phys.*, **10**, 4932.

30 Stähler, J., Mehlhorn, M., Bovensiepen, U., Meyer, M., Kusmierek, D.O., Morgenstern, K., and Wolf, M. (2007) *Phys. Rev. Lett.*, **98**, 206105.

31 Stähler, J., Meyer, M., Kusmierek, D.O., Bovensiepen, U., and Wolf, M. (2008) *J. Am. Chem. Soc.*, **130** (27), 8797.

32 Petrenko, V.F. and Whitworth, R.W. (2002) *Physics of Ice*, Oxford University Press.

33 Devlin, J.P. and Buch, V. (1995) *J. Chem. Phys.*, **99**, 16534.

34 Furukawa, Y. and Nada, H. (1997) *J. Phys. Chem. B*, **101**, 6167.

35 Smith, R.S. and Kay, B.D. (1997) *Surf. Rev. Lett.*, **4**, 781.

36 Bovensiepen, U., Gahl, C., Stähler, J., Bockstedte, M., Meyer, M., Baletto, F., Scandolo, S., Zhu, X.-Y., Rubio, A., and Wolf, M. (2009) *J. Phys. Chem. C*, **113** (3), 979.

37 Klar, D., Ruf, M.W., Fabrikant, I., and Hotop, H. (2001) *J. Phys. B*, **34**, 3855.

38 Bertin, M., Meyer, M., Stähler, J., Gahl, C., Wolf, M., and Bovensiepen, U. (2009) *Faraday Discuss.*, **141**, 293.

39 Gahl, C., Bovensiepen, U., Frischkorn, C., and Wolf, M. (2002) *Phys. Rev. Lett.*, **89**, 107402.

Color Plates

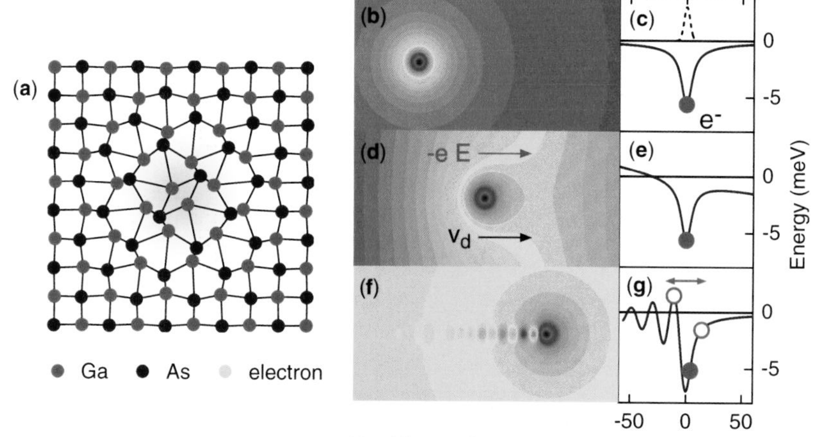

Figure 1.1 (a) Lattice distortion of the Fröhlich polaron in GaAs. Self-induced polaron potential (b, contour plot and solid line in (c)) and electron wave function (dashed line in (c)) of a polaron at rest. (d and e) Linear transport: For low applied electric fields, the total potential is the sum of the applied potential and the zero-field polaron potential. (f and g) Nonlinear transport: In a strong DC field (which has been subtracted from the potentials shown in (f) and (g)), the drifting electron (red dot) is displaced from the minimum of the LO phonon cloud and generates coherent phonon oscillations in its stern wave. As the amplitude of coherent LO phonons exceeds a certain threshold, the polaron potential eventually causes electron oscillations (shown as open circles) along the relative coordinate *r* on top of the drift motion of the entire quasiparticle. (This figure also appears on page 4.)

Dynamics at Solid State Surface and Interfaces Vol. 1: Current Developments
Edited by Uwe Bovensiepen, Hrvoje Petek, and Martin Wolf

ISBN: 978-3-527-40937-2

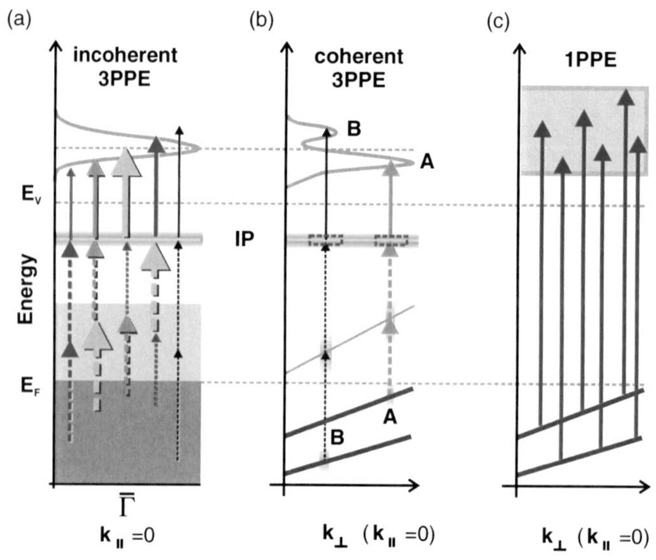

Figure 2.4 Principle of detecting coherent bulk band structure resonances in 3PPE via an image potential state on Cu(001). (a) Nonresonant incoherent excitation of the IP state from unspecific bulk bands at normal emission ($k_{||} = 0$). The initial states are averaged over $k_{\perp}$ illustrating the loss of $k_{\perp}$ information in the sequential one-photon pathways due to the lack of a $k_{\perp}$ selection rule in the IP to final-state one-photon transition ($k_{\perp}$ is undefined for surface states). The colors of the arrows correspond to photon frequencies provided by the excitation pulse, their thickness corresponds to the respective spectral weight. The photons can act in any combination for sequential incoherent excitations. The peak position of the IP state is determined by the central photon energy of the pulse (thickest arrow) and the measured width of the IP state is determined by the width of the pulse spectrum. (b) Coherent 3PPE resonances in the bulk band structure coupled to the image potential state produce specific peaks A and B in the spectrum, providing $k_{\perp}$ information. A fixed photon energy defines each resonant pathway. The width of features A and B is not limited by the spectral pulse width. (c) The corresponding 1PPE process for the same level scheme using a single-photon energy three times as large as that in the corresponding 3PPE process is not k-selective if no resonance to a dispersing final-state band is present. (This figure also appears on page 43.)

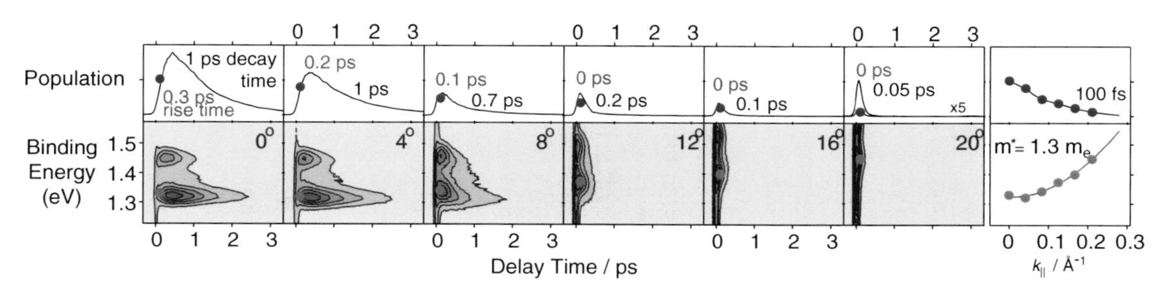

Figure 4.5 Angle-resolved dynamic scans of 3 ML of *p*-xylene/Ag(111). Dynamic population traces for each angle are given in the top panels, and contour plots of the kinetic energy versus time for each angle are given in the bottom panels. The bottom right panel shows the dispersion and effective mass that remain constant at all times, and the top right panel shows the 100 fs population distribution versus $k_{//}$. The red dots correspond to the main peak's center energy and the blue dots are the population after 100 fs. Reproduced with permission from [24]. (This figure also appears on page 85.)

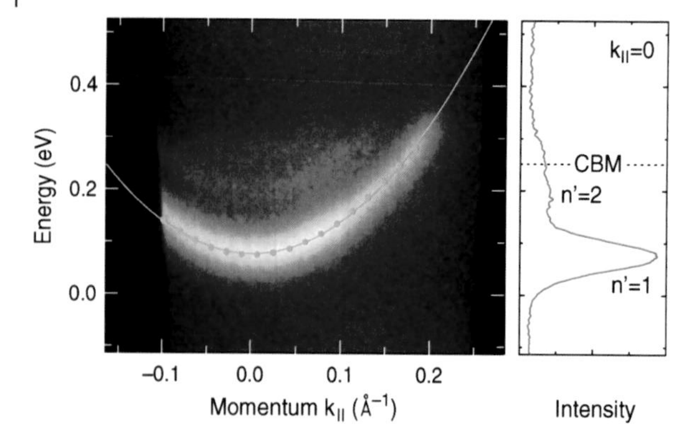

Figure 5.7 (a) $E(k_{\|})$ 2PPE spectrum for 25 ML Ar/Cu(100) acquired with a display-style electron analyzer. The dots along the dispersion parabola of the $n' = 1$ interface state mark the points used for evaluation of the time-resolved pump–probe traces. The faint parabola of the $n' = 2$ state is also visible. (b) 2PPE energy spectrum for $k_{\|} = 0$ obtained from a cut through the 2d spectrum with a width of about 0.018 Å^{-1}. Reproduced from Ref. [29]. Copyright 2005, IOP Publishing Ltd. and Deutsche Physikalische Gesellschaft. (This figure also appears on page 109.)

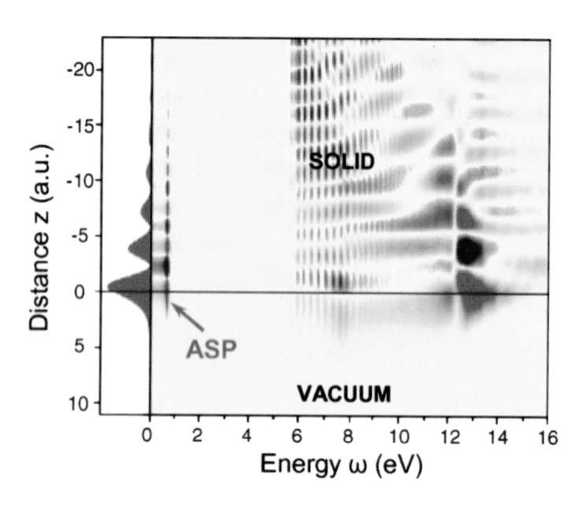

Figure 8.10 2D plot of the imaginary part of the induced charge density, Im $[n_{\mathrm{ind}}(z, q_{\|}, \omega)]$ at $q_{\|} = 0.05$ a.u.$^{-1}$ versus energy ω and coordinate z, at the Be(0001) surface, as obtained from Eq. (8.8) in response to an external potential of Eq. (8.13). The solid (vacuum) is located at negative (positive) distances z. Positive (negative) values of Im $[n_{\mathrm{ind}}(z, q_{\|}, \omega)]$ are represented by red (blue) color. More intense color corresponds to larger values. The surface state charge density is shown on the left-hand side of the figure. The charge density oscillations corresponding to the acoustic surface plasmon at this momentum are marked by arrow. Oscillations at energies above approximately 5.8 eV are due to interband transitions. The strong oscillations with z features around energy of 13 eV correspond to the conventional surface plasmon that produce a peak "SP" in the surface loss function of Figure 8.4. Adapted from Ref. [38]. (This figure also appears on page 182.)

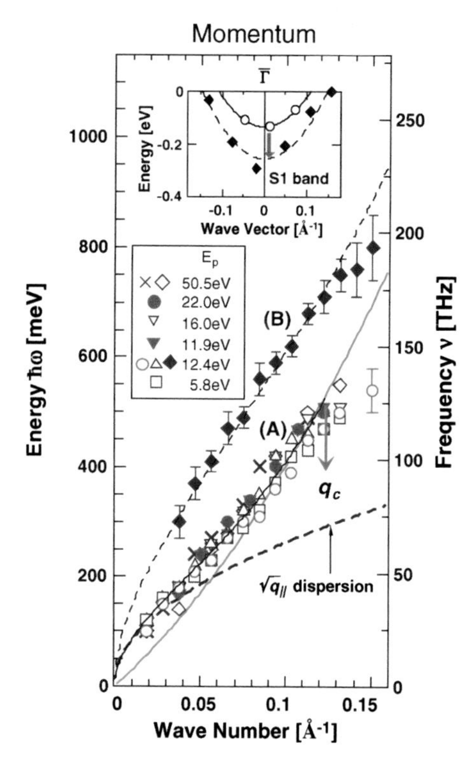

Figure 9.9 Plasmon dispersion curve determined by the momentum-resolved EELS spectra. Experimental curve (A) shows the data before the deposition of donor adatoms. The bold solid curve at curve (A) is the best fit to the nonanalytic full RPA dispersion by Stern [21]. Bold dashed curve below the curve (A) shows the $\sqrt{|\mathbf{q}|}$-type dispersion by Eq. (9.4). The thin solid curve just beneath the curve (A) is the upper edge of the single-particle excitation continua. Curve (B) is the data taken after 0.15 ML of donor Ag adatoms are deposited. Inset shows the "band sinkage" of this 2DES upon additional Ag deposition. (This figure also appears on page 199.)

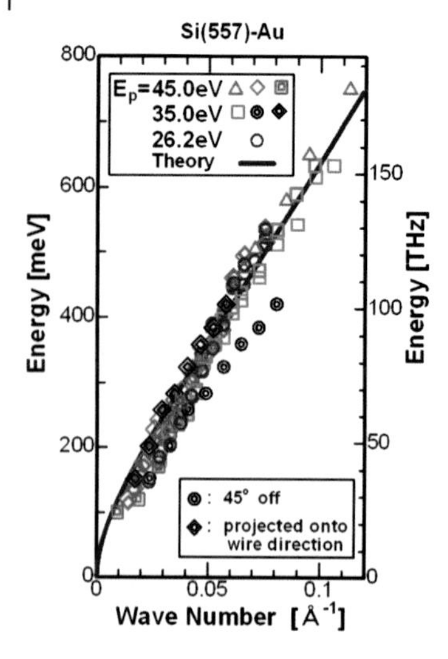

Figure 9.14 Energy dispersion curve determined from the peak positions in the momentum-resolved EELS spectra. Bold curve is the theoretical curve, including spin–orbit coupling (SOC) and exchange correlation effects. The precision in energy is roughly represented by the size of each data point. Below $q_{||} = 0.02\,\text{Å}^{-1}$, the precision is ± 30 meV due to the high tail from the quasielastic peak. Double circles and double diamonds are the data points taken at $\theta = 45°$ and plotted against $q = |\mathbf{q}|$ and against $q \cos(45°)$, respectively (θ is defined in Figure 9.11b, upper right). (This figure also appears on page 205.)

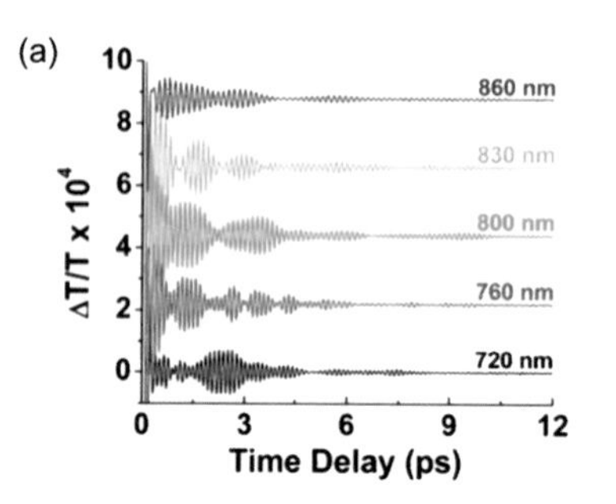

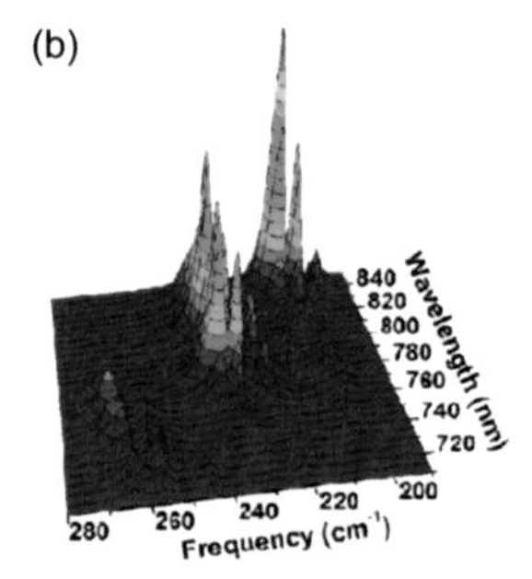

Figure 10.12 (a) Oscillatory part of the transient transmittance of SWNTs in a suspension excited and measured with 50 fs pulses at different photon energies. (b) A 3D plot of its fast FT obtained over a photon energy range of 1.746–1.459 eV (wavelength of 710–850 nm) (from Ref. [35]). (This figure also appears on page 225.)

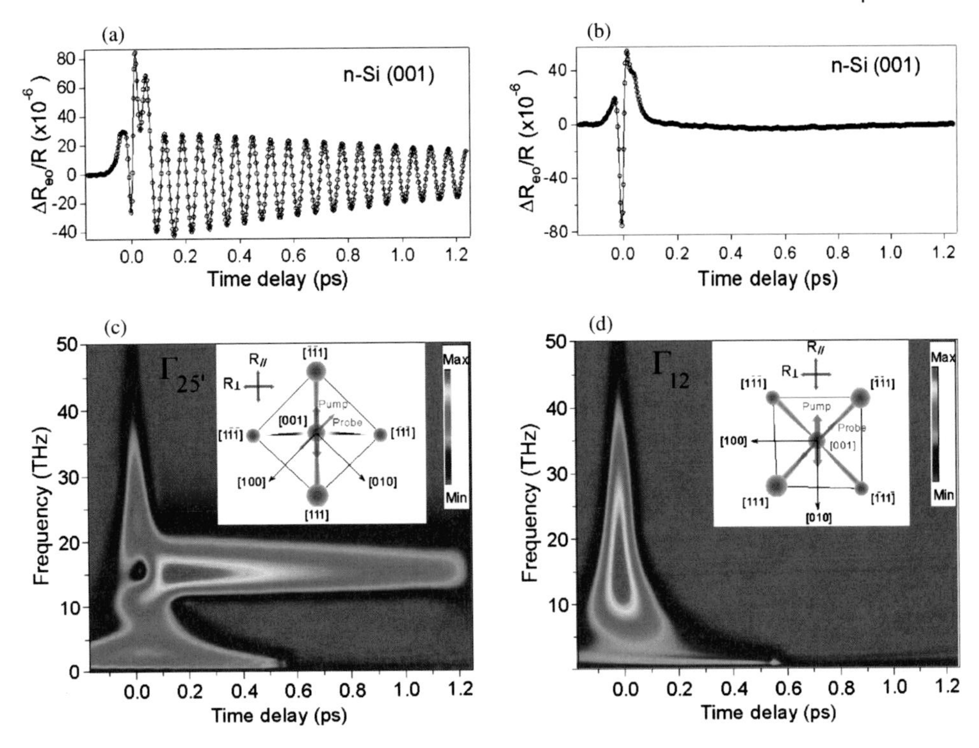

Figure 10.14 Transient anisotropic reflectivity change for Si(001) in $\Gamma_{25'}$ (a) and Γ_{12} (b) geometry. (c and d) Continuous wavelet transform of (a) and (b), respectively. Insets define the polarization of the pump and probe light relative to the crystalline axes (from Ref. [41]). (This figure also appears on page 226.)

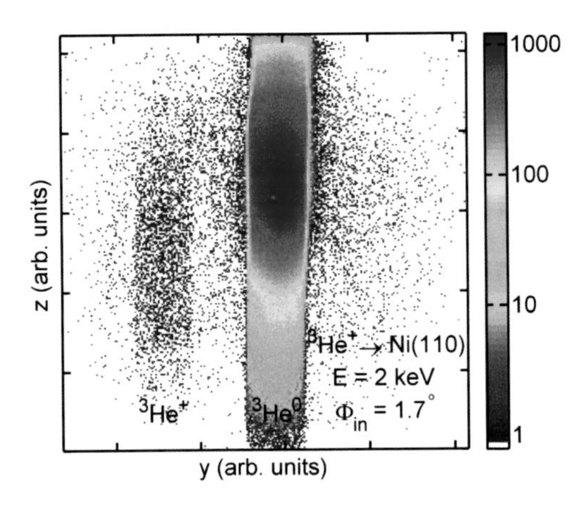

Figure 13.9 Intensity pattern (color scale) as recorded with position-sensitive microchannel plate detector for scattering of 2 keV $^3He^+$ ions from Ni(110) surface under $\Phi_{in} = 1.7°$. Reflected beam is dispersed normal to scattering plane by means of pair of electric field plates. *Intense bar in middle*: signal from neutral atoms; *faint bar on left-hand side*: signal from singly charged ions that have survived complete scattering event. (This figure also appears on page 295.)

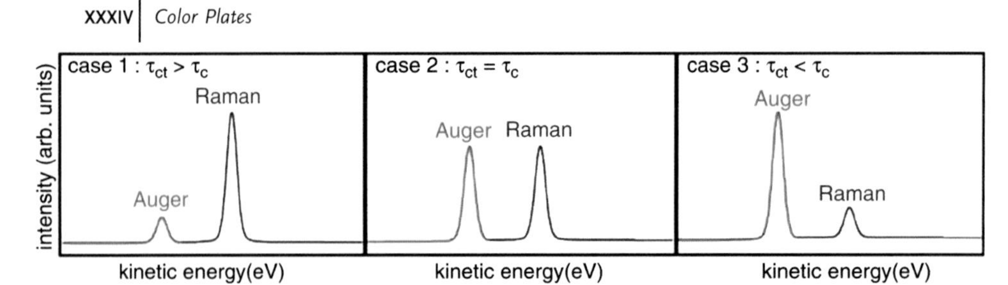

Figure 14.2 The cartoons show qualitatively the relative intensities for Raman and Auger channels for different relations of the charge transfer time τ_{CT} to the core hole lifetime τ_{Γ}. (This figure also appears on page 307.)

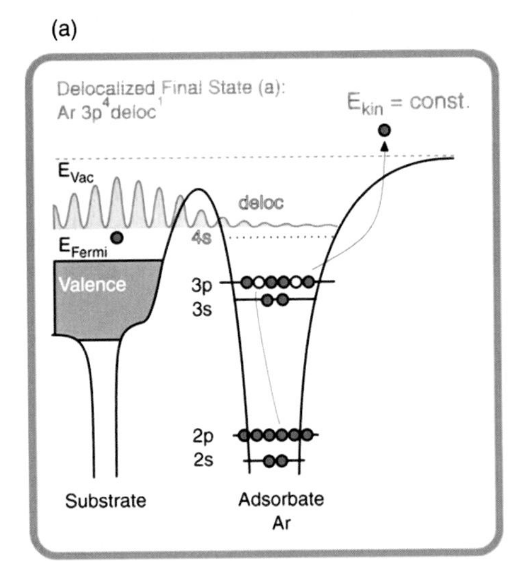

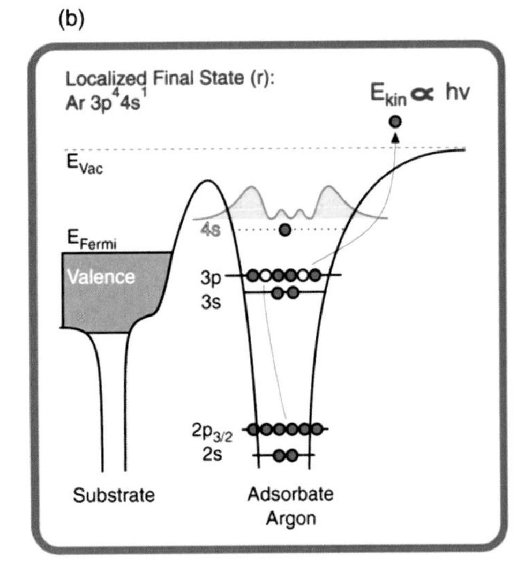

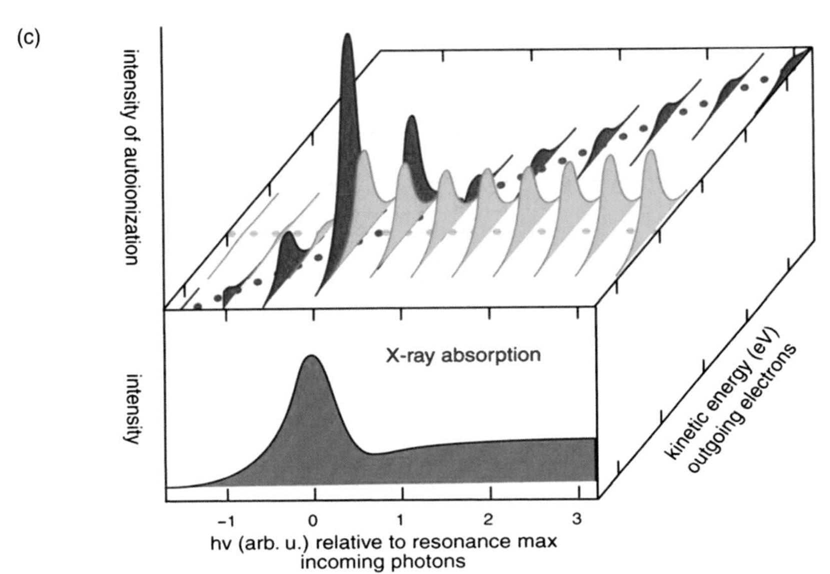

Figure 14.4 The figure (c) shows schematically the photon energy dependence of the kinetic energy of the outgoing electrons when the photon energy is tuned across the absorption resonance for a Raman channel where the excited electron is localized at the site of the excitation ((b), green) and for an Auger channel where the excited electron is delocalized ((a), red). (This figure also appears on page 308.)

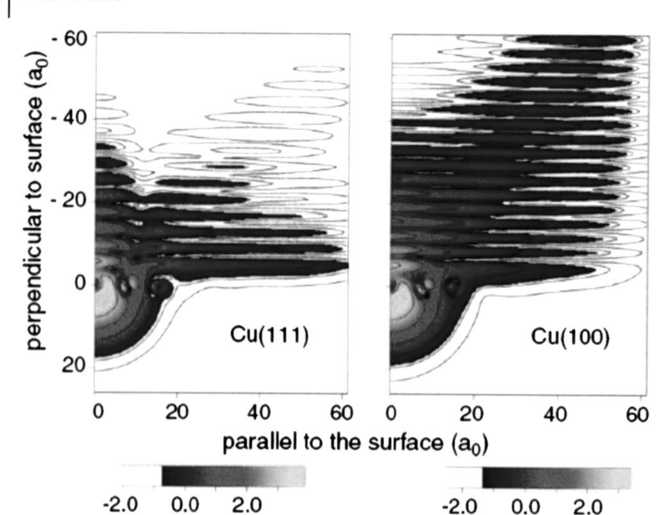

Figure 14.6 Contour plot of a cut of the resonant 4s wave packet in the Ar $2p_{3/2}^{-1}4s^{+1}$/Cu systems. The plot presents the logarithm of the electron density in cylindrical coordinates parallel (ϱ) and perpendicular (z) to the surface. The z-coordinate is positive in vacuum. The adsorbate is centered at the origin. The ϱ-axis is in a plane that contains one of the nearest Ar neighbors. The coordinates are in atomic units $a_0 = 0.529$ Å. The electron density decreases when going from yellow to dark red. White corresponds to very small densities. The contour lines are separated by a factor e (taken from Ref. [17]). (This figure also appears on page 310.)

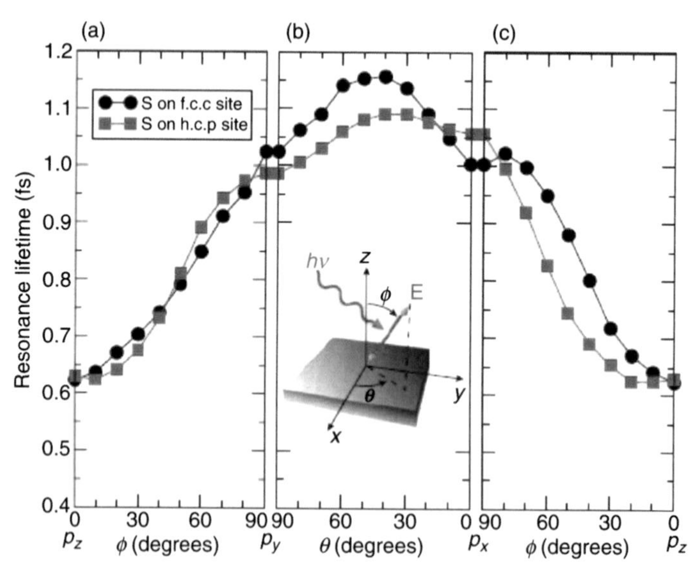

Figure 14.8 Calculated values for the electron transfer time from the 3p state of S on Ru(001) for two different S adsorption sites (fcc versus hcp) and different orientations of the 3p orbitals (taken from Ref. [19]). (This figure also appears on page 312.)

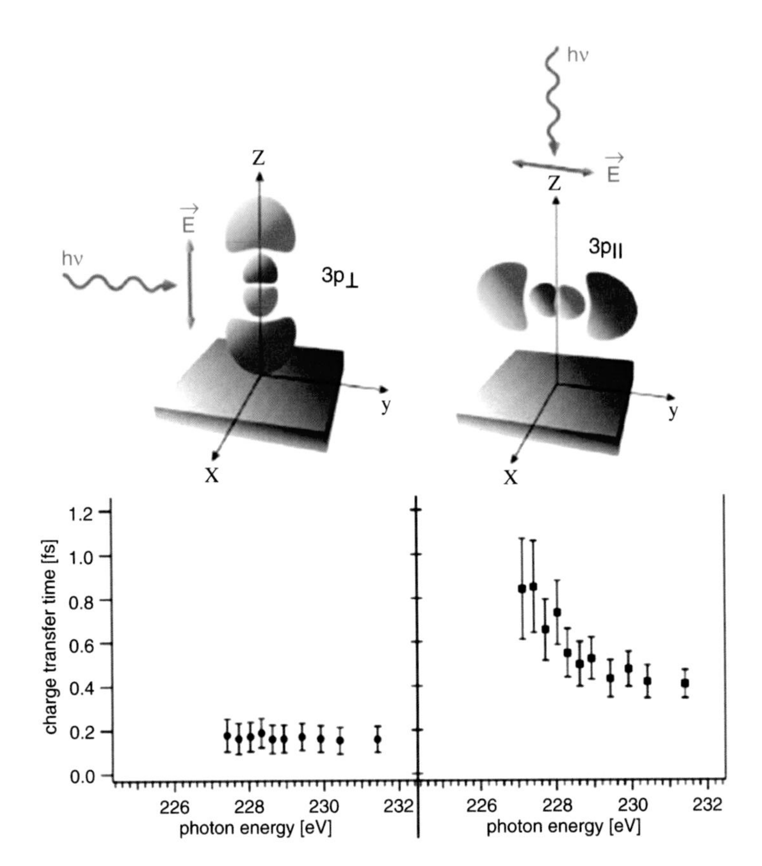

Figure 14.9 The figure shows the electron transfer time obtained for electrons excited to the 3p orbitals of S adsorbed on Ru(001). On the left side the polarization of the exciting photons is chosen such that the 3p orbital perpendicular to the surface is populated, while on the right side the 3p orbitals oriented parallel to the surface are selected in the excitation process. (This figure also appears on page 313.)

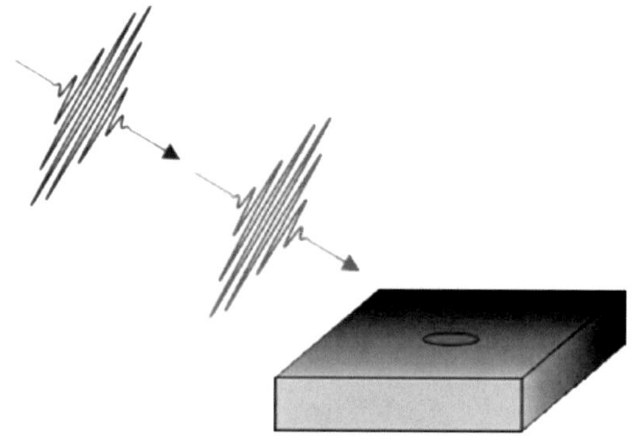

Figure 14.11 A two-color pump–probe experiment is schematically shown, where two pulses with different photon energies and a fixed delay in time impinge on a surface. In our experiment, the first optical pulse photoexcites the system, while the second soft X-ray pulse probes the system using core-level photoelectron spectroscopy. (This figure also appears on page 316.)

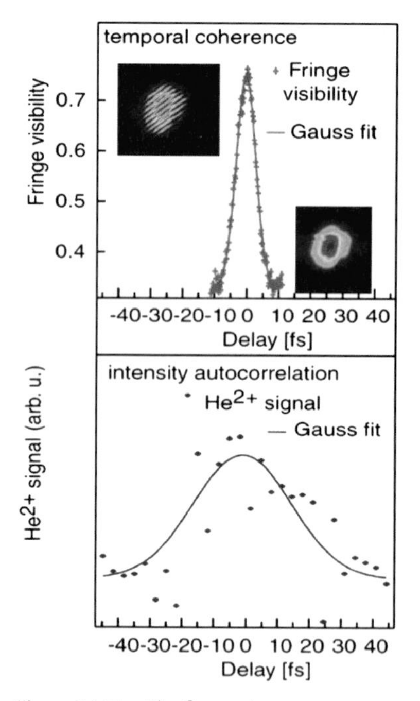

Figure 14.12 The figure shows a measurement of the temporal coherence (top) and the pulse length (bottom) of pulses from the free electron laser FLASH performed with our XUV split and delay unit based on a Mach–Zehnder interferometer [26]. The temporal coherence was obtained from the visibility of interference fringes recorded at 32 nm, while the intensity autocorrelation was obtained monitoring two-photon double ionization of He atoms at 47 eV. (This figure also appears on page 317.)

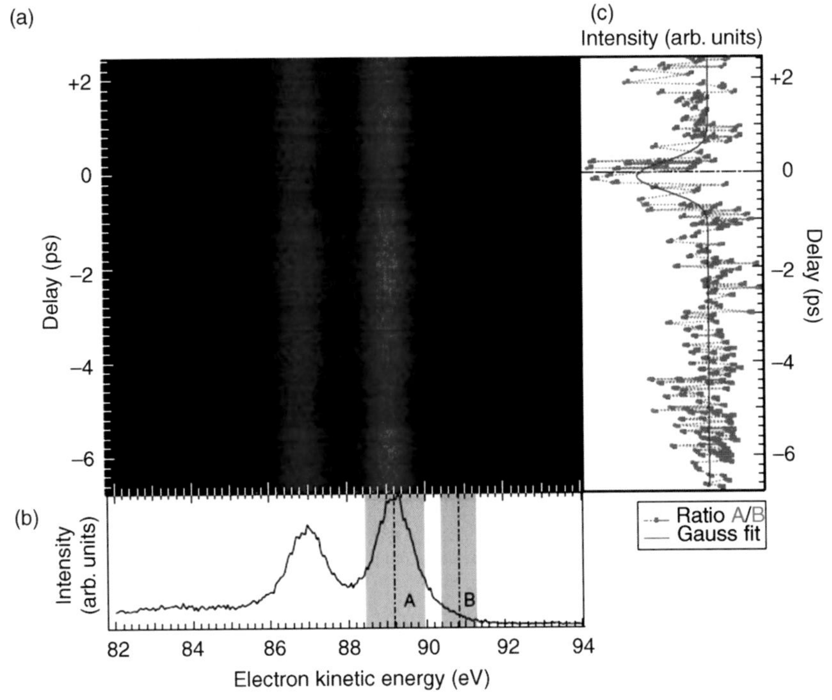

Figure 14.13 The figure shows a density plot of the spin–orbit split W4f photoemission lines as a function of the delay between an optical (800 nm $\hat{=}$ 1.55 eV) laser pulse and a XUV pulse (118.5 eV) from FLASH impinging on a W(110) surface. At the bottom, a horizontal cut showing a W4f photoemission spectrum is given where two regions of interest are highlighted. Region A includes the unperturbed $W4f_{7/2}$ line, while Region B is shifted to higher energies by the photon energy of the optical laser. To the right, the intensity ratio A/B is shown as a function of delay between the two pulses. Around zero delay, a deep can be seen with a width of about 400 fs indicative of the appearance of sidebands when both pulses are present (taken from Ref. [33]). (This figure also appears on page 319.)

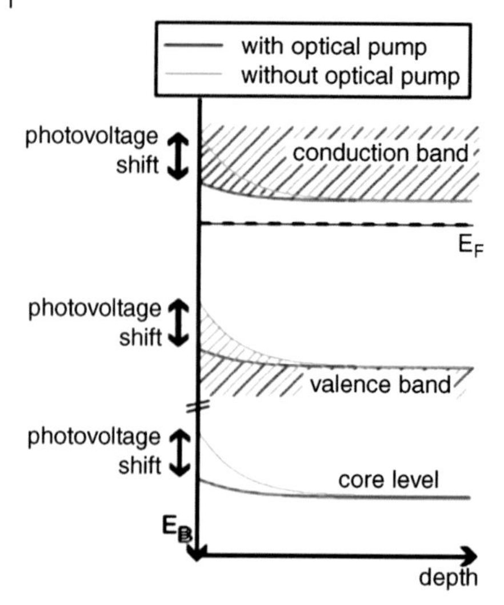

Figure 14.14 The figure shows schematically the transient change in surface potential, that is, the so-called surface photovoltage, induced by the creation of electron–hole pairs by an optical laser pulse. Note that as a result all electronic levels, including core levels, are shifted with respect to the Fermi level. (This figure also appears on page 320.)

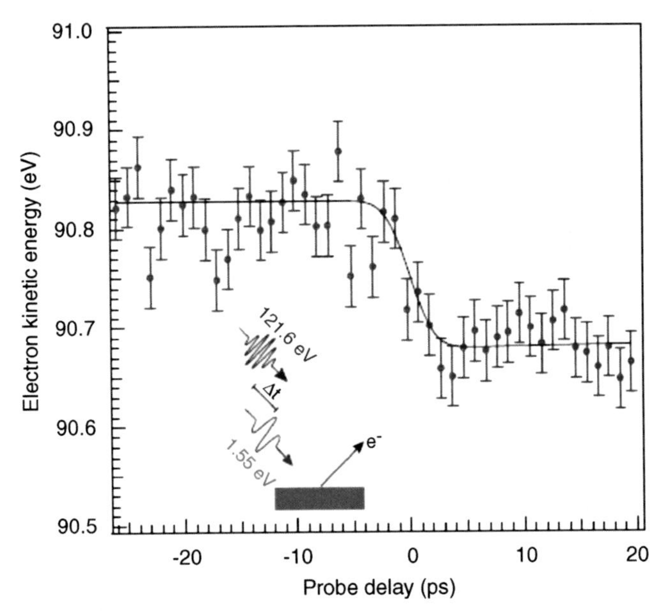

Figure 14.15 Femtosecond time-resolved ESCA of the Ge 3d photoemission lines of an n-doped Ge crystal using XUV pulses (photon energy 121.6 eV) from the free electron laser FLASH at DESY in Hamburg. The optical pump pulse (1.55 eV) creates mobile charge carriers within the surface depletion layer. The difference in electron and hole mobility leads to an ultrafast lifting of the band bending at the surface, which is probed through the binding energy shift of the photoemitted Ge 3d electrons relative to the Fermi level. (This figure also appears on page 321.)

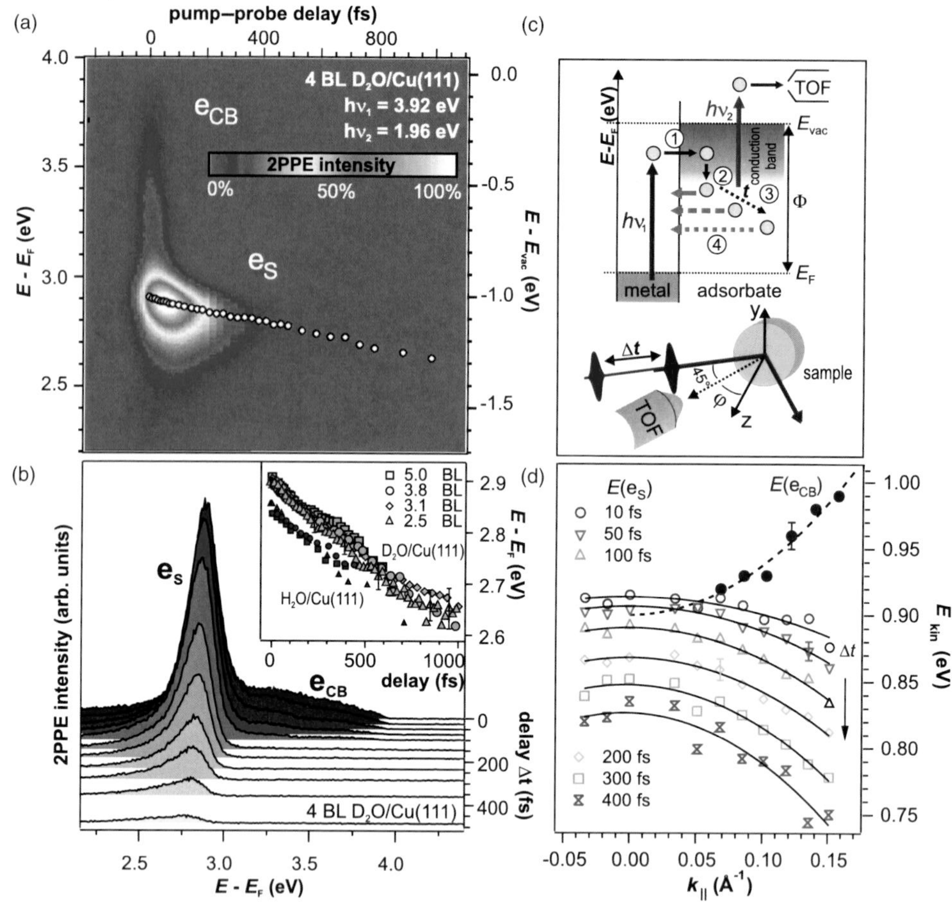

Figure 16.1 (a) Time-resolved 2PPE intensity of amorphous multilayers D_2O/Cu(111) in false color representation. The energy shift of the solvated electrons is indicated by white markers (modified from Ref. [19]). (b) 2PPE spectra of the same data set (modified from Ref. [39]). *Inset*: Energy shift of e_S for various coverages of H_2O and D_2O on Cu(111). (c) Time- and angle-resolved 2PPE scheme. (d) Dispersion of conduction band and solvated electrons at a series of time delays (modified from Ref. [19]). (This figure also appears on page 362.)

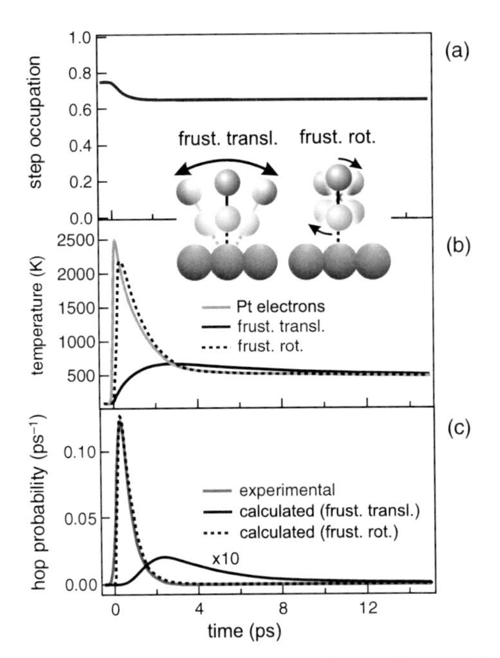

Figure 18.16 (a) Time-dependent fractional occupation of the step sites as a function of time after excitation. (b) Time-dependent electron temperature, as calculated by the two temperature model, along with the transient temperatures associated with the frustrated translation, and frustrated rotation. (c) Experimentally observed and calculated probability for a CO molecule to hop from a step to a terrace site as a function of time. The calculation assuming coupling to the frustrated translation is too slow and has too low an amplitude; the calculation assuming coupling to the frustrated translation works remarkably well. (This figure also appears on page 437.)

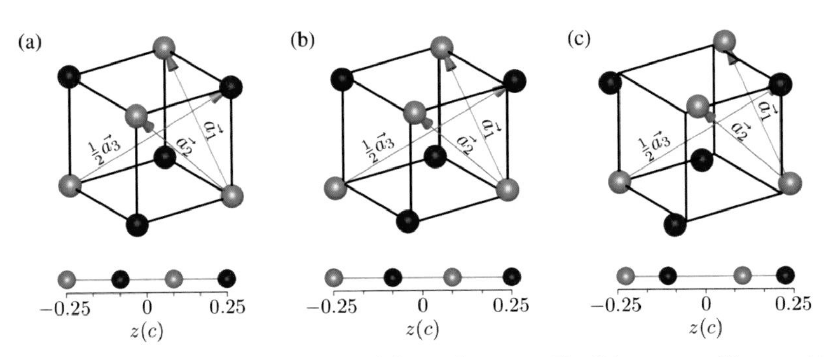

Figure 19.2 Relation between the A7 and the simple cubic structures. (a) The simple cubic structure. The lattice vectors $\mathbf{a}_1$, $\mathbf{a}_2$, and $\mathbf{a}_3$ belong to the A7 structure (the simple cubic structure is a special case of the A7 structure). (b) Intermediate structure. Compared to (a), the unit cell of the simple cubic lattice is elongated along the vector $\mathbf{a}_3$ keeping the volume per atom constant. The new positions of the atoms are shown together with a cube of the simple cubic lattice. (c) The A7 structure of As at ambient pressure. Compared to (b) the atomic planes perpendicular to $\mathbf{a}_3$ are displaced alternatingly in the $\mathbf{a}_3$ and $-\mathbf{a}_3$ directions due to a Peierls distortion. The cube still represents the simple cubic structure of (a), for reference. Below (a)–(c) projections of the atomic planes onto $\mathbf{a}_3$ are shown. (This figure also appears on page 459.)

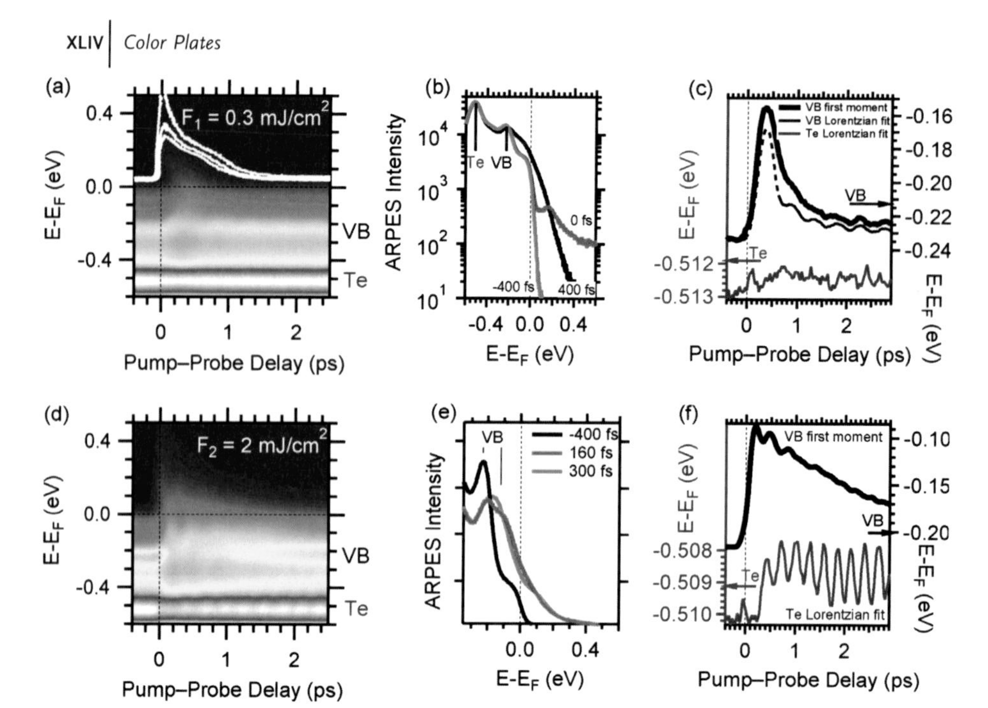

Figure 20.8 (a)/(d) Photoelectron intensity as a function of energy $E - E_F$ and pump–probe delay taken at $k = k_F$, $T = 100$ K, and an incident fluence of $F_1 = 0.3$ mJ/cm^2 (upper row) and $F_2 = 2$ mJ/cm^2 (lower row). The Te band and the VB are indicated. (b)/(e) Spectra from (a)/(d) for selected delays. (b) The peak at $E - E_F = 0.15$ eV in the unoccupied part of the band structure evidences the photodoping at zero delay. (e) The VB peak shifts and splits due to the intense excitation. (c)/(f) The transient binding energies of the Te band (left) and the VB (right) reveal oscillations. The Te and VB states were analyzed by Lorentzian line profiles and the first statistical moment for the VB. Modified from [25]. (This figure also appears on page 488.)

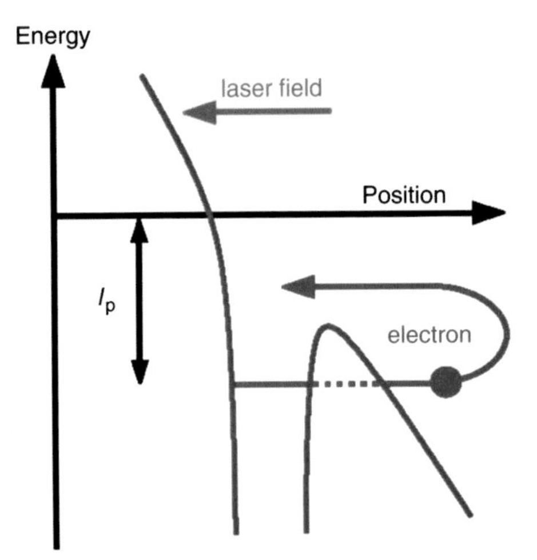

Figure 21.1 Illustration of the three-step model for high harmonic generation. An electron tunnels out with the help of a strong laser field suppressing the Coulomb potential inside the atom. The free electron is then accelerated by the laser field and can come back and recombine with the parent ion, emitting a high-energy photon. The photon energy is a sum of the ionization potential I_p and the electron's kinetic energy obtained in the laser field. (This figure also appears on page 503.)

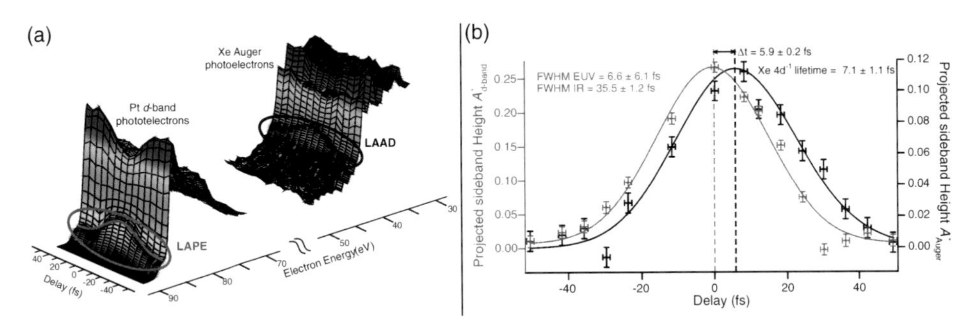

Figure 21.21 (a) Photoemission spectra from Xe/Pt(111) as a function of the time delay between the IR and the XUV pulse. The time evolution of the redistribution of electrons due to the IR dressing at the Pt d-band and at the Xe Auger photoelectron peaks can be observed. (b) The sideband height for the Pt d-band (gray curve) corresponds to a cross correlation between the XUV and the IR pulses. The Auger sideband height (black curve) is shifted and broadened with respect to the Pt d-band curve. A value of 7.1 ± 1.1 fs for the Xe 4d core-hole lifetime was extracted using these data [116]. (This figure also appears on page 528.)

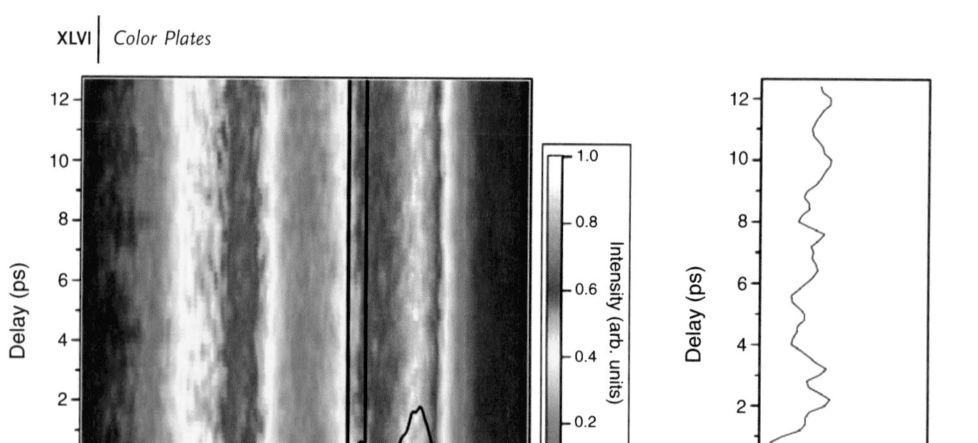

Figure 21.22 (a) Time-resolved Ta 4f core-level spectra as a function of the pump–probe delay between −5 and + 10 ps. (b) Integrated intensity at a CDW-split core level (black box in Figure 21.2a] as a function of the pump–probe delay. (This figure also appears on page 529.)

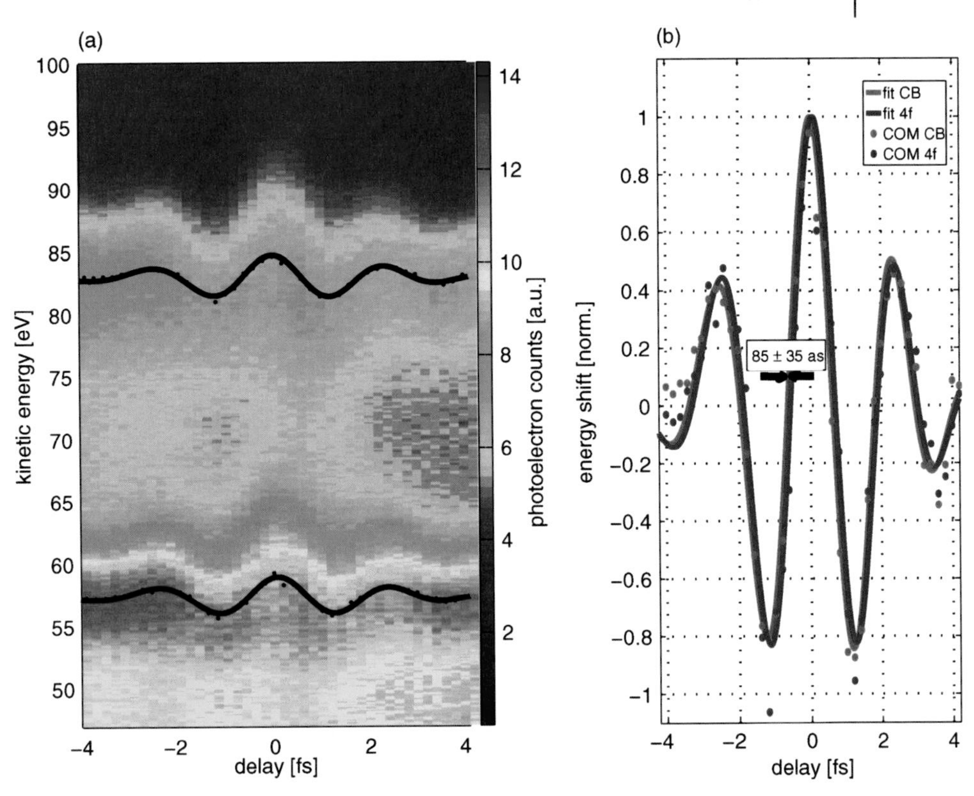

Figure 22.4 (a) Raw ATR spectrogram of a tungsten (110) surface. Photoelectron intensities are given in arbitrary units. The 5d/6sp and 4f/5p peaks are fitted with a Gaussian function and the peak positions at every time point are indicated by dots. The solid line shows the transient shifts of these peaks as obtained from a global fit equivalent to the center-of-mass (COM) analysis. (b) COM analysis of the spectrogram. The COM of both emission lines as measured are given as dots, a global fit to both COM traces is shown in solid lines. See text for details. (This figure also appears on page 547.)

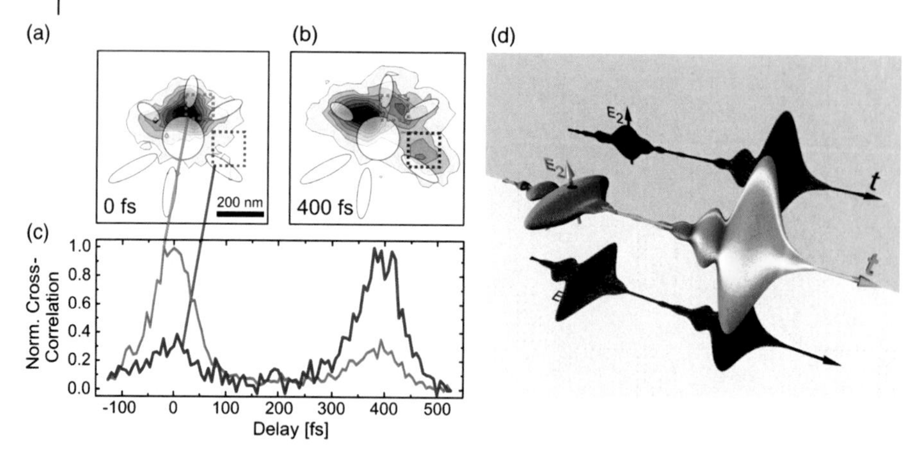

Figure 23.9 Time-resolved cross-correlation PEEM measurements [15]. (a and b) Two-photon PEEM emission patterns for temporal overlap of the circularly polarized probe pulse with the first and second subpulse of the excitation pulse, respectively, from the same nanostructure. The gray solid lines indicate the position of the nanostructure and the rectangles show the regions of interest used for the cross-correlation plot shown in (c). (c) Time-resolved averaged cross-correlation signals for different regions of the nanostructure as indicated by the lines. (d) Reconstructed excitation pulse shape in intuitive 3D representation. The subpulses centered at delay $t = 0$ fs and $t = 400$ fs correspond to polarization directions with respect to the incidence plane and as seen in direction of the propagating beam of 45° and −45°, respectively. (This figure also appears on page 572.)

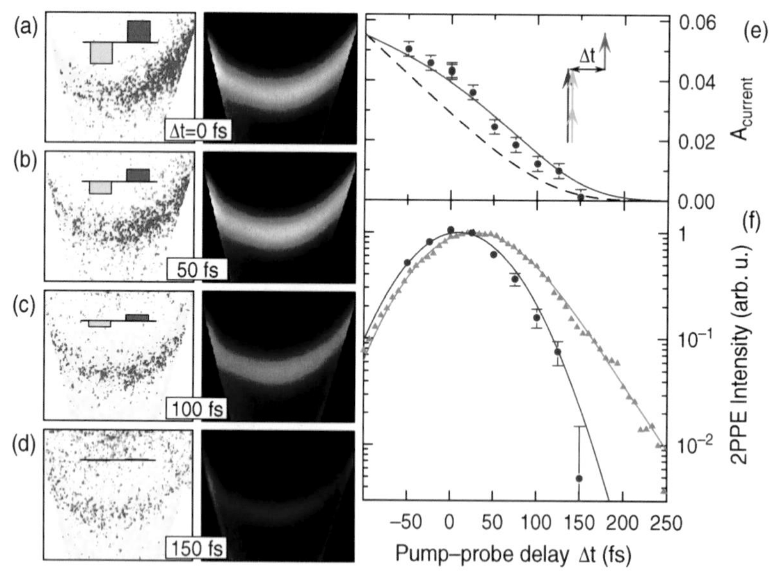

Figure 24.5 Decay of the current in comparison with the average population decay. ((a)–(d)) Difference $E(k_{\|})$ spectra (left) and phase-averaged $E(k_{\|})$ spectra (right) for different time delays Δt of the photoemission laser pulse with respect to the current generation. (e) Relative contribution of the electron population that carries the current with respect to the total population as a function of Δt evaluated at $k_{\|} = \pm 0.15\ \text{Å}^{-1}$. (f) Population decay at the band bottom compared to the current decay as determined from e). Reprinted with permission from *Science* (http://www.aaas.org) Ref. [6]. Copyright 2007. American Association of the Advancement of Science. (This figure also appears on page 588.)

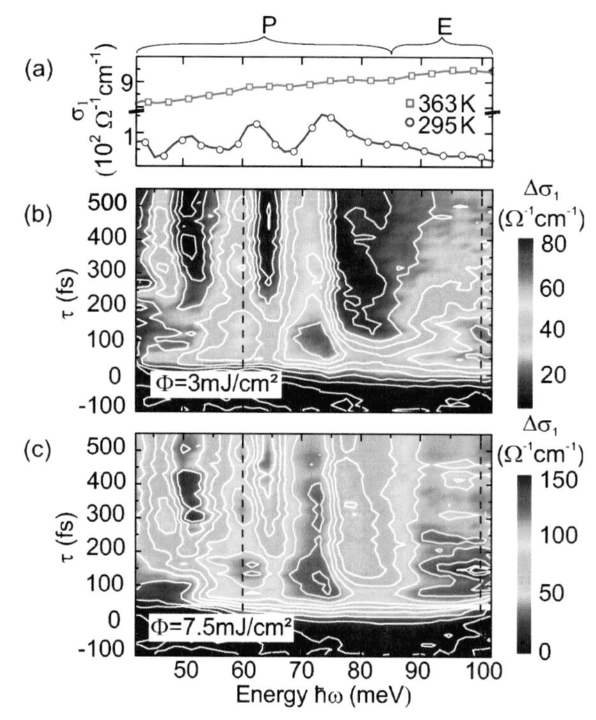

Figure 25.8 2D optical pump–THz probe data. (a) Equilibrium conductivity of metallic (squares) and insulating (circles) VO_2. Color plot of the pump-induced *changes* of the conductivity $\Delta\sigma_1(\omega, \tau)$ for pump fluences (b) $\Phi = 3$ mJ/cm^2 and (c) $\Phi = 7.5$ mJ/cm^2 at $T_L = 250$ K. The broken vertical lines indicate the frequency positions of cross sections reproduced in Figure 25.9. The spectral domain P comprises predominantly changes of the phonon resonances, while the region E reflects the electronic conductivity. (This figure also appears on page 602.)

Part Four
Photoinduced Modification of Materials and Femtochemistry

Dynamics at Solid State Surface and Interfaces Vol.1: Current Developments
Edited by Uwe Bovensiepen, Hrvoje Petek, and Martin Wolf

ISBN: 978-3-527-40937-2

17
Theory of Femtochemistry at Metal Surfaces: Associative Molecular Photodesorption as a Case Study

Peter Saalfrank, Tillmann Klamroth, Tijo Vazhappilly, and Rigoberto Hernandez

17.1
Introduction

Reactions on surfaces are of great scientific interest because of their diverse applications. Well-known examples are the production of ammonia on metal surfaces for fertilizers and reduction of poisonous gases from automobiles using catalytic converters [1]. For these applications, metal surfaces – most notably, those consisting of transition metals – are particularly interesting as they offer a great variety of electronic structure and, therefore, "tunability" toward specific targets.

Photoinduced reactions at surfaces were also studied in detail [2–5]. These are useful, for example, for photocatalysis and nanostructuring of surfaces. Some of these reactions are occurring on femtosecond ($1\,\mathrm{fs} = 10^{-15}\,\mathrm{s}$) timescale [6]. The advent of femtosecond lasers – that is, ultrashort and intense laser pulses – offers the possibility to monitor these motions in real time, for example, by two-pulse correlation techniques. Moreover, femtochemistry at metal surfaces can be very distinct – on quantitative and qualitative scales – from more traditional surface photochemistry driven by conventional light sources [5, 6].

Photochemistry at metal surfaces may or may not proceed *indirectly*, that is, by initial excitation of the substrate electrons rather than by direct excitation of the adsorbate. One way to distinguish between the direct or the substrate-mediated route is through a dependence (direct) [7], or lack thereof (substrate mediated) [8, 9], of the reaction cross section on the polarization of the incoming light. In particular for initiating photons in the UV/vis regime, which can deeply penetrate into metal surfaces [10], indirect mechanisms have been suggested to be operative in many cases. Further, when femtosecond lasers (FLs) are used, hot metal electrons that are attached to the adsorbate are believed to often trigger the reaction rather than phonons [6, 11] (see below). For the reaction of interest below, photodesorption of H_2 from Ru(0001), experiments suggest a substrate-mediated, hot-electron-driven mechanism [12].

The indirect, substrate-mediated excitation can also be categorized according to its fluence dependence. For the prototypical case of photodesorption, for example,

Dynamics at Solid State Surface and Interfaces Vol. 1: Current Developments
Edited by Uwe Bovensiepen, Hrvoje Petek, and Martin Wolf

ISBN: 978-3-527-40937-2

low-fluence, nanosecond lasers enforce so-called DIET, desorption induced by electronic transitions, for which the desorption probability Y scales *linearly* with laser fluence [2–5]. DIET is a result of infrequent and uncorrelated electronic excitations of the adsorbate–substrate complex. In the desorption of adsorbates from metals, the excited state is often a *negative ion resonance* formed by electron attachment of excited electrons in the substrate to the adsorbate. The resonance is short-lived with an electronic lifetime τ_{el} typically as short as femtoseconds [6]. Desorption occurs in the ground state after electronic quenching on this timescale. The adsorbate also relaxes vibrationally, by vibration-electron coupling, with a vibrational lifetime τ_{vib} typically in the order of picoseconds [13]. DIET occurs if the average time between two subsequent electronic excitations, t_{exc}, is long as compared to these lifetimes.

In contrast, in the presence of intense femtosecond laser pulses, one observes so-called DIMET, desorption induced by multiple electronic transitions [14–17]. The short, but intense, laser pulses can cause more than one photon collision with the substrate during the course of an event leading to multiple excitations of the adsorbate on the timescales of electronic and vibrational relaxation. The pulse creates "hot electrons" in the metal surface, which enforce "ladder climbing" of the adsorbate on the electronic ground state into the desorption continuum. The FL pulse may also heat surface and bulk phonons, thus leading to thermal desorption. Experimentally, one can discriminate between phonon and electronic mechanisms by two-pulse correlation (2PC) traces [16]. Accordingly, one records observables such as the desorption yield Y, as a function of the delay time $\Delta\tau$ between two laser pulses: Correlations on ultrashort timescales (approximately picoseconds [6]) are indicative of an electronic mechanism.

A "hallmark" of DIMET is its *superlinear* increase of the desorption yield with laser fluence, F, frequently fitted as a power law

$$Y = A\,F^n \tag{17.1}$$

with $n > 1$ (typically 2–10). An example – see below – is associative desorption of H_2 from Ru(0001) by FL pulses, where $n \approx 3$ has been observed experimentally [12, 18, 19]. A second key difference between DIMET and DIET is that DIMET typically leads to higher yields but – due to electronic quenching – they are still low as compared to yields for photoreactions at insulators. A third "hallmark" of DIMET is that also its energy distribution into various degrees of freedom of the desorbing particles is fluence dependent. For example, for 2D:Ru(0001), the translational energy of desorbing D_2 changes from about 300–500 meV in the fluence range $F \in \{60,\ 120\}$ J/m^2 according to experiment [19]. In general, photoreactions such as DIET and DIMET can show a nonequal and nonthermal energy distribution into various degrees of freedom. For DIMET of H_2 from Ru(0001), for example, translation is typically "hotter" than vibration by a factor of about 4 [19]. The precise value depends on the laser fluence.

This chapter focuses on a comparison of various methods and models to treat FL-induced photochemistry at metal surfaces theoretically. In particular, we will consider DIMET of H_2 and D_2 from Ru(0001), which has been studied experimentally in

Refs [12, 18, 19], and theoretically in Refs [20] and [21, 22]. While our analysis is presented only for a specific system and reaction, it is expected to be relevant also for other FL-induced reactions at metal surfaces [23].

17.2 Theory of Femtochemistry at Surfaces

In this section, several theoretical models to rationalize femtochemistry at surfaces will be reviewed. We will distinguish, somewhat arbitrarily, between "weakly nonadiabatic" and "strongly nonadiabatic" models. We further distinguish classical (or quasiclassical) approaches in which the nuclear motion is described by classical mechanics from quantum dynamical approaches that account properly for quantum mechanical effects such as zero-point motion and tunneling.

17.2.1 Weakly Nonadiabatic Models

We use the term "weakly nonadiabatic" in cases when the dynamics is dominated by nuclear motion on a single (ground-state) potential energy surface, with electronic transitions to excited states serving merely to drive the dynamics on the ground state. Besides many hot-electron-driven reactions, a further physical example of this type is the vibrational relaxation of adsorbates at metal surfaces by vibration-electron coupling [13].

17.2.1.1 Two- and Three-Temperature Models

Hot-electron-mediated femtochemistry at metals can be treated by two- or three-temperature models in conjunction with classical Langevin dynamics, master equation approaches, or with Arrhenius-type rate equations. All of these are based on the concept of "electronic friction," and the assumption that temperatures can be assigned to various subsystems (electrons, phonons, and adsorbate vibrations) of the adsorbate–substrate complex. Accordingly, in a typical experiment, the laser pulse is absorbed by the metal thereby heating the electrons to a time-dependent (and generally also coordinate-dependent) electron temperature, $T_{\mathrm{el}}(t)$. By electron–phonon coupling, phonons are also excited giving rise to a phonon temperature $T_{\mathrm{ph}}(t)$. This can be described by the two-temperature model (2TM), where one solves two coupled equations [24, 25]

$$C_{\mathrm{el}}\frac{\partial T_{\mathrm{el}}}{\partial t} = \frac{\partial}{\partial z}\varkappa\frac{\partial}{\partial z}T_{\mathrm{el}} - g(T_{\mathrm{el}} - T_{\mathrm{ph}}) + S(z,t) \tag{17.2}$$

$$C_{\mathrm{ph}}\frac{\partial T_{\mathrm{ph}}}{\partial t} = g(T_{\mathrm{el}} - T_{\mathrm{ph}}) \tag{17.3}$$

for the electron and phonon temperatures T_{el} and T_{ph} within the metal. In Eqs. (17.2) and (17.3), C_{el} and C_{ph} are electron and lattice specific heat constants, respectively, which can be calculated from the electron specific heat constant, the Debye temperature, and the density of the solid. According to Eq. (17.2), $T_{el}(t)$ is affected by thermal diffusion along a coordinate z in direction perpendicular to the surface, as indicated in the first term on the rhs. Here, $\varkappa = \varkappa_0 \, T_{el}/T_{ph}$ is the thermal conductivity and $\varkappa_0$ an empirical parameter. $T_{el}(t)$ is also affected by electron–phonon coupling (second term, with g = electron–phonon coupling constant), and the external laser pulse that gives rise to a time- and coordinate-dependent source term (third term). Assuming a laser pulse of Gaussian temporal shape, the source term is given by

$$S(z,t) = \frac{Fe^{-\frac{z}{\zeta}}}{\zeta} \frac{e^{-\frac{t^2}{2\sigma^2}}}{\sqrt{2\pi}\sigma}. \tag{17.4}$$

Here, ζ is the optical penetration depth calculated from $\zeta = \frac{\lambda}{4\pi n_I}$, where λ is the wavelength of the femtosecond laser and n_I the imaginary part of the refractive index. Further, σ is the width parameter for the Gaussian. For a Gaussian pulse, the fluence is $F = E_0^2 \varepsilon_0 c \sigma \frac{\sqrt{\pi}}{2}$, where E_0 is the field amplitude, ε_0 the vacuum permittivity, and c the speed of light. In summary, for the calculation of T_{el} and T_{ph} in the 2TM a set of internal (material) and external (laser) parameters is required.

As an example, we show results of the 2TM for a Ru surface in Figure 17.1a, when excited by various femtosecond laser pulses to be used for modeling DIMET of H_2

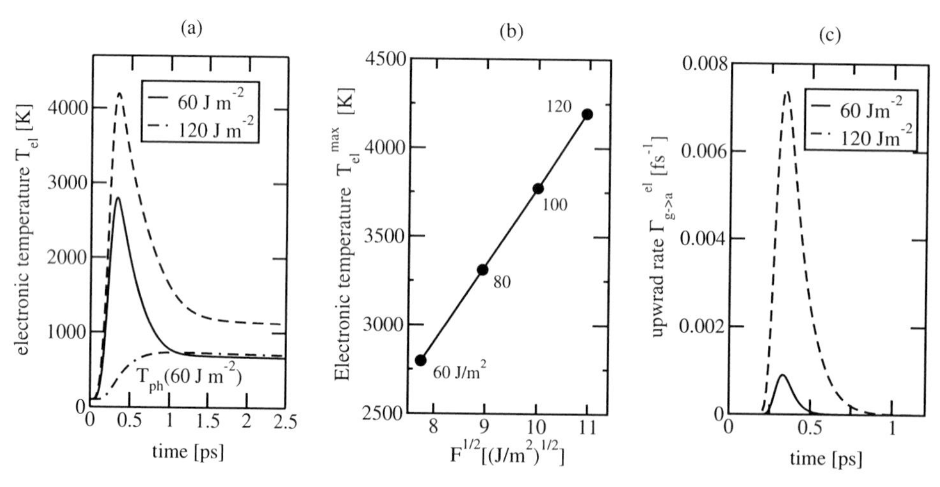

Figure 17.1 (a) Electronic temperatures calculated for Ru from the 2TM with two different laser fluences. The dashed-dotted curve represents the phonon temperature, T_{ph}, for a fluence of 60 J/m^2. Gaussian pulse parameters: FWHM $= \sqrt{8 \ln 2}\,\sigma = 130$ fs, $\lambda = 800$ nm. The material parameters entering the 2TM are Electron–phonon coupling constant, $g = 185 \times 10^{16}$ W/m^3/K; electron specific heat constant, $\gamma = 400$ J/m^3/K^2; thermal conductivity, $\varkappa_0 = 117$ W/m/K; Debye temperature, $\theta = 600$ K; density, $\varrho_0 = 12370$ kg/m^3; optical penetration depth at $\lambda = 800$ nm, $\varsigma = 15.6$ nm. (b) Maximum electronic temperature T_{el}^{max} calculated from the 2TM, plotted as a function of the square root of laser fluence, $\sqrt{F}$, showing an approximate relation $T_{el}^{max} \propto \sqrt{F}$. (c) Dependence of the upward rate $\Gamma_{g\to a}^{el}$ for 2H:Ru(0001) on the applied laser fluence calculated from Eq. (17.20). Parameters: $\Gamma_{a\to g}^{el} = (2\text{ fs})^{-1}$, $\Delta V = 1.55$ eV.

from Ru(0001). The 2TM predicts a maximum electronic temperature of several thousand Kelvin around a few hundred femtosecond, before equilibration with the lattice occurs. The maximum electronic temperature increases with increasing fluence roughly according to $T_{el}^{max} \propto \sqrt{F}$ as indicated in Figure 17.1b, in agreement with earlier findings [26, 27]. Thus, the laser fluence can be used as a control parameter to enhance reaction yields. Another efficient control parameter that also affects T_{el}^{max} is the thickness d of a metal film, if films are used rather than bulk metals [10, 27].

17.2.1.2 One-Dimensional Classical, Arrhenius-Type Models

In the simplest theoretical models of FL-induced photodesorption, the 2TM is often extended to a three-temperature model (3TM). Here, one also assigns a temperature $T_{ads}(t)$ to the adsorbate-surface vibration (or to an internal mode), which arises from the coupling of this mode to the electrons or phonons of the metal. This can be described by an *electron-vibration coupling constant*, η_{el}, and a *vibration-phonon coupling constant*, η_{ph}, respectively. Often the electronic channel dominates. If the concept of thermal equilibration of the active adsorbate is accepted, it can be shown [28] from a Langevin model for the electron heat bath that

$$\frac{\partial T_{ads}}{\partial t} = \eta_{el}\left(T_{el} - T_{ads}\right). \tag{17.5}$$

Equation (17.5) must be solved in addition to Eqs. (17.2) and (17.3), resulting in the 3TM. The basic concepts of the 3TM are illustrated in Figure 17.2a.

Given $T_{ads}(t)$, in a classical, one-dimensional limit [28] where the reaction proceeds along a reaction coordinate, q (e.g., the adsorbate–surface distance), the desorption rate is given as

$$R_{des} = \frac{D\,\eta_{el}}{k_B T_{ads}(t)} \exp\left\{-\frac{D}{k_B T_{ads}(t)}\right\}. \tag{17.6}$$

Here, k_B is Boltzmann's constant, and D an activation energy that can be interpreted, formally, as the adsorption energy of the molecule (see Figure 17.2b). In practice, D and η_{el} are frequently used as model parameters to fit experimental data. Related Arrhenius-type expressions have been suggested elsewhere [16, 17].

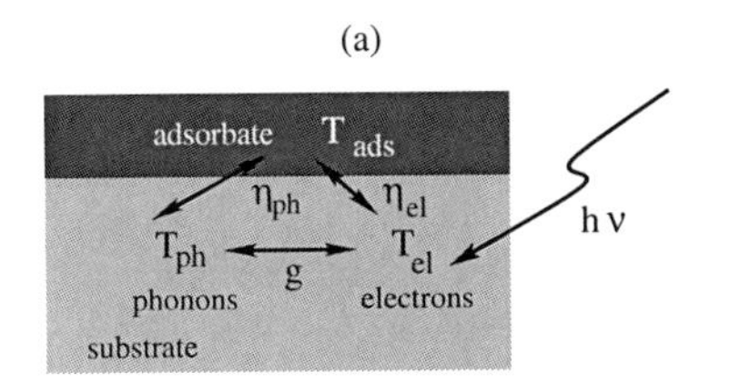

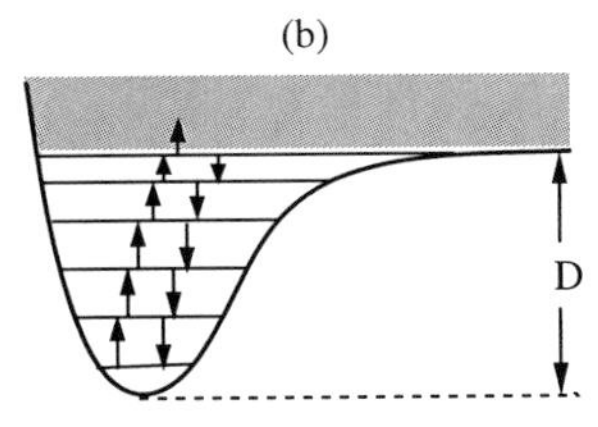

Figure 17.2 (a) Illustration of the three-temperature model, see text. (b) Illustration of a substrate-mediated, photochemical "ladder climbing" process into the continuum of a ground-state potential. Alternatively, Arrhenius-type expressions such as Eq. (17.6) describe the desorption process, with D playing the role of an activation energy (see text). Adapted in part with permission from [23]. Copyright 2006 American Chemical Society.

Arrhenius models are limited to one-dimensional reaction coordinates, treat nuclear motion classically, and assume the existence of electronic and vibrational temperatures even though the conditions are not at equilibrium. The concept of a thermal electron bath is questionable at least within the first few 100 fs or so, after the pulse, according to experimental [29] and theoretical evidence [30, 31]. The concept of an adsorbate temperature can be inaccurate for even much longer timescales [32, 33].

17.2.1.3 Langevin Dynamics with Electronic Friction

Some of the restrictions discussed above can be overcome by Langevin molecular dynamics with electronic friction [34]. Here, nuclear motion is still classical, and the electronic degrees of freedom are expressed in the form of a friction and fluctuating forces. If q is the only degree of freedom considered (e.g., the molecule-surface distance Z with associated reduced mass, m_q), the equation of motion is

$$m_q \frac{d^2 q}{dt^2} = -\frac{dV}{dq} - \eta_{qq} \frac{dq}{dt} + R_q(t). \tag{17.7}$$

Here, V is the ground state potential, and η_{qq} is related to the electronic friction coefficient of above through $\eta_{qq} = m_q \eta_{el}$. $R_q(t)$ is a fluctuating force that obeys a fluctuation–dissipation theorem, and depends on the electronic temperature, $T_{el}(t)$, obtained from the 2TM. When modeled as Gaussian white noise, the random force has the properties

$$\langle R_q(t) \rangle = 0 \tag{17.8}$$

$$\langle R_q(t) | R_q(t') \rangle = 2k_B T_{el} \eta_{qq}\, \delta(t-t'). \tag{17.9}$$

$R(t)$ can be calculated from the Box–Müller algorithm [35] as

$$R_q(t) = \left[\frac{2k_B T_{el}(t) \eta_{qq})}{\Delta t} \right]^{\frac{1}{2}} (-2 \ln b)^{\frac{1}{2}} \cos 2\pi c \tag{17.10}$$

where Δt is the time step, and b and c are random numbers uniformly distributed on the interval [0,1]. This equation shows clearly that if the electronic temperature is zero, or if electronic friction vanishes, we have no random force. Increasing the electronic temperature at finite friction, on the other hand (e.g., by a laser pulse according to the 2TM), leads to large random forces that may drive a photoreaction.

Various methods, based, for example, on Hartree–Fock cluster calculations [34] or on periodic density functional theory (DFT) [36], have been suggested to calculate the friction coefficient from first principles. All of these methods are based on a perturbative, golden rule-type treatment of the vibration-electron coupling. For example, the diagonal electronic friction coefficients are determined from (periodic) DFT as

$$\eta_{qq} = 2\pi\hbar \sum_{\alpha,\beta} \left| \left\langle \psi_\alpha | \frac{\delta v}{\delta q} | \psi_\beta \right\rangle \right|^2 \delta(\varepsilon_\alpha - \varepsilon_F)\, \delta(\varepsilon_\beta - \varepsilon_F), \tag{17.11}$$

where $\psi_{\alpha(\beta)}$ and $\varepsilon_{\alpha(\beta)}$ are Kohn–Sham orbitals and energies for combined band and k-point index $\alpha(\beta)$. Further, ε_F is the Fermi energy and $\frac{\delta v}{\delta q}$ the functional derivative of the exchange–correlation functional with respect to nuclear coordinate q. The electronic friction coefficient is also directly related to the vibrational relaxation rate, and thus to the vibrational lifetime τ_{vib}. At zero temperature, the vibrational relaxation rate $\Gamma^{vib}_{1\to 0}$ of the first excited vibrational level of the mode of interest is simply

$$\Gamma^{vib}_{1\to 0} = \tau^{-1}_{vib} = \eta_{el}. \tag{17.12}$$

Equation (17.7) can be easily generalized to more degrees of freedom by interpreting the friction coefficient as a tensor with elements $\eta_{qq'}$ [34, 36] and extending the equation of motion (17.7) by a coupling term $-\sum_{q'\neq q}\eta_{qq'}\frac{dq'}{dt}$. In general, there are F^2 tensor matrix elements $\eta_{qq'}$ each of which depends on F degrees of freedom $q_1, \ldots, q_F$ for an F-dimensional system. This can be extremely demanding computationally, which is why one often neglects off-diagonal elements $\eta_{qq'}$ for $q \neq q'$ and/or makes additional approximations regarding the coordinate dependence of η_{qq} [36].

17.2.1.4 **Quantum Treatment with Master Equations**

The Langevin approach is multidimensional and free of the assumption of vibrational temperatures, but still classical. The classical approximation can be overcome with the help of master equations,

$$\frac{dP_\alpha}{dt} = \sum_\beta W_{\beta\to\alpha}P_\beta(t) - \sum_\beta W_{\alpha\to\beta}P_\alpha(t), \tag{17.13}$$

giving the population P_α of state $|\alpha\rangle$ of the ground state potential. The latter can be derived from (multidimensional) system Hamiltonians. The populations are determined by interlevel transition rates,

$$W_{\alpha\to\beta} = \Gamma^{vib}_{\alpha\to\beta} + W^{ex}_{\alpha\to\beta}. \tag{17.14}$$

Here, $\Gamma^{vib}_{\alpha\to\beta}$ is a vibrational relaxation rate or re-excitation rate if the temperature is finite. $W^{ex}_{\alpha\to\beta}$ is a rate caused by an external energy source that drives the "ladder climbing." For example, for FL-induced, hot-electron-mediated reactions driven by electron attachment to form a negative ion resonance, the transition rates calculated perturbatively depend on the electron temperature $T_{el}(t)$, and the energetic position ε_a and width Δ_a of the adsorbate resonance as shown elsewhere [33, 37].

By solving Eq. (17.13), a reaction probability can be defined by analyzing those populations that reach the desorption continuum, with an energy $E_\alpha > D$ where D is again the binding energy of the adsorbate if "desorption" is the reaction of interest. Such a damped ladder climbing process is schematically illustrated in Figure 17.2b. As a simplification, in the so-called "truncated harmonic oscillator" (THO) model and when additionally assuming that the P_α is a Boltzmann distribution associated with a vibrational temperature T_{ads}, the master equation (17.13) simplifies to reaction rates R of the Arrhenius type as in Eq. (17.6) [37].

17.2.2
Strongly Nonadiabatic Dynamics

In contrast to weakly nonadiabatic dynamics, "strongly nonadiabatic" dynamics requires the explicit inclusion of electronically excited states. Examples where this is necessary are resonant charge transfer during atom-surface scattering and reactions proceeding through long-lived excited states, such as adsorbate excitations at insulators. Of course, any photoreaction at a metal surface with short-lived excited states can also be described in this way.

17.2.2.1 Multistate Time-Dependent Schrödinger Equation

The most general approach to treating nonadiabatic dynamics is by a coupled, multistate time-dependent Schrödinger equation (TDSE). As the initial condition, we assume that the system is initially in a product of its electronic ground state $|g\rangle$ and an arbitrary nuclear wave function $\psi_g(R,t)$ (where R denotes a set of nuclear coordinates). Assuming that only a single, excited adsorbate state $|a\rangle$ is of importance, the TDSE reads

$$i\hbar\frac{\partial}{\partial t}\begin{pmatrix}\psi_a\\ \psi_g\\ \psi_{k_1}\\ \psi_{k_2}\\ \vdots\end{pmatrix} = \begin{pmatrix}\hat{H}_a & V_{ag} & V_{ak_1} & V_{ak_2} & & \cdots\\ V_{ga} & \hat{H}_g & V_{gk_1} & V_{gk_2} & & \cdots\\ V_{k_1a} & V_{k_1g} & \hat{H}_{k_1} & V_{k_1k_2} & V_{k_1k_3} & \cdots\\ V_{k_2a} & V_{k_2g} & V_{k_2k_1} & \hat{H}_{k_2} & V_{k_2k_1} & \cdots\\ \vdots & \vdots & \vdots & \vdots & \vdots & \ddots\end{pmatrix}\begin{pmatrix}\psi_a\\ \psi_g\\ \psi_{k_1}\\ \psi_{k_2}\\ \vdots\end{pmatrix}. \tag{17.15}$$

In Eq. (17.15), a discretized continuum of substrate-excited states $|k_i\rangle$ was included to model a metal surface. The diagonal elements of the Hamiltonian matrix are $\hat{H}_n = \hat{T}_{\text{nuc}} + V_n(R)$ with $\hat{T}_{\text{nuc}}$ being the nuclear kinetic energy operator, and $V_n(R)$ the potential energy curve for state $|n\rangle$. The off-diagonal elements $V_{nm}(R,t)$ stand for general couplings between different electronic states, for example, for direct dipole, spin-orbit, or non-Born–Oppenheimer couplings, the latter expressed here as potential couplings in a diabatic representation. The multistate model is illustrated schematically in Figure 17.3a. Equation (17.15) can hardly ever be solved in practice. In particular, for metals with a continuum of electronic excitations, the basis of the latter is often far too large to be computationally tractable.

17.2.2.2 Open-System Density Matrix Theory

An alternative is to map the closed-system, N-state model (17.15) on an open-system, *reduced* density matrix model with only a few relevant states. In particular, two-state models are common, and retain only the ground $|g\rangle$ and excited $|a\rangle$ states. Here, the excited state (e.g., the negative ion state) is treated as a nonstationary *resonance* with a finite energetic width Δ_a. The energy is generally not conserved in contrast to the unitary approach using the TDSE – see Figure 17.3b for illustration. The resonance width, Δ_a, can in principle be calculated from the non-Born–Oppenheimer coupling

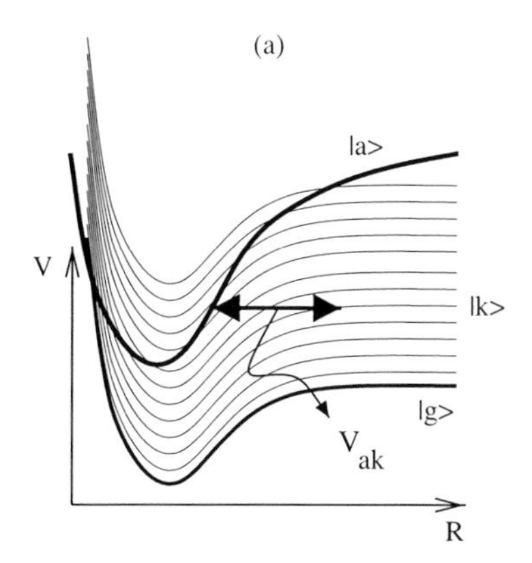

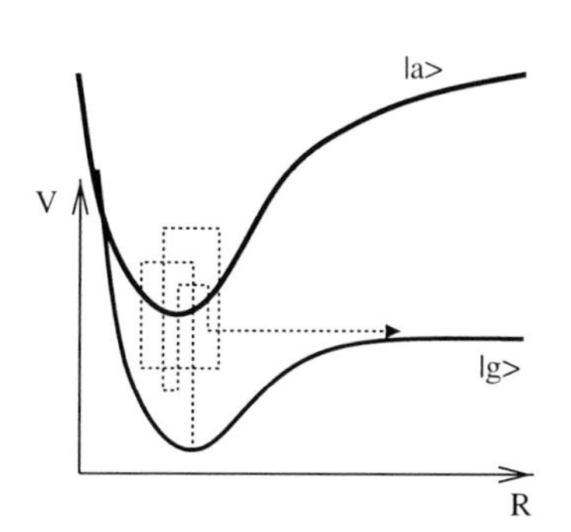

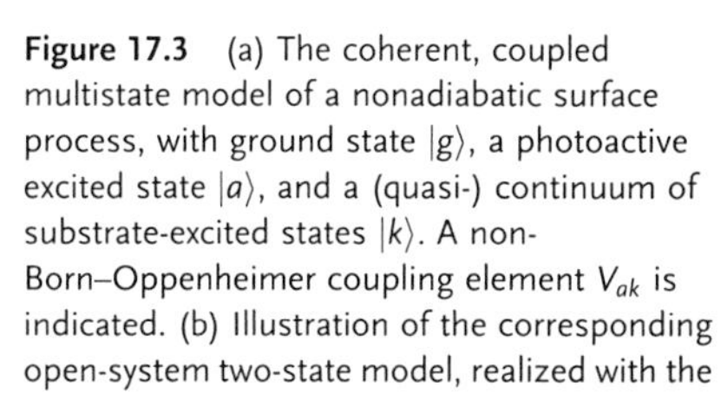

Figure 17.3 (a) The coherent, coupled multistate model of a nonadiabatic surface process, with ground state $|g\rangle$, a photoactive excited state $|a\rangle$, and a (quasi-) continuum of substrate-excited states $|k\rangle$. A non-Born–Oppenheimer coupling element V_{ak} is indicated. (b) Illustration of the corresponding open-system two-state model, realized with the MCWP approach or its classical variant. The dashed curve indicates a (quantum) trajectory schematically, with multiple transitions between the ground state $|g\rangle$ and resonance state $|a\rangle$. The latter has a finite energetic width, $\Delta_a = \hbar\Gamma^{\mathrm{el}}_{a\to g}$. Adapted in part with permission from [23]. Copyright 2006 American Chemical Society.

matrix elements V_{ak} that couple metal continuum states $|k\rangle$ to the excited state $|a\rangle$, using, for instance, a Newns–Anderson approach [38, 39]. Often, the resonance width is treated as an empirical parameter instead. In general, the resonance width Δ_a depends on the nuclear coordinates R. If this dependence is neglected, the resonance decays strictly exponentially with an ultrashort (approximately femtoseconds) electronic lifetime

$$\tau_{\mathrm{el}} = \frac{\hbar}{\Delta_a} = 1/\Gamma^{\mathrm{el}}_{a\to g}, \tag{17.16}$$

Here, $\Gamma^{\mathrm{el}}_{a\to g}$ is the electronic quenching rate.

The suggested equation of motion for the open-system two-state model of DIMET has been cast in the form of a Liouville–von Neumann (LvN) equation [40, 41]

$$\begin{aligned}\frac{\partial}{\partial t}\begin{pmatrix}\hat{\varrho}_a & \hat{\varrho}_{ag}\\ \hat{\varrho}_{ga} & \hat{\varrho}_g\end{pmatrix} &= -\frac{i}{\hbar}\left[\begin{pmatrix}\hat{H}_a & V_{ag}\\ V_{ga} & \hat{H}_g\end{pmatrix},\begin{pmatrix}\hat{\varrho}_a & \hat{\varrho}_{ag}\\ \hat{\varrho}_{ga} & \hat{\varrho}_g\end{pmatrix}\right]\\ &\quad + \frac{\partial}{\partial t}\begin{pmatrix}\hat{\varrho}_a & \hat{\varrho}_{ag}\\ \hat{\varrho}_{ga} & \hat{\varrho}_g\end{pmatrix}_{\mathrm{rel}} + \frac{\partial}{\partial t}\begin{pmatrix}\hat{\varrho}_a & \hat{\varrho}_{ag}\\ \hat{\varrho}_{ga} & \hat{\varrho}_g\end{pmatrix}_{\mathrm{exc}}.\end{aligned} \tag{17.17}$$

In Eq. (17.17), $\hat{\varrho}_i$ and $\hat{\varrho}_{ij}$ are the diagonal and off-diagonal reduced density operators in the vibrational space of the ground and excited states.

The first term on the rhs of Eq. (17.17) describes the coherent evolution of the system, with possible direct couplings. In Eq. (17.17), "rel" accounts for energy relaxation of the resonance state $|a\rangle$ with an electronic relaxation rate $\Gamma^{\mathrm{el}}_{a\to g}$ according to

$$\frac{\partial}{\partial t}\begin{pmatrix} \hat{\varrho}_a & \hat{\varrho}_{ag} \\ \hat{\varrho}_{ga} & \hat{\varrho}_g \end{pmatrix}_{\mathrm{rel}} = \Gamma^{\mathrm{el}}_{a\to g}\begin{pmatrix} -\hat{\varrho}_a & -\dfrac{\hat{\varrho}_{ag}}{2} \\ -\dfrac{\hat{\varrho}_{ga}}{2} & +\hat{\varrho}_a \end{pmatrix} \tag{17.18}$$

and "exc" describes the hot-electron-, substrate-mediated excitation of the adsorbate–substrate complex [40, 41], which depopulates the ground state:

$$\frac{\partial}{\partial t}\begin{pmatrix} \hat{\varrho}_a & \hat{\varrho}_{ag} \\ \hat{\varrho}_{ga} & \hat{\varrho}_g \end{pmatrix}_{exc} = \Gamma^{el}_{g\to a}(t)\begin{pmatrix} +\hat{\varrho}_g & -\dfrac{\hat{\varrho}_{ag}}{2} \\ -\dfrac{\hat{\varrho}_{ga}}{2} & -\hat{\varrho}_g \end{pmatrix}. \tag{17.19}$$

For DIMET, the initial condition is $\hat{\varrho}_0 = |g\rangle\langle g| \otimes |\phi_0^g\rangle\langle\phi_0^g|$ if $T = 0$, that is, the system is in the ground vibrational state at the ground-state surface. Further,

$$\Gamma^{\mathrm{el}}_{g\to a}(t) = \Gamma^{\mathrm{el}}_{a\to g}\exp\left\{-\frac{V_a - V_g}{k_B\, T_{\mathrm{el}}(t)}\right\} \tag{17.20}$$

is a time-dependent upward rate that obeys detailed balance. The assumption of an electronic temperature, which can be calculated from the 2TM, is also made in Eq. (17.20). Since both the excited- and ground-state potentials V_a and V_g depend on all nuclear coordinates, $\Gamma^{\mathrm{el}}_{g\to a}(t)$ is generally also coordinate-dependent, even for coordinate-independent quenching. Frequently, the approximation is made of $V_a - V_g$ being a constant, given by the potential difference at the Franck–Condon point that is assumed to coincide with the energy of the exciting laser. For 2H:Ru (0001), for example, the energy difference is 1.55 eV, corresponding to the excitation wavelength of 800 nm. For this system, the upward rates for laser fluences and $T_{\mathrm{el}}(t)$ curves of Figure 17.1a are indicated in Figure 17.1c, where an excited-state lifetime of $\tau_{\mathrm{el}} = 2$ fs has been assumed.

In passing we note that the two-state DIMET model of above can be recast in more compact notation as

$$\frac{\partial\hat{\varrho}_s}{\partial t} = -\frac{i}{\hbar}\left[\hat{H}_s, \hat{\varrho}_s\right] + \sum_{k=1}^{2}\left(\hat{C}_k\hat{\varrho}_s\hat{C}_k^\dagger - \frac{1}{2}\left[\hat{C}_k^\dagger\hat{C}_k, \hat{\varrho}_s\right]_+\right) \tag{17.21}$$

where $\hat{\varrho}_s = \hat{\varrho}_g|g\rangle\langle g| + \hat{\varrho}_a|a\rangle\langle a| + \hat{\varrho}_{ag}|a\rangle\langle g| + \hat{\varrho}_{ga}|g\rangle\langle a|$ is the system density operator and $\hat{H}_s = \hat{H}_g|g\rangle\langle g| + \hat{H}_a|a\rangle\langle a| + V_{ag}|a\rangle\langle g| + V_{ga}|g\rangle\langle a|$ is the system Hamiltonian. Further,

$$\hat{C}_1 = \sqrt{\Gamma^{\mathrm{el}}_{a\to g}}|g\rangle\langle a| \tag{17.22}$$

$$\hat{C}_2 = \sqrt{\Gamma^{\mathrm{el}}_{g\to a}(t)}|a\rangle\langle g| \tag{17.23}$$

are the so-called Lindblad operators for the relaxation ($k = 1$) and excitation ($k = 2$) process, respectively. In Eq. (17.21), $[\,]_+$ refers to an anticommutator. It is easy to extend the formalism to account for further "dissipation channels," excitations with more than one laser pulse, and finite initial temperature [32].

17.2.2.3 Stochastic Wave Packet Approaches

Open-system density matrix theory is a quantum approach that is quadratically more costly than wave packet methods since matrices, rather than state vectors have to be propagated forward in time. Thus, the direct numerical solution of Eq. (17.21) is limited to low-dimensional models in which only a few, often only one, system modes are included. An alternative to the direct solution is stochastic wave packet approaches [42–44], which are easily applicable to LvN equations of the Lindblad form such as Eq. (17.21). Here, a matrix propagation is replaced by a set of stochastically selected wave packet (state vector) propagations. Thus, though costly, higher-dimensional problems can be treated quantum mechanically.

In particular, we refer to the Monte Carlo wave packet (MCWP) method [45], which has also been adopted for photodesorption problems [46–49]. For concreteness, suppose that $T = 0$ K of an arbitrary number K of dissipative channels, the algorithm works as follows:

1) We start with an initial wave function $|\psi(0)\rangle$, typically the vibrational ground state on the ground-state surface at $T = 0$ K.
2) An auxiliary wave function is created at every time step as $|\psi'(t+\Delta t)\rangle = \exp(-i\hat{H}_s^{'}\Delta t/\hbar)|\psi(t)\rangle$, by propagating $|\psi(t)\rangle$ under the influence of a non-hermitian Hamiltonian, $\hat{H}_s^{'}$, given by

$$\hat{H}'_s = \hat{H}_s - \frac{i\hbar}{2}\sum_k^K \hat{C}_k^\dagger \hat{C}_k \tag{17.24}$$

where, $\hat{C}_k$ are the Lindblad operators from Eq. (17.21). Due to the non-Hermiticity of the Hamiltonian, $|\psi'(t+\Delta t)\rangle$ is not normalized. The loss of norm Δp in each time step is

$$\Delta p = 1 - \langle\psi'(t+\Delta t)|\psi'(t+\Delta t)\rangle. \tag{17.25}$$

3) In each time step, a random number ε is drawn from a uniform distribution between 0 and 1. If $\Delta p > \varepsilon$, a quantum jump occurs, and the new wave function is chosen among the different possibilities $\hat{C}_k|\psi(t)\rangle$ with a probability $p_k = \Delta p_k/\Delta p$. Here, $\Delta p_k = \Delta t\langle\psi(t)|\hat{C}_k^\dagger\hat{C}_k|\psi(t)\rangle \geq 0$ is the approximate loss in dissipative channel k up to first-order. The new wave function is then calculated in normalized form as

$$|\psi(t+\Delta t)\rangle = \frac{\hat{C}_k|\psi(t)\rangle}{\langle\psi(t)|\hat{C}_k^\dagger\hat{C}_k|\psi(t)\rangle^{\frac{1}{2}}}. \tag{17.26}$$

For $\Delta p \leq \varepsilon$, no quantum jump occurs and the new wave function $|\psi(t+\Delta t)\rangle$ is obtained by renormalizing $|\psi'(t+\Delta t)\rangle$ as

$$|\psi(t+\Delta t)\rangle = \frac{|\psi'(t+\Delta t)\rangle}{(1-\Delta p)^{\frac{1}{2}}}. \tag{17.27}$$

4) This procedure is repeated N times, that is, there are N "quantum trajectories." For each realization n, the expectation value of an operator $\hat{A}$ is calculated as

$$A_n(t) = \langle \psi_n(t)|\hat{A}|\psi_n(t)\rangle \tag{17.28}$$

and final expectation values of an operator $\hat{A}$ are obtained by averaging over N trajectories:

$$\langle \hat{A}\rangle(t) = \frac{1}{N}\sum_{n}^{N} A_n(t). \tag{17.29}$$

When applied to DIMET, the algorithm specializes as follows. Initially, the wave packet is on the ground-state surface, $|g\rangle$. The term $-\frac{i\hbar}{2}\hat{C}_2^{\dagger}\hat{C}_2 = -\frac{i\hbar}{2}\Gamma^{\text{el}}_{g\to a}(t)|g\rangle\langle g|$ acts as a negative imaginary potential on the ground-state potential, which leads to a loss of norm of $\Delta p = 1-\text{e}^{-\Gamma^{\text{el}}_{g\to a}\Delta t} \leq 1$ during time interval Δt. If $\Delta p > \varepsilon$, the wave packet jumps as a whole to the excited state $|a\rangle$, where it propagates under the influence of V_a and an imaginary absorber, $-\frac{i\hbar}{2}\Gamma^{\text{el}}_{a\to g}|a\rangle\langle a|$. The wave packet jumps back to the ground state $|g\rangle$ as soon as $\Delta p > \varepsilon$, where Δp is now given as $\Delta p = 1-e^{-\Gamma^{el}_{a\to e}\Delta t}$. Once in the ground state again, the wave packet can be re-excited provided the upward rate $\Gamma^{\text{el}}_{g\to a}(t)$ is still large. As a consequence, *multiple excitations* and de-excitations are possible during the action of the femtosecond laser pulse. In fact, the algorithm allows one to quantify the notion of *multiple* in DIMET by simply counting the number of quantum jumps from $|g\rangle$ to $|a\rangle$ and vice versa [22]. Again, one has to average over a large ensemble of quantum trajectories to arrive at meaningful results. For illustration, the desorption dynamics within the two-state MCWP model is illustrated schematically for a single trajectory in Figure 17.3b.

17.2.2.4 **Quantum–Classical Hopping Schemes**

The MCWP method is a quantum approach, and thus computationally demanding. One may, however, replace the quantum trajectories by classical trajectories while using the same procedure as in MCWP, that is, with transition probabilities calculated from the quenching rate $\Gamma^{\text{el}}_{a\to g}$ (for the relaxation process) and the excitation rates $\Gamma^{\text{el}}_{g\to a}(t)$ (for the FL-induced excitation process). This idea has been followed in the early, one-dimensional models of DIMET of diatomic molecules from metal surfaces [15]. Since more than one potential surface is involved, the method is a mixed quantum–classical approach with dissipation. Because nuclear dynamics is treated classically, the method can easily be extended to multidimensions, provided the relevant excited state potential energy surface(s) is (are) known.

There is also a quantum–classical approach to solve the nondissipative, coupled multistate Schrödinger equation (17.15), namely, the surface hopping method of Tully and coworkers [50]. In the latter, at each instant of time, the nuclear dynamics is described by a classical trajectory in one electronic state only. Hops or *switches*, between different electronic states $|n\rangle$ and $|m\rangle$ occur due to nonadiabatic couplings V_{nm}. Many ways exist of how to realize the switches in practice [50–54], and Tully's fewest switching algorithm [50] is a very popular one. The method has also been applied for nonadiabatic dynamics at surfaces [55, 56].

17.3 Femtosecond-Laser Driven Desorption of H_2 and D_2 from Ru(0001)

17.3.1 Experimental Facts

Several of the concepts discussed above will now be illustrated with the specific example of associative desorption of H_2 and D_2, from H- or D-covered Ru(0001) surfaces. The reaction has been studied by Frischkorn *et al.* using femtosecond lasers [12, 18, 19]. Their key experimental observations are as follows:

- Using 120–130 fs laser pulses at a wavelength of 800 nm and with fluences ranging from 60 to 120 J/m^2, associative desorption of H_2 and D_2 from a fully (1×1) H- or D-covered Ru(0001) surface has been observed.
- The recombinative desorption shows a large isotope effect with a desorption probability for H_2 typically 3–10 times higher than that of D_2 depending on fluence.
- The desorbing molecules have high translational energy compared to their vibrational energy.
- The desorption probability increases nonlinearly with laser fluence. At the same time, the isotope effect in the desorption probabilities goes down and the energy partitioning into different degrees of freedom is also fluence dependent.
- Two-pulse correlation experiments suggest that the reaction is probably due to a hot-electron-mediated mechanism with a short response time of $\sim$1 ps.

17.3.2 Potentials, Electronic Lifetime, and Friction

For theoretical modeling, both Langevin molecular dynamics with electronic friction and the quantum Monte Carlo wave packet two-state model have been used effectively [22].

17.3.2.1 The Ground-State Potential and Vibrational Levels

In both models, the same ground-state potential was adopted. We used a modified version of the six-dimensional (6D) potential of Luppi *et al.* [57, 58] calculated by periodic DFT for two H atoms per cell corresponding to a (2×2) coverage of H on Ru (0001). The DACAPO code [59] was adopted, using a plane wave basis, a three-layer relaxed slab model, and the so-called RPBE exchange–correlation functional [60]. The potential was fitted to an analytic form using the corrugation reducing procedure (CRP) [61]. From the 6D potential, a two-dimensional (2D) reduced dimensionality minimal model was constructed in which only the two most important coordinates for associative desorption are included. These are the distance of the center of mass of the molecule from the surface, Z, and the internal vibrational coordinate (H–H distance) of the desorbing molecule, r. The other four degrees of freedom of a diatomic molecule relative to a rigid surface, namely, lateral coordinates X and Y and

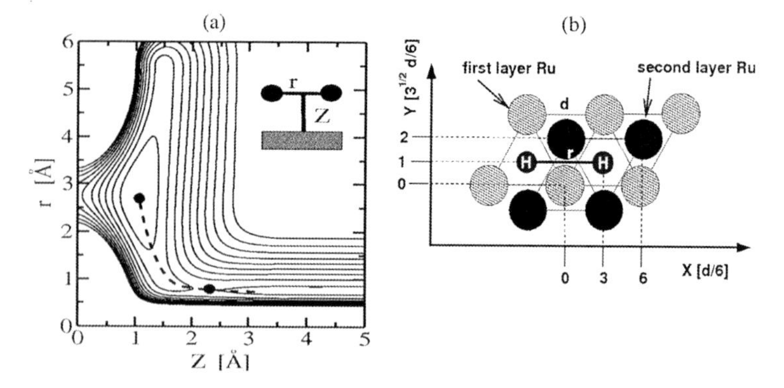

Figure 17.4 (a) Ground-state potential $V_g(r, Z)$ used in this work. Contours have an increment of 0.5 eV, starting at 0.5 eV. Bullets indicate the minimum at $Z_0 = 1.06$ Å and $r_0 = 2.75$ Å and the transition state at $Z^{\ddagger} = 2.24$ Å and $r^{\ddagger} = 0.77$ Å. The dotted line is the approximate minimum energy path S for the associative desorption. (b) Sketch of the coordinate system used. The H atoms reside in a plane defined by the interatomic axis, and the surface normal. Initially, they are in fcc (face-centered cubic) sites of the Ru(0001) surface. Adapted in part with permission from [22]. Copyright 2009 American Chemical Society.

azimuthal and polar angles φ and θ were fixed at their equilibrium values as $X_0 = 0$, $Y = \sqrt{3}d/6$, $\theta = 90°$, and $\varphi = 0°$. Here, $d = 2.75$ Å is the smallest distance between the two Ru atoms in the first layer of the Ru(0001) surface. These chosen coordinates are shown in Figure 17.4. The original potential was modified by extending it to regions that were not calculated in Ref. [62], and by adding smooth repulsive walls behind the transition states toward diffusion and subsurface adsorption. This allows one to model a full-coverage situation H-(1 × 1) as used in Ref. [12, 18], and to exclude subsurface adsorption after excitation [22].

The resultant two-dimensional PES, which is also shown in Figure 17.4(a), predicts an adsorption minimum at $Z_0 = 1.06$ Å and $r_0 = 2.75$ Å (= d), with a binding energy of 0.85 eV (i.e., 0.42 eV per H atom). Further, in the 2D model, an incoming H_2 molecule feels a classical barrier of 0.18 eV. The barrier is located "early" on the reaction path for dissociative adsorption, or "late" on the reaction path for associative desorption.

The 2D ground-state potential supports vibrational states. The lowest of these can be classified according to the number of nodes along r and Z. There are 32 bound vibrational levels for 2H:Ru(0001) and 64 for 2D:Ru(0001). The calculated vibrational energies [22] are $\hbar\omega_r = 94$ meV ($758\,\text{cm}^{-1}$) for the r-, and $\hbar\omega_Z = 136$ meV ($1097\,\text{cm}^{-1}$) for the Z-mode. The corresponding experimental values [63] are $\hbar\omega_r = 85$ meV and $\hbar\omega_Z = 140$ meV at a coverage of (1 × 1). For 2D:Ru(0001), theory gives $\hbar\omega_r = 67$ meV and $\hbar\omega_Z = 97$ meV.

17.3.2.2 The Excited-State Potential and Electronic Lifetime

In the excitation–de-excitation models, relevant excited state potentials are needed. For 2H:Ru(0001) we adopt a single excited state potential $V_a(r, Z)$ with an associated excited-state lifetime, τ_{el}. Reliable *ab initio* calculations of (adsorbate) excited states at metal surfaces and lifetimes are not yet available. Recently, attempts have been made to calculate excited states of metal/adsorbate systems using cluster approaches in combination with time-dependent DFT (TD-DFT) [64]. Calculations using clusters H_2Ru_n give already with $n = 3$ a multitude of excited states [22]. Most of these have a topology very similar to the ground-state surface. Several of them, however, contain a minimum simultaneously shifted along the r and Z modes by an amount Δ approximately according to

$$V_a(r, Z) = V_g(r-\Delta, Z+\Delta) + E_{ex}. \tag{17.30}$$

All calculations below are for a "representative excited state," with $\Delta = 0.2$ Å. The electronic excitation energy, $E_{ex} = 1.55$ eV, is chosen to match the excitation energy in the experiments.

The excited-state lifetime τ_{el} was varied between 1 and about 10 fs, and thus treated as an empirical parameter. All calculations below refer to $\tau_{el} = 2$ fs. The dependence of the results on different choices for τ_{el} and the potential shift parameter Δ was discussed in Refs [21, 22].

17.3.2.3 **Frictional Surfaces and Vibrational Relaxation**

In the Langevin model, excited states are not explicitly included but enter the formalism indirectly through the friction coefficients $\eta_{qq'}(r, Z)$ (with $q, q' = r, Z$). Below, we neglect the off-diagonal terms η_{rZ} such that the Langevin equations in the 2D model are

$$m_r\ddot{r} = -\frac{\partial V_g(r, Z)}{\partial r} - \eta_{rr}\dot{r} + R_r(t), \tag{17.31}$$

$$m_Z\ddot{Z} = -\frac{\partial V_g(r, Z)}{\partial Z} - \eta_{ZZ}\dot{Z} + R_Z(t). \tag{17.32}$$

Thus, mode coupling enters only through the potential term. In the equations above, m_r and m_Z are the reduced masses for motion along r and Z, respectively, for 2H:Ru(0001) given as $m_r = m_H/2$ and $m_Z = 2m_H$, respectively (m_H is the hydrogen mass).

The friction coefficients were represented by analytical forms for $\eta_{ZZ}(r,Z)$ and $\eta_{rr}(r,Z)$, which are based on generic models [65] with input from the DFT study on 2H:Ru(0001) of Ref. [20]. These functions are shown in Figure 17.5 along a minimum energy path of the ground-state potential $V_g(r, Z)$, which connects the potential minimum at (r_0, Z_0) with a point corresponding to desorbed H_2 (see also the dotted line in Figure 14.4(a)). The functional forms and parameters are given in the figure caption. From the figure, we note that around the equilibrium position (r_0, Z_0) both coefficients are similar and finite, and both of them vanish when the molecule is away from the surface and cannot couple to the metal electron–hole pairs any more. In between, in particular the η_{ZZ} values vary significantly with the position of the adsorbate relative to the surface.

The frictional terms give rise to vibrational relaxation in the ground electronic state due to vibration-electron coupling. This is demonstrated in Figure 17.6, where the

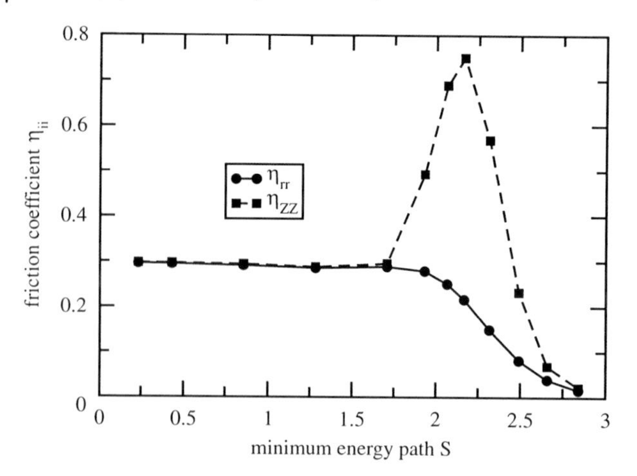

Figure 17.5 Electronic friction coefficients η_{ii} in units of meV ps Å^{-2} along the minimum energy path (S), given in units of Å. $S = 0$ Å corresponds to hydrogen adsorbed at Ru(0001) in the equilibrium position (r_0, Z_0); $S = 2.1$ Å is the transition state and $S = 3$ Å corresponds to the situation where the molecule is about to leave the Ru surface. The functional forms are $\eta_{rr}(r,Z) = a_2 \frac{1}{1+e^{c(Z-Z_2)}} + b_2 \exp\left[\frac{-(r-r_2)^2}{2\sigma_r^2} - \frac{(Z-Z_2)^2}{2\sigma_r^2}\right]$ and $\eta_{ZZ}(r,Z) = a_1 \frac{1}{1+e^{c(Z-Z_1)}} + b_1 \exp\left[\frac{-(r-r_1)^2}{2\sigma_Z^2} - \frac{(Z-Z_1)^2}{2\sigma_Z^2}\right]$ with the following parameters: $a_1 = a_2 = 0.3$ meV ps Å^{-2}, $b_1 = 0.6$ meV ps Å^{-2}, $b_2 = 0.1$ meV ps Å^{-2}, $c = 4$ Å^{-1}, $Z_1 = 2.36$ Å, $Z_2 = 2.25$ Å, $r_1 = r_2 = 0.74$ Å, $\sigma_Z = 0.2$ Å, $\sigma_r = 0.2$ Å [22].

energy loss (upper panels) and the coordinates r and Z (lower panels) for 2H:Ru (0001) are shown, both as a function of time, for the cases when either the r mode (a) or the Z mode vibration (b) was excited initially. Specifically, single trajectories were run using Eqs. (17.31) and (17.32), with the initial conditions that kinetic energies of $\hbar\omega_r = 94$ meV or $\hbar\omega_Z = 136$ meV were deposited in the respective modes r (a) and Z (b). The initial positions were chosen as the equilibrium values of $r_0 = 2.75$ Å and $Z_0 = 1.06$ Å, respectively. Furthermore, the choice $T_{el} = 0$ K was made. This is equivalent to the assumption that only the r- or Z-excited systems can relax and that electronic re-excitation is impossible because the fluctuating forces vanish. The equations of motion were solved with the help of an algorithm due to Ermak and Buckholz [66].

From an exponential fit of the energies such as

$$E_{tot}(t) = E_{tot}^q(0)\, e^{-t/\tau_{vib}^q}, \tag{17.33}$$

vibrational lifetimes can be estimated for the r and Z modes ($q = r$ and $q = Z$), with $E_{tot}^q(0) = \hbar\omega_q$. Figure 17.6 shows that the exponential form holds well, particularly for the Z mode. One obtains $\tau_{vib}^r = 215$ fs and $\tau_{vib}^Z = 685$ fs. The vibrational relaxation is accompanied by damped oscillations around the equilibrium values r_0 and Z_0.

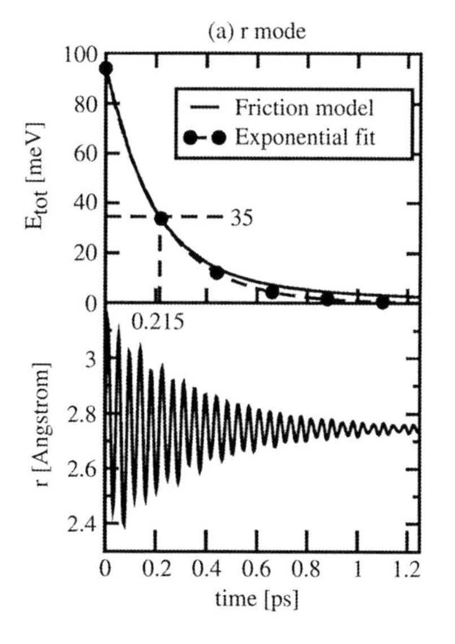

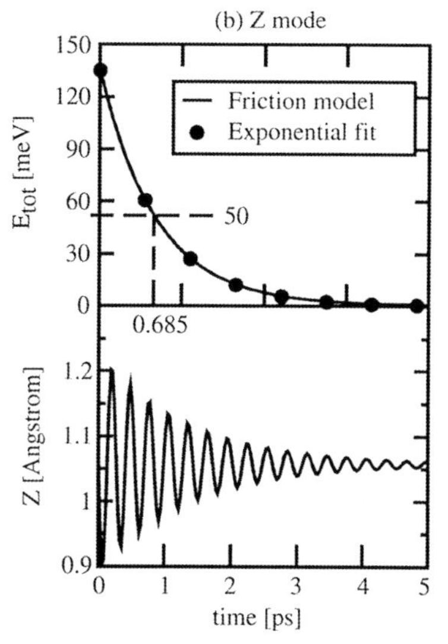

Figure 17.6 The vibrational relaxation of the r (a) and Z mode (b) for 2H:Ru(0001) is shown. Upper panels: The total energy of the system, E_{tot}, as a function of time. From an exponential fit to the data, the vibrational lifetime τ_{vib} can be estimated for each mode as the time when the energy has dropped to $1/e$th of its initial value, $E_{tot}(0)$. Lower panel: The position expectation value along r or Z as a function of time, showing damped oscillations around the equilibrium positions.

17.3.3 DIMET at a Single Laser Fluence

17.3.3.1 Langevin Dynamics Approach

To treat DIMET of H_2 or D_2 for H/D:Ru(0001), the 2D Langevin equations were propagated for many stochastic trajectories. All the trajectories are initialized on the ground-state surface at the adsorption minimum (r_0, Z_0). Initial velocities are chosen from a Boltzmann distribution with a kinetic energy in each mode of $k_B T_{surf}$, with the surface temperature $T_{surf} = 100$ K chosen in accordance with experiment [19]. The trajectories were propagated for 1 ps to equilibrate the system at this temperature. After the initial thermalization period, a visible laser pulse is applied, which heats the metal electrons and the lattice according to the 2TM. The visible pulse was chosen to have the form of a Gaussian temporal shape with 800 nm wavelength and 130 fs FWHM.

When the laser pulse has a fluence of 120 J/m^2, it leads to the electronic temperature as shown in Figure 17.1a, dashed curve. At finite electron temperature, the fluctuating forces R_r and R_Z are different from zero, and the adsorbed H atoms have a chance to associate and desorb as H_2. This is demonstrated in Figure 17.7, where a "representative trajectory" that leads to desorption is shown.

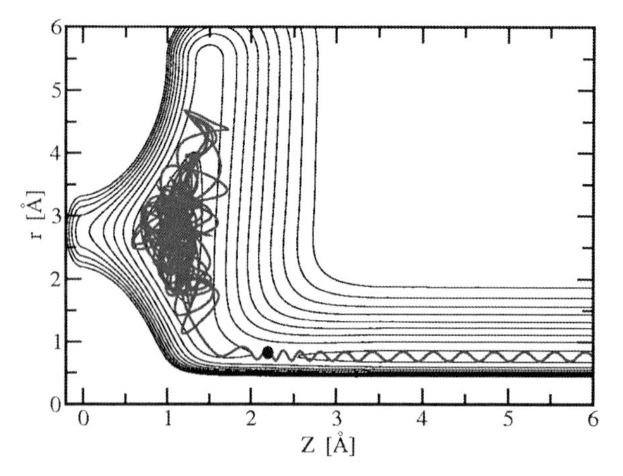

Figure 17.7 An example for a typical trajectory desorbing as molecular hydrogen is overlaid on the ground state PES. The laser fluence was 120 J/m^2. The transition state for associative desorption is indicated by a bullet.

To compare to experiments, one has to do statistics by counting and analyzing the desorbing trajectories for an ensemble of propagations. A trajectory is classified as desorbed, when it is beyond a certain line Z_{des} far from the surface at the end of the propagation time (here, $Z_{des} = 9.5$ Å). Typically, several ten thousands of trajectories are required for desorption of H_2 or D_2 from Ru(0001). When using $N = 60\,000$ for H_2 and $N = 80\,000$ for D_2, one obtains a desorption yield of $Y = 2.55 \times 10^{-2}$ for H_2 and $Y = 5.46 \times 10^{-3}$ for D_2. The desorption probability was calculated as $Y = N_d/N$, where N_d is the number of desorbing trajectories. The yields are *desorption probabilities per laser pulse*, and not very high. Relatively low reaction yields are typical for FL-induced processes at metal surfaces, because quenching and/or electronic friction is very efficient. Various strategies have been suggested how to enhance reactive cross sections, also for metal surfaces [23] – a topic that we will not discuss in this chapter. The larger desorption yield for H_2 compared to D_2 is consistent with experimental observations [18], and also a consequence of the low desorption yields [23]. Most of the desorptions take place immediately after the peaking of the electronic temperature T_{el} [22], when the fluctuating forces are still large but vibrational relaxation was not yet efficient.

One can also determine average translational and vibrational energies of the desorbing hydrogen molecules. The translational energies are 363 and 324 meV for H_2 and D_2 with a vibrational energy of 218 and 163 meV, respectively. This unequal energy partitioning is in good agreement with experimental results. The propensity for translation is due to the "late" barrier along the desorption path. It favors translationally rather than vibrationally excited species according to Polanyi's rules. In fact, associative desorption is the reverse reaction to dissociative adsorption. In the latter, translational rather than vibrational energy helps to overcome "early barriers" [67]. The desorption characteristics for H_2 and D_2 at $F = 120$ J/m^2 according to the Langevin model is summarized in Table 17.1, left half.

Table 17.1 DIMET with 800 nm, 130 fs laser pulses.

	Langevin, $F = 120$ J/m^2				MCWP, $F = 100$ J/m^2		
	Y	E_{trans} (meV)	E_{vib} (meV)		Y	E_{trans} (meV)	E_{vib} (meV)
H_2	2.55×10^{-2}	363	218	H_2	3.78×10^{-3}	454	161
D_2	5.46×10^{-3}	324	163	D_2	7.59×10^{-4}	403	146

a) Left half: Langevin dynamics, at a fluence of 120 J/m^2. Right half: MCWP dynamics, for a fluence of 100 J/m^2. Y is the desorption yield, E_{trans} and E_{vib} are the average translational and vibrational energies of desorbing H_2 and D_2 molecules.

17.3.3.2 **Stochastic Wave Packet Approach**

A similar analysis can be carried out by using the Monte Carlo wave packet method to solve the two-state LvN equation (17.21) for DIMET. To do so, we use the same ground-state potential and the same 2TM as in the Langevin dynamics. The excited state was given by Eq. (17.30), and the electronic lifetime chosen as $\tau_{el} = 2$ fs as described earlier. Starting from the ground vibrational state as initial state, and choosing a laser fluence of $F = 100$ J/m^2 (with otherwise identical laser parameters as above), one needs about 8000 "quantum trajectories" for H_2 and 10 000 for D_2 to converge results to an accuracy of a small percentage [22]. The propagation of the wave packets was done by a split-operator propagator and by representing the wave functions on a two-dimensional grid. The time–energy method was used to analyze the properties of the desorbing molecules [22].

Desorption arises in this model because the nonvanishing excitation rates $\Gamma^{el}_{g \to a}(t)$ enable population to transfer temporarily from the ground state $|g\rangle$ to the excited state $|a\rangle$. Since the latter is displaced from the ground state by the displacement parameter Δ according to Eq. (17.30), the excited wave packet is nonstationary in the excited state. After return to the ground state within an average lifetime τ_{el}, the wave packet continues to move on $V_g(r, Z)$ until a part of the wave packet can reach the asymptotic region with $Z \to \infty$. Multiple excitations are possible with FL pulses, thereby enhancing the chance of desorption.

In Figure 17.8, the excitation probability is shown not to be very large: The averaged excited-state population $\langle P_a \rangle$ does not exceed 1% or so when a 130 fs laser pulse with $F = 100$ J/m^2 is applied. As a consequence, in this model, the desorption probability is also small as demonstrated in the right half of Table 17.1. In fact, in the MCWP model, the desorption yield is lower than in the Langevin model. However, it is hard to compare both calculations quantitatively. This is so partly because the Langevin approach is classical. Also, a lower fluence was used in the quantum excitation–de-excitation model and, finally, the results depend somewhat on the excited state potential and lifetimes parameters that are not known accurately [22]. On the positive side, both the classical Langevin and the quantum MCWP models agree in the most important findings: (i) The desorption cross sections are small, (ii) a clear isotope effect is found with H_2 desorbing easier than D_2, and (iii) the desorbing particles are

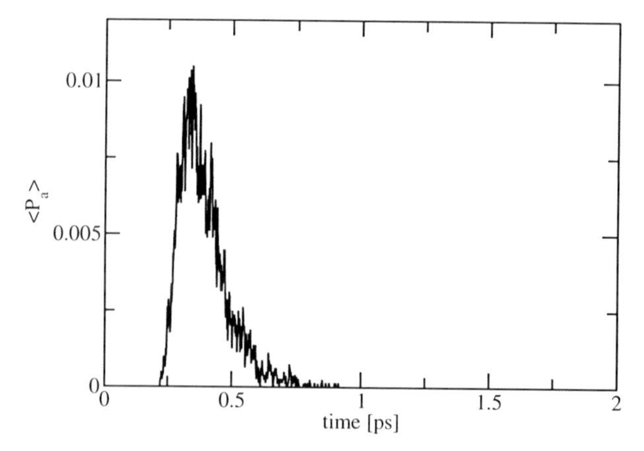

Figure 17.8 Time-dependent electronic excited-state population $\langle P_a \rangle$ for 2H:Ru(0001), for a fluence of 100 J/m^2 in the DIMET model. The populations are averaged over 8000 trajectories obtained from the MCWP method.

much more excited translationally than vibrationally. These similarities are due to the fact that the *desorption process is dominated by the dynamics in the electronic ground state*, even in the excitation–de-excitation models, because the excited-state lifetime is so short [22].

17.3.4
Scaling of DIMET with Laser Fluence

It is therefore also not surprising that both models explain equally well the experimental facts regarding the scaling of desorption yields and other properties of the desorbing molecules with laser fluence. This is demonstrated for H/D:Ru (0001) in Figure 17.9, which shows the desorption yield for H_2 and D_2 as a function of laser fluence F, and in Figure 17.10, which shows the dependence of energy distributions of the desorbates, also as a function of F [22]. In the classical Langevin case, fluences in the range $\{80, 160\}$ J/m^2 were considered while in the quantum MCWP model we apply fluences within the interval of $\{60, 120\}$ J/m^2. Apart from the different fluence range, the same computational protocols were followed as in Chapter 17.3.3.

From Figure 17.9, it is noted that the desorption yield increases nonlinearly as predicted by Eq. (17.1), both for H_2 and D_2, and for both models – Langevin (a) and MCWP (b). The computed exponents are somewhat higher than experimentally observed (in particular for MCWP), that is, $n = 2.8$ for H_2 and $n = 3.2$ for D_2 [18], but still in general agreement. In further agreement with experiment, the isotope ratio $I = Y(H_2)/Y(D_2)$ decreases with increasing F. In the MCWP case, this ratio changes from ~10 to 5 in the fluence range $F \in \{60, 120\}$ J/m^2. In the Langevin model, I changes from about 8 to 3 in the range $F \in \{60, 120\}$ J/m^2 [22]. The nonlinear dependence of the desorption probability on fluence is best explained in the MCWP model by the upward rate $\Gamma^{el}_{g \to a}$, which increases dramatically with F as shown in

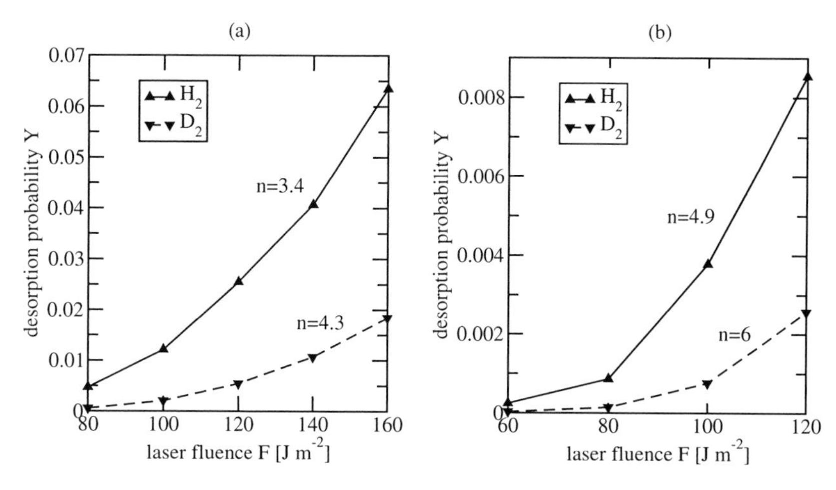

Figure 17.9 Dependence of the desorption yields for H_2 and D_2 on the applied laser fluence calculated with the classical Langevin method (a) and the quantum MCWP case (b). In the former, simulations are performed for $F \in \{60, 120\}$ J/m^2, and an excited-state lifetime of 2 fs. In the latter, fluences F are taken in the range, $\{80, 160\}$ J/m^2. All curves are associated with a different value of the power law constant, n according to Eq. (17.1).

Figure 17.1c. As a consequence, increasing the fluence greatly enhances the electronic excitation probability. This can also be seen from the average number of electronic excitations that increase with fluence as quantitatively demonstrated in Ref. [22]. In that study, it is found that at $F = 60$ J/m^2, the average number of excitation–de-excitation cycles per pulse is 0.13, and that the probability of multiple excitations is vanishingly small. At $F = 120$ J/m^2, on the other hand, the average

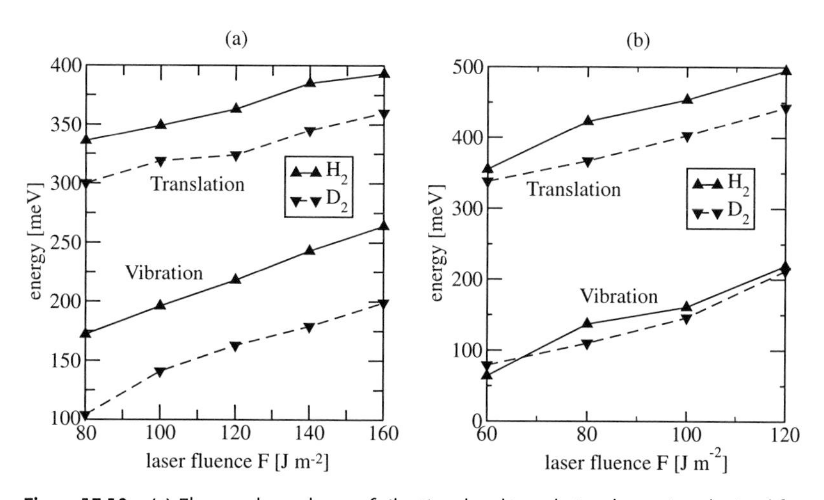

Figure 17.10 (a) Fluence dependence of vibrational and translational energies obtained from the Langevin simulations for different fluences for H_2 and D_2, in the fluence range $F \in \{80, 160\}$ J/m^2. (b) The same for the quantum MCWP case, and a fluence range of $F \in \{60, 120\}$ J/m^2.

number of excitation–de-excitation cycles is 1.5, and the probability of multiple excitations is high – up to sixfold excitations were observed in Ref. [22]. As a result, more molecules reach higher vibrational levels on the ground electronic state after the final electronic relaxation when high fluences are used, and this promotes desorption. The process is highly nonlinear, thus resulting in a nonlinear scaling as in Eq. (17.1).

Figure 17.10 illustrates the nearly linear increase with fluence found in the computed translational and vibrational energies. The increasing energy content in the product channels with increasing fluence arises because the latter leads to more energy absorption and hence more energy can be released into the desorbates. The translation remains hotter compared to vibration and H_2 acquires more energy than D_2 in both modes. These two findings are in good agreement with experimental observations [19]. For example, the translational energy of D_2 increases from about 300 to 500 meV in the fluence range $F \in \{60, 120\}$ J/m^2 according to the experiment studied in Ref. [19]. Our corresponding theoretical values are 338 meV (at 60 J/m^2), and 443 meV (at 120 J/m^2), respectively. In further agreement with experiment, the ratio between translational and vibrational energies, E_{tr}/E_{vib}, decreases from 5.6 to 2.3 for H_2 and from 4.3 to 2.1 for D_2, respectively. This is due to the fact that the additional energy gained by the desorbing molecule at higher fluences is approximately equally distributed between translation and vibration, that is, the difference between energies remains nearly constant while their ratio changes.

17.4
Conclusions

Theoretical concepts for the treatment of femtochemistry at solid surfaces can be categorized as effective single-state ("weakly nonadiabatic") or nonadiabatically coupled multistate models ("strongly nonadiabatic"), respectively. At metal surfaces, the electronically excited states that enforce the nuclear motion toward new products are typically very-short lived ($\sim$ fs). As a consequence, the nuclear dynamics proceeds in the electronic ground state for most of the time. This explains why, if the same ground-state potentials are used, semiquantitative agreement is obtained for both sets of theories. The importance of quantum effects in the nuclear coordinates is not quite clear because classical treatments of the nuclear motion is often sufficient to explain the major effects. Note, however, that the explicit inclusion of electronically excited states is, by definition, at least partially quantum in nature. FL-induced dynamics at surfaces, notably at metal surfaces, in practice requires system-bath-type models because coherent, multistate models cannot be realized computationally. In the Langevin approach, the (electron–hole) "bath" enters through friction coefficients, while in the excitation–de-excitation models the excited electronic states are short-lived. For the laser excitation of metal electrons, the concept of a time-dependent electron temperature is useful, but limitations are known.

Despite their similarities, both the "weakly nonadiabatic" friction and the "strongly nonadiabatic" excitation–de-excitation models of femtosecond laser chemistry rely on different assumptions. Therefore, there are also clear differences that may lead to different interpretations of the desorption dynamics [32]. For example, the two- and multistate models account for dynamical details arising from the characteristic topology of the excited state potential, for example, the postexcitation “inward motion” in the Antoniewicz models of photodesorption [4]. In the friction models, the excited states enter only implicitly, for example, through the friction coefficients $\eta_{qq'}$ and their coordinate dependence. Furthermore, in the friction model, vibrational relaxation and desorption are deeply interconnected: Without the former, the latter cannot happen. In contrast, in the excitation–de-excitation models, vibrational relaxation is merely a small correction while electronic relaxation is decisive [32].

These differences are due to different approximations for the underlying physics. Still, both sets of models are capable of describing femtochemistry at surfaces semiquantitatively. It should now be the goal to extend the approaches toward computationally tractable multidimensional simulation tools, and to include hereto neglected effects such as coupling to surface phonons. Ideally, these models should use input only from *first principles* electronic structure theory. It will then be interesting to apply the tools also to other FL-induced elementary processes at adsorbate-covered metal surfaces, such as photodissociation [68], photodiffusion [69, 70], or more complex photoreactions [11].

Acknowledgment

We acknowledge the support of this work by the Deutsche Forschungsgemeinschaft through *Sonderforschungsbereich 450*, project C7, and by the Fonds der Chemischen Industrie. Rigoberto Hernandez thanks the Alexander-von Humboldt foundation for the support.

References

1 Ertl, G. (1990) *Angew. Chem. Int. Ed.*, **29**, 1219.
2 Menzel, D. and Gomer, R. (1964) *J. Chem. Phys.*, **41**, 3311.
3 Redhead, P.A. (1980) *Can. J. Phys.*, **42**, 886.
4 Antoniewicz, P.R. (1980) *Phys. Rev. B*, **21**, 3811.
5 Ho, W. (1996) *Surf. Sci.*, **363**, 166.
6 Frischkorn, C. and Wolf, M. (2006) *Chem. Rev.*, **106**, 4207.
7 Vondrak, T. and Zhu, X.-Y. (1999) *Phys. Rev. Lett.*, **82**, 1967.
8 Buntin, S.A., Richter, L.J., Cavanagh, R.R., and King, D.S. (1988) *Phys. Rev. Lett.*, **61**, 1321.
9 Kidd, R.T., Lennon, D., and Meech, S.R. (1999) *J. Phys. Chem. B*, **103**, 7480.
10 Hohlfeld, J., Wellershoff, S.-S., Güdde, J., Conrad, U., Jähnke, V., and Matthias, E. (2000) *Chem. Phys.*, **251**, 237.
11 Bonn, M., Funk, S., Hess, Ch., Denzler, D.N., Stampfl, C., Scheffler, M., Wolf, M., and Ertl, G. (1999) *Science*, **285**, 1042.
12 Denzler, D.N., Frischkorn, C., Hess, C., Wolf, M., and Ertl, G. (2003) *Phys. Rev. Lett.*, **91**, 226102.
13 Morin, M., Levinos, N.J., and Harris, A.L. (1992) *J. Chem. Phys.*, **96**, 3950.
14 Prybyla, J.A., Heinz, T.F., Misewich, J.A., Loy, M.M.T., and Glownia, J.H. (1990) *Phys. Rev. Lett.*, **64**, 1537.

15 Misewich, J.A., Heinz, T.F., and Newns, D.M. (1992) *Phys. Rev. Lett.*, **68**, 3737.
16 Budde, F., Heinz, T.F., Loy, M.M.T., Misewich, J.A., deRougemont, F., and Zacharias, H. (1991) *Phys. Rev. Lett.*, **66**, 3024.
17 Budde, F., Heinz, T.F., Kalamarides, A., Loy, M.M.T., and Misewich, J.A. (1993) *Surf. Sci.*, **283**, 143.
18 Denzler, D.N., Frischkorn, C., Wolf, M., and Ertl, G. (2004) *J. Phys. Chem. B*, **108**, 14503.
19 Wagner, S., Frischkorn, Ch., Wolf, M., Rutkowski, M., Zacharias, H., and Luntz, A. (2005) *Phys. Rev. B*, **72**, 205404.
20 Luntz, A.C., Persson, M., Wagner, S., Frischkorn, C., and Wolf, M. (2006) *J. Chem. Phys.*, **124**, 244702.
21 Vazhappilly, T., Beyvers, S., Klamroth, T., Luppi, M., and Saalfrank, P. (2007) *Chem. Phys.*, **338**, 299.
22 Vazhappilly, T., Klamroth, T., Saalfrank, P., and Hernandez, R. (2009) *J. Phys. Chem. C*, **113**, 7790.
23 Saalfrank, P. (2006) *Chem. Rev.*, **106**, 4116.
24 Kaganov, M.I., Lifshitz, I.M., and Tanatarov, L.V. (1957) *Sov. Phys. JETP*, **4**, 173.
25 Anisimov, S.I., Kapeliovich, B.L., and Perel'man, T.L. (1974) *Zh. Eksp. Teor. Fiz.*, **66**, 776.
26 Corkum, P.B., Brunel, F., Sherman, N.K., and Srinivasan-Rao, T. (1988) *Phys. Rev. Lett.*, **61**, 2886.
27 Nest, M. and Saalfrank, P. (2004) *Phys. Rev. B*, **69**, 235405.
28 Brandbyge, M., Hedegård, P., Heinz, T.F., Misewich, J.A., and Newns, D.M. (1995) *Phys. Rev. B*, **52**, 6042.
29 Fann, W.S., Storz, R., Tom, H.W.K., and Bokor, J. (1992) *Phys. Rev. Lett.*, **68**, 2834.
30 Weik, F., de Meijere, A., and Hasselbrink, E. (1993) *J. Chem. Phys.*, **99**, 682.
31 Knorren, R., Bouzerar, G., and Bennemann, K.-H. (2001) *Phys. Rev. B*, **63**, 094306.
32 Nest, M. and Saalfrank, P. (2002) *J. Chem. Phys.*, **116**, 7189.
33 Gao, S., Lundquist, B.I., and Ho, W. (1995) *Surf. Sci.*, **341**, L1031.
34 Tully, J.C., Gomez, M., and Head-Gordon, M. (1993) *J. Vac. Sci. Technol. A*, **11**, 1914.
35 Press, W.H., Teukolsky, S.A., Vetterling, W.T., and Flannery, B.P. (2003) *Numerical Recipes*, 2nd edn, Cambridge University Press, Cambridge.
36 Luntz, A.C. and Persson, M. (2005) *J. Chem. Phys.*, **123**, 074704.
37 Gao, S. (1997) *Phys. Rev. B*, **55**, 1876.
38 Anderson, P.W. (1961) *Phys. Rev.*, **124**, 41.
39 Newns, D.M. (1969) *Phys. Rev.*, **178**, 1123.
40 Saalfrank, P., Baer, R., and Kosloff, R. (1994) *Chem. Phys. Lett.*, **230**, 463.
41 Saalfrank, P. and Kosloff, R. (1996) *J. Chem. Phys.*, **105**, 2441.
42 Zoller, P., Marte, M., and Walls, D.F. (1987) *Phys. Rev. A*, **35**, 198.
43 Dum, R., Zoller, P., and Ritsch, H. (1992) *Phys. Rev. A*, **45**, 4879.
44 Dalibard, J., Castin, Y., and Mølmer, K. (1992) *Phys. Rev. Lett.*, **68**, 580.
45 Mølmer, K., Castin, Y., and Dalibard, J. (1993) *J. Opt. Soc. Am. B*, **10**, 524.
46 Saalfrank, P. (1996) *Chem. Phys.*, **211**, 265.
47 Saalfrank, P., Boendgen, G., Finger, K., and Pesce, L. (2000) *Chem. Phys.*, **251**, 51.
48 Gao, S., Strömquist, J., and Lundqvist, B.I. (2001) *Phys. Rev. Lett.*, **86**, 1805.
49 Guo, H. (1997) *J. Chem. Phys.*, **106**, 1967.
50 Tully, J.C. (1990) *J. Chem. Phys.*, **93**, 1061.
51 Blais, N.C. and Truhlar, D.G. (1983) *J. Chem. Phys.*, **79**, 1334.
52 Coker, D.F. and Xiao, L. (1995) *J. Chem. Phys.*, **102**, 496.
53 Krylov, A.I., Gerber, R.B., and Coalson, R.D. (1996) *J. Chem. Phys.*, **105**, 4626.
54 Müller, U. and Stock, G. (1997) *J. Chem. Phys.*, **107**, 6230.
55 Groß, A. and Bach, C. (2001) *J. Chem. Phys.*, **114**, 6396.
56 Carbagno, C., Groß, A., and Rohlfing, M. (2007) *Appl. Phys. A*, **88**, 579.
57 Luppi, M., Oslen, R.A., and Baerends, E.J. (2006) *Phys. Chem. Chem. Phys.*, **8**, 688.
58 Vincent, J.K., Olsen, R.A., Kroes, G.-J., Luppi, M., and Baerends, E.-J. (2005) *J. Chem. Phys.*, **122**, 044701.
59 DACAPO code, http://www.fysik.dtu.dk/campos/Dacapo.
60 Hammer, B., Hansen, L.B., and Nørskov, J.K. (1999) *Phys. Rev. B*, **59**, 7413.

61 Busnengo, H.F., Salin, A., and Dong, W. (2000) *J. Chem. Phys.*, **112**, 7641.
62 Vincent, J.K., Olsen, R.A., Kroes, G.-J., Luppi, M., and Baerends, E.-J. (2005) *J. Chem. Phys.*, **122**, 044701.
63 Shi, H. and Jacobi, K. (1994) *Surf. Sci.*, **313**, 289.
64 Besley, N.A. (2004) *Chem. Phys. Lett.*, **390**, 124.
65 Diekhöner, L., Hornekær, L., Mortensen, H., Jensen, E., Baurichter, A., Petrunin, V.V., and Luntz, A.C. (2002) *J. Chem. Phys.*, **117**, 5018.
66 Allen, M.P. and Tildesley, D.J. (1987) *Computer Simulation of Liquids*, Oxford University Press, New York.
67 Halstead, D. and Holloway, S. (1990) *J. Chem. Phys.*, **93**, 2859.
68 Camillone, N. III, Khan, K.A., Lasky, P.J., Wu, L., Moryl, J.E., and Osgood, R.M. Jr. (1998) *J. Chem. Phys.*, **109**, 8045.
69 Bartels, L., Wang, F., Möller, D., Knoesel, E., and Heinz, T.F. (2004) *Science*, **305**, 648.
70 Stépán, K., Güdde, J., and Höfer, U. (2005) *Phys. Rev. Lett.*, **94**, 236103.

18
Time-Resolved Investigation of Electronically Induced Diffusion Processes

Jens Güdde, Mischa Bonn, Hiromu Ueba, and Ulrich Höfer

18.1
Introduction

Hopping of atoms or molecules between different adsorption sites on a surface is one of the most elementary steps in surface reactions such as catalysis and crystal growth. Usually, surface diffusion is a thermally activated process that is initiated by heating the substrate. Electronic excitations, however, can play an even more important role, particularly at metal surfaces. This has been studied in detail for the process of desorption induced by electronic transitions (DIET) where the reaction is initiated by single electronic transitions [1, 2]. With the introduction of intense femtosecond laser pulses as an excitation source, a new regime of desorption induced by multiple electronic transitions (DIMET) [3] has been discovered, where the high density of optical excited electrons causes repetitive electronic transitions between the ground and the excited states of the adsorbate–metal system on the timescale of nuclear motion. The repetitive excitation and de-excitation cycles result in reaction yields that are many orders of magnitude higher than in conventional photochemical reactions at metal surfaces [4–7]. In addition, the DIMETregime is characterized by a nonlinear dependence of the reaction yield on laser fluence. The use of ultrashort laser pulses, however, further offers unique capabilities, both for initiating and for analyzing such types of surface reactions. Femtosecond lasers not only allow to study the energy transfer dynamics on the timescale of nuclear motion directly in the time-domain but also hold potential for controlling adsorbate motion before the excitation energy is thermalized within the whole adsorbate–substrate system.

For experimental reasons, femtochemistry at surfaces has so far been investigated mainly for desorption since desorbing atoms or molecules can be detected easily and efficiently in the gas phase with quadrupole mass spectrometry (QMS). From an energetic point of view, lateral motion is easier to excite than desorption since diffusion barriers are generally much lower than chemisorption energies. Similarly, one can expect the amount of electronic excitation required to initiate diffusion to be less than the amount required for desorption. In fact, the concept of electronic friction to describe the electronic coupling of an adsorbate with a metal surface was

Dynamics at Solid State Surface and Interfaces Vol.1: Current Developments
Edited by Uwe Bovensiepen, Hrvoje Petek, and Martin Wolf

ISBN: 978-3-527-40937-2

first discussed in the context of diffusion [8, 9]. Therefore, models that have been developed for the analysis of femtosecond desorption experiments should be transferable to the case of laser-induced diffusion. In many studies of desorption, a basic understanding of the electronic excitation of adsorbate motion has been achieved by models that do not consider the multitude of individual transitions but describe the nonadiabatic coupling between the electronic system of the metal substrate and the adsorbate degrees of freedom by electronic friction [10–17].

The first observation of molecular motion on a surface induced by femtosecond laser pulses has been made in the group of Tony Heinz, where the adsorbate motion has been detected by scanning tunneling microscopy [18]. Since then only two experiments have been reported investigating the dynamics of electronically induced diffusion at surfaces on a femtosecond timescale [19, 20]. In this section, we will review these two experiments and their analysis using different variants of the electronic friction model. It will be shown that the coupling of different vibrational modes of the adsorbate is particularly important for the initiation of diffusive motion by electronic excitation. This coupling can be described by a recently developed theory for the energy transfer between adsorbates and ultrafast laser-heated hot electrons [17, 21]. This theory represents an extension of the standard electronic friction model that considers a general anharmonic oscillator potential for the adsorbate motion and coupling between different vibrational modes of the adsorbate.

In the following sections, we will start with a description of different experimental methods for the observation of the dynamics of electronically induced diffusion. Then, we will briefly discuss the different variants of the electronic friction models for the description of electronically induced surface reactions, with emphasis on the recently developed extension that considers the coupling of different vibrational modes. In the last section, we will review the two experiments on electronically induced diffusion of atomic oxygen and carbon monoxide on vicinal Pt(111) samples and their analysis within the electronic friction models.

18.2
Detection of Electronically Induced Diffusion

Experimentally, lateral motion following electronic excitation is more difficult to detect than desorbing atoms or molecules where quadrupole mass spectroscopy is typically employed. Early reports of electron-stimulated diffusion or migration of adsorbates were based on the changes of low-energy electron diffraction (LEED) patterns and changes of angular distributions of desorbing species as a function of electron irradiation [22, 23]. Direct spectroscopic evidence could be obtained for CO/Pt(335) where the migration of CO molecules from terrace to step sites induced by an electron gun was detected using infrared spectroscopy [24].

The observation of single hopping events of adsorbates on surfaces obviously falls into the domain of scanning tunneling microscopy (STM). Its capability to manipulate adsorbates on surfaces have led to a growing interest in electron-stimulated rearrangement processes of adsorbates. Several groups have shown that inelastic tunneling of electrons from an STM tip can lead to rotational and diffusive motion of adsorbed molecules [25–28]. Actually, the first observation of femtosecond laser-induced diffusion used the STM to detect adsorbate motion. Bartels and co-workers [18] were able to show for the system CO/Cu(110) that optical excitation of the substrate electrons by ultrashort laser pulses gives rise to diffusion of CO not only parallel but also perpendicular to the close-packed rows, while thermal excitation leads to diffusion only along the rows [18]. The STM, however, cannot readily be used to monitor laser-induced diffusion simultaneously with laser irradiation, which makes the determination of quantitative diffusion rates as a function of the parameters of an external excitation source, such as a laser, very time consuming.

Optical techniques for the detection of adsorbate diffusion, on the other hand, cannot observe the motion of individual atoms. They can nevertheless provide ensemble-averaged information about diffusion on a nanometer scale if the optical signal can discriminate between different adsorption sites. One example, infrared reflection absorption spectroscopy (IRAS), has already been mentioned. The inherent averaging of the optical signal over many individual events is a clear advantage compared to the observation of single events in terms of statistics.

18.2.1 Time-Resolved Techniques

The application of ultrashort laser pulses for the excitation of adsorbate motion makes it straightforward to implement time-resolved detection schemes. In general, one has to distinguish between ultrashort transient and time-integrated techniques to probe the dynamics of the adsorbate motion. If the probe is an ultrashort laser pulse that is synchronized with the pump pulse, as is the case in time-resolved sum-frequency generation (SFG), pump–probe techniques, which measure pump-induced changes as a function of the time delay between pump and probe pulses, can be used. In this way, the time-evolution of a specific optical property of the system is monitored with respect of the excitation. This technique can be universally applied even for DIET-type reactions where the reaction yield depends linearly on laser fluence. If the reaction yield has a nonlinear dependence on laser fluence, as is typical for DIMET-type reactions, time-resolved information about the excitation process can be also obtained by the application of a two-pulse correlation (2PC) scheme. This technique, which is well established in nonlinear optics, was first applied to laser-induced desorption by Budde *et al.* [29]. In a 2PC scheme, the pump pulse is split into two pulses with a variable time delay (Figure 18.1). Only at pulse–pulse delay times for which the first excitation is not already decayed the second pulse can generate an enhanced yield compared to independent excitations. A 2PC scheme can be combined with all, even nonoptical, detection techniques since it requires neither that the probe is an ultrashort laser pulse nor that the probe is synchronized with the pump pulses.

In principle, the time resolution of both techniques is only limited by the pulse durations of pump and probe pulses or the two pump pulses, respectively. However, the time resolution with which a specific reaction can be observed is in the end

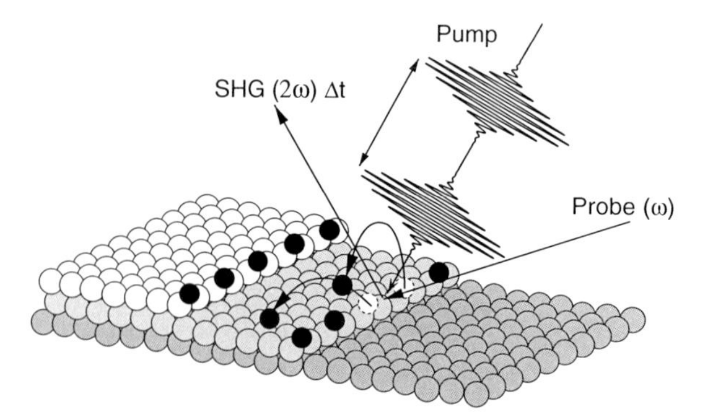

Figure 18.1 Observation of electronically induced diffusion by SHG. Step sites are decorated selectively by dissociative adsorption. Intensive ultrashort laser pulses induce diffusion onto the terraces. Adsorption as well as laser-induced diffusion are monitored by SHG of a probe beam. Two-pulse correlations are taken by measuring the hopping rate as a function of time delay between two pump laser pulses.

limited by the time response of the initial excitation. This applies for pump–probe schemes equally as for 2PC schemes because although the pump pulse is short, the effective start signal for the reaction is broadened. In addition, the initial excitation, for example, of metal electrons, might affect an optical probe in a complicated way until the initial excitation has decayed even if no reaction takes place. Pump–probe techniques, however, have the specific advantage that they provide the possibility to observe transient intermediate steps of the reaction. In the case of SFG, for example, this is the transient heating of the observed vibrational modes. On the other hand, 2PC techniques can often detect even very low reaction rates, since the probe can average over a long time independent on the optical excitation. Pump–probe techniques typically require a minimum reaction rate per laser shot in order to observe a transient change of the probe signal. Therefore, pump–probe and 2PC techniques give complementary information about the dynamics of a reaction.

18.2.2
Second Harmonic Generation

Even-order nonlinear optical techniques such as second harmonic generation (SHG) intrinsically have a high surface sensitivity owing to the symmetry breaking at the crystal surface. Regularly spaced steps on a vicinal surface result in an additional symmetry breaking parallel to the surface, which makes them into very efficient sources of SHG [31, 32]. This makes it possible to use SHG as a sensitive monitor of the occupation of step sites with adsorbates, where the simultaneous and continuous observation of a large number of sites allows the determination of even very small hopping rates [33].

In a typical SHG experiment, a relatively weak probe beam is focused onto the sample. The probe fluence has to be kept low enough to prevent the initiation of diffusion by the probe. If a strongly focused probe beam and a much larger pump beam spot are used, the probed area is illuminated by a nearly constant pump fluence. This makes it possible to correlate the change of the SH signal due to the pump beam directly with a certain pump fluence and yield averaging of the laser fluence is not necessary in contrast to laser-induced desorption experiments [34].

Even for low-index surfaces, the number of independent components of the second-order nonlinear susceptibility $\chi^{(2)}$ that describes the SH response of a surface is rather large [35] and in general unknown. For stepped surfaces the symmetry of the surface is reduced and the number of unknown components is even larger [31]. Therefore, it is in general not possible to predict quantitatively neither the contribution of steps to the total SH signal nor its change with adsorbate coverage. For this reason, the step sensitivity of the SH signal has typically to be optimized by varying the polarization of the fundamental and the detected second harmonic (Figure 18.2). For the case of oxygen adsorption at the steps of a vicinal P(111) surface, we have found that the polarization combination $p \rightarrow P$ (p-polarized fundamental and analyzing P-polarized second harmonic) gives the highest sensitivity to oxygen absorption [36]. With the step edges parallel to the plane of incidence the SH signal is, in this case, sensitive to the perpendicular contributions of the step edges to the nonlinear susceptibility, which dominates the total yield when the threefold rotational response of the in-plane components of the (111)-terraces has its minimum.

In order to relate the second-harmonic signal to the step coverage one can often express the nonlinear susceptibility in first approximation as a sum of a coverage-dependent step contribution $\chi_s^{(2)}$ proportional to the relative number of free step sites $(1-\theta_s)$ and a coverage independent (or only weakly coverage-dependent) background $\chi_{bg}^{(2)}$ that includes a remaining response of the terraces and possible bulk quadrupole contributions [37].

$$\chi^{(2)}(\theta_s) = \chi_s^{(2)}(1-\theta_s) + \chi_{bg}^{(2)} \tag{18.1}$$

The detected second-harmonic intensity $I(2\omega)$ is proportional to the square of the induced nonlinear polarization and thus to the square of the nonlinear susceptibility

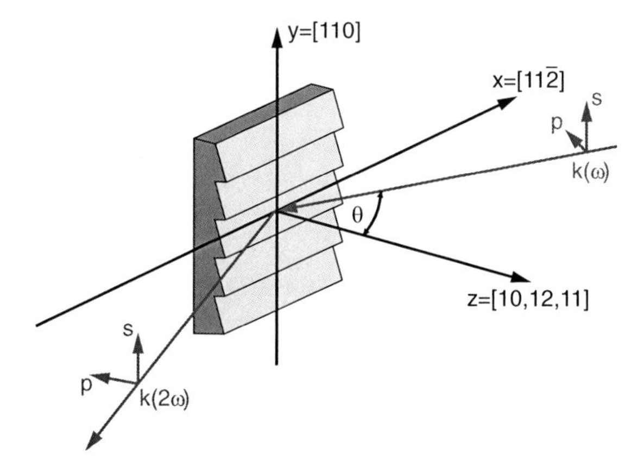

Figure 18.2 Scheme of the geometry used in the second-harmonic measurements.

$I(2\omega) \propto |\chi_s^{(2)} + \chi_{bg}^{(2)}|^2$. In general, the nonlinear susceptibility $\chi^{(2)}$ is a complex quantity and a phase shift between $\chi_s^{(2)}$ and $\chi_{bg}^{(2)}$ has to be considered. This phase shift can be determined by measuring the phase of the generated SH as a function of coverage. In UHV applications, this can be done by frequency domain second-harmonic interferometry [38, 39]. If the phase does not depend on coverage, $\chi_s^{(2)}$ and $\chi_{bg}^{(2)}$ have the same phase and the relative step coverage θ_s can be simply determined by taking the square root of the measured SH signal and by applying Eq. (18.1) with real quantities $\chi_s^{(2)}$ and $\chi_{bg}^{(2)}$.

18.2.3
Time-Resolved Sum-Frequency Generation

In the nonlinear technique of sum-frequency generation, the sample is irradiated with two different laser pulses. In a vibrational SFG experiment, an infrared photon (e.g., wavelength $\lambda = 5000$ nm corresponding to 2000 cm^{-1}) and a visible photon (e. g., $\lambda = 800$ nm corresponding to 12 500 cm^{-1}) are used. The resulting polarization emits light with the sum frequency of the two incoming beams, for example, 14 500 cm^{-1} ($\lambda = 690$ nm), a different frequency than the incoming beams. As a second-order nonlinear optical technique, SFG is surface sensitive, similar to its degenerate counterpart SHG. More importantly, however, when the infrared radiation is close to a vibrational resonance, the second-order susceptibility is enhanced, resulting in a peak in the spectrum; inversely, the spectrum reflects the vibrational properties of the surface molecules. The vibrational spectrum of the surface molecules contains detailed information about their local environment; for adsorbates on metal surfaces, one can generally readily distinguish adsorbates on different adsorption sites.

The time-resolved version of this spectroscopy is illustrated in Figure 18.3, which depicts transient spectra for the C–O stretch of adsorbed CO on a Pt surface, following optical excitation of the surface.

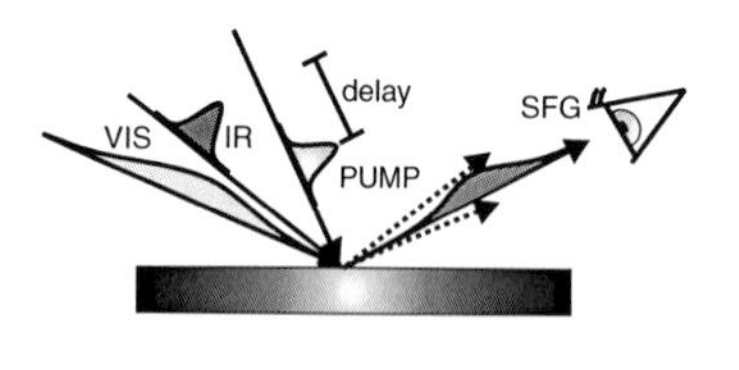

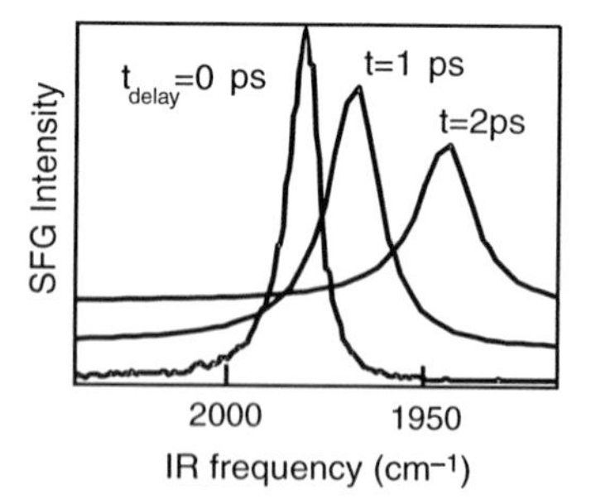

Figure 18.3 Scheme for time-resolved pump–SFG probe spectroscopy to map the time evolution of the vibrational properties of the adsorbates following an optical excitation pulse. The optical pulse excites electrons and, indirectly, phonons in the metal surface, both of which can couple to the molecular vibrations of the adsorbate. The rate and magnitude of the line shift reflects the coupling between the substrate and the adsorbate.

In these experiments, a broadband (~200 cm^{-1}, femtosecond (~100 fs) infrared laser pulse is incident on the surface, together with a narrowband visible pulse (bandwidth typically ~10 cm^{-1}, with ~ps duration). The cross correlation between the IR pulse and the femtosecond excitation ("pump") pulse determines the time resolution of the experiment, whereas the picosecond visible pulse determines the spectral resolution. Despite the fact that femtosecond laser pulses have inherently large bandwidths, frequency-resolved SFG is still possible, as demonstrated in Figure 18.3. The essence of femtosecond broadband SFG is that the vibrational resonance will select from the IR light pulse only those colors that are resonant with the vibrational transition. The IR field drives the coherent polarization (the constructively interfering sum of the oscillating microscopic dipoles) of the vibration under investigation, and the frequency of the polarization (and frequency distribution, i.e., linewidth) is determined by the system, and not by the driving IR field. Therefore, if the coherent polarization is upconverted by a spectrally narrow VIS-pulse, the width of the resulting SFG spectrum (right panel) reflects that of the vibrational transition, and is independent of the spectral shape of the IR pulse.

Within the limit in which the spectral parameters such as central frequency and linewidths are varying slowly with respect to the inverse of the vibrational linewidth, the system can be described by a quasisteady state, further facilitating the analysis. However, in the example presented below, the time-dependence of the spectral changes has to be taken into account explicitly.

18.3 Description of Electronically Induced Motion by Electronic Friction

18.3.1 Electronic Friction Models

Surface reactions of adsorbates are typically triggered by an excitation of specific degrees of freedom of the adsorbate such as vibrational or rotational modes. The mechanism of an electronic excitation of these modes should be similar for different kinds of reactions such as diffusion or desorption. The main differences are related to the relevance of the different degrees of freedom. Desorption, for example, typically requires vibrational excitations of the adsorbate against the substrate, whereas diffusion requires an excitation of lateral vibrational modes such as the frustrated translation. This suggests that the detailed reaction pathways for these two processes may differ and depend on the degree of electronic versus internal coupling of the different vibrational modes. The electronic excitation of the vibrational modes, however, can be described by similar models.

On metal surfaces, femtosecond laser-induced surface reactions are usually described as an indirect excitation of the adsorbate by laser-excited hot electrons of the substrate. Direct photoexcitation of the adsorbate, in particular for near-infrared excitation, typically plays a minor role. Such an indirect hot-electron excitation process is usually analyzed in two steps. First, the dynamics of the laser-excited hot-electron distribution is described by applying the well-known two-temperature model [40]. This model describes the energy flow between the electronic and the phononic system of the substrate by two coupled heat diffusion equations, which allows the assignment of time-dependent electronic and ionic temperatures $T_e(t)$ and $T_i(t)$ assuming instantaneous equilibration in each subsystem. This assumption is reasonable in most cases, since an indirect excitation of an adsorbate by hot electrons typically proceeds on the timescale of at least several hundred femtoseconds. This is also the typical time until the initial nonequilibrium energy distribution of the excited electrons transforms into a hot Fermi–Dirac distribution by electron–electron scattering [41–43].

Two principal models have been employed for the description of the second step, the energy transfer to the nuclear coordinates of the adsorbate. One is the DIMET model [3] that is derived from the MGR model [1, 2] of the usual DIET process. The DIMET model describes the excitation of the adsorbate by multiple electronic transitions between potential energy curves with high-lying adsorbate levels and emphasizes the dynamics of the adsorbate on the excited potential energy surface (PES). A description in the DIMET-model is most advantageous if the adsorbate dynamics proceed most of the time on one of the PES, and excitations are relatively rare.

For an increasing number of excitation and deexcitation cycles by Franck–Condon transitions at a different distance to the surface, the electronic excitations lead to an effective vibrational heating in the ground state. Within this limit, the DIMET model makes contact to a second one, the electronic friction model [10, 13–15] that treats the coupling between the adsorbate and the electrons via an effective electronic friction. It is based on the theory of vibrational damping of adsorbates by the creation of electron–hole pairs [44, 45], which is the inverse process of electronic excitation of adsorbate motion. It emphasizes the dynamics on the ground state PES perturbed by the excited electrons and is most suitable in the presence of low-lying excited states of the adsorbate–substrate system. However, both descriptions are closely related and describe the same physical processes from different points of view.

In the empirical friction model [11, 12], the adsorbate is treated as a harmonic oscillator that is coupled to the heat bath of the electrons via an electronic friction coefficient η_e. By applying a master equation formalism, a relation for the temporal evolution of the average vibrational energy of the adsorbate of the form

$$\frac{d}{dt} U_a = \eta_e [U_e - U_a] \tag{18.2}$$

has been derived, where U_e and U_a are the Bose–Einstein distributed mean vibrational energies of an oscillator in equilibrium with the electron temperature T_e and an adsorbate temperature T_a, respectively, and

$$U_x = h\nu_a \left(e^{h\nu_a/k_B T_x} - 1\right)^{-1}. \tag{18.3}$$

Here, ν_a is the frequency of the adsorbate vibration along the reaction coordinate. The friction coefficient η_e is normalized to the adsorbate mass M and can be regarded as the inverse of an electron–adsorbate energy transfer time $\tau_{ea} = 1/\eta_e$. It is related to the friction coefficient f of a classical velocity proportional friction force $F_e = -fv$ by $f = \eta_e M$ [10]. Coupling to the phononic system has been considered by adding a corresponding term to Eq. (18.2), which introduces an ion–adsorbate energy transfer time [34]. This term can describe contributions to the reaction dynamics that proceeds on a timescale of several picoseconds [34, 46].

In the high-temperature limit ($h\nu_a/k_B T_x \ll 1$ and therefore $U_x \approx k_B T_x$), Eq. (18.2) can be simplified to

$$\frac{d}{dt} T_a(t) = \eta_e [T_e(t) - T_a(t)], \tag{18.4}$$

which represents the conventional heat transfer equation for the calculation of the transient adsorbate temperature $T_a(t)$ for an electronic friction.

In the empirical friction model, the friction coefficient η_e is constant, and the reaction rate $R(t)$, for first-order kinetics, is given by an Arrhenius-type expression as

$$R(t) = -\frac{d}{dt}\theta(t) = \theta(t)\nu_a\, e^{-E_a/k_B T_a(t)}, \tag{18.5}$$

where θ denotes the relative coverage of the available sites. Thus, the initial reaction probability for a single laser shot at small changes of the coverage ($\theta \approx 1$) is simply given by $p = \int \nu_a \exp(-E_a/k_B T_a(t))\, dt$.

Brandbyge *et al.* have shown that both DIMET and friction regime can be covered within a generalization of the electronic friction model [14]. They derived an analytic result for a spatially independent friction and a truncated harmonic oscillator potential of depth E_a for the description of the adsorbate–surface interaction in the electronic ground state. This leads to the same heat transfer equation as in the empirical friction model (Eq. (18.4)), but with a friction coefficient that depends in general on time and space and that needs to be calculated by microscopic theories [13, 16, 47]. It has been shown that the friction coefficient has a weak dependence on electron temperature and therewith on time if the adsorbate resonance is broad and close to the Fermi level. This case corresponds to the theory of adsorbate vibrational damping. The DIMET limit of a high-lying and well-defined affinity level, on the other hand, is characterized by a friction coefficient, which is small at low electron temperatures and increases strongly if the electron temperature is large enough to populate significantly the adsorbate resonance [14]. For thermal energies $k_B T_a$ that are small compared to E_a, the probability to overcome the barrier E_a is given in this formalism as the time integral over a rate

$$R(t) = \eta_e(t) \frac{E_a}{k_B T_a(t)}\, e^{-E_a/k_B T_a(t)}. \tag{18.6}$$

For the laser-induced diffusion experiment, E_a is given by the diffusion barrier E_{dif}, and the hopping probability per laser shot is given by $p_{dif} = \int R(t)dt$.

18.3.2
Generalized Description of Heat Transfer at Surfaces

Recently, Ueba and Persson have developed a theory for the energy transfer between adsorbates and ultrafast laser-heated hot electrons using a newly introduced heat transfer coefficient that depends on the adsorbate rather than on the electron temperature [17, 21]. This theory extends the standard electronic friction model by considering a general anharmonic oscillator potential for the adsorbate motion as well as a coupling between different vibrational modes of the adsorbate. The latter was motivated by real-time studies on lateral motion of atomic oxygen and CO molecules on vicinal Pt surfaces [19, 20] as described in this chapter.

For the system oxygen on vicinal Pt(111), Stépán *et al.* have found that a consistent modeling of both the observed 2PC and the fluence dependence of the hopping probability requires the introduction of an empirical temperature-dependent electronic friction coefficient $\eta_e(T_e(t))$ in the conventional heat transfer Eq. (18.4) [19]. This result was interpreted in terms of an indirect excitation mechanism [48]. It has been proposed that the substrate electrons primarily excite O-Pt vibrations that then couple anharmonically to the frustrated translation (FT) mode required to overcome the barrier for lateral motion. Such a temperature-dependent friction model has also been used to explain the 2PC and the fluence dependence observed for femtosecond laser-induced desorption of oxygen molecules from a Pd(111) surface [49].

Employing time-resolved SFG spectroscopy to monitor the hopping of CO molecules from step onto terrace sites on Pt(533), Backus *et al.* unexpectedly found that the excitation of the frustrated rotation (FR) mode is essential in the CO hopping process [20]. Excitation of the FT mode is significantly too slow to be primarily responsible for the hopping motion of the CO on subpicosecond timescale observed experimentally. It has been shown that the reaction pathway involves translational motion, which is already thermally populated in their experiment performed at 100 K. This precursor state couples to the rotational motion excited by ultrafast coupling to hot substrate electrons. The hopping motion is such as a dance in which the CO molecules execute concerted rocking and translational steps [50]. It should be mentioned here that long time before this experiment was performed, Dobbs and Doren [51] have demonstrated using classical molecular dynamics that the bending and the lateral translational motions are strongly coupled near the transition state for diffusion in a model system of CO/Ni(111). This mechanism permits excitation of the FT mode by the FR mode. The FR mode then mediates energy transfer between the surface and the reaction coordinate modes. The subtleties of the coupling of low frequency modes to reaction coordinate modes in surface–adsorbate systems may be generally more complex than we imagine. The idea behind such anharmonic coupling between different modes whereby only one of them needs to be initially excited has been originally used to rationalize hopping experiments of CO molecules on a Cu(111) surface by electron attachment during vertical manipulation with a scanning tunneling microscope (STM) [52].

These experimental studies, and the elementary processes suggested thereby, motivated Ueba and Persson [21, 53, 54] to explore energy transfer in the presence of the coupling between two different vibrational modes excited by femtosecond laser pulse heated hot electrons in the substrate. The vibrational mode coupling has been demonstrated to play a key role in hopping of a single CO molecule on Pd(110) [26] and a NH_3 molecule on Cu(100) [28], where energy transfer from the C−O (N−H) stretch vibration (excited by tunneling electron injected using a scanning tunneling microscope) to the FT mode induces hopping [55].

In the model of Ueba and Persson, the energy (heat) transfer between the adsorbed layer and the substrate, which is excited to high temperature $k_B T_e \gg \hbar\omega_0$ (where $\hbar\omega_0$ is the vibrational quantum) is determined by a heat transfer coefficient α. For small adsorbed molecules, the different vibrational modes will exhibit different temperatures, so that each molecular vibrational mode ν may be characterized by a different temperature T_ν. In this case, the energy Q_ν in a mode ν satisfies

$$\frac{dQ_\nu}{dt} = \alpha_\nu (T_e - T_\nu), \tag{18.7}$$

assuming that the coupling between the adsorbate vibrations and the substrate occurs primarily via the electronic excitations in the substrate. Experiment and model calculations have shown that laser pulses in the femtosecond range can generate very high surface temperatures on subpicosecond timescales at metal surfaces [7, 12]. The laser photons are absorbed in the substrate within the optical skin depth (typically 10 nm) and may lead to an effective electronic temperature of the order of several thousands of Kelvin. At these high temperatures, the adsorbate modes may be highly excited, and in this case strong intramolecular anharmonic coupling may lead to transfer of energy between different adsorbate modes in order to conserve the energy involving some energy exchange with the substrate. Because of heat transfer to phonons and heat diffusion into the bulk, the high electronic temperature only exists for few picoseconds. Because of the relatively small heat capacity of a degenerate electron system, the increase in the phonon temperature is quite small. For this reason, laser-induced adsorbate reactions are often caused by the interaction between the adsorbates and the hot electron system, rather than interaction with phonons. Experimental data of the type discussed above are usually analyzed by assuming instantaneous equilibrium in each subsystem so that the surface region of the solid exhibits time dependent electronic and phonon temperatures $T_e(t)$ and $T_{ph}(t)$. The population of vibrational levels in the adsorbate system is also assumed to be in (local) equilibrium so that we can speak about an adsorbate temperature $T_{ad}(t)$ that may differ for different adsorbate vibrational modes. We are interested in reactions (e.g., desorption or diffusion) that are going over a reaction energy barrier. However, before going over the barrier, an adsorbate will in general perform many jumps between its vibrational levels. Thus, the ensemble of adsorbates will be nearly thermalized, except for the levels close to the barrier top, where the population is nonequilibrium because of the transfer over the barrier.

We define the time dependent adsorbate temperature $T_{\mathrm{ad}}(t)$. For $k_{\mathrm{B}}T_{\mathrm{ad}} \gg \hbar\omega_0$ and $k_{\mathrm{B}}T_{\mathrm{e}} \gg \hbar\omega_0$ the adsorbate temperature $T_{\mathrm{ad}}(t)$ is usually calculated using the heat transfer equation with a constant friction coupling η_{e} in Eq. (18.4).

Let P_{n} be the probability that an adsorbate is in the vibrational excited state $|n\rangle$. The function $P_{\mathrm{n}}(t)$ satisfies the rate equation

$$\frac{\mathrm{d}P_n}{\mathrm{d}t} = \sum_m P_m w_{m\to n} - \sum_m P_n w_{n\to m}, \tag{18.8}$$

where $w_{m\to n}$ is the transition rate from the vibrational state $|m\rangle$ to the state $|n\rangle$, caused by the interaction with hot electrons in the metal. We assume local thermal equilibrium on the adsorbate so that

$$P_n = Z_{\mathrm{a}}^{-1}\mathrm{e}^{-\beta_{\mathrm{ad}}E_n}, \tag{18.9}$$

where $\beta_{\mathrm{ad}} = 1/k_{\mathrm{B}}T_{\mathrm{ad}}$ and E_n $(n = 0, 1, \ldots)$ are the vibrational energy levels (for a harmonic oscillator $E_n = n\hbar\omega$) and $Z_a = \sum_n \mathrm{e}^{-\beta_{\mathrm{ad}}E_n}$. Substituting Eq. (18.9) in Eq. (18.8), multiplying with E_n and summing over n give an evolution of the vibrational energy $Q_{\mathrm{ad}} = \sum_n E_n P_n$,

$$\frac{dQ_{\mathrm{ad}}}{\mathrm{d}t} = \sum_{nm}\left[\mathrm{e}^{-\beta_{ad}(E_m - E_n)}W_{m\to n} - W_{n\to m}\right]E_n P_n. \tag{18.10}$$

Now we consider an adsorbate with two different vibrational **a** and **b** modes, excited by the hot electrons via the friction couplings η_{a} and η_{b}, respectively. We assume an anharmonic coupling between the modes excited by energy (heat) transfer from hot electrons so that energy can be directly transferred between the two modes. In order to conserve the total energy, the vibrational transitions are accompanied by the emission or absorption of substrate elementary excitation such as electron–hole pair excitation. We assume the Hamiltonian to be of the form

$$H = H_{\mathrm{a}} + H_{\mathrm{b}} + V(u), \tag{18.11}$$

where H_{a} and H_{b} are related to the vibrational a and b modes (normal mode coordinate u_{a} and u_{b}), respectively. The anharmonic mode coupling $V(u)$ between a and b modes is assumed to be of the form

$$V(u) = \sum_q \lambda_q (b_q^+ + b_q) f(u_{\mathrm{a}}, u_b), \tag{18.12}$$

where the coupling constant λ_q is assumed to be temperature independent and function f determines the order of the coupling through its expansion in coordinates u_{a} and u_{b}. The nucleus–electron interaction is, in general, of the form $\sum_{\alpha,\beta} V_{\alpha,\beta}(u_1, u_2 \ldots) c_\alpha^+ c_\beta$ (where u_1, u_2, are, e.g., normal mode or reaction coordinates of the nucleus and $c_\alpha^+ c_\beta$ represents the electronic excitation from level ε_β to level ε_α in the substrate). We treat the electronic excitations as bosons (b_q^+, b_q) so that we get terms of the form $\sum_q V_q(u_1, u_2 \ldots)(b_q^+ + b_q)$. The lifetimes of the low frequency adsorbate modes are determined by the same coupling [56], but with $V_{\alpha,\beta}(u_1, u_2 \ldots)$ expanded to linear order in $u_1, u_2, \ldots$. The linear order terms, however, cannot give rise to the process of energy transfer between different adsorbate vibrational modes. We therefore consider the leading nonlinear term in the expansion of $V(u)$ that can give rise to mode–mode coupling and energy transfer from one mode to another.

Let us now calculate the probability rate $W_{n_a \to m_a}$s that a vibrational quantum is transferred between a and b mode, while mode a makes a transition from state n_a to m_a accompanied with an emission or absorption of a bulk elementary excitation ω_q. We get

$$W_{n_a \to m_a} = -\int d\omega[\varrho_\lambda(\omega) - \varrho_\lambda(-\omega)]n(-\omega)w_{n_a m_a}(\omega), \tag{18.13}$$

where

$$\varrho_\lambda(\omega) = \sum_q |\lambda_q|^2 \delta(\omega_q - \omega), \tag{18.14}$$

is the spectral density of states of the substrate elementary excitations and

$$\begin{aligned} w_{n_a m_a}(\omega) &= \frac{2\pi}{\hbar}\sum_{n_b m_b} |\langle m_a, m_b | f(u_a, u_b) | n_a, n_b\rangle|^2 \\ &\quad \times \delta(E_{m_a} + E_{m_b} + \omega - E_{n_a} - E_{n_b})P(n_b). \end{aligned} \tag{18.15}$$

In a similar way, we also obtain

$$W_{m_a \to n_a} = \int d\omega[\varrho_\lambda(\omega) - \varrho_\lambda(-\omega)]n(\omega)w_{m_a n_a}(-\omega), \tag{18.16}$$

where

$$\begin{aligned} w_{m_a n_a}(-\omega) &= \frac{2\pi}{\hbar}\sum_{n_b m_b} |\langle m_a, m_b | f | n_a, n_b\rangle|^2 \\ &\quad \delta(E_{n_a} + E_{n_b} - \omega - E_{m_a} - E_{m_b})P(m_b). \end{aligned} \tag{18.17}$$

Substituting Eq. (18.13) and (18.16) in Eq. (18.10) gives a heat transfer equation for the a-mode at the high temperatures $n(\omega) \approx 1/\beta_e\omega$ so that $[\varrho_\lambda(\omega) - \varrho_\lambda(-\omega)]\, n(\omega) \approx 2\lambda^2\varrho^2(\varepsilon_F)/\beta_e$, where $\varrho(\varepsilon_F)$ is the density of states at the Fermi level ε_F, we finally obtain (see [21, 53] for more details),

$$\frac{dQ_a(t)}{dt} = \alpha_a(T_a)(T_e - T_a) + k_B S\,[(T_e - T_a)M_{aa}\beta_a + (T_e - T_b)\,M_{ab}\beta_b], \tag{18.18}$$

$$M_{aa} = 1/2\sum_{n_a m_a}\sum_{n_b m_b} |\langle m_a, m_b | \bar{f} | n_a, n_b\rangle|^2(-E_{m_a n_a})E_{n_a}P(n_a)P(n_b), \tag{18.19}$$

and

$$M_{ab} = 1/2 \sum_{n_a m_a} \sum_{n_b m_b} |\langle m_a, m_b | \bar{f} | n_a, n_b \rangle|^2 (-E_{m_b n_b})\, E_{n_a} P(n_a) P(n_b). \qquad (18.20)$$

where $S = 8\pi\, [u_{a0} u_{b0} \lambda \varrho\, (\varepsilon_F)]^2$ (where u_{a0} and u_{b0} are the zero point vibration amplitudes of the a and b modes, respectively), $\bar{f} = f/(u_{a0} u_{b0})$, $E_{m_a n_a} = E_{m_a} - E_{n_a}$, $E_{m_b n_b} = E_{m_b} - E_{n_b}$. In Eq. (18.18) we have added the contribution $\alpha_a\,(T_a)\,(T_e - T_a)$ due to the direct coupling between a-mode and hot electrons in the substrate, Here the heat transfer coefficient $\alpha_a\,(T_a)$ is defined by

$$\alpha\,(T_a) = k_B \eta_e \beta_a \sum_n M_n E_n P_n, \qquad (18.21)$$

where

$$M_n = \sum_m |\langle m | \bar{f}(u) | n \rangle|^2 \frac{E_{mn}^2}{2 E_n \hbar \omega_a}, \qquad (18.22)$$

takes into account that when the adsorbate temperature increases, the adsorbate will probe regions in u far away from the equilibrium position $u = 0$, and this will result in a modified overlap between the adsorbate orbitals and the metal electrons. This picture is particularly clear when u corresponds to the vertical displacement of the adsorbate. In this case, as $u \to \infty$ the overlap between the adsorbate and the substrate vanishes, that is, $f(u) = 0$, and in this limit $M_n = 0$. It has been shown before for a Morse potential that M_n decreases with increasing vibrational energy [17, 21]. This simply reflects that the (average) position of the adsorbate moves further away from the surface as the vibrational excitation energy increases, which reduces the adsorbate–surface interaction.

If we assume that the adsorbate mode can be treated as a harmonic oscillator and if we assume a linear coupling $f(u) = u = u_0\,(b + b^+)$, we get $M_n = 1$ from Eq. (18.22). Using this at high temperatures, $\sum_n E_n P_n = k_B T_a$ for a harmonic oscillator, Eq. (18.21) gives $\alpha(T_a) = k_B \eta_e$, so that Eq. (18.18) reduces to Eq. (18.4) without mode coupling. This indicates that Eq. (18.4) is valid for a harmonic oscillator with linear electron-vibration coupling. In addition, if the adsorbate temperature becomes so high that the vibrational levels in the anharmonic part of the potential well are strongly populated, we can expect that $\alpha(T_a)$ exhibits a strong temperature dependence and gets modified as a result of anharmonicity. This is likely to be the case for a very short time in many high-fluency ultrafast laser spectroscopy experiments. It is also noted that $\alpha(T_a)$ determines not only the heat transfer from the substrate to the adsorbate when the substrate is hotter than the adsorbate but also the reverse process when the adsorbate is hotter than the substrate.

In the same way, we also obtain

$$\frac{dQ_b(t)}{dt} = \alpha_b (T_e - T_b) + k_B S\, [(T_e - T_b) M_{bb} \beta_b + (T_e - T_a) M_{ba} \beta_a], \qquad (18.23)$$

where $M_{\mathrm{ba}} = M_{\mathrm{ab}}$ and

$$M_{bb} = 14 \sum_{n_a m_a} \sum_{n_b m_b} |\langle m_a, m_b | \bar{f} | n_a, n_b \rangle|^2 E^2_{m_b n_b} P(n_a) P(n_b). \tag{18.24}$$

Note that M_{aa}, M_{ab}, and M_{bb} depend on T_{a} and T_{b}, but not on the substrate temperature T_{e}. It is also remarked that if $T_{\mathrm{e}} \neq T_{\mathrm{a}}$ (or $T_{\mathrm{e}} \neq T_{\mathrm{b}}$) the second terms of Eqs. (18.18) and (18.23) do not vanish even for two identical modes ($\omega_{\mathrm{a}} = \omega_{\mathrm{b}}$) with the same temperatures $T_{\mathrm{a}} = T_{\mathrm{b}}$. This can be understood as follows: Assume first that there are no direct frictional coupling between a and b modes and the substrate, that is, $\eta_{\mathrm{a}} = \eta_{\mathrm{b}} = 0$ (so that $\alpha_{\mathrm{a}} = \alpha_{\mathrm{b}} = 0$ in Eqs. (18.18) and (18.23)). But, we will assume that higher order terms in the expansion of $f(u_{\mathrm{a}}, u_{\mathrm{b}})$, which couple a-mode to b-mode, do not vanish. Now, assume that $T_{\mathrm{a}} = T_{\mathrm{b}} \neq T_{\mathrm{e}}$. In this case, if the new terms we derived (second terms in Eqs. (18.18) and (18.23)) would vanish when $T_{\mathrm{a}} = T_{\mathrm{b}}$ then $\mathrm{d}Q_{\mathrm{a}}/dt = 0$, and we would *never* reach thermal equilibrium in spite of the fact that there is a coupling between the adsorbed molecule and the electronic excitations of the substrate. However, the terms we calculated do not vanish unless $T_{\mathrm{e}} = T_{\mathrm{a}} = T_{\mathrm{b}}$ so that the system, after long enough time, reaches thermal equilibrium as expected because of the coupling to the substrate electrons.

We now consider that two different vibrational modes in the harmonic potential interact with each other through the lowest-order coupling $f(u_{\mathrm{a}}, u_{\mathrm{b}}) = u_{\mathrm{a}} u_{\mathrm{b}}$. Substituting this in Eqs. (18.19) and (18.20) gives

$$M_{\mathrm{aa}} = \frac{\omega_{\mathrm{a}}}{\omega_{\mathrm{b}}} (\beta_{\mathrm{a}} \beta_{\mathrm{b}})^{-1}, \; M_{\mathrm{bb}} = \frac{\omega_{\mathrm{b}}}{\omega_{\mathrm{a}}} (\beta_{\mathrm{a}} \beta_{\mathrm{b}})^{-1}, \; M_{\mathrm{ab}} = \omega_{\mathrm{a}} \omega_b, \tag{18.25}$$

so that

$$\frac{\mathrm{d}T_{\mathrm{a}}(t)}{\mathrm{d}t} = \eta_{\mathrm{a}}(T_{\mathrm{e}} - T_{\mathrm{a}}) + S\omega_{\mathrm{a}} \left[\frac{(T_{\mathrm{e}} - T_{\mathrm{a}})}{\omega_{\mathrm{b}} \beta_{\mathrm{b}}} + (T_{\mathrm{e}} - T_{\mathrm{b}}) \omega_{\mathrm{b}} \beta_{\mathrm{b}} \right], \tag{18.26}$$

$$\frac{\mathrm{d}T_{\mathrm{b}}(t)}{\mathrm{d}t} = \eta_{\mathrm{b}}(T_{\mathrm{e}} - T_{\mathrm{b}}) + S\omega_{\mathrm{b}} \left[\frac{(T_{\mathrm{e}} - T_{\mathrm{b}})}{\omega_{\mathrm{a}} \beta_{\mathrm{a}}} + (T_{\mathrm{e}} - T_{\mathrm{a}}) \omega_{\mathrm{a}} \beta_{\mathrm{a}} \right]. \tag{18.27}$$

We define indirect friction parameters $\eta_{\mathrm{ab}} = S\omega_{\mathrm{a}}$ and $\eta_{\mathrm{ba}} = S\omega_{\mathrm{b}}$ so that $\eta_{\mathrm{ab}}/\eta_{\mathrm{ba}} = \omega_{\mathrm{a}}/\omega_{\mathrm{b}}$. Notice that the second terms in the [...] brackets in Eqs. (18.26) and (18.27) are in general negligible compared to the first terms. Then, we obtain

$$\frac{\mathrm{d}T_{\mathrm{a}}(t)}{dt} = \left[\eta_{\mathrm{a}} + \eta_{\mathrm{ab}} \frac{k_{\mathrm{B}} T_{\mathrm{B}}}{\hbar \omega_{\mathrm{b}}} \right] (T_{\mathrm{e}} - T_{\mathrm{a}}), \tag{18.28}$$

$$\frac{\mathrm{d}T_{\mathrm{b}}(t)}{\mathrm{d}t} = \left[\eta_{\mathrm{b}} + \eta_{\mathrm{ba}} \frac{k_{\mathrm{B}} T_{\mathrm{a}}}{\hbar \omega_{\mathrm{a}}} \right] (T_{\mathrm{e}} - T_{\mathrm{b}}), \tag{18.29}$$

where the effective friction

$$\eta^{\mathrm{eff}}_{\mathrm{a(b)}} = \eta_{\mathrm{a(b)}} + \eta_{\mathrm{ab(ba)}} (k_{\mathrm{B}} T_{\mathrm{b}(a)} / \hbar \omega_{\mathrm{b}(a)}), \tag{18.30}$$

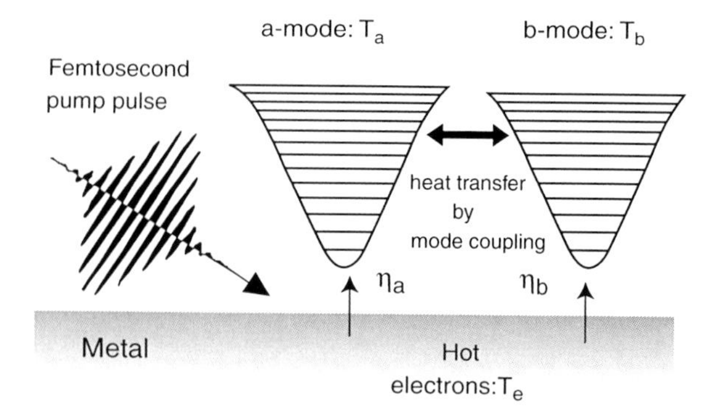

Figure 18.4 Schematic illustration of the heat transfer between the a and the b modes via η_{ab} where both modes couple to the heat source of hot electrons with η_a and η_b, respectively.

depends on the temperature of the coupling partner mode. Because the energies of the low-frequency modes involved in adsorbate motions (hopping, rotation, and desorption) are, in general, quite low (e.g., the FT mode and the FR mode of a CO molecule on a Pt surface are few meV and several tens of meV (1 meV = 11.6 K), respectively), the additional heat transfer coefficients $\eta_{ab(ba)}\ (k_B T_{b(a)}/\hbar\omega_{b(a)})$ become extremely large. This is the most important ingredient for heating by mode coupling.

It is now clear that the mode coupling $V = u_a u_b \sum_q \lambda_q (b_q + b_q^+)$ can give a strong temperature increase and may give the dominant contribution to the heating of a-mode when the direct coupling η_a between the a-mode and the hot electrons is small, as it is often the case for parallel adsorbate vibrations. Figure 18.4 illustrates the heat transfer between the a and b modes via mode coupling η_{ab} where both modes are coupled to the heat source of hot electrons with η_a and η_b, respectively.

Prior to the application of this heat transfer/mode-coupling model, we will first demonstrate how it works. Figure 18.5 shows the time dependence of T_a and T_b

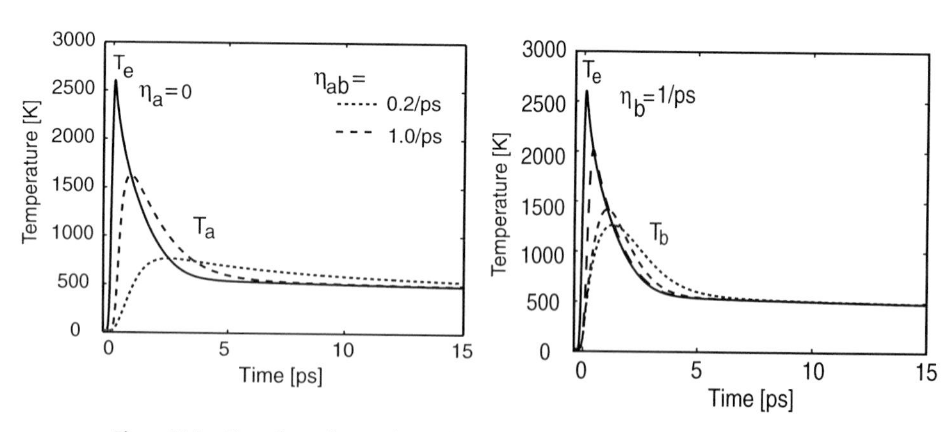

Figure 18.5 Time dependence of T_e (solid curve), T_a (left) for $\eta_{ab} = 0.2$/ps (dotted curve) and 1.0/ps (dashed curve) and T_b (right) for $\eta_b = 1$/ps (dotted: without mode coupling) and $\eta_{ab} = 0.2$/ps (short-dashed curve), 1.0/ps (long-dashed curve). See the text for the parameters used herein.

calculated using Eqs. (18.28) and (18.29) for the a-mode $\hbar\omega_a = 20$ meV with $\eta_a = 0$ (no direct heating by hot electrons and remains at $T_a = T_0$ without a-mode coupling) and the b-mode $\hbar\omega_b = 50$ meV with $\eta_b = 1$/ps for the mode coupling $\eta_{ab} = 0.2$ and 1.0/ps, where T_e (the solid curve) is calculated for a laser pulse width (FWHM) of 0.13 ps and wavelength of 800 nm, an absorbed laser fluence of 70 J/m^2 irradiated on a Pt metal at the initial temperature $T_0 = 20$ K using a set of the material parameters (Debye temperature = 240 K, electron heat capacity = 750 J/m^{3}K^2, heat conductivity = 72 W/mK, skin depth = 13 nm, and the electron–phonon coupling = 0.5×10^{18} W/m^{3}K). The laser pulses excite the electrons in the substrate, and due to the small heat capacity of the electrons compared to the lattice, $T_e(t)$ rises within the subpicosecond pulse width to levels far above the melting point of the lattice. This electronic excitation energy is then dissipated either by electron diffusion into the bulk or by energy transfer into the phonon heat bath via electron–phonon interaction. This causes an increase of the phonon temperature T_{ph} on much slower timescales of picoseconds than the electronic response. $T_e(t)$ and $T_{ph}(t)$ then gradually reach thermal equilibrium on a timescale of approximately the electron–phonon coupling time. More details of the energy transfer processes in the metal substrate are described in Ref. [7].

As shown in Figure 18.6, T_a and T_b increase with η_{ab}. The a-mode without direct heating can be heated up through the coupling to the b-mode excited by hot electrons. Here, the following list of parameters is used: the initial temperature $T_0 = 60$ K, the fluence $F = 15$ J/m^2, $\hbar\omega_a = 4.4$ meV, $\eta_a = 0.25$/ps, $\hbar\omega_b = 51$ meV, $\eta_b = 0.3$/ps, and $\eta_{ab} = 1.25$/ps.

It is also found that $T_a(t)$ exhibits a peak after $T_e(t)$ has already reached its maximum. This is also true for slower phonon-mediated processes characterized by

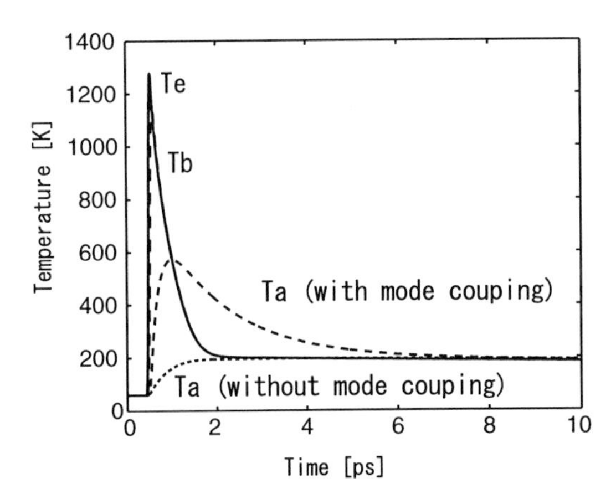

Figure 18.6 Time dependence of T_e (solid curve), T_a (dashed curve) and T_b calculated using Eqs. (18.28) and (18.29). The dotted curve shows T_a without mode coupling. Here, T_e and T_b are hard to be distinguished because of a strong coupling η_b. See the text for the rest of parameters.

prolonged cooling. What Luntz *et al.* found with a three-dimensional model based on molecular dynamics is that this time separation between the maximums of T_e and T_a is essentially the time a molecule needs to climb out of the adsorption well. In terms of the heat path picture, this delayed rise and peak of the adsorbate temperature with respect to T_e reflects the time that is needed to heat up the energy reservoir of the relevant adsorbate motion leading to reaction [57].

18.4 Results

18.4.1 O/Pt(111)

The choice of atomic oxygen on a vicinal Pt(111) surface for the study of laser-induced diffusion was motivated by its model character for diffusion of a strongly chemisorbed atomic adsorbate and for experimental reasons. Due to the importance of Pt as a catalyst in oxidation reactions the dissociative adsorption of O_2 on Pt has been well characterized by a variety of methods (see, for example, Refs [58, 59] and references therein). Steps increase the reactivity of the substrate dramatically [59–61] and STM investigations at temperatures lower than 160 K, where atomic oxygen is immobile, have shown that the dissociative adsorption takes place directly at the step edges [62]. This makes it possible to easily generate a well-defined initial distribution of oxygen atoms that occupy all available step sites. Hopping from step sites onto the initially empty terraces was induced by absorption of femtosecond laser pulses as sketched in Figure 18.1. The relatively large diffusion barrier ensured that diffusion was only induced by the laser pulses and not initiated by thermal activation at a temperature of 80 K.

Figure 18.7a shows typical raw SHG data sets for various absorbed fluences of the pump laser pulses as a function of time during dosage and induced diffusion of oxygen. For each run, the sample was first kept at 160 K and exposed to constant flux of molecular oxygen. At this temperature, chemisorbed O_2 is not stable on the terraces. It desorbs or it diffuses to the step edges where it preferentially dissociates and forms strongly bound atomic oxygen on top of the step edges [62]. Filling of the step sites with atomic oxygen leads to a strong reduction of the SHG signal until the steps are saturated ("dosing" regime in Figure 18.7). This strong reduction with oxygen coverage demonstrates the high sensitivity of the SHG detection even for a low step density of 1/12. The monotonous decrease of the SHG signal makes it possible to relate the SHG signal to the relative coverage of the step sites θ_s (right scale in Figure 18.7) using a simple model for the coverage dependence of the nonlinear susceptibility [36]. Following step decoration of the sample with atomic oxygen, the sample was cooled down to 80 K where oxygen is immobile even on the terraces. Partial depletion of the steps by thermal diffusion has been observed for sample temperatures exceeding 260 K. After reaching 80 K, the sample was irradiated with femtosecond laser pulses at a repetition rate of 1 kHz ("laser-induced diffusion"

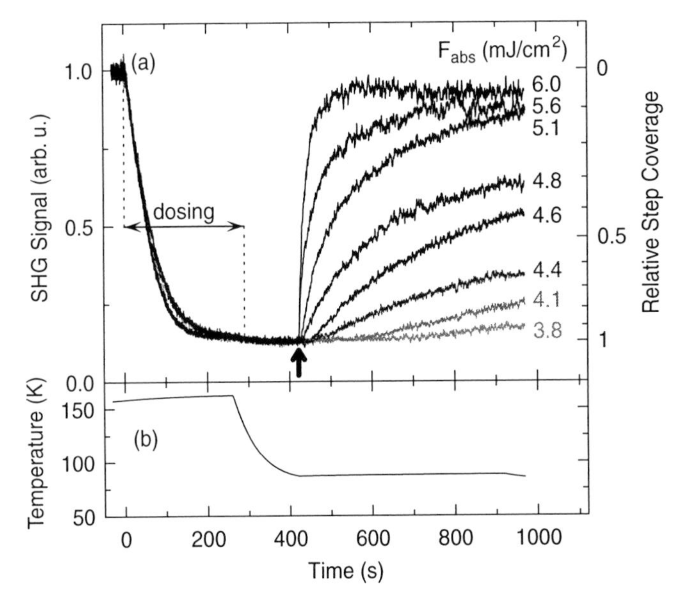

Figure 18.7 (a) Second-harmonic response of the Pt sample during dissociative adsorption of O_2 at the steps (dosing) and diffusion of atomic oxygen induced by femtosecond laser pulses of various absorbed fluences F_{abs}. The pump pulses with a repetition rate of 1 kHz were switched on at the time marked by the vertical arrow. The right y-scale gives the conversion of the SHG signal into the relative step coverage [36]. (b) Typical temperature progression. Reprinted with permission from [19]. Copyright 2005 by the American Physical Society.

regime in Figure 18.7). A continuous recovery of the SHG signal has been observed for pulses that exceed an absorbed fluence of 3.5 mJ/cm^2. The recovery of the SHG signal is due to the depletion of the step sites by oxygen diffusion onto the terraces and not due to desorption as has been checked by subsequent temperature-programmed desorption (TPD). By comparison of the thermal barriers for diffusion and desorption, it can be estimated that laser pulses exceeding the damage threshold of the sample would be required to induce desorption of atomic oxygen. In order to exploit the excellent statistics of the whole data sets, the hopping probability p_{dif} per laser shot for migration from the completely filled step sites onto the empty terraces has been determined by describing the diffusion kinetics using a simple one-dimensional rate equation model [36]. p_{dif} corresponds roughly to the initial slope of the step coverage as a function of the number of laser shots when the pump laser starts to depopulated the step sites.

The hopping probability p_{dif} depends in a strongly nonlinear way on the fluence of the pump laser as can be even seen in the raw data of Figure 18.8. Variation of the laser fluence by only 50% covers the whole accessible dynamic range of the diffusion rate. In the investigated fluence range, it can be described by a power law of the form $p_{\text{dif}} \propto F^{15}$. This is a much stronger nonlinearity compared to previous laser-induced desorption experiments, where typical exponents were in the range 3–8 [34, 63–65]. In laser-induced desorption experiments, the nonlinearity of the fluence dependence

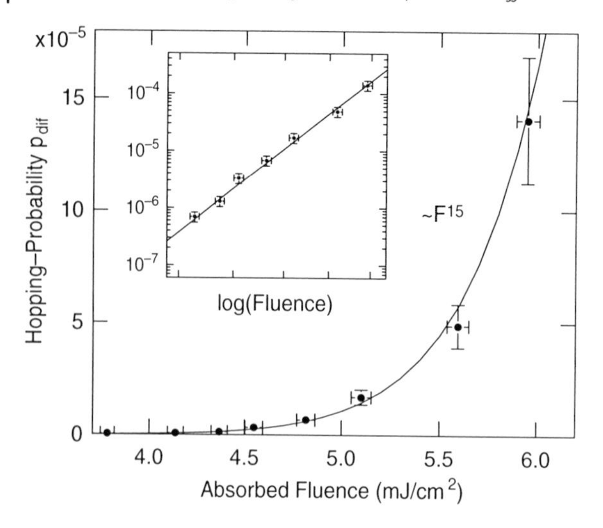

Figure 18.8 Hopping probability per laser shot $p_{\rm dif}$ for the diffusion of oxygen from the filled step sites onto the empty terraces as a function of absorbed laser fluence F. The inset shows the data in a logarithmic scale. The solid line shows a power law $\propto F^{15}$. Reprinted from Ref. [48]. Copyright 2006, IOP Publishing Ltd.

makes it necessary to apply yield averaging of the laser fluence [34] since adsorbates that are desorbing from the whole illuminated laser spot are detected. In contrast to this, the SHG detection of the adsorbate coverage makes it possible to use a much smaller spot size for the probe beam compared to the pump beam. In the laser-induced diffusion experiments discussed here, a ratio of the spot sizes of 1: 10 has been used. This ensures that even for a nonlinearity as strong as $\propto F^{15}$, the yield in the center of the pumped area is spatially uniform over the diameter of the probe beam within 10%. The nonlinear detection further narrows the effective probe diameter by a factor of $\sqrt{2}$, which finally reduces the nonuniformity to 5%.

The nonlinear dependence of $p_{\rm dif}$ on laser fluence makes it possible to apply a two-pulse correlation scheme that provides information about the energy transfer time from the initial excitation of the electrons to the adsorbate motion (see Section 18.2). For these experiments, the pump pulse is split and the step depletion rate is determined as a function of the time delay between the two pulses (Figure 18.9). The two-pulse correlation displayed in Figure 18.9 has a high contrast between $p_{\rm dif}$ at zero and large delays, which is related to the high nonlinearity of the fluence dependence. The width of 1.45 ps (FWHM) is much larger than the cross-correlation of the two laser pulses and has the value of a typical electron–phonon coupling time. This unambiguously shows that diffusion is driven by the laser-excited electrons of the metallic substrate. An energy transfer that is mediated by substrate phonons would typically be slower by one order of magnitude [30].

The two-pulse correlation and fluence dependence of the hopping rate has been modeled with the different variants of the electronic friction model. These results are shown in Figure 18.10 in comparison with the experimental data. The time-

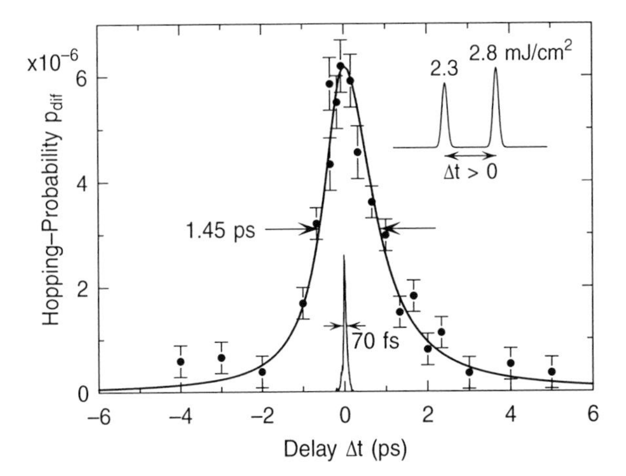

Figure 18.9 Hopping probability per laser shot p_{dif} as a function of delay between the p- and the s-polarized pump beams with absorbed fluences of 2.3 and 2.8 mJ/cm^2, respectively (symbols). The solid line is a guide to the eye. The time resolution of 70 fs is depicted by the black line that shows the SHG cross-correlation of the two pump pulses. Reprinted with permission from Ref. [19]. Copyright 2005 by the American Physical Society.

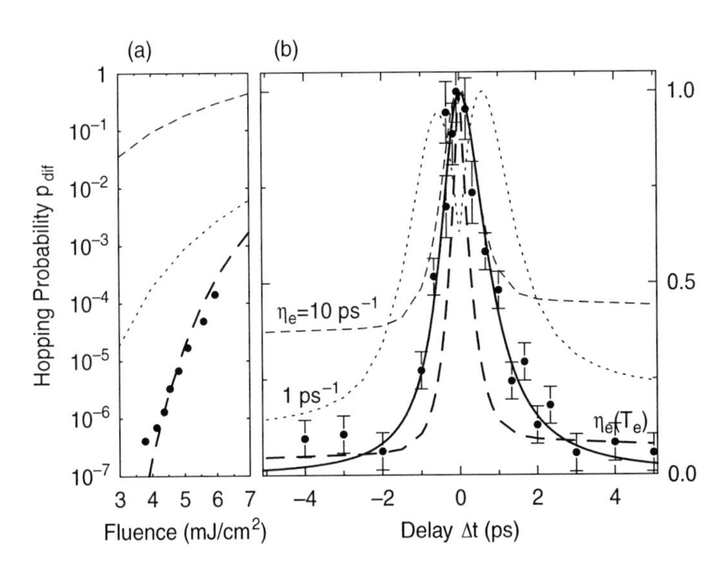

Figure 18.10 Calculated hopping probability p_{dif} (a) as a function of absorbed laser fluence and (b) as a function of time delay between two pump pulses for two constant friction coefficients (thin dotted and dashed lines) and an electron temperature dependent η_e (thick dashed line). Experimental values are indicated by symbols and the solid line. Reprinted with permission from Ref. [19]. Copyright 2005 by the American Physical Society.

dependence of T_e and T_i have been calculated with the two-temperature model [40] using material parameters for Pt reported in Ref. [43] (electron–phonon coupling parameter $g = 6.76 \times 10^{17}$ $WK^{-1}m^{-3}$, electronic specific heat $\gamma = 748$ $JK^{-2}m^{-3}$, thermal conductivity at 77 K $\varkappa_0 = 71.6$ $WK^{-1}m^{-1}$, Debye temperature $T_D = 240$ K, and skin depth at 800 nm $\lambda_s = 12.56$ nm). The optical excitation has been modeled using Gaussian shape laser pulses at a central wavelength of 800 nm with a pulse width of 55 fs (FWHM) and fluences of 23 and 28 mJ/cm^2, respectively. From thermal diffusion experiments, the diffusion barrier from step to terrace sites has been estimated to $E_{dif} \approx 0.8$ eV that is ≈ 0.3 eV larger than the barrier for hopping between terrace sites [66]. As shown by the thin dashed and dotted lines, the fluence dependence and the two-pulse correlation cannot be described simultaneously with a constant η_e. A narrow two-pulse correlation can only be modeled with strong friction. However, this results in a hopping probability that is too large by several orders of magnitude and in a poor contrast between zero and large delays. A strong fluence dependence and a high contrast of the two-pulse correlation can only be obtained with a small η_e. This, however, results in a two-pulse correlation that is much broader than measured.

A much better agreement with the experimental data could be achieved by assuming an empirical dependence of η_e on electron temperature, $\eta_e(T_e) = \eta_e^0 T_e^2$ with $\eta_e^0 = 1.8 \times 10^5$ $K^{-2}ps^{-1}$. In particular, the high nonlinearity of the fluence dependence can be well reproduced with this assumption (Figure 18.10a). It should be noted that the best parameterization of $\eta_e(T_e)$ depends on the diffusion barrier E_{dif}. With a higher value, which is more than 0.8 eV, the required exponent of T_e drops below 2 but does not vanish. For all reasonable parameterizations of $\eta_e(T_e)$, the width of the two-pulse correlation remains somewhat smaller than the experimental one (Figure 18.10b).

A dependence of the electronic friction on electron temperature, however, is supported by neither the electronic structure of O on Pt(111) nor *ab initio* calculations [48]. For this reason, it has been suggested that the electron-temperature dependence of η_e is only an effective one and appears due to the neglect of the coupling between different vibrational modes. An indirect excitation mechanism has been proposed that is based on a primary excitation of the O-Pt vibration by the hot electrons, which then couples anharmonically to the FT mode required to initiate lateral motion [19, 48]. Such indirect excitation due to vibrational mode coupling has been found in inelastic scanning tunneling experiments where hopping of CO on Pd(110)[26] and NH_3 on Cu(100)[28] has been induced by excitation of the internal C–O (N–H) stretch vibration. The lateral motion has been shown to be initiated by anharmonic coupling between the high-frequency internal stretch and the low-frequency FT mode. The existence of such mode coupling for O/Pt(111) has been deduced from the temperature dependence of the O-Pt stretch mode observed by infrared absorption spectroscopy [67].

Recently, Ueba *et al.* have modeled the electronic excitation of the oxygen atoms under explicit consideration of such mode coupling using their generalization of the heat transfer at surfaces [54]. For this purpose, Eqs. (18.28) and (18.29) were

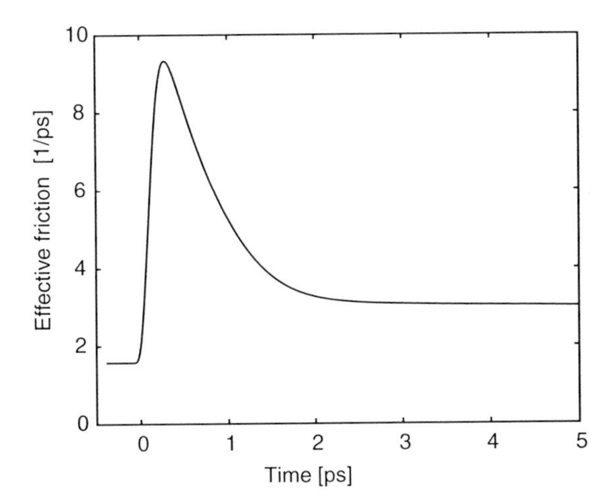

Figure 18.11 Transients of the effective friction $\eta_a^{eff}(t)$. See the text for the parameters used herein.

combined with the two-temperature model [40] to calculate $T_e(t)$, $T_a(t)$, and $T_b(t)$ for atomic oxygen on a Pt surface at $T=80$ K. The optical excitation and the dynamics of the electron temperature of the Pt substrate have been calculated with the same material parameters for Pt from Ref. [43] as have been described above. The present calculation has been done using a laser pulse width of 55 fs and a fluence of 23 mJ/cm^2. For the vibrational energies of the a and the b modes, we have used $\hbar\omega_a = 50$ meV and $\hbar\omega_b = 60$ meV, which, respectively, correspond to the atomic oxygen FT mode (the reaction coordinate mode for hopping) and the O-Pt stretch mode on a Pt (111) surface [67]. We fixed $1/\eta_b = 1$ ps as has been estimated from the low temperature limit of the linewidth of the O-Pt stretch mode [67].

Figure 18.11 shows the time dependent effective friction $\eta_a^{eff}(t) = \eta_a + \eta_{ab}(k_B T_b/\hbar\omega_b)$ of the a-mode due to direct ($\eta_a = 1$/ps) and indirect friction coupling ($\eta_{ab} = 5$/ps), where $1/\eta_b = 1$ ps is used to calculate $T_e(t)$, $T_a(t)$, and $T_b(t)$. The transient behavior of $\eta_a^{eff}(t)$ is the same as that of $T_b(t)$. Because of the efficient coupling of the b-mode to hot electrons, $T_b(t)$ immediately follows $T_e(t)$.

Figure 18.12 shows the temperature transients of the electron $T_e(t)$ (solid curve) and a-mode $T_a(t)$ (dashed curve) for a pulse pair at a time delay of 3 ps, a fluence of 23 mJ/cm^2 in each pulse, an initial temperature of $T_0=80$ K and friction coefficients $\eta_a = 0.3$/ps, $\eta_b = 1.0$/ps, and $\eta_{ab} = 6$/ps. For comparison, $T_a(t)$ calculated for $\eta_{ab} = 0$ is shown as dotted curve. It is found that $T_a(t)$ can be substantially heated up to higher temperature via the mode coupling compared to the calculation using Eq. (18.4) for a constant as well as for a $T_e(t)$ dependent friction.

Knowing the time-dependent adsorbate temperature $T_a(t)$ describing the vibrational excitation of the adsorbate, the reaction rate is calculated using Eq. (18.5). Güdde *el al.* [48] compared the transient behaviors of $T_a(t)$ and $R(t)$ calculated using Eqs. (18.5) and (18.6) for the generalized friction model [14]. They found that the

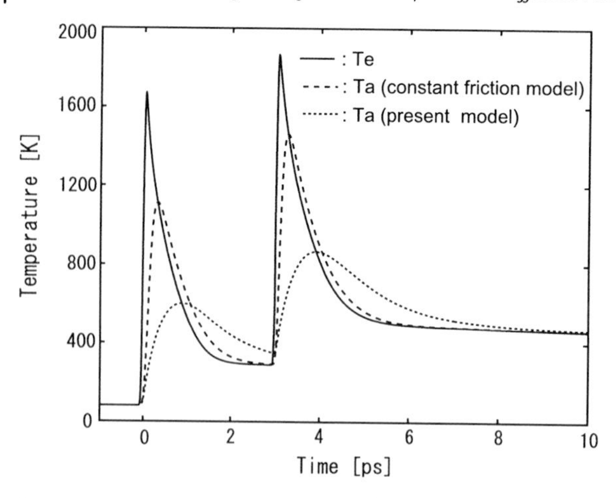

Figure 18.12 Temperature transients of the electron $T_e(t)$ (solid curve) and a-mode $T_a(t)$ with (dashed curve) and without (dotted curve) a coupling to the b-mode. See the text for the parameters used herein.

empirical and the generalized friction model showed nearly identical results not only for $T_a(t)$ but also for $R(t)$. Here, we use Eq. (18.5) to calculate the 2PC for atomic oxygen hopping. Figure 18.13 shows the normalized 2PC for atomic oxygen hopping as a function of time delay (t_d) between the two pump pulses of the equal fluence. The

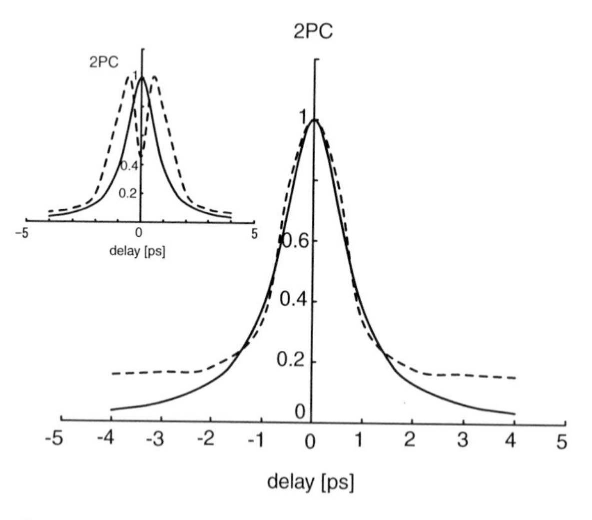

Figure 18.13 Normalized 2PC for the oxygen hopping probability as a function of time delay between two pump pulses. The dashed curve is calculated using the present model and compared with the experimental result (solid curve) from Ref. [19]. In the inset the dashed curve is calculated with constant friction. See text for the parameters used.

calculated 2PC nicely reproduces the experimental (solid curve). The 2PC exhibits a peak at $t_d = 0$, while a dip is observed when we use a constant friction $\eta_a = 1/\text{ps}$ as shown in the insert (dotted curve). Here, we used $E_a = 1.4$ eV in Eq. (18.5). When we assume a very large friction coupling time $1/\eta_a = 0.1$ ps in Eq. (18.4), we obtain a peak at $t_d = 0$ accompanied with too large tails compared to the experimental result at longer t_d. In our model $T_a(t_d)$ shows a peak at $t_d = 0$ due to the strong vibrational energy transfer mediated by the friction coupling to the substrate and bears a close resemblance to $T_e(t_d)$. We also found that a peak evolves to a dip in the 2PC($t_d = 0$) with a decrease in η_{ab}. It is remarked that the 2PC($t_d = 0$) exhibits a peak even when there is no direct heating of the a-mode, that is, $\eta_a = 0$, as far as we assume an efficient friction η_b and mode coupling η_{ab}.

As has been found before [48], the width of the 2PC becomes narrower with increasing E_a. In the analysis using a T_e-dependent friction model the best agreement with the experimental results was achieved for $E_a = 1.2$ eV, which is, however, much larger than that estimated for the thermal activated diffusion. A choice of $E_a = 1.2$ eV nicely reproduces the experimental 2PC except in the tail at longer time delay. The 2PC calculated using $E_a = 0.8$ eV in Eq. (18.4) also exhibits a peak at $t_d = 0$, but a quite large deviation from the experimental result is found in the tail. The width of a 2PC provides information about the timescale of the energy transfer from the initial excitation of the electrons to the adsorbate motion and allows the distinction between an electron- and a phonon-mediated processes [30]. We obtain the full width at half maximum (FWHM) $\tau = 1.53$ ps (the $\eta_a(T_e)$ model gives FWHM = 0.68 ps), which is very close to the experimentally observed FWHM = 1.45 ps. This seems to support an indirect excitation mechanism via the mode coupling mediated by laser-heated hot electrons.

The choice of the activation barrier height is one of the key issues in the analysis of adsorbate dynamics induced by ultrafast laser pulses. Using $T_e(t)$ dependent friction Szymanski *et al.* [49] attempted to reproduce the 2PC for the desorption of molecular oxygen from Pd(111). They found that the reproduction of their experimental 2PC is significantly improved with the use of a desorption barrier much higher than that determined from thermal desorption experiments. A question remains open whether the activation barrier deduced from the femtosecond photoinduced motions is the same as that from the thermal-induced ones [30]. Perhaps, an even more important and fundamental issue is how appropriate a description of the reaction rate in terms of Arrhenius behavior is, which assumes a thermal equilibrium of a system, for adsorbate dynamics taking place on ultrafast timescale.

We may be able to relate the $T_e(t)$-dependent friction model to the present model. When we assume strong coupling η_b for the accepting mode, $T_b(t)$ (the frustrated O-Pt mode for oxygen hopping) immediately follows $T_e(t)$ and gives rise to a delayed rapid increase in $T_a(t)$ of the reaction coordinate mode (the FT mode). Even when a weak direct coupling η_a is very small for the parallel motion, the intramode heat transfer $\eta_{ab}(k_B T_b/\hbar\omega_b)$ becomes extremely large because $k_B T_b/\hbar\omega_b \gg 1$ so that $T_b(t) \simeq T_e(t)$ for a strong η_b.

18.4.2
CO/Pt(533)

In the research described in the preceding section, information about the ultrafast dynamics of molecular motion over surface was obtained using changes in the electronic response of the substrate, following molecular motion over the surface. In this section, we demonstrate that one can also monitor the adsorbate response using time-resolved vibrational SFG. Combined with a stepped surface, SFG can provide "snapshots" of the surface with subpicosecond time resolution while diffusion is occurring [20]. These experiments rely on the fact that small diatomic molecules such as carbon monoxide exhibits different internal vibrational frequencies depending on where precisely they are adsorbed on the surface.

Here, we use a stepped platinum (533) crystal, which consists of 4-atom wide terraces that are 7 Å in width, separated by monatomic steps. A schematic picture of the surface can be found in Figure 18.14. As also shown in this figure, the internal C–O stretch vibration of carbon monoxide amounts to 2080 and 2100 cm^{-1} on step and terrace platinum atoms, respectively, when CO is adsorbed in atop position. This frequency difference allows one to readily distinguish CO adsorbed on the two different sites, and hence also the motion of CO from one to the other site.

The diffusion of CO molecules is initiated by excitation of platinum electrons using ultrashort (~100 fs) laser excitation. The motion is time resolved using femtosecond time-resolved vibrational SFG, monitoring the distribution of site-

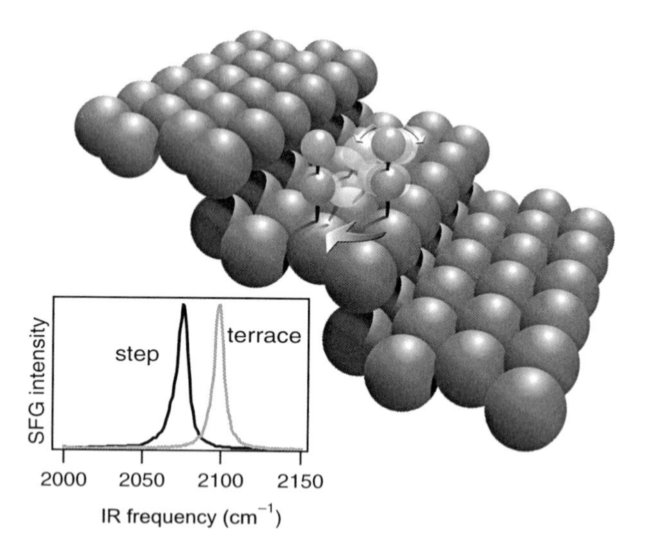

Figure 18.14 Schematic representation of the platinum Pt(533) surface, which consists of three-atom wide terraces separated by monatomic steps. Carbon monoxide absorbs strongly to the under coordinated step sites, but can be moved to a terrace site by ultrafast laser excitation. This process can be followed using sum-frequency generation, because the SFG spectra of carbon monoxide adsorbed on step and terrace sites reveal distinct vibrational frequencies for the internal C–O stretch vibration, as shown in the lower left panel.

specific CO stretch vibrations as a function of time between the excitation and the SFG pulses.

In the experimental approach, we employ the fact that CO binds appreciably more strongly on the step sites than on the terrace sites (binding energies: 2.0 and 1.5 eV, respectively [20]). The desorption temperatures of CO from the two sites are correspondingly different, and as a result, heating the CO-saturated surface to 420 K leads to CO molecules to be preferentially removed from the terrace sites, leaving only the step sites occupied. The resulting SFG spectra (Figure 18.14) reveal one resonance at 2080 cm^{-1}, associated with these step-CO molecules. For a fully covered platinum 533 surface, the majority of CO molecules is situated on terrace atop sites, characterized by a vibrational frequency of 2100 cm^{-1} (Figure 18.14). Although the step sites are still occupied, the vibrational intensity at 2080 cm^{-1} has vanished, as a result of dipole–dipole coupling. This results in a vibrational intensity transfer from step to terrace site.

Steady-state SFG spectra (not shown here) revealed that irradiation of the CO-covered platinum surface leads to the transfer of vibrational intensity from the peak associated with CO on step sites to that associated with CO on terrace sites. This is direct evidence that the CO molecules can undergo laser-induced motion from step to terrace sites: laser excitation can bring the system temporarily out of the equilibrium situation where the step sites are predominantly occupied. It was found that for sufficiently high excitation densities, typically 10% of the CO molecules located on step sites could be made to move onto the terrace sites. Repopulation of step sites was found to occur on a timescale of seconds at liquid nitrogen temperatures, which is consistent with the low diffusion constant at these low temperatures.

The real-time dynamics of the motion of CO over the surface could therefore be followed by recording transient SFG spectra as a function of delay time between the pump and the SFG probe pair. One set of such measurements is shown in Figure 18.15. The spectrum at long negative delay times corresponds to the situation where 75% of the step and 20% of the terrace sites were covered, respectively. This *initial* nonequilibrium situation is due to the fact that the pump pulse excites the surface at a repetition rate of 83 Hz. The diffusion of CO molecules back to the step sites occurs on a timescale of seconds, so that a steady-state depletion of step site occupation is reached.

It is clear from Figure 18.15 that the ratio between the intensities of the step and terrace peaks changes on very fast (subpicosecond) timescales: the stepped peak disappears almost completely within a picosecond. This is strong qualitative evidence for the involvement of platinum electrons in the hopping process.

The transient SFG spectra contain the details of the time-dependent occupation of the two types of surface sites. Here, it is important to note that the transient SFG spectra are changing not only because of the relative occupation of steps and terrace sites is changing but also because of the “trivial” heating of the surface, which also affects the internal C–O stretch vibration. Specifically, the low-frequency frustrated translation mode is known to couple to the platinum electrons. This low-frequency mode, in turn, is coupled to the C–O internal stretch vibration. It is easy to correct for this effect by simply recording transient spectra at fluences where hopping does not

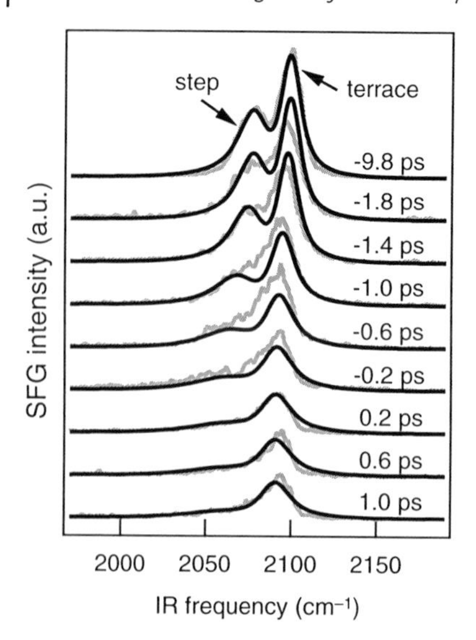

Figure 18.15 Experimental (gray) and calculated (black) time-resolved sum-frequency generation spectra of the C—O stretch of carbon monoxide adsorbed on the Pt(533) surface, after excitation of the platinum surface with an intense near-IR pulse at $t = 0$ ps. Excitation leads to an intensity transfer from CO on step sites to CO on terrace sites, evidencing laser-induced motion of carbon monoxide over the surface.

yet significantly occur. It turns out that the timescale on which energy flows between electrons and the low-frequency frustrated translation mode is rather slow, amounting to 2.5 and 4 ps for terrace and steps sites, respectively.

Including this correction, it is possible to quantitatively retrieve the fractional occupation of steps sites as a function of time after the excitation pulse. The calculations confirm what is evident from the raw data: step-to-terrace motion occurs on subpicosecond timescales. The step occupation as a function of time is plotted in Figure 18.16a.

The question that now comes to mind concerns the mechanism of this laser-induced ultrafast surface hopping process. Figure 18.16b shows the time-dependent electron and phonon temperature of the platinum substrate. It is evident from the very fast timescale associated with molecular motion, the coupling to the electrons must predominate the hopping process.

It has often been stated [19, 26, 28] that the vibrational mode most strongly coupled to the reaction coordinate responsible for diffusion, should be the frustrated translation. The inset in Figure 18.16 shows the atomic motion associated with this mode. It is clear from the raw data, however, that hopping occurs much too fast to be explained solely by coupling to the frustrated translation. As stated above, independent measurements of the coupling rate between the hot-surface electrons and the frustrated translation mode reveal coupling times in excess of 2 ps, whereas the surface motion is finished within 1 ps. Indeed, attempts to model the time-dependent hopping probability assuming only coupling to the frustrated translation mode are

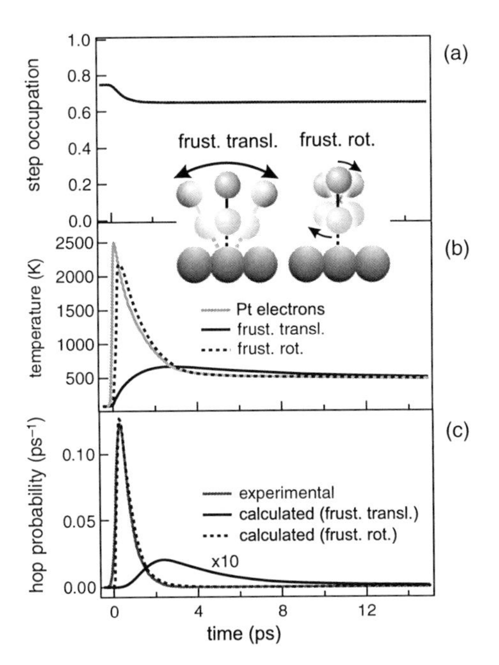

Figure 18.16 (a) Time-dependent fractional occupation of the step sites as a function of time after excitation. (b) Time-dependent electron temperature, as calculated by the two temperature model, along with the transient temperatures associated with the frustrated translation, and frustrated rotation. (c) Experimentally observed and calculated probability for a CO molecule to hop from a step to a terrace site as a function of time. The calculation assuming coupling to the frustrated translation is too slow and has too low an amplitude; the calculation assuming coupling to the frustrated translation works remarkably well. (Please find a color version of this figure on the color plates.)

unsuccessful, as evidenced by Figure 18.16c. The theoretically predicted rate is too low, as is the overall time-integrated hopping probability.

As noted above, however, Dobbs and Doren [51] have previously theoretically demonstrated that the frustrated translation mode and the frustrated rotational mode are strongly coupled near the transition state for diffusion in a model system of CO/Ni(111). This is, in fact, intuitive. As is evident from the inset of Figure 18.16, the frustrated translation involves significant center-of-mass motion across the surface, but at the expense of a rotating molecule. It is evident that this rotation has to be lifted when going to the next adsorption site, to be able to adsorb there. Indeed, quantitative agreement between the model and the data can be obtained when one assumes that the rate-limiting step in the hopping process is the excitation of the frustrated translation mode by the hot metal electrons, as evidenced by the trace in Figure 18.16c. The frustrated rotational mode couples efficiently to the hot-electron distribution, with the characteristic time constant on the order of 100 fs [68, 69]. The data therefore reveal that laser-induced excitation of rotational motion rather than translational motion is the rate-limiting step of the motion of CO over the stepped platinum surface.

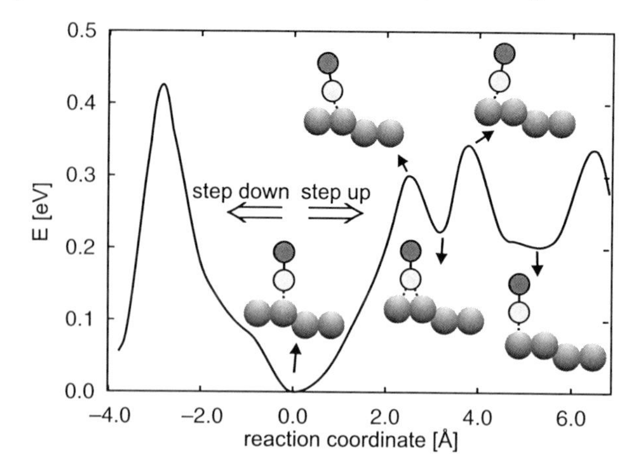

Figure 18.17 Reaction pathways for the diffusion of CO from the step sites to the lower or upper terrace obtained from DFT calculations. The barrier for diffusion to the upper terrace is lower by 0.1 eV. For diffusion to the upper terrace the crucial step over the transition state includes excitation of the frustrated rotation.

It is important to note that these experiments were performed at 100 K. There will therefore be substantial thermal excitation of the very low-frequency frustrated translation mode ($\sim$50 cm^{-1}), and not of the frustrated rotation, which has a substantially higher mode frequency ($\sim$400 cm^{-1}) [70]. The picture thus emerges is that both modes are required for the hopping process, but at 100 K, excitation of the rotational motion is rate limiting.

The experimental conclusions about the hopping mechanism were corroborated by density functional theory (DFT) calculations. The results of these calculations are shown in Figure 18.17. This figure shows reaction pathways for the diffusion of CO initially adsorbed on the step sites to both the lower (left) and the upper (right) terrace sites. The calculated value for the diffusion barrier ($\sim$0.3 eV) is in reasonable agreement with the 0.4 eV barrier that has been reported experimentally for CO on a stepped platinum surface [71]. It is further evident from the DFT calculations that the CO molecules will have a propensity to move up, rather than down the step, given the reduced barrier height in the direction. The molecular structures that are shown in the figure (obtained from the calculations) reveal that, at the two transition states, the molecule has performed a translational motion and must perform a rotational motion to overcome the barrier and move to the next adsorption site: the CO molecules first perform a translational motion but for the Pt–C bond to be broken one requires rotation of the molecule with the carbon and oxygen atoms moving in opposite directions. It is therefore apparent that the frustrated rotation plays a crucial role for the diffusion, in agreement with the experimental results.

Hence, it is clear that an indispensable role exists of the FR mode in addition to the FT mode, for the laser-activated hopping of CO molecules on a stepped Pt surface. This conclusion was based on $T_a(t)$ calculated for the FT and FR mode using Eq. (18.4) with the coupling time $\tau_a = 1/\eta_a = 4$ ps and $\tau_b = 1/\eta_b = 0.1$ ps,

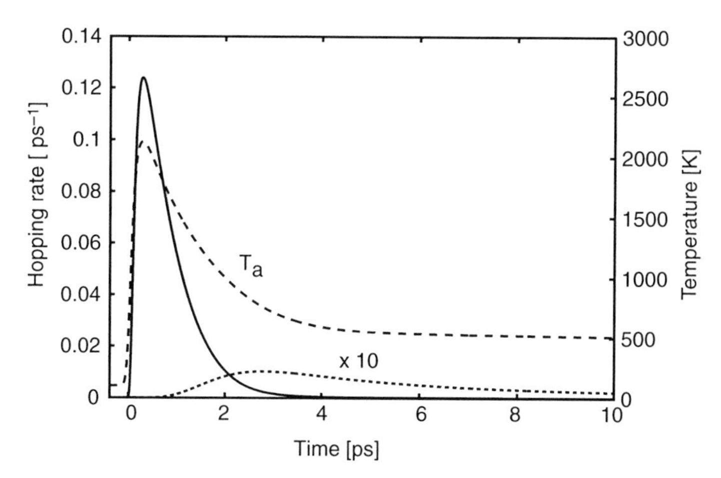

Figure 18.18 CO hopping probability calculated for the FT mode temperature $T_a(t)$ (dashed curve) with mode coupling to the FR mode ($1/\eta_{ab} = 0.8$ ps). The dotted curve is calculated for the FT mode with $1/\eta_a = 4.0$ ps $1/\eta_{ab} = 0$ ps. See the text for the parameters used herein.

respectively. Note that in the experiment the substrate temperature was at $T_0 = 100$ K so that the FT mode is thermally excited, although not high enough temperature for hopping. We now apply Eqs. (18.28) and (18.29) to this system. It should be noted that the FT and FR mode of CO on metal surfaces are diagonal modes in a sense that the energy cannot flow from one mode to another. When we include the coupling to the substrate electronic degrees of freedom, it becomes possible that the energy can flow from one mode to another via substrate electronic excitations.

18.4.2.1 Numerical Results of CO Hopping on a Pt (553) Surface

The hopping probability of CO on a Pt(553) surface at $T_0 = 100$ K is calculated using an Arrhenius type expression $R(t) = k_0 \exp\left[-U_0/k_B T_a(t)\right]$ for the thermally activated process based on the well-known Kramer's theory. To go over the barrier, the system must be highly vibrationally excited along the reaction coordinate, and in this case, a classical treatment should be a very good approximation. The calculated hopping rate $R(t)$ shown in Figure 18.18 (calculated using the prefactor $k_0 = 1.3 \times 10^{12}$ s^{-1}, the barrier height $U_0 = 0.4$ eV, ω_a (FT mode) = 4.4 meV, $1/\eta_a = 4.0$ ps, ω_b (FR mode) = 51 meV, $1/\eta_b = 0.2$ ps, and $1/\eta_{ab} = 0.8$ ps) is in good agreement with the experimental result. Here, $T_a(t)$ for the FT mode with mode coupling to the FR mode is also depicted together with $R(t)$ without mode coupling. Although adsorbate substrate energy exchange solely via excitation of the FT mode is significantly too slow to cause the hopping motion of CO, the strong coupling to the FR mode heats up the FT mode high enough for hopping. It is also very important to emphasize that the excitation of the FR mode by hot electrons is absolutely required in order for the FT reaction coordinate mode to get excited within a picosecond for the hopping motion.

Now one may ask if the excitation of the FT mode by hot electrons is required no matter how weak it is. Because of the quite low energy ($\hbar\omega a = 4.4$ meV) the FT mode is thermally excited at $T_0 = 100$ K in the experiments, and can be a precursor state for hopping [20, 50]. In order to examine this idea we have made the whole calculation with $\eta_a = 0$ (no direct heat transfer from hot electrons). It is found that results shown above remain unchanged. The energy needed to activate the FT mode above the barrier is supplied from the partner FR mode excited by hot electrons. This is reminiscent of CO hopping on Pd(110) [26] and NH_3 hopping on Pd(110) on Cu (100) [28] using a STM, where excitation of the C–O (N–H) stretch mode (these vibrational energies are larger than the barrier for hopping) by a single tunneling electron activates the FT mode above the barrier via the anharmonic mode coupling. Observation of the FT and FR modes of CO on Cu(100) by STM-IETS (inelastic electron tunneling spectroscopy) [72] provides clear evidence of the fact that a direct excitation of these modes by tunneling electrons does not lead to any motion of a CO molecule. Because of the quite low energies of these frustrated modes compared to the hopping barrier and the high vibrational damping rate compared to tunneling current rates, it is unlikely for the FT mode to climb up the vibrational ladders by a multielectron process. The vibrational line shapes of the C–O stretch mode exhibit an increase in the width of the infrared absorption peak with increasing temperature, thereby suggesting vibrational dephasing via anharmonic coupling to the low-frequency FT or FR modes. Schweizer *et al.* [73] reported detailed measurements of the lineshapes and the intensities of the C–O stretch mode in the CO/Pt(111) system using infrared absorption spectroscopy. The change of the linewidth and frequency as a function of temperature has been explained in terms of a vibrational dephasing model. They found that the anharmonic coupling to the FT mode is very small (only 2 cm^{-1}) corresponding to $\eta_a = 0.36$/ps for the on-top CO molecules. The coupling to the FR mode was not observed and is likely negligible small. This makes an efficient heat transfer unlikely from the C–O stretch mode to the low-frequency frustrated modes involved in laser-induced hopping of CO on Pt(111) surface. However, this was not the case for CO on a Ru(001). A transient redshift, a broadening and a decrease in intensity of the C–O stretch mode observed by time-resolved SFG during femtosecond near-IR laser excitation leading to desorption have been explained in terms of the weak (strong) anharmonic mode coupling to the FT (FR) [74].

In summary, it is evident that the reaction coordinate for diffusion involves both the frustrated translation and the frustrated rotation mode. Although the frustrated translation mode appears to be weakly coupled to the electron bath, it can be excited thermally at elevated temperatures (such as the 100 K here), or indirectly through the frustrated rotation mode, which is strongly coupled to the electron bath. At low temperatures, we expect this indirect coupling path to dominate the dynamics.

18.4.2.2 Two-Pulse Correlation Measurements

Recently, we have studied laser-induced diffusion of CO on vicinal Pt(111) also by applying the 2PC scheme combined with SHG detection [75]. In these experiments, a

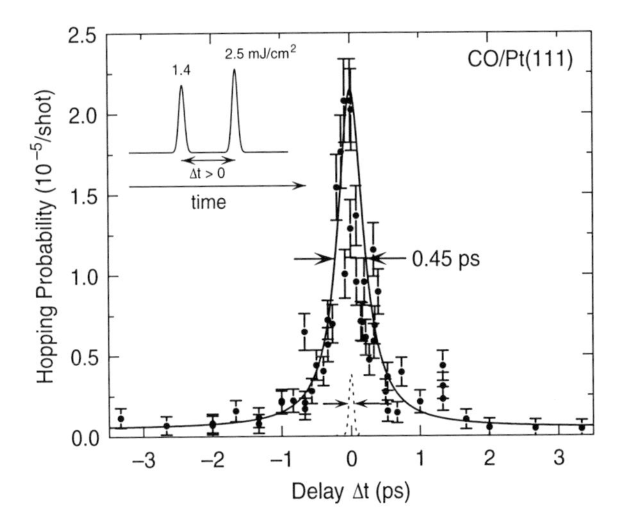

Figure 18.19 Hopping probability per laser shot p_{dif} for diffusion of CO from step to terrace sites on Pt(111) as a function of delay between the p- and the s-polarized pump beams with absorbed fluences of 1.4 and 2.5 mJ/cm^2, respectively (symbols).The solid line is a guide to the eye. The time resolution of 70 fs is depicted by the dashed black line that shows the SHG cross-correlation of the two pump pulses.

Pt(111) surface with low step density but with the same step orientation as compared to the experiments on Pt(533) has been used. Figure 18.19 shows a 2PC for a total absorbed fluence of $F_{abs} = 3.9$ mJ/cm^2. The width of the 2PC is in good agreement with the time constant of about 0.5 ps observed for the hopping of CO on Pt(533) [20] and confirms the fast electronic energy transfer from the optically excited substrate electrons to the lateral motion. One should have in mind, however, that it is not possible to identify the energy transfer time directly with the width of the 2PC. It has been shown that the width of the 2PC depends on laser fluence [75]. For weaker excitation, the 2PC becomes even narrower. This dependence is a general phenomenon in laser-induced substrate-mediated surface reactions and can be qualitatively described even within the empirical friction model.

18.5 Summary

In summary, we have shown that diffusion of adsorbates on metal surfaces can be induced electronically at low substrate temperatures by femtosecond laser excitation of the metal electrons. The application of nonlinear optical techniques made it possible to sensitively detect hopping of the adsorbates from step onto terrace sites on vicinal surfaces and to study the dynamics of the energy transfer from the optically excited metal electrons to the lateral motion of the adsorbate on a subpicosecond timescale. It has been shown that this energy transfer can be described by models that treat the coupling of the optically excited hot electron to the adsorbate degrees of freedom in terms of an electronic friction. The experiments presented here have revealed that in contrast to most of the electronically induced desorption experiments, the coupling between different vibrational modes plays an important role for the initiation of lateral motion by electronic excitation. This coupling has been included in the description of the energy transfer by developing an extension of the electronic friction model. This generalization of the electronic friction model is also capable to consider the anharmonicity of the potential for the adsorbate motion, which is an important issue for the comprehensive description of an activated process such as diffusion, since the adsorbate particularly probes the anharmonic part of the potential when it goes over the barrier.

Acknowledgments

We are grateful to E. Backus, A. Eichler, M. Lawrenz, K. Stépán and A. Kleyn for many helpful discussions. This work was supported by the Deutsche Forschungsgemeinschaft through grant No. HO 2295/1–4, and GK 790, the German-Israel Science Foundation, and the Marburg Center for Optodynamics. It is part of the research program of the Stichting Fundamenteel Onderzoek der Materie (Foundation for Fundamental Research on Matter) with financial support from the Nederlandse Organisatie voor Wetenschappelijk Onderzoek (Netherlands Organization for the Advancement of Research).

References

1 Menzel, D. and Gomer, R. (1964) Desorption from metal durfaces by low-energy electrons. *J. Chem. Phys.*, **41**, 3311.

2 Redhead, P.A. (1964) Interaction of slow electrons with chemisorbed oxygen. *Can. J. Phys.*, **42**, 886.

3 Misewich, J.A., Heinz, T.F., and Newns, D.M. (1992) Desorption induced by multiple electronic-transitions. *Phys. Rev. Lett.*, **68**, 3737–3740.

4 Zhou, X.L., Zhu, X.Y., and White, J.M. (1991) Photochemistry at adsorbate/metal interfaces. *Surf. Sci. Rep.*, **13**, 73–220.

5 Feulner, P. and Menzel, D. (1995) Electronically stimulated desorption of neutrals and ions from adsorbed and condensed layers, in *Laser Spectroscopy and Photo-Chemistry on Metal Surfaces Part II* (eds H.L. Dai and W. Ho), World Scientific, Singapore, pp. 627–684.

6 Hasselbrink, E. (1990) Coupling of the rotational and translational degrees of freedom in molecular DIET: a classical trajectory study. *Chem. Phys. Lett.*, **170**, 329–334.

7 Frischkorn, C. and Wolf, M. (2006) Femtochemistry at metal surfaces: nonadiabatic reaction dynamics. *Chem. Rev.*, **106**, 4207–4233.

8 Bohnen, K.P., Kiwi, M., and Suhl, H. (1975) Friction coefficient of an adsorbed H-atom on a metal surface. *Phys. Rev. Lett.*, **34**, 1512–1515.

9 Nourtier, A. (1977) Friction coefficient of atoms near a metal surface. *J. Phys-Paris*, **38**, 479–502.

10 Newns, D.M., Heinz, T.F., and Misewich, J.A. (1991) Desorption by femtosecond laser pulses: an electron–hole effect? *Prog. Theor. Phys. Supp.*, **106**, 411.

11 Budde, F., Heinz, T.F., Kalamarides, A., Loy, M.M.T., and Misewich, J.A. (1993) Vibrational distributions in desorption induced by femtosecond laser pulses: coupling of adsorbate vibration to substrate electronic excitation. *Surf. Sci.*, **283**, 143–157.

12 Misewich, J.A., Heinz, T.F., Weigand, P., and Kalamarides, A. (1996) Femtosecond surfaces science: the dynamics of desorption, in *Laser Spectroscopy and Photo-Chemistry on Metal Surfaces Part II* (eds H.L. Dai and W. Ho), World Scientific, Singapore, pp. 764–826.

13 Head-Gordon, M. and Tully, J.C. (1995) Molecular dynamics with electronic frictions. *J. Chem. Phys.*, **103**, 10137–10145.

14 Brandbyge, M., Hedegård, P., Heinz, T.F., Misewich, J.A., and Newns, D.M. (1995) Electronically driven adsorbate excitation mechanism in femtosecond-pulse laser desorption. *Phys. Rev. B*, **52**, 6042–6056.

15 Springer, C. and Head-Gordon, M. (1996) Simulations of the femtosecond laser-induced desorption of CO from Cu(100) at 0.5mL coverage. *Chem. Phys.*, **205**, 73–89.

16 Luntz, A.C. and Persson, M. (2005) How adiabatic is activated adsorption/associative desorption? *J. Chem. Phys.*, **123**, 074704.

17 Persson, B.N.J. and Ueba, H. (2007) Heat transfer at surfaces exposed to short-pulsed laser fields. *Phys. Rev. B*, **76**, 125401.

18 Bartels, L., Wang, F., Moller, D., Knoesel, E., and Heinz, T.F. (2004) Real-space observation of molecular motion induced by femtosecond laser pulses. *Science*, **305**, 648–651.

19 Stépán, K., Güdde, J., and Höfer, U. (2005) Time-resolved measurement of surface diffusion induced by femtosecond laser pulses. *Phys. Rev. Lett.*, **94**, 236103.

20 Backus, E.H.G., Eichler, A., Kleyn, A.W., and Bonn, M. (2005) Real-time observation of molecular motion on a surface. *Science*, **310**, 1790–1793.

21 Ueba, H. and Persson, B.N.J. (2008) Heat transfer between adsorbate and laser-heated hot electrons. *J. Phys. Condes. Matter*, **20**, 224016.

22 Fedorus, A.G., Klimenko, E.V., Naumovets, A.G., Zasimovich, E.M., and Zasimovich, I.N. (1995) Electron-stimulated mobility of adsorbed particles. *Nucl. Instrum. Methods Phys. Res. B*, **101**, 207–215.

23 Madey, T.E., Netzer, F.P., Houston, J.E., Hanson, D.M., and Stockbauer, R. (1983) *Springer Series in Chemical Physics*, vol. **24**, Springer, Heidelberg, p. 120.

24 Jänsch, H.J., Xu, J.Z., and Yates, J.T. (1993) Electron-stimulated surface migration of CO on Pt(335): 1st spectroscopic evidence for a new phenomenon. *J. Chem. Phys.*, **99**, 721–724.

25 Ho, W. (2002) Single-molecule chemistry. *J. Chem. Phys.*, **117**, 11033–11061.

26 Komeda, T., Kim, Y., Kawai, M., Persson, B.N.J., and Ueba, H. (2002) Lateral hopping of molecules induced by excitation of internal vibration mode. *Science*, **295**, 2055–2058.

27 Morgenstern, K. and Rieder, K.H. (2002) Formation of the cyclic ice hexamer via excitation of vibrational molecular modes by the scanning tunneling microscope. *J. Chem. Phys.*, **116**, 5746–5752.

28 Pascual, J.I., Lorente, N., Song, Z., Conrad, H., and Rust, H.P. (2003) Selectivity in vibrationally mediated single-molecule chemistry. *Nature*, **423**, 525–528.

29 Budde, F., Heinz, T.F., Loy, M.M.T., Misewich, J.A., de Rougemont, F., and Zacharias, H. (1991) Femtosecond time-resolved measurement of desorption. *Phys. Rev. Lett.*, **66**, 3024–3027.

30 Bonn, M., Funk, S., Hess, C., Denzler, D.N., Stampfl, C., Scheffler, M., Wolf, M., and Ertl, G. (1999) Phonon- versus electron-mediated desorption and oxidation of CO on Ru(0001). *Science*, **285**, 1042–1045.

31 Lüpke, G., Bottomley, D.J., and van Driel, H.M. (1994) Second- and third-harmonic generation from cubic centrosymmetric crystals with vicinal faces: phenomenological theory and experiment. *J. Opt. Soc. Am. B*, **11**, 33–44.

32 Kratzer, P., Pehlke, E., Scheffler, M., Raschke, M.B., and Höfer, U. (1998) Highly site-specific adsorption of molecular hydrogen on vicinal Si(001)

surfaces. *Phys. Rev. Lett.*, **81**, 5596–5599.

33 Raschke, M.B. and Höfer, U. (1999) Investigations of equilibrium and non-equilibrium hydrogen coverages on vicinal Si(001) surfaces: energetics and diffusion. *Phys. Rev. B*, **59**, 2783–2789.

34 Struck, L.M., Richter, L.J., Buntin, S.A., Cavanagh, R.R., and Stephenson, J.C. (1996) Femtosecond laser-induced desorption of CO from Cu(100): comparison of theory and experiment. *Phys. Rev. Lett.*, **77**, 4576–4579.

35 Sipe, J.E., Moss, D.J., and van Driel, H.M. (1987) Phenomenological theory of optical second- and third-harmonic generation from cubic centrosymmetric crystals. *Phys. Rev. B*, **35**, 1129–1141.

36 Stépán, K., Dürr, M., Güdde, J., and Höfer, U. (2005) Laser-induced diffusion of oxygen on a stepped Pt(111) surface. *Surf. Sci.*, **593**, 54–66.

37 Höfer, U. (1996) Nonlinear optical investigations of the dynamics of hydrogen interaction with silicon surface. *Appl. Phys. A*, **63**, 533–548.

38 Wilson, P.T., Jiang, Y., Aktsipetrov, O.A., Mishina, E.D., and Downer, M.C. (1999) Frequency-domain interferometric second-harmonic spectroscopy. *Opt. Lett.*, **24**, 496–498.

39 Conrad, U., Güdde, J., Jähnke, V., and Matthias, E. (2001) Phase effects in magnetic second-harmonic generation on ultrathin Co and Ni films on Cu(001). *Phys. Rev. B*, **6314**, 144417.

40 Anisimov, S.I., Kapeliovich, B.L., and Perelman, T.L. (1974) Electron emission from metal surfaces exposed to ultrashort laser pulses. *Sov. Phys. JETP*, **39**, 375.

41 Fann, W.S., Storz, R., Tom, H.W.K., and Bokor, J. (1992) Direct measurement of nonequilibrium electron-energy distributions in subpicosecond laser-heated gold films. *Phys. Rev. Lett.*, **68**, 2834.

42 Fann, W.S., Storz, R., Tom, H.W.K., and Bokor, J. (1992) Electron thermalization in gold. *Phys. Rev. B*, **46**, 13592.

43 Lei, C., Bauer, M., Read, K., Tobey, R., Liu, Y., Popmintchev, T., Murnane, M.M., and Kapteyn, H.C. (2002) Hot-electron-driven charge transfer processes on O_2/Pt(111) surface probed by ultrafast extreme-ultraviolet pulses. *Phys. Rev. B*, **66**, 245420.

44 Persson, B.N.J. and Persson, M. (1980) Vibrational lifetime for CO adsorbed on Cu(100). *Solid State Commun.*, **36**, 175.

45 Persson, B.N.J. and Ryberg, R. (1985) Vibrational phase relaxation at surfaces: CO on Ni(111). *Phys. Rev. Lett.*, **54**, 2119.

46 Misewich, J.A., Heinz, T.F., Kalamarides, A., Höfer, U., and Loy, M.M.T. (1994) Femtosecond surface chemistry of O_2/Pd (111): vibrational assisted electronic desorption. *J. Chem. Phys.*, **100**, 736–739.

47 Tully, J.C., Gomez, M., and Head-Gordon, M. (1993) Electronic and phonon mechanisms of vibrational relaxation: CO on Cu(100). *J. Vac. Sci. Technol. A*, **11**, 1914–1920.

48 Güdde, J. and Höfer, U. (2006) Dynamics of femtosecond laser-induced lateral motion of an adsorbate: O on vicinal Pt (111). *J. Phys. Condens. Matter*, **18**, S1409.

49 Szymanski, P., Harris, A.L., and Camillone, N. (2007) Temperature-dependent electron-mediated coupling in subpicosecond photoinduced desorption. *Surf. Sci.*, **601**, 3335–3349.

50 Ueba, H. and Wolf, M. (2005) Lateral hopping requires molecular rocking. *Science*, **310**, 1774–1775.

51 Dobbs, K.D. and Doren, D.J. (1993) Dynamics of molecular surface diffusion: energy distributions and rotation-translation coupling. *J. Chem. Phys.*, **99**, 10041.

52 Bartels, L., Wolf, M., Meyer, G., and Rieder, K.H. (1998) On the diffusion of 'hot' adsorbates: a non-monotonic distribution of single particle diffusion lengths for CO/Cu(111). *Chem. Phys. Lett.*, **291**, 573–578.

53 Ueba, H. and Persson, B.N.J. (2008) Heating of adsorbate by vibrational-mode coupling. *Phys. Rev. B*, **77**, 035413.

54 Ueba, H., Hayashi, M., Paulsson, M., and Persson, B.N.J. (2008) Adsorbate hopping via vibrational-mode coupling induced by femtosecond laser pulses. *Phys. Rev. B*, **78**, 113408.

55 Ueba, H. (2007) Adsorbate motions induced by vibrational mode coupling. *Surf. Sci.*, **601**, 5212.

56 Persson, B.N.J. and Gadzuk, J.W. (1998) Comments on vibrational dynamics of low-frequency adsorbate motion. *Surf. Sci.*, **410**, L779.

57 Frischkorn, C. (2008) Ultrafast reaction dynamics of the associate hydrogen desorption from Ru(001). *J. Phys. Condens. Matter*, **20**, 313002.

58 Wang, H., Tobin, R.G., Lambert, D.K., DiMaggio, C.L., and Fisher, G.B. (1997) Adsorption and dissociation of oxygen on Pt(335). *Surf. Sci.*, **372**, 267–278.

59 Gee, A.T. and Hayden, B.E. (2000) The dynamics of O_2 adsorption on Pt(533): step mediated molecular chemisorption and dissociation. *J. Chem. Phys.*, **113**, 10333–10343.

60 Gland, J.L., Sexton, B.A., and Fisher, G.B. (1980) Oxygen interactions with the Pt(111) surface. *Surf. Sci.*, **95**, 587–602.

61 Winkler, A., Guo, X., Siddiqui, H.R., Hagans, P.L., and Yates, J.T. (1988) Kinetics and energetics of oxygen-adsorption on Pt(111) and Pt(112): a comparison of flat and stepped surfaces. *Surf. Sci.*, **201**, 419–443.

62 Gambardella, P., Šljivăncanin, Ž., Hammer, B., Blanc, M., Kuhnke, K., and Kern, K. (2001) Oxygen dissociation at Pt steps. *Phys. Rev. Lett.*, **87**, 056103.

63 Prybyla, J.A., Heinz, T.F., Misewich, J.A., Loy, M.M.T., and Glownia, J.H. (1990) Desorption induced by femtosecond laser-pulses. *Phys. Rev. Lett.*, **64**, 1537–1540.

64 Kao, F.J., Busch, D.G., Cohen, D., da Costa, D.G., and Ho, W. (1993) Femtosecond laser-desorption of molecularly adsorbed oxygen from Pt(111). *Phys. Rev. Lett.*, **71**, 2094–2097.

65 Deliwala, S., Finlay, R.J., Goldman, J.R., Her, T.H., Mieher, W.D., and Mazur, E. (1995) Surface femtochemistry of O_2 and CO on Pt(111). *Chem. Phys. Lett.*, **242**, 617–622.

66 Wintterlin, J., Schuster, R., and Ertl, G. (1996) Existence of a "hot" atom mechanism for the dissociation of O_2 on Pt(111). *Phys. Rev. Lett.*, **77**, 123–126.

67 Engström, U. and Ryberg, R. (1999) Atomic oxygen on a Pt(111) surface studied by infrared spectroscopy. *Phys. Rev. Lett.*, **82**, 2741–2744.

68 Backus, E.H.G., Forsblom, M., Persson, M., and Bonn, M. (2007) Highly efficient ultrafast energy transfer into molecules at surface step sites. *J. Phys. Chem. C*, **111**, 6149–6153.

69 Fournier, F., Zheng, W.Q., Carrez, S., Dubost, H., and Bourguignon, B. (2004) Vibrational dynamics of adsorbed molecules under conditions of photodesorption: pump–probe SFG spectra of CO/Pt(111). *J. Chem. Phys.*, **121**, 4839–4847.

70 Germer, T.A., Stephenson, J.C., Heilweil, E.J., and Cavanagh, R.R. (1993) Picosecond measurement of substrate-to-adsorbate energy transfer: the frustrated translation of CO/Pt(111). *J. Chem. Phys.*, **98**, 9986–9994.

71 Ma, J.W., Xiao, X.D., DiNardo, N.J., and Loy, M.M.T. (1998) Diffusion of CO on Pt (111) studied by an optical diffraction method. *Phys. Rev. B*, **58**, 4977–4983.

72 Pascual, J.I. (2005) Single molecule vibrationally mediated chemistry: towards state-specific strategies for molecular handling. *Eur. Phys. J. D*, **35**, 327–340.

73 Schweizer, E., Persson, B.N.J., Tüshaus, M., Hoge, D., and Bradshaw, A.M. (1989) The potential energy surface, vibrational phase relaxation and the order–disorder transition in the adsorption system Pt (111)-CO. *Surf. Sci.*, **213**, 49.

74 Bonn, M., Hess, C., Funk, S., Miners, J.H., Persson, B.N.J., Wolf, M., and Ertl, G. (2000) Femtosecond surface vibrational spectroscopy of CO adsorbed on Ru(001) during desorption. *Phys. Rev. Lett.*, **84**, 4653–4656.

75 Lawrenz, M., Stépán, K., Güdde, J., and Höfer, U. (2009) Time-domain investigation of laser-induced diffusion of CO on vicinal Pt(111). *Phys. Rev. B*, **80**, 075429.

19
Laser-Induced Softening of Lattice Vibrations

Eeuwe S. Zijlstra and Martin E. Garcia

Intense femtosecond laser pulses bring solids to a nonequilibrium state in which the electrons acquire an extremely high temperature while the lattice remains cold. The hot electrons modify dramatically the potential energy surface acting on the ions. As a consequence, phonon softening, phonon destabilization, cold melting, or solid–solid structural phase transitions can occur. In this chapter, we discuss the physical picture underlying these effects and present an accurate theoretical description for selected examples.

19.1
Introduction

The interaction of intense femtosecond laser pulses with materials gives rise to different kinds of interesting phenomena that have been experimentally studied by means of pump–probe measurements since the early 1980s [1]. Ultrafast structural changes constitute a particular class of ultrafast phenomena in solids and include coherent phonons [2–6], solid-to-solid structural changes [7], melting [8–10], and ablation [11, 12].

When an ultrashort laser pulse excites a solid, an extreme nonequilibrium state is created, which lives for a couple of picoseconds. This state is characterized by electrons that, after absorbing a large amount of photons from the laser pulse, acquire an extremely high temperature while the lattice still remains cold. This occurs because intense pulses can excite a considerable fraction of electron–hole pairs [13], which thermalize very rapidly. The excitation and thermalization processes that the electrons undergo produce dramatic changes in the potential energy of the ions. Due to their much slower motion, ions “feel” a sudden change of the potential surface and start moving. This motion is in most cases cooperative and can be sometimes even coherent. Most of the experimentally observed laser-induced ultrafast structural changes can be analyzed in the framework of the above description.

In this chapter, we present a detailed theoretical description of the physical picture underlying laser-excited coherent phonons and structural changes. We also derive

Dynamics at Solid State Surface and Interfaces Vol.1: Current Developments
Edited by Uwe Bovensiepen, Hrvoje Petek, and Martin Wolf

ISBN: 978-3-527-40937-2

mathematical expressions that support the qualitative arguments of the previous paragraph. We start in Section 19.2 from the Born–Oppenheimer approximation and show how it can be generalized to the case of a solid immediately after having been excited by a femtosecond laser pulse. We describe not only the mechanisms leading to the excitation of coherent phonons but also how their frequency depends on the laser fluence. In Section 19.3, we discuss how the theory presented in Section 19.2 can be applied, making also use of density functional theory calculations, to explain a variety of experiments performed on different materials.

Our results from Section 19.3 illustrate the interplay between laser excitations and crystal structure. To name some examples, we show how the softening of lattice vibrations in the A7 structure can be employed to study the ultrafast process of electron–hole thermalization in bismuth and how it precedes a solid–solid transition in arsenic under pressure. For the zinc-blende structure, we show how the phonon softening of selected modes leads to an acceleration of the ions, which subsequently disorder. This process is called ultrafast melting.

19.2 Theoretical Framework

In the following, we introduce the concept of potential energy surface starting from the Hamiltonian of the solid and using the adiabatic approximation.

19.2.1 The Hamiltonian of the Solid

We consider a solid consisting of N_e electrons (each one having a charge $-e$) and N_I ions of masses M_i and charges $Z_i e$ ($i = 1, N_\mathrm{I}$). Due to charge neutrality, it must hold that $N_e = \sum_{i=1}^{N_\mathrm{I}} Z_i$. Neglecting relativistic effects, the corresponding Hamiltonian can be written as

$$\hat{H} = \hat{T}_\mathrm{I} + \hat{T}_e + V_{ee} + V_{e\mathrm{I}} + V_\mathrm{II}. \tag{19.1}$$

Here, the first term that corresponds to the kinetic energy of the ions is given by

$$\hat{T}_\mathrm{I} = -\sum_{i=1}^{N_I} \frac{\hbar^2}{2M_i} \nabla^2_{\vec{R}_i}, \tag{19.2}$$

where $\vec{\nabla}_{\vec{R}_i}$ refers to the gradient with respect to the coordinate $\vec{R}_i$ of the ith ion.

The second term in Eq. (19.1) describes the kinetic energy of the electrons, which is given by

$$\hat{T}_e = -\sum_{j=1}^{N_e} \frac{\hbar^2}{2m} \nabla^2_{\vec{r}_j}, \tag{19.3}$$

where $\vec{\nabla}_{\vec{r}_j}$ refers to the gradient with respect to the coordinate $\vec{r}_j$ of the jth electron.

The third term in Eq. (19.1) accounts for the Coulomb interactions between electrons, which is given by

$$V_{ee} = \sum_{j<l}^{N_e} \frac{e^2}{|\vec{r}_j - \vec{r}_l|}. \tag{19.4}$$

Finally, the last two terms of the Hamiltonian $\hat{H}$ refer to the attractive interaction between electrons and ions and to the ion–ion repulsion, respectively. Both interaction terms consist of pair potentials and can be written as $V_{eI} = \sum_{j,i}^{N_e,N_I} v_{eI}(|\vec{r}_j - \vec{R}_i|)$ and $V_{II} = \sum_{i,k}^{N_I} v_{II}(|\vec{R}_i - \vec{R}_k|)$. Note, however, that the form of the pair potentials $v_{eI}(|\vec{r}_j - \vec{R}_i|)$ and $v_{II}(|\vec{R}_i - \vec{R}_k|)$ depends on how many electrons are taken as degrees of freedom of the problem. For example, if only the valence electrons are considered, the ions do include not only the atomic nuclei but also the core electrons. Then, both pair potentials must include not only the Coulomb interactions but also the Pauli principle, and they acquire a more complex form. If, however, all electrons are considered as degrees of freedom, both potentials $v_{eI}(|\vec{r}_j - \vec{R}_i|)$ and $v_{II}(|\vec{R}_i - \vec{R}_k|)$ can be written as pure Coulomb interactions, and the corresponding terms of the Hamiltonian have the form

$$V_{eI} = \sum_{j,i}^{N_e,N_I} \frac{-Z_i e^2}{|\vec{r}_j - \vec{R}_i|},$$

$$V_{II} = \sum_{i,k}^{N_I} \frac{Z_i Z_k e^2}{|\vec{R}_i - \vec{R}_k|}. \tag{19.5}$$

Since ions are considerably heavier and consequently much slower than electrons ($m/M_i \sim 10^{-4}$) their kinetic energy is much smaller than that of electrons. Moreover, the term $\hat{T}_I$ is much smaller than the other terms of the Hamiltonian $\hat{H}$ [14]. In many physical situations, it is possible to assume a separation of the timescales for electrons and ions, and perform the adiabatic (or Born–Oppenheimer) approximation.

19.2.2 Born–Oppenheimer Approximation

Following the above discussion, one can treat $\hat{T}_I$ as a perturbation and write $\hat{H} = \hat{H}_0 + \hat{T}_I$. Note that $\hat{H}_0$ is the zeroth order of $\hat{H}$ in $\hat{T}_I$ and is a purely electronic Hamiltonian, with the operators $\vec{\nabla}_{\vec{r}_j}$ and V_{ee} acting on the electron wave functions and the terms V_{eI} and V_{II} depending on the ionic coordinates $\{\vec{R}_i\}$ as external parameters. Thus, for fixed ionic coordinates $\hat{H}_0$ describes N_e electrons in a static potential. The corresponding Schrödinger equation is expressed as

$$\begin{aligned} \hat{H}_0\, \phi_\alpha(\{\vec{r}_j\},\{\vec{R}_i\}) &= [\hat{T}_e + V_{ee} + V_{eI} + V_{II}]\, \phi_\alpha(\{\vec{r}_j\},\{\vec{R}_i\}) \\ &= [\varepsilon_\alpha(\{\vec{R}_i\}) + V_{II}(\{\vec{R}_i\})]\, \phi_\alpha(\{\vec{r}_j\},\{\vec{R}_i\}), \end{aligned} \tag{19.6}$$

where the wave function ϕ_α is an eigenstate of $\hat{H}_0$, which describes all N_e electrons considered as degrees of freedom and can therefore be expressed as a linear combination of $N_e \times N_e$ Slater determinants of Bloch functions. ε_α is the corresponding eigenvalue. Since $V_{\rm II}$ does not depend on the electronic coordinates, we write it explicitly as an additive constant in Eq. (19.6). Note that both ε_α and $V_{\rm II}$ contain the ionic coordinates, which act as $3N_{\rm I}$ external parameters. $\{\alpha\}$ is a complete set of quantum numbers and $\{\phi_\alpha\}$ a complete set of N_e-electron wave functions.

19.2.2.1 General Solution of $\hat{H}$

We can now seek for the complete solution of $\hat{H}$, which will be given by

$$\hat{H}\,\psi(\{\vec{r}_j\},\{\vec{R}_i\}) = E\,\psi(\{\vec{r}_j\},\{\vec{R}_i\}), \tag{19.7}$$

where ψ is the complete wave function of the solid describing the N_e electrons and $N_{\rm I}$ ions. For fixed ionic coordinates, and due to completeness, we can express ψ as

$$\psi(\{\vec{r}_j\},\{\vec{R}_i\}) = \sum_\alpha \chi_\alpha(\{\vec{R}_i\})\,\phi_\alpha(\{\vec{r}_j\},\{\vec{R}_i\}). \tag{19.8}$$

Inserting Eq. (19.8) in Eq. (19.7), one obtains

$$\sum_\alpha [\hat{H}_0 + \hat{T}_{\rm I} - E]\,\chi_\alpha(\{\vec{R}_i\})\,\phi_\alpha(\{\vec{r}_j\},\{\vec{R}_i\}) = 0. \tag{19.9}$$

Here, the coefficients $\{\chi_\alpha\}$ of the linear expansion depend on the ionic coordinates. For different sets of ionic coordinates $\{\vec{R}_i\}$, one obtains different χ_α values.

Taking into account that the operator $\hat{T}_{\rm I}$ affects both the ϕ_α values and the χ_α values, multiplying Eq. (19.9) by $\phi^*_\beta(\{\vec{r}_j\},\{\vec{R}_i\})$ and integrating over all $3N_e$ electronic coordinates, after some steps and after interchanging the labels α and β, one obtains

$$\begin{aligned}&[\hat{T}_{\rm I} + \varepsilon_\alpha(\{\vec{R}_i\}) + V_{\rm II}(\{\vec{R}_i\})]\,\chi_\alpha(\{\vec{R}_i\})\\ &\quad + \sum_\beta B_{\beta\alpha}(\{\vec{R}_i\})\,\chi_\beta(\{\vec{R}_i\}) = E\chi_\alpha(\{\vec{R}_i\}),\end{aligned} \tag{19.10}$$

where $B_{\beta\alpha}$ are operators acting on the ionic coordinates and therefore on the coefficients $\chi_\alpha(\{\vec{R}_i\})$ (see Ref. [14] for the detailed expressions).

19.2.2.2 Adiabatic Approximation

One can estimate the magnitude of the terms contained in the operators $B_{\beta\alpha}$ [14]. It turns out that the terms scale roughly as $(m/M_i)^{3/4}$ and (m/M_i) and that they can be neglected if the velocity of the ions is small. Thus, assuming that $B_{\beta\alpha} \to 0$, one obtains an eigenvalue equation fulfilled by the coefficients χ_α given by

$$[\hat{T}_{\rm I} + U(\{\vec{R}_i\})]\chi_\alpha(\{\vec{R}_i\}) = E\chi_\alpha(\{\vec{R}_i\}), \tag{19.11}$$

where

$$U(\{\vec{R}_i\}) = \varepsilon_\alpha(\{\vec{R}_i\}) + V_{\text{II}}(\{\vec{R}_i\}). \tag{19.12}$$

Note that Eq. (19.11) has the form of a Schrödinger equation for the coefficients χ_α, which are functions of the ionic coordinates $(\{\vec{R}_i\})$. Therefore, the $\chi_\alpha(\{\vec{R}_i\})$ values can be interpreted as the wave functions of the ions, which move under the influence of an effective potential $U(\{\vec{R}_i\})$, which is called potential energy surface (PES).

The PESs of the form in Eq. (19.12) are called adiabatic because they result from first solving the electron problem for fixed ionic coordinates. The different states of the N_e electronic system are labeled with α, and for each N_e-electron state there is a particular potential energy surface. $\alpha = 0$ corresponds to the ground-state potential energy $U_0(\{\vec{R}_i\})$, which is obtained from the ground state of $\hat{H}_0$. There is a huge amount of PESs, since already each single-electron excitation leads to a different PES. The PESs are multidimensional surfaces in a space of $3N_{\text{I}}$ dimensions. In the adiabatic approximation, there is no connection between the different PESs. They can cross with each other but there is no term connecting them. However, as soon as the factors $B_{\beta\alpha}$ (Eq. (19.10)) are considered to be nonzero, hoppings between the potential surfaces merely induced by the motion of the ions are possible.

In the adiabatic approximation the ions move on the multidimensional potential energy surface $U(\{\vec{R}_i\})$, which can be obtained by solving the electronic Schrödinger equation (19.6) and adding the ion–ion repulsion V_{II}. For each electronic excitation, there is a different PES. Hoppings between PESs may be induced by the motion of the ions.

19.2.3 Properties of $\hat{H}_0$

As mentioned above, the eigenstates of $\hat{H}_0$ are N_e-electron wave functions that can be written as linear combinations of Slater determinants of Bloch functions. In some cases, it is possible to reduce $\hat{H}_0$ to an effective single-particle Hamiltonian $\hat{H}_0^{\text{sp}}$ by performing some mean field treatment of the Coulomb interactions (such as Hartree–Fock, local density approximation, etc.). In this case, the electronic Hamiltonian becomes $\hat{H}_0^{\text{sp}} = \sum_{j=1}^{N_e} \hat{h}_0^{(j)}(\vec{r}_j, \{\vec{R}_i\})$, that is, a sum of single-particle operators $\hat{h}_0^{(j)}$, which satisfy $\hat{h}_0^{(j)}(\vec{r}_j, \{\vec{R}_i\}) = \hat{h}_0^{(j)}(\vec{r}_j + \vec{R}_{\vec{n}}, \{\vec{R}_i\})$, where $\vec{R}_{\vec{n}}$ is a vector of the Bravais lattice of the investigated solid. If $\hat{H}_0^{\text{sp}}$ is an effective single-particle operator, then its eigenvalues are given by $\varepsilon_\beta(\{\vec{R}_i\}) = \sum_{j=1}^{N_e} \varepsilon(\vec{k}_j, \nu_j, \sigma_j)$, where $\varepsilon(\vec{k}_j, \nu_j, \sigma_j)$ are the eigenvalues of the operators $\hat{h}_0^{(j)}$ and $\vec{k}_j$, ν_j and σ_j refer to the wave number, band index, and spin projection of the jth electron, respectively.

The complete $\hat{H}_0$, written in second quantization, is given by

$$\hat{H}_0 = \hat{H}_0^{\mathrm{sp}} + \hat{V}_{ee} = \sum_{\nu,\vec{k}\sigma} \varepsilon_\nu(\vec{k})(\{R_i\})\, c^+_{\nu\vec{k}\sigma}\, c_{\nu\vec{k}\sigma} + \sum_{1234} v^{1234}_{ee}\, c_1^+\, c_2^+\, c_3 c_4, \tag{19.13}$$

where the first term is the energy of noninteracting electrons in the solid, containing the electron band structure $\varepsilon_\nu(\vec{k})$. The second term refers to the electron–electron interactions, contained in the matrix elements v_{1234} of the Coulomb operator. The generic indexes 1, 2, 3, and 4 contain the wave vector $\vec{k}$ (which runs over the first Brillouin zone), the band index ν, and the spin projection σ.

19.2.4
Phonons in the Ground-State Potential Energy Surface

When the electrons are in the ground state of $\hat{H}_0$, the minimum of the ground-state PES is obtained for the set of ionic positions $\{\vec{R}_i^0\}$ that correspond to the stable crystal structure of the solid. It holds that $\partial U/\partial \vec{R}_i|_{\{\vec{R}_i^0\}} = 0$ for all ions. Thus, if we expand the PES in powers of the ionic displacements with respect to the ground-state configuration $\delta \vec{R}_i$, the first order will be zero. Such an expansion is expressed as

$$U(\{\vec{R}_i\}) = U(\{\vec{R}_i^0\}) + \frac{1}{2}\sum_{i',i'',a,a'} \left.\frac{\delta^2 U}{\delta R_{i'a}\,\delta R_{i''a'}}\right|_{\{R_i^0\}} \delta R_{i'a}\,\delta R_{i''a'} + \mathcal{O}[(\delta R_i)^3], \tag{19.14}$$

with $a, a' = x, y, z$. This means that for small displacements of the ions with respect to their equilibrium positions the PES is harmonic. The harmonic part is used to construct the dynamical matrix

$$D_{ia,i'a'} = \frac{1}{\sqrt{M_i M_{i'}}}\frac{\delta^2 U}{\delta R_{ia}\,\delta R_{i'a'}}, \tag{19.15}$$

which in solids satisfies different properties due to the translational symmetry. From the diagonalization of the Fourier transform of $D_{ia,i'a'}$, one obtains the phonon spectrum $\omega(\vec{q},\lambda)$, where $\vec{q}$ refers to the phonon momentum within the first Brillouin zone and λ labels the phonon branches. Note that the terms written as $\mathcal{O}[(\delta R_i)^3]$ in Eq. (19.14), which are neglected in the harmonic approximation, are those that will contribute to the anharmonicities of the lattice vibrations.

So far, we have shown that the ground-state PES $U_0(\{\vec{R}_i\})$, that is, the potential acting on the ions when the electrons are in the ground state, gives rise to the crystal structure and to the phonon modes of the solid. For each set of ionic coordinates $\{\vec{R}_i\}$, the ground-state PES is well defined, since it is determined from the lowest electronic state. PESs of excited electronic states can undergo crossings with other surfaces. In each of such crossings, nonadiabatic effects, that is, transitions from one surface to the other due to the surface coupling coefficients $B_{\beta\alpha}$ described in Eq. (19.10) are very likely to occur.

19.2.5
Electron–Phonon Coupling and Excitation of Coherent Phonons in the Ground-State Potential Energy Surface

The interaction between the lattice vibrations and the electrons is essentially a nonadiabatic effect, which should be contained in the coefficients $B_{\beta\alpha}$ described in Eq. (19.10). However, one can determine the form and magnitude of the electron–phonon interactions without explicitly calculating the $B_{\beta\alpha}$ values. For the derivation of Eqs. (19.6) and (19.9) we have assumed, making use of the separation of timescales for electronic and ionic motion, that the ionic coordinates remain fixed. However, the separation of timescales should not be considered very strictly when ions move over very short distances, since this could happen on timescales not far away from that for the electronic motion. Therefore, the ion–electron interaction terms in Eq. (19.6) must reflect this fact. For this purpose, one makes the expansion $v_{el}(|\vec{r}_j - \vec{R}_i|) = v_{el}(|\vec{r}_j - \vec{R}_i^0|) + \vec{\nabla}_{\vec{R}_i} v_{el}(|\vec{r}_j - \vec{R}_i^0|) \cdot \delta\vec{R}_i$, which holds for small displacements of the ions, like those occurring when thermal phonons are excited. The first term of this expansion was treated before and contributes to the PES. The second one corresponds to the electron–phonon interaction. By performing a Fourier transformation of it, and expressing the ionic displacements $\delta\vec{R}_i$ in terms of the phonon creation and destruction operators (b_q^+, b_q), one obtains the widely used expression for the electron–phonon interactions [14]. In the case of two electronic bands (valence and conduction states) and considering only one particular phonon branch, the expression for electron–phonon interactions is given by

$$\hat{V}_{\text{el-ph}} = \sum_{\nu,\vec{k},\vec{q}} M_{\nu\vec{k},\vec{q}} (b_{\vec{q}} + b_{-\vec{q}}^+)\, c_{\nu\vec{k}}^+\, c_{\nu\vec{k}+\vec{q}}. \tag{19.16}$$

As mentioned in Section 19.1, when the system becomes excited by a light pulse of considerably shorter duration than the phonon period, coherent lattice vibrations can be excited. In addition to the phenomenological theory proposed by Zeiger *et al.* [2], a microscopic approach based on the electron–phonon interaction expressed in Eq. (19.16) was used by Kuznetsov and Stanton [15]. They defined the coherent amplitude of the $\vec{q}$th phonons as the time-dependent statistical average $D_{\vec{q}} = \langle b_{\vec{q}} + b_{-\vec{q}}^+ \rangle$. The equation of motion for $D_{\vec{q}}$ can be obtained from the commutator of b_q^+ and b_q with the Hamiltonian $\hat{H}_{\text{el}} + \hat{H}_{\text{ph}} + \hat{V}_{\text{el-ph}}$, where $\hat{H}_{\text{el}} = \hat{H}_0^{\text{sp}}$ and $\hat{H}_{\text{ph}} = \hat{T}_I + U(\{\vec{R}_i\})$. One obtains [15]

$$\partial^2 \frac{D_{\vec{q}}}{\partial t^2} + \omega_{\vec{q}} D_{\vec{q}} = -2\omega_{\vec{q}} \sum_{\nu\vec{q}} M_{\nu,\vec{k},\vec{q}}\, \hat{\varrho}_{\nu\vec{k},\vec{k}+\vec{q}}, \tag{19.17}$$

where $\hat{\varrho}_{\nu\vec{k},\vec{k}+\vec{q}} = c_{\nu\vec{k}}^+\, c_{\nu\vec{k}+\vec{q}}$ is the electron-reduced density operator. Note that Eq. (19.17) describes the motion of a harmonic oscillator under the action of an external driving force. If one assumes a spatially homogeneous excitation, the density matrix is diagonal in the momentum representation, indicating that the driving force depends on the electronic occupations. The driving force is also time dependent, since the electron occupations are changed by the laser pulse. Integration of Eq. (19.17) yields, in particular for the case $\vec{q} = 0$, a harmonic motion for $D_0(t)$ with frequency ω_0. This motion is of displacive character ("cosine"-like), in contrast to the case of impulsive Raman excitation of coherent phonons. This was also the result of the phenomenological theory by Zeiger *et al.* [2].

Merlin and coworkers [3, 4] pointed out that the displacive excitation of coherent phonons is not entirely a new phenomenon, but can also be described applying the theory for Raman excitations to nontransparent (i.e., absorbing) materials. By starting from the phenomenological potential for stimulated Raman scattering but treating the susceptibility, and therefore the Raman tensors, within a microscopic theory, they have shown that the driving force for the coherent phonons is of displacive character as soon as the system is no longer transparent or, more generally, as soon as the decay rates of the electronic states are nonzero [3, 4].

It is worth mentioning that the theories mentioned above are able to describe the existence of laser-excited coherent lattice oscillations and are correct for low laser intensities. However, they cannot explain experimentally the observed effects such as laser-induced softening of coherent phonons [5, 6, 16] or laser-induced phonon–phonon interactions [5]. These effects can only be taken into account by describing the changes of the PES due to the excitation of electron–hole pairs by the laser pulse.

19.2.6 "Laser Excited" Potential Energy Surfaces

The ground-state PES $U_0(\{\vec{R}_i\})$ is used to calculate mechanical and chemical properties of solids in thermodynamical equilibrium. For this purpose, $U_0(\{\vec{R}_i\})$ is usually fitted using two- or three-body model potentials, such as Lennard-Jones [17], Tersoff [18], or Stillinger–Weber [19] potentials, which consist of functions of the interatomic distances $\vec{R}_{ii'} = \vec{R}_i - \vec{R}_{i'}$.

For the description of laser-induced structural changes, however, such model potentials are not useful, since they depend only on the ionic coordinates and do not contain the electrons as degrees of freedom. Therefore, they do not allow any reasonable description of the excitation process, since, as pointed out before, the laser pulse interacts first with the electrons. Hence, for the studies to be presented in this chapter, a many-body PES $U(\{\vec{R}_i\}, \{n_{\nu\vec{k}}(t)\})$ depending on the electronic occupations $\{n_{\nu\vec{k}}(t)\}$, which themselves are dependent on time, must be derived from a microscopic electronic Hamiltonian.

The interaction of the electrons with a laser pulse modifies the Hamiltonian $\hat{H}_0$ (Eq. (19.6)) by a term proportional to the scalar product of the time-dependent vector potential corresponding to the electromagnetic field of the laser and the momentum $\hat{\vec{p}}_j$ of each electron [20]. This term, which we call $\hat{H}_{\text{laser}}$, is a single-particle operator that creates electrons above the Fermi level and holes below it. This means that it is responsible for the rapid creation of electron–hole pairs, as long as the pulse is acting.

The dynamics of electrons is governed by the equation of motion for the reduced density matrix $\hat{\varrho}$, which is expressed as

$$i\hbar\frac{\partial\hat{\varrho}}{\partial t}=\left[\hat{H}_0^{sp}+\hat{H}_{laser},\hat{\varrho}\right]+\left.\frac{\partial\hat{\varrho}}{\partial t}\right|_{collisions}, \tag{19.18}$$

where $\hat{H}_0^{sp}$ is, as described before, the effective single-particle part of $\hat{H}_0$ and the last term describes the collisions due to the electron–electron Coulomb interactions. The first term describes the coherent motion of the electrons, which is driven by the laser field and involves optical transitions between the energy levels of the Hamiltonian $\hat{H}_0^{sp}$. The coherence is reflected in the fact that due to the action of the laser pulse off-diagonal elements of the density matrix become nonzero. The density matrix keeps this nondiagonal form only during a very short time τ_{deph}, since the system undergoes a rapid dephasing due to the action of the second, dissipative term in Eq. (19.18). For high laser fluences, the dephasing time τ_{depth} is so short that off-diagonal elements can be assumed to be zero as soon as the laser pulse is over. The second very fast step is the electron thermalization, which occurs on a thermalization timescale τ_{therm}. For very low fluences, τ_{therm} is known to be of the order of 100 fs [21]. For high fluences, a carrier thermalization time faster than 25 fs in GaAs has been reported [22]. Therefore, one can assume that the diagonal elements of the density matrix, which represent a nonequilibrium electron distribution created by the laser, will rapidly converge to a Fermi–Dirac distribution, that is,

$$\varrho_{\nu\vec{k},\nu,\vec{k}}=n_{\nu\vec{k}}=\frac{1}{\exp\{\beta(\varepsilon_\nu(\vec{k})-\mu)+1\}},\quad \text{with } \beta=\frac{1}{k_BT_{el}}. \tag{19.19}$$

Here, T_{el} is the electron temperature and μ is the chemical potential. Note that T_{el} is very high (usually of the order of many thousand Kelvin) due to the large amount of energy pumped into the electronic system by the laser.

As mentioned before, the ground-state PES is obtained after solving the electron problem, that is, from the ground-state energy of the electronic system. Now, in the laser-excited solid and for times larger than τ_{therm}, the electrons will have a well-defined temperature T_{el}. Thus, solving the electron problem in this case means determining the function that uniquely describes electrons at a given temperature, which is no longer the ground-state energy but the Helmholtz free energy of the electrons $F_{el}(\{\vec{R}_i\},t)$, which is given by

$$\begin{aligned}U(\{\vec{R}_i\},t) &= F_{el}(\{\vec{R}_i\},t)=\sum_{\nu\vec{k}}n_{\nu\vec{k}}(t)\,\varepsilon_\nu(\vec{k})+\\ &\quad +V_{II}(\{\vec{R}_i\})-T_{el}(t)S_{el}(t).\end{aligned} \tag{19.20}$$

The assumption of fast electron thermalization is justified for laser pulse intensities that excite a significant percentage of the valence electrons to the conduction band. The third term in Eq. (19.20) contains the electronic temperature T_{el} and the electronic entropy S_{el}, which is given by

$$S_{\text{el}} = -k_{\text{B}} \sum_{\nu\vec{k}} [n(\varepsilon_\nu(\vec{k}), t) \log(n(\varepsilon_\nu(\vec{k}), t)) + (1 - n(\varepsilon_\nu(\vec{k}), t)) \log(1 - n(\varepsilon_\nu(\vec{k}), t))]. \tag{19.21}$$

Equation (19.20) represents a generalization of the Born–Oppenheimer approximation. Note that the usual Born–Oppenheimer approximation is recovered for $T_{\text{el}} = 0$ [13].

It is important to point out that the functional dependence of $U(\{\vec{R}_i\}, t)$ on the atomic positions $\{\vec{R}_i\}$ is strongly dominated by the electronic occupations $n_{\nu\vec{k}}(t)$ present in the first and third terms in Eq. (19.20). If the electronic occupations undergo strong changes, the PES $U(\{\vec{R}_i\}, t)$ will change significantly [13, 23–25].

In view of the above discussion, one can understand the mechanism for laser-induced structural changes by using a simple physical picture: Before laser heating the solid is in thermodynamical equilibrium at a low initial (lattice and electronic) temperature ($T_i \leq 300$ K). This means that electrons fill mainly the states below the Fermi level. The resulting ground-state potential landscape $U_0(\{\vec{R}_i\})$ shows minima at the crystal lattice sites. Atoms are sitting on these sites and therefore no forces act on them. However, the situation changes dramatically when the solid gets heated up by a laser pulse, which creates a considerable amount of electron–hole pairs. This means that the electronic occupations $n_{\nu\vec{k}}(t)$ change. Consequently, the functional form of $U(\{\vec{R}_i\})$ undergoes quantitative and qualitative changes, which results in the shifting of the minima, or their disappearance for very high fluences [26]. Since these modifications occur on a timescale that is much shorter than the reaction time of the lattice, during the laser pulse the atoms remain in their positions, most of them no longer minima of the new potential landscape. Therefore, forces now act on the atoms, which start to move [27]. As a consequence of the laser excitation, the system is in an extreme nonequilibrium state in which electrons are very hot while the lattice remains cold. This may lead to a variety of laser-induced structural events.

It is important to point out that the approach described above for obtaining the laser-excited PES is not the only possible one. The assumption that electrons acquire a well-defined temperature during the first stages of the laser-induced lattice dynamics was used for some molecular dynamics simulations [28, 29] and by us for the construction of excited PESs [30, 31]. Fahy and coworker use for semiconductors and semimetals a constraint model in which it is assumed that electrons and holes thermalize separately and acquire two different chemical potentials [32]. We discussed in Refs [33] and [16] the drawbacks of this assumption for semimetals in general and also for semiconductors excited by very intense laser pulses. We have also compared the behavior of phonon softening assuming one and two chemical potentials. A more detailed discussion on this is presented in Section 19.3.2.

Another possible starting point for the calculation of the laser-excited PES is to assume that the electronic entropy remains constant. This idea was implemented by Stampfli and Bennemann [13, 23, 34] and by us [16, 33, 35]. In that case, the thermodynamical function that is relevant for the construction of the laser-excited PES is the total energy.

Finally, molecular dynamics simulations were performed in which the absorbed energy is the quantity that does not change during the first stages of the laser induced lattice dynamics. Those simulations take the laser pulse explicitly into account. Examples are the works by Jeschke *et al.* [24–26, 36, 37].

19.2.7
Displacive Excitation of Coherent Phonons. Phonon Softening

The main changes that excitation of electron–hole pairs can produce on the PES are (i) a shift of the minimum with respect to the ground-state PES U_0 in clusters and solids with more than one atom per unit cell [32, 33, 38, 39] and (ii) a deformation of the shape near the minimum due to the presence of anharmonicities [24–26, 31, 34]. In most nanostructures, and particularly in covalent systems, the minimum is shifted to larger interatomic distances. This is due to the fact that the laser excitation creates holes in bonding states and electrons in antibonding states. In general, when electrons are heated up to very high temperatures, the system becomes weakly bonded. An interesting example of this shift occurs in laser-excited carbon nanotubes. Dumitrica *et al.* [38, 39] and Romero *et al.* [40] have shown that this effect may lead to exotic nonequilibrium structural changes. In solids, such a shift of the minimum to larger interatomic distances is only possible if the total volume remains conserved due to inertial confinement, that is, because volume expansions are governed by the sound velocity and occur on timescales that are much longer than those that play a major role in coherent phonons and ultrafast structural changes. An example of such a shift occurs in Bi, which exhibits a lattice structure with two atoms per unit cell, which is distorted due to a Peierls transition. The effect of laser pulses on solid Bi will be discussed in detail in Sections 19.3.1–19.3.4, where we also address the laser-induced deformations of the PES near the minimum and the importance of anharmonicities.

The mechanism for the displacive excitation of coherent phonons is illustrated in Figure 19.1. Note that it looks very similar to the scheme used to describe light excitation of molecules, although the underlying physics is considerably different. The femtosecond laser excitation in solids can be considered as a vertical transition due to the ultrashort pulse duration and the previously discussed inertial confinement.

19.2.8
Comparison to Other Theories for the Generation of Coherent Phonons

The previous model for the displacive excitation of coherent phonons has some advantages with respect to the ones proposed by Zeiger *et al.* [2] and Stanton *et al.* [15] in the sense that it does not only explain the "cosine"-like time dependence of reflectivity changes but also describes, even quantitatively, both the laser-induced softening and phonon–phonon interactions. Moreover, it does not rely on a particular basis to define the creation and destruction phonon operators, such as the one discussed in Ref. in [15].

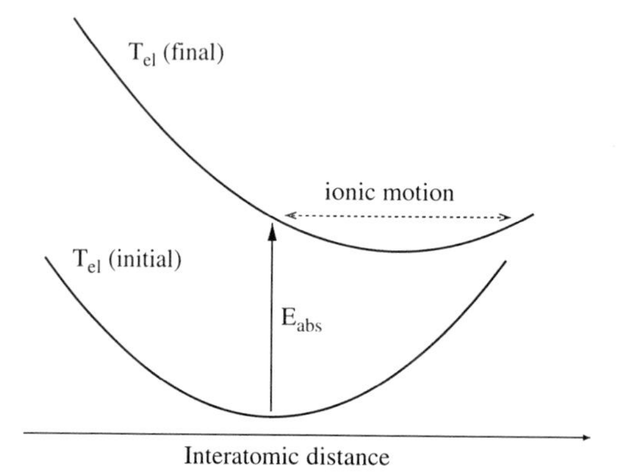

Figure 19.1 Illustration of the displacive excitation of coherent phonons. Before the laser excitation, ions are on the global minimum of the ground-state PES at an initial electronic temperature T_{el}(initial) close to zero. The laser pulse acts on an extremely short time in which the ions cannot move. As a consequence, atoms suddenly find themselves on the laser-excited PES at a very high final temperature T_{el}(final). Due to bond weakening, the laser-excited PES can be shifted to larger interatomic distances and be deformed (phonon softening). The laser excitation is represented by the vertical arrow. E_{abs} is the energy that the electrons absorb from the light pulse. As a consequence of the excitation, ions move coherently on the new PES.

Although our model does not include Raman-like excitations, it shows that the displacive excitation of coherent phonons is produced by a sudden change in the PES, which could be produced either by ultrashort light pulses or by other types of ultrashort excitation, for example, ion bombardment. Therefore, the conclusions by Merlin *et al.* [3, 4], that the displacive mechanism constitutes a special case of Raman excitation for opaque materials, is incomplete.

In Section 19.3, we discuss different laser-induced structural events on the basis of the construction of laser-excited PESs using accurate all-electron density functional theory calculations.

19.3 Laser-Induced Events Involving Phonon Softening

When phonons soften under the influence of a femtosecond laser pulse, different phenomena, which typically involve ultrafast structural changes, may be observed. Events that should be distinguished are the following:

- **Coherent phonons**: [41]. These may be induced displacively when the equilibrium coordinates of the atoms are shifted by an ultrashort laser pulse and impulsively by Raman scattering.
- **Solid-to-solid transitions**: [7]. These occur when the symmetry of a solid is affected by a laser, for example, the undoing of a Peierls distortion.

- **Ultrafast melting**: [42]. This transition to a liquid-like state typically takes place through the softening of transverse acoustic phonon modes. The atoms accelerate on a repulsive potential and subsequently disorder. The liquid that is formed after ultrafast melting may have different properties than the ordinary liquid phase, where the electrons and ions are in thermodynamic equilibrium [28].

Due to the complex interplay between the atomic and electronic structures of materials, it is often useful to explicitly compute the PES of laser-excited systems, which can then be used to understand phenomena that have already been observed and to make novel predictions. Here, we present several examples. Because of the importance of the atomic structure for these selected cases, we also provide a description of the crystal lattices involved, namely, the A7 and the zinc-blende structures.

19.3.1 The A7 Structure

The group V semimetals arsenic, antimony, and bismuth crystallize in the A7 structure, which can be derived from a simple cubic atomic packing in two steps. This is illustrated for the case of arsenic in Figure 19.2. The steps are as follows: (1) A simple cubic lattice is deformed by elongating it along one of the body diagonals. (2) A Peierls instability causes the atoms to be displaced along the same diagonal, in opposite directions. In the A7 structure, the Peierls parameter has a one-to-one relation to the only atomic coordinate that is not fixed by symmetry. We will show that

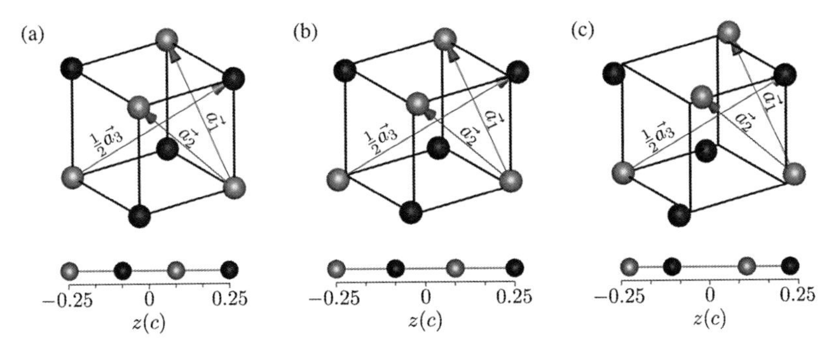

Figure 19.2 Relation between the A7 and the simple cubic structures. (a) The simple cubic structure. The lattice vectors $\mathbf{a}_1$, $\mathbf{a}_2$, and $\mathbf{a}_3$ belong to the A7 structure (the simple cubic structure is a special case of the A7 structure). (b) Intermediate structure. Compared to (a), the unit cell of the simple cubic lattice is elongated along the vector $\mathbf{a}_3$ keeping the volume per atom constant. The new positions of the atoms are shown together with a cube of the simple cubic lattice. (c) The A7 structure of As at ambient pressure. Compared to (b) the atomic planes perpendicular to $\mathbf{a}_3$ are displaced alternatingly in the $\mathbf{a}_3$ and $-\mathbf{a}_3$ directions due to a Peierls distortion. The cube still represents the simple cubic structure of (a), for reference. Below (a)–(c) projections of the atomic planes onto $\mathbf{a}_3$ are shown. Reprinted from [30]. Copyright 2009 IOP Publishing Ltd. (Please find a color version of this figure on the color plates.)

this circumstance can be used to excite large-amplitude coherent phonons, and that, if a laser pulse is intense enough, the Peierls distortion may be undone, leading to a solid-to-solid phase transition.

In general, the magnitude of a Peierls distortion is determined by the detailed interactions between the ions and the electronic structure. When the electrons are heated with a laser, the Peierls parameter is affected and the corresponding equilibrium positions of the atoms change. If the duration of the laser pulse is short in comparison with the timescale of the atomic motion, the ions are lifted at their original equilibrium positions to the PES that is created by the hot electrons, from where they start to swing about their new equilibrium positions [2]. For Bi, this vertical transition is illustrated in Figure 19.3. In the A7 structure, this displacive motion of the atoms corresponds to the coherent excitation of the A_{1g} phonons. Coherent E_g phonons, which correspond to the motion of the atoms in the plane perpendicular to the elongated body diagonal, can be excited impulsively through Raman scattering [43].

In the following subsections, we shall look at a variety of ultrafast events that may be induced by a femtosecond laser in bismuth and arsenic. The first three examples involve coherent phonons in bismuth and are presented in order of increasing fluence. For low laser fluences, we shall look at the effect of electron–hole thermalization immediately following laser excitation. We then consider the phonon beatings

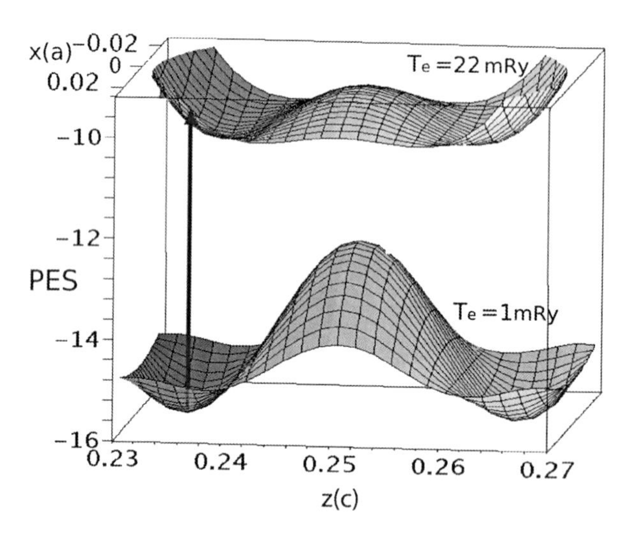

Figure 19.3 PESs of Bi for the ground state ($T_e = 1$ mRy) and for an excited electronic state ($T_e = 22$ mRy ≈ 3500 K). The arrow shows the vertical transition of the atoms from the ground state to the excited energy surface. After excitation the atoms will start to move in the z direction, which is the degree of freedom associated with the A_{1g} phonons. Note that this figure is a practical example of what has been qualitatively sketched in Figure 19.1. It can be clearly seen how the PES of the hot electron system is shifted and flattened, so that the A_{1g} coherent phonons that are excited displacively have a lower frequency than the A_{1g} phonons in the ground-state PES. The E_g phonons (the direction of the corresponding atomic motion is indicated by x) are not excited displacively due to the symmetry of the PES.

that have experimentally been observed above a certain threshold fluence. For even higher fluences, we show that the coupling between coherent phonons of different symmetries becomes relevant. Our last example of a laser-induced event in a material that assumes the A7 structure concerns the possibility to induce a solid–solid phase transition in arsenic under pressure with an ultrashort laser pulse.

19.3.2
Ultrafast Electron–Hole Thermalization in Bismuth

When bismuth is excited with an ultrashort laser pulse electron–hole pairs are created. As explained in Section 19.2.6, for the electrons and the holes, it is reasonable to assume an instant intraband thermalization [44]. However, when the laser fluence is low, the electrons and holes need a finite time to equilibrate with each other. On the basis of density functional theory computations and their comparison with depth-dependent ultrafast X-ray diffraction measurements, we have estimated that the electrons and holes thermalize on a timescale of ~ 260 fs [16]. Here, we give an account of our density functional theory calculations and of how we have arrived at our main conclusion, that fully thermalized electrons and holes lead to a greater softening of the phonons as compared to a scenario where the number of electron–hole pairs is fixed. Some additional technical details may also be found in Ref. [33].

Frequencies and anharmonic constants of the atomic potential describing the A_{1g} phonon mode were obtained by fitting calculated total energies for $z = 0.2320$, $0.2321, \cdots, 0.2390$ (These are the curves with $x = 0$ in Figure 19.3) to the function

$$E_{\text{tot}} = E_{\text{min}} + 4373.0\,\nu_{A_{1g}}^2 (z - z_{\text{min}})^2 + \gamma (z - z_{\text{min}})^3, \tag{19.22}$$

where $\nu_{A_{1g}}$ is the harmonic A_{1g} phonon frequency in THz, E_{tot} is the total energy in mRy/atom, and z is the z-coordinate of Bi in units of the lattice parameter c (11.80 Å).

Total energies at elevated electronic temperatures T_e were calculated from

$$E_{\text{tot}}(T_e) = E_{\text{tot}}(\text{gs}) + \Delta E_{\text{band}}, \tag{19.23}$$

where $E_{\text{tot}}(\text{gs})$ was the self-consistent total energy of the electronic ground state and $\Delta E_{\text{band}} = E_{\text{band}}(T_e) - E_{\text{band}}(\text{gs})$. This approach was based on the interpretation of the Kohn–Sham energies as single-electron excitation energies. In standard temperature-dependent density functional theory the electronic occupation numbers are incorporated in the self-consistent cycle to take into account possible screening effects. We also performed such standard temperature-dependent density functional theory calculations and found that the differences between the predictions of both approaches were less than 0.020 mRy/atom for the electronic temperatures considered in this chapter. As these differences showed no clear trend, we concluded that they were probably caused by rounding errors.

For the electron–hole recombination, first two limiting cases are considered, namely, instantaneous recombination and no electron–hole recombination, which we will call fast and slow electron–hole recombination in the following. For the case of fast electron–hole recombination, PESs for laser-excited bismuth were obtained by calculating E_{tot} for a range of initial electronic temperatures (defined at $z = 0.2341$), namely, $T_e = 1, 2, \cdots, 20$ mRy, using a single Fermi–Dirac distribution for the occupation numbers of the valence and the conduction bands. The electronic response to the atomic coordinates of bismuth, which were treated as external parameters, was derived using the microcanonical ensemble, meaning that the electronic entropy S_e, not the temperature T_e, was a constant of motion. The electronic entropy-dependent fitting parameters E_{min}, z_{min}, $\nu_{A_{1g}}$, and γ [Eq. (19.22)] were then fitted to functions of the form

$$\nu_{A_{1g}} = \nu_0 + \nu_1 E_{min} + \nu_2 E_{min}^2, \quad (19.24)$$

$$z_{min} = z_0 + z_1 E_{min} + z_2 E_{min}^2, \quad (19.25)$$

$$\gamma = \gamma_0 + \gamma_1 E_{min} + \gamma_2 E_{min}^2, \quad (19.26)$$

where E_{min} was shifted by a constant to make it zero for the electronic ground-state calculation ($T_e = 1$ mRy).

For the case of slow electron–hole recombination, PESs were obtained by calculating E_{tot} for $n_{el-h} = 0, 0.01, \ldots, 0.08$ electron–hole pairs per unit cell. In this case, the entropies of the electrons and the holes S_e and S_h as well as the number of electron–hole pairs per unit cell n_{el-h} were constants of motion. We obtained the entropies of the electrons and the holes by first creating electron–hole pairs at k points where the energy difference between unoccupied and occupied levels was $E \approx E_\gamma = 0.114$ Ry, the energy of one 800 nm photon. For the probability of each transition, we used Gaussian weights

$$w = \exp\left[-\left(\frac{E - E_\gamma}{\varepsilon}\right)^2\right], \quad (19.27)$$

where $\varepsilon = 0.001$ Ry was small compared to E_γ. Then, a very fast thermalization of the electrons and of the holes separately was assumed, leading to two separate Fermi–Dirac distributions for the electrons and the holes with well-defined electron and hole entropies.

The fitting parameters in Eqs. (19.24–19.26) for the two limiting cases of fast and slow electron–hole recombinations are given in Table 19.1. Data points and fits of the parameters defining the harmonic part of the potential energy curve are shown in Figure 19.4. In the following, we neglect the anharmonicity of the interatomic potential, because we estimated it to contribute as little as at most 3% to the softening of the A_{1g} phonon frequency for the fluences considered, that is, $E_{min} \leq 5$ mRy/atom. The energy absorbed from the laser can for an ultrashort laser pulse easily be calculated from Eq. (19.22) by assuming that the atoms do not move during the laser pulse. It turned out that the values thus obtained deviated less than $\sim 3\%$ from E_{min}. Therefore we could safely interpret E_{min} as the energy absorbed from the laser. Under these assumptions, it is apparent from Figure 19.4 that the phonon softening in laser-excited bismuth depends critically on the electronic occupation numbers and that

Table 19.1 Fitting parameters for Eqs. (19.24–19.26) for fast and slow electron–hole recombination. The ground-state parameters ν_0, z_0, and γ_0 were constrained to be equal in both cases.

Parameter	Electron–Hole recombination	
	Fast	Slow
ν_0	2.9907 ± 0.0014	
ν_1	-0.1325 ± 0.0018	-0.0431 ± 0.0018
ν_2	0.0066 ± 0.0004	-0.0010 ± 0.0004
z_0	0.234386 ± 0.000001	
z_1	0.000489 ± 0.000002	0.000251 ± 0.000002
z_2	$-(4.1 \pm 0.4)10^{-6}$	$(7.7 \pm 0.5)10^{-6}$
γ_0	$-(1.030 \pm 0.003)10^{6}$	
γ_1	$-(36 \pm 5)10^{3}$	$(38 \pm 5)10^{3}$
γ_2	$(9.2 \pm 1.1)10^{3}$	$-(2.2 \pm 1.4)10^{3}$

for a given excitation energy $E_{\min}$ the fast electron–hole thermalization scenario gives the greatest phonon softening. As already mentioned above, on the basis of X-ray diffraction measurements we have estimated that the electron–hole thermalization time is $\sim$ 260 fs in Bi for moderate excitation densities [16].

We now discuss the physical implications for the cases of high and low excitation densities. First of all, when the laser fluence is low, we expect that the electron–hole thermalization time may be longer than 260 fs. In this case, our calculations predict that the time-dependent A_{1g} frequency in Bi will further soften during the thermalization process after laser excitation, before hardening. When this takes more than

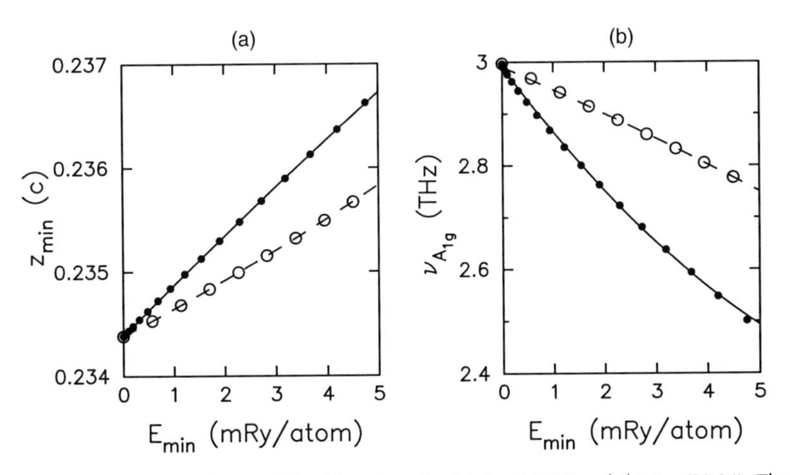

Figure 19.4 Data points and fitted functions for (a) Eq. (19.25) and (b) Eq. (19.24). The closed symbols and solid curves are for the case of fast electron–hole recombination and the open symbols and dashed curves are for slow electron–hole recombination.

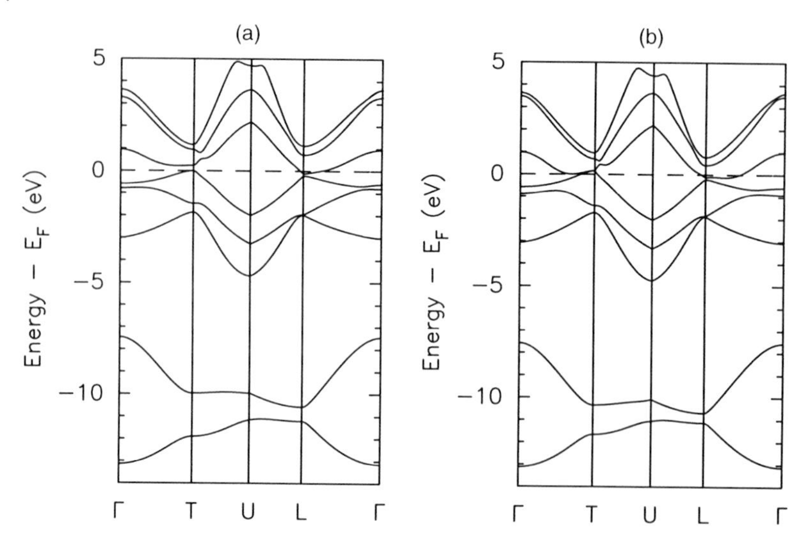

Figure 19.5 Band structure of Bi calculated for (a) $z = 0.234$ (the equilibrium structure of Bi) and (b) $z = 0.239$ (the value where the valence and conduction bands start to cross at the T point). In both figures, the Fermi energy is indicated by a dashed line.

two phonon periods (one period is ≈ 300 fs) we expect that this further lattice softening can most readily be observed in optical reflectivity measurements.

For the case of high excitation densities, it is important to note that when the amplitude of the excited coherent A_{1g} phonons is large, the conduction and valence bands cross (Figure 19.5). In this case, the electrons and the holes thermalize very fast with each other, so the occupation numbers of the conduction and valence bands can be described by a single Fermi–Dirac distribution function. According to Figure 19.5, this situation occurs when the A_{1g} phonons have an amplitude $\Delta z > 0.005c$, where $c = 11.80$ Å is a lattice parameter of Bi. At this amplitude, the A_{1g} phonon frequency $\nu(A_{1g}) < 2.45$ THz [33], implicating that, for example, the experimental works in Refs [5] and [6] should be described assuming fast electron–hole recombination.

19.3.3
Amplitude Collapse and Revival of Coherent A_{1g} Phonons in Bismuth: A Classical Phenomenon?

Very intense laser excitations are extremely interesting, because new phenomena based on anharmonicities and phonon–phonon couplings may be observed. In order to study this domain for bismuth, we have parameterized the PES using accurate full-potential linearized augmented plane wave calculations [45]. Anharmonic contributions up to the fifth power in the A_{1g} phonon coordinate have been tabulated as a function of the absorbed laser energy. The resulting parameterization is similar to the one presented in Table 19.1, but for a greater range of validity. For details regarding this parameterization, the interested reader is referred to Ref. [45].

Using a model including effects of electron–phonon coupling and carrier diffusion [16], we have obtained the time-dependent PES for any given laser pulse shape and duration. We have then performed quantum dynamical simulations to study the experimentally observed amplitude collapse and revival of coherent A_{1g} phonons in bismuth [46]. On the basis of these simulations, we have reached the conclusion that the observed beatings are not related to quantum effects and are most probably of classical origin [45].

19.3.4
Laser-Induced Phonon–Phonon Interactions in Bismuth

When large-amplitude coherent phonons of different symmetries are simultaneously excited by a laser pulse, the interactions between these different modes may become significant and modulations of the individual modes may be observed. We have studied and analyzed this possibility for the case of bismuth [33]. In our simulation, after an ultrashort, strong laser pulse, the electrons were heated to 2500 K. The fully symmetric A_{1g} phonon was softened from 2.9 THz to a frequency of 2.5 THz. In Figure 19.6, we show the x and z coordinates of the Bi atoms in the first picosecond after the laser pulse. The z (x) coordinate describes the A_{1g} (E_g) phonon mode. From Figure 19.6, it is clear that the excited A_{1g} and E_g phonons cannot be described by a single cosine or sine function, but that, instead, there is a significant amount of interaction between these modes, which leads to modulations. Another observation is that the modulations are more pronounced for the E_g phonons than for the A_{1g} phonon mode.

The role of the phonon–phonon coupling is also illustrated in Figure 19.7, where we have plotted the trajectory of the atoms during the first picosecond after the laser pulse without and with coupling. Again, the role of the phonon–phonon interactions

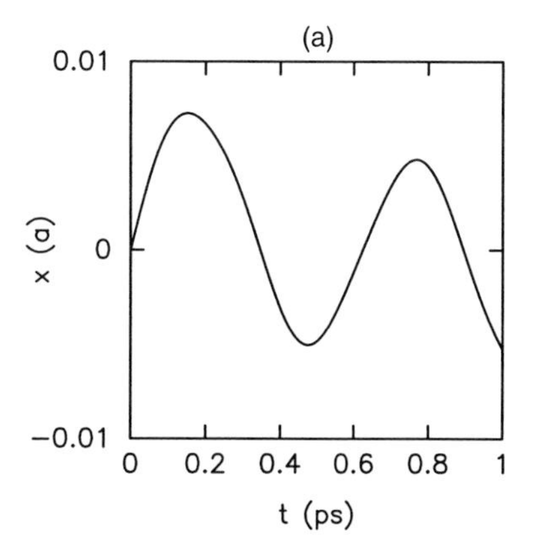

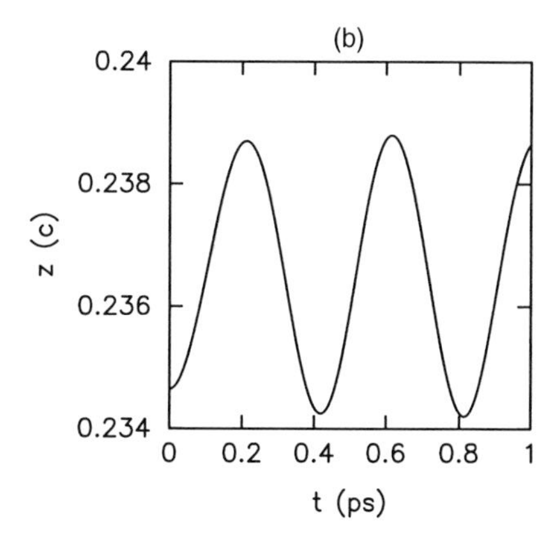

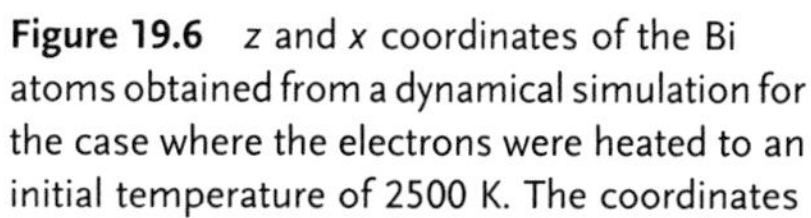

Figure 19.6 z and x coordinates of the Bi atoms obtained from a dynamical simulation for the case where the electrons were heated to an initial temperature of 2500 K. The coordinates are given in units of the lattice parameters $a = 4.53$ and $c = 11.80$ Å, respectively. The effect of coupling between both degrees of freedom shown leads to a modulation of x and z.

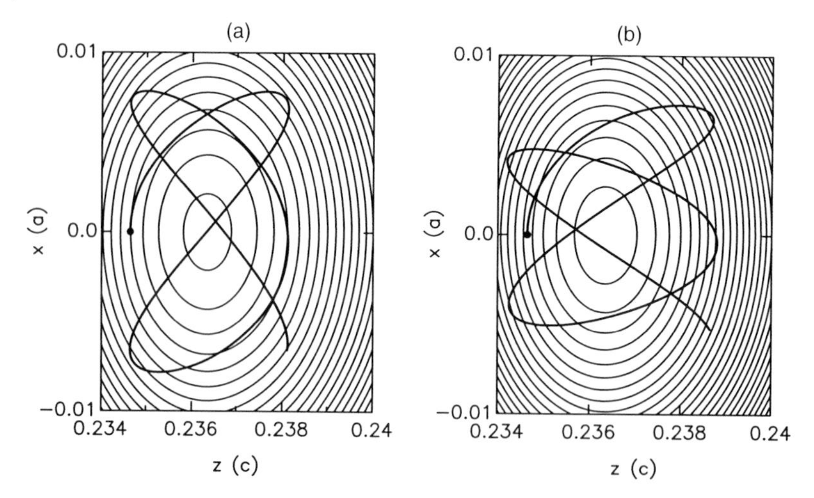

Figure 19.7 Calculated PESs in laser-excited bismuth (a) without and (b) with phonon–phonon coupling. The distance between the contour lines equals 0.025 mRy/atom. Trajectories for the A_{1g} (z-direction) and E_g (x-direction) degrees of freedom are shown for the first picosecond after the laser pulse. The initial coordinates are indicated by a dot. The initial velocity of the atoms is in the x-direction.

is very clearly visible. Without coupling, the x and z coordinates evolve independently (Figure 19.7a). In this case, the maximal amplitude of the E_g phonon mode is proportional to the initial velocity of the atom in the x direction, induced by the laser, and the maximal amplitude of the A_{1g} phonon is reached when the velocity in the z direction becomes zero, which is at the point where the potential energy equals the initial potential energy for this mode. When the coupling is introduced in Figure 19.7b, we see that the trajectory of the atoms is changed quite dramatically. On the one hand, a very fast energy transfer from the E_g phonons to the A_{1g} phonons leads to a larger amplitude of the coherent A_{1g} phonons in comparison to the case without coupling, on the other hand, it leads to a reduced E_g amplitude. From the fact that the phonon–phonon coupling is a term of higher order in the effective phonon potential than the terms describing the harmonic phonon modes, it is immediately clear that these coupling effects become stronger at higher amplitudes, which can be obtained by using laser pulses that have higher fluences and/or shorter pulse durations.

19.3.5
Ultrafast Laser-Induced Solid-to-Solid Transition in Arsenic Under Pressure

In arsenic a series of pressure-induced phase transitions exists. At ambient conditions As crystallizes in the A7 structure. At 25 GPa there is a transition to the simple cubic phase, followed by transitions to the so-called As(III) structure at 48 GPa and to body-centered cubic As at 97 GPa. On the basis of density functional theory

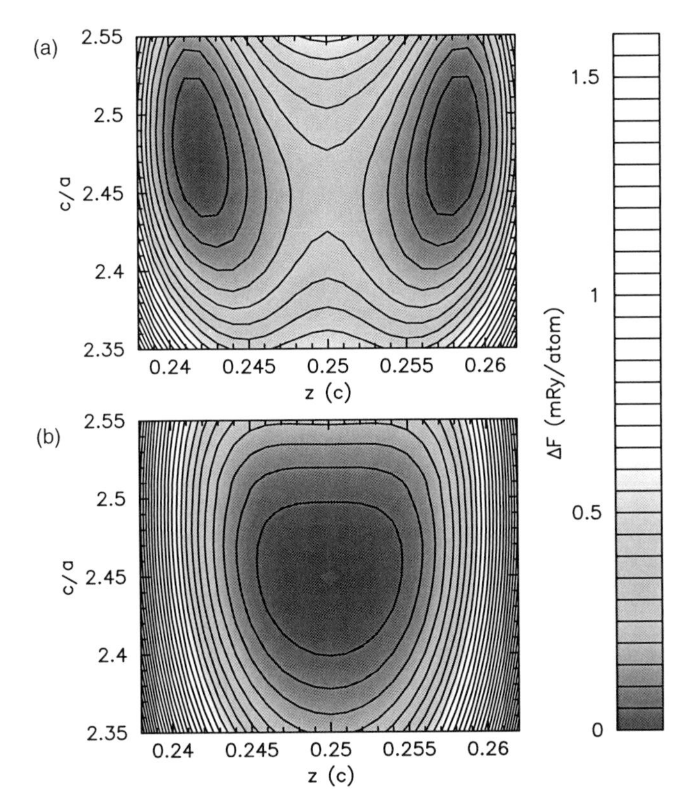

Figure 19.8 PES contour plot of As (a) before and (b) after laser-excitation as a function of the internal parameters z and c/a at the volume $V = 114.0\, a_0^3/\text{atom}$, which corresponds to an applied pressure of 23.8 GPa in (a). The two equivalent minima before laser excitation merge into a single minimum after the laser excitation, indicating that the Peierls distortion of the A7 structure can be undone by a femtosecond laser pulse. The minimum in (b) is located at the simple cubic values of z and c/a (0.25 and $\sqrt{6} \approx 2.45$, respectively). Reprinted from [30]. Copyright 2009 IOP Publishing Ltd.

calculations, we have predicted that the first pressure-induced phase transition can also be induced by an ultrashort laser pulse in As under pressure [30].

Figure 19.8 shows the computed PES of arsenic under an applied pressure of 23.8 GPa before and after the excitation with a laser pulse with a fluence of 2 mJ/cm^2 [30]. These PESs demonstrate that one may induce a solid–-solid phase transition from the A7 to the simple cubic structure in As under pressure using an ultrashort laser pulse. Further computations [47] have shown that for higher fluences, the applied pressure may be reduced according to Table 19.2. Unfortunately, these computations do not include the degrees of freedom that are needed to describe competing processes, such as, laser ablation or nonthermal melting, as it is to be expected that for very high fluences or very low applied pressures the system would either melt or simply evaporate.

Table 19.2 Computed required minimal fluence to induce a solid–solid transition in arsenic as a function of applied pressure.

Pressure (GPa)	Fluence (mJ/cm^2)
23.8	2
12.4	10
4.0	23

19.3.6
The Zinc-Blende Structure. Ultrafast Melting of InSb

The group III–V semiconductors GaAs, InP, and InSb crystallize in the zinc-blende structure. Here, the atomic positions are identical to those of diamond, except that the two face-centered cubic sublattices that form the diamond structure are occupied by different types of atoms as is shown in Figure 19.9. In the zinc-blende structure, each atom is tetragonally surrounded by four atoms. Hence, all atomic coordinates are fixed by symmetry and there is no displacive excitation of coherent phonons. Nevertheless, when, for example, InSb is excited with an ultrashort laser pulse lattice vibrations soften [9, 48]. However, it is important to notice that the softening is not more than 25% for most phonon modes, even for very high fluences due to a saturation in the frequency change [48]. Below, we indicate how the extreme softening of a few selected phonon modes in InSb by a femtosecond laser pulse can lead to ultrafast melting.

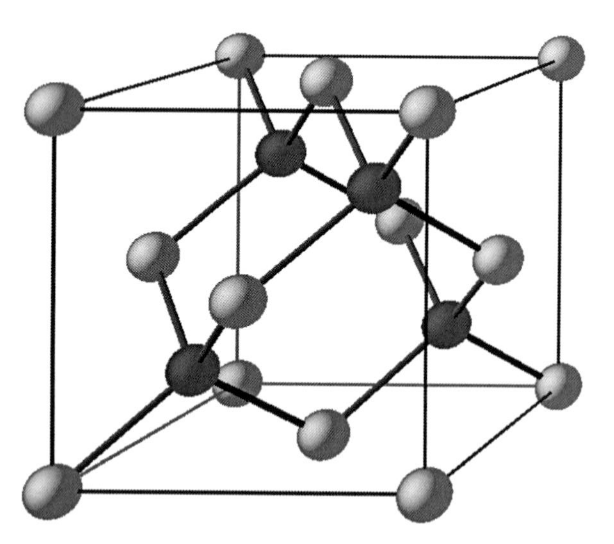

Figure 19.9 The zinc-blende structure consists of two face-centered cubic sublattices, which are shifted by (1/4,1/4,1/4) with respect to each other. In the case of InSb, one sublattice is occupied by indium and the other sublattice is occupied by antimony atoms.

This process of laser-induced ultrafast melting has recently been studied on the basis of a time-resolved "optical pump"–"X-ray probe" experimental setup by Lindenberg *et al.* [9], who have measured the time evolution of the (111) and (220) Bragg-peak heights of InSb after intense laser excitation. They found that the intensity of the diffraction peaks follows a Gaussian decay with an ultrashort time constant [400 fs for the (111) peak]. Assuming that the Debye–Waller theory is valid for time-dependent phenomena, they described the decay of the Bragg-peak intensity $I(\vec{Q}, t)$ by [9]

$$I(\vec{Q}, t) = \mathrm{e}^{-Q^2 \langle x^2(t) \rangle / 3}, \tag{19.28}$$

where $\vec{Q}$ is the reciprocal lattice vector corresponding to the Bragg peak considered, and $\langle x^2(t) \rangle$ refers to the mean square displacements of the ions. Since $I(\vec{Q}, t)$ has a Gaussian time behavior, inversion of Eq. (19.28) leads to a quadratic dependence of $\langle x^2(t) \rangle$, meaning that the root mean square (rms) displacements exhibit, quite remarkably, within the experimental error bars, a linear dependence on time during the first 400 fs after the excitation by the pump pulse. The other remarkable feature of the mentioned experimental results is that within these 400 fs the ions cover in average a distance of about 1 Å, which is extremely large compared to the constrained motion of ions in a solid. Moreover, the almost constant slope of the rms displacements, which the authors interpret as a velocity, roughly corresponds to the thermal velocity of atoms at 300 K (i.e., at the temperature before excitation), which is $\langle v \rangle_{300} = 2.5$ Å/ps.

Based on accurate, all-electron density functional theory calculations of the laser-excited PES of InSb, we have provided a quantitative explanation of the experimental results of Ref. [9], which is already on a qualitative level different from the original inertial dynamics model used by the authors of Ref. [9]. For an explanation of this inertial dynamics model, which is inadequate, the reader is referred to the original publication [9] and Ref. [31].

From our computations, we have seen that above a certain threshold fluence, the transverse acoustic phonons at the zone boundary point X and L become unstable (Figure 19.10) [48]. On the basis of dynamical simulations on the computed laser-excited PES, we have further found that when the potential becomes repulsive the atoms accelerate in the corresponding directions. In the harmonic approximation, this leads to an exponential increase of the root mean square displacement of the atoms. When one also takes into account the anharmonicity of the phonons, the increase becomes roughly linear (Figure 19.11) with a slope of 2.3 Å/ps [31], in good agreement with the above-mentioned experimental result [9].

Based on our calculations, we now discuss what happens during nonthermal melting. First of all, an ultrashort laser pulse creates a high density of electron–hole pairs, which thermalize on a very short time and find a common chemical potential. As a consequence, a very high electronic temperature is achieved, which reflects the large amount of energy absorbed from the laser pulse. This high T_{el} leads to dramatic changes in the total energy and in the entropy of the electronic system, which produces a dramatic change of the PES for the lattice motion. However, this change

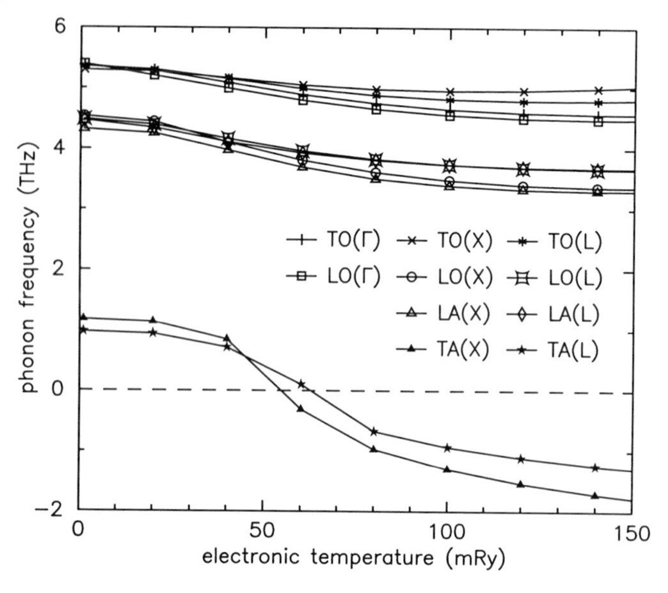

Figure 19.10 Phonon frequencies at the Γ, X, and L points as a function of the electronic temperature. The labels indicate whether the phonons are longitudinal (L) or transverse (T) and optical (O) or acoustic (A). The negative values represent repulsive potentials, that is, imaginary frequencies. The lines are a guide to the eye.

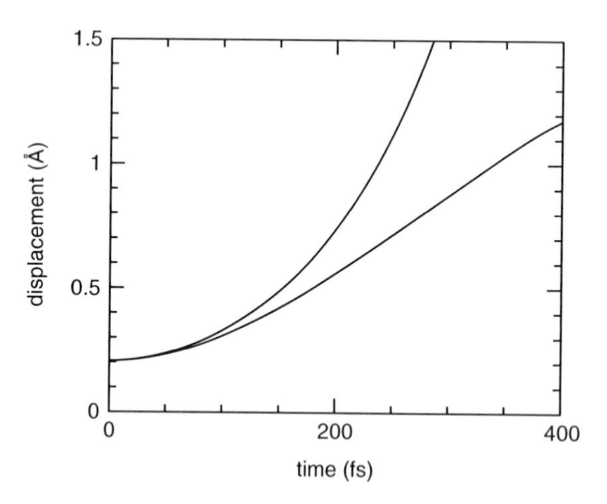

Figure 19.11 Root mean square atomic displacement versus time for InSb. The electronic temperature $T_e = 100$ mRy. Whereas the top curve shows the displacement within the harmonic approximation, the bottom curve includes anharmonic effects. The discrepancy between both curves demonstrates the importance of anharmonicity for times greater than ~ 100 fs.

affects only a fraction of the lattice degrees of freedom, and in particular, the transverse acoustic modes at the boundary of the Brillouin zone, while most of the other lattice modes are only slightly softened, that is, at most 25%. In the few modes affected, the PES is so strongly changed into a repulsive potential, that the ions experience large forces that make them to cover average distances as large as 1 Å within the first 400 fs after excitation. Our results confirm the original physical picture proposed by Stampfli and Bennemann for the ultrafast melting of silicon [23].

19.4 Conclusion

In this chapter, we have considered laser-induced events involving phonon softening from a theoretical perspective. In Section 19.2, we have presented several approaches that have been developed to explain these events. In particular, we have shown how the Born–Oppenheimer approximation can be generalized to laser-excited solids and how this leads to the concept of a laser-excited potential energy surface. We have indicated that this constitutes a general theoretical framework that can explain the excitation by a femtosecond laser pulse of displacive coherent phonons, changes in bond lengths, equilibrium structures, and so on.

In Section 19.3, we have used the general concepts of Section 19.2 to describe selected examples of ultrafast events in more detail based on computed potential energy surfaces. Among other things, we have discussed the mechanism leading to the displacive excitation of coherent phonons, the laser-induced softening of these phonon modes, the thermalization of laser-excited electrons and holes in a semimetal, the coupling between laser-excited phonons of different symmetries, the possibility to induce solid-to-solid phase transitions by a femtosecond laser pulse, and last but not least, ultrafast melting. All these processes take place on, and can be understood through, laser-excited potential energy surfaces.

References

1 Harris, C.B., Ippen, E.P., Morou, G.A., and Zeweil, A.H. (eds) (1990) *Ultrafast Phenomena VII*, Series in Chemical Physics, Springer-Verlag, Berlin.

2 Zeiger, H.J., Vidal, J., Cheng, T.K., Ippen, E.P., Dresselhaus, G., and Dresselhaus, M.S. (1992) Theory for displacive excitation of coherent phonons. *Phys. Rev. B*, **45**, 768–778.

3 Merlin, R. (1997) Generating coherent THz phonons with light pulses. *Solid State Commun.*, **2–3**, 207–220.

4 Stevens, T.E., Kuhl, J., and Merlin, R. (2002) Coherent phonon generation and the two stimulated Raman tensors. *Phys. Rev. B*, **65**, 144304.

5 Hase, M., Kitajima, M., Nakashima, S.I., and Mizoguchi, K. (2002) Dynamics of coherent anharmonic phonons in bismuth using high density photoexcitation. *Phys. Rev. Lett.*, **88**, 067401.

6 Sokolowski-Tinten, K., Blome, C., Blums, J., Cavalleri, A., Dietrich, C., Tarasevitch, A., Uschmann, I., Förster, E., Kammler, M., Horn-von-Hoegen, M., and von der Linde, D. (2003) Femtosecond X-ray measurement of coherent lattice

vibrations near the Lindemann stability limit. *Nature*, **422**, 287–289.
7 Cavalleri, A., Tóth, Cs., Siders, C.W., Squier, J.A., Ráksi, F., Forget, P., and Kieffer, J.C. (2001) Femtosecond structural dynamics in VO_2 during an ultrafast solid–solid phase transition. *Phys. Rev. Lett.*, **87**, 237401.
8 Johnson, S.L., Heimann, P.A., Lindenberg, A.M., Jeschke, H.O., Garcia, M.E., Chang, Z., Lee, R.W., Rehr, J.J., and Falcone, R.W. (2003) Properties of liquid silicon observed by time-resolved X-ray absorption spectroscopy. *Phys. Rev. Lett.*, **91**, 157403.
9 Lindenberg, A., Larsson, J., Sokolowski-Tinten, K., Gaffney, K.J., Blome, C., Synnergren, O., Sheppard, J., Caleman, C., MacPhee, A.G., Weinstein, D., Lowney, D.P., Allison, T.K., Matthews, T., Falcone, R.W., Cavalieri, A.L., Fritz, D.M., Lee, S.H., Bucksbaum, P.H., Reis, D.A., Rudati, J., Fuoss, P.H., Kao, C.C., Siddons, D.P., Pahl, R., Als-Nielsen, J., Duesterer, S., Ischebeck, R., Schlarb, H., Schulte-Schrepping, H., Tschentscher, Th., Schneider, J., von der Linde, D., Hignette, O., Sette, F., Chapman, H.N., Lee, R.W., Hansen, T.N., Techert, S., Wark, J.S., Bergh, M., Huldt, G., van der Spoel, D., Timneanu, N., Hajdu, J., Akre, R.A., Bong, E., Krejcik, P., Arthur, J., Brennan, S., Luening, K., and Hastings, J.B. (2005) Atomic-scale visualization of inertial dynamics. *Science*, **308**, 392–395.
10 Harb, M., Ernstorfer, R., Hebeisen, Ch.T., Sciaini, G., Peng, W., Dartigalongue, T., Eriksson, M.A., Lagally, M.G., Kruglik, S.G., and Miller, R.J. (2008) Electronically driven structure changes of Si captured by femtosecond electron diffraction. *Phys. Rev. Lett.*, **100**, 155504.
11 Sokolowski-Tinten, K., Bialkowski, J., Cavalleri, A., von der Linde, D., Oparin, A., Meyer-ter-Vehn, J., and Anisimov, S.I. (1998) Transient states of matter during short pulse laser ablation. *Phys. Rev. Lett.*, **81**, 224–227.
12 Glover, T.E., Ackerman, G.D., Belkacem, A., Heimann, P.A., Hussain, Z., Lee, R.W., Padmore, H.A., Ray, C., Schoenlein, R.W., Steele, W.F., and Young, D.A. (2003) Metal–insulator transitions in an expanding metallic fluid: particle formation kinetics. *Phys. Rev. Lett.*, **90**, 236102.
13 Stampfli, P. and Bennemann, K.H. (1990) Theory for the instability of the diamond structure of Si, Ge, and C induced by a dense electron–hole plasma. *Phys. Rev. B*, **42**, 7163–7173.
14 Czycholl, G. (2000) *Theoretische Festkörperphysik*, Vieweg-Verlag, Braunschweig/Wiesbaden.
15 Kuznetsov, A.V. and Stanton, C.J. (1994) Theory of coherent phonon oscillations in semiconductors. *Phys. Rev. Lett.*, **73**, 3243–3246.
16 Johnson, S.L., Beaud, P., Milne, C.J., Krasniqi, F.S., Zijlstra, E.S., Garcia, M.E., Kaiser, M., Grolimund, D., Abela, R., and Ingold, G. (2008) Nanoscale depth-resolved coherent femtosecond motion in laser-excited bismuth. *Phys. Rev. Lett.*, **100**, 155501.
17 Halgren, T.A. (1992) Representation of van der Waals (vdW) interactions in molecular mechanics force fields: potential form, combination rules, and vdW parameters. *J. Am. Chem. Soc.*, **114**, 7827–7843.
18 Tersoff, J. (1988) New empirical approach for the structure and energy of covalent systems. *Phys. Rev. B*, **37**, 6991–7000.
19 Stillinger, F. and Weber, T.A. (1985) Computer simulation of local order in condensed phases of silicon. *Phys. Rev. B*, **31**, 5262–5271.
20 Jackson, J.D. (1975) *Classical Electrodynamics*, John Wiley & Sons, New York.
21 Elsaesser, T., Shah, J., Rota, L., and Lugli, P. (1991) Initial thermalization of photoexcited carriers in GaAs studied by femtosecond luminescence spectroscopy. *Phys. Rev. Lett.*, **66**, 1757–1760.
22 Knox, W.H., Chemla, D.S., Livescu, G., Cunningham, J.E., and Henry, J.E. (1988) Femtosecond carrier thermalization in dense Fermi seas. *Phys. Rev. Lett.*, **61**, 1290–1293.
23 Stampfli, P. and Bennemann, K.H. (1992) Dynamical theory of the laser-induced lattice instability of silicon. *Phys. Rev. B*, **46**, 10686–10692.

24 Jeschke, H.O., Garcia, M.E., and Bennemann, K.H. (2002) Time-dependent energy absorption changes during ultrafast lattice deformation. *J. Appl. Phys.*, **91**, 18–23.

25 Jeschke, H.O., Garcia, M.E., Lenzner, M., Bonse, J., Krüger, J., and Kautek, W. (2002) Laser ablation thresholds of silicon for different pulse durations: theory and experiment. *Appl. Surf. Sci.*, **197–198**, 839–844.

26 Jeschke, H.O., Garcia, M.E., and Bennemann, K.H. (1999) Microscopic analysis of the femtosecond graphitization of diamond. *Phys. Rev. B*, **60**, R3701–R3704.

27 Jeschke, H.O. and Garcia, M.E. (2003) Ultrafast structural changes induced by femtosecond laser pulses, in *Nonlinear Optics, Quantum Optics and Ultrafast Phenomena with X-Rays* (ed. B.W. Adams), Kluwer Academic Publishers, Boston/Dordrecht/London, pp. 175–214.

28 Silvestrelli, P.L., Alavi, A., Parrinello, M., and Frenkel, D. (1996) *Ab initio* molecular dynamics simulation of laser melting of silicon. *Phys. Rev. Lett.*, **77**, 3149–3152.

29 Zickfeld, K., Garcia, M.E., and Bennemann, K.H. (1999) Theoretical study of the laser induced femtosecond dynamics of small Si_n clusters. *Phys. Rev. B*, **59**, 13422–13430.

30 Zijlstra, E.S., Huntemann, N., and Garcia, M.E. (2008) Laser-induced solid–solid phase transition in As under pressure: a theoretical prediction. *New J. Phys.*, **10**, 033010.

31 Zijlstra, E.S., Walkenhorst, J., and Garcia, M.E. (2008) Anharmonic noninertial lattice dynamics during ultrafast nonthermal melting of InSb. *Phys. Rev. Lett.*, **101**, 135701.

32 Tangney, P. and Fahy, S. (1999) Calculations of the A_1 phonon frequency in photoexcited tellurium. *Phys. Rev. Lett.*, **82**, 4340–4343.

33 Zijlstra, E.S., Tatarinova, L.L., and Garcia, M.E. (2006) Laser-induced phonon–phonon interactions in bismuth. *Phys. Rev. B*, **74**, 220301(R).

34 Stampfli, P. and Bennemann, K.H. (1994) Time dependence of the laser-induced femtosecond lattice instability of Si and GaAs: role of longitudinal optical distortions. *Phys. Rev. B*, **49**, 7299–7305.

35 Zijlstra, E.S., Tatarinova, L.L., and Garcia, M.E. (2006) *Ab-initio* study of coherent phonons excited by femtosecond laser pulses in bismuth. *Proc. SPIE*, **6261**, 62610.

36 Jeschke, H.O., Garcia, M.E., and Bennemann, K.H. (1999) Theory for laser induced ultrafast phase transitions in carbon. *Appl. Phys. A*, **69** (Suppl.), S49–S53.

37 Jeschke, H.O., Garcia, M.E., and Bennemann, K.H. (2001) Theory for the ultrafast ablation of graphite films. *Phys. Rev. Lett.*, **87**, 015003.

38 Dumitrica, T., Garcia, M.E., Jeschke, H.O., and Yakobson, B.I. (2004) Selective cap opening in carbon nanotubes driven by laser-induced coherent phonons. *Phys. Rev. Lett.*, **92**, 117401.

39 Dumitrica, T., Garcia, M.E., Jeschke, H.O., and Yakobson, B.I. (2006) Breathing coherent phonons and caps fragmentation in carbon nanotubes following ultrafast laser pulses. *Phys. Rev. B*, **74**, 193406.

40 Romero, A.H., Garcia, M.E., Valencia, F., Terrones, H., Terrones, M., and Jeschke, H.O. (2005) Femtosecond laser nanosurgery of defects in carbon nanotubes. *Nano Lett.*, **5**, 1361–1365.

41 Hunsche, S., Wienecke, K., Dekorsy, T., and Kurz, H. (1995) Impusive softening of coherent phonons in tellurium. *Phys. Rev. Lett.*, **75**, 1815–1818.

42 Tom, H.W.K., Aumiller, G.D., and Brito-Cruz, C.H. (1988) Time-resolved study of laser-induced disorder of Si surfaces. *Phys. Rev. Lett.*, **60**, 1438–1441.

43 Garrett, G.A., Albrecht, T.F., Whitaker, J.F., and Merlin, R. (1996) Coherent THz phonons driven by light pulses and the Sb problem: what is the mechanism? *Phys. Rev. Lett.*, **77**, 3661–3664.

44 Tangney, P. and Fahy, S. (2002) Density-functional theory approach to ultrafast laser excitation of semiconductors: application to the A_1 phonon in tellurium. *Phys. Rev. B*, **65**, 054302.

45 Diakhate, M.S., Zijlstra, E.S., and Garcia, M.E. (2009) Quantum dynamical study of the amplitude collapse and revival of coherent A_{1g} phonons in bismuth: a

classical phenomenon? *Appl. Phys. A*, **96**, 5–10.

46 Misochko, O.V., Hase, M., Ishioka, K., and Kitajima, M. (2004) Observation of an amplitude collapse and revival of chirped coherent phonons in bismuth. *Phys. Rev. Lett.*, **92**, 197401.

47 Huntemann, N., Zijlstra, E.S., and Garcia, M.E. (2009) Fluence dependence of the ultrafast transition from the A7 to the simple cubic structure in arsenic. *Appl. Phys. A*, **96**, 19–22.

48 Zijlstra, E.S., Walkenhorst, J., Gilfert, C., Sippel, C., Töws, W., and Garcia, M.E. (2008) *Ab initio* description of the first stages of laser-induced ultra-fast nonthermal melting of InSb. *Appl. Phys. B*, **93**, 743(R)–747(R).

20
Femtosecond Time- and Angle-Resolved Photoemission as a Real-time Probe of Cooperative Effects in Correlated Electron Materials

Patrick S. Kirchmann, Luca Perfetti, Martin Wolf, and Uwe Bovensiepen

20.1
Introduction

Many-body and correlation effects in solid-state physics manifest in specific signatures of the electronic band structure. The interaction of electrons with elementary excitations of the solid results in renormalization of the single-particle band structure due to, for example, electron–phonon, electron–magnon, and electron–electron interactions. This is explained in more detail in Volume 2 (Chapter 2) and this volume (Chapter 8). In particular for highly correlated materials such as charge (or spin) density wave materials, Mott insulators, and superconductors, such many-body effects are key to understanding macroscopic properties, for example, the electrical conductivity. Due to competition of these correlations with thermal fluctuations, phase transitions to ordered ground states with broken symmetry occur with decreasing temperature. Below the respective critical temperatures, the electronic system gains stability by opening of a bandgap (see Volume 2, Chapter 1). Understanding these many-particle effects is one goal of present efforts in solid-state physics.

From an experimental point of view, angle-resolved photoelectron spectroscopy (ARPES) is the method of choice to analyze the electronic band structure of a solid [1, 2]. It retrieves the information on the electronic states from the energy and the momentum of the photoelectrons emitted by a monochromatic photon source (see Volume 2, Chapter 3). This conversion is possible because the kinetic energy of the photoelectron is related to the binding energy of the photohole left behind. Furthermore, the translational symmetry of the surface guaranties that the momentum component of the photoelectron and of the photohole parallel to the surface are identical. However, the photoelectron wave vector perpendicular to the surface is not conserved due to the photoemission process. This limitation can be overcome if the material has a quasi-two-dimensional band structure. In this case, the angle-dependent photoelectron intensity represents the spectral function $A_k(\omega)$ multiplied by the Fermi–Dirac distribution $f(\omega)$. The sample volume that is probed in ARPES is determined by the mean free path of the excited electrons. For photon energies in the

Dynamics at Solid State Surface and Interfaces Vol.1: Current Developments
Edited by Uwe Bovensiepen, Hrvoje Petek, and Martin Wolf

ISBN: 978-3-527-40937-2

VUV range, as used in conventional ARPES experiments employing He lamps or synchrotrons as light sources, photoelectrons are emitted from a region very close to the surface. For kinetic energies of photoelectrons in the few eV regime, as they are generated in laser-induced ARPES [3, 4], the mean free path increases to several nanometers. Thus, laser-induced ARPES should be very bulk sensitive. However, due to matrix element effects (see Volume 2, Chapter 5) the method can still be surface sensitive. Under favorable conditions, surface and bulk phenomena can be compared directly, as will be shown in this article.

In ARPES, the electronic band dispersion can be represented by a two-dimensional map of the photoelectron intensity as a function of binding energy and electron wave vector. As an example, Figure 20.1 displays a photoelectron intensity map acquired for the high-temperature superconductor $Bi_2Sr_2CaCu_2O_{8+\delta}$ along the ΓY direction using femtosecond laser pulses at $h\nu = 6$ eV photon energy. The ultrashort 6 eV pulses are generated by quadrupling the fundamental output of an amplified Ti:Sa laser operating at 30 kHz. The electronic band disperses as a function of binding energy and is occupied for wave vectors smaller than the Fermi momentum k_F. The intrinsic broadening of the electronic band is due to an interaction between the removed electron and the remaining system. Accordingly, linewidth measurements are a well established way of studying many-particle effects (see Volume 2, Chapter 3) and Chapter 8 of this volume.

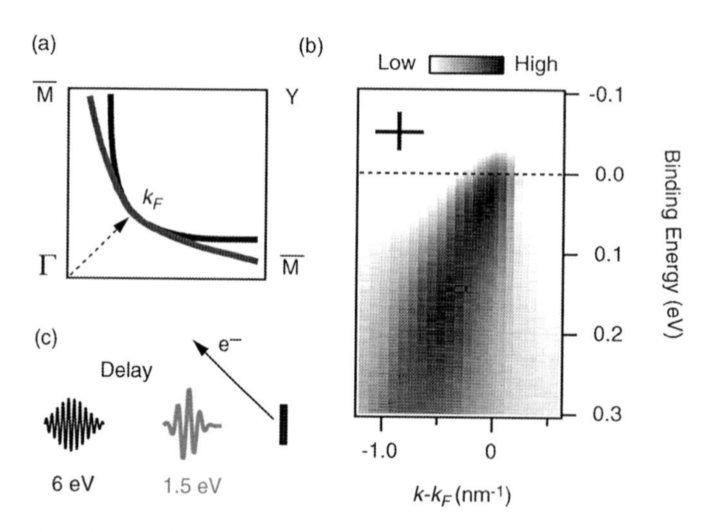

Figure 20.1 (a) First Brillouin zone of the high temperature superconductor $Bi_2Sr_2CaCu_2O_{8+\delta}$ at optimal doping. The solid lines depict the Fermi surface and k_F designates the Fermi wave vector along the ΓY direction. (b) Photoelectron intensity as a function of binding energy acquired along the ΓY direction. An electronic band crosses the Fermi level at k_F. The cross indicates the energy and momentum resolution. (c) Sketch of the tr-ARPES experiment. An infrared pulse excites the sample while a delayed UV pulse generates the photoelectrons; Reprinted with permission from [5]. Copyright 2007 by the American Physical Society.

In terms of energy and momentum resolution, the photoelectron map shown in Figure 20.1 cannot compete with high-resolution setups that are dedicated to *frequency* domain spectroscopy [2–4]. However, our approach is to analyze modifications of the electronic structure in the *time* domain, which becomes accessible if laser-based ARPES is conducted as a pump–probe experiment. For this, a promising compromise of spectrometer transmission, acceptance angle, momentum, and energy resolution has to be achieved. The working principle of time-resolved ARPES (tr-ARPES) is straightforward: a first femtosecond laser pulse excites the sample while a second pulse in the UV spectral range generates photoelectrons after a variable time delay τ (see Figure 20.1c). The photoelectrons are analyzed in a spectrometer that determines their energy at a defined emission angle. In this way, the time-dependent relaxation of a perturbed system is monitored, which reveals essential aspects of the many-body interactions. While a hypothetical gas of non-interacting electrons would remain in the photoexcited state for an unlimited amount of time, processes such as electron–electron, electron–phonon, and phonon–phonon scattering determine the relaxation dynamics back to the equilibrium conditions in a real material. Since each of these processes acts on a specific timescale, the relaxation of electronic energy generates a sequence of transient states that are coupled by the respective many-particle interaction under investigation. It is this relaxation through sequential or competing channels that is analyzed in order to disentangle contributions of purely electronic correlations and electron–phonon interactions to the ordered phases.

20.2
Hot Electron Relaxation

The relaxation mechanisms of the photoexcited state are qualitatively different in metals and insulators. When the material displays a bandgap, electron scattering processes that involve electron–hole pair excitation must overcome the gap energy, which strongly limits the available phase space for electron–electron scattering. In contrast, in photoexcited metals such an energy boundary is not imposed. As a consequence, relaxation of the metallic phase is much faster in comparison to the insulating counterpart. In the following discussion, we focus on the scattering channels leading to the relaxation of a photoexcited metal. In the subsequent sections, we discuss highly correlated materials.

When a femtosecond pump pulse interacts with a solid, it induces a coherent polarization of the material. In a density matrix description of the system, the optical coherence of this photoexcited state is retained as long as the off-diagonal elements have not decayed. In metals, quasielastic scattering induces very rapid decay of the coherence and dephasing within few femtoseconds [6] (see Chapter 2). As a consequence, we will discuss in the following the population dynamics of electron–hole pair excitations using the concept of a nonequilibrium electron distribution function, that is, the evolution of the diagonal elements of the density matrix. This is justified because the employed pump and probe pulses of 50–100 fs are considerably

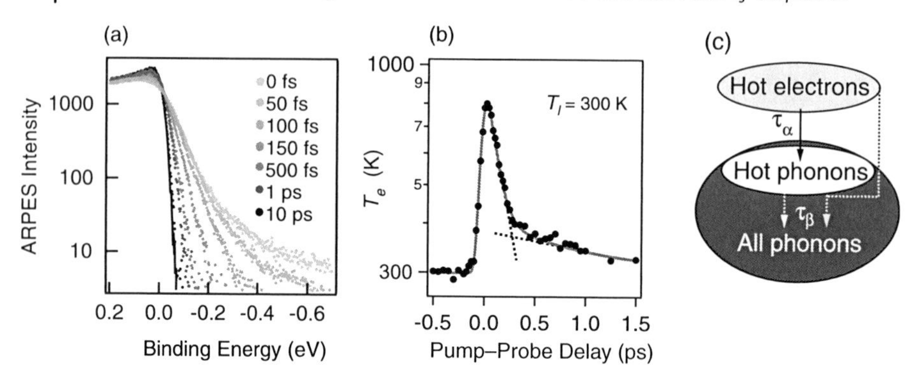

Figure 20.2 (a) Time-resolved spectra collected in $Bi_2Sr_2CaCu_2O_{8+\delta}$ near k_F for several pump–probe delays after photoexcitation. The equilibrium sample temperature is 300 K. (b) Temporal evolution of the electronic temperature T_e on a logarithmic temperature axis for $T_l = 300$ K. The initial decay for $\tau < 300$ fs originates from coupling of hot electrons to a subset of hot phonons, whereas the subsequent relaxation ($\tau > 300$ fs) represents decay of these hot phonons. (c) Illustration of the electronic relaxation ((a) and (c) are reprinted with permission from [5] Copyright 2007 by the American Physical Society).

longer than the pure dephasing time, which has been reported to be < 10 fs [6]. On the basis of these arguments, we assume that the temporal evolution of the single-particle excitations is given by a transient spectral function $A_k(\omega, \tau)$ multiplied by a time-dependent nonequilibrium distribution function $f(\omega, \tau)$. Supposing that the analytical form of one of these factors is known, $A_k(\omega, \tau)$ and $f(\omega, \tau)$ can be separated. Then, the dynamics of the electronic band structure can be distinguished from the dynamics of the distribution function under optically excited conditions.

Now, we shift our attention to the mechanisms that redistribute the energy of photoexcited electron–hole pairs among each other and other degrees of freedom, for example, phonons. On the one hand, the electron–electron scattering mediates the thermalization of $f(\omega, \tau)$ from a nonthermal toward a Fermi–Dirac distribution function. On the other hand, the electrons transfer energy to the lattice by electron–phonon scattering. If the thermalization of the hot electron gas is faster than the energy dissipation to the lattice, the electrons reach an effective electron temperature T_e that is above the temperature of the lattice T_l before complete thermalization is established and $T_e = T_l$. This situation is expected in most materials with strong electron–electron interaction and has been experimentally verified for various systems such as Au [7], Ru(0001) [8], Ni(111) [9], Gd(0001) [10], 1T-TaS_2 [11], and $Bi_2Sr_2CaCu_2O_{8+\delta}$ [5]. The last system is discussed here as an example.

Figure 20.2a shows time-resolved photoemission spectra of $Bi_2Sr_2CaCu_2O_{8+\delta}$ acquired at k_F for several pump–probe delays.[1] After optical excitation, the pronounced signature of hot electrons above the Fermi level can be clearly identified. At $\tau = 0$ fs, the ARPES intensity can be described by a constant spectral function and

1) Details about the experimental setup can be found in Ref. [8].

a Fermi distribution down to 2.5% of the maximum intensity at a binding energy of −0.25 eV. Evaluation of the respective excess energy residing in the thermalized and nonthermalized part of the distribution shows that 93.5% of the excess energy is described by a thermalized electron system. As can be recognized from Figure 20.2, the thermalized fraction increases further for longer delays. Hence, we can safely describe the dominant part of the optically excited system by a Fermi distribution function. Furthermore, the spectra in Figure 20.2 show that $T_e(\tau)$ decreases until the electrons reach an equilibrium with the lattice. We readily extract the temporal evolution of T_e, which is depicted in Figure 20.2b and shows a nearly biexponential decay. We describe the fast initial decay of the electronic temperature by a rate that is proportional to the second momentum $\lambda\Omega^2$ of the Eliashberg coupling function [12]. This parameter can be extracted by solving the rate equation

$$\frac{\partial T_e}{\partial \tau} = \frac{3\hbar\lambda\Omega^2}{\pi k_B T_e}(T_l - T_e). \tag{20.1}$$

According to the standard notation, λ is the momentum averaged electron–phonon coupling and Ω^2 is the averaged squared energy of the coupled phonon modes. Figure 20.2b shows the temporal evolution of the electronic temperature in optimally doped $Bi_2Sr_2CaCu_2O_{8+\delta}$. We identify two decay times $\tau_a = 110$ fs and $\tau_b = 900$ fs. The initial decay of T_e leads to $\lambda\Omega^2 = 360 \pm 30$ meV^2 [5]. Buckling and stretching modes of the copper–oxygen planes, which are potential mediators of the pairing interaction, have $\Omega > 40$ meV. From the extracted value of $\lambda\Omega^2$, it follows that these modes can contribute to the electronic pairing at most with an average electron–phonon coupling of $\lambda < 0.25$. Such a moderate momentum-averaged value of the electron–phonon coupling suggests that either specific and strongly coupled phonon modes drive the pairing in cuprates or another Boson is responsible. For a more detailed discussion, see Ref. [5].

Now we turn to the second slower decay in T_e, which embeds also indirect information on the dynamics of the phonon modes. The large specific heat of the lattice modes usually guaranties a small increase in their occupation number and hence a small increase in T_l. The occurrence of nonequilibrium phonons becomes evident only when a minor subset of phonons are coupled to the electrons. This situation can be expected in materials with anisotropic conduction and directional bonding like in $Bi_2Sr_2CaCu_2O_{8+\delta}$. Supposing that a subset of phonons is strongly coupled to hot electrons, thermalization of the phonon system requires that these hot phonons transfer their excess energy to the remaining cold phonon bath as illustrated in Figure 20.2c. On a similar timescale, the remaining hot electrons can couple to the cold boson bath.

The ratio of electronic temperature measured before and just after the generation of nonequilibrium phonons, provides a rough estimate on the number of lattice modes that are preferentially excited. In the case of $Bi_2Sr_2CaCu_2O_{8+\delta}$, our analysis indicates that only 20% of the lattice modes are more strongly coupled to the electrons than the average. This result confirms the conjecture of nonuniform coupling in these layered materials.

20.3
Photoinduced Insulator–Metal Transitions

Upon photoexcitation of a band insulator such as silicon, the electrons undergo optical transitions from the valence band to the conduction band. The resulting electron–hole plasma increases by a large amount the conductance of the nonequilibrium state. Despite the transient changes in conductance, we cannot consider this process as an insulator–metal transition. Indeed, the electronic band structure of silicon and in particular the electronic bandgap are not appreciably modified by optical pumping. In the following discussion, we will consider as insulator–metal transitions only those excitations leading to a photoinduced collapse of the electronic bandgap. Such drastic changes in the electronic structure occur in materials in which electrons experience strong interactions and correlation effects. Typical examples are Mott insulators and Peierls insulators. In the case of Mott insulators, the charge gap originates from the mutual repulsions of electrons in a half filled electronic band. Instead, the Peierls compound display a Fermi surface instability arising from the strong coupling between the conducting electrons and the specific lattice modes. In this section, we investigate the transient response of electronic states of these two different insulating ground states following optical excitation. As an example of a Mott–Hubbard insulator, we discuss the dichalcogenide 1T-TaS_2, in which the transition to a metallic state is found to occur simultaneously with the optical excitation. These results are contrasted to the response of the charge density wave material $TbTe_3$, which is discussed later in this chapter. Here a delay in the closing of the energy gap has been resolved and is attributed to nuclear motion required to close the bandgap in this Peierls-like system.

20.3.1
Response of 1T-TaS_2 to Optical Excitation

In the following, we discuss the dichalcogenide compound 1T-TaS_2 that already forms a charge density wave (CDW) at temperatures far above room temperature. In general, CDW materials are among the well-established model systems that have had considerable impact on our understanding of quantum many-body problems [14]. In brief, a CDW is a broken symmetry ground state with a spatially modulated electron density, which can form at low temperature in low-dimensional materials with high densities of states at the Fermi level due to a strong and anisotropic electron–electron and electron–phonon coupling. The driving force of this coupled electron–lattice instability is the minimization of the system's total energy. The increase in the elastic energy due to a displacement of the nuclei is overcompensated by the reduction of the kinetic electron energy by opening of a bandgap at the Fermi surface. Such low-dimensional electronic systems are characterized by a strong Lindhard response function close to the Fermi wave vector.

1*T*-TaS_2 is a two-dimensional material with layered structure. At far above the room temperature, this material forms a charge density wave. It shows a first-order metal–insulator phase transition with a change in conductivity by a factor of 10 at

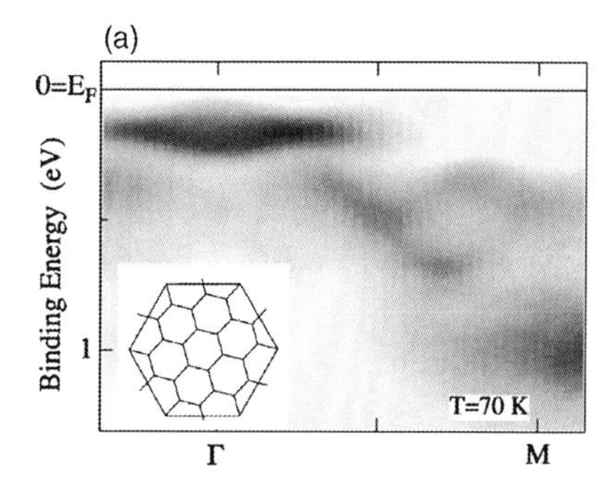

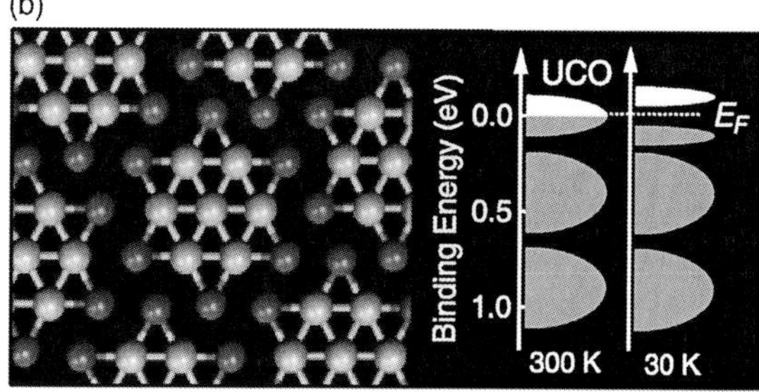

Figure 20.3 (a) ARPES intensity of 1T-TaS_2 below the metal–insulator transition temperature (reproduced from Ref. [13]). (b): Sketch of the commensurate charge density wave structure (Ta ions are shown) and electronic density of the states for the metallic (300 K) and the insulating state (30 K); Reprinted with permission from [5]. Copyright 2007 by the American Physical Society.

a transition temperature of 170 K, which is accompanied by a structural transition from an incommensurate to a commensurate CDW. This CDW breaks the symmetry of the Ta layers and the Ta ions are rearranged into a star-like shape depicted in Figure 20.3. The 13 valence electrons per Ta star form the electronic structure of these clusters, which is seen in the ARPES intensity shown in Figure 20.3a. The respective upper cluster orbital (UCO) is occupied by one electron, which is responsible for the metallic behavior along the planes at high temperature. At temperatures below this transition, the electronic band structure close to the Fermi energy is characterized by localized UCO with 200 meV binding energy.

Since in Mott–Hubbard insulators the kinetic energy of valence electrons, which is given by the bandwidth W, competes with the Coulomb repulsion U, a certain degree of localization is characteristic of such an insulator (for more information, see Volume 2, Chapter 1). On the other hand, these materials do not require a good angular resolution in a tr-ARPES measurement, as the lower Hubbard band covers a considerable part of the first Brillouin zone around the $\overline{\Gamma}$-point, see Figure 20.3a.

Figure 20.4a shows a phase diagram for a one-band Hubbard model calculated as a function of U/W and T_e for a half filled electron band [11]. The first order Mott transition takes place when varying U/W at an electronic temperature below $T_e \sim 90$ K across the solid line in Figure 20.4a. In contrast, above $k_B T_e = 0.02\ W$, the metallic and the insulating regimes are connected by the so-called crossover phase. An optical excitation can change the electron temperature on an ultrafast timescale as shown in Section 20.1.3. Then, the transition can be photoinduced. If the energy deposited by the pump pulse exceeds a given threshold, the charge fluctuations become large enough to disrupt the Mott insulating phase. Supposing that the collapse of electronic correlations and the internal thermalization occurs on a timescale faster than the probing pulse, the phase diagram might be extended to a description of the photoexcited state. Note that in the absence of nuclear motion the photoexcitation leaves the bandwidth, the Coulomb repulsion, and the filling

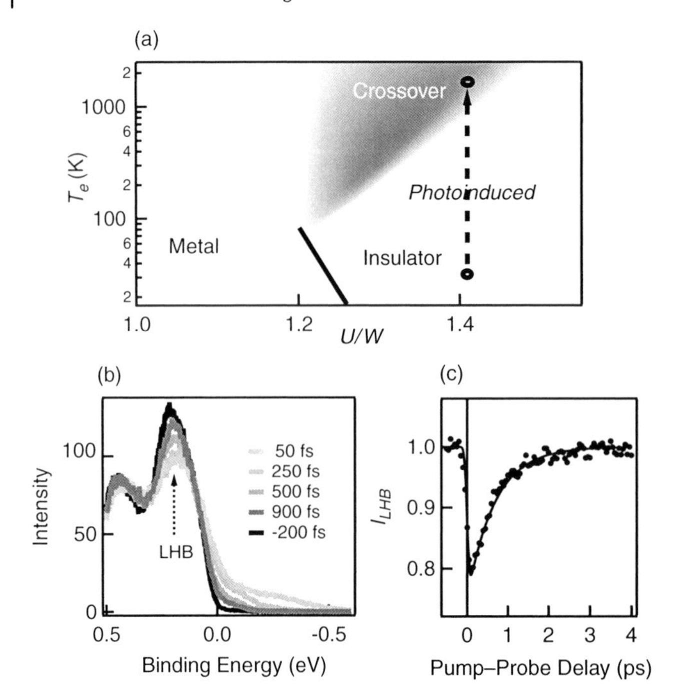

Figure 20.4 (a) Phase diagram of the Hubbard model as a function of the U/W ratio and of the electronic temperature. The diagram refers to a half filled system with bandwidth $W = 0.35$ eV. A dashed arrow shows that photoexcitation of the Mott insulator leads to a vertical transition into the crossover phase. (b) Time-resolved ARPES spectra of the Mott insulator 1T-TaS_2 at 30 K equilibrium temperature for several pump–probe delays. Due to photoexcitation spectral weight is transferred from the Lower Hubbard Band to the gapped energy region. (c) Normalized intensity of the LHB as a function of the pump probe delay. The solid line is a model calculation assuming an instantaneous collapse and an exponential recovery of the electronic correlations.

factor of the material almost constant. Hence, as shown in Figure 20.4a, a vertical transition of the electronic temperature induces a transition from the insulating to the crossover phase.

Figure 1.4b reports a tr-ARPES experiment on 1T-TaS_2. If this Mott insulator is in equilibrium at $T_e = T_l = 30$ K, the spectral function displays the UCO, which is considered as the Lower Hubbard Band (LHB) at an energy $\hbar\omega \simeq U/2$ and vanishing density of electronic states at the Fermi level. In the atomic limit, the position of the LHB coincides with the energy necessary to remove an electron from the single occupied site. The chemical potential is pinned at the edge of the LHB and the spectral function holds an energy gap in the interval $-0.1\,\text{eV} < \hbar\omega < 0\,\text{eV}$. A sufficiently large photoinduced increase in electronic temperature leads to the collapse of the Mott insulating phase. Figure 20.4b shows the photoinduced transfer of spectral weight from the LHB to the Fermi level. The electronic gap is partially filled by transient electronic states and recovers along with the energy relaxation of the hot electrons. This is not a result of mere electron population. In fact, the electronic structure did change upon excitation as in equilibrium an energy gap is found at E_F. The LHB intensity I_{LHB} in Figure 20.4c is an empirical parameter measuring the transfer of spectral weight. Due to the pulse duration of the 6 eV probe beam of 90 fs, the intensity of the LHB displays a quasi-instantaneous drop upon photoexcitation. In reality, the collapse of the Mott insulator should take place within the characteristic timescale of the charge fluctuations of about $\sim \hbar/W \approx 1$ fs [15]. Assuming a temporal profile of the laser pulse, the dynamics occurring within the laser pulse duration can be analyzed by convoluting the laser pulse profile with the anticipated evolution, for example, exponential decay as considered for the model calculation of the solid line in Figure 20.4c. In this way, we determine that the intensity minimum in Figure 20.4c has a maximum delay with respect to the laser excitation of 10 fs. Thus, a scenario where nuclear motion drives the transition leading to a delayed response can be safely excluded for the present case since nuclear motion occurs typically on a > 100 fs timescale (see below). The encountered shift of the intensity minimum with respect to $\tau = 0$ is due to the laser pulse duration, whereas the monotonic recovery is due to the gradual decrease in electronic temperature via phonon emission. Therefore, the recovery time of the Mott insulating phase is governed by cooling of the photoexcited electron distribution due to electron–phonon coupling.

Beside the photoinduced transition, the tr-ARPES spectra of 1T-TaS_2 exhibit periodic modulations of the binding energy. The photoelectron intensity map in Figure 20.5a shows pronounced oscillations of $A_k(\omega, \tau)$ that start at zero pump–probe delay and endure up to several 10 ps. This phenomenon is due to the coherent response of the lattice at a specific mode. At early pump–probe delays, the electrons of the photoexcited state occupy different electronic states from the equilibrium state. This time-dependent evolution of the electronic density leads to a change in the potential energy landscape of the ions and thus to an instantaneous force on the atoms of the lattice. As a consequence, the nuclei oscillate around their potential minimum until several scattering mechanisms lead to damping and dephasing of the oscillation, bringing the systems back to the initial conditions. The coherent lattice motion, which is identified as the lateral breathing mode of the star-shaped clusters, modulates the binding energy of the electronic states, causing an oscillatory behavior in the photoelectron spectra. From Figure 20.5c, it is seen that this binding energy variation contains two closely spaced frequencies, which present a beating behavior with a node at 10 ps. The lower frequency of 2.45 THz has been observed by time-resolved linear reflectivity studies before [16] and is attributed to the breathing mode in the bulk of the material [17]. Due to the combined surface and bulk sensitivity of low energy photoemission, we attribute the higher frequency to a stiffer mode at the surface. The higher frequency of the surface mode may be attributed to a different screening of the ion motion in the surface layer. Dephasing of this coherent lattice motion occurs during several 10 ps. This can be explained by damping through anharmonic coupling of the coherent vibration with thermal phonons. Decay of the coherent mode by electron–hole pair excitation is blocked because the system has returned to its insulating state after about 2 ps (Figure 20.4c). An analysis of the coherent mode at room temperature, which can be found in Refs [16, 18], showed that

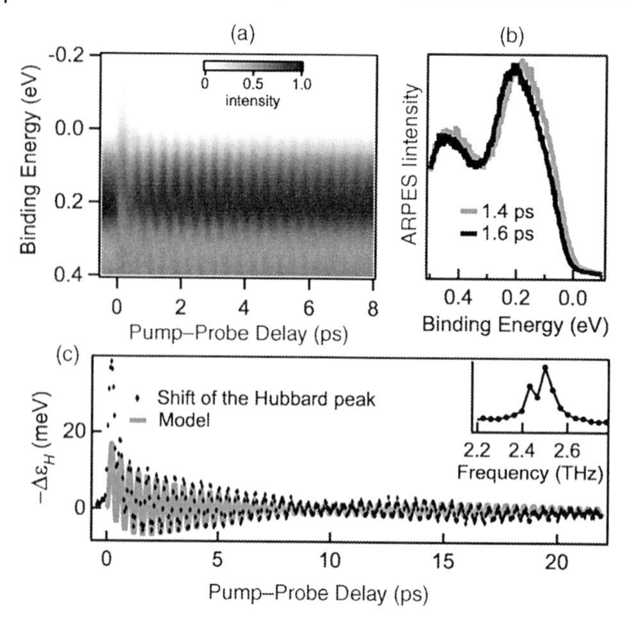

Figure 20.5 (a) Photoelectron intensity map acquired for 1T-TaS_2 as a function of pump–probe delay in normal emission direction of photoelectrons. The equilibrium temperature is 30 K. (b) Two spectra acquired on a maximum ($\tau = 1.4$ ps) and on a minimum ($\tau = 1.6$ ps) of one oscillation period display a rigid shift with respect to each other. (c) Time-dependent shift of the spectra as a function of pump–probe; Reprinted with permission from [11]. Copyright 2006 by the American Physical Society.

here the dephasing is considerably more efficient. It is almost completed after 2 ps due to the presence of low-energy electronic excitations.

To conclude the analysis of the dynamics in TaS_2, we would like to emphasize the following points: (i) The shift of spectral weight from the LHB peak to the range of the energy gap, which unambiguously shows the transition from an insulating ground to a transiently metallic state, proceeds on an ultrafast timescale with a delay of < 10 fs. This vividly demonstrates the Mott–Hubbard nature of the ground state in 1T-TaS_2: By excitation the systems is put into a nonequilibrium. According to the observed photoemission spectra, this state can be identified as the crossover phase introduced in Figure 20.4a. Recovery of the insulating ground state proceeds on the timescale of the hot electron decay, which can be described by reduction of the hot electron temperature, through electron–phonon scattering, that is, by lattice heating. (ii) The coherent lattice mode has a different excitation and decay mechanism. It is driven by a photoinduced modification of the spatial charge distribution and state filling in the star-like structures leading to the breathing mode. Its damping continues after the transient metallic state has decayed to the insulating state. Hence, phonon–phonon scattering remains as a decay channel for the coherent lattice mode. (iii) In this way we can separate the CDW and the Mott insulating character of this material, which is still a challenge for experiments performed in thermodynamic equilibrium. As we found that the response of the CDW mode and the Mott insulating gap to optical excitation occurs on different timescales, we propose that to first order the CDW and Mott insulating behavior can be treated independently.

20.3.2
Dynamics of a Photoinduced Melting of a Charge Density Wave in $TbTe_3$

Now, we turn to the photoinduced dynamics of a second CDW system and start with a brief discussion of a nesting-driven CDW formation and address the underlying microscopic interactions. In a nesting-driven CDW, the opening of a bandgap at the Fermi surface is a consequence of a spontaneous periodic lattice modulation Δz (see Figure 20.6b) with an appropriate wave vector q_{CDW} (see Figure 20.6c) that nests parts of the Fermi surface [14]. The coupled electron–lattice dynamics of CDW systems are characterized by a well-defined excitation mode: The CDW amplitude mode is a collective vibration that modulates the lattice distortion. It hence changes the CDW amplitude and the size of the CDW bandgap periodically in time [14]. In view of the electronic structure, this collective mode exists in the nesting region close to k_{F} below the CDW phase transition temperature T_{CDW} and exhibits a characteristic mode softening upon approaching T_{CDW}. It can be anticipated that the signature of the amplitude mode is directly observable in tr-ARPES experiments by a modulation of the bandgap near k_{F}.

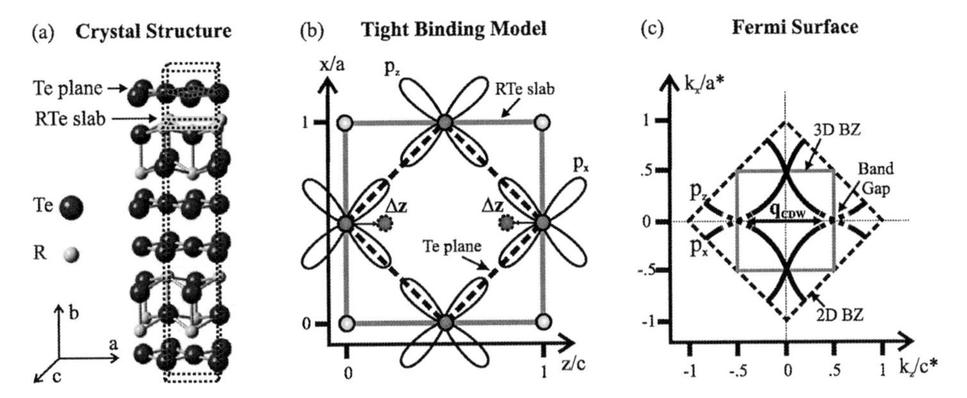

Figure 20.6 (a) The crystal structure of RTe_3 [19] is composed of square Te planes and buckled RTe slabs. Note that the *b*-axis in the space group Cmcm of RTe_3 is perpendicular to the Te plane with $b = 25.7\,\text{Å} \gg a \approx c = 4.3, \ldots, 4.5\,\text{Å}$ [19, 20]. The crystals cleave between two RTe slabs. (b) Real-space model of the square-planar Te sheet with 2D (solid line) and 3D (dashed line) unit cells of the Te planes and the RTe slab, respectively. The Te p_x and p_z orbitals as well as the lattice distortion Δz due to the CDW are depicted. (c) Fermi surface calculated in a tight binding model [21]. The extended 2D (dashed) and reduced 3D Brillouin zone (solid) are indicated. The small perpendicular coupling between the p_y orbitals results in two sets of quasi-1D bands and an almost rectangular Fermi surface. The nesting of the phonon mode with nesting vector q_{CDW} opens two bandgaps in the Fermi surface along the z-direction. Crystal Structure reprinted with permission from [19]. Copyright 2006 by the American Physical Society.

Recent efforts have identified the RTe_3 (R = rare-earth element) family of compounds to be a quasi-2D model system to study Fermi surface nesting-driven CDW formation [21–23]. The RTe_3 unit cell as depicted in Figure 20.6a is constructed from the basic building block of a square Te plane and a buckled RTe slab, where one RTe slab resides between two largely decoupled Te planes. Two of these building blocks are stacked up along the *b*-axis. The crystal structure and band dispersion of RTe_3 near E_F are closely related and well described in a tight binding model [21] with weakly hybridized $5p_x$- and $5p_z$-orbitals of the Te plane as shown in Figure 20.6b. Since the planar square net of Te atoms is rotated by 45° with respect to the RTe slab, two different unit cells in real space are appropriate: The larger 3D unit cell is defined by the Tb atoms of the buckled slab (solid line in Figure 20.6b), whereas the smaller 2D unit cell is given by the square Te plane (dashed line). For the band structure in reciprocal space (see Figure 20.6c), this results in a larger 2D Brillouin zone (dashed line) corresponding to the Te square, which is rotated by 45° with respect to the smaller 3D Brillouin zone (solid line). The underlying RTe bilayer with its finite 3D character reduces the Brillouin zone to $\sqrt{2} \times \sqrt{2}$-$R45°$ symmetry and determines the diamond-like shape of the Fermi surface [21, 23, 24]. Below the CDW phase transition temperature T_{CDW}, the Fermi surface nesting with a nesting vector q_{CDW} opens up a bandgap at the tip of the Fermi surface along the *c*-axes, as depicted by dashed curves in Figure 20.6c. In real space, this leads to a freezing of the associated lattice vibration, which is equivalent to a displacement Δz of the ion positions, see schematic sketch in Figure 20.6b.

20.3.2.1 Electronic Band Structure of $TbTe_3$ in Thermal Equilibrium

Before turning to the photoinduced CDW dynamics in $TbTe_3$, the band structure in thermal equilibrium is analyzed in Figure 20.7 using ARPES at photon energies of $h\nu = 23$ eV and 6 eV from a synchrotron and Ti:sapphire laser-based light source, respectively. The two sets of ARPES measurements are compared for temperatures of 300 K and 100 K corresponding to the ungapped metallic and the gapped CDW state. As the transition temperature for $TbTe_3$ is $T_{CDW} = 335$ K [20], the CDW gap at 300 K is almost closed and the Fermi surface in Figure 20.7a exhibits the characteristic diamond-shaped topology that is well described by a tight binding model including the back-folded shadow bands [21]. Figure 20.7d shows the Fermi surface of $TbTe_3$ for a temperature of 100 K, where the CDW gap extends throughout a significant part of the Brillouin zone. The CDW gap is further analyzed by taking photoemission spectra as a function of momentum indicated by the solid arc in Figure 20.7d.

The ARPES intensity maps in Figure 20.7b/e were taken at $h\nu = 23$ eV and show a weakly dispersing p-like Te band at $E - E_F \approx -0.45$ eV and a p-like Te valence band (VB). Near T_{CDW} the VB disperses from $E - E_F \approx -0.2$ eV through E_F and the bandgap extends down to 200 meV below E_F. Figure 20.7c/f shows the same spectral sections as in (b)/(e), but these were taken with the tr-ARPES setup using 6 eV fs laser pulses without optical pumping. In general, both ARPES measurements agree well, although some details of the lower lying Te band differ. Nevertheless, the dispersion of the VB and the opening of the CDW gap below T_{CDW} is captured correctly by the laser-based ARPES measurement. The good agreement of both experiments ensures

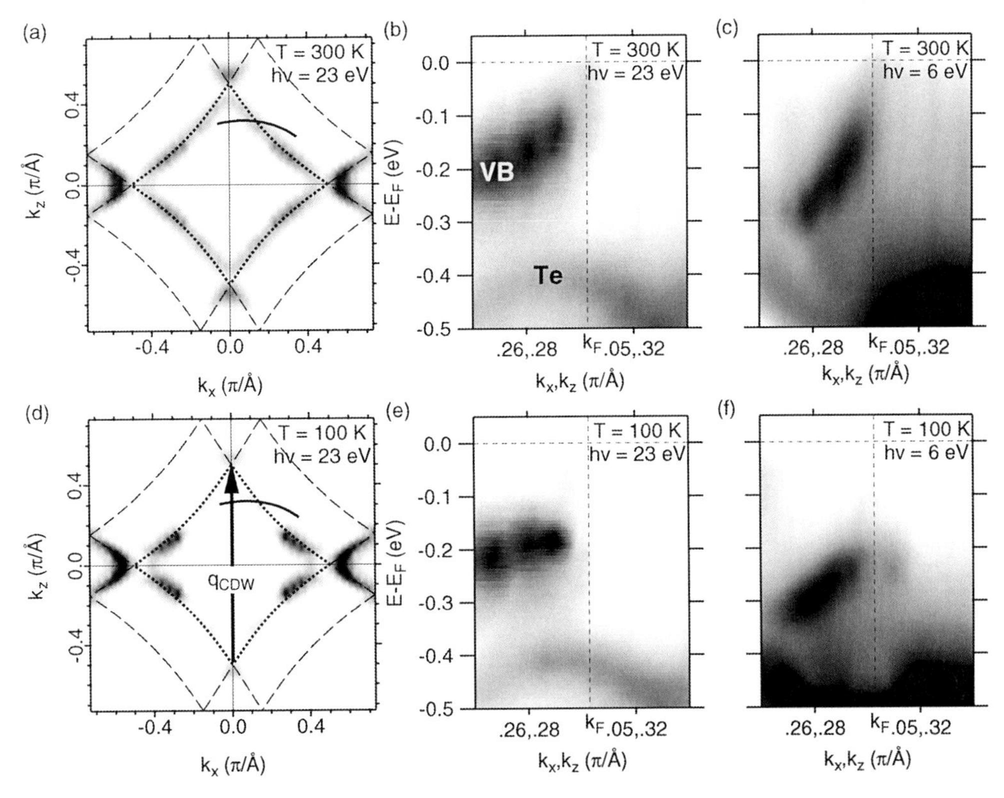

Figure 20.7 Comparison of ARPES measurements of $TbTe_3$ in thermal equilibrium. The data were taken at $T = 300$ K (upper row) and $T = 100$ K (lower row). (a)/(d) The Fermi surface maps along the in-plane directions k_x and k_z were obtained from synchrotron-based ARPES at $h\nu = 23$ eV by integrating an energy window of 20 meV centered at E_F. Also plotted is a tight binding model [21] (bold dotted lines), the backfolded shadow bands (thin dashed lines), and the CDW nesting vector q_{CDW} (arrow) along the nesting direction k_z. (b)/(e) Spectral sections along the in-plane directions as indicated in (a)/(d). In (b) the localized Te band and the VB are indicated. The positions of E_F and k_F are given by horizontal and vertical lines, respectively. (c)/(f) Identical spectral sections using the laser-based tr-ARPES system at $h\nu = 6$ eV. Adapted from [25]. Reprinted with permission from AAAS.

that the same Brillouin zone position is probed in tr-ARPES as in conventional ARPES. Thus, the transient electronic band structure can be safely analyzed with tr-ARPES. To monitor the time evolution of the CDW bandgap after optical excitation, we focus on $k = k_F$ in the following section.

20.3.2.2 **Weak Perturbation Regime**

We begin the discussion of the photoinduced dynamics with the tr-ARPES data in Figure 20.8, which were acquired at an incident pump fluence of $F_1 = 0.3$ mJ/cm^2 in the IR pump laser beam with a photon energy of $h\nu = 1.5$ eV. In Figure 20.8a, the photoelectron yield at the Fermi vector $k = k_F$ is given as function of binding energy

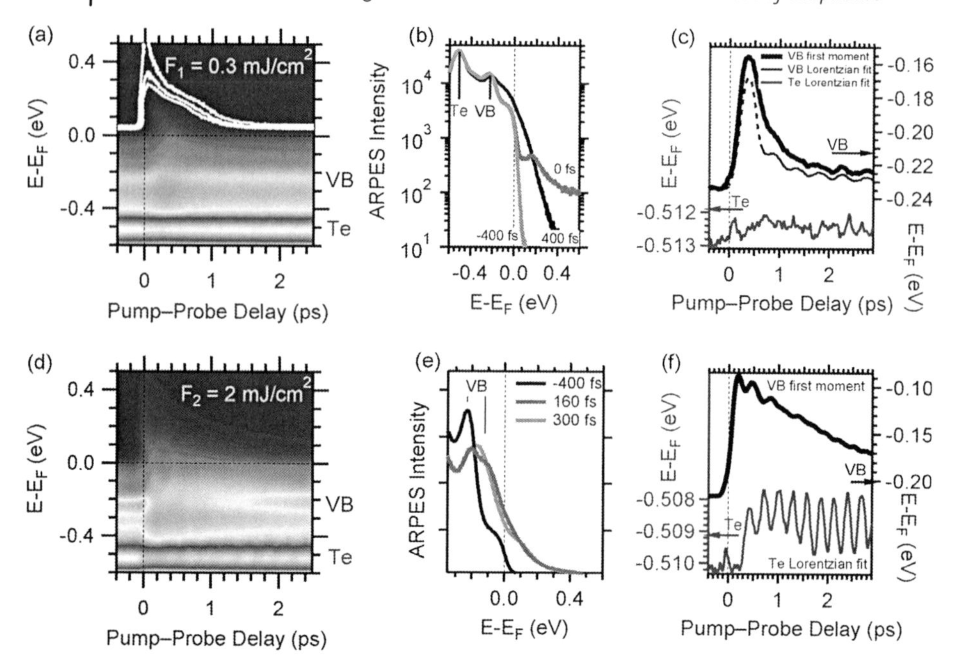

Figure 20.8 (a)/(d) Photoelectron intensity as a function of energy $E-E_F$ and pump–probe delay taken at $k = k_F$, $T = 100$ K, and an incident fluence of $F_1 = 0.3$ mJ/cm^2 (upper row) and $F_2 = 2$ mJ/cm^2 (lower row). The Te band and the VB are indicated. (b)/(e) Spectra from (a)/(d) for selected delays. (b) The peak at $E-E_F = 0.15$ eV in the unoccupied part of the band structure evidences the photodoping at zero delay. (e) The VB peak shifts and splits due to the intense excitation. (c)/(f) The transient binding energies of the Te band (left) and the VB (right) reveal oscillations. The Te and VB states were analyzed by Lorentzian line profiles and the first statistical moment for the VB. Adapted from [25]. Reprinted with permission from AAAS. (Please find a color version of this figure on the color plates.)

$E-E_F$ and pump–probe delay t for a temperature of 100 K well below $T_{CDW} = 335$ K. Figure 20.8b shows spectra before, at, and after the optical excitation on a logarithmic intensity scale. Before the arrival of the pump pulse, the spectrum exhibits two peaks from the Te band and the VB at $E-E_F = -0.50$ eV and -0.25 eV, respectively, and a well-defined Fermi–Dirac distribution around E_F. At zero time delay, the optical excitation with the IR pulse generates a nonthermal hot electron distribution and populates an unoccupied state, which is identified in panel (b) by a peak at $E-E_F = 0.15$ eV. At $t = 0$ fs, the optical pumping has only a minor influence on the occupied states and the well-defined Fermi–Dirac distribution remains essentially unchanged around E_F accounting for 90% of the electron population. The remaining 10% of the electron population consists of hot electrons above E_F, which populate orbitals that might have a different character than in the CDW ground state. The associated change in the electronic screening and the electron–phonon coupling due to optical excitation of electron–hole pairs is the essence of photodoping and the driving force of the lattice dynamics observed at later delays.

This photodoping is evidenced by the discrete peak appearing at $E - E_F = 0.15$ eV at $t = 0$ fs.

The continuous population of states above E_F occurs through electron–electron scattering within the experimental time resolution of 100 fs, as under the excitation conditions used, inelastic electron–electron scattering is known to be the dominant relaxation process within the first few 100 fs [7, 8, 10] as discussed in Section 20.3.1. These efficient scattering processes ensure that a transient electronic population is observed. After 400 fs, the initially nonthermal electron distribution has thermalized and the broadened Fermi–Dirac distribution indicates a significantly enhanced electron temperature $T_e \gg 100$ K. As seen in Figure 20.8a, the VB shows a considerable shift in its spectral weight toward E_F after pumping, which reaches its maximum at 400 fs.

This brings us to the transient binding energies of the Te band and the VB presented in Figure 20.8c, which are analyzed[2)] by a Lorentzian line fit, and the analysis of the first moment of the time-dependent spectra. Within 400 fs the VB shifts toward E_F. After 400 fs, the VB binding energy starts to recover and reveals an oscillatory behavior persisting for several picoseconds. Initially, the binding energy of the VB decreases because the optically deposited excess energy resides mainly in the electronic system during the first ~ 400 fs. At these delays, T_e is considerably higher than the lattice temperature and we attribute the lowering of the VB binding energy to a modified screening that affects the electronic part of the CDW. Subsequently, electrons and phonons equilibrate through electron–phonon scattering within roughly 2 ps.

In the next step, the oscillatory behavior is further analyzed[2)] by a Fourier transformation of the transient binding energies for delays > 500 fs avoiding complications due to the initial binding energy shift. For the VB mode, this results in a frequency of $\Omega_{VB} = 2.3(5)$ THz. The binding energy changes of 80 meV in the VB are much larger than the ~ 0.5 meV shift of the Te band (Figure 20.8c), which is less susceptible to the excitation. Nevertheless, a Fourier transformation of the transient Te binding energy is able to resolve a periodic contribution at $\Omega_{Te} = 3.6(5)$ THz [26]. Notably, we observe two distinct collective modes Ω_{VB} and Ω_{Te} for these two different electronic bands.

This result poses the question of the origin of these two different collective modes. Focusing on the VB mode, we consider the CDW amplitude mode as the fundamental collective excitation mode of a CDW system. The VB peak shifts toward E_F (top row of Figure 20.8) due to the photodoping, which is consistent with a reduction in the bandgap. This initiates nuclear motion to compensate for the instantaneous change in electronic screening. Since the pulse duration is shorter than the oscillation period, a phase relation with respect to time zero is established [10]. Thus, a coherent oscillation of the ions is launched, which, in turn, modulates the electronic part of the CDW. The signature of the CDW amplitude mode is very specific since it exists only below T_{CDW} and close to k_F due to the anisotropic electronic response of the nested

2) Details of the binding energy analysis are discussed in the supplemental material of Ref. [25].

Fermi surface [14]. Employing low fluences that result in a weak perturbation is another prerequisite for observation of the amplitude mode: These excitation densities are sufficient to excite the amplitude mode but yet below the threshold to drive a transition from the gapped CDW to the ungapped metallic state. In the following section, we show that the 2.3 THz VB mode is observed only near k_F, only for $T < T_{CDW}$ and only for low fluences. On the basis of this characteristic momentum, temperature dependence, and fluence dependence, the 2.3 THz VB oscillation is attributed to the CDW amplitude mode.

20.3.2.3 Strong Perturbation Regime

We now proceed to a strongly perturbative excitation regime, where the excitation is strong enough to melt the CDW on an ultrafast timescale. The lower row of Figure 20.8 shows tr-ARPES data that have been taken at an increased IR pump fluence of $F_2 = 2$ mJ/cm^2. Clearly, the oscillations of the electronic states have increased in amplitude compared to F_1. First, focusing on the Te band in Figure 20.8f, the oscillations are now well resolved and the Fourier analysis yields a frequency of $\Omega_{Te} = 3.6(1)$ THz. Although the increased pump fluence leads to an increased amplitude of the Te mode, its frequency is unaffected.

In contrast to the Te band, the response of the VB has changed qualitatively compared to the weakly perturbative excitation (top row of Figure 20.8). The spectral function of the VB depends on the oscillation phase as seen from the spectra given in Figure 20.8e. For a delay of 160 fs, the VB peak is significantly broadened, whereas at 300 fs an additional splitting of the line is observed. These spectra suggest not a simple shift and broadening of the line but a more complex transient state, which is under further investigation at present. Due to a significant deviation from a Lorentzian line profile, the transient binding energy of the VB, as shown in Figure 20.8f, is analyzed simply by determining the center of mass. Its time evolution seems to be separated into two regimes: The first two oscillations display a different period of 280 fs and 360 fs and subsequently vanish for delays > 1 ps. This unique behavior of the VB can be understood in terms of an ultrafast melting of the frozen CDW state: The intense photodoping at this fluence initiates an ion displacement that is large enough to lift the periodicity initially given by the CDW state. In turn, the electronic system responds to the rearranged ion positions by a collapse of the bandgap. Thus, the charge ordered state melts. Consequently, the amplitude mode in the VB first is severely softened or even is no longer defined. We observe a transient softening by the increasing oscillation periods before the amplitude ceases to exist after a few oscillation periods. After 1 ps, when the melting of the CDW phase is complete and the amplitude mode no longer exists, only the 3.6 THz Te oscillation is observed. However, before discussing this ultrafast CDW transition in more detail, it is instructive to turn to the assignment of the observed frequencies.

20.3.2.4 Assignment of the Collective Modes

The temperature dependence of the Te and VB frequencies is analyzed in Figure 20.9 as a function of fluence F. At $T = 300$ K, the Te-derived 3.6 THz mode is observed.

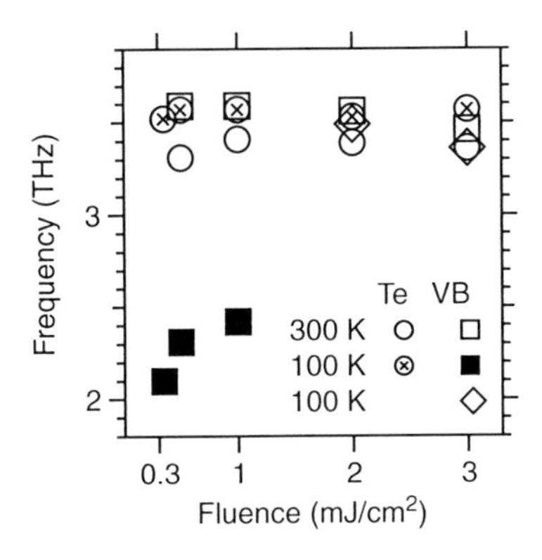

Figure 20.9 Dependence of frequencies of the collective modes on the incident fluence for 300 and 100 K. Circles represent the Te band, squares the VB. Frequencies have been determined after 500 fs from a Lorentzian line fit of spectra acquired at $k = k_F$. Diamonds give the frequency of the first statistical moment of the VB. Adapted from [25]. Reprinted with permission from AAAS.

Since the gap in the VB is almost closed in $TbTe_3$ at 300 K, the absence of the 2.3 THz mode at 300 K points to the lower frequency mode as being the amplitude mode.[3)]

In contrast, for $T = 100$ K $\ll T_{CDW}$ two modes are identified for fluences below 1 mJ/cm^2. The high frequency mode with 3.6 THz observed in the Te band is robustly present in all measurements, regardless of the position in the Brillouin zone, the temperature, or the pump power [25]. Thus, we assign the 3.6 THz mode to an excitation that is independent of the CDW state and is associated with the lower lying Te band. The low-frequency mode at 2.3 THz is observed by transient modifications of the VB. For pump fluences above 1 mJ/cm^2 this mode is no longer present (see Figure 20.8f), which is explained by the strongly perturbative regime in which the CDW melts. Since the 2.3 THz mode is found only if the system forms the CDW and is weakly excited, we identify the 2.3 THz mode with the amplitude mode. Further support of this assignment is available from the momentum dependence discussed in the next section.

20.3.2.5 Ultrafast Melting of the CDW State

Now, we return to the discussion of the impact of the intense photodoping on the CDW phase. We performed a momentum-dependent analysis of the electron dynamics in the CDW nesting region in momentum space, which provides an explicit link between the ion core motion underlying the amplitude mode and the transient behavior of the electronic bandgap. Considering the nesting-driven CDW

3) Strictly speaking, the amplitude mode is not absent at $T = 300$ K$< T_{CDW} = 335$ K. However, as T is close to T_{CDW}, the spectral weight might be too weak to be detected.

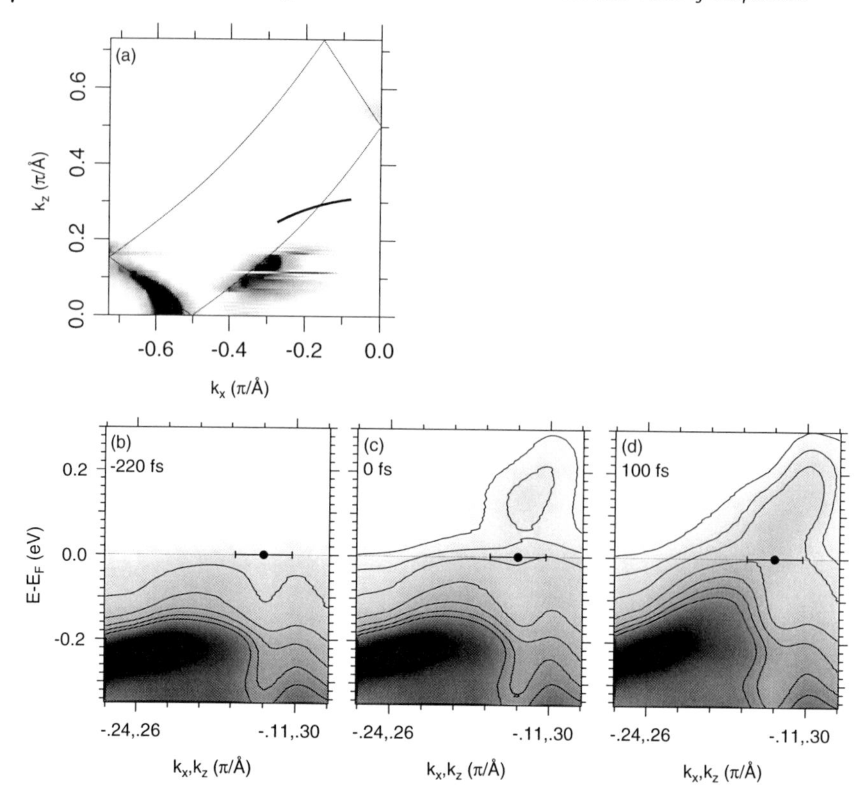

Figure 20.10 (a) Detail of the Fermi surface plot from Figure 20.7(d). (b)–(d) Photoelectron intensity as a function of energy and momentum along the arc in (a) for the delays indicated: (a) before excitation, (b) during optical excitation, and (c) after the system has recovered the high-temperature state featuring a quasifree electron dispersion $E(k)$ ($T = 100$ K, $F_2 = 2$ mJ/cm^2). k_F is marked in panels (b)–(d) by a solid dot. The contour lines emphasize the photoelectron intensity above E_F. Adapted from [25]. Reprinted with permission from AAAS.

formation in the RTe_3 materials, the electrons at E_F are very susceptible to scattering by the q_{CDW} wave vector and the respective phonon mode. The Lindhard response function accounts for this behavior and shows that the electronic response increases upon approaching k_F (for details see [14]). Therefore, the CDW amplitude mode is expected to influence the electronic structure strongest at k_F.

In Figure 20.10, tr-ARPES data of the CDW nesting region are presented. The arc in Figure 20.10a defines the position in the Brillouin zone where the data sets of Figure 20.10b–d have been recorded under strongly perturbative conditions of $F_2 = 2$ mJ/cm^2. In Figure 20.10b–d, the photoelectron intensity is shown as a function of momentum and energy for selected delays to resolve the transient electronic band structure in the nesting region of the Fermi surface. This allows a momentum-specific analysis of the electronic excitation and the response of the lattice revealing the process of the ultrafast CDW melting.

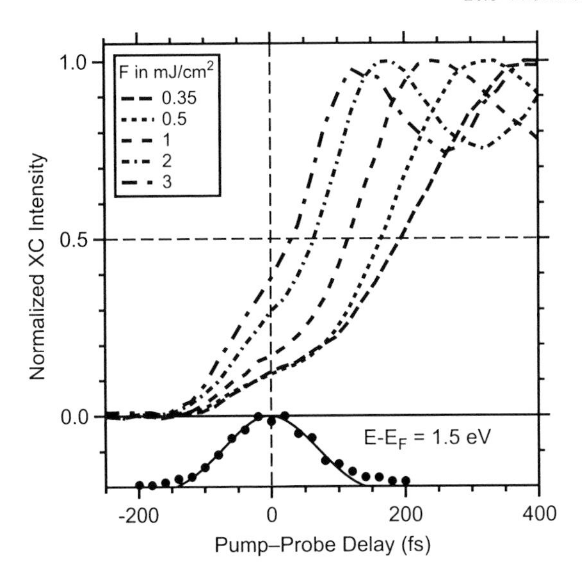

Figure 20.11 The time-dependent photoemission intensities are acquired at $E = E_F$, $k = k_F$ and a temperature of $T = 100$ K for various incident fluences by integrating over an energy window of ± 50 meV width centered around E_F. The lower panel indicates the transient photoemission intensity at $E - E_F = 1.5$ eV defining zero pump–probe delay as well as the convoluted temporal profile of pump and probe pulses [8].

These data feature a pronounced momentum-dependent response of the electronic structure. At $t = -220$ fs, the unperturbed equilibrium band structure with the gapped CDW state is equivalent to the equilibrium band structure (Figure 20.7f). At a delay of 0 fs, where the leading edge of the IR pump pulse has already excited the electronic system by electron–hole pair creation, the bandgap is still preserved and the flat dispersion of the VB represents a localized state as prior to pumping. However, in the vicinity of k_F (dots in Figure 20.10b–d), an intensity increase at $E - E_F = 0.15$ eV is encountered. This feature coincides with the unoccupied state observed in Figure 20.8b and manifests a quasi-instantaneous photodoping of the system. This drives the amplitude mode at low excitation densities and initiates the melting of the CDW phase at higher fluences.

After 100 fs and hence delayed with respect to the excitation and the occurrence of the photodoped state above E_F, (i) the bandgap has closed and (ii) the dispersion of the VB recovers the quasifree electron dispersion known from the metallic state, compared to Figure 20.7c. Note that the VB dispersion can be followed well above E_F due to the photoexcited hot electron population. The recurrence of a gapless VB dispersion, resembling the high-temperature situation, vividly demonstrates the transition from the gapped CDW state near k_F to a continuous dispersion through E_F. Thereby, we have demonstrated that the VB responds preferably at k_F.

In the next step, we analyze the timescale of the CDW melting in more detail. Considering that CDW formation is based on nuclear rearrangement, the CDW melting process cannot proceed faster than the respective nuclear motion. Hence, one might expect the nuclear rearrangement to limit the timescale of the CDW melting. To analyze the respective timescale, we evaluate the photoemission intensities at E_F and k_F as a function of time delay. The onset of the photoemission intensity within the initially gapped region as function of time reveals the closing of the bandgap and the timescale of the response.

In Figure 20.11, the relative change in the photoelectron intensity at E_F is plotted as a function of pump–probe delay for various incident fluences. The lower panel shows the intensity at $E - E_F = 1.5$ eV indicating the convoluted temporal envelope of pump and probe pulses [8]. We compare this instrument response function to the experimentally determined intensities and find that the gap closing, that is, the response of the CDW phase to optical excitation, is clearly delayed compared to time zero. We attribute this delay to the nuclear motion as part of the CDW melting process. A detailed understanding of the electronic and nuclear contributions to the dynamics of the gap closing is being developed. The observed fluence dependence might provide valuable information toward this goal.

20.4 Discussion

We have studied the transient electronic structure of TaS_2 and $TbTe_3$, which are considered model systems for Mott insulators and charge density wave materials, respectively. Using time-resolved photoemission spectroscopy, we investigate the characteristic timescale of the collapse of the electronic bandgaps in these insulators after sufficiently intense optical excitation. In a Mott insulator, purely electronic correlations lead to electron localization and formation of Mott–Hubbard gap around E_F. Hence, the fastest timescale at which the gap is filled is determined by electronic hopping among the sites. The respective timescale is on the order of $\hbar/W$ equal to few femtoseconds (see above). This is beyond the available time resolution in the present experimental setup. However, this picture is consistent with the observation that states within the gap are populated quasi-instantaneously within the laser pulse profile. The fact that in Figure 20.4c no delay between the temporal profile of the laser pulse and the material response is detected is consistent with the conclusion that TaS_2 is indeed a Mott insulator and its response to optical excitation is purely dominated by electronic correlations. However, we cannot exclude that the pronounced signature of nuclear vibrations in the breathing mode may lead to small variations in the electrical conductivity. Such variations are, however, too weak to drive the transition into a metallic state.

In a charge density wave material, nuclear motion is a prerequisite for CDW formation and thus the comparably slow nuclear motion is expected to delay the bandgap collapse and "melting" of the electronic structure. This is consistent with the observed delayed filling of the bandgap in $TbTe_3$ (see Figure 20.11). The bandgap filling and the recurrence of the quasifree electron dispersion $E(k)$ indicate that the electronic system has transiently locked in the high-temperature state. However, since a CDW involves nuclear motion collective excitations of the CDW system require more time to equilibrate, as can be concluded from the transient softening in the CDW amplitude mode observed under the highly perturbative excitation conditions. This time can be estimated to be 1 ps because after this delay no reminiscence of the amplitude mode is encountered in Figure 20.8f. The observations of the formation of a quasifree electron dispersion within 100 fs and the vanishing of the amplitude mode derived vibration within 1 ps suggest that the electronic and nuclear constituents of the CDW evolve differently in time. Future investigations will focus on this aspect and are expected to clarify the question how and on what timescale the transient potential energy landscape, in which the nuclear motion proceeds, is formed.

20.5 Conclusions and Outlook

The presented investigations of ultrafast dynamics in materials with strong electron correlation have shown that the information available from tr-ARPES is valuable to achieve a microscopic understanding of relaxation processes on ultrafast timescales and interaction of low-energy excitations. In particular, tr-ARPES can access both the populations dynamics of single quasiparticle states and the coupling of collective (boson) modes to the electronic structure. The technique provides thus new insights into electron–phonon coupling in general as well as in the context of photoinduced phase changes of (correlated) materials.

In this chapter, complementary ways to analyze electron–phonon interaction have been presented. In $Bi_2Sr_2CaCu_2O_{8+\delta}$, the energy transfer between one electron and two phonon heat baths – the latter describe one subset of phonons that is strongly coupled to electrons and one that accounts for the remaining phonons – has been modeled. The momentum-averaged electron–phonon coupling constant has been determined for a given phonon frequency. In addition to these energy transfer processes, coherent phonons are encountered in the transient response of the electronic structure in 1T-TaS_2 and $TbTe_3$. This signature contains very detailed information on how the electron system couples to particular phonons. However, quantitative information on the coupling strength requires further theoretical input as well, as has been shown for Gd(0001) surfaces (see Chapter 12). Nevertheless, the results obtained for $TbTe_3$ demonstrate the electron momentum-dependent coupling of the CDW amplitude mode in this CDW material. The electron–phonon coupling of the amplitude mode of the CDW-ordered state is qualitatively different from a generic coherent phonon, which lacks the characteristic dependence on electron momentum and temperature encountered here for the amplitude mode.

The timescale at which optical excitation of materials with temperature-dependent insulator-to-metal transitions leads to a filling of the bandgap contains information uniquely available from time-resolved studies. Here, we have demonstrated in a proof-of-principle experiment that in case of a Mott insulator the bandgap is closed on an intrinsic electronic timescale, which is expected to be few femtoseconds and has not been resolved due to the present experimental limitations. In CDW materials, where nuclear motion is intimately linked to opening or closing of the gap, a delayed closing of the bandgap has been observed, which is attributed to the nuclear rearrangement required during the bandgap collapse.

In the employed tr-ARPES setup, the accessible region of the Brillouin zone is limited by the low kinetic energy of photoelectrons generated with $h\nu = 6$ eV using nonlinear optical crystals for frequency quadrupling of the fundamental laser pulse. In the future, schemes that overcome these limitations may utilize high-harmonic generation in polarizable gases (see Chapters 21 and 22). Such a development would open up the use of higher photon energies to enlarge the accessible region of the Brillouin zone. Simultaneously, the broad bandwidth of high harmonic light sources should support much shorter laser pulses for a better sensitivity to high frequency modes. However, the time–bandwidth product of the laser pulses employed in tr-ARPES needs to be adapted to the particular problem of interest to maintain the required energy resolution and sufficiently short laser pulses.

Finally, we point out that tr-ARPES has the ability to characterize phonons through their coupling to the electron system and can separate phonons that modify the single-particle energy in the time domain from phonons that influence the electronic structure under certain restrictions in, for example, temperature and momentum. Such phonons – the amplitude mode in $TbTe_3$ is the example that has been discussed in this chapter – can be an essential part of cooperative phenomena and hence their understanding is essential for the ground-state properties of materials. In this respect, tr-ARPES is complementary to various established phonon spectroscopies.

Acknowledgments

This work is based on several very fruitful collaborations that have been essential to obtain the results presented here. We are grateful to H. Berger, S. Biermann, P. S. Cornaglia, H. Eisaki, A. Georges, M. Krenz, M. Lisowski, P. A. Loukakos, R. G. Moore, L. Rettig, F. Schmitt, and Z. X. Shen.

References

1 Hüfner, S. (2003) *Photoelectron Spectroscopy*, 3rd edn) Springer.
2 Damascelli, A., Hussain, Z., and Shen, Z.-X. (2003) *Rev. Mod. Phys.*, **75**, 473.
3 Koralek, J.D., Douglas, J.F., Plumb, N.C., Sun, Z., Fedorov, A.V., Murnane, M.M., Kapteyn, H.C., Cundiff, S.T., Aiura, Y., Oka, K., Eisaki, H., and Dessau, D.S. (2006) *Phys. Rev. Lett.*, **96**, 017005.
4 Kiss, T., Kanetaka, F., Yokoya, T., Shimojima, T., Kanai, K., Shin, S., Onuki, Y., Togashi, T., Zhang, C., Chen, C.T., and Watanabe, S. (2005) *Phys. Rev. Lett.*, **94**, 057001.
5 Perfetti, L., Loukakos, P.A., Lisowski, M., Bovensiepen, U., Eisaki, H., and Wolf, M. (2007) *Phys. Rev. Lett.*, **99**, 197001.
6 Petek, H. and Ogawa, S. (1997) *Prog. Surf. Sci.*, **56**, 239.
7 Fann, W.S., Storz, R., Tom, H.W.K., and Bokor, J. (1992) *Phys. Rev. B*, **46**, 13592.
8 Lisowski, M., Loukakos, P.A., Bovensiepen, U., Stähler, J., Gahl, C., and Wolf, M. (2004) *Appl. Phys. A*, **78**, 165.
9 Rhie, H.-S., Dürr, H.A., and Eberhardt, W. (2003) *Phys. Rev. Lett.*, **90**, 247201.
10 Bovensiepen, U. (2007) *J. Phys. Cond. Matter*, **19**, 083201.

11 Perfetti, L., Loukakos, P.A., Lisowski, M., Bovensiepen, U., Berger, H., Biermann, S., Cornaglia, P.S., Georges, A., and Wolf, M. (2006) *Phys. Rev. Lett.*, **97**, 067402.

12 Allen, P.B. (1987) *Phys. Rev. Lett.*, **59**, 1460–1463.

13 Perfetti, L., Gloor, T.A., Mila, F., Berger, H., and Grioni, M. (2005) *Phys. Rev. B*, **71**, 153101.

14 Grüner, G. (1994) *Density Waves in Solids*, vol. 89, Frontiers in Physics, Addison-Wesley.

15 Eckstein, M. and Kollar, M. (2008) *Phys. Rev. B*, **78** (24), 245113.

16 Demsar, J., Forró, L., Berger, H., and Mihailovic, D. (2002) *Phys. Rev. B*, **66**, 041101.

17 Sugai, S. (1985) *Phys. Status Solidi B*, **129**, 13.

18 Perfetti, L., Loukakos, P.A., Lisowski, M., Bovensiepen, U., Wolf, M., Berger, H., Biermann, S., and Georges, A. (2008) *New J. Phys.*, **10**, 053019.

19 Ru, N. and Fisher, I.R. (2006) *Phys. Rev. B*, **73**, 033101.

20 Ru, N., Condron, C.L., Margulis, G.Y., Shin, K.Y., Laverock, J., Dugdale, S.B., Toney, M.F., and Fisher, I.R. (2008) *Phys. Rev. B*, **77**, 035114.

21 Brouet, V., Yang, W.L., Zhou, X.J., Hussain, Z., Ru, N., Shin, K.Y., Fisher, I.R., and Shen, Z.-X. (2004) *Phys. Rev. Lett.*, **93**, 126405.

22 DiMasi, E., Aronson, M.C., Mansfield, J.F., Foran, B., and Lee, S. (1995) *Phys. Rev. B*, **52** (20), 14516.

23 Laverock, J., Dugdale, S.B., Major, Z., Alam, M.A., Ru, N., Fisher, I.R., Santi, G., and Bruno, E. (2005) *Phys. Rev. B*, **71**, 085114.

24 Gweon, G.-H., Denlinger, J.D., Clack, J.A., Allen, J.W., Olson, C.G., DiMasi, E., Aronson, M.C., Foran, B., and Lee, S. (1998) *Phys. Rev. Lett.*, **81**, 886.

25 Schmitt, F., Kirchmann, P.S., Bovensiepen, U., Moore, R.G., Rettig, L., Krenz, M., Chu, J.-H., Ru, N., Perfetti, L., Lu, D.H., Wolf, M., Fisher, I.R., and Shen, Z.-X. (2008) *Science*, **321**, 1649.

26 Schmitt, F. *et al.* (to be published).

Part Five
Recent Developments and Future Directions

Dynamics at Solid State Surface and Interfaces Vol. 1: Current Developments
Edited by Uwe Bovensiepen, Hrvoje Petek, and Martin Wolf

ISBN: 978-3-527-40937-2

21
Time-Resolved Photoelectron Spectroscopy at Surfaces Using Femtosecond XUV Pulses

Stefan Mathias, Michael Bauer, Martin Aeschlimann, Luis Miaja-Avila, Henry C. Kapteyn, and Margaret M. Murnane

21.1
Introduction

Laser-based femtosecond light sources in the infrared, visible, and the near-ultraviolet spectral regime have been successfully employed for a large number of experimental studies of real-time dynamics of ultrafast surface processes, as is discussed at length in the preceding chapters. Visible and near-ultraviolet wavelengths, however, are capable of probing only a relatively restricted range of energies in the surface electronic structure – specifically electronic states within a few electron volts around the Fermi energy. Most of the surface-related time-resolved photoemission studies to date are, for instance, based on a two-photon photoemission process (the first (static) two-photon photoemission work was performed by Teich *et al.* [1]) and focus on the ultrafast dephasing and decay of electronic excitations in this optical energy range [2]. The use of femtosecond pulsed light sources in the extreme ultraviolet (XUV)[1] spectral region, in contrast, allows direct access to much more deeply bound electronic levels. Furthermore, the range of electron momentum that can be studied can be substantially expanded by the use of these light sources. Both aspects are highly attractive in the study of ultrafast processes since they provide new and relevant system information, for instance, on the chemical or magnetic state, or the structure, of a surface – information that is often not accessible in time-resolved experiments employing sources at lower photon energy.

The first experimental work showing that ultrafast light sources in the XUV and soft X-ray region of the spectrum were possible were conducted in the late 1980s. Studies by McPherson *et al.* [3] and Ferray *et al.* [4] were the first to clearly show the

1) The extreme ultraviolet regime covers the wavelength between 50 and 1 nm (20–1200 eV).

Dynamics at Solid State Surface and Interfaces Vol.1: Current Developments
Edited by Uwe Bovensiepen, Hrvoje Petek, and Martin Wolf

ISBN: 978-3-527-40937-2

potential for nonlinear optical high-order upconversion by observing harmonic radiation at wavelengths well below 100 nm and with a large range of harmonic orders of comparable brightness (i.e., in a *nonperturbative* interaction). The average flux, and thus the utility, of high-order sources improved dramatically with the use of very short-duration pulses (i.e., <10 optical cycles or ~ 25 fs) to drive the process [5]. State-of-the-art high harmonic sources can deliver coherent radiation from the XUV into the soft X-ray region of the spectrum [6, 7], and even into the keV regime [8], with a pulse duration of at most a few tens of a femtosecond and as short as 80 attoseconds [9–11]. Alternative sources of ultrashort short-wavelength pulses include femtosecond laser plasma sources (which first directly demonstrated the generation of ultrafast X-ray pulses [12–14]), free-electron laser facilities [15–17], femtosecond slicing of a synchrotron electron bunch [18–20], and X-ray lasers. The use of these ultrafast sources in a surface science photoemission experiment employing a stroboscopic (pump–probe) scheme has the potential to provide insight into the ultrafast dynamics of surface processes at a level of detail not possible heretofore.

In the early 1980s, picosecond light pulses in the far ultraviolet regime were used for the first time in photoemission experiments as a probe of carrier relaxation processes in semiconductors [21, 22]. The use of ultrashort-pulse XUV light for photoemission was pioneered by R. Haight who showed that high-order harmonic generation can be used as a light source for valence- and core-level photoelectron spectroscopy [23]. This work also recognized the ability to perform real-time experiments on a femtosecond timescale [24, 25], providing a new route toward the real-time study of ultrafast processes at surfaces.

This chapter is organized as follows: We will start with a brief introduction to the technique of high harmonic generation (HHG) and the basic principles of operation of free electron lasers (FEL) (Section 21.2). HHG and FEL are the two most appropriate short-pulse XUV sources for time-resolved photoemission experiments by virtue of their ability to generate ultrashort XUV pulses continuously at a relatively high repetition rate; however, X-ray laser sources with subpicosecond pulse duration have also been used for photoemission spectroscopy where a high-peak flux source is required [26, 27], and laser plasma sources have been used for photoemission spectroscopy [28, 29]. In Section 21.3, further important experimental aspects of the time-resolved photoemission technique will be discussed. This section includes the description of the basic scheme of the setup, the choice of suitable wavelength-selective optics, and the characterization of the XUV pulse length available in the experiment. Also addressed will be the effect of space charging, an issue that can become critical in photoemission experiments using low repetition rate light sources and that particularly limits the use of the intense light pulses from FEL sources. In Section 21.4, we will review pioneering and recent photoemission experiments using short-pulse XUV sources. Time-resolved studies are the primary focus of this section, but selected results from static photoemission experiments that illustrate general capabilities are also included. We will conclude the article with the discussion of some future perspectives of the XUV-based time-resolved photoelectron spectroscopy.

21.2
Femtosecond XUV Sources

21.2.1
High Harmonic Generation

In high harmonic generation, atoms exposed to a strong laser field can emit harmonic radiation at photon energies in the extreme ultraviolet regime. Using intense visible or infrared femtosecond light pulses to drive the process, a broad spectrum of harmonics extending into the "water window" and keV energies can be generated [6–8]. At first glance, this process is the extension of traditional nonlinear processes to very high orders, making use of extremely intense pump lasers. However, the spectrum of HHG emission is qualitatively different from that observed in the past work on harmonic generation in gases using nanosecond or picosecond lasers [30] in that a large number of (odd) harmonic orders are observed, with a large number of these orders exhibiting comparable intensity, that is, a "plateau" region, followed by a rather abrupt cutoff. This plateau behavior of the harmonic spectrum indicates that in contrast to past work where the generation process can be described through perturbative nonlinear optics, the HHG process is nonperturbative.

An intuitive physical interpretation of HHG, developed in quantum simulations by Kulander *et al.* [31, 32], and formulated in a classical trajectory pictured by Corkum [33], explains HHG as the consequence of ionization of the medium in a three-step "rescattering" model (Figure 21.1). First, the Coulomb barrier binding

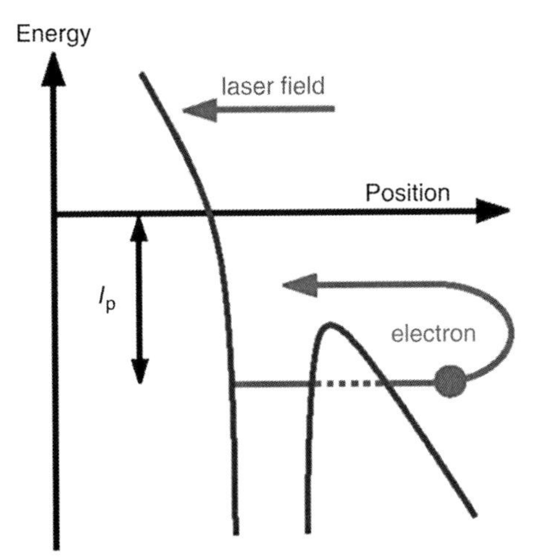

Figure 21.1 Illustration of the three-step model for high harmonic generation. An electron tunnels out with the help of a strong laser field suppressing the Coulomb potential inside the atom. The free electron is then accelerated by the laser field and can come back and recombine with the parent ion, emitting a high-energy photon. The photon energy is a sum of the ionization potential I_p and the electron's kinetic energy obtained in the laser field. (Please find a color version of this figure on the color plates.)

the electron to the atom is suppressed by the strong laser field, resulting in quantum mechanical tunnel ionization. Once free, the electron then accelerates in the laser field, gaining kinetic energy. When the laser field reverses, the electron can re-encounter the parent ion, and in the third “recollision” step, there is a finite probability that it will recombine with the ion and emit a high-energy photon.

Whether or not the electron will return to the parent ion and at which kinetic energy depends on the phase of the light field when the electron tunnels out. Solving the equation of motion for the electron in the oscillating electric field of the laser yields a maximum electron kinetic energy on return of $3.17\,U_{\mathrm{p}}$. The ponderomotive energy $U_{\mathrm{p}} = I/4\omega^2$ (atomic units) is the mean kinetic energy of a free electron oscillating in the laser field and I and ω are the intensity and frequency of the laser field, respectively. Thus, the energy of the highest harmonic emitted is

$$\hbar\omega|_{\max} = I_{\mathrm{p}} + 3.17\,U_{\mathrm{p}}.$$

Ionization of the medium occurs in bursts near the times within the cycle when the field peaks, that is, twice per cycle. The Fourier relationship thus dictates spectral peaks spaced by 2ω, that is, ω, 3ω, 5ω. Furthermore, because of the extreme nonlinearity of the HHG process, the HHG light pulse duration is considerably shorter than the driving laser light pulse [34, 35]. This fact means that the Fourier “harmonic” picture is only a first-order approximation. The “intrinsic” atomic phase [36, 37] (i.e., the phase advance of the electron wave function during its path between ionization and recombination, which can be $\gg\pi$) and ionization effects can substantially alter the spectrum and shift the peak positions. This alteration of the spectrum must be considered in any experimental setup. The atomic phase can be used to manipulate the spectrum of HHG in useful ways [37–39]. For sufficiently short driving pulse duration, the emission at high energy can be confined primarily to a single attosecond burst [35, 40].

The development of few optical-cycle solid-state lasers using Ti:sapphire was instrumental in the practical application of HHG [41–43]. The utility of HHG as a light source dramatically advanced with the use of very short-duration pulses for two reasons: first, since HHG is the result of field ionization and will terminate when the medium is fully ionized, the efficiency improves with increasing ratio of peak-to-total pulse duration, that is, with shorter driver pulse duration. This observation provided a direct confirmation of the basic validity of the ionization mechanism of HHG [5]. Second, phase-matching conditions are dramatically improved with the use of very short-duration pulses, resulting in optimally efficient generation of harmonic light (Figure 21.2). Phase matching means that the driving laser and the generated harmonics travel with the same phase velocity, resulting in constructive buildup of the signal field over an extended propagation distance. Since the (real part of) refractive index in the extreme ultraviolet wavelength region is less than unity, high harmonics travel at phase velocities faster than c (speed of light in vacuum). As a result, free space propagation of the laser light will not travel at the same speed as the harmonics. However, the medium gets ionized by the laser as it propagates. Since the index of refraction of the neutral gas is positive and that of the ionized plasma is

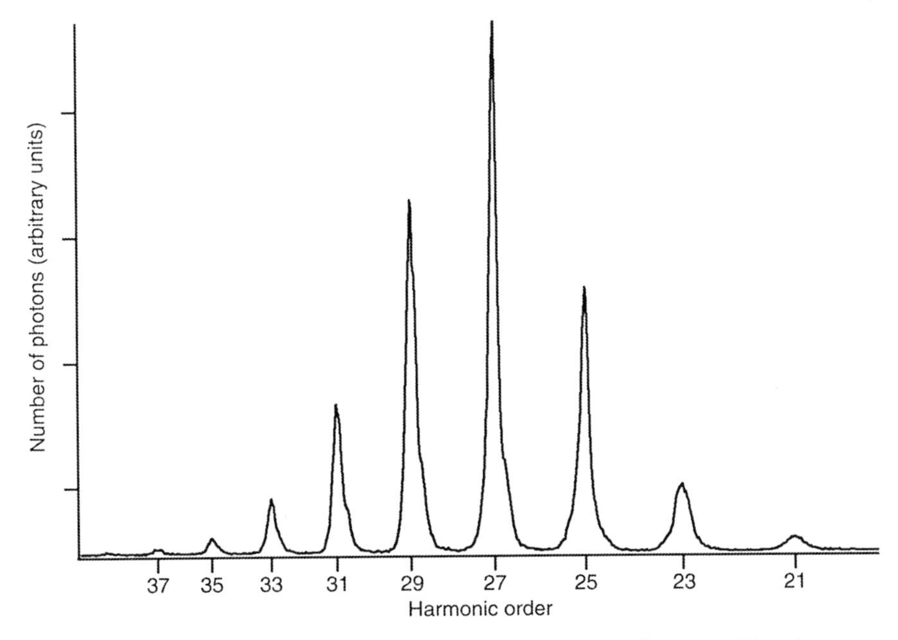

Figure 21.2 Phase-matched HHG spectrum generated in approximately 20 torr of Ar using a waveguide geometry.

negative, the index of refraction of the gas for the driving laser is dynamically changing, and phase matching of the emission can occur provided the required $I_p + 3.17\,U_p$ intensity is reached while the medium is still only weakly ionized, that is, using very short duration pulses. This method of phase matching was first realized in the case of HHG in a guided wave geometry [44–46]. For a sufficiently short pulse, light of ionizing intensity can be reliably guided in a gas-filled hollow capillary without damage to the structure. In this case, the k-vector of propagation is

$$k_{\text{laser}} \approx \frac{2\pi}{\lambda} + \frac{2\pi N_a \delta(\lambda)}{\lambda} - N_e r_e \lambda - \frac{u_{nm}^2 \lambda}{4\pi a^2},$$

where the first term corresponds to simple vacuum propagation and the second and third terms result from dispersion of the gas and the plasma created by ionization. Here, N_a is the atom density, N_e is the electron density, λ is the vacuum wavelength, and δ depends on the dispersive characteristics of the atom. The last term is due to the waveguide, where a is the radius of the waveguide, and u_{nm} is a constant corresponding to the propagation mode of the fiber. Phase matching corresponds to $\Delta k = qk_{\text{laser}} - k_{\text{EUV}} = 0$, where q is the harmonic order. The gas-filled hollow waveguide provides a stable mode over an extended propagation distance and allows for careful control to engineer the phase-matched conditions including the wavelength, gas pressure, gas species, waveguide size, and the spatial mode. It also allows pressure tuning where the neutral gas index balances the waveguide dispersion. More generally, however, this phase matching occurs even in free space plane wave propagation of the driving laser as the level of ionization passes through a critical ionization value. In the general case of a laser focused in free space, the waveguide term in the above expression must be replaced by the Guoy phase, which varies through the focus. The free focus geometry was considered, neglecting neutral gas terms, in Ref. [47], and resulted in a spatial profile driven by local phase-matching conditions that vary through the focus. The rapidly varying intrinsic phase as the intensity varies must also be considered, and the result is that optimal generation in a free focus geometry requires a gas medium that is well confined to a region of propagation corresponding to a small fraction of the confocal parameter that is optically "thick" for the desired wavelength, that is, corresponds to a density length product of a few absorption depths. Where the efficiency is severely limited by absorption of the XUV (such as using neon to generate light in the approximately $<$ 100 eV region), sometimes this limit can be approximated in a free gas jet geometry (although care must still be taken ensure the gas medium ends *before* the beam intensity falls below what is necessary to generate the desired harmonic order). In practice, most gas jet or gas cell setups are not optimized to near maximum conversion. More generally, for example, for argon gas in the region of approximately 40–50 eV, or for HHG using helium gas, the required density-length product is not practical to engineer in a gas jet. Using argon, optimal efficiency can be realized using a hollow waveguide and approximately 0.3 mJ laser energy [44], or in a gas cell with a very long focus geometry ($f = 5$ m) using 5–10 mJ [48]. Although laser pulse energy in the 0.5–5 mJ range can be used in a free focus geometry even for HHG in argon (at suboptimal conversion efficiency), the waveguide geometry makes it possible to obtain optimal conversion efficiency in a wide range of conditions in a compact geometry (i.e., $\sim$1 m beamline) using pulse energies readily obtainable using kHz and higher repetition rate lasers, resulting in an efficiency improvement of up to a 100-fold at these pulse energies. In addition, the waveguide is an optimal configuration for differential pumping of the gas, with a typical gas load many orders of magnitude smaller than for a gas jet and several orders of magnitude lower than for a typical static cell (and corresponding reduced gas usage, which can be significant for some noble gases). Thus, ultrahigh-vacuum (UHV) implementation can be done by pumping an intermediate chamber with a small ($\sim$70 l/s) turbo pump coupled to the UHV system through a thin metal filter (which also serves to reject the copropagating laser light).

Expanding the useful range of photon energy that can be generated using HHG relies on continued progress in understanding how to phase match the process. Recent years have seen rapid progress in this area, with the development of new techniques for quasiphase matching using periodic structures in the medium [49–51], or counter-propagating light pulses [39] that periodically modulate the emission and can thus facilitate coherent signal buildup at higher energies than is possible through simple phase matching. Other recent work has shown that driving HHG using longer wavelengths can shift the region of phase matching to higher photon energy, with conversion efficiencies into the water window and keV region of the spectrum predicted to be comparable to those obtained in the XUV [52]. These future developments will continue to expand the capabilities of laser-based photoemission.

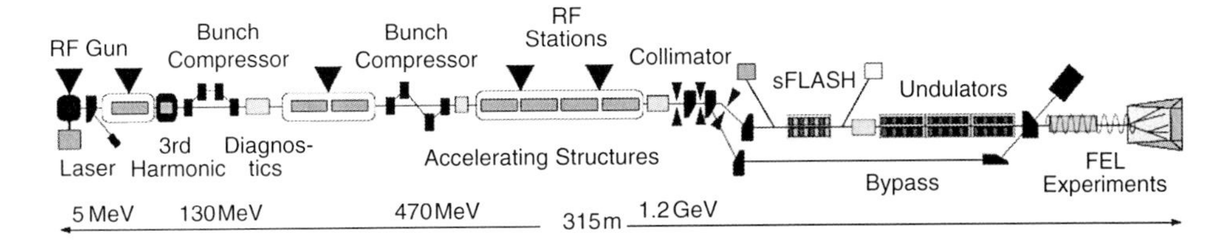

Figure 21.3 Schematic layout of the VUV-FEL at DESY, Hamburg, in an upgrade configuration as expected to be in operation by spring 2010 (from Ref. [53]).

21.2.2
Free Electron Laser

The principle of the free electron laser has been known since the early 1970s [15]. The first FEL operation, emitting at a wavelength of 3.4 μm, was demonstrated only a few years later [54]. However, only in recent years has technical progress made it possible to implement an FEL source operating in the XUV spectral regime [16]. Because of their high brilliance,[2)] these sources are also referred to as fourth-generation synchrotron sources. Furthermore, the generated XUV light is emitted in pulses that are predicted to be as short as 10 fs.

The configuration of the FEL source FLASH operated by the synchrotron facility DESY in Hamburg, Germany, is shown in Figure 21.3 and described, for instance, in Ref. [55]. Three main sections of the FEL can be distinguished: a photocathode inside a radiofrequency (RF) cavity and driven by a picosecond UV laser generates short electron bunches – the so-called macrobunches – at a repetition rate of 2.25 MHz. These bunches are injected into a linear accelerator and accelerated to kinetic energies of up to 1 GeV (at the time this chapter was written). The intense and highly coherent XUV pulses are then generated in the following 30 m long undulator, a periodic array of magnets that is also used in a much shorter configuration in conventional synchrotron radiation sources for the generation of high brilliance monochromatic X-ray light. However, one main difference in the operational mode of the undulator at the FEL is that the light emission arises from small electron packets, the microbunches, that evolve along the undulator line within the macrobunch. The microbunches are confined to a size that is shorter than the XUV wavelength. The resulting emission of a single microbunch is fully coherent, and the light intensity increases with the number N of microbunch electrons as N^2.

In more detail, incoherent XUV radiation, generated because of spontaneous emissions by the wiggling electrons at the undulator entrance, interacts with the propagating macrobunch. This interaction gives rise to the evolution of longitudinal electron density modulations (see Figure 21.4). The distance between these micro-bunches is identical to the wavelength of the emitted XUV light, so that the coherent emission from a single microbunch can now constructively interact with the

2) The term brilliance is defined as the number of photons per unit time emitted into a given bandwidth interval per angle of beam spread and source area.

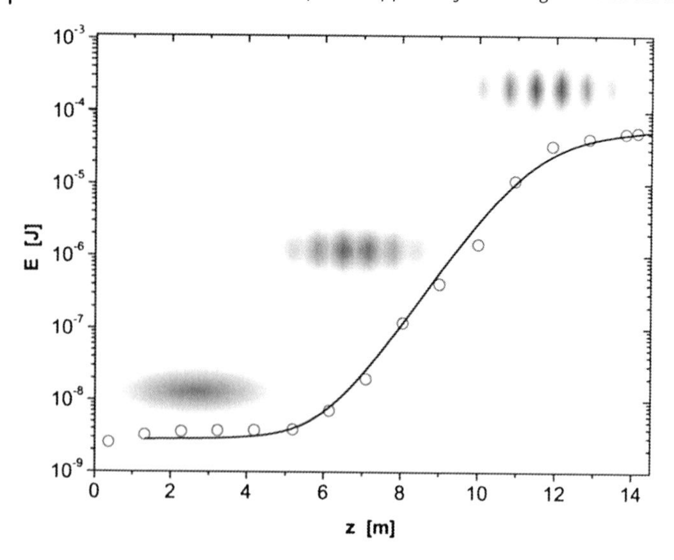

Figure 21.4 Increase in the XUV pulse energy along the undulator measured at a wavelength of 98mm at FLASH. Also shown is an illustration of the microbunch evolution in feedback to the interaction with the emitted light. Reprinted from [56] with kind permission of Springer Science & Business Media.

emission of the neighboring microbunches. As shown in Figure 21.4, this interaction gives rise to an exponential increase in the radiated XUV intensity until a complete bunch modulation is achieved and saturation sets in.

As the amplification starts from radiation noise at the undulator entrance, this FEL mode is referred to as self-amplified spontaneous emission (SASE) operation. An alternative seeding of the amplification process by a laser-based coherent high harmonic source is planned for implementation in future FEL light sources. This configuration will significantly reduce the pulse-to-pulse fluctuations in intensity and timing that are characteristic for the SASE process starting from noise.

The XUV light pulses emitted by HHG sources and FELs exhibit differing characteristics that make these sources complimentary for different types of experiments. Table 21.1 compares the XUV output of a state-of-the-art HHG source with

Table 21.1 Characteristic source parameters at a photon energy of about 50 eV for a laboratory-based state-of-the-art HHG system and the FLASH at DESY.

	HHG	FEL FLASH
Photon energy	45 eV	52 eV
Photon flux/pulse	$\sim 10^7$–10^8	$\sim 10^{12}$–10^{13} [53]
Pulse length	Typical: 5–50 fs demonstrated: 80 as	10–50 fs
Timing accuracy	<1 fs	~200 fs [57]
Repetition rate	3 kHz	≤ 500 pulses/s
Availability	Permanent	Restricted by available beam times

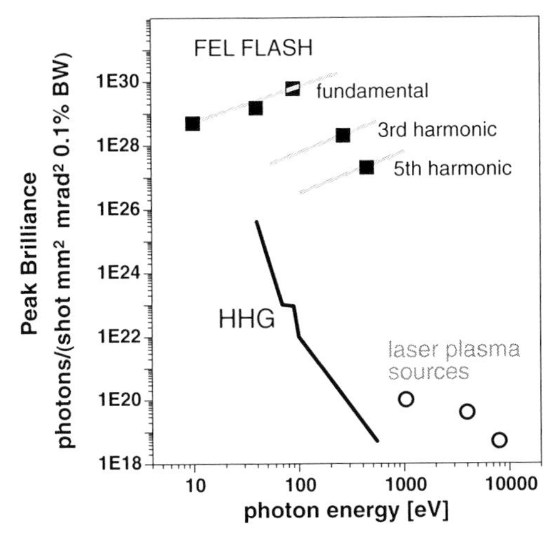

Figure 21.5 Comparison of the peak brilliance of HHG sources and the FEL FLASH as achievable over a wide range in the XUV regime. In addition, typical peak brilliance values of laser-based femtosecond XUV plasma sources are included. Data are taken from Refs [16, 58].

corresponding values from the FEL FLASH at a photon energy of about 50 eV. With respect to the photon flux per pulse, the FEL beats the laser-based source by several orders of magnitude. However, particularly in this photon energy regime, HHG sources have quite successfully proven that the delivered photon flux is adequate to perform even time-resolved photoemission experiments, which require many spectra at a varying pump–probe delay in a reasonable acquisition time. The per-pulse flux available with a HHG source is often, in fact, appropriate to avoid space charge effects. In photoemission experiments on solid surfaces, the FEL pulse energy is likely to require attenuation to avoid this effect. The HHG source repetition rate is dictated by the driver laser, with rapid technological advance in this area immediately translating into higher repetition rate [59] and overall flux. Furthermore, in terms of pulse width and timing accuracy in a visible–XUV pump–probe experiment, HHG is clearly superior to the FEL. However, the photon energy range available with FELs is much larger, with the LCLS FEL at SLAC recently operating at $h\nu = 8$ keV. As can be seen from Figure 21.5, the intensity from current HHG sources rapidly decreases beyond approximately 120 eV, and reasonable photoemission experiments are not feasible, yet. Thus, FEL sources are likely to provide unique experimental access for time-resolved deep core-level experiments. Furthermore, the very large per-pulse energy of the FEL makes it possible, in principle, to spectrally filter the emission for high-energy resolution studies of picosecond or subpicosecond dynamics. For completeness, we also include peak brilliance values obtainable from laser-based plasma sources, which can provide X-ray pulses up to several keV photon energy. The relatively low yield has dissuaded implementation of time-resolved photoemission experiments with these sources to date, although plasma sources using ns-duration drive lasers have been used for photoemission [28, 29].

To summarize, HHG sources in time-resolved photoemission are ideally suited for the study of ultrafast processes taking place in the sub-100 fs time regime by monitoring transient changes in the valence electron states and the shallow core levels. On the other hand, FEL sources enable the access to a much broader spectrum of core levels and are particularly useful for time-resolved core-level and valence-level studies that rely on a high-energy resolution, or seek to study dynamics in dilute targets such as clusters, as has been done using photoionization studies employing long-pulse XUV laser sources [60].

21.3
Photoelectron Spectroscopy Using XUV Pulses: Some Technical Aspects

The technical realization of an XUV time-resolved photoemission experiment designed to study ultrafast surface dynamics combines femtosecond optics and sophisticated ultrahigh-vacuum surface science technology. At the interface between an XUV light source and a UHV analysis chamber, important issues such as wavelength selection at minimum pulse temporal distortion, appropriate XUV-focusing conditions, and isolation of the gas load of the XUV light source from the UHV must be considered. Furthermore, to facilitate stroboscopic pump–probe studies with femtosecond resolution, the setup must provide appropriate tools for obtaining temporal and spatial overlap between the pump and the probe pulses. Some of these technical aspects will be addressed in Section 21.3.1.

The quantitative analysis of the time-resolved photoemission data requires knowledge of the temporal profiles of pump and probe pulses. Particularly, the experimental characterization of XUV pulse profiles can be a challenging task. The laser-assisted photoelectric effect (LAPE) has emerged as a powerful tool in this context and will be discussed in Section 21.4.2.3. We will also present promising developments devoted to the specific demands in the characterization of FEL-generated XUV pulses.

Strong electromagnetic fields absorbed by a surface can generate intense bunches of photoemitted electrons, which become significantly distorted during their propagation (i.e., toward the electron energy analyzer) by Coulomb repulsion between the electrons. Such parasitic space charge effects can result in loss of both energy and angular resolution, and will be topic of Section 21.3.3.

21.3.1
Time-Resolved PES and Angle-Resolved PES

21.3.1.1 Experimental Setup

The extreme time resolution offered by the femtosecond-to-attosecond temporal structure of XUV pulses can be exploited only using all-optical pump–probe schemes. An example of an experimental setup for time-resolved PES is shown in Figure 21.6. The XUV source used here is a hollow waveguide high harmonic source driven by a state-of-the-art Ti:sapphire multipass amplifier or regenerative amplifier, ideally

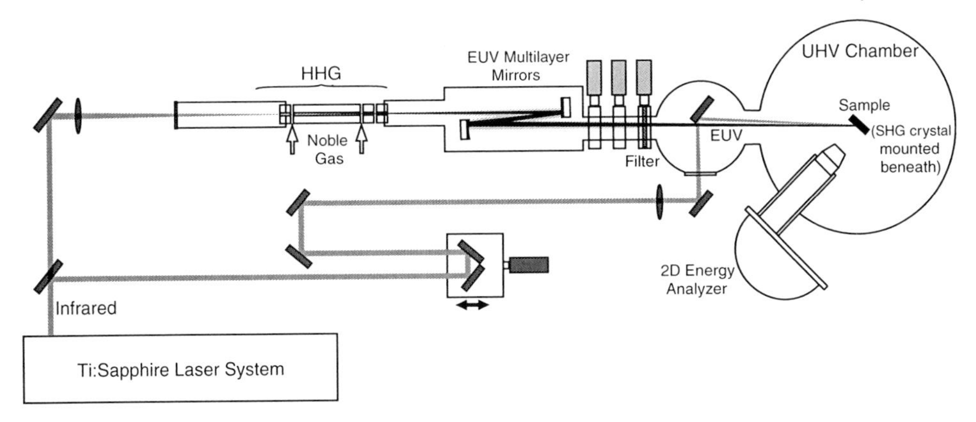

Figure 21.6 Experimental setup, taken from Ref. [62]. The setup consists of a Ti:sapphire amplifier laser system, an HHG light source, a multilayer double-mirror monochromator, a 2D energy analyzer for parallel energy and momentum detection, or a TOF detector. For time-resolved pump–probe spectroscopy, part of the fundamental laser beam is separated with a beam splitter and directly focused onto the sample after passing a computer-controlled optical delay stage. Reprinted with permission from [62]. Copyright 2007, American Institute of Physics.

delivering sub-30 fs light pulses at an energy in the mJ regime and a repetition rate of several kHz [42]. In this setup, an optical beam splitter creates separate optical pump and probe beams prior to the XUV generation. An alternative technique, which is particularly useful for attosecond studies, is to generate high-order harmonics using the full laser output, and then subsequently separating XUV and a fraction of the remaining infrared light using a concentric mirror arrangement [61]. XUV multilayer mirrors are used in either setup to select a specific harmonic order, or a range of photon energies, to be used in the experiment. In the setup of Figure 21.6, one of the multilayer mirrors is concave to focus the XUV light onto the sample in the center of the UHV chamber. A small focal spot for the XUV light is of particular relevance to angular-resolved photoelectron spectroscopy (ARPES), which provides additional information about the electron momentum distribution. Residual copropagating light from the laser fundamental and low harmonics that are still reflected in part by the multilayers are blocked by ultrathin metal filters (typical thickness 100–500 nm), made, for example, of aluminum or zirconium (for experiments with, for example, 42 or 90 eV, respectively) at the exit of the monochromator. As a side effect, the metal filter is also very effective as a differential pump stage. In the present geometry, the pressure in the UHV chamber rises to a value of about 5×10^{-10} mbar when the HHG light source is in operation.

A helpful tool to determine temporal and spatial overlap for pump–probe measurements is a second harmonic crystal mounted inside the UHV chamber that can be positioned *in situ* at the sample position. Cross-correlation second harmonic (SHG) signal measurements between the optical pump pulse and the transmitted fundamental light pulse used for HHG provide information on the quality of the lateral adjustment of pump and probe light with respect to each other and the position of time zero. The temporal overlap between the optical pump and the XUV probe pulse can be determined from a photoemission measurement based on LAPE. LAPE allows for a quantitative temporal characterization of the XUV light, as detailed in Section 21.3.2.

21.3.1.2 Wavelength Selection

Photoelectron spectroscopy relies on a photon source exhibiting a limited bandwidth, that is, ideally significantly narrower than the electronic spectral signature to be probed. Therefore, selection of a finite energy window from the broad XUV source spectrum is necessary. On the other hand, the time–bandwidth limit dictates that narrowing the spectrum limits the shortest time duration of the XUV pulse and, therefore, the obtainable time resolution in a photoemission experiment. As an illustration, Figure 21.7 shows the energy broadening as a function of the temporal width for a photon energy of $h\nu = 50$ eV. A compromise between these two constraints (energy resolution and temporal resolution) must be found.

Using reflection gratings is the most common approach for wavelength selection in the XUV regime. In a typical monochromator design, the gratings are arranged in a toroidal configuration that focuses the XUV light onto the sample. The particular advantage of a grating monochromator in comparison to alternative selection techniques is the flexibility in tunability over a wide photon energy range. However, the diffraction of a specific wavelength by the grating results in a tilt of the wave front with respect to the propagation direction and gives rise to a temporal broadening of the pulse [63]. This effect can be quite critical for sub-100 fs pulses. The use of gratings in an optimized configuration is, however, a reasonable compromise if one is interested in processes on timescales of 100 fs and longer [64]. More recently, double-monochromator designs have been demonstrated that allow both energy selection and very short (< 10 fs) near-time–bandwidth limited pulse outputs – at the cost of complexity of the optical system [65, 66].

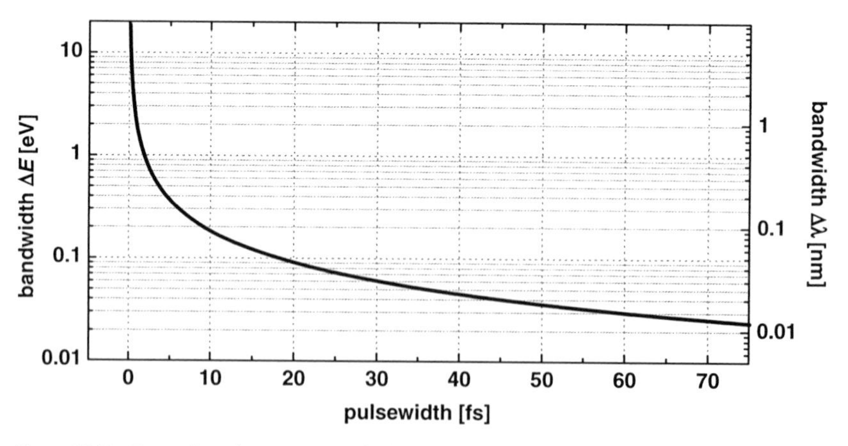

Figure 21.7 Energy broadening (ΔE and $\Delta\lambda$) as function of the temporal width of a Gaussian centered at $E = 50$ eV ($\lambda = 24.8$ nm).

An alternative approach to restrict the spectral bandwidth of the harmonics is the use of XUV multilayer mirrors [62, 67–69]. These optical components are equivalent to dielectric filters in the visible wavelength regime. They consist of a layered structure in which the optical constants vary periodically with depth, resulting in selective reflection of electromagnetic waves in a particular wavelength range. In contrast to diffraction gratings, wavelength selection results from interaction with the periodicity perpendicular to the surface plane and reflection into the zeroth diffraction order (specular reflection). Therefore, the outgoing ray does not exhibit the wave front tilting seen with conventional gratings, and temporal broadening of the reflected XUV pulse is not present. An example where multilayer mirrors have been used for the selection of a single harmonic from a HHG spectrum is given in Ref. [62]. Multilayer mirrors consisting of a stack of 20 alternate layers of molybdenum and silicon are used. The thicknesses of the layers have been chosen to maximize the reflectivity at a wavelength of about 30 nm for near-normal incidence [70]. These values fit rather closely to the 27th harmonic of the 800 nm laser light used for harmonic generation. The selectivity of the monochromator is further enhanced by a *Z*-fold double mirror setup (see Figure 21.6). The reflectivity as function of photon energy of this monochromator arrangement calculated from the reflectivity data of the individual mirrors is shown in the vicinity of the 27th harmonic in Figure 21.8. A maximum selectivity is achieved for an incidence angle of about 85°. The monochromator arrangement exhibits an extinction ratio of 2000 with respect to the 25th harmonic and an extinction ratio of 200 with respect to the 29th harmonic. The expected bandwidth of the transmittance is about 1.9 eV (full width at half maximum, FWHM), a value that is capable of supporting pulses shorter than 1 fs.

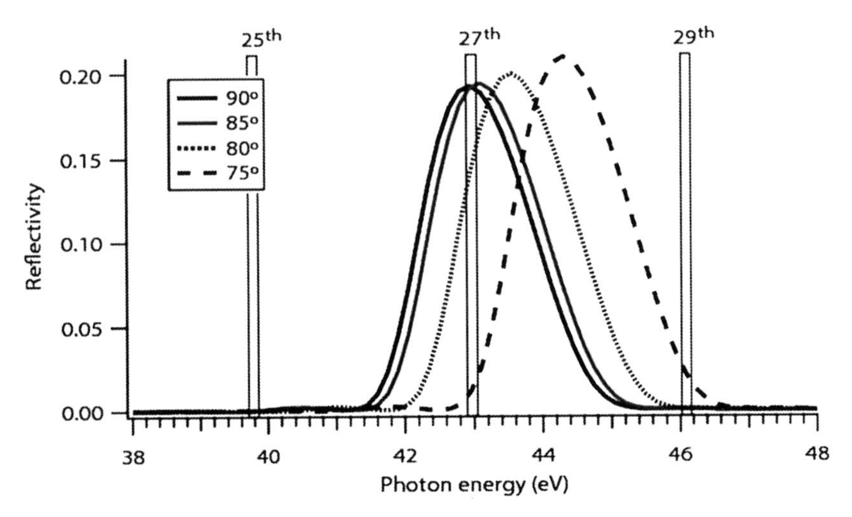

Figure 21.8 Calculated reflectivity of the double-mirror monochromator setup. Reprinted with permission from [62]. Copyright 2007, American Institute of Physics. The bright gray line corresponds to the experimental configuration (incidence angle of the harmonic light of 85°).

21.3.2
XUV Pulse Profile Characterization

For a quantitative analysis of time-resolved photoemission traces, an accurate knowledge of temporal XUV pulse profiles is highly desirable. Pulse characterization by a nonlinear XUV autocorrelation technique that is based on a direct two-photon ionization process in rare gases has been proven to be successful up to a maximum photon energy of 54 eV [71–73]. However, this technique relies intrinsically on high XUV intensities not necessarily available with conventional laboratory-based HHG sources.

Cross-correlation measurements between an XUV pulse and a well-characterized pulse in the optical regime have proven to be an effective alternative approach for XUV pulse characterization. In particular, LAPE can be used to provide femtosecond to attosecond resolution. LAPE was first used for XUV pulse characterization in a gas-phase photoemission experiment by Glover *et al.* [74]. Miaja-Avila and coworkers later reported the observation of LAPE in photoemission from surfaces [75–77]. In a laser-assisted process, illuminating an atom simultaneously with an XUV pulse and an optical pulse gives rise to characteristic sidebands in the photoemission spectrum. These sidebands arise from the simultaneous absorption and emission of visible photons along with the XUV photon. Consequently, they are present only when the two beams overlap in time. Figure 21.9 shows an experimental cross-correlation LAPE trace between an infrared pump (35 fs) and an XUV probe pulse from a Pt(111) surface. The trace was generated by plotting the intensity of the LAPE sidebands next to the platinum 5d-lines as a function of temporal delay between the XUV and light pulses. From this trace, the temporal width of the XUV pulse could be determined to about 10 ± 1fs. A more detailed discussion on LAPE follows in Section 21.4.2.3.

Another promising cross-correlation technique was recently presented by Gahl *et al.* [78]. This technique employs a transient change in reflectivity of a GaAs surface

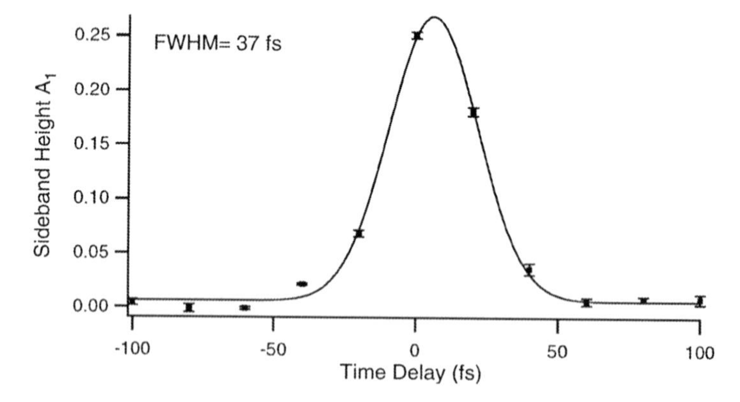

Figure 21.9 Measured cross correlation of an infrared (35–40 fs) and approximately 10 fs XUV pulse that was determined by plotting the height of LAPE sidebands in the photoemission spectra as a function of time delay between XUV and IR pulses. Reprinted with permission from [75]. Copyright 2006 by the American Physical Society; the FWHM of the cross correlation trace is mainly governed by the temporal width of the infrared pulse.

in the visible spectral regime induced by an XUV pulse. This technique was used to determine the temporal characteristics of the pump–probe setup at the FEL facility FLASH. The measured width of the XUV-induced drop in reflectivity of about 160 fs was limited by the pulse length of the long optical laser pulse (120–150 fs). A quantitative measure of the (much shorter) XUV pulse width was not yet possible in this experiment.

21.3.3 Efficient Detection Schemes for XUV Time- and Angle-Resolved Photoemission

The extension of conventional photoelectron spectroscopy into momentum space by means of ARPES allows unique access to the bulk or surface electronic band structure, $E(k)$, of solids. ARPES has emerged as a leading experimental technique in identifying static key properties of complex phenomena such as high-temperature superconduction [79], spin–orbit splitting [80–82], or condensed matter properties associated with reduced dimensionality of electronic systems [83, 84]. Femtosecond time- and angle-resolved photoemission spectroscopy in the ultraviolet regime has been applied so far to the study of ultrafast dynamics associated with transient charge density wave melting in $TbTe_3$ [85]. In this work, described in more detail in chapter IV.7 of this book, the fourth harmonic (6 eV) of a high-repetition rate ($\sim$200 kHz) Ti: sapphire laser system was used as the probing light. The application of photon energies in the XUV regime for these kinds of studies offers intriguing prospects for the investigation of ultrafast dynamics in complex electron systems. Next to the extension of the binding energy range, the use of XUV pulses will also substantially extend the accessible momentum range in a photoemission experiment compared to pulses in the vacuum ultraviolet (VUV).[3] This effect is illustrated in Figure 21.10, which displays ARPES data from ultrathin silver films grown on Cu(111), recorded at photon energies of 6 eV (fourth harmonic of the Ti:sapphire oscillator, Figure 21.11a) and 21.22 eV (He I line of the discharge VUV lamp, Figure 21.11b), respectively. The data were taken at identical settings of the used electron energy analyzer. In particular, the acceptance angle for the photoemitted electrons was kept constant at a value of $\pm 7°$. The plots map the electron kinetic energy distribution as function of $k_{\|}$, the electron momentum parallel to the surface. The series of dispersive states visible in both maps arise from the Shockley surface state (topmost feature) and quantum well states (higher binding energy states) confined in the silver film. While the VUV pulse probes only a momentum range in the very vicinity of the brillouin zone center ($k_{\|} = 0\,Å^{-1}$), the data recorded with the XUV pulse cover a momentum range equivalent to $\pm 0.25\,Å^{-1}$of the brillioiun zone for a detection angle of $\pm 7°$. This extension is a direct consequence of the relation between the emission angle θ, the kinetic electron energy E_{kin}, and $k_{\|}$, which is given by

3) In the definition used here, the term vacuum ultraviolet covers the wavelength regime between 100 and 200 nm.

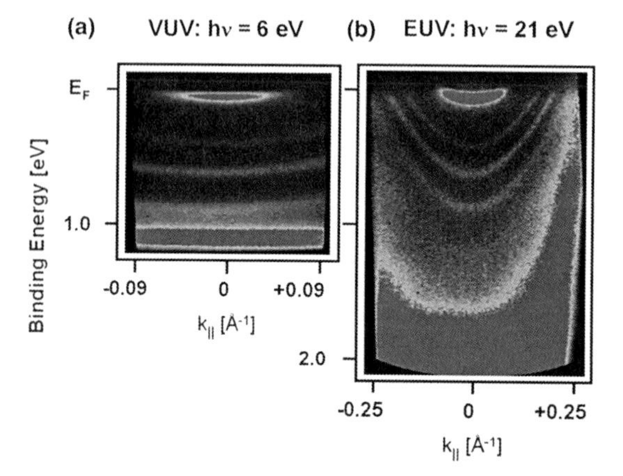

Figure 21.10 Angle-resolved photoemission spectra of the Shockley surface state and a series of quantum well states of 40 ML Ag/Cu(111) recorded at photon energies of (a) 6 eV (fourth harmonic of the Ti:sapphire oscillator) and (b) 21.22 eV (He I line of a discharge VUV lamp).

$$k_{||} = \sqrt{\frac{2m_e}{\hbar^2} E_{kin}} \cdot \sin(\theta). \tag{21.1}$$

Because of the relatively low pulse repetition rates available from high harmonic sources and FEL facilities, reasonable time-resolved XUV ARPES experiments rely on detection schemes that enable an efficient collection of the available information on energy, momentum, and time. For instance, in this approach hemispherical energy analyzers equipped with imaging detectors have been used in conventional ARPES for more than a decade and allow a parallel detection of the electron kinetic energy and one momentum component of the emitted electrons [86]. The development of new electron energy analyzer schemes based on multidimensional detectors is a promising approach in this direction. For example, a more recent scheme is based

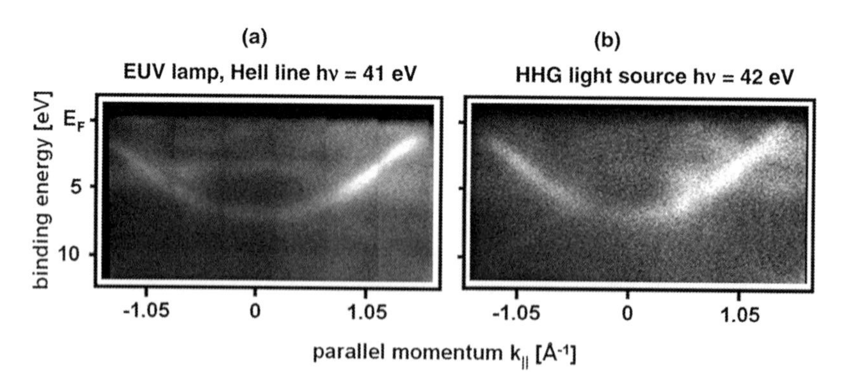

Figure 21.11 Photoemission maps of the Pt(111) surface band structure at a finite angle with the He II line (a) and the approximately 7 fs pulsed HHG light source, following selection of the 27th harmonic (b).

on the electron time-of-flight (eToF) technique and makes use of the advantages of imaging delay-line detectors [87, 88]. This configuration even allows for parallel access to the kinetic energy and both $k_{\|}$ components.

First studies using these types of analyzers in photoemission with femtosecond XUV light pulses were discussed by S. Mathias *et al.* [62, 89]. In their work, they used an imaging detector-type hemispherical analyzer and compared photoemission data recorded using 7 fs XUV pulses ($h\nu = 41.85$ eV) from a high harmonic source and corresponding data recorded with the He II line of a gas discharge lamp ($h\nu = 40.81$ eV). Figure 21.11 shows corresponding photoemission intensity maps from a Pt (111) surface recorded with the He discharge lamp (Figure 21.11a) and the HHG source (Figure 21.11b). Both maps were recorded at identical analyzer settings with an energy resolution of 400 meV and an angular resolution of 0.2°. The data cover an overall $k_{\|}$-range of about ± 1.5 Å^{-1}. Clearly visible is a substantial energy broadening of the spectral signatures in the HHG spectrum compared to the He II spectrum. This energy broadening is due to the larger bandwidth of the ultrashort HHG light pulses. A quantitative analysis shows, however, that the achieved momentum resolution is not affected by the use of HHG light at all.

21.3.4
Space Charge Effects

The Coulomb repulsion within a short electron bunch as, for instance, generated in a photoemission process by an ultrashort light pulse, can distort the energy distribution of the electrons and result in a considerable loss in both spectral resolution and angular resolution [90–95]. Figure 21.12 illustrates the extent to which the effect of Coulomb interaction, also referred to as space charging, can affect

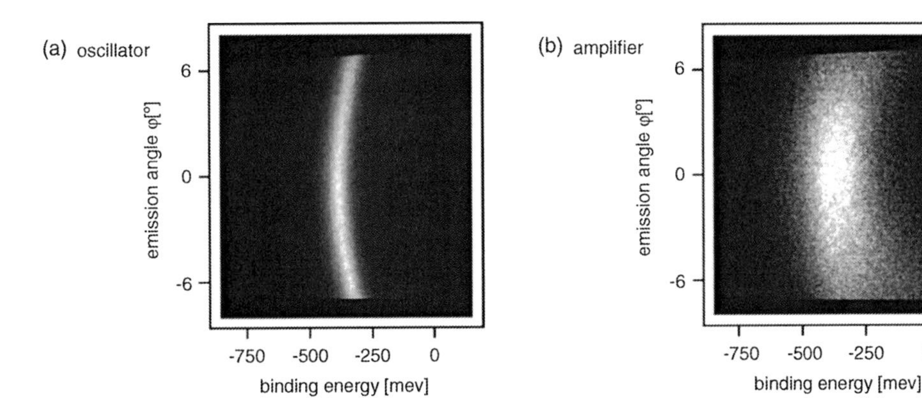

Figure 21.12 (a) 2PPE $E(k)$ photoemission map of the Cu(111) surface recorded using the second harmonic of a 80 MHz laser system. The main signal is due to the dispersing Shockley surface state. (b) Equivalent 2PPE map recorded with the second harmonic of a 1 kHz laser system. A total of about 80 000 e/incident laser pulse are emitted from the surface in this experiment. Clearly visible is a drastic distortion of the spectral distribution in comparison to (a). Reprinted with permission from [96]. Copyright 2006, American Institute of Physics.

the (angle-resolved) photoelectron spectral distribution. The data are taken from Ref. [96] and show the angle-resolved two-photon photoemission signal from the Cu (111) Shockley surface state recorded using the second harmonic light of an amplified 1 kHz Ti:sapphire laser system (Figure 21.12b, 1 kHz spectrum). This signal is compared to results achieved with the second harmonic light of a 80 MHz Ti:sapphire laser system (Figure 21.12b, 80 MHz spectrum). In the 80 MHz spectrum, the photoemitted electron bunch has been generated on a surface area of about 150 μm and contains an average of about three electrons. For the high-intensity spectrum 80 000 e/pulse are emitted from a surface area of about 3.1 mm^2. The high charge density of this electron bunch gives rise to drastic energy broadening and a slight energy shift of the dispersing spectral signature associated with the surface state band. Additional evidence for space charge effects is the appearance of ghost peaks in the spectrum [90], and the observation of extremely high electron kinetic energies, which exceed the original photon energies by several orders of magnitude [91, 92].

Because of increasing interest in the use of low-repetition rate, and intense pulsed light sources in photoemission experiments, distortions of the photoelectron spectra due to the space charge effect have been analyzed from a quantitative point of view by different groups in the recent past [95–98]. For instance, space charge distortions at the Fermi-edge signal from a polycrystalline gold sample at illumination with picosecond XUV pulses from a synchrotron source were observed by Zhou *et al.* and quantitatively analyzed using a Monte Carlo-based technique [95]. Passlack *et al.* systematically studied the spectral-broadening effects and peak shifts in the photoelectron spectra at excitation by amplified sub-50 fs laser pulses at varying intensities in the visible spectral regime [96]. Recent XUV photoemission results obtained at the FEL source FLASH by Pietzsch *et al.* also exhibited distinct distortions that became particularly strong at high light pulse intensities [98]. In a comprehensive theoretical study by Hellmann *et al.*, space charge distorted photoemission spectra have been calculated using a molecular dynamic full *N*-electron simulation incorporating a modified tree-code algorithm [97]. The authors showed that this approach is very successful in consistently describing various experimental results that were obtained under different conditions.

Some general rules and implications regarding space charge effects in photoemission can be deduced from these results. For example, the critical parameters responsible for a space charge-induced energy broadening and energy shift are the number of electrons in a single electron bunch and the source diameter of the electron bunch at the sample surface. The pulse duration and the kinetic energy of the photoelectrons are of much less importance. The emergence of positive mirror charges induced in the substrate by the negatively charged photoemitted electron cloud significantly counteracts the space charge effect and reduces the energy broadening and energy shift by about 30–50%. As a rule of thumb for a core-level experiment, one finds that the energy broadening can be limited to a value of 50 meV with an electron bunch exhibiting no more than approximately 10 000 electrons per mm (e/mm) spot diameter. With respect to valence-band spectroscopy, a 5 meV resolution limit corresponds to a value of approximately 3000 e/mm. For angle-resolved PES experiments, momentum broadening is a critical issue. The simulations by Hellmann *et al.* show that the momentum distribution of the photoemitted electrons is much less sensitive to space charging than the electron energy distribution [97]. However, this momentum distribution exhibits a clear dependence on the electron kinetic energy ($\Delta k \propto \sqrt{E_{\text{kin}}}$). For instance, the critical value for a momentum broadening of 0.01 Å^{-1} is 10 000 electrons/$\text{eV}^{1/2}$ per 1 mm of illumination diameter.

21.4
Review of Pioneering Experiments

21.4.1
Static Photoelectron Spectroscopy Using High Harmonic Sources

The tunability of high-order harmonic sources over a rather broad photon energy range is a specific feature not offered by conventional XUV light sources in laboratory use. This feature is particularly useful in photoelectron spectroscopy as photoemission cross sections in the XUV-range depend critically on photon energy. For example, close to the absorption threshold of a core-level resonance, photoemission can be used to enhance certain valence band characteristics or to discriminate between the different excitation channels responsible for an emission process [101]. Another example are the so-called shape resonances of adsorbed molecules that can be used as monitors of the adsorption geometry [102].

Figure 21.13a shows photoemission spectra of the $c(4 \times 2)$-2CO/Pt(111) surface recorded at different photon energies using a high harmonic source [100]. For these measurements, not only wavelength tuning was achieved by the selection of an individual harmonic but also the tuning within the spectral width of a single harmonic was allowed. In these spectra, a clear resonance-like behavior in the peak intensities of the 4σ, 5σ, and 1π orbitals of the adsorbed CO molecule is observed, as shown in Figure 21.13b. The 4σ and 5σ resonance energies of 37 and 28 eV, respectively, had been reported in an earlier synchrotron-based work [99] and can be assigned to shape resonances of the CO molecule [103]. In contrast, the 1π shape resonance had not been previously observed. A comparison with theoretical calculations [104] indicates that this feature arises from an autoionization process involving the 3σ and 2π^* orbitals of the CO molecule.

In another example by Haight and Peale, the flexibility of the high harmonic source was used to distinguish between the bulk and the surface contributions to the signal in a photoemission study of an As-terminated Ge(111) surface [106]. As bulk bands disperse in general in k perpendicular to the surface plane, their contribution to the spectra could be identified by their distinct energy dispersion with the excitation photon energy. Other peaks in the spectrum did not disperse at all. The comparison of these data with calculations of the surface band structure showed that these features are surface states derived from As atomic orbitals.

The first photoemission core-level spectra with a high harmonic source were recorded by Haight and Seidler [107]. In their study, they clearly resolved the 3/2–5/2

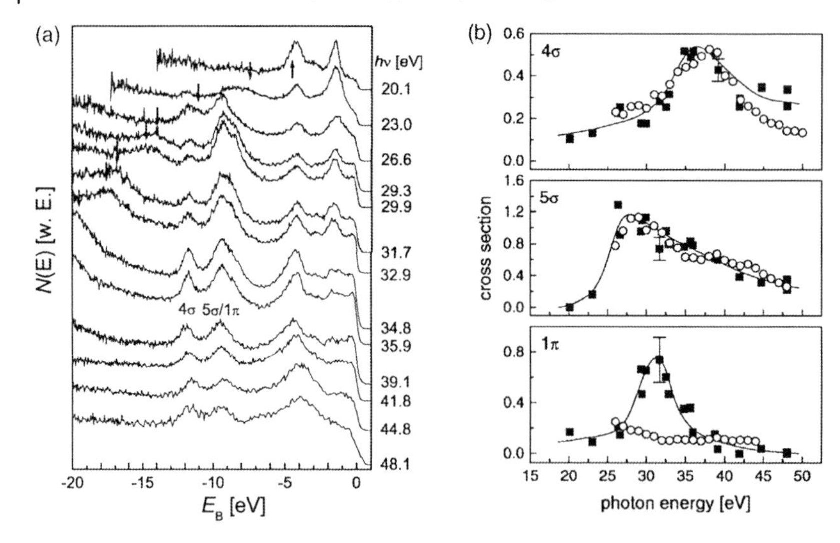

Figure 21.13 (a) Ultraviolet-PES spectra for $c(4 \times 2)$–2CO Pt (111) in the $h\nu = 20$–50 eV range. The 4σ, 5σ, and 1π valence resonances of the adsorbed CO exhibit a clear dependence on photon energy. (b) Photoemission cross section of the CO-induced 4σ, 5σ, and 1π states for $c(4 \times 2)$–2CO Pt (111) using p-polarized high-order harmonic radiation (solid squares). The experimental results of Ref. [99] measured on Pt (110) have been added (open circles). Reprinted from [100]. With kind permission of Springer Science & Business Media.

splitting of the Ga 3d core level of approximately 400 meV (Figure 21.14). Another example is the spectroscopy of the Pb 5d core level where a significant cross-sectional enhancement of the photoemission signal with increasing photon energy was observed [107].

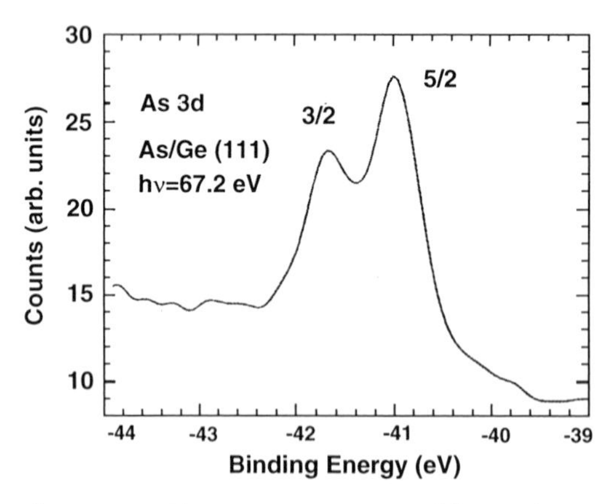

Figure 21.14 Photoemission spectrum of the As 3*d* 3/2 and 5/2 spin–orbit split core states from a single monolayer of As deposited on Ge(111) recorded using an HHG source at a photon energy of 67.2 eV. Reprinted with permission from [105]. Copyright 1996 The Optical Society.

The preceding examples illustrate the prospects for high harmonic generation as a relatively inexpensive table-top laboratory source for static photoelectron spectroscopy. The data show that this kind of source provides a high degree of flexibility in the XUV range that is offered at present only by synchrotron sources. Even though the HHG source flux is orders of magnitude smaller than what is possible with modern synchrotrons, the typical integration times in photoemission experiments are rather convenient (a few minutes integration time per spectrum in our examples). For various future applications, high harmonic generation will be a promising alternative to, or even complementary to, synchrotron-based XUV sources.

21.4.2 Time-Resolved Photoemission Using XUV Pulses

21.4.2.1 Probing Electron Excitations

The first time-resolved photoemission experiments using short-pulse XUV sources focused on the thermalization dynamics of hot carriers in semiconductor such as InP (110), GaAs, Au/GaAs(110), and As-terminated Ge(111) (for an overview of these studies, see Ref. [109]). In principle and as discussed by several contributors to this book, photon energies in the visible and near-UV regime are suitable for probing electron excitations and decays at surfaces. However, at increasing excitation densities, thermo-, multiphoton, and tunnel emission induced by the excitation pump pulse become relevant. They will, at some point, dominate the low-energy part of the photoemission spectrum as recorded by the probe pulse. Figure 21.15 illustrates the emergence of this parasitic background with increasing pump laser fluence from an IR-pump XUV-probe study of a Pd(111) surface [108]. Riffe *et al.* provide a quantitative discussion of the different effects involved in this type of process [94].

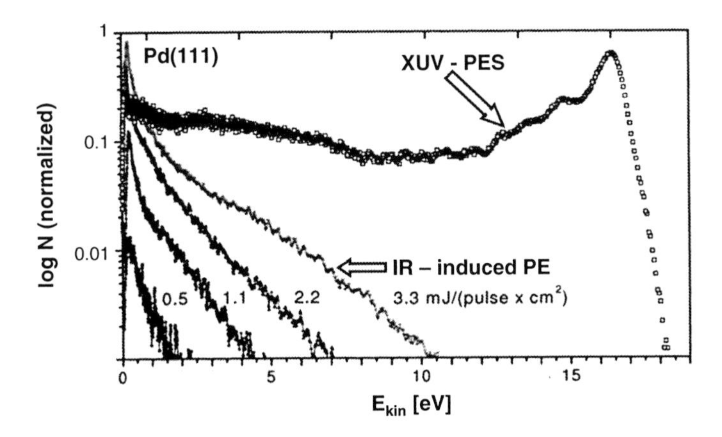

Figure 21.15 Example of high kinetic-energy photoemission from a Pd(111) surface induced by the 800 nm pump pulse at different intensities. In comparison to the XUV probe spectrum (boxes, $h\nu = 23$ eV), photoemission of up to several eV is observed from the IR pulse. This IR photoemission can exceed the count rates achieved with the harmonic light in particular for low kinetic energy. Reprinted from [108]. Copyright 2005 IOP Publishing Ltd.

In summary, the pump pulse-induced electron emission background can extend to quite high kinetic energies up to several eV. Such high kinetic energies can result in the photoemission spectrum from a probe pulse with low-photon energy becoming completely blurred. Any information on the excited state distribution and its decay will be lost. The higher energy probe photons in the XUV regime can remove the photoelectrons from this background to give a clear spectral distribution.

Figure 21.16 summarizes the results of a time-resolved electron relaxation study from a GaAs(110) surface using high harmonic light of 10.7 eV [110]. For semiconductors, the phase space for inelastic carrier decay in the excitation energy regime close to the conducting band minimum, as addressed in this work, is quite limited and governed by the electron–lattice interaction (electron–phonon scattering). Typical timescales for the carrier energy relaxation are therefore on the order of several picoseconds; thus, reflective grating optics, as used in this work, are not critical. In this example, the angular resolution mode of the employed time-of-flight electron spectrometer was efficiently used to probe the dynamics of population exchange between two separate points in momentum space resulting from momentum transfer due to electron–phonon interactions. Photoemission spectra recorded at different emission angles showed that the population of the GaAs $\bar{X}$-valley after absorption of the 700 nm pump beam is delayed by approximately 0.9 picoseconds with respect to the population of the $\bar{\Gamma}$- valley. It is this delay that is related to intervalley scattering from the $\bar{\Gamma}$-point to the $\bar{X}$-point, which was induced by

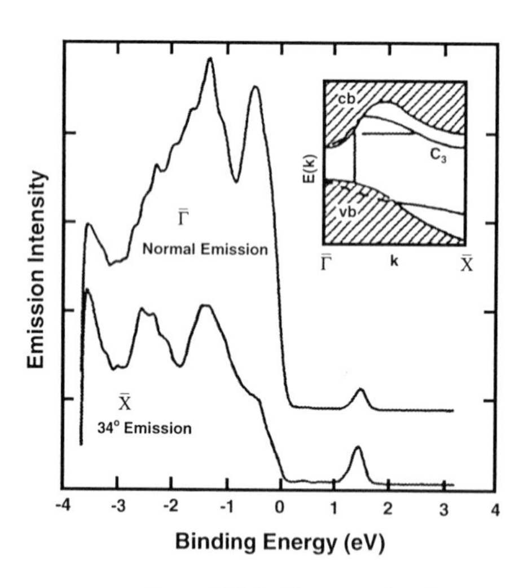

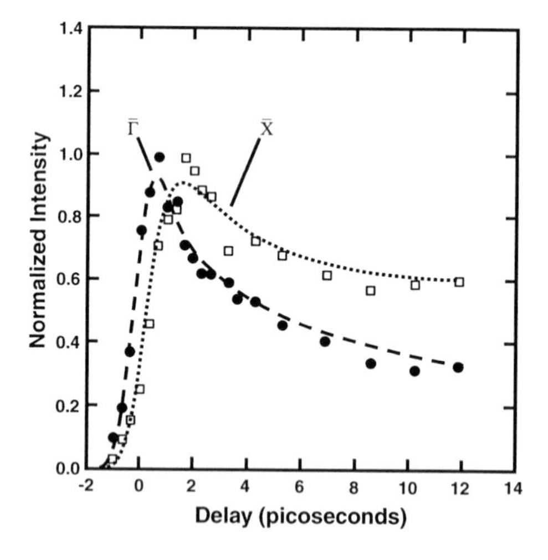

Figure 21.16 Pump–probe photoemission spectra from cleaved GaAs(110). The bottom spectrum was collected at an emission angle of 34° relative to the surface normal in the $\bar{\Gamma}-\bar{X}$ direction. The top spectrum was collected along the surface normal corresponding to emission from the $\bar{\Gamma}$-point. The peak appearing at a binding energy of about 1.5 eV is due to photoemission from electrons excited by the pump pulse. *Inset*: The relevant surface and bulk-projected band structures (hatched regions); (b) pump–probe delay curves for emission from $\bar{\Gamma}$ (filled circle) and from $\bar{X}$ (open squares). Fits from the model calculations described in the text are shown as dashed ($\bar{\Gamma}$) and dot-dashed ($\bar{X}$) curves.

electron–phonon scattering. A rate equation model was fitted to the data and revealed that the characteristic scattering time was approximately 0.4 ps.

21.4.2.2 Time-Resolved Valence and Core-Level Spectroscopy

Femtosecond-pulsed XUV sources make it possible to study photoinduced phase transitions (PIPTs) in real time by monitoring the transient changes of the valence band and core-level spectrum of a solid during the transition. This aspect is highly attractive as these states are sensitive monitors for a variety of surface and bulk properties, including the chemical state, magnetic moments, collective electronic phenomena, and structural properties.

The chemical sensitivity of valence-state photoemission to adsorbate–surface interaction has been used in a work by Bauer *et al.* on the transient response of molecular oxygen adsorbed onto platinum to an intense surface excitation with 800 nm laser pulses (see Figure 21.17) [25, 111]. In the experiment, a saturation layer of molecular oxygen was first adsorbed onto the Pt(111) surface at 77 K. These conditions result in the predominant occupation of oxygen sites in a superoxo configuration. The characteristic signature for the oxygen chemical state is the occupied 1π orbital of the molecular oxygen. It appears at a binding energy of about 6 eV with respect to E_F in the photoemission spectrum. Ultrafast laser heating of the platinum substrate induces a chemical transition (nuclear motion) of the molecular oxygen that is metastable on a picosecond timescale. Specifically, the authors observe the instantaneous creation of a nonthermal electron distribution in the platinum substrate (not shown) that decays on a timescale of approximately 200 fs. Simultaneously, a change in the spectral characteristics of the oxygen 1π orbital appears in the

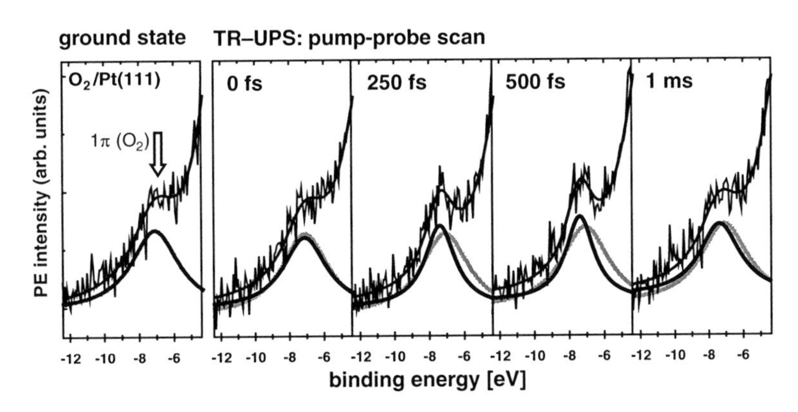

Figure 21.17 Time-resolved photoemission results from a saturation layer of molecular oxygen adsorbed on a Pt(111) surface. Photoemission spectra were obtained for the ground state of the system and at four different temporal delays. The shoulder (or peak) appearing in the spectra in the energy region between 6 and 8 eV binding energy is due to photoemission from the 1π orbital of the oxygen. Transient changes observed in this region can be assigned to an excitation from the superoxo ground state of the oxygen into a peroxo excited state. The full black lines are Lorentzian fits to the data. The gray line is a Lorentzian fit of the ground-state spectrum.

photoemission spectrum where this change is significantly delayed with respect to the pump excitation and does not appear before approximately 100 fs have passed. A Lorentzian fit of this feature for different time delays indicates a continuous shift of the oxygen 1π feature to higher binding energies and a narrowing of its line width as the time delay increases to 1 ps. For longer times (several picoseconds), an almost complete recovery of the original (time zero) signal is observed, indicating that the observed chemical modification, that is, the nuclear motion from one adsorption side to another, is mainly of a transient nature. However, this recovery is not fully complete as a permanent (nonreversible) change of the spectral characteristics of the oxygen is observed for prolonged interaction times. This change is an indication that, in addition to relaxation into the original ground-state configuration (superoxo oxygen), an alternative decay channel into a different, stable oxygen configuration is possible for the transient chemical oxygen state. Comparing this result to static photoemission spectra obtained for molecular oxygen adsorption at low coverages shows that this second relaxed state corresponds to the peroxo state of oxygen on Pt(111).

In a time-resolved core-level photoemission experiment, Siffalovic and co-workers monitored surface photovoltage transients on *p*-GaAs after photoexcitation of electrons from the valence into the conduction band using an intense 400 nm femtosecond pump pulse [112]. For these experiments, they used $<$100 fs XUV pulses with a photon energy of about 70 eV. The carrier transport dynamics between the bulk and the surface regions and carrier recombination at the surface were probed by monitoring transient shifts of the Ga-3d core levels (binding energy $\approx$ 20 eV). The corresponding kinetic energy of the photoemitted electrons ($\approx$50 eV) guarantees an exceptionally high surface sensitivity, as this energy is located right at the minimum of the mean free path of electrons in solids [113] and has, therefore, an enhanced sensitivity to the surface photovoltaic effect. Furthermore, the relevant spectral probe feature is by this means sufficiently retracted from the photoemission background induced by the intense pump pulse (10 GW/cm^2) extending to kinetic energies of up to 10 eV, which would otherwise significantly blur the signal (see Figure 21.15).

To understand the dynamic processes in the *p*-GaAs surface region induced by the absorption of the pump pulse, the static electronic structure in the vicinity of the surface has to be considered (Figure 21.18a). The excess charge associated with the existence of GaAs surface states located within the bandgap is compensated by a thin space charge region of a few nanometers. The space charge region leads to band bending in this area compared to the bulk band. The absorption of light results in the creation of free carriers (electrons in the conduction band and holes in the valence band) that are then accelerated by the surface potential toward the surface (for the electrons) and into the bulk (for the holes), respectively. This charge redistribution in the surface layer results in a transient reduction in the band bending (flat band transition) that can be monitored by tracking the corresponding energy shift in the binding energy of the 3d core levels of the Ga atoms located inside the space charge layer. In Figure 21.19b, the corresponding photoelectron spectra are displayed as a function of temporal delay between the pump and XUV-probe pulse. The observed shifts in the center of gravity of the Ga-3d peak reflect the dynamics of the discharging and recharging of the surface layer due to the transient charge redistribution. A fast

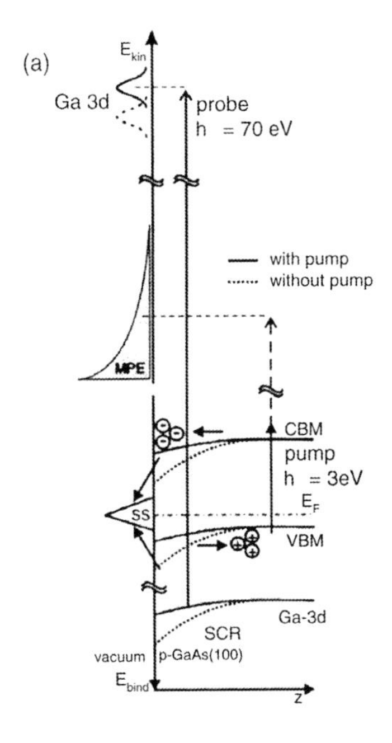

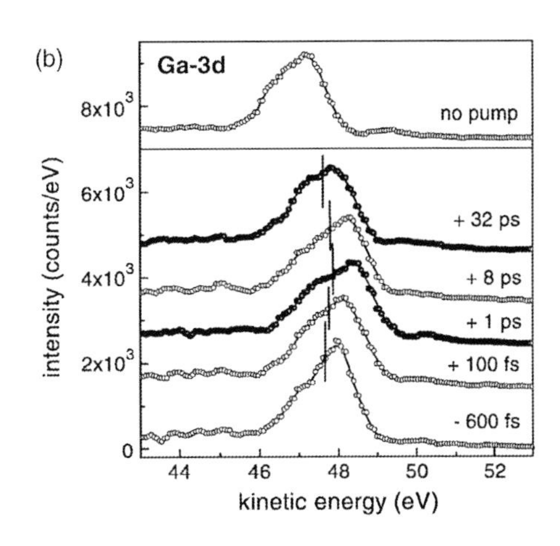

Figure 21.18 (a) Principle of the band-bending variation after photoexcitation of *p*-GaAs detected by measuring the kinetic energy of the photoelectrons from the Ga-3d shell (for details see text). (b) Surface photovoltage transient in *p*-GaAs(100) revealed by a Ga-3d core-level shift as a function of the pump–probe delay within a TR-UPS experiment. The thin vertical lines show the peak's center of gravity. Reprinted from [108] Copyright 2002 EDP Sciences.

carrier transport to the surface layer is indicated by a fast peak shift (within approximately 500 fs) to lower binding energies. However, the relaxation into the surface layer ground state occurs on a much longer timescales (approximately 15 ps) and is driven by recombination processes and trapping of the electrons in surface states.

21.4.2.3 **Laser-Assisted Photoelectric Effect From Surfaces**

The examples presented so far are based on a classical pump–probe scheme. An intense (infrared) pump pulse generates a nonequilibrium state at the surface, and the XUV pulse probes in photoemission the femtosecond dynamics associated with the relaxation back into equilibrium. In time-resolved experiments based on LAPE, the infrared pulse probes the XUV photoemission (or another XUV-induced electron emission process) and consequently probes the dynamics associated with the excitation state prepared by the XUV absorption.

LAPE was first reported in gas-phase photoemission [74, 114]. In this process, an atom is simultaneously irradiated by XUV and intense infrared light. The presence of the infrared laser modifies the normal XUV photoelectron spectrum. The effect can be described as the simultaneous absorption or stimulated emission of one (or several) infrared photons with the XUV photon. For many-cycle infrared pulses, the

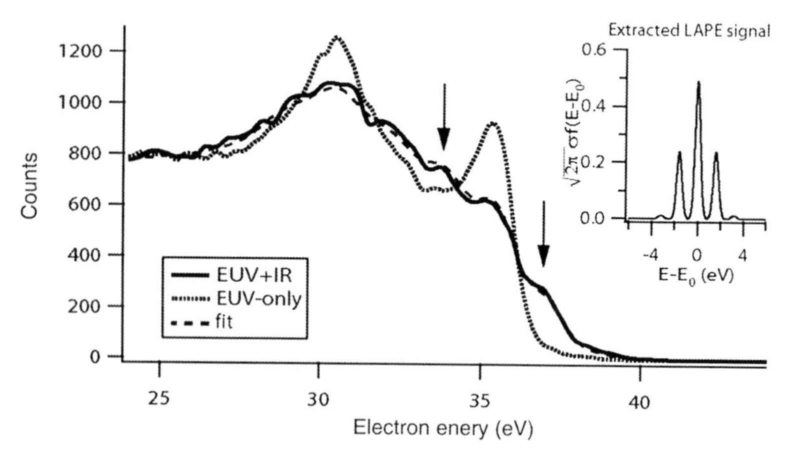

Figure 21.19 Experimental observation of laser-assisted photoemission from a Pt(111) surface using HHG light. The XUV + IR spectrum (solid line) shows a strong distortion around the Fermi edge compared to the XUV-only spectrum (dotted line). This change is due to the appearance of sidebands in the spectrum. The dashed line shows the fit of the XUV + IR spectrum by convolving a LAPE response function with the XUV-only spectrum. The inset shows the LAPE response function associated with the fit. The energy separation between the sidebands is equal to the IR photon energy, as expected for surface LAPE. Reprinted with permission from [75]. Copyright 2006 by the American Physical Society.

LAPE process leads to discrete sidebands in the XUV photoelectron spectrum at energies corresponding to multiples of the infrared photon energy. Few-cycle infrared pulses give rise to the streaking of the electrons photoemitted by the XUV pulse by the electric field of the laser light [10, 115]. Because of the instantaneous character of LAPE, this effect is particularly useful for temporal characterization of XUV pulses (see also Section 21.3.2).

Miaja-Avila and coworkers reported the first observation of LAPE in a time-resolved photoemission experiment from surfaces [75, 76]. In their work, a clean Pt(111) single crystal was used as the solid surface, since it exhibits a large density of d-states at the Fermi edge, with a characteristic peak-like structure in the XUV photoelectron spectrum. In the presence of an intense laser field and at temporal overlap with the XUV pulse, sidebands appeared in the Pt(111) photoemission spectrum that could unambiguously be assigned to a LAPE process (see Figure 21.19). From a model fit to the photoelectron spectrum, it was possible to reconstruct the LAPE response function, which is shown in the inset of Figure 21.19 (for details of the data analysis see Refs [75, 76]). In comparison to LAPE in the gas phase, surface LAPE benefits from the high density of target atoms on a surface and can therefore be used to characterize lower flux and higher energy XUV sources. This enhanced sensitivity can be further increased by the use of longer wavelength infrared dressing beams. The use of 1300 nm infrared pulses instead of 800 nm infrared pulses enhances the amplitude of the surface LAPE signal by a factor of 7 [77].

In a similar manner, the emission of an Auger electron from an atom can be modified by an intense infrared laser pulse [114, 116]. In this process, an Auger electron is first generated by the decay of a core hole created by the absorption of

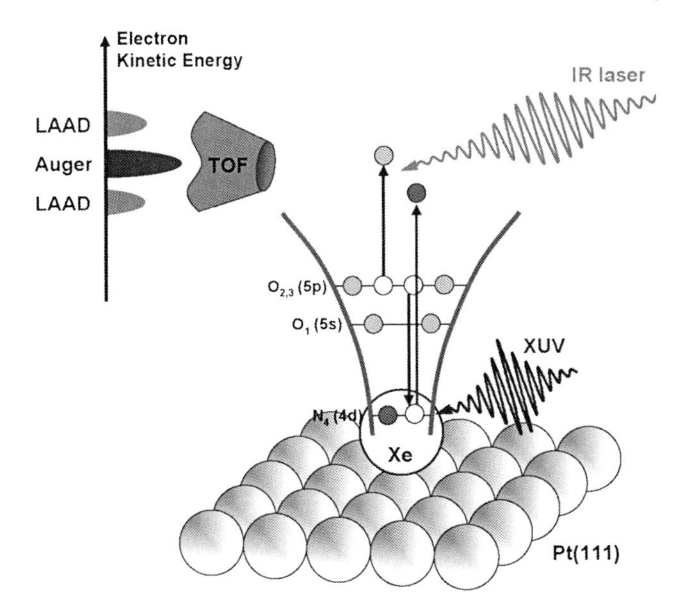

Figure 21.20 Ultrafast XUV and IR pulses are focused onto a Xe/Pt(111) surface, where the kinetic energy of the photoemitted electrons is measured by a time-of-flight detector. An electron from the Xe N_4 (4d) shell is ejected by the XUV, followed by filling of the core hole by an $O_{2,3}$ (5p) shell electron and ejection of a secondary (Auger) electron from the $O_{2,3}$ shell. In the presence of the IR beam, sidebands appear in the photoelectron spectrum of both Xe and Pt. By comparing the sideband amplitude versus time delay for the Pt and the Xe Auger photoelectron peaks, the lifetime of the Xe 4d core hole can be extracted. Reprinted with permission from [116]. Copyright 2008 by the American Physical Society.

a XUV photon (see Figure 21.20). Subsequent absorption of an infrared photon by this electron gives rise to sidebands, this time located next to the Auger peak characteristic for the probed atom species. Surface laser-assisted Auger decay (LAAD) was first demonstrated for the Auger decay of a Xenon adsorbate on Pt(111) [116]. By comparing the time evolution of the LAPE response from the Pt substrate with the LAAD signal from the adsorbed Xe, Miaja-Avila and coworkers measured the lifetime of the 4d core hole initially created by the XUV pulse (see Figure 21.21). This measurement was possible despite the many complexities that result from using the metallic substrate such as above-threshold photoemission, hot electrons, and space charge effects. The extracted lifetime of 7 ± 1 fs for the Xe/Pt 4d core hole was consistent with high-resolution linewidth measurements of Xe in the gas phase [117], clearly validating the success of this experimental approach.

In another work by Cavalieri *et al.* [118], the few-cycle infrared LAPE technique (streaking technique) was applied to study the subfemtosecond dynamics associated with the photoemission process from delocalized (valence) states and localized (core) states of a tungsten surface. Differences on the order of 100 attoseconds in the photoemission signal response from these different states could be resolved (for more details about this work, readers may refer to Chapter 22).

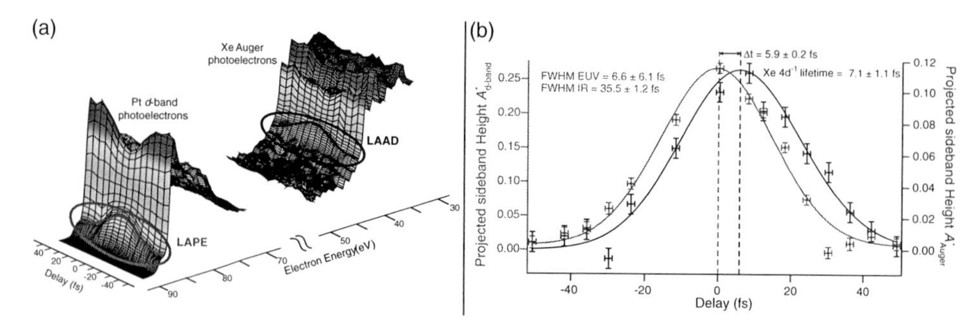

Figure 21.21 (a) Photoemission spectra from Xe/Pt(111) as a function of the time delay between the IR and the XUV pulse. The time evolution of the redistribution of electrons due to the IR dressing at the Pt d-band and at the Xe Auger photoelectron peaks can be observed. (b) The sideband height for the Pt d-band (gray curve) corresponds to a cross correlation between the XUV and the IR pulses. The Auger sideband height (black curve) is shifted and broadened with respect to the Pt d-band curve. A value of 7.1 ± 1.1 fs for the Xe 4d core-hole lifetime was extracted using these data. Reprinted with permission from [116]. Copyright 2008 by the American Physical Society. (Please find a color version of this figure on the color plates.)

The results of Miaja-Avila *et al.* and Cavalieri *et al.* clearly demonstrate that surface LAPE can be used to study ultrafast femtosecond-to-attosecond timescale electron dynamics in solids and in surface adsorbate systems where complex, correlated, electron relaxation processes are expected.

21.4.2.4 **FEL-Based Time-Resolved Photoemission: First Results**

Time-resolved studies using a free electron laser XUV light source pose some serious challenges to an experimentalist. However, the extreme intensities in the XUV regime and the accessible broad energy range (e.g., up to 1 keV from the 5th harmonic light at the free electron laser facility, FLASH) make this source highly attractive for future time-resolved photoemission experiments. In the context of ultrafast surface studies, these challenges include the timing synchronization and temporal jitter reduction between an excitation laser pulse and the FEL XUV-probing pulse, the handling of the (intrinsically large) pulse-to-pulse fluctuations, severe space charging at high photon fluxes, and the restricted experimental time available at FEL facilities. Despite these challenges, first proof-of-principle experiments have been accomplished illustrating the potential of these sources in the context of time-resolved photoelectron spectroscopy.

Pietzsch *et al.* [98] have performed femtosecond time-resolved core-level photoelectron spectroscopy at the W(100) 4f line (binding energy $E_B \approx 31$ eV). They used third harmonic pulses from FLASH with a photon energy of 118.5 eV. One objective of their work was the study of space charge effects in photoemission due to illumination with high-intensity FEL light, as discussed in Section 21.3.4. Furthermore, they succeeded in recording the first surface photoemission infrared-XUV cross-correlation signal at an FEL source. LAPE sideband formation could be monitored next to the tungsten 4f core lines at temporal overlap between the

XUV pulse and an amplified 800 nm laser pulse. The measured width of this cross-correlation signal was determined to approximately 440 fs. This width is mainly governed by the temporal jitter of the FEL pulses with respect to the optical laser pulses and the temporal drifts within the FLASH macrobunch trains caused by electronic feedback systems of the accelerator structure [96]. The temporal width of the optical pulse (120–150 fs) and the XUV pulse are negligible in this case, but their cross correlation can be retrieved when the jitter is determined independently through electrooptical sampling [92–94].

In a more recent FEL study, time-resolved core-level PES has been applied to the correlated layered compound 1*T*-TaS_2 [119]. The choice of the sample is motivated by the very clear valence band photoemission results reported by Perfetti *et al.* in a time-resolved study addressing the dynamics of a laser-induced metal–insulator transition in this system [120]. The results clearly demonstrate that core-level dynamics from the femtosecond up to the nanosecond timescale can be investigated in photoelectron spectroscopy using the high-intensity XUV beam at FLASH. Figure 21.22 presents the observed ultrafast effects occurring on a sub-1 ps timescale. It shows the evolution of the Ta 4f core-level doublet of 1*T*-TaS_2 as a function of the pump–probe delay in response to excitation with a 800 nm laser pulse. Note that the Ta $4f_{7/2}$ and $4f_{5/2}$ peaks are split into peak doublets because of the presence of a charge density wave (CDW) at the experimental sample temperature of 10 K. Figure 21.22 displays the transient evolution of the integral photoemission signal from the low energetic Ta $4f_{7/2}$ core peak (see vertical solid lines in Figure 21.22). For this scan, the infrared pump and XUV-probe intensities were kept at a sufficiently low level so that the space charge

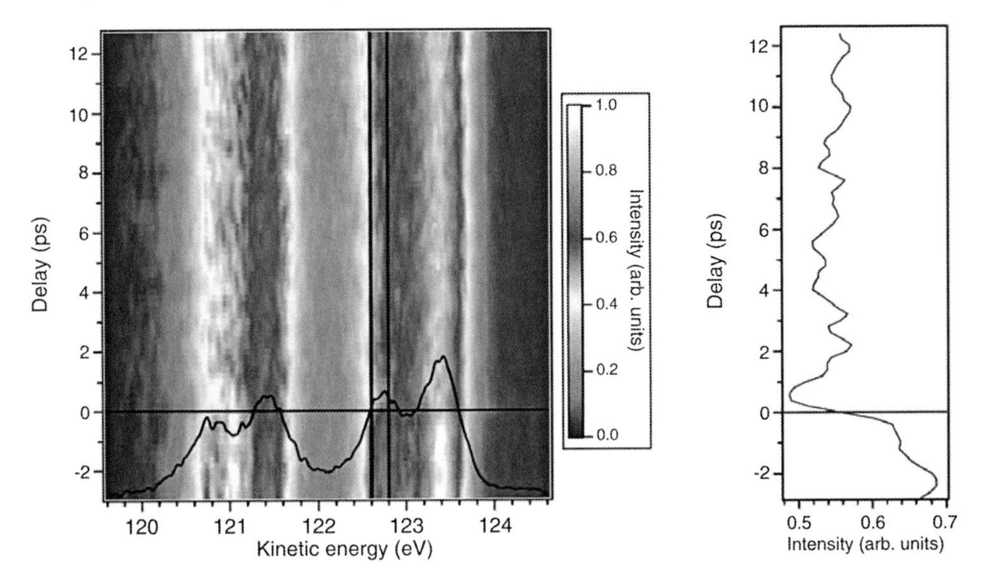

Figure 21.22 (a) Time-resolved Ta 4f core-level spectra as a function of the pump–probe delay between −5 and + 10 ps. (b) Integrated intensity at a CDW-split core level (black box in Figure 21.2a] as a function of the pump–probe delay. (Please find a color version of this figure on the color plates.)

effects do not affect the probed ultrafast dynamics. A fast-transient depression within about 800 fs and a slow recovery of the intensity can be identified. A detailed analysis of these data to disentangle the different physical mechanisms responsible for the observed effects occurring on different timescales is in progress.

21.5
Conclusions and Outlook

In probing the electronic structure of solids and their surfaces, photoelectron spectroscopy using XUV and X-ray light sources has emerged as a leading experimental technique in identifying key properties of phenomena associated with surface chemistry, electronic correlation, or bulk and surface magnetism. Extension of this technique into the ultrafast time domain will make it possible to address the fundamental processes on an atomic scale, exploring the dynamics of these phenomena from a new perspective. In this context, we expect that time-resolved photoemission at short wavelengths will contribute to a more comprehensive understanding of condensed matter. Recent important technical advances will routinely enable these kinds of experiments in the future. These advances include the development of high-flux pulsed XUV sources, progress in the detection efficiency of electron energy analyzers, and the development of suitable schemes for XUV-pulse profile analysis.

Further developments will certainly extend the applicability of time-resolved XUV photoemission. Circularly polarized XUV pulses are highly attractive for the study of the magnetization dynamics in solids on the femtosecond timescale. Circular polarization can be generated with sophisticated XUV sources [121] or quarter-wave plates specifically designed for the XUV regime [122]. A substantial increase in the repetition rate of the XUV sources will also further benefit research in this field. In a very promising approach, which uses field enhancement effects due to light coupling to plasmonic nanostructures, high harmonics up to the 17th were recently generated from a Ti:sapphire laser system without amplification [123]. In another experiment, a high-power Yb-doped fiber chirped pulse amplification system was used for XUV pulse generation and harmonics up to 31st order at a 1 MHz repetition rate were reported [124]. High-repetition sources are also particularly valuable for circumventing limitations due to space charge effects. Furthermore, high-repetition sources would help facilitate multidimensional detection techniques such as $E(k)$ mapping or photoemission microspectroscopy.

References

1 Teich, M.C., Schroeer, J.M., and Wolga, G.J. (1964) *Phys. Rev. Lett.*, **13**, 611.
2 Petek, H. and Ogawa, S. (1997) *Prog. Surf. Sci.*, **56**, 239.
3 McPherson, A., Gibson, G., Jara, H., Johann, U., Luk, T.S., McIntyre, I.A., Boyer, K., and Rhodes, C.K. (1987) *J. Opt. Soc. Am. B*, **4**, 595.

4 Ferray, M., Lhuillier, A., Li, X.F., Lompre, L.A., Mainfray, G., and Manus, C. (1988) *J. Phys. B At. Mol. Opt.*, **21**, L31.

5 Zhou, J., Peatross, J., Murnane, M.M., and Kapteyn, H.C. (1996) *Phys. Rev. Lett.*, **76**, 752.

6 Chang, Z.H., Rundquist, A., Wang, H.W., Murnane, M.M., and Kapteyn, H.C. (1997) *Phys. Rev. Lett.*, **79**, 2967.

7 Spielmann, C., Burnett, N.H., Sartania, S., Koppitsch, R., Schnurer, M., Kan, C., Lenzner, M., Wobrauschek, P., and Krausz, F. (1997) *Science*, **278**, 661.

8 Seres, E., Seres, J., Krausz, F., and Spielmann, C. (2004) *Phys. Rev. Lett.*, **92**, 163002.

9 Paul, P.M., Toma, E.S., Breger, P., Mullot, G., Auge, F., Balcou, P., Muller, H.G., and Agostini, P. (2001) *Science*, **292**, 1689.

10 Hentschel, M., Kienberger, R., Spielmann, C., Reider, G.A., Milosevic, N., Brabec, T., Corkum, P., Heinzmann, U., Drescher, M., and Krausz, F. (2001) *Nature*, **414**, 509.

11 Goulielmakis, E., Schultze, M., Hofstetter, M., Yakovlev, V.S., Gagnon, J., Uiberacker, M., Aquila, A.L., Gullikson, E.M., Attwood, D.T., Kienberger, R., Krausz, F., and Kleineberg, U. (2008) *Science*, **320**, 1614.

12 Murnane, M.M., Kapteyn, H.C., and Falcone, R.W. (1989) *Phys. Rev. Lett.*, **62**, 155.

13 Murnane, M.M., Kapteyn, H.C., Rosen, M.D., and Falcone, R.W. (1991) *Science*, **251**, 531.

14 Kmetec, J.D., Gordon, C.L., Macklin, J.J., Lemoff, B.E., Brown, G.S., and Harris, S.E. (1992) *Phys. Rev. Lett.*, **68**, 1527.

15 Madey, J.M.J. (1971) *J. Appl. Phys.*, **42**, 1906.

16 Ackermann, W., Asova, G., Ayvazyan, V., Azima, A., Baboi, N., Bahr, J., Balandin, V., Beutner, B., Brandt, A., Bolzmann, A., Brinkmann, R., Brovko, O.I., Castellano, M., Castro, P., Catani, L., Chiadroni, E., Choroba, S., Cianchi, A., Costello, J.T., Cubaynes, D., Dardis, J., Decking, W., Delsim-Hashemi, H., Delserieys, A., Di Pirro, G., Dohlus, M., Dusterer, S., Eckhardt, A., Edwards, H.T., Faatz, B., Feldhaus, J., Flottmann, K., Frisch, J., Frohlich, L., Garvey, T., Gensch, U., Gerth, C., Gorler, M., Golubeva, N., Grabosch, H.J., Grecki, M., Grimm, O., Hacker, K., Hahn, U., Han, J.H., Honkavaara, K., Hott, T., Huning, M., Ivanisenko, Y., Jaeschke, E., Jalmuzna, W., Jezynski, T., Kammering, R., Katalev, V., Kavanagh, K., Kennedy, E.T., Khodyachykh, S., Klose, K., Kocharyan, V., Korfer, M., Kollewe, M., Koprek, W., Korepanov, S., Kostin, D., Krassilnikov, M., Kube, G., Kuhlmann, M., Lewis, C.L.S., Lilje, L., Limberg, T., Lipka, D., Lohl, F., Luna, H., Luong, M., Martins, M., Meyer, M., Michelato, P., Miltchev, V., Moller, W.D., Monaco, L., Muller, W.F.O., Napieralski, O., Napoly, O., Nicolosi, P., Nolle, D., Nunez, T., Oppelt, A., Pagani, C., Paparella, R., Pchalek, N., Pedregosa-Gutierrez, J., Petersen, B., Petrosyan, B., Petrosyan, G., Petrosyan, L., Pfluger, J., Plonjes, E., Poletto, L., Pozniak, K., Prat, E., Proch, D., Pucyk, P., Radcliffe, P., Redlin, H., Rehlich, K., Richter, M., Roehrs, M., Roensch, J., Romaniuk, R., Ross, M., Rossbach, J., Rybnikov, V., Sachwitz, M., Saldin, E.L., Sandner, W., Schlarb, H., Schmidt, B., Schmitz, M., Schmuser, P., Schneider, J.R., Schneidmiller, E.A., Schnepp, S., Schreiber, S., Seidel, M., Sertore, D., Shabunov, A.V., Simon, C., Simrock, S., Sombrowski, E., Sorokin, A.A., Spanknebel, P., Spesyvtsev, R., Staykov, L., Steffen, B., Stephan, F., Stulle, F., Thom, H., Tiedtke, K., Tischer, M., Toleikis, S., Treusch, R., Trines, D., Tsakov, I., Vogel, E., Weiland, T., Weise, H., Wellhoffer, M., Wendt, M., Will, I., Winter, A., Wittenburg, K., Wurth, W., Yeates, P., Yurkov, M.V., Zagorodnov, I., and Zapfe, K. (2007) *Nat. Photonics*, **1**, 336.

17 Cho, A. (2002) *Science*, **296**, 1008.

18 Zholents, A.A. and Zolotorev, M.S. (1996) *Phys. Rev. Lett.*, **76**, 912.

19 Schoenlein, R.W., Chattopadhyay, S., Chong, H.H.W., Glover, T.E., Heimann, P.A., Shank, C.V., Zholents, A.A., and Zolotorev, M.S. (2000) *Science*, **287**, 2237.

20 Khan, S., Holldack, K., Kachel, T., Mitzner, R., and Quast, T. (2006) *Phys. Rev. Lett.*, **97**, 074801.

21 Williams, R.T., Royt, T.R., Rife, J.C., Long, J.P., and Kabler, M.N. (1982) *J. Vac. Sci. Technol.*, **21**, 509.
22 Haight, R., Bokor, J., Stark, J., Storz, R.H., Freeman, R.R., and Bucksbaum, P.H. (1985) *Phys. Rev. Lett.*, **54**, 1302.
23 Haight, R. and Peale, D.R. (1994) *Rev. Sci. Instrum.*, **65**, 1853.
24 Probst, M. and Haight, R. (1997) *Appl. Phys. Lett.*, **71**, 202.
25 Bauer, M., Lei, C., Read, K., Tobey, R., Gland, J., Murnane, M.M., and Kapteyn, H.C. (2001) *Phys. Rev. Lett.*, **87**, 25501.
26 Nelson, A.J., Dunn, J., van Buuren, T., and Hunter, J. (2004) *Appl. Phys. Lett.*, **85**, 6290.
27 Nelson, A.J., Dunn, J., Hunter, J., and Widmann, K. (2005) *Appl. Phys. Lett.*, **87**, 154102.
28 Joyce, J.J., Arko, A.J., Cox, L.E., and Czuchlewski, S. (1998) *Surf. Interface Anal.*, **26**, 121.
29 Arko, A.J., Joyce, J.J., Morales, L., Wills, J., Lashley, J., Wastin, F., and Rebizant, J. (2000) *Phys. Rev. B*, **62**, 1773.
30 Reintjes, J.F. (1984) *Nonlinear Optical Parametric Processes in Liquids and Gases*, Academic Press, Orlando, FL.
31 Krause, J.L., Schafer, K.J., and Kulander, K.C. (1992) *Phys. Rev. Lett.*, **68**, 3535.
32 Kulander, K.C., Schafer, K.J., and Krause, J.L. (1993) In: *Super Intense Laser-Atom Physics*, (eds B. Piraux, A. L'Huillier, and K. Rzazewski), NATO ASI Series, Plenum Press, New York.
33 Corkum, P.B. (1993) *Phys. Rev. Lett.*, **71**, 1994.
34 Christov, I.P., Zhou, J., Peatross, J., Rundquist, A., Murnane, M.M., and Kapteyn, H.C. (1996) *Phys. Rev. Lett.*, **77**, 1743.
35 Christov, I.P., Murnane, M.M., and Kapteyn, H.C. (1997) *Phys. Rev. Lett.*, **78**, 1251.
36 Lewenstein, M., Balcou, P., Ivanov, M.Y., Lhuillier, A., and Corkum, P.B. (1994) *Phys. Rev. A*, **49**, 2117.
37 Chang, Z., Rundquist, A., Wang, H., Christov, I., Kapteyn, H.C., and Murnane, M.M. (1998) *Phys. Rev. A*, **58**, R30.
38 Bartels, R., Backus, S., Zeek, E., Misoguti, L., Vdovin, G., Christov, I.P., Murnane, M.M., and Kapteyn, H.C. (2000) *Nature*, **406**, 164.
39 Zhang, X.H., Lytle, A.L., Popmintchev, T., Zhou, X.B., Kapteyn, H.C., Murnane, M.M., and Cohen, O. (2007) *Nat. Physics*, **3**, 270.
40 Baltuska, A., Udem, T., Uiberacker, M., Hentschel, M., Goulielmakis, E., Gohle Ch, H.R., Yakovlev, V.S., Scrinzi, A., Hansch, T.W., and Krausz, F. (2003) *Nature*, **421**, 611.
41 Zhou, J.P., Huang, C.P., Shi, C.Y., Murnane, M.M., and Kapteyn, H.C. (1994) *Opt. Lett.*, **19**, 126.
42 Backus, S., Durfee, C.G., Murnane, M.M., and Kapteyn, H.C. (1998) *Rev. Sci. Instrum.*, **69**, 1207.
43 Backus, S., Bartels, R., Thompson, S., Dollinger, R., Kapteyn, H.C., and Murnane, M.M. (2001) *Opt. Lett.*, **26**, 465.
44 Rundquist, A., Durfee, C.G., Chang, Z.H., Herne, C., Backus, S., Murnane, M.M., and Kapteyn, H.C. (1998) *Science*, **280**, 1412.
45 Durfee, C.G., Rundquist, A.R., Backus, S., Herne, C., Murnane, M.M., and Kapteyn, H.C. (1999) *Phys. Rev. Lett.*, **83**, 2187.
46 Constant, E., Garzella, D., Breger, P., Mevel, E., Dorrer, C., Le Blanc, C., Salin, F., and Agostini, P. (1999) *Phys. Rev. Lett.*, **82**, 1668.
47 Balcou, P., Salieres, P., Lhuillier, A., and Lewenstein, M. (1997) *Phys. Rev. A*, **55**, 3204.
48 Tamaki, Y., Itatani, J., Obara, M., and Midorikawa, K. (2000) *Phys. Rev. A*, **62**, 063802.
49 Gibson, E.A., Paul, A., Wagner, N., Tobey, R., Gaudiosi, D., Backus, S., Christov, I.P., Aquila, A., Gullikson, E.M., Attwood, D.T., Murnane, M.M., and Kapteyn, H.C. (2003) *Science*, **302**, 95.
50 Seres, J., Yakovlev, V.S., Seres, E., Streli, C., Wobrauschek, P., Spielmann, C., and Krausz, F. (2007) *Nat. Physics*, **3**, 878.
51 Lytle, A.L., Zhang, X.S., Sandberg, R.L., Cohen, O., Kapteyn, H.C., and Murnane, M.M. (2008) *Opt. Express*, **16**, 6544.
52 Popmintchev, T., Chen, M.-C., Bahabad, A., Gerrity, M.R., Sidorenko, P., Cohen, O., Christov, I., Murnane, M.M., and

Kapteyn, H.C. (2009) *Proc. Natl. Acad. Sci. U.S.A.*, **106**, 10516.

53 http://flash.desy.de/.

54 Deacon, D.A.G., Elias, L.R., Madey, J.M.J., Ramian, G.J., Schwettman, H.A., and Smith, T.I. (1977) *Phys. Rev. Lett.*, **38**, 892.

55 Ayvazyan, V., *et al.* (2006) *Eur. Phys. J. D*, **37**, 297.

56 Ayvazyan, V., *et al.* (2002) *Eur. Phys. J. D*, **20**, 149.

57 Cavalieri, A.L., *et al.* (2005) *Phys. Rev. Lett.*, **94**, 114801.

58 Pfeifer, T., Spielmann, C., and Gerber, G. (2006) *Rep. Prog. Phys.*, **69**, 443.

59 Chen, M.-C., Gerrity, M.R., Backus, S., Popmintchev, T., Zhou, X., Arpin, P., Zhang, X.H., Murnane, M.M., and Kapteyn, H.C. (2009) *Opt. Express*, **17**, 17376.

60 Dong, F., Heinbuch, S., Rocca, J.J., and Bernstein, E.R. (2006) *J. Chem. Phys.*, **124**, 224319.

61 Drescher, M., Hentschel, M., Kienberger, R., Tempea, G., Spielmann, C., Reider, G.A., Corkum, P.B., and Krausz, F. (2001) *Science*, **291**, 1923.

62 Mathias, S., Miaja-Avila, L., Murnane, M.M., Kapteyn, H., Aeschlimann, M., and Bauer, M. (2007) *Rev. Sci. Instrum.*, **78**, 083105.

63 Villoresi, P. (1999) *Appl. Opt..*, **38**, 6040.

64 Nugent-Glandorf, L., Scheer, M., Samuels, D.A., Bierbaum, V., and Leone, S.R. (2002) *Rev. Sci. Instrum.*, **73**, 1875.

65 Poletto, L., Villoresi, P., Benedetti, E., Ferrari, F., Stagira, S., Sansone, G., and Nisoli, M. (2007) *Opt. Lett.*, **32**, 2897.

66 Poletto, L., Villoresi, P., Benedetti, E., Ferrari, F., Stagira, S., Sansone, G., and Nisoli, M. (2008) *Opt. Lett.*, **33**, 140.

67 Siffalovic, P., Drescher, M., Spieweck, M., Wiesenthal, T., Lim, Y.C., Weidner, R., Elizarov, A., and Heinzmann, U. (2001) *Rev. Sci. Instrum.*, **72**, 30.

68 Braun, S., Mai, H., Moss, M., Scholz, R., and Leson, A. (2002) *Jpn. J. Appl. Phys. 1*, **41**, 4074.

69 Bauer, M., Lei, C., Tobey, R., Murnane, M.M., and Kapteyn, H. (2003) *Surf. Sci.*, **532**, 1159.

70 www.cxro.lbl.gov.

71 Benis, E.P., Charalambidis, D., Kitsopoulos, T.N., Tsakiris, G.D., and Tzallas, P. (2006) *Phys. Rev. A*, **74**, 051402.

72 Nikolopoulos, L.A.A., Benis, E.P., Tzallas, P., Charalambidis, D., Witte, K., and Tsakiris, G.D. (2005) *Phys. Rev. Lett.*, **94**, 113905.

73 Sekikawa, T., Kosuge, A., Kanai, T., and Watanabe, S. (2004) *Nature*, **432**, 605.

74 Glover, T.E., Schoenlein, R.W., Chin, A.H., and Shank, C.V. (1996) *Phys. Rev. Lett.*, **76**, 2468.

75 Miaja-Avila, L., Lei, C., Aeschlimann, M., Gland, J.L., Murnane, M.M., Kapteyn, H.C., and Saathoff, G. (2006) *Phys. Rev. Lett.*, **97**, 113604.

76 Saathoff, G., Miaja-Avila, L., Aeschlimann, M., Murnane, M.M., and Kapteyn, H.C. (2008) *Phys. Rev. A*, **77**, 022903.

77 Miaja-Avila, L., Yin, J., Backus, S., Saathoff, G., Aeschlimann, M., Murnane, M.M., and Kapteyn, H.C. (2009) *Phys. Rev. A*, **79**, 030901.

78 Gahl, C., Azima, A., Beye, M., Deppe, M., Dobrich, K., Hasslinger, U., Hennies, F., Melnikov, A., Nagasono, M., Pietzsch, A., Wolf, M., Wurth, W., and Fohlisch, A. (2008) *Nat. Photonics*, **2**, 165.

79 Damascelli, A., Hussain, Z., and Shen, Z.X. (2003) *Rev. Mod. Phys.*, **75**, 473.

80 Nicolay, G., Reinert, F., Hufner, S., and Blaha, P. (2002) *Phys. Rev. B*, **65**, 033407.

81 Koroteev, Y.M., Bihlmayer, G., Gayone, J.E., Chulkov, E.V., Blugel, S., Echenique, P.M., and Hofmann, P. (2004) *Phys. Rev. Lett.*, **93**, 046403.

82 Ast, C.R., Henk, J., Ernst, A., Moreschini, L., Falub, M.C., Pacile, D., Bruno, P., Kern, K., and Grioni, M. (2007) *Phys. Rev. Lett.*, **98**, 186807.

83 Chiang, T.C. (2000) *Surf. Sci. Rep.*, **39**, 181.

84 Milun, M., Pervan, P., and Woodruff, D.P. (2002) *Rep. Prog. Phys.*, **65**, 99.

85 Schmitt, F., Kirchmann, P.S., Bovensiepen, U., Moore, R.G., Rettig, L., Krenz, M., Chu, J.H., Ru, N., Perfetti, L., Lu, D.H., Wolf, M., Fisher, I.R., and Shen, Z.X. (2008) *Science*, **321**, 1649.

86 Martensson, N., Baltzer, P., Bruhwiler, P.A., Forsell, J.O., Nilsson, A., Stenborg,

A., and Wannberg, B. (1994) *J. Electron. Spectrosc.*, **70**, 117.

87 Schicketanz, M., Oelsner, A., Morais, J., Mergel, V., Schmidt-Bocking, H., and Schonhense, G. (2001) *Nucl. Instrum. Meth. A*, **467**, 1519.

88 Kirchmann, P.S., Rettig, L., Nandi, D., Lipowski, U., Wolf, M., and Bovensiepen, U. (2008) *Appl. Phys. A: Mater. Sci. Process.*, **91**, 211.

89 Mathias, S., Wiesenmayer, M., Deicke, F., Ruffing, A., Miaja-Avila, L., Murnane, M.M., Kapteyn, H., Bauer, M., and Aeschlimann, M. (2009) *J. Phys. Conference Series*, **148**, 012042.

90 Gilton, T.L., Cowin, J.P., Kubiak, G.D., and Hamza, A.V. (1990) *J. Appl. Phys.*, **68**, 4802.

91 Farkas, G. and Toth, C. (1990) *Phys. Rev. A*, **41**, 4123.

92 Girardeaumontaut, C. and Girardeaumontaut, J.P. (1991) *Phys. Rev. A*, **44**, 1409.

93 Clauberg, R. and Blacha, A. (1989) *J. Appl. Phys.*, **65**, 4095.

94 Riffe, D.M., Wang, X.Y., Downer, M.C., Fisher, D.L., Tajima, T., Erskine, J.L., and More, R.M. (1993) *J. Opt. Soc. Am. B*, **10**, 1424.

95 Zhou, X.J., Wannberg, B., Yang, W.L., Brouet, V., Sun, Z., Douglas, J.F., Dessau, D., Hussain, Z., and Shen, Z.X. (2005) *J. Electron. Spectrosc.*, **142**, 27.

96 Passlack, S., Mathias, S., Andreyev, O., Mittnacht, D., Aeschlimann, M., and Bauer, M. (2006) *J. Appl. Phys.*, **100**, 024912.

97 Hellmann, S., Rossnagel, K., Marczynski-Buhlow, M., and Kipp, L. (2009) *Phys. Rev. B*, **79**, 035402.

98 Pietzsch, A., Fohlisch, A., Beye, M., Deppe, M., Hennies, F., Nagasono, M., Suljoti, E., Wurth, W., Gahl, C., Dobrich, K., and Melnikov, A. (2008) *New J. Phys.*, **10**, 033004.

99 Bare, S.R., Griffiths, K., Hofmann, P., King, D.A., Nyberg, G.L., and Richardson, N.V. (1982) *Surf. Sci.*, **120**, 367.

100 Tsilimis, G., Kutzner, J., and Zacharias, H. (2003) *Appl. Phys. A: Mater. Sci. Process.*, **76**, 743.

101 Hüfner, S. (1995) *Photoelectron Spectroscopy*, Springer, Berlin.

102 Gustafsson, T. (1980) *Surf. Sci.*, **94**, 593.

103 Davenport, J.W. (1976) *Phys. Rev. Lett.*, **36**, 945.

104 Stener, M. and Decleva, P. (2000) *J. Chem. Phys.*, **112**, 10871.

105 Haight, R. (1996) *Appl. Optics.*, **35**, 6445.

106 Haight, R. and Peale, D.R. (1993) *Phys. Rev. Lett.*, **70**, 3979.

107 Haight, R. and Seidler, P.F. (1994) *Appl. Phys. Lett.*, **65**, 517.

108 Bauer, M. (2005) *J. Phys. D: Appl. Phys.*, **38**, R253.

109 Haight, R. (1995) *Surf. Sci. Rep.*, **21**, 277.

110 Haight, R. and Silberman, J.A. (1989) *Phys. Rev. Lett.*, **62**, 815.

111 Lei, C., Bauer, M., Read, K., Tobey, R., Liu, Y., Popmintchev, T., Murnane, M.M., and Kapteyn, H.C. (2002) *Phys. Rev. B*, **66**, 245420.

112 Siffalovic, P., Drescher, M., and Heinzmann, U. (2002) *Europhys. Lett.*, **60**, 924.

113 Seah, M.P. and Dench, W.A. (1979) *Surf. Interface Anal.*, **1**, 2.

114 Schins, J.M., Breger, P., Agostini, P., Constantinescu, R.C., Muller, H.G., Grillon, G., Antonetti, A., and Mysyrowicz, A. (1994) *Phys. Rev. Lett.*, **73**, 2180.

115 Kienberger, R., Goulielmakis, E., Uiberacker, M., Baltuska, A., Yakovlev, V., Bammer, F., Scrinzi, A., Westerwalbesloh, T., Kleineberg, U., and Heinzmann, U. (2004) *Nature*, **427**, 817.

116 Miaja-Avila, L., Saathoff, G., Mathias, S., Yin, J., La-o-vorakiat, C., Bauer, M., Aeschlimann, M., Murnane, M.M., and Kapteyn, H.C. (2008) *Phys. Rev. Lett.*, **101**, 046101.

117 Jurvansuu, M., Kivimaki, A., and Aksela, S. (2001) *Phys. Rev. A*, **64**, 012502.

118 Cavalieri, A.L., Muller, N., Uphues, T., Yakovlev, V.S., Baltuska, A., Horvath, B., Schmidt, B., Blumel, L., Holzwarth, R., and Hendel, S. (2007) *Nature*, **449**, 1029.

119 Hellmann, S., Sohrt, C., Rohwer, T., Marczynski-Buhlow, M., Bauer, M., Kipp, L., Rossnagel, K., Beye, M., Sorgenfrei, F., Fohlisch, A., Berglund, M., Schlotter, W., and Wurth, W. (2008) DESY Annual Reports.

120 Perfetti, L., Loukakos, P.A., Lisowski, M., Bovensiepen, U., Berger, H., Biermann, S., Cornaglia, P.S., Georges, A., and Wolf, M. (2006) *Phys. Rev. Lett.*, **97**, 067402.

121 Tong, X.M. and Chu, S.I. (1998) *Phys. Rev. A*, **58**, R2656.

122 Viefhaus, J., Avaldi, L., Snell, G., Wiedenhoft, M., Hentges, R., Rudel, A., Schafers, F., Menke, D., Heinzmann, U., Engelns, A., Berakdar, J., Klar, H., and Becker, U. (1996) *Phys. Rev. Lett.*, **77**, 3975.

123 Kim, S., Jin, J.H., Kim, Y.J., Park, I.Y., Kim, Y., and Kim, S.W. (2008) *Nature*, **453**, 757.

124 Boullet, J., Zaouter, Y., Limpert, J., Petit, S., Mairesse, Y., Fabre, B., Higuet, J., Mevel, E., Constant, E., and Cormier, E. (2009) *Opt. Lett.*, **34**, 1489.

22
Attosecond Time-Resolved Spectroscopy at Surfaces

Adrian L. Cavalieri, Ferenc Krausz, Ralph Ernstorfer, Reinhard Kienberger, Peter Feulner, Johannes V. Barth, and Dietrich Menzel

22.1
Overview

An area of physics exhibiting rapid progress and strongly focused interest is that of extremely fast phenomena; both the development of suitable methods and the understanding of such processes are at issue. While the term "ultrafast" was originally used for processes occurring in the (first upper, then lower) femtosecond range, today timescales of 1 fs or fractions thereof are accessible. This does not represent a marginal improvement, but encompasses a qualitative change: from the range where in principle the Born–Oppenheimer approximation, albeit with diabatic extensions, can be applied, to that of true nonadiabatic response, where the motion of individual electrons becomes the focus of attention. Therefore, the opportunity now exists to investigate the motions of electrons within many-electron systems, including the development of coupled many-body responses like the formation of band structure and the buildup of screening responses. First attempts to push into this range by looking at, for example, the ultrafast charge transfer at surfaces have been made successfully in the last 15 years with the core hole clock method. However, the latest developments utilizing attosecond laser pulses promise to provide a general tool to investigate in detail, and with few restrictions, this realm of electronic motions. While significant methodological development remains necessary for complete utilization and interpretation of measurements made on the attosecond timescale, the results obtained so far are extremely promising. It is timely, therefore, to give an overview of some past approaches, the most recent developments, and to identify the types of problems that should be attacked in the future using this exciting new tool. We will focus in particular on processes occurring at surfaces, because they contain *in nuce* the ingredients operative in atoms and molecules together with the important effects of solids.

Dynamics at Solid State Surface and Interfaces Vol. 1: Current Developments
Edited by Uwe Bovensiepen, Hrvoje Petek, and Martin Wolf

ISBN: 978-3-527-40937-2

With this contribution, a short overview of surface-relevant processes proceeding on the attosecond timescales will be given. In its first part, we present a selection of ultrafast phenomena and show how they can be accessed under continuous wave (cw) conditions, that is, in experiments where no temporal resolution is provided either by attosecond excitation pulses or by ultrafast detection. We use this part to introduce key phenomena of ultrafast dynamics on surfaces and to illustrate the limits of conventional methods. In the second part (Chapter 22.3), we demonstrate how the application of laser-based attosecond metrology can be used to overcome these limits.

22.2 Examples for Ultrafast Dynamics on Solid Surfaces

22.2.1 Electronic Response

When comparing electron binding energies from photoemission experiments on isolated atoms with one-electron energy levels of the unperturbed particle obtained from one-electron calculations, significant differences appear, which are larger for more localized and compact photoionized orbitals [1, 2]. This effect, known as intraatomic relaxation, is due to a rearrangement of the ion's electron distribution in response to the creation of the vacancy. As a result, the total energy of the ion is decreased and the kinetic energy of the photoelectron is increased. Coulombic and exchange energy also contribute to the observed increase in kinetic energy. For incomplete relaxation, part of the excess energy remains in the ion as a bound or continuum excitation, and additional shake-up or shake-off satellites are observed in addition to the main lines in the photoelectron spectrum [1, 3]. For molecules or solid compounds, this satellite structure can be very rich due to contributions of local and nonlocal excitations (e.g., plasmons) and charge transfer between the atomic constituents of the compound [1]. To gain perspective, we will examine a simple case: an atom adsorbed onto a surface that is photoionized. We assume that charge transfer is slow. In this case, only the polarization of the surrounding matrix will contribute the additional, extraatomic part to the total relaxation energy. It is largest for metallic substrates with delocalized conduction electrons (typically $\sim$1–2 eV) and smallest for insulators with low ε [1, 4]. How fast does this simple screening process proceed (and all the other, more complicated processes mentioned above)? Calculations indicate that the dynamics can be divided into two parts: For *short* times, defined by less than one half of a plasma frequency oscillation period, the electrons react ballistically as independent particles. For *long* times, collective oscillations with the plasma frequency are encountered [5]. For all metals, these dynamics proceed in the sub-femtosecond regime due to their corresponding plasma frequencies. So far, time-resolved measurements have only been possible for semiconductors with very low electron density and correspondingly small plasma frequency [6]. Do cw experiments allow access to these dynamics? One could consider photoemission experiments on identical atoms bound to different surfaces with very different plasma frequencies. Using different photon energies resulting in low and high kinetic photoelectron energies, that is, slow or fast photoelectron release, one could hope to see differences in the response of the matrix showing up as shifts of the experimentally obtained binding energies. Unfortunately, such an experiment fails. Due to the temporally extended photon pulses from conventional cw sources, the photoemission process will occur for a long time and the experimentally obtained spectrum will be dominated by the fully screened final state in both cases. Since the adiabatic limit is never reached, there will also be lines from less or unscreened final states even if the excitation process is extended by using extremely narrowband radiation [3]. The spectrum sums over all possible final states and for practical situations, it is always close to the sudden limit. Therefore, feasible cw approaches to screening dynamics are difficult.

One way to access the screening dynamics would be to monitor a process in which two electrons are emitted sequentially, separated by a time interval comparable to the critical screening time noted above. The analysis of postcollision interaction (PCI) line shapes [7] falls into this category and consequently seems promising. During PCI, near-threshold ionization of inner shell levels results in slow photoelectrons that are still within reach of the ion's Coulomb potential when the core hole decays by a subsequent Auger process. This doubles the ion's positive charge and increases the attractive force on the threshold photoelectron. Because of the statistical nature of core decay, broadened and red (blue) shifted energy profiles result for the photo (decay) electrons for such near-threshold ionization of core levels [7]. For an adsorbate on a surface, screening by extraatomic polarization reduces both PCI effects, that is, level broadening as well as energy shift [8]. We propose, therefore, that PCI profiles from very short-lived core levels of adsorbates contain information on the dynamics of this screening process. Because of the small ionization cross sections of such levels [9] and the large widths of their spectral lines [10], related experiments are difficult, which is probably the reason why no successful attempts have been made so far. Access to the dynamics of more complex satellites or to that of matrix-induced energy losses like plasmons or electron–hole excitations will be even more difficult to observe without the use of true attosecond metrology.

There is, however, an interesting exception. The spin-dependent screening dynamics of magnetic impurities in 3d ferromagnets have been successfully investigated by inelastic resonant Raman scattering (IRRS) [11]. Using the lifetime of the 2p core holes as a time ruler (see the description of the core hole clock method below), magnetic screening time constants ranging from 1.5 fs for Ni to less than 180 as for Fe have been obtained [11].

22.2.2
Charge Transfer Dynamics and Resonant Photoemission

Lifetimes of inner shell levels can be very short. For those with large binding energies or those decaying via Coster–Kronig or super-Coster–Kronig processes, lifetimes are well below 1 fs (see Ref. [10] for a compilation). Lifetime-broadened maxima are observed in kinetic energy spectra for the photoelectrons as well as for the decay (Auger) electrons. We note, however, that these homogeneous broadenings are correlated. For all decay processes with a final state of well-defined energy (e.g., the ground state of a doubly charged ion) and no further inelastic channels, energy conservation can only be violated for the (short) lifetime of the intermediate core-ionized state; it has to apply between initial (ground) and final (doubly ionized) states. In other words, a positive energy shift ($+\Delta E$) of the photoelectron is compensated by $-\Delta E$ of the decay electron. It has been convincingly demonstrated by coincidence experiments that excitation and de-excitation are coherent processes [12]. Unfortunately, coincidence experiments are very difficult to carry out, due to the very large background counts from solids. It is much easier to start with a *bound* core *resonance* instead and excite it with narrowband radiation. If the resonantly excited electron does not exchange energy with its environment during the lifetime of the core hole, which is typically the case for an isolated particle, due to energy conservation the energy of the decay electron (or decay photon) disperses linearly with the excitation energy. The decay electron spectrum resembles the off-resonant photoemission spectrum, with one-hole states (participator lines, corresponding to the main line in photoemission) and one-hole two-electron states (spectator lines, corresponding to shake-up satellites), however with (often very) different relative amplitudes [13]. For molecules, which may be vibrationally excited in the intermediate as well as in the final states, this dispersion rule holds strictly only within one set of vibrational quanta [13]; interesting redistribution processes among the vibrations are observed by detuning the initial excitation within the resonance. This coherent process, which is termed resonant photoemission or Auger resonant Raman effect (ARRE), is – besides other interesting applications in molecular physics – also a very powerful tool for investigation of ultrafast charge delocalization dynamics on surfaces [14–16].

Particles bound to surfaces or other molecules are not isolated. They can exchange charge and energy. If the status of the resonantly excited electron is changed during the lifetime of the core hole by interaction with the environment, the coherence of excitation and deexcitation noted above is lost. Particularly, delocalization of the resonance electron into a continuum, for example, into unoccupied electronic states of the substrate, converts the shape of the decay spectrum from resonant into normal Auger (Figure 22.1).

In many cases, both contributions can be well separated. Combined with the known lifetime of the core hole, the ratio of the integrated intensities of these two channels yield the charge delocalization time:

$$\tau_{\text{delocalization}} = \tau_{\text{core hole}} \cdot \frac{I_{\text{resonant}}}{I_{\text{nonresonant}}},$$

that is, $\tau_{\text{delocalization}}$ is measured in units of $\tau_{\text{core hole}}$ [14]. The CHC method looks at processes that are different from those observable by lasers in the optical range, but it also has some clear limitations. A fundamental complication is that the

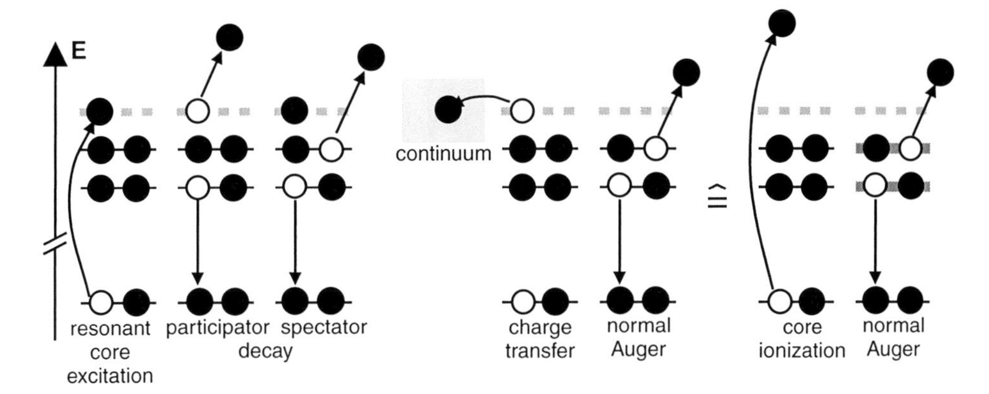

Figure 22.1 Charge delocalization measured by the core hole clock method. Inner shell levels and occupied outer shell/valence levels are indicated as solid black lines and unoccupied levels as dashed gray lines. Core excitons (left) decay via participator or spectator processes if the lifetime of the resonance is longer than the lifetime of the core hole. If the delocalization of the resonantly excited electron, that is, its transfer to a continuum, for example, into unoccupied levels of the substrate, is faster than core decay, normal Auger is observed (middle) as for primary core ionization (right).

core ionization increases the effective atomic number from Z to $Z + 1$ and therefore changes the chemical behavior of the atom during the lifetime of the core hole ("equivalent core approximation" [1]). The excited electron whose transfer is observed is therefore initially localized on the atom with the core hole, while for laser excitation in the optical range, the electron wave packet is delocalized in the adsorbate layer, and the hole is in the substrate. This does not necessarily mean that one method, either the CHC or the optical laser pump–probe, provides a more accurate depiction of the systems dynamics. In fact, in the case of the CHC, the clear initial localization may well be closer to actual charge transfer processes, but these differences must be taken into account in interpretation and comparison. A severe limitation is that core levels of appropriate lifetime [10] may not be available for the surface particles of specific interest, or that the resonance energy may not be suitable for such measurements [14]. In addition, in molecular adsorbates with spatially extended valence state wave functions, it is difficult to distinguish intramolecular charge delocalization from heterogeneous electron transfer, without further geometrical and/or energetic information. Finally, the CHC method supplies only a number, namely, the ratio of both channels, albeit as a function of energy through the resonance, but this number does not yield information about the temporal evolution of the charge distribution. We will show that future experiments with attosecond pulses can avoid these limitations. We finally note that extracting charge delocalization times simply from the widths of resonance lines fails in most cases because other effects, particularly from inhomogeneity of the samples and also from vibrational envelopes, will cause additional line broadening, which in most cases cannot be discriminated from the lifetime effect, even by sophisticated data evaluation.

22.2.3
Scattering Experiments and Band Structure Buildup

Atomic and molecular orbitals of particles condensed into the solid-state interact, forming energy bands with energy values dependent on electron momentum. However, as for the collective behavior in the screening example, a few scattering events are necessary for the electrons to probe the structure of their environment in order to form bands. On a very short timescale, we can assume that the electrons "do not know" about the solid structure or the corresponding electronic band structure. This is confirmed by theoretical investigations and experiments on charge transfer during scattering of fast ions at surfaces. Neutralization of negative ions probes the unoccupied, while that of positive ions the occupied density of states at the surface. For crystalline materials, the density of states varies strongly with energy showing maxima, minima, and gaps. As a result, charge transfer rates should vary depending on the energetic positions of the projectile ionization or affinity levels and indeed such effects have been found using the CHC method outlined above [14]. Theory [17, 18] and experiment [18] show that the signature of the band structure in the charge transfer rates vanishes for increasing particle energies, corresponding to decreasing interaction times. We will see that this aspect is also of fundamental importance for experiments with attosecond light pulses and their interpretation.

22.3
Attosecond Experiments at Surfaces

22.3.1
General Remarks

The measurement of attosecond dynamics relies on the reproducible generation of events with attosecond duration and on probing techniques with commensurate resolution [19]. Until recently, the briefest reproducible events have been pulses of near-infrared (NIR) laser light, with durations of around 5 fs [20, 21]. At present, high-energy NIR pulses as short as 3.4 fs can be produced [22]. Traditionally, the fastest measurement techniques have used the envelope of these laser pulses for sampling [23]. However, these techniques have not yet been capable of resolving the time structure of subfemtosecond transients.

With the arrival of isolated attosecond pulses in the extreme ultraviolet (XUV) [24–27] and of carrier-envelope phase-stabilized intense few-cycle laser pulses [28], an apparatus was developed allowing reconstruction of atomic processes with a time resolution within the Bohr orbit time of $\sim$150 as. In this apparatus, an accurately controlled few-cycle wave of visible light is used to collect "tomographic images" of the time–momentum distribution of electrons ejected from atoms following sudden excitation, for example, by XUV pulses of attosecond duration. From these images, the temporal evolution of both the emission intensity and initial momentum of released electrons can be retrieved on a subfemtosecond timescale. As a first application, primary photo and secondary Auger electrons were probed. In general, transients can be triggered by an isolated attosecond electron or photon burst synchronized with the probing light field oscillations.

22.3.2
Principles of Surface-Related Attosecond Metrology

The technique draws on the basic operation principle of a streak camera [29–33]. Deflection of the electrons in a rapidly varying electric field allows reconstruction of the temporal profile of the electron bunch. In our case, the intense few-cycle pulse of a laser supplies this field (see below). By measuring the evolution of the emission intensity and momentum distribution of photoelectrons, this attosecond transient recorder (ATR) [34] has already provided insight into the temporal rearrangement of the electronic shell of excited atoms on a subfemtosecond scale. This technique has now been extended to the investigation of solid surfaces.

22.3.3
Generation of Isolated Attosecond Pulses and Principle of Attosecond Spectroscopy of Solids

In a gas target, the electric field of intense linearly polarized femtosecond laser pulses induces – in a highly nonlinear interaction – gigantic dipole oscillations by pulling an electron out of the atom and subsequently forcing it back toward the core in the next half-cycle of the electric field. These oscillations contain high-frequency components extending into the extreme ultraviolet and soft-X-ray regimes [20]. In a long laser pulse containing many field oscillation cycles, the giant dipole oscillations are repeated quasiperiodically, resulting in the emission of a series of high-energy bursts of subfemtosecond duration that appear in the frequency domain as high-order harmonics of the drive laser. In contrast, for an ultrashort few-cycle laser pulse, only a few dipole oscillations can occur, and furthermore these few oscillations vary strongly in amplitude. The field oscillation with the largest amplitude produces a single burst of radiation in the spectral range of the highest emitted photon energies [35].

With waveform-controlled few-cycle laser pulses [28], the giant atomic dipole oscillations can be precisely controlled and reproduced from one laser shot to the next. As a result, the XUV burst parameters, including duration, energy, and timing with respect to the laser field, are well reproduced from one shot to the next. Because the XUV burst is precisely synchronized to the field oscillations of the driving laser pulse, we can use the XUV burst *in combination with* the oscillating laser field for attosecond spectroscopy. This is essential, as these laser-produced XUV bursts are currently too weak to be used for *both* triggering *and* probing electron dynamics. Instead, the oscillating laser field, which changes its strength from minimum to maximum within only ~1200 as for a wavelength of 750 nm, must take the role of the probing attosecond pulse. Experiments have proven that precise control over the waveform of few-cycle laser pulses makes it possible to *control* and *observe* atomic processes on a subfemtosecond timescale.

Corkum and coworkers [32] proposed the basic concept of ATR metrology in 2002, which was further developed with a comprehensive quantum theory by Brabec and coworkers [33]. Today, ATR based on photoemission from noble gases is an established technique for characterizing the duration of attosecond XUV pulses [22, 34, 36]. The principle of ATR is based on generating a photoelectron by an attosecond XUV burst in the presence of an intense, linearly polarized, few-cycle laser field $E_L(t) = E_0(t)\cos(\omega_L t + \varphi)$. The momentum of the released electrons is changed by $\Delta p = eA_L(t)$ along the laser field vector, where $A_L(t_r) = \int_{t_r}^{\infty} E_L(t)\mathrm{d}t$ is the vector potential of the laser field, e stands for the rest mass of the electron, and t_r is the instant of release of the photoelectron. This momentum transfer Δp maps the temporal profile of the electron emission onto a similar distribution of final momenta $p_f = p_i + \Delta p$ within a time window of $T_0/2 = \pi/\omega_L$. If the electron's initial momentum p_i is constant in time and their emission terminates within $T_0/2$, the temporal evolution of the electron emission can be unambiguously retrieved from a single "streaked" momentum distribution.

In the case of multiple, distinct emission lines in the photoelectron spectrum, the ATR can be used to compare the characteristics of the photoelectrons originating from the different states. For example, the photoelectron spectrum of a (110)-oriented tungsten surface obtained with the attosecond XUV pulses (Figure 22.2) shows two distinct peaks originating from 5d/6sp valence band electrons and 4f/5p core levels. By employing the ATR, the relative timing of the photoelectrons emerging through the surface can be determined. As illustrated in Figure 22.3, photoelectrons traversing the surface at different instants of time are subject to different phases of the streaking field. It is possible to extract an effective emergence time of the electrons using the waveform-controlled streaking field even though it penetrates the bulk due to the optical index of refraction. As a result, the effective delay in photoemission can be reconstructed from the spectrograms.

22.3.4
Hardware Requirements

At present, solid-state and surface experiments are performed in an attosecond beamline consisting of a HHG generation chamber, an XUV spectrometer, and several differential pumping stages in order to ensure UHV vacuum conditions in the measurement chamber. The measurement chamber contains a two-part multilayer mirror for the XUV and IR pulses as described in Ref. [23]. The mirror acts as a spectral filter for the generation of single, isolated attosecond pulses from 80 to 300 as duration and is – with its two parts – used to introduce the temporal delay between the XUV and the IR pulses. Emitted electrons are recorded with a time-of-flight spectrometer. The analyzer section of the measurement chamber is rotatable, allowing independent settings of the angles of detection and light incidence; the latter is important to meet the Brewster criterion for the NIR, which minimizes the effect of the phase-shifted reflected wave in the streaking experiment. The sample is

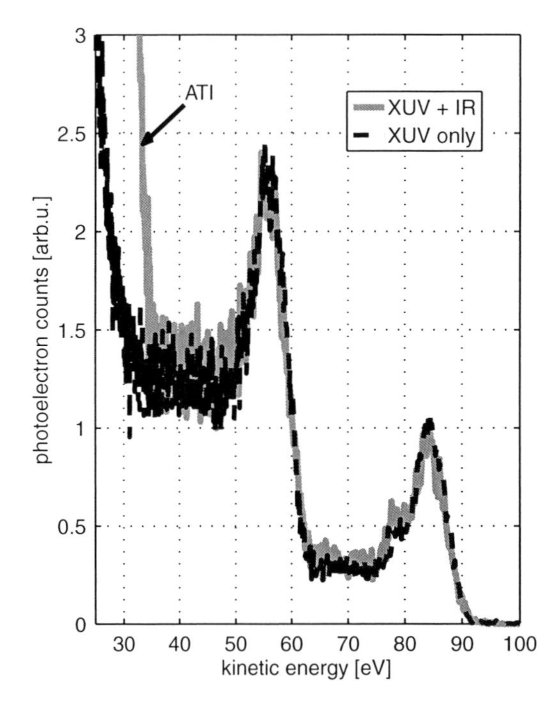

Figure 22.2 Raw photoelectron spectra of tungsten (110) measured with (gray curve) and without (dashed black curve) the presence of the probe NIR streaking field using XUV photons of ~91 eV. The spectra show two distinct peaks originating from 5d/6sp valence electrons and 4f/5p core levels at ~83 and ~56 eV, respectively. In the presence of the NIR probe field, there is an intense photoelectron signal below 35 eV induced by above-threshold ionization (ATI). Each spectrum was obtained by integration over 60 000 laser pulses.

mounted on a manipulator that moves it between the measurement chamber and a separate preparation chamber for cleaning, conditioning, and characterization purposes.

22.3.5
First Experimental Results

A tungsten (110) surface was the sample used in the first proof-of-principle experiments [38]. From clean W(110), two parallel spectrograms are observed that originate from different electronic states, 5d valence and 6sp conduction band electrons, and 4f, 5p core electrons, which are emitted with different kinetic energies (Figure 22.4a). The large bandwidth of the excitation pulse does not allow separation of the 4f/5p and the 5d/6sp intensities, respectively. But due to their comparatively small ionization cross sections and densities of states, the 5p and 6sp contributions are minor and the two emission peaks are dominated by 5d valence and 4f core contributions, respectively. Both spectrograms show the change in electron energy corresponding to the evolution of the electric field of the NIR streaking pulse. Figure 22.4b shows a

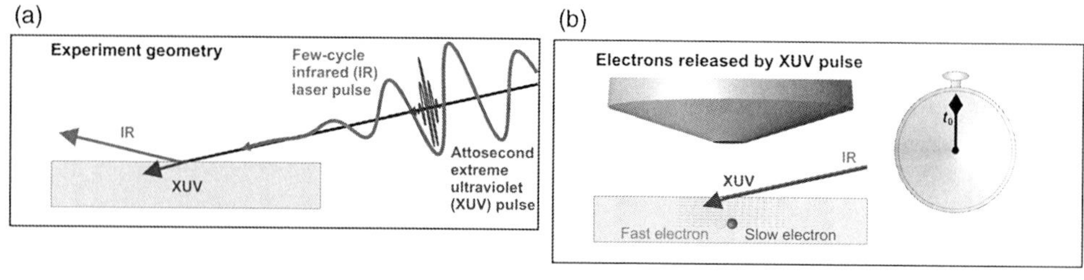

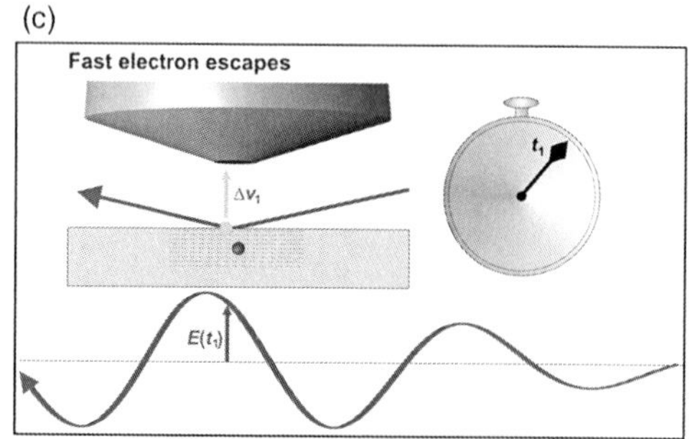

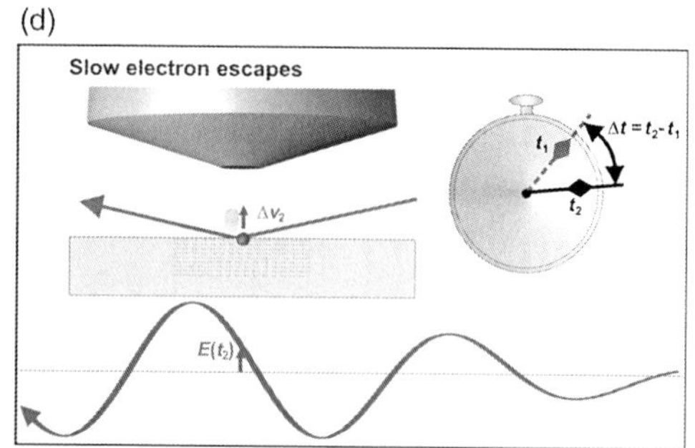

Figure 22.3 Attosecond spectroscopy on solids: Electrons arriving at the surface at different instants of time are subject to different phases of the streaking field outside the metal. (In this figure, we neglect the streaking field inside the solid, which due to the metal's optical properties is weak.) In panel (a), an isolated attosecond XUV pulse and a delayed few-cycle waveform-controlled streaking field are incident on a solid surface. In panel (b), at time t_0, the XUV pulse is absorbed in the solid and two types of photoelectrons are born, for simplicity one photoelectron is called "slow" and the other "fast." In panel (c), at time t_1, the fast electron has propagated to the surface and is now subject to the strong streaking field, which modulates its outgoing kinetic energy depending on the instant of release. In panel (d), at time t_2, the slow electron has reached the surface and feels the strong streaking field at the vacuum side; since it has emerged at a different time, the modulation of its kinetic energy will vary depending on the precise delay in emission. By evaluation of the full streaking spectrograms, collected as a function of relative delay between the attosecond XUV pulse and the streaking field, the delay in photoemission can be determined. Compared to streaking experiments at isolated particles, detailed models of electron localization, and electron and photon transport and interaction are necessary for the evaluation of such spectrograms.

center-of-mass (COM) analysis of the spectrograms. For this analysis, the time-dependent COM of both emission lines were calculated in a global fit by

$$\mathrm{COM}_{\mathrm{CB}}(t) = a_1 \mathrm{e}^{-4\ln(2)\frac{(t-t_0)^2}{\mathrm{FWHM}^2}} \sin(\omega t + \phi_0) + \mathrm{offset}_{\mathrm{CB}},$$

$$\mathrm{COM}_{4\mathrm{f}}(t) = a_2 \mathrm{e}^{-4\ln(2)\frac{(t-t_0-\Delta t)^2}{\mathrm{FWHM}^2}} \sin(\omega t + \phi_0 - \omega\Delta t) + \mathrm{offset}_{4\mathrm{f}},$$

where a_1, a_2, $\mathrm{offset}_{\mathrm{CB}}$, and $\mathrm{offset}_{4\mathrm{f}}$ denote the streaking amplitudes and the time-independent positions of the emission lines, respectively. t_0 and FWHM denote center and full width at half maximum of the Gaussian-shaped envelope of the streaking field, and ϕ_0 gives its carrier envelope phase. The parameter Δt accounts for a temporal shift between the spectrograms of both emission lines. The fit results are shown as solid lines in Figure 22.4b and a temporal shift in the streaking of 85 ± 35 as corresponding to the same delay in emission of the photoelectrons from the tungsten surface is extracted. This result is in good agreement with the initial study, where the valence electrons were found to be emitted approximately 100 as earlier than their tightly bound core-state counterparts [38].

These results demonstrate the technical capability of measuring photon-induced electron release, electron transport through the topmost atomic layers of a solid sample, and emission into the vacuum in real time, with attosecond temporal resolution. However, explaining the state-dependent differences of these emission dynamics as seen in the streaking experiment is, unfortunately, a much more challenging task than its equivalent for isolated particles [34]. Four different theoretical approaches to describe the dynamics seen in this W(110) experiment exist so

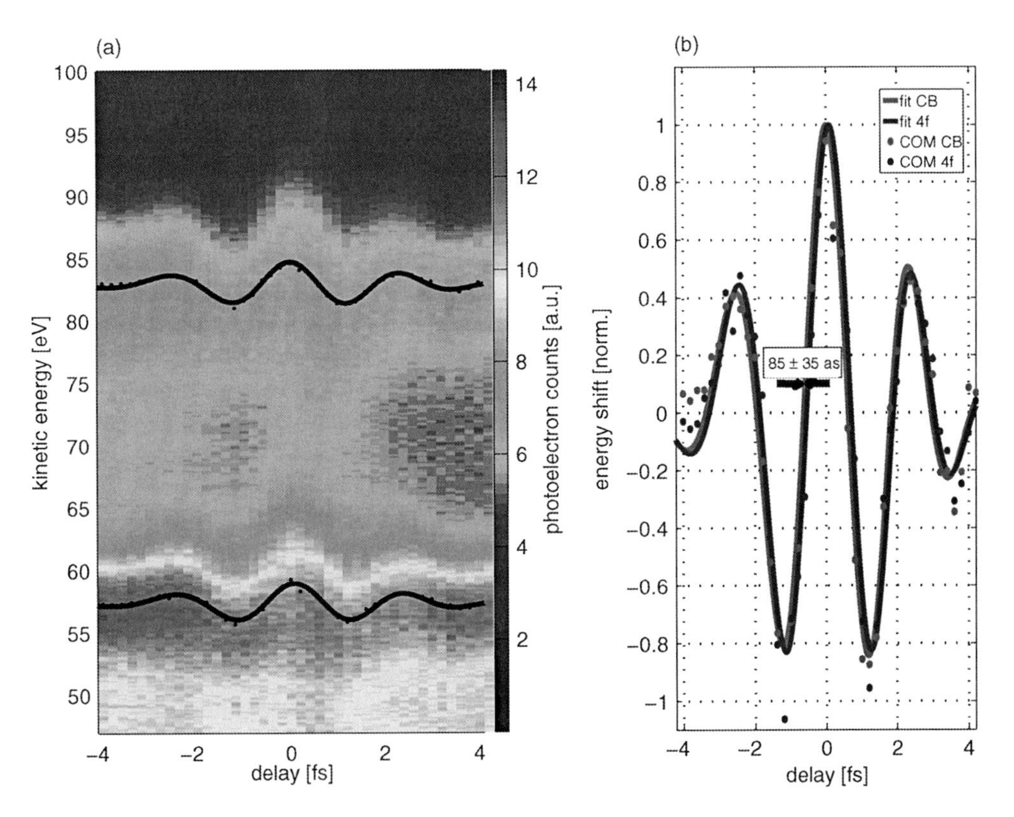

Figure 22.4 (a) Raw ATR spectrogram of a tungsten (110) surface. Photoelectron intensities are given in arbitrary units. The 5d/6sp and 4f/5p peaks are fitted with a Gaussian function and the peak positions at every time point are indicated by dots. The solid line shows the transient shifts of these peaks as obtained from a global fit equivalent to the center-of-mass (COM) analysis. (b) COM analysis of the spectrogram. The COM of both emission lines as measured are given as dots, a global fit to both COM traces is shown in solid lines. See text for details. (Please find a color version of this figure on the color plates.)

far, all yielding delayed emission of the core versus valence electrons from 42 to 110 as. The first theoretical approach by Echenique and coworkers explained the delayed emission of the core electrons by different group velocities of the final states [38]. Assuming validity of the static band structure picture, the authors showed that for the photon energy of the experiment (91 eV), the final state bands of the valence electrons exhibit stronger dispersion than those of the core electrons. As a result, a smaller effective mass, larger group velocity, and more rapid transport were concluded for the valence electrons, explaining the observed timing. A critical point of this model was the application of the static band structure. As indicated in the introduction, at least a few scattering events are expected to be required for the formation of delocalized states, and attosecond time intervals may be too short for these processes to occur. Considering this, Kazansky and Echenique have investigated in a second study the relative contributions of final and initial state effects to the observed dynamics [39]. In their revised quantum model, they treat the core electrons as localized and the valence electrons as completely delocalized. Attenuation by inelastic scattering is taken into account as well as electron–hole interaction for the inner shell levels, but not for the valence states. The streaking field inside the metal is assumed to be zero. Pseudopotentials obtained for copper are used to model electron transport. Calculations based on this one-dimensional theory also reproduce the experimentally obtained result quite well. Compared with the first approach, however, the relative magnitudes of final and initial state effects are reversed. The authors show that different final state energies contribute only a Δt of 10 as to the total delay; the main effect can now be attributed to the localized nature of the core electrons in contrast to the valence electrons that are treated as delocalized.

A third study by Zhang and Thumm also assumes localized core and delocalized valence electrons in the jellium approximation [40]. Compared to Ref. [39], the authors include the streaking field inside the solid: they argue that the skin depth (~100 Å) is much larger than the electron's mean free path (~5 Å). While the effect of streaking is treated nonperturbatively, the photoemission by the XUV pulse is treated by first-order perturbation theory. The authors introduce interfering contributions from different lattice layers to the photoemission dipole matrix element for the localized core states but not for the valence states, which according to their assumption are completely delocalized (jellium approximation). According to their model, this interference is the main source for the delayed emission of the 4f core electrons. For this delay they obtain 110 as, a value compatible with the experiment.

A fourth theoretical study by Lemell *et al.* models the attosecond streaking experiment by classical transport [41]. Quantum effects enter this classical calculation via a stochastic force $F_{\mathrm{stoc}}(t)$ containing elastic and inelastic scatterings with tungsten cores, as well as with conduction electrons. Other forces changing the electron's momenta result from the NIR streaking field, which penetrates ~85 layers of the solid in their model, and from the potential barrier at the surface. Rather detailed assumptions are made concerning the properties of the various involved electronic states. They discriminate 5p and 4f core electrons and 5d and 6s valence electrons, and treat only the 6s band as delocalized. Elastic scattering cross sections are calculated with the ELSEPA package [42]. Inelastic scattering cross sections and angular distributions of inelastically scattered electrons are obtained from the momentum distribution and energy-dependent dielectric constant of tungsten. Two limiting cases are considered for the final states: (a) a free particle dispersion relation and (b) the group velocity distribution of Silkin *et al.* (Ref. [38], supplementary material). Depending on these two alternatives, delayed emission of the core electrons from 42 (case a) to 110 as (case b) is obtained; the authors point out, however, that the group velocity distribution from Ref. [38] had to be blueshifted by 8 eV to obtain the maximum effect. For the limiting case (a), larger emission depth of the core electrons and the valence electrons scattered inelastically into the energy region of the core photoelectrons are the main sources of the observed delay. We note that the authors address possible extensions of their model, particularly inclusion of local field enhancements at the surface (plasmons), which might affect emission and transport of localized and delocalized states differently.

22.3.6 Future Experiments

22.3.6.1 Improving the Theoretical Description of Streaking Experiments on Solid Targets

The fact that four different theoretical approaches based on rather different assumptions yield contradicting explanations for the experimentally observed electron emission dynamics from a metal surface, yet numerical values for the expected delay, which all are within the experimental scatter, clearly indicates that the streaking process at solid surfaces is not well understood. Many processes and properties, such as primary photoemission, screening by itinerant electrons, transport, including elastic and inelastic scatterings in the bulk and across the solid/vacuum interface, the influence of the streaking field inside the material, and the localization of the individual electronic states contribute to the effect. Future experiments must help to solve this puzzle. A homogeneous sample like tungsten with its overlapping localized (5d) and delocalized (6sp) valence electrons and core levels of different symmetry (5p, 4f) that are energetically not resolvable under the conditions of an attosecond photoemission experiment may be too complex for this purpose. Well-tailored samples and experiments performed on these samples with different photon energies are required to disentangle the individual contributions of the above processes to the overall delay observed in the experiment. Such samples could be thin crystalline layers heteroepitaxially grown on top of substrates from different materials (Figure 22.5).

By varying thickness and elemental composition of these layered samples, transport and electronic properties could be varied independently and disentangled. Experiments with different photon energies would test the influence of final state effects.

22.3.6.2 Dynamics of Band Structure Formation, Screening, and Magnetic Effects

It is clear from the considerations given above that a comprehensive description of the streaking process must encompass information about the dynamics of band

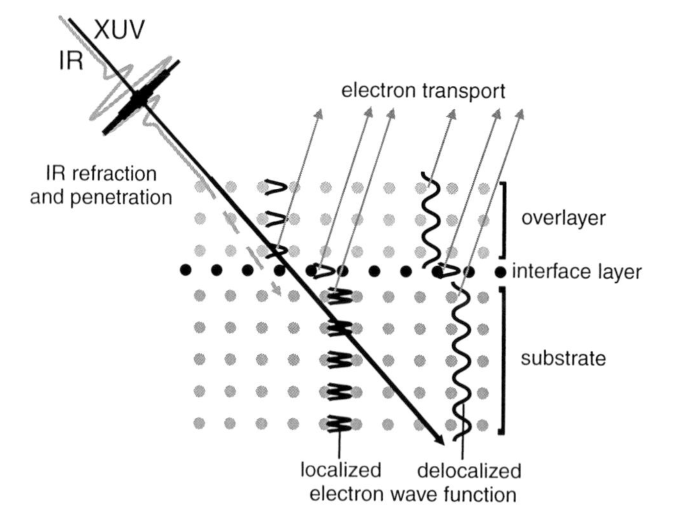

Figure 22.5 Sketch of an exemplary future sample design. Achieving a better understanding of the streaking process requires discriminating the influences of electron localization, electron transport surface and interface barriers, and refraction and penetration of the IR streaking field. Appropriately tailored crystalline sandwich structures of different materials, metallic and insulating, can help to solve this task.

structure formation and screening. To attain this, it will be particularly important to compare experiments with different photon energies, resulting in electron propagation in different final states, with different effective masses in the static band structure picture. We expect the maximum contrast for flat final states, for example, states with d or f symmetry, where the deviation from free particle dispersion is largest. Exciting electrons into such states, while varying the duration of the time interval during which the photoemission process occurs by tuning the spectral width of the XUV pulse, could reveal the transition from the free particle (for short pulses) to the static band regime (for long pulses), similar to the variation of the projectile energy in ion scattering experiments mentioned in Section 22.2.3.

On a short timescale, the concept of screening encompasses scattering and redistribution of electrons, and as such its dynamics must enter a detailed theory of streaking, as does that of the dynamics of band formation. Following the considerations above, well-designed samples can help to isolate the related phenomena. For many materials, photoemission is accompanied by simultaneous excitation of intrinsic plasmons, that is, collective excitations of the electron gas. Because of their many-body character, time for at least a few scattering events is required for their formation. A goal of future experiments will be the investigation of the emission dynamics of plasmon satellites and of features resulting from the inelastic electron–hole excitations. We also expect attosecond-related deviations from the static picture in line shape and line energy for polarization screening of photoemission from adsorbates, for example, from inert layers like monolayers of rare gases, and for charge transfer screening satellites in photoemission from compounds, for example, Mott insulators like nickel oxide. Of particular interest will be the class of valence band screening satellites obtained for 3(4,5)d transition metals, 4f rare earth metals, and 5f actinides. For these materials, vacancies in the 3(4,5)d or 4(5)f shells can be screened either by localized d or f electrons or by delocalized conduction electrons [1]. The latter screening process is less efficient, leading to photoemission satellites with binding energies larger than that of the main line by a few eV. It is of particular interest that these satellites can also be resonantly excited by the decay of [3(4,5)p]3(4,5)d or [4(5)d]4(5)f transitions, giving rise to interference between direct photoemission (with different screening scenario) and autoionization channels with emission delayed by the lifetime of the core hole. It is clear that by varying the material, the thickness of the active material (by layered samples), the photon energy, and the XUV pulse duration that many routes to study these different processes selectively will be opened. In particular, analysis of the temporal evolution of Fano line profiles resulting from interference phenomena will exclusively be possible with the attosecond spectroscopy approach [43].

The investigation of ultrafast magnetic phenomena is another field that has attracted considerable interest during the last years. Recently, coherent coupling of the polarization induced by the photon field of a femtosecond laser pulse with the spins of ferromagnetic samples has been demonstrated [44]. The origin of this coupling mechanism has been explained by relativistic quantum electrodynamics, beyond the spin–orbit coupling involving the ionic potential [44]. Intense, few-cycle IR pulses in combination with ultrashort XUV probe pulses and spin-sensitive detection will enable the investigation of such ultrafast magnetic phenomena with unprecedented temporal resolution. These studies will also include the microscopic details of the incoherent thermalization of excited spin and allow comparison with the corresponding charge processes.

22.3.6.3 **Charge Transport**

Ultrafast charge transport and adsorbate-to-substrate charge transfer are processes of key importance for the dynamics of bond formation and bond breaking, modification of materials by electronic excitations, electrochemistry, including processes in dye-sensitized solar cells, spectroscopy, and molecular electronics. The core hole clock method has clear limitations, as mentioned in Section 22.2.3. In particular, it is not possible to follow the evolution of the charge distribution in real time. Attosecond streaking, however, can achieve this task. After a resonant inner shell excitation by an attosecond XUV pulse, it can follow the entire evolution of the charge redistribution, starting with autoionization spectra immediately after excitation and ending with nonresonant core decay after complete charge delocalization. This transition between the two regimes will be evident in both the energetics and the line shapes of the streaking spectrogram of the decay electrons. This method will reach its full potential as soon as the XUV photon energies in the K-shell excitation region of the chemically important elements C, N, and O become available (from 280 to 540 eV).

All the experiments mentioned so far, apart from the magnetization studies, are common in that the excitation does not disturb the system under investigation. That is, the excitation does not drive the population of states far from equilibrium, as is the case, for example, in classical pump–probe experiments with femtosecond laser. At present, the intensity of the attosecond XUV pulses is too low to promote a large enough fraction of electrons into an excited state to be probed by a subsequent laser pulse.

Once they become feasible, such pump–probe experiments, with two attosecond pulses, will be powerful despite their perturbative nature. The second pulse could be used for ultrafast time-resolved XPS, where time-dependent chemical shifts would make it possible to monitor the electronic system's evolution. Tracer atoms with well-separated spectral features along the trajectory of charge transport would allow a site-selective view. At present, this detection scheme is applicable only for excitation with (low) harmonics of the NIR pulse. Despite these limitations, it is an extremely promising approach. The progress in light wave synthesis will enable detailed control of pulse shapes in time, making excitation of selected electronic states possible by coherent control of electron wave packets, the motion of which can then be followed by time-resolved XPS. We foresee for this method a level of potential comparable to multipulse NMR, which has revolutionized stereochemistry. We expect an analogous impact on the investigation of electron dynamics in solids and at interfaces and all related phenomena.

Acknowledgments

Help during the experiments by N. Müller, E. Magerl, S. Neppl, M. Stanislawski, E. Bothschafter, and N. Karpowicz is gratefully acknowledged. We thank Martin Wolf for helpful discussions. We gratefully acknowledge support by the Deutsche Forschungsgemeinschaft through the Excellence Cluster Munich Center for Advanced Photonics, Research Area A1 and C1. R.K. acknowledges support from the Sofja Kovalevskaja Award by the Alexander-von-Humboldt Foundation and the ERC Starting Grant.

References

1 Hüfner, S. (2003) *Photoelectron Spectroscopy*, Springer, Berlin, and references therein.
2 Shirley, D.A. (1972) *Chem. Phys. Lett.*, **16**, 220;Gelius, U. (1974) *J. Electron Spectrosc.*, **5**, 985.
3 Manne, R. and Åberg, T. (1970) *Chem. Phys. Lett.*, **7**, 282.
4 Chiang, T.C., Kaindl, G., and Mandel, T. (1986) *Phys. Rev. B.*, **33**, 695.
5 Borisov, A., Sánchez-Portal, D., Díez Muiño, R., and Echenique, P.M. (2004) *Chem. Phys. Lett.*, **397**, 95.
6 Huber, R., Tauser, F., Brodschelm, A., Bichler, M., Abstreiter, G., and Leitensdorfer, A. (2001) *Nature*, **414**, 286.
7 Thomas, T.D., Hall, R.I., Hochlaf, M., Kjeldsen, H., Penent, F., Lablanquie, P., Lavollée, M., and Morin, P. (1996) *J. Phys.*, **B29**, 3245.
8 Kassühlke, B., Romberg, R., Averkamp, P., and Feulner, P. (1998) *Phys. Rev. Lett.*, **81**, 2771.
9 Yeh, J.J. and Lindau, I. (1985) *Atom. Data Nucl. Data Tables*, **32**, 1.

10 Campbell, J.L. and Papp, T. (2001) *Atom. Data Nucl. Data Tables*, **77**, 1.

11 Braicovich, L. and van der Laan, G. (2008) *Phys. Rev.*, **B78**, 174421, and references therein.

12 Jensen, E., Bartynski, R.A., Hubert, S.L., Johnson, E.D., and Garrett, R. (1989) *Phys. Rev. Lett.*, **62**, 71.

13 Gel'mukhanov, F. and Ågren, H. (1999) *Phys. Rep.*, **312**, 87;Piancastelli, M.N. (2001) *J. Electron Spectrosc.*, **107**, 1, and references therein.

14 Menzel, D. (2008) *Chem. Soc. Rev.*, **37**, 2212;Brühwiler, P.A., Karis, O., and Mårtensson, N. (2002) *Rev. Mod. Phys.*, **74**, 703, and references therein.

15 Föhlisch, A., Feulner, P., Hennies, F., Fink, A., Menzel, D., Sanchez-Portal, D., Echenique, P.M., and Wurth, W. (2005) *Nature*, **436**, 373;Deppe, M., Föhlisch, A., Hennies, F., Nagasono, M., Beye, M., Sanchez-Portal, D., Echenique, P.M., and Wurth, W. (2007) *J. Chem. Phys.*, **127**, 174708.

16 Wurth, W. and Menzel, D. (2000) *Chem. Phys.*, **251**, 141.

17 Borisov, A.G., Kazansky, A.K., and Gauyacq, J.P. (1998) *Phys. Rev. Lett.*, **80**, 1996;Borisov, A.G., Kazansky, A.K., and Gauyacq, J.P. (1999) *Phys. Rev.*, **B59**, 10935.

18 Canário, A.R., Borisov, A.G., Gauyacq, J.P., and Esaulov, V.A. (2005) *Phys. Rev.*, **B71**, 121401(R).

19 Zewail, A. (2000) *J. Phys. Chem.*, **A104**, 5660–5694.

20 Brabec, T. and Krausz, F. (2000) *Rev. Mod. Phys.*, **72**, 545–591.

21 Keller, U. (2003) *Nature*, **424**, 831–838.

22 Goulielmakis, E. *et al.* (2008) *Science*, **320**, 1614.

23 Drescher, M. *et al.* (2001) *Science*, **291**, 1923–1927.

24 Paul, P.M., Toma, E.S., Breger, P., Mullot, G., Augé, F., Balcou, Ph., Muller, H.G., and Agostini, P. (2001) *Science*, **292**, 1689–1692.

25 Mairesse, Y. *et al.* (2003) *Science*, **302**, 1540–1543.

26 Tzallas, P. *et al.* (2003) *Nature*, **426**, 267–271.

27 Hentschel, M. *et al.* (2001) *Nature*, **414**, 509–513.

28 Baltuska, A. *et al.* (2003) *Nature*, **421**, 611–615.

29 Wheatstone, C. (1835) *Philos. Mag.*, **6**, 61.

30 Bradley, D.J., Liddy, B., and Sleat, W.E. (1971) *Opt. Commun.*, **2**, 391.

31 Schelev, M. *et al.* (1971) *Appl. Phys. Lett.*, **18**, 354.

32 Itatani, J. *et al.* (2002) *Phys. Rev. Lett.*, **88**, 173903.

33 Kitzler, M., Milosevic, N., Scrinzi, A., Krausz, F., and Brabec, T. (2002) *Phys. Rev. Lett.*, **88**, 173904.

34 Kienberger, R. *et al.* (2004) *Nature*, **427**, 817–821.

35 Christov, I.P., Murnane, M.M., and Kapteyn, H.C. (1997) *Phys. Rev. Lett.*, **78**, 1251–1254.

36 Goulielmakis, E. *et al.* (2004) *Science*, **305**, 1267.

37 Bandrauk, A.D. *et al.* (2004) *Int. J. Quantum Chem.*, **100**, 834; Niikura, H. *et al.* (2006) *Phys. Rev. A*, **73**, 021402; Yudin, G.L. *et al.* (2006) *Phys. Rev. A*, **72**, 051401.

38 Cavalieri, A.L. *et al.* (2007) *Nature*, **449**, 1029.

39 Kazansky, A.K. and Echenique, P.M. (2009) *Phys. Rev. Lett.*, **102**, 177401.

40 Zhang, C.-H. and Thumm, U. (2009) *Phys. Rev. Lett.*, **102**, 123601.

41 Lemell, C., Solleder, B., Tökési, K., and Burgdörfer, J. (2009) *Phys. Rev.*, **A79**, 062901.

42 Salvat, F., Jablonski, A., and Powell, C. (2005) *Comput. Phys. Commun.*, **165**, 157.

43 Wickenhauser, M., Burgdörfer, J., Krausz, F., and Drescher, M. (2005) *Phys. Rev. Lett.*, **94**, 023002.

44 Bigot, J.-Y., Vomir, M., and Beaurepaire, E. (2009) *Nat. Phys.*, **5**, 515;see also Bovensiepen, U. (2009) *Nat. Phys.*, **5**, 461.

23
Simultaneous Spatial and Temporal Control of Nanooptical Fields

Walter Pfeiffer and Martin Aeschlimann

23.1
Introduction

The fundamental understanding and flexible control of the light–matter interaction on very small length and ultrashort timescales is the key to innovations in a broad range of disciplines, including physics, biology, chemistry, communication and computation technology, as well as energy conversion technologies. The spatiotemporal dynamics of optical excitations govern the function of a wide range of natural and artificial nanodevices ranging from light-harvesting complexes [1], photovoltaic cells [2], and light-emitting devices [3] to more recently proposed surface plasmon-polariton-based nanosensors, nanowaveguides [4], or optical switches [5]. In all these systems, excitations evolve on exceedingly small nanometer length scales and on ultrafast femtosecond timescales. Strategies to manipulate the field evolution on the relevant time and length scales will thus provide a better understanding of the underlying mechanisms in nanophotonic devices and in addition will give ample possibilities to develop new applications in nanophotonics.

The spatiotemporal control of optical near-field distributions is based on the excitation of suitably designed nanostructures with coherent broadband radiation, that is, it combines the two up to now rather separated research fields of nanooptics and ultrafast laser spectroscopy. This combination has recently opened a new realm for probing and manipulating the time evolution of optical excitations on the nanoscale. The achievements in this emerging field include tip-enhanced second harmonic microscopy [6] and tip-enhanced electron emission microscopy [7] with 10 nm spatial resolution, simultaneous control of the spatial and temporal evolution of the optical near-field distribution [8–12], and the prospect of directly monitoring the field evolution with attosecond resolution [13] and low laser fluence high harmonic generation [14]. In the present context, the prediction that laser pulse shaping provides a handle to tailor the spatial and temporal evolution of the local field in the vicinity of a nanostructure [8, 9] is most relevant. This nanooptical control scheme [9] was recently demonstrated experimentally using polarization-shaped laser pulses in combination with two-photon photoemission electron microscopy (PEEM) [11, 15].

Dynamics at Solid State Surface and Interfaces Vol. 1: Current Developments
Edited by Uwe Bovensiepen, Hrvoje Petek, and Martin Wolf

ISBN: 978-3-527-40937-2

In the present review, we concentrate on the recent progress to tailor simultaneously the spatial and temporal degrees of freedom of the optical near-field distribution in the vicinity of a metal nanostructure. One perspective of this research is the development of innovative space–time-resolved spectroscopies [9]. However, this is only one possibility. Beyond mere spectroscopic applications, the tremendous flexibility of field control that is achievable in optical near fields driven by coherent broadband radiation offers an improved controllability of nanosystems. In comparison to transverse electromagnetic fields, such optical near fields exhibit more degrees of freedom and thus open new possibilities to control nanooptical excitations [16]. Nanooptical excitations exhibit fascinating properties, such as subwavelength variation of the field, local field enhancement, and local fields with vector components perpendicular to those of the incident field. We are not presenting details on these interesting phenomena here and refer the reader to excellent recent publications covering the state of the art in this field [17–23].

Nonlinear optics and in particular ultrafast laser spectroscopy of metallic nanostructures has attracted considerable interest over the years [24–28]. The interplay between the dynamics of collective and single-particle excitations was studied showing that time-resolved spectroscopy provides means to disentangle both dynamics although they occur on the same timescale [29]. However, these investigations were restricted purely to spectroscopy until Stockman *et al.* proposed to utilize the spectral phase of the optical near-field response to actually manipulate the spatiotemporal near-field distribution [8]. This stimulated the application of adaptive control schemes in nanooptics and thus also initiated the results presented in this review.

The basic idea of this control of nanooptical excitations is the following: The local electromagnetic field in the vicinity is in first approximation described by a linear response function that is driven by an external field. The actual spectral phase and amplitude of this response function determine the local fields and, therefore, the manipulation of the spectral phase and amplitude of the incident field independently for both polarization components can be used to tailor the local field evolution. The local response of the nanostructure is determined via Maxwell's equations by the material dielectric properties and the geometry of the nanostructure. Therefore, even the degree of controllability that can be achieved in the optical near field is limited by the actual shape and composition of the nanostructure. It is one of the driving forces in this field to understand and expand the limitations for near-field control.

The control demonstrated here should not be confused with "coherent control" usually restricted to the manipulation of quantum mechanic systems using the phase relationship between the excitations in a multiphoton process. Although the electrons in the nanostructure interacting with light represent a complex quantum mechanical many-body system, it is more conveniently described by Maxwell's equations using the linear susceptibility to characterize the behavior of the electrons. Consequently, the effects considered here are related to linear effects such as absorption of radiation. One possible control goal could therefore be to manipulate the local population of excited electrons by influencing whether locally light is absorbed at a given time or not.

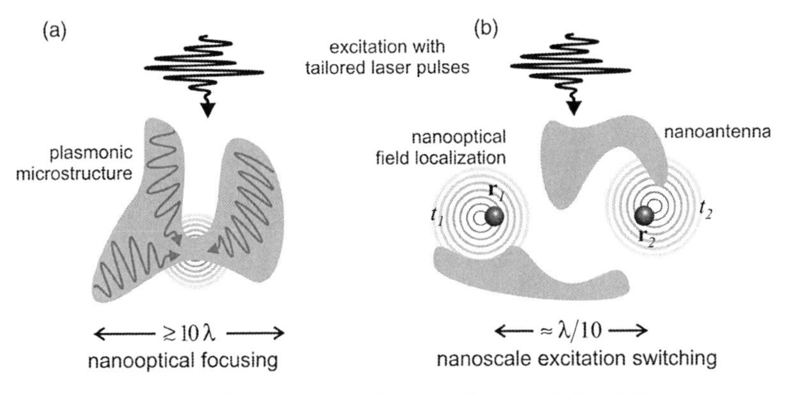

Figure 23.1 Scenarios for spatiotemporal control of nanooptical excitations.

The ability to manipulate the spatiotemporal evolution of nanooptical excitations is in itself fascinating. However, it is primarily only a tool to achieve better control over the nanoscale light–matter interaction and thus opens new prospects for optical driven quantum systems. We conceive primarily two different scenarios, as schematically depicted in Figure 23.1, for which the simultaneous spatial and temporal control of nanooptical excitations will be relevant. Propagating electromagnetic modes localized at the interface of the nanostructure are important for objects large compared to the optical wavelength. The resulting optical near-field distribution depends critically on the relative phase of the various such modes that interfere locally with each other. In Figure 23.1a, three different propagating modes excited in different parts of the nanostructure are shown that interfere at the constriction of the structure. The momentary frequency, amplitude, and polarization of the incident field determine the local phase and amplitude of these propagating modes. For example, the spatiotemporal focusing of propagating modes has been demonstrated for tapered plasmonic waveguides [30] and wedges [31]. The concept is not restricted to plane wave excitation of the plasmonic structure, but remains applicable for focused illumination of waveguides [32, 33]. Spatiotemporal control of such modes with subwavelength spatial resolutions could become essential in anticipated applications of plasmonic waveguides for information processing and nanooptical interconnects [34]. An even more interesting possibility arises on much shorter length scale, as depicted in Figure 23.1b. A suitably designed nanostructure exhibits highly localized optical near-field enhancements at various spatial locations. The actual spatial distribution depends on the incident field (wavelength and polarization) and the temporal evolution of the optical near-field distribution can be controlled by the momentary incident polarization, amplitude, and polarization [16]. For example, a switching of the nanolocalized excitation between $\mathbf{r}_1$ and $\mathbf{r}_2$ is achieved [9]. Methods based on nonlinear nanooptical processes such as local field-assisted scanning probe two-photon fluorescence microscopy [6] or nanooptical high harmonic generation [14] would significantly benefit from a signal enhancement because of an improved field localization that can be achieved by coherent control methods [9, 30, 35–37]. The applications are not limited to signal enhancements, but a flexible control over spatial evolution of nanooptical fields [9] provides means to control excitations in quantum systems in an unprecedented way. As will be discussed below in more detail, the new strategies for nanooptical switching and space–time-resolved spectroscopies are conceivable [9].

In Section 23.2, the basic concept of near-field control via polarization pulse shaping is presented. Based on a theoretical model, the importance of the polarization degree of freedom in the incident field is shown. The first experimental demonstration of this control scheme is based on two-photon photoemission (2PPE) electron microscopy, a polarization pulse shaper, and adaptive pulse shaping methods. These methods are studied in detail in Sections 23.3.1–23.3.4. The experimental results demonstrating simultaneous spatial and temporal control of optical near-field distributions are given in Sections 23.3.5 and 23.3.6. Finally, the anticipated future prospects for applications of the near-field control scheme are discussed in Section 23.4.

23.2 Optical Near-Field Control via Polarization Pulse Shaping

Polarization pulse shaping can be used to simultaneously control the spatial and temporal evolution of optical near-field distributions. Figure 23.2 summarizes the first study that demonstrated that local switching can be achieved by polarization-shaped laser pulses, whereas only marginal switching is achieved by phase shaping alone [9]. As a model nanostructure, a sharp metal tip positioned above a metal sphere (Figure 23.2a) is considered and two locations $\mathbf{r}_1$ and $\mathbf{r}_2$ in the vicinity of the structure are chosen at which the optical near-field evolution shall be controlled independently. The choice of the appropriate time-dependent polarization of the illuminating laser is not trivial and hence a genetic algorithm [38] is applied for

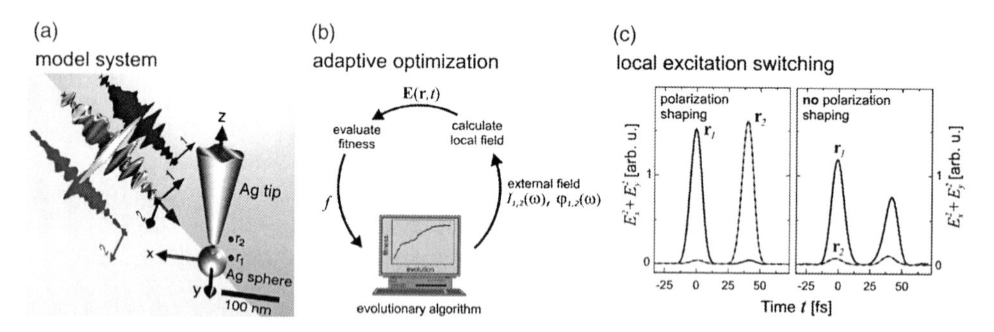

Figure 23.2 Ultrafast local excitation switching in the optical near field of a metal nanostructure using adaptively optimized polarization-shaped laser pulses. (a) Model nanostructure excited with a polarization-shaped laser pulse. (b) Evolutionary algorithm used to adaptively optimize a time-dependent field distribution with respect to a suitably defined fitness function f. (c) Comparison of optimal local switching efficiency for the spatial locations defined in (a) achieved with (left panel) and without (right panel) polarization shaping. In this case, a Au nanostructure was used for the calculations.

automated optimization of the incident field (Figure 23.2b). Based on the theoretically derived response function of the nanostructure and spectral intensities $I_i(\omega)$ $(i = 1, 2)$ and phases $\varphi_i(\omega)$ of the two incident polarization components E_i of the incident electric field $\mathbf{E}(\omega)$, the local field $\mathbf{E}(\mathbf{r}, t)$ is derived after Fourier transformation. The goal of an actual optimization is translated into a fitness function and together with the calculated local field, this provides a feedback for the genetic algorithm used for optimization. Switching of intensity of the x- and y-components of the electric field is defined as target and for polarization pulse shaping, a significant switching of the local intensities is observed (Figure 23.2c, left part), whereas pure phase shaping, that is, incident pulses with time-independent polarization, allows only a marginal switching (Figure 23.2c, right part). This demonstrates that polarization shaping is crucial to achieve simultaneous spatial and temporal control of an optical near-field distribution. This is also confirmed by a recent theoretical investigation of time-reversed nanooptical fields that shows that complex polarization-shaped fields are required to localize a nanooptical excitation in a random nanostructure simultaneously in space and time [39]. In the present case, the dominant control mechanism is based on two-pathway interference between the modes excited by the two incident polarization components [35]. The demonstrated ultrafast nanoscale switching of an optical near-field distribution opens a route toward space–time-resolved spectroscopies below the diffraction limit, in which pump and probe excitations can be located simultaneously and independently on a nanometer length scale and femtosecond timescale using a single laser pulse. Using this strategy to tailor nanooptical fields provides means to control quantum systems in an unprecedented manner and could significantly improve the possibilities of coherent control [16] that are already enhanced for far-field light–matter interaction by the polarization degree of freedom [40].

Any modifications in the incident field, such as altered spectral phase, amplitude, or polarization influence the optical near-field distribution. Using broadband coherent radiation, that is, the superposition of many different modes of the electromagnetic field, leads to a tremendously large parameter space. Thus, it is important to identify both the dominant control mechanisms and the critical degrees of freedom of the incident radiation for achieving a particularly tailored optical near field. We consider three different basic control mechanisms depicted schematically in Figure 23.3 that allow manipulation of the optical near-field distribution: (i) local mode interference, (ii) polarization matching to local dipoles, and (iii) local pulse compression by compensation of the spectral phase of the response function. These mechanisms are further discussed in the following. Note that all mechanisms are based on linear response theory and thus no nonlinear optical phenomena are considered here.

The mechanism of near-field control via local mode interference is depicted in Figure 23.3a. The nanostructure is illuminated with an incident far-field radiation. In the far field, two orthogonal polarization components $E_1(t)$ and $E_2(t)$ are required to describe the general polarization state of the incident wave and the phase between them determines the polarization state. In the far field, these two components do not interfere. However, in the near field, this is no longer the case. The local field in the

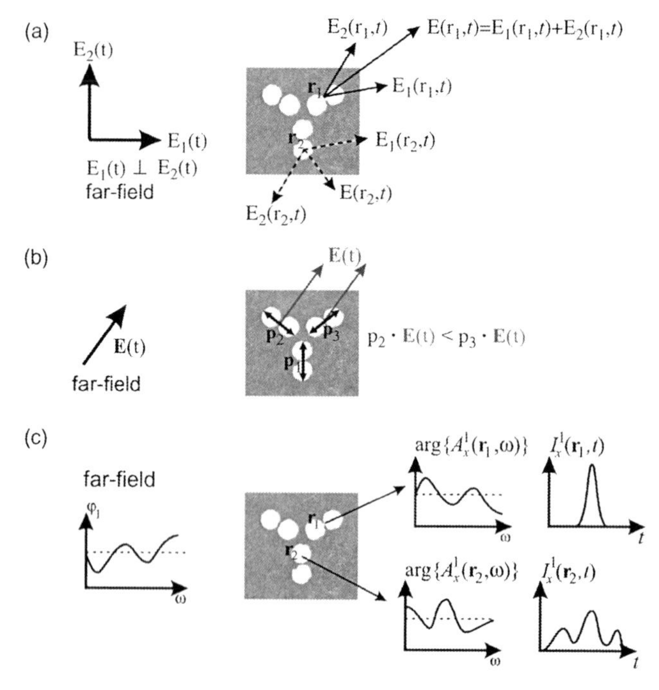

Figure 23.3 Mechanisms for optical near-field control. The left part shows the far-field degrees of freedom and the right part shows the effect for a model nanostructure. (a) Local interference of optical near-field modes. The two orthogonal far-field polarizations $\mathbf{E}_1(t)$ and $\mathbf{E}_2(t)$ excite local fields $\mathbf{E}_j(\mathbf{r},t)(j=1,2)$ that are no longer orthogonal and interfere with each other. The phase between $\mathbf{E}_1(t)$ and $\mathbf{E}_2(t)$ determines the local field. (b) Polarization matching to local dipoles. The response of the model nanostructure is approximated by three differently oriented local dipole moments $\mathbf{p}_k$. These moments are best excited if the far-field polarization is aligned to the local moments. (c) Local pulse compression. The spectral phase $\varphi_1(\omega)$ compensates the spectral phase of the x-component of the local response function $\arg\{A_x^1(\mathbf{r}_1,\omega)\}$ and therefore the corresponding local field is bandwidth limited. For a different location with different local response, this is not fulfilled and the local field is not bandwidth limited.

vicinity of the nanostructure can be expressed as convolution of the external field $\mathbf{E}(t)$ with a linear local response function $\mathbf{A}(\mathbf{r},t)$. In contrast to the far field, the local fields $\mathbf{E}_1(\mathbf{r},t)$ and $\mathbf{E}_2(\mathbf{r},t)$ generated by the two independent far-field components are not necessarily orthogonal to each other and can interfere. This gives a handle to manipulate the local field since the relative phase $\Delta\varphi$ of the incident field components $E_1(t)$ and $E_2(t)$, that is, the incident polarization state, determines the local interference of both local fields $\mathbf{E}_1(\mathbf{r},t)$ and $\mathbf{E}_2(\mathbf{r},t)$. If the local field amplitude is maximal for a given phase difference $\Delta\varphi$, the amplitude is minimal for a phase difference $\Delta\varphi+\pi$, since a phase shift of π shifts between "constructive" and "destructive" interference of the local fields. Note that the interference is only partial since the local fields $\mathbf{E}_1(\mathbf{r},t)$ and $\mathbf{E}_2(\mathbf{r},t)$ are not necessarily parallel. The scalar product of $\mathbf{E}_1(\mathbf{r},t)$ and $\mathbf{E}_2(\mathbf{r},t)$ represents a quantitative measure of the degree of control that can be achieved via the relative phase of the corresponding incident field components.

Even without interference of the local fields generated by the two incident polarization components, the spatial optical near-field distribution can be tailored at a subdiffraction length scale. If the nanostructure exhibits localized resonances that are approximated by local dipole moments $\mathbf{p}_k(\mathbf{r},\omega)$ with different orientations, it is possible to match the momentary polarization of the incident field to the orientation of one particular dipole moment $\mathbf{p}_1(\mathbf{r},\omega)$ and then later to another differently oriented dipole moment $\mathbf{p}_2(\mathbf{r},\omega)$. This mechanism is depicted schematically in Figure 23.3b. Thus, an ultrafast switching of the incident polarization will also result in a switching of the nanoscale field distribution. Both the bandwidth of the incident radiation and the spectral "linewidth" of the excited dipole moment limit the switching time. The influence of the local response function on the local field evolution is closely related to the third control mechanism discussed in the following.

The local response can be represented either in the time domain or in the spectral domain. "Local pulse compression," the third mechanism, is best considered in the spectral domain and schematically represented in Figure 23.3c. Fourier transform of the local response function $\mathbf{A}^j(\mathbf{r},t)$ (with j being the index of the polarization component) yields the response in the frequency domain $\mathbf{A}^j(\mathbf{r},\omega)$ that can for each Cartesian component be separated in spectral amplitude $\left|A_i^j(\mathbf{r},\omega)\right|$ $(i = x, y, z)$ and spectral phase $\varphi_{A_i}^j(\mathbf{r},\omega) = \arg\{A_i^j(\mathbf{r},\omega)\}$. The convolution of incident field and response function yielding the local field in the time domain corresponds to a multiplication of $A_i^j(\mathbf{r},\omega)$ and $E_j(\omega)$ with $E_j(\omega)$ being one component $j = 1, 2$ of the incident field in the spectral domain. A local field component $E_i^j(\mathbf{r},\omega)$ with the shortest possible duration, that is, a bandwidth-limited local field, is then achieved, if the sum of the spectral phase of the incident field $\varphi^j(\omega)$ and spectral phase of the local response $\varphi_{A_i}^j(\mathbf{r},\omega)$ gives a constant value. An accordingly chosen incident field $\mathbf{E}(\omega)$ leads to the highest local field strength of the particular local field component $E_i^j(\mathbf{r},\omega)$ and, therefore, also maximizes the local nonlinear response of a given nanostructure. Note that it is straightforward to compress one local field component. However, the local nonlinear response can well depend on all field components and thus the optimum incident field cannot be determined analytically. Adaptive optimization algorithms overcome this limitation. The condition for a bandwidth-limited local field is a local property and thus the compensation of the spectral phase of the response function at a given location $\mathbf{r}$ leads to local fields at other locations of the nanostructure that are not necessarily bandwidth limited.

These three control mechanisms are utilized to achieve the desired spatiotemporal evolution of the optical near field for a given nanostructure. The degrees of freedom are defined by the response function, that is, the geometry and material properties of the nanostructure, and the possibilities to manipulate the incident field. The recent experimental demonstrations of optical near-field control [11, 15] discussed in Section 23.3 allow only a rather limited assessment of the controllability that can in principle be achieved by the proposed scheme of optical near-field control via polarization pulse shaping of coherent broadband radiation. However, theoretical modeling provides valuable information. Various numerical techniques such as the boundary element method, multiple elastic scattering of the multipole expansions [41], or the finite difference time domain method provide solutions of

Maxwells's equations for complex nanostructures and allow calculating the response function. Based on the so determined local response, it is possible to investigate the near-field controllability in simulations. From these, the following conclusions are drawn:

- Simultaneous control over temporal and spatial degrees of freedom of the optical near field is significantly improved for polarization shaping in comparison to phase shaping alone [9, 35].
- The optical near-field controllability persists under illumination in a tight focus [42].
- Polarization pulse shaping also affects the local spectral distribution and thus opens a route to ultrafast nanooptical spectral multiplexing [35].
- The scheme of spatiotemporal field control is also applicable to propagating nanooptical modes in larger nanostructures that are locally excited with a tightly focused incident field [32, 33].
- In general, the external field required to achieve a particular optical near-field evolution is accessible via a direct solution and thus adaptive optimization strategies are applied to find the optimal incident field. However, for some targets, analytical solutions for the external field compare well with adaptively optimized fields [32]. The comparison of both solutions provides insight into the dominating field control mechanisms.
- The nanostructure defines and limits the possible optical near-field control and, therefore, it is conceived as a light focusing tool or nanoantenna. In addition to adaptive optimizations of nanooptical fields, an adaptive optimization of the nanostructure itself increases the flexibility of field control even further [43].
- Spatial control over nanooptical fields via polarization pulse shaping provides means for force field control and nanomechanical manipulation in optical tweezers [44].
- Two-pathway interference as it was used to achieve nanoscale field switching in the vicinity of a metal nanotip [9] is also applicable to control the longitudinal fields in a tightly focused Gaussian beam [45].

23.3 Experimental Demonstration of Spatiotemporal Control

Simultaneous spatiotemporal control of optical near fields is based on the identification of the required excitation pulse shape. Provided the complex response function of the used nanostructure is known, the incident pulse can in some cases be determined analytically [32]. However, in general, even for a known response of the nanostructure, the incident pulse required to achieve a particular spatiotemporal evolution of the local field cannot be derived analytically. Adaptive optimization as it has also been applied in theoretical studies to determine the incident field [9, 12, 35] provides means to determine the optimal pulse shape for a given target. Optimization is based on the definition of a suitable fitness function that captures the desired evolution of the optical near field. To minimize the chance of trapping in a local extremum of the fitness landscape, genetic algorithms are applied. In an experimental realization of the proposed near-field control scheme, the ability to determine the incident field in a closed-loop optimization creates constraints concerning the long-term stability and the speed with which the actually achieved local field can be determined. Multiphoton photoemission electron microscopy [10, 46–52] as a nonlinear method is sensitive to the local intensity of the electromagnetic field distribution. In addition, the lateral distribution of emitted electrons is detected spatially resolved and typically within a short integration time of several seconds, an emission pattern with sufficient statistics is recorded. This allows a fast feedback in the adaptive optimization of polarization shape of the incident laser pulses.

23.3.1
PEEM

A photoemission electron microscope is a versatile tool for mapping the lateral distribution of photoelectrons generated at a solid sample surface (as an overview, see Ref. [53]). Unlike all common types of electron microscopes, the PEEM is based on a "hybrid technique": it combines optical excitation with electro-optical imaging as shown in Figure 23.4. The present lateral resolution limit in PEEM is typically 5–20 nm, depending on the sample and illumination source. Since the first PEEM experiments of Brüche and Johannson [54], the method has been demonstrated for different applications, particularly in surface physics, chemistry, material science, magnetism, and nanooptics. The main advantage of the PEEM technique is its fast parallel image acquisition, similar to that of an optical microscope, which ensures a high degree of comparability of data taken at different locations inside the field of view. This is especially an important advantage in the time-resolved mode on a femto- and subfemtosecond timescale as it will be shown below. In combination with

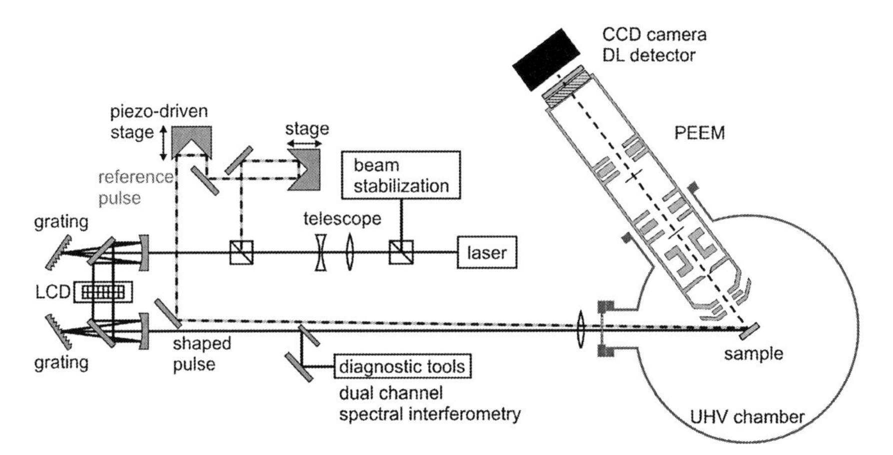

Figure 23.4 Experimental setup for the demonstration of simultaneous spatial and temporal control of optical near-field distributions.

various methods of spectroscopy, a number of different contrast mechanisms can be exploited to gain information on the local electronic, magnetic, and chemical structure of the surface [55].

23.3.2
PEEM as a Near-Field Probe

In the past years, femtosecond laser-induced nonlinear photoemission used in combination with PEEM has attracted considerable attention as a highly sensitive tool as a near-field probe. For instance, Cinchetti *et al.* used two-photon photoemission to map the lateral distribution of optical near fields in the vicinity of plasmon-resonant moon-shaped nanostructures [47]. Conventional threshold photoemission (off-resonant) showed a completely different picture of the surface mapping the nanostructure topography. Two examples illustrating the potential of the PEEM technique to map the local near field associated with a plasmon excitation at a nanometer resolution are shown in Figure 23.5. One-photon and two-photon

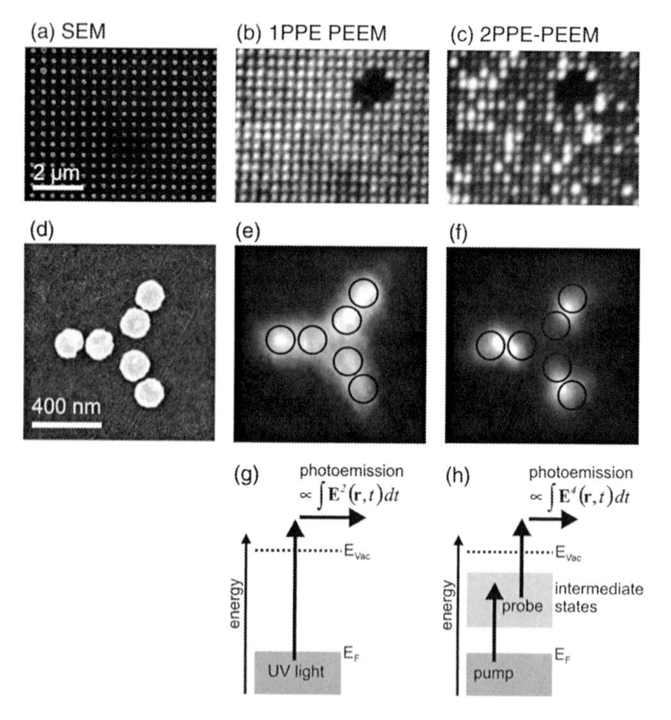

Figure 23.5 Comparison of one-photon and two-photon photoemission patterns from the planar Ag nanostructures on ITO. SEM images of Ag nanodot array (a) and a single Ag nanostructure (d) fabricated by electron beam lithography. (b and e) One-photon photoemission patterns obtained under mercury lamp UV illumination from the samples shown in (a) and (d). (c and f) Corresponding two-photon photoemission. (g and h) Excitation schemes for one-photon and two-photon photoemission processes, respectively.

photoemissions for two different nanostructures, a silver nanodot array (dot diameter: 200 nm, dot height: 50 nm) (Figure 23.5a–c), and an individual planar star-like arrangement of Ag nanodisk dimers (height 50 nm) (Figure 23.5d and f) are compared. Figure 23.5b and e represents the 1PPE images obtained under illumination at off-resonance conditions using a mercury vapor lamp ($h\nu = 4.9$ eV). The photon is so large that photoelectrons are directly emitted from the surface in a one-photon process (Figure 23.5g). The yield is proportional to the time-integrated local intensity and thus represents a measure of the local absorbed laser fluence. The silver particles are the dominating photoemitters, whereas the photoemission yield from the ITO interstitial areas is of the order of the background signal (MCP and CCD noise) or below. The one-photon photoemission pattern recorded for the dimer star (Figure 23.5e) reveals the complete nanostructure and under close inspection, the gap between the dimers is also detected in the emission pattern as a local dip providing an estimate for the spatial resolution of about 40 nm. Figure 23.5c and f represents the corresponding two-photon photoemission images obtained at resonant excitation for the nanodot array ($h\nu = 3.1$ eV) and at off-resonant excitation for the star-like structure ($h\nu = 1.5$ eV), respectively. For the nanodot array, a strongly increased photoemission signal from the silver nanoparticles is observed at the plasmon resonance frequency of the silver nanoparticles. The inhomogeneity in the local two-photon photoemission yield from different dots is related to statistical variations in the resonance frequency and in the density of structural imperfections (lattice defects) of the nanoparticles. Such imperfections or other scattering processes act as momentum source for nonradiative damping processes (decay of plasmons due to the creation of electron–hole pairs) and support intraband excitations required for the two-photon photoemission from noble metal particles [29]. For the star-like nanostructure, the work function was reduced just below 3 eV by dosing the surface with Cs to allow a two-photon photoemission process at an excitation wavelength of 800 nm. The two-photon emission is strongest from the dimer gaps (Figure 23.5f) indicating that the field enhancement in dimer gaps [56] is the reason for the enhanced electron emission. The gap field is also enhanced for an excitation wavelength far off the resonance and thus can also affect the emission pattern in two-photon photoemission at 800 nm excitation. In two-photon photoemission, the yield is proportional to the time integral over the fourth power of the local field (Figure 23.5h) and thus is influenced by the intensity of the local field. A stronger field enhances the yield nonlinearly and thus the emission pattern serves as a qualitative measure of the local field enhancement.

Near-field mapping as an alternative contrast mechanism in PEEM has the potential to open up complementary insights into the physics of nanostructured surfaces. At present, in these experiments, the two-photon PEEM is used to qualitatively monitor the local near-field distribution. The interpretation of these images is based on the assumption that the highest local field intensity will produce the highest photoemission yield in a two-photon photoemission process. However, for a quantitative interpretation of such data, an improved understanding of the nonlinear photoemission process at nanostructured surfaces is required where not only the local optical field amplitude is considered but also the electronic band structure of the emitting material. The response of the highly mobile conduction band electrons to an intense time-varying light field can open up additional pathways for photoemission that are not easily caught in the framework of standard one-step or three-step models. The two-photon photoemission yield can be influenced by additional parameters like the intermediate-state lifetime in the metal or the field component perpendicular to the surface. Only a detailed modeling of the emission process and the comparison with the experimentally observed emission pattern will provide an improved understanding of two-photon photoemission from nanostructured objects.

23.3.3
Time-Resolved PEEM

Due to the increasing interest in the specific hot-electron dynamics of spatially heterogeneous systems such as metallic nanostructures, an extension to a time-resolved PEEM (TR-PEEM) has been established. The potential of time-resolved PEEM as a versatile tool for mapping the electron dynamics of metallic nanoparticles has been introduced by Schmidt *et al.* [46]. A review article about the TR-PEEM method can be found in Ref. [57]. In contrast to the setup shown in Figure 23.4, typically a collinear path of the pump and probe beam is chosen, and either a Mach–Zehnder or a Michelson interferometer is employed to generate a sequence of two excitation pulses with defined temporal delay. This delay is achieved by a different path length of the two interferometer arms. The path length difference has to be controlled to better than 150 nm when a time step resolution of 1 fs is to be achieved. For each step, a PEEM image is taken resulting in a series of images containing a correlation trace for each pixel. A lifetime map is generated by plotting the FWHM of the cross-correlation traces deduced from a pixel-wise analysis of the TR-PEEM scan (p-polarized pump and s-polarized probe pulse). The lifetime map contains information on the dynamical behavior of the electron system at the sample surface (decay time T_1 of the intermediate state) with the spatial resolution of the PEEM. When the two pulses are collinearly overlapped, interferometric correlation signals with a periodicity of the optical wavelength are obtained. For 400 nm light, a much smaller step size (20 nm) and higher stability and reproducibility are required. Then, instead of using conventional motorized translation stages, piezoelectrically driven ones and interferometers actively controlled by a frequency-stabilized HeNe laser beam are employed [58]. It is also of importance to control the polarization – s, p, or circular – of both beams applied to the sample.

Figure 23.6 shows a PEEM image of a Ag nanodot array (as shown in Figure 23.5a–c) at resonance excitation. The 400 nm laser light used for the TR two-photon photoemission experiment couples almost resonantly to the in-plane mode of the particles. As discussed above, the two-photon photoemission image shows distinct interparticle brightness variations. The lifetime map as shown in Figure 23.6b visualizes the lateral variations in the electron dynamics in gray scale coding. Bright tones correspond to small FWHM values of the cross-correlation traces, indicating short decay times, while dark tones are associated with slow decay.

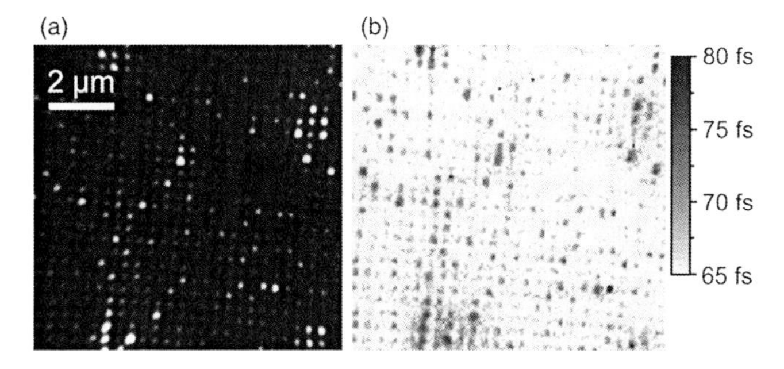

Figure 23.6 Time-resolved two-photon PEEM of a Ag nanodot array. (a) The two-photon photoemission yield ($h\nu = 3.1$ eV) from the nanodot array mapped with PEEM. (b) Corresponding FWHM lifetime map deduced from a local analysis of the cross-correlation traces recorded in a time-resolved PEEM scan.

Interestingly, the width of the cross-correlation measured for the individual nanoparticles shows some variation. A histogram analysis of the lifetime map and statistical analysis yield an average FWHM width of the cross-correlation of 67 fs with a standard deviation of 4 fs. The average width is significantly longer than the second-order autocorrelation width of about 50 fs of the used laser pulses. This increase reflects the lifetime of excited electrons in the nanoparticles that is probed here by TR-2PPE via indirect transitions. The standard deviation of 4 fs is larger than the statistical uncertainty and reflects a variation cross-correlation signal between different nanoparticles. However, this variation does not mean that the excited electrons in the various particles have different relaxation lifetimes, rather it is an indication of a variation of the individual plasmon resonances from particle to particle. As we have demonstrated in interferometrical time-resolved 2PPE from supported nanoparticles, the 2PPE signal is influenced by the lifetime of the collective excitation, that is, the plasmon, and the lifetime of excited electron in the intermediate state [29]. Consequently, we attribute the 4 fs standard deviation of the cross-correlation width for different nanoparticles to a variation of the plasmon dephasing time. Slow dephasing corresponds to a narrow plasmon resonance giving rise to a higher local field enhancement and longer lifetime of the collective excitation. In contrast, a fast dephasing leads to broad resonances and relatively weak local fields. The thus anticipated correlation between lifetime map and 2PPE yield is indeed observed in the comparison of Figure 23.6a and b.

The dynamic processes associated with a plasmon excitation in a single particle can be studied in further detail when the phase-resolved setup is employed. The image sequence in Figure 23.7 shows the plasmon dynamics in a single particle. The time interval between two images is 0.13 fs. A clear variation in the contrast within the area of the nanoparticle in the subfemtosecond timescale is seen. The result can be explained in the following way: The incident field drives the local polarization and the local electric field amplitude is determined from the interference of both fields. In case of pulsed illumination, the center frequencies of the driving laser light field and of the local polarization field are in general not identical. This gives rise to a time-dependent local phase between both fields, and thus at some times during the

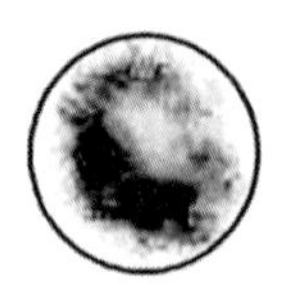

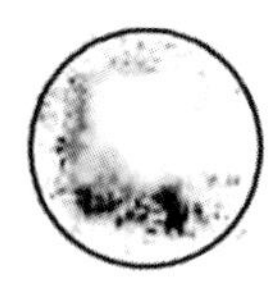

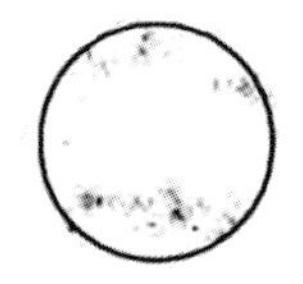

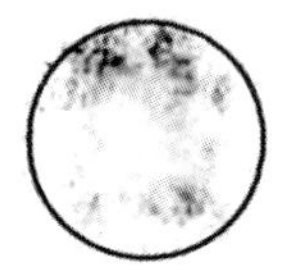

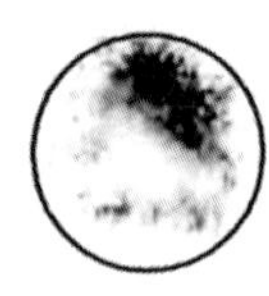

Figure 23.7 Spatiotemporal femtosecond dynamics of a silver nanoparticle. The particle has a diameter of 200 nm. A modulation of the lateral photoemission distribution here shown normalized to the emission at $t = 0$ fs as a function of phase delay between two identical exciting femtosecond laser pulses for two-photon photoemission is observed. This is assigned to a phase propagation of a plasmon through the nanoparticle. Reprinted from [49]. With kind permission of Springer Science & Business Media.

evolution of the pulse, the fields interfere constructively and at others destructively, that is, a local beating of incident field and induced polarization. In a single-pulse 2PPE pattern, this beating is averaged out and the nanoparticle appears as a more or less homogeneous emitter. However, in an interferometric time-resolved 2PPE experiment, the phase delay $\Delta\varphi(\tau)$ between pump and probe laser pulses affects the spectral phase and amplitude of the incident radiation via spectral interference of both pulses. This also affects the beating between incident field and local polarization giving rise to a lateral variation of the emission pattern of an individual nanoparticle as $\Delta\varphi(\tau)$ is varied. From the known incident field (illumination of the surface under an incidence angle of 65°) and the phase delay dependence of the emission pattern, the local field evolution can in principle be reconstructed. Effectively, the 2PPE snapshots recorded for different values of the phase delay $\Delta\varphi(\tau)$ reveal the "oscillation" of the local field via a variation of the local interference between incident field and induced polarization. Thus, the particle internal structure visible in a single PEEM image of Figure 23.7 is a residual of the varying interference between the external light field and the particle internal LSP field, directly connected to the plasmon phase.

These results are in good agreement with the findings by Kubo *et al.* [52]. They also combined the ultrafast laser spectroscopy and electron microscopy in order to image the quantum interference of localized SP-polariton waves with subwavelength spatial resolution and subfemtosecond temporal precision [10]. By scanning the time delay between identical, phase-correlated pump and probe in successive 330 as steps (corresponding to $\pi/2$ steps at 400 nm wavelength) and recording the resulting change in the polarization interference pattern, a movie of the surface plasmon-polariton (SPP) propagation wave packet at the Ag/vacuum interface can be created. They observed propagation of the plasmon with ~60% of the speed of light as a systematic shift of the beat pattern as a function of the delay time between pump and probe pulses [52]. The SPP wave packet propagation length and, hence, coherent control studies can even be improved by using single-crystal nanostructures as shown by Chelaru *et al.* [48].

23.3.4
Polarization Pulse Shaping

An important technique in this context is the shaping of the phase and amplitude of femtosecond laser pulses. Initially, femtosecond laser pulse shaping was developed for telecommunication purposes, but it is now a well-established technique [59] with applications in many branches of the ultrafast science community. Especially for coherent control purposes, shaped pulses provide a very flexible tool for influencing the dynamical evolution of quantum systems. An overview and also detailed discussion of the different developments and applications is provided in several review articles [60–63].

As discussed in Section 23.2, the polarization degree of freedom is essential to achieve simultaneous spatial and temporal control of nanooptical fields. The first demonstration of polarization shaping was achieved in Keith Nelson's group [64], although limited by the unbalanced diffraction efficiencies for the two orthogonal polarization components. Brixner and coworkers have eliminated this limitation and have developed a method of polarization shaping by which the polarization state can be modulated within a single femtosecond pulse [65].

The femtosecond polarization pulse shaper (Figure 23.4) contains a two-layer 128-pixel liquid crystal display (LCD) spatial light modulator in the Fourier plane of a zero-dispersion compressor in folded 4f configuration [65, 66]. The first and second LCD layers modulate the spectral phase of two transverse polarization components, leading to a spectral variation of polarization state and phase. Hence, in the time domain, the intensity, momentary oscillation frequency, and polarization state (i.e., elliptical eccentricity and orientation) can all be made to vary within a single laser pulse. In the present improved experimental setup, we use volume-holographic gratings to achieve high diffraction efficiency for both polarization components, avoiding efficiency-compensating Brewster windows employed earlier [65, 66]. Experimental characterization of the polarization-shaped laser pulses proceeds directly in the beam path of the experiment via dual-channel spectral interferometry with an unshaped reference laser pulse [66] whose spectral phase is determined separately by SPIDER [67].

23.3.5
Adaptive Optimization of Nanoscale Nonlinear Photoemission Patterns

As discussed in Section 23.2, the incident pulse shape required to achieve a particular optical near-field distribution can in general not be directly determined. Consequently, the adaptive optimization applied in the theoretical study of simultaneous spatial and temporal control of nanooptical excitations [9] is also used in an experimental demonstration. Figure 23.8 summarizes the first successful control of nanooptical fields by adaptive polarization pulse shaping [11]. Here the sample is illuminated only with the polarization-shaped laser beam, that is, the dashed reference beam depicted in Figure 23.4 is blocked. Two-photon PEEM is applied to monitor qualitatively the field distribution achieved by excitation of the nano-

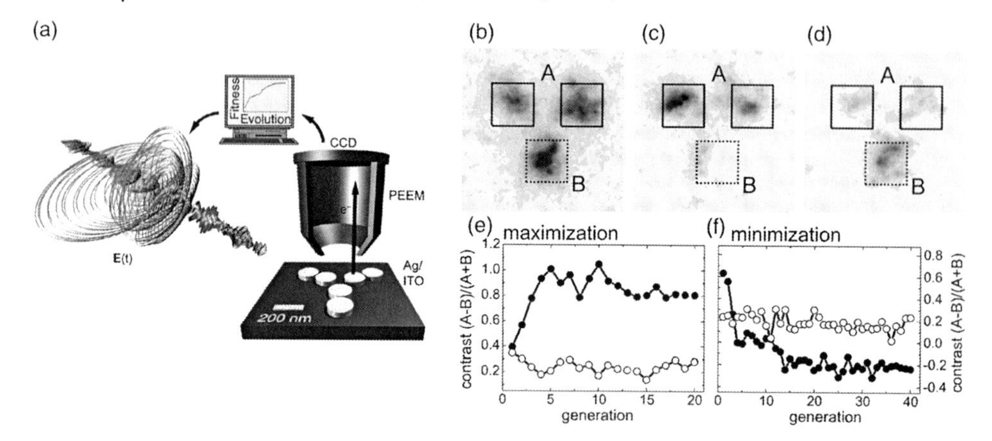

Figure 23.8 Adaptive nanoscopic field control. (a) A polarization shaper for ultrashort laser pulses controls the temporal evolution of the vectorial electric field **E**(t). These pulses illuminate a planar nanostructure and the two-photon photoemission is recorded by a photoemission electron microscope (PEEM). A CCD camera records the photoemission image and provides a feedback signal for an evolutionary learning algorithm. (b) The photoelectron distribution (PEEM pattern) for an individual nanostructure is displayed for p-polarized excitation. This emission pattern serves as a reference for the optimization experiments with polarization-modulated laser pulses. (c–f) Adaptive optimization of the A/B photoemission ratio leads to increased (open circle in (e)) and decreased (open circle in (f)) contrast compared to unshaped laser pulses recorded as a reference (filled circle in e and f). (c and d) The experimental PEEM images after adaptive contrast maximization and minimization using complex polarization-shaped laser pulses, respectively. The contour plots in (b–d) are shown on the same scale.

structure by polarization-shaped laser pulses. A fitness value is derived from the measured two-photon photoemission pattern and used as feedback in a genetic algorithm [38] to optimize the pattern according to the defined goal. The star-like nanostructure consists of three nanodisk dimers oriented under 120° to each other as shown in Figure 23.5d. For illumination with p-polarized unshaped laser pulses (center wavelength 790 nm, spectral full-width at half maximum 24 nm, repetition rate 75 MHz, maximum flux 25 μJ/cm^2, angle of incidence 65°, projected direction of incidence k parallel to the vertical boundary of the contour plot), the emission pattern (Figure 23.8b) consists of three strongly emitting areas. As discussed in Section 23.3.2, the emission induced by a two-photon process is localized in gap regions of the dimers. The p-polarized light of the laser pulses couples better to the bottom dimer oriented parallel to the projected k-vector of the incident pulse and therefore a stronger two-photon photoemission is observed. Note that the diameter of the whole star-like nanostructure is about 700 nm and hence the emission pattern observed reveals a field distribution with structure well below the diffraction limit. In the present case, two different optimization goals are defined. The contrast between the top two maxima (region of interest ROI A) and the bottom maximum (ROI B) is adaptively maximized (Figure 23.8c and e) and minimized (Figure 23.8d and f), respectively. After five generations, the maximized contrast saturates and the emission pattern clearly demonstrates the successful optimization, that is, the emission from ROI B is strongly suppressed. The emission pattern for unshaped incident pulses recorded as control signal shows no significant variation of the time required to run the optimization. In case of the minimization, the contrast saturates only after 20 generations, however, a reversal of the contrast is achieved and the emission from ROI B outweighs the emission from ROI A.

The successful adaptive optimization of a two-photon photoemission pattern proves that polarization pulse shaping is indeed a suitable tool to manipulate optical near-field distributions. Interestingly, it was not possible to achieve a contrast reversal as it is observed in the contrast minimization by "simple" polarized pulse shapes (compare supplementary information in Ref. [11]). Thus, the flexible pulse shaping capability is essential for spatial optical near-field control.

The emission patterns reflect not only the spatial distribution of the excited optical near field but are also influenced by the temporal evolution of the fields. The two-photon photoemission is proportional to the fourth power of the local field. Assuming the same deposited energy per pulse, intense local fields lead to stronger two-photon photoemission compared to weaker local fields of an optical near-field evolution that is spread out over longer times. The control of the emission pattern shown in Figure 23.8 thus reflects a control of the local excitation and/or a control of the local field evolution. As we have already mentioned in the introduction, linear quantities are controlled although a nonlinear optical process is used to monitor the control. A local excitation in the present case is, for example, the density of excited electrons that reflects the local absorption in the nanostructure and the local 2PPE yield is proportional to this quantity. From the time-integrated emission pattern alone, it cannot be determined how temporal evolution and spatial distribution of the local field and the associated local excitation are controlled. Therefore, time-resolved spectroscopy is required to indeed demonstrate spatiotemporal control (see Section 23.3.6).

The limited stability of the pulse shaper setup and laser system does not yet allow a direct comparison of several successive field optimizations with the same goal. Such an experiment would give evidence about both the robustness of the method and the uniqueness and reproducibility of the determined optimal pulse shape. With improved experimental stability, this is an important task for coming experiments and will provide information about the actual near-field control mechanisms as well. However, in theoretical adaptive optimizations of optical near-field distributions, we have always observed a rather robust behavior, that is, repeated optimizations starting from random initial pulses using the same genetic algorithm as it is applied in the experimental demonstration yielded always the same optimal pulse shape. This indicates that the genetic algorithm indeed finds the global extremum of the fitness landscape. For a stable surface condition and stable laser operation, we thus expect a robust behavior of the adaptive optimization scheme in the experiment.

23.3.6 Simultaneous Spatial and Temporal Control of Optical Near Fields

As mentioned in Section 23.3.4, the adaptive control of two-photon photoemission patterns does not yet demonstrate simultaneous spatial and temporal control over nanooptical excitations since the photoemission yield is integrated over the duration of the polarization-shaped laser pulse. Time-resolved two-photon PEEM [46] provides a way out. The cross-correlation between excitation with a polarization-shaped laser pulse and an unshaped reference pulse yields information about the evolution of the local excitation. The setup used for TR two-photon PEEM is depicted in Figure 23.4. The solid black line indicates the beam of the polarization-shaped laser pulse. The reference pulse follows the dashed path. In the present experiments, both pulses are centered at 800 nm wavelength. The delay between both pulses is adjusted by two delay stages. One stage allows large optical path differences and is used to overlap both pulses in time and to vary the delay in cross-correlation measurements. The second stage is piezo driven and introduces a fast phase jitter of about 2π in the path of the reference pulse. This cancels interferences between both pulses. The latter is particularly important since the interference between both fields also influences the optical near-field distribution.

Figure 23.9 shows the first demonstration of ultrafast optical near-field switching [15]. The polarization-shaped pulse is shown in an intuitive 3D representation in Figure 23.9d. The almost linearly polarized first subpulse has its polarization oriented parallel to the $\mathbf{E}_1$direction, whereas the second subpulse has its largest component parallel to the $\mathbf{E}_2$direction. This corresponds to an illumination of the sample with two almost linearly polarized pulses with 45° and −45° polarization with respect to the incidence plane. Figure 23.9a and b shows photoemission patterns

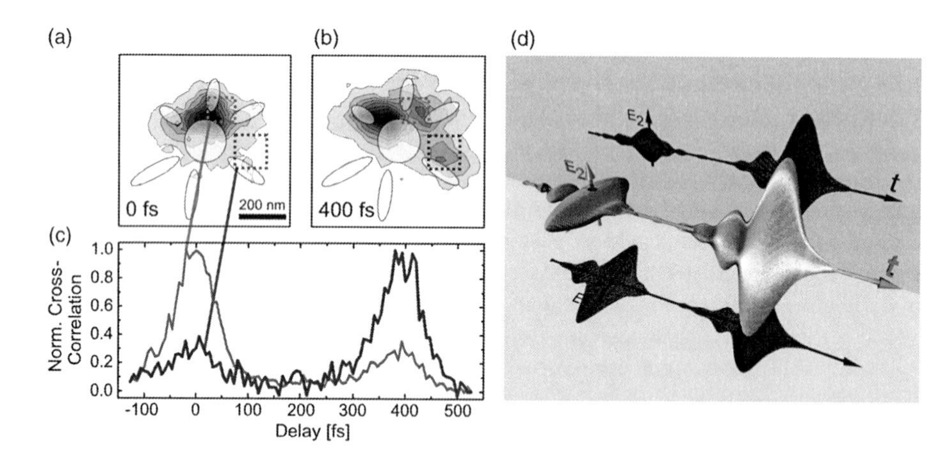

Figure 23.9 Time-resolved cross-correlation PEEM measurements [15]. (a and b) Two-photon PEEM emission patterns for temporal overlap of the circularly polarized probe pulse with the first and second subpulse of the excitation pulse, respectively, from the same nanostructure. The gray solid lines indicate the position of the nanostructure and the rectangles show the regions of interest used for the cross-correlation plot shown in (c). (c) Time-resolved averaged cross-correlation signals for different regions of the nanostructure as indicated by the lines. (d) Reconstructed excitation pulse shape in intuitive 3D representation. The subpulses centered at delay $t = 0$ fs and $t = 400$ fs correspond to polarization directions with respect to the incidence plane and as seen in direction of the propagating beam of 45° and −45°, respectively. (Please find a color version of this figure on the color plates.)

recorded for overlapping polarization-shaped pump pulse and circularly polarized probe pulse after background subtraction for a delay of 0 and 400 fs, respectively. Pump and probe pulses also generate photoelectrons via a two-photon photoemission process and this background contribution that contains no information about the evolution of the nanooptical excitation is subtracted. The emission patterns exhibit a striking variation – the maximum of the emission is "rotated" from the topmost arm of the nanostructure to the next arm on the left side. This corresponds to a lateral spatial shift of the excitation maximum of about 200 nm. In addition, the shape of the entire emission pattern also changes. To monitor the evolution of the emission pattern over time regions of interest are defined and the normalized yield is shown as function of the delay between pump and probe pulses in Figure 23.9c. The time-resolved cross-correlation traces for different regions of the nanostructure show a clear variation of their relative intensities in the time span of 400 fs. The emission patterns contain information about the interaction of pump and probe pulses with the nanostructure. Interferences between the two pulses are canceled by the phase jitter. This and the fact that the only parameter varying with the delay is the momentary polarization state of the pump pulse indicates that the observed pattern variation directly reflects the excitation pattern generated by the polarization-shaped pump pulse. The polarization-shaped incident laser pulse indeed switches ultrafast between two different excitation patterns. In the present example, the switching time is 400 fs since the separation between both subpulses of the polarization-shaped laser pulse was set to this value.

23.4 Future Prospects and Conclusions

So far the theoretical and experimental investigations serve primarily as proof of principle for the novel nanooptical control scheme that is based on suitably designed nanostructures acting as nanoantennas for polarization-shaped laser pulses. We have demonstrated that this scheme facilitates ultrafast spatiotemporal control of nanooptical fields and excitations. In addition, adaptive nanooptics has been demonstrated. This constitutes a first step toward fascinating new possibilities to control the light–matter interaction in nanophotonics. Starting from the successful demonstration of simultaneous spatial and temporal control of nanooptical excitations, it will now be essential to investigate the limitations, that is, identification of the ultimate spatial and temporal resolution, as well as the flexibility of the novel nanooptical control scheme. In addition, it is important to apply the scheme in a new spectroscopic method that reveals information about ultrafast processes that are not accessible with common methods.

Limitations and flexibility of field control are primarily defined by the optical response of the nanostructure. The investigation of different nanostructures is therefore essential to develop an understanding of the dominating control mechanism and will provide design rules for particular control targets. The use of PEEM as a tool for optical near-field characterization restricts us to planar nanostructures and thus limits the possibilities to investigate complex three-dimensional nanophotonic aggregates. Other spectroscopic tools such as fluorescence microscopy in combination with functionalized nanostructures might be of help in that respect.

A fascinating goal in photochemistry is the spatially and temporally resolved addressing of individual atoms in a molecule to achieve the ultimate control over the light-driven reaction [68]. The demonstrated resolution of optical near-field control is still far from addressing individual atoms. However, for artificial molecules formed from interacting quantum nanodots, the addressing of individual dots seems feasible.

Acknowledgments

We thank the PEEM team (M. Bauer, D. Bayer, T. Brixner, S. Cunovic, F. Dimler, A. Fischer, M. Rohmer, C. Schneider, C. Spindler, F. Steeb, C. Strüber, and D.V. Voronine) for the excellent experimental work. The theoretical support and the many discussions with F.J. García de Abajo have greatly contributed to the success of the project. This work was supported by the German Science Foundation (DFG) within the SPP 1391.

References

1 Engel, G.S., Calhoun, T.R., Read, E.L., Ahn, T.-K., Mancal, T., Cheng, Y.-C., Blankenship, R.E., and Fleming, G.R. (2007) Evidence for wavelike energy transfer through quantum coherence in photosynthetic systems. *Nature*, **446**, 782–786.

2 Brabec, C.J., Sariciftci, N.S., and Hummelen, J.C. (2001) Plastic solar cells. *Adv. Funct. Mater.*, **11** (1), 15–26.

3 Baldo, M.A., Thompson, M.E., and Forrest, S.R. (2000) High-efficiency fluorescent organic light-emitting devices using a phosphorescent sensitizer. *Nature*, **403**, 750–753.

4 Maier, S.A., Kik, P.G., Atwater, H.A., Meltzer, S., Harel, E., Koel, B.E., and Requicha, A.A.G. (2003) Local detection of electromagnetic energy transport below the diffraction limit in metal nanoparticle plasmon waveguides. *Nat. Mater.*, **2** (4), 229–232.

5 MacDonald, K.F., Samson, Z.L., Stockman, M.I., and Zheludev, N.I. (2009) Ultrafast active plasmonics. *Nat. Photonics*, **3** (1), 55–58.

6 Sanchez, E.J., Novotny, L., and Xie, X.S. (1999) Near-field fluorescence microscopy based on two-photon excitation with metal tips. *Phys. Rev. Lett.*, **82** (20), 4014–4017.

7 Ropers, C., Solli, D.R., Schulz, C.P., Lienau, C., and Elsaesser, T. (2007) Localized multiphoton emission of femtosecond electron pulses from metal nanotips. *Phys. Rev. Lett.*, **98** (4), 043907.

8 Stockman, M.I., Faleev, S.V., and Bergman, D.J. (2002) Coherent control of femtosecond energy localization in nanosystems. *Phys. Rev. Lett.*, **88** (6), 067402.

9 Brixner, T., García de Abajo, F.J., Schneider, J., and Pfeiffer, W. (2005) Nanoscopic ultrafast space–time-resolved spectroscopy. *Phys. Rev. Lett.*, **95** (9), 093901.

10 Kubo, A., Onda, K., Petek, H., Sun, Z., Jung, Y.S., and Kim, H.K. (2005) Femtosecond imaging of surface plasmon dynamics in a nanostructured silver film. *Nano Lett.*, **5** (6), 1123–1127.

11 Aeschlimann, M., Bauer, M., Bayer, D., Brixner, T., García de Abajo, F.J., Pfeiffer, W., Rohmer, M., Spindler, C., and Steeb, F. (2007) Adaptive subwavelength control of nano-optical fields. *Nature*, **446**, 301–304.
12 Sukharev, M. and Seideman, T. (2006) Phase and polarization control as a route to plasmonic nanodevices. *Nano Lett.*, **6** (4), 715–719.
13 Stockman, M.I., Kling, M.F., Kleineberg, U., and Krausz, F. (2007) Attosecond nanoplasmonic-field microscope. *Nat. Photonics*, **1**, 539–544.
14 Kim, S., Jin, J., Kim, Y.-J., Park, I.-Y., Kim, Y., and Kim, S.-W. (2008) High-harmonic generation by resonant plasmon field enhancement. *Nature*, **453**, 757–768.
15 Aeschlimann, M., Bauer, M., Bayer, D., Brixner, T., Cunovic, S., Dimler, F., Fischer, A., Pfeiffer, W., Rohmer, M., Schneider, C., Steeb, F., Strüber, C., Voronine, D.V. (2010) Spatiotemporal control of nanooptical excitations. *P. Natl. Acad. Sci. USA.*, **107**(12) 5329–5333.
16 Brixner, T., Pfeiffer, W., and García de Abajo, F.J. (2004) Femtosecond shaping of transverse and longitudinal light polarization. *Opt. Lett.*, **29** (18), 2187–2189.
17 Barnes, W.L., Dereux, A., and Ebbesen, T.W. (2003) Surface plasmon subwavelength optics. *Nature*, **424** (6950), 824–830.
18 Krenn, J.R. and Weeber, J.-C. (2004) Surface plasmon polaritons in metal stripes and wires. *Philos. Trans. R. Soc. Lond. A*, **362** (1817), 739–756.
19 Girard, C. (2005) Near fields in nanostructures. *Rep. Prog. Phys.*, **68** (8), 1883–1933.
20 Novotny, L. and Hecht, B. (2006) *Principles of Nano-Optics*, Cambridge University Press, Cambridge.
21 Novotny, L. and Stranick, S.J. (2006) Near-field optical microscopy and spectroscopy with pointed probes. *Annu. Rev. Phys. Chem.*, **57**, 303–331.
22 Ozbay, E. (2006) Plasmonics: merging photonics and electronics at nanoscale dimensions. *Science*, **311** (5758), 189–193.
23 Murray, W.A. and Barnes, W.L. (2007) Plasmonic Materials. *Adv. Mater.*, **19** (22), 3771–3782.
24 Klar, T., Perner, M., Grosse, S., von Plessen, G., Spirkl, W., and Feldmann, J. (1998) Surface-plasmon resonances in single metallic nanoparticles. *Phys. Rev. Lett.*, **80** (19), 4249–4252.
25 Lehmann, J., Merschdorf, M., Pfeiffer, W., Thon, A., Voll, S., and Gerber, G. (2000) Surface plasmon dynamics in silver nanoparticles studied by femtosecond time-resolved photoemission. *Phys. Rev. Lett.*, **85** (14), 2921–2924.
26 Link, S., Burda, C., Mohamed, M.B., Nikoobakht, B., and El-Sayed, M.A. (2000) Femtosecond transient–absorption dynamics of colloidal gold nanorods: shape independence of the electron–phonon relaxation time. *Phys. Rev. B*, **61** (9), 6086–6090.
27 Lamprecht, B., Schider, B., Lechner, R.T., Ditlbacher, H., Krenn, J.R., Leitner, A., and Aussenegg, F.R. (2000) Metal nanoparticle gratings: influence of dipolar particle interaction on the plasmon resonance. *Phys. Rev. Lett.*, **84** (20), 4721–4724.
28 Pfeiffer, W., Kennerknecht, C., and Merschdorf, M. (2004) Electron dynamics in supported metal nanoparticles: relaxation and charge transfer studied by time-resolved photoemission. *Appl. Phys. A*, **78** (7), 1011–1028.
29 Merschdorf, M., Kennerknecht, C., and Pfeiffer, W. (2004) Collective and single-particle dynamics in time-resolved two-photon photoemission. *Phys. Rev. B*, **70** (19), 193401.
30 Stockman, M.I. (2004) Nanofocusing of optical energy in tapered plasmonic waveguides. *Phys. Rev. Lett.*, **93** (13), 137404.
31 Durach, M., Rusina, A., Stockman, M.I., and Nelson, K.A. (2007) Toward full spatiotemporal control on the nanoscale. *Nano Lett.*, **7** (10), 3145–3149.
32 Tuchscherer, P., Voronine, D.V., Rewitz, C., García de Abajo, F.J.,

Pfeiffer, W., and Brixner, T. (2009) Analytic control of plasmon propagation in nanostructures. *Opt. Express*, **17** (16), 14235–14259.

33 Huang, J.S., Voronine, D.V., Tuchscherer, P., Brixner, T., and Hecht, B. (2009) Deterministic spatiotemporal control of optical fields in nanoantennas and plasmonic circuits. *Phys. Rev. B*, **79** (19), 195441.

34 Brongersma, M.L., Zia, R., and Schuller, J.A. (2007) Plasmonics: the missing link between nanoelectronics and microphotonics. *Appl. Phys. A*, **89** (2), 221–223.

35 Brixner, T., García de Abajo, F.J., Schneider, J., Spindler, C., and Pfeiffer, W. (2006) Ultrafast adaptive optical near-field control. *Phys. Rev. B*, **73** (12), 125437.

36 Stockman, M.I. (2008) Ultrafast nanoplasmonics under coherent control. *New J. Phys.*, **10**, 025031.

37 Rusina, A., Durach, M., Nelson, K.A., and Stockman, M.I. (2008) Nanoconcentration of terahertz radiation in plasmonic waveguides. *Opt. Express*, **16** (23), 18576–18589.

38 Baumert, T., Brixner, T., Seyfried, V., Strehle, M., and Gerber, G. (1997) Femtosecond pulse shaping by an evolutionary algorithm with feedback. *Appl. Phys. B*, **65** (6), 779–782.

39 Li, X.T. and Stockman, M.I. (2008) Highly efficient spatiotemporal coherent control in nanoplasmonics on a nanometer–femtosecond scale by time reversal. *Phys. Rev. B*, **77** (19), 195109.

40 Brixner, T., Krampert, G., Pfeifer, T., Selle, R., Gerber, G., Wollenhaupt, M., Graefe, O., Horn, C., Liese, D., and Baumert, T. (2004) Quantum control by ultrafast polarization shaping. *Phys. Rev. Lett.*, **92** (20), 208301.

41 García de Abajo, F.J. (1999) Multiple scattering of radiation in clusters of dielectrics. *Phys. Rev. B*, **60** (8), 6086–6102.

42 Brixner, T., García de Abajo, F.J., Spindler, C., and Pfeiffer, W. (2006) Adaptive ultrafast nano-optics in a tight focus. *Appl. Phys. B*, **84** (1–2), 89–95.

43 Yelk, J., Sukharev, M., and Seideman, T. (2008) Optimal design of nanoplasmonic materials using genetic algorithms as a multiparameter optimization tool. *J. Chem. Phys.*, **129** (6), 064706.

44 García de Abajo, F.J., Brixner, T., and Pfeiffer, W. (2007) Nanoscale force manipulation in the vicinity of a metal nanostructure. *J. Phys. B At. Mol. Opt. Phys.*, **40** (11), S249–S258.

45 Spindler, C., Pfeiffer, W., and Brixner, T. (2007) Field control in the tight focus of polarization-shaped laser pulses. *Appl. Phys. B*, **89** (4), 553–558.

46 Schmidt, O., Bauer, M., Wiemann, C., Porath, R., Scharte, M., Andreyev, O., Schönhense, G., and Aeschlimann, M. (2002) Time-resolved two photon photoemission electron microscopy. *Appl. Phys. B*, **74** (3), 223–227.

47 Cinchetti, M., Gloskovskii, A., Nepjiko, S.A., Schönhense, G., Rochholz, H., and Kreiter, M. (2005) Photoemission electron microscopy as a tool for the investigation of optical near fields. *Phys. Rev. Lett.*, **95** (4), 047601.

48 Chelaru, L.I., Horn von Hoegen, M., Thien, D., and Meyer zu Heringdorf, F.-J. (2006) Fringe fields in nonlinear photoemission microscopy. *Phys. Rev. B*, **73** (11), 115416.

49 Bauer, M., Wiemann, C., Lange, J., Bayer, D., Rohmer, M., and Aeschlimann, M. (2007) Phase propagation of localized surface plasmons probed by time-resolved photoemission electron microscopy. *Appl. Phys. A*, **88** (3), 473–480.

50 Wiemann, C., Bayer, D., Rohmer, M., Aeschlimann, M., and Bauer, M. (2007) Local 2PPE-yield enhancement in a defined periodic silver nanodisk array. *Surf. Sci.*, **601** (20), 4714–4721.

51 Kubo, A., Jung, Y.S., Kim, H.K., and Petek, H. (2007) Femtosecond microscopy of localized and propagating surface plasmons in silver gratings. *J. Phys. B At. Mol. Opt. Phys.*, **40** (11), S259–S272.

52 Kubo, A., Pontius, N., and Petek, H. (2007) Femtosecond microscopy of surface plasmon polariton wave packet evolution at the silver/vacuum interface. *Nano Lett.*, **7** (2), 470–475.

53 Schönhense, G. (1999) Imaging of magnetic structures by photoemission

electron microscopy. *J. Phys. Condens. Matter*, **11** (48), 9517–9547.

54 Brüche, E. and Johannson, H. (1932) Some new cathode experiments with the electric electron microscope. *Physikalische Zeitschrift*, **33**, 898–899.

55 Merkel, M., Escher, M., Settemeyer, J., Funnemann, D., Oelsner, A., Ziethen, C., Schmidt, O., Klais, M., and Schönhense, G. (2001) Microspectroscopy and spectromicroscopy with photoemission electron microscopy using a new kind of imaging energy filter. *Surf. Sci.*, **480** (3), 196–202.

56 Hao, E. and Schatz, G.C. (2004) Electromagnetic fields around silver nanoparticles and dimers. *J. Chem. Phys.*, **120** (1), 357–366.

57 Schönhense, G., Elmers, H.J., Nepijko, S.A., and Schneider, C.M. (2006) Time-resolved photoemission electron microscopy. *Adv. Imag. Elect. Phys.*, **142**, 159–323.

58 Lange, J., Bayer, D., Rohmer, M., Wiemann, C., Gaier, O., Aeschlimann, M., and Bauer, M. (2006) Probing femtosecond plasmon dynamics with nanometer resolution. *Proc. SPIE*, **6195**, 61950.

59 Weiner, A.M. (2000) Femtosecond pulse shaping using spatial light modulators. *Rev. Sci. Instrum.*, **71** (5), 1929–1960.

60 Rabitz, H., de Vivie-Riedle, R., Motzkus, M., and Kompa, K.-L. (2000) Chemistry: whither the future of controlling quantum phenomena? *Science*, **288** (5467), 824–828.

61 Brixner, T., Damrauer, N.H., and Gerber, G. (2001) Femtosecond quantum control. *Adv. At. Mol. Opt. Phys.*, **46**, 1–54.

62 Dantus, M. and Lozovoy, V.V. (2004) Experimental coherent laser control of physicochemical processes. *Chem. Rev.*, **104** (4), 1813–1859.

63 Silberberg, Y. (2009) Quantum coherent control for nonlinear spectroscopy and microscopy. *Annu. Rev. Phys. Chem.*, **60**, 277–292.

64 Wefers, M.M. and Nelson, K.A. (1995) Generation of high-fidelity programmable ultrafast optical waveforms. *Opt. Lett.*, **20** (9), 1047–1049.

65 Brixner, T. and Gerber, G. (2001) Femtosecond polarization pulse shaping. *Opt. Lett.*, **26** (8), 557–559.

66 Brixner, T., Krampert, G., Niklaus, P., and Gerber, G. (2002) Generation and characterization of polarization-shaped femtosecond laser pulses. *Appl. Phys. B*, **74** (S1), S133–S144.

67 Iaconis, C. and Walmsley, I.A. (1998) Spectral phase interferometry for direct electric-field reconstruction of ultrashort optical pulses. *Opt. Lett.*, **23** (10), 792–794.

68 Yang, N., Tang, Y.Q., and Cohen, A.E. (2009) Spectroscopy in sculpted fields. *Nano Today*, **4** (3), 269–279.

24
Coherently Controlled Electrical Currents at Surfaces

Jens Güdde, Marcus Rohleder, Torsten Meier, Stephan W. Koch, and Ulrich Höfer

24.1
Introduction

During the last decade, great progress has been made by investigating electron transport through single molecules, metallic point contacts, or chains of single atoms by using scanning tunneling microscopy or break junctions [1, 2]. Many of these works have focused on the coherent regime where inelastic scattering of the electrons due to vibronic or electronic excitation within the junction can be neglected. In this regime, nanoscale wires show no ohmic behavior and a quantization of the conductance has been observed as can be described by the Landauer theory [3]. While these experiments observe electron transport on the atomic scale, information about the dynamics of electron transport on the timescale of the relevant scattering mechanisms is scarce. The microscopic understanding of the mechanisms of electron transport, however, is not only of great interest for the development of new small-scale electronic devices, the connection of the electric conductivity σ of a material with the microscopic scattering processes of individual charge carriers is also a fundamental issue, central to solid state physics [4]. This question goes back to Paul Drude [5] who has connected the electric conductivity σ of a metal to an empirical scattering time τ of free electrons that are accelerated in an external electric field, resulting in the well known Drude formula $\sigma = e^2 n_e \tau / m_e$, where e, m_e, and n_e are the electron charge, mass and density, respectively. Even if this model is oversimplified as it does not consider the quantum nature and the many-body aspects of electron–electron interaction, it gives the correct order of magnitude for the timescale of electron scattering processes. For Cu at room temperature, for example, it gives a scattering time of 27 fs [4], which shows that usual conductivity measurements cannot resolve the individual scattering events because available electronic equipment can neither produce trigger signals nor detect transients that are shorter than tens of picoseconds.

Recently, we have introduced a new experimental technique that is capable of accessing the dynamics of electrical currents on the femtosecond timescale [6]. This contact-free technique combines the pure optical generation of electric

Dynamics at Solid State Surface and Interfaces Vol.1: Current Developments
Edited by Uwe Bovensiepen, Hrvoje Petek, and Martin Wolf

ISBN: 978-3-527-40937-2

currents by methods of coherent control with time- and angle-resolved two-photon photoemission (2PPE). Our technique can directly and sensitively detect electric currents by measuring energy and momentum of excited electrons with high time resolution.

The optical excitation scheme that we apply has been used to induce electric [7–9] and spin currents [10, 11] in bulk direct-bandgap semiconductors and to generate THz-radiation [12]. It is a variant of the Brumer–Shapiro scheme of coherent control [13]. Two phase-locked laser fields with frequencies ω_a and $\omega_a/2$ coherently excite electrons from an occupied into an unoccupied state, that is, from the valance band into the conduction band of the semiconductor. The quantum mechanical interference of the two different excitation pathways, the one- and the two-photon transition, allows to control the transfer of electrons at different points in momentum space (k-space). In this way, the direction as well as the magnitude of a current density can be controlled by changing the relative phase between the two laser fields.

In the first semiconductor experiments, the photoinduced currents have been detected by measuring a voltage drop between the two contacts [7, 9]. While this straightforward method clearly showed that the optical excitation resulted in macroscopic electrical current, the method not suitable for pump–probe investigations with ultrafast time resolution. For this purpose, a contact-free laser-based method is required not only for the generation but also for the detection of the current. Time- and angle-resolved photoelectron spectroscopy is an ideal experimental tool for this purpose because it makes it possible to map the momentum distribution of excited electrons with femtosecond time resolution. This is achieved by photoemission of the excited electrons with a time-delayed probe pulse combined with an angle-resolved detection as sketched in Figure 24.1. For the surface of a well-ordered solid, the momentum of the photoemitted electrons parallel to the surface $k_{\|}$ is conserved and the angle-resolved photoemission yield can be directly related to the momentum distribution of the density of the excited electrons $n_e(k_{\|})$. The asymmetry of the electron density with respect to the surface normal $n_e(+k_{\|})-n_e(-k_{\|})$ represents the momentum-resolved contribution of the excited electrons to a total lateral current density of the electrons

$$j_{\|} = -e\int_{0}^{k_{\max}} \left[n_e(+k_{\|})-n_e(-k_{\|})\right] \frac{\hbar k_{\|}}{m_e}\, \mathrm{d}k_{\|} \qquad (24.1)$$

where e is the electron charge, m_e the electron mass, and $k_{\max}$ is the maximum value of $k_{\|}$. Therefore, the combination of the optical current generation with time-resolved photoelectron spectroscopy not only is capable of detecting the total electron current with high temporal resolution but also, with the time-resolved observation of the momentum distribution, makes it possible to visualize the individual contributions of the excited electrons to the current. In this way, we gain information on the generation and decay of ultrashort current pulses in unprecedented detail. In particular, this technique makes it possible to observe the scattering processes of excited electrons leading to current decay in terms of an incoherent population dynamics in momentum space. In general, such optically excited electron current is

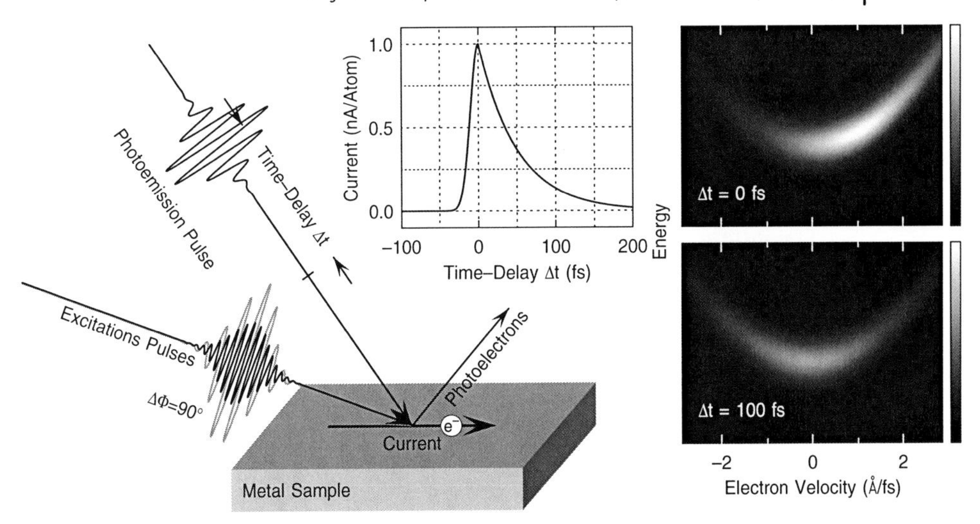

Figure 24.1 Sketch of the coherently controlled generation and time-resolved detection of ultrashort current pulses at surfaces. Two phase-locked laser pulses with carrier frequencies ω_a and $\omega_a/2$ excite electrons into a parabolic surface band. The coherent two-color excitation generates an asymmetry of the momentum distribution in the surface band, which corresponds to an electric current parallel to the surface. The magnitude as well as the direction of the current can be controlled by the relative phase $\Delta\phi = \phi(\omega_a) - 2\phi(\omega_a/2)$ between the excitation pulses. The time evolution of the current is observed by time- and angle-resolved photoemission spectroscopy using a third time-delayed laser pulse with carrier frequency ω_b, which maps the momentum distribution within the surface band. The graphs show synthetic data which are representative for an excitation of the first image-potential state on Cu(100) but with an enhanced asymmetry. The magnitude of the current is a rough estimation from the experimental results (see text).

accompanied by a hole current that is not observed in the present photoelectron detection scheme.

We have applied the optical current generation and detection scheme to the first ($n = 1$) image-potential state on a Cu(100) surface, which represents a prototype of a lateral delocalized electronic surface state. In these normally unoccupied states, excited electrons are bound only perpendicular to the metal surface by the Coulombic image potential and exhibit a free-electron-like dispersion parallel to the surface [14, 15]. Therefore, an excitation of electrons with specific momenta parallel to the surface into these normally unoccupied states generates a ballistic current flowing within a tiny sheet in front of the surface, which has a thickness of only a few angstrom.

24.2 Observation of Coherently Controlled Currents by Photoelectron Spectroscopy

The excitation of the first image-potential state on the Cu(100) surface using the coherent two-color excitation scheme requires two phase-locked ultrashort laser

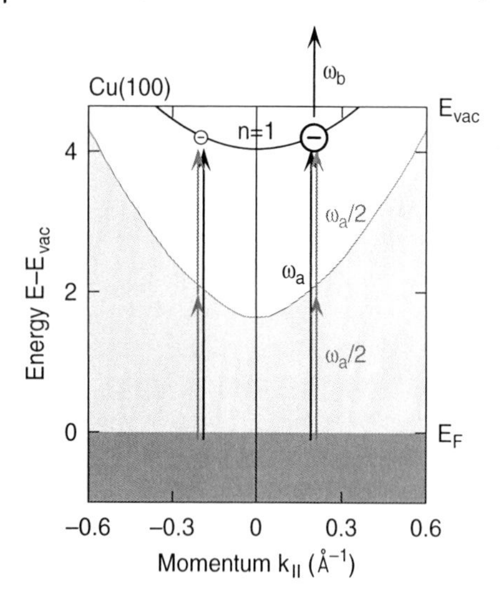

Figure 24.2 Projected surface band structure of Cu(100) including the first ($n = 1$) image-potential band together with the optical excitation scheme for the generation and detection of a surface current within the image-potential band. Light (dark) gray areas indicate empty (filled) projected bulk bands. Adapted from [6]. Reprinted with permission from AAAS.

pulses in the UV and in the visible spectral range as depicted in the projected surface band structure in Figure 24.2. Visible laser pulses in the wavelength range between 540 and 570 nm with a pulse duration of 60 fs served as excitation pulses with frequency $\omega_a/2$ and were generated using a high repetition rate Ti:sapphire amplifier system combined with an optical parametric amplifier. The excitation pulses with frequency ω_a were generated by frequency doubling of the visible pulses. Locking and tuning of the relative phase $\Delta\phi = \phi(\omega_a) - 2\phi(\omega_a/2)$ between the excitation pulses was achieved by using an actively stabilized two-color Mach–Zehnder interferometer [16]. A part of the fundamental output of the Ti:sapphire amplifier at a wavelength of 800 nm served as time-delayed probe pulses with frequency ω_b. Further details of the optical setup are described in detail in Refs [16, 17].

The laser beams were focused onto a Cu(100) single crystal sample that was located in a UHV chamber with a base pressure below 5×10^{-11} mbar at room temperature. The sample was prepared by standard sputtering and annealing procedures. Surface cleanliness and order were verified by X-ray photoelectron spectroscopy (XPS), low energy electron diffraction (LEED), and by linewidth measurements of the image-potential state with 2PPE [18].

Photoelectrons emitted from the sample by either the visible laser pulses with frequency $\omega_a/2$ or by the time-delayed 800 nm fundamental laser pulses were detected by a hemispherical analyzer equipped with an angle-resolved lens mode and a two-dimensional (2D) charge-coupled-device (CCD) detector, which allowed for single-shot $E(k_{||})$ measurements [19]. At a pass energy of 10 eV, a kinetic energy range of 1.3 eV was visible along the energy-dispersive axis of the spectrometer with an energy resolution of 14 meV. The full acceptance angle of the electron lens was $\pm 13^\circ$ with an angular resolution of better than 0.4° that corresponds to a momentum resolution of about 0.005 $\mathring{A}^{-1}$ in the investigated energy range.

As will be described in detail in the next section, not all excited electrons contribute to the coherently controlled current. Partly this results from the possibility to populate the image-potential band by excitation with only the UV pulses, which neither produces a current nor does it lead to a coherent control. In addition, the break of the inversion symmetry at the metal surface opens up a two-color excitation pathway that depends on the relative phase between the two light fields, but produces also a symmetric population and therefore no current in the band. For this reason, the coherently controlled current can be best visualized by subtracting a phase-averaged background from the phase-dependent energy and momentum distribution of the photoemission intensity. Figure 24.3 shows such difference plots of the measured energy and momentum distribution of the photoemission intensity ($E(k_{\|})$ spectra) for various relative phases $\Delta\phi$ where a phase-averaged distribution has been subtracted. These data show clearly the coherent control of the current. $\Delta\phi = 90^\circ$ represents the case where the current in $+k_{\|}$ direction has its maximum. Changing $\Delta\phi$ from 90° to 270° clearly reverses the direction of the current. As shown below, up

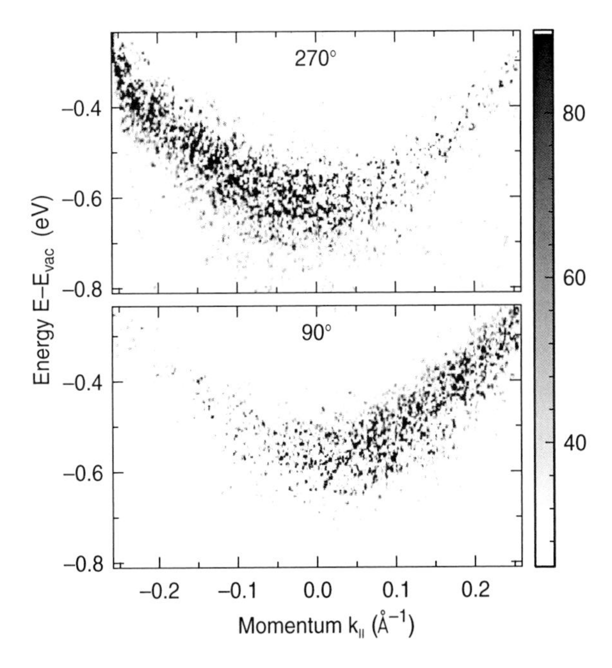

Figure 24.3 Coherent control of the momentum distribution in the $n = 1$ image-potential band. The figures show the energy and momentum distribution of the photoemission intensity for two different relative phases $\Delta\phi$ between the ω_a and the $\omega_a/2$ excitation pulses. A phase-averaged momentum distribution has been subtracted in order to emphasize the phase-dependent contribution to the photoemission intensity. Note that only positive values of the intensity differences are shown.

to 5% of all excited electron contribute to the current. With an estimated excitation density of $n_e \approx 10^{-6}$ per surface atom, a mean momentum $k_{\parallel} = 0.1\,\text{Å}^{-1}$ that corresponds to a velocity of $v_{\parallel} \approx 1$ Å/fs, we estimate the generated current density to $j_{\parallel} \approx 1$ nA per surface atom.

24.3
Modeling of the Coherent Excitation

The generation of electric currents by a coherent two-color excitation scheme can be described by different models for the light–matter interaction.

In a macroscopic view, the generation of a DC current using optical light fields with frequencies ω_a and $\omega_a/2$ can be understood as third-order nonlinear optical rectification that leads to a current injection that has for electrons as well as for holes the form [8]

$$\frac{\mathrm{d}\boldsymbol{j}(t)}{\mathrm{d}t} = \nu^{(3)}\boldsymbol{E}_1(\omega_a)\boldsymbol{E}_2(-\omega_a/2)\boldsymbol{E}_2(-\omega_a/2). \tag{24.2}$$

The fourth-rank tensor $\nu^{(3)}$ can be connected to the third-order nonlinear optical susceptibility $\chi^{(3)}$ [20]. Its symmetry properties make it possible to generate photocurrents in this way even in unbiased centrosymmetric materials [21].

Microscopically the acceleration of the excited electrons or holes by the electric fields of the laser pulses can be described by taking into account that optical dipole transitions in solids include contributions of *intraband* transitions within electronic bands, which change the electron momentum [21–23]. This can be seen by evaluating the matrix element of the dipole operator $\boldsymbol{d} = -e\boldsymbol{r}$ for Bloch states of an electron with momentum $\boldsymbol{k}$ in a band λ as

$$\langle \lambda, \boldsymbol{k}|\boldsymbol{d}|\lambda', \boldsymbol{k}'\rangle = e\boldsymbol{r}_{\boldsymbol{k}}^{\lambda\lambda'}\delta(\boldsymbol{k}-\boldsymbol{k}') + ie\,\delta_{\lambda\lambda'}\nabla_{\boldsymbol{k}}\delta(\boldsymbol{k}-\boldsymbol{k}'), \tag{24.3}$$

where $\boldsymbol{r}_{\boldsymbol{k}}^{\lambda\lambda'}$ describes the optical *interband* transitions and $\nabla_{\boldsymbol{k}}$ is the *intraband* acceleration. Intraband accelerations are the standard components for the description of the electronic response of solids to static and low-frequency electric fields [4]. This process, however, can often safely be neglected when light–matter interaction is analyzed since optical frequencies do not resonantly generate intraband transitions that are associated with small energy differences. For this reason, optical excitations are typically described only by vertical momentum-conserving interband transitions between different electronic bands. The optical generation of an electrical current, however, requires a redistribution of the electrons in momentum space, which can only be provided by the intraband acceleration.

For semiconductor nanostructures, a microscopic theory has been developed that provides a detailed description of the ultrafast dynamics of the optical generation and the decay of photocurrents including many-body effects and scattering processes [23–25]. This approach is based on the semiconductor Bloch equations [26] that are extended by the electric field induced intraband acceleration [22, 24]. This theory, however, cannot be directly applied for the description of photoexcited electron currents at metal surfaces where the electronic structure as well as the decay processes of excited electrons are different compared to bulk semiconductors. Whereas the initial state in a bulk system or a quantum well or wire with translational symmetry is given by a single or few discrete valence bands, the projection of a metal bulk band onto the surface results in a continuum of initial states for the photoexcitation [27, 28] (compare Figure 24.2). In addition, the break of inversion symmetry at the surface opens up additional excitation pathways because interband transitions are allowed between all bands even in dipole approximation.

In a first attempt to describe the optical excitation of currents at metal surfaces, we have developed a model system that is able to describe some of the most important aspects of the photoexcitation at the Cu(100) surface including intraband excitations. As shown in Figure 24.4d, the model approximates the continuum of initial bulk

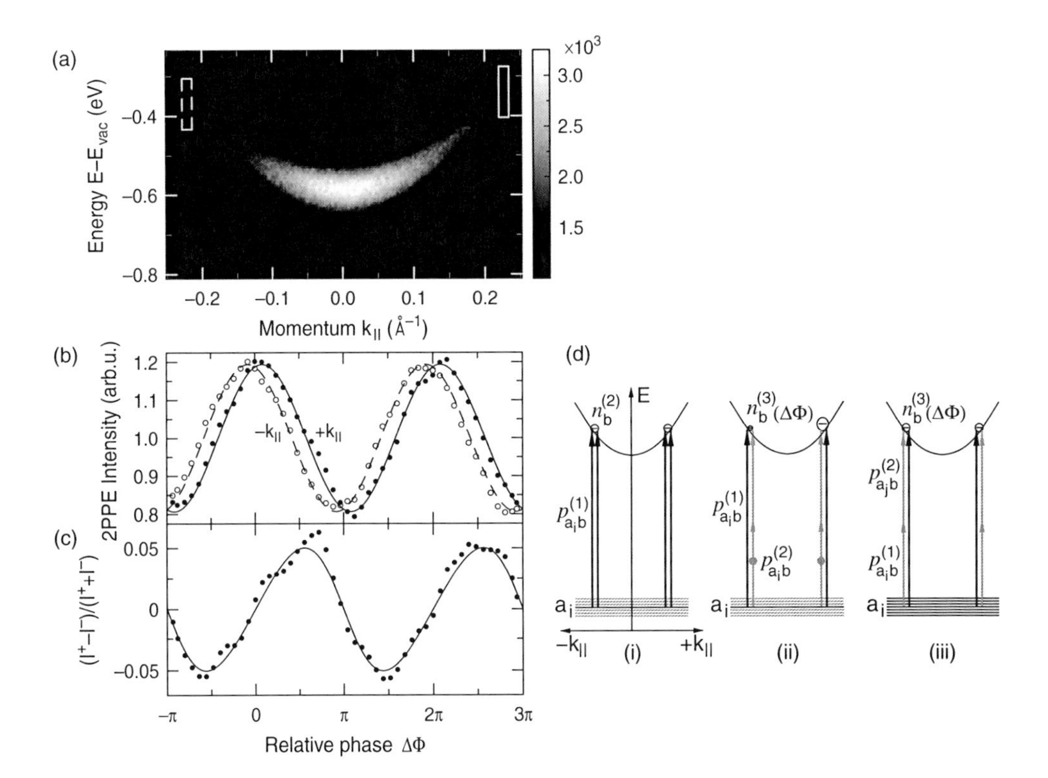

Figure 24.4 (a) Energy and momentum distribution of the photoemission intensity for a relative phase of $\Delta\phi = 90°$. (b) Normalized experimental photoemission intensity integrated over the dashed and solid rectangles in (a) (open and filled data points, respectively) together with the results of the model calculation (dashed and solid lines). (c) Relative intensity contrast for opposite parallel momenta from b) corresponding to the relative contribution of the coherently excited current to the total photoemission signal. (d) Sketch of the multiband model and the three considered excitation pathways (see text for details). Adapted from Ref. [6]. Reprinted with permission from AAAS.

states by a set of narrowly spaced dispersionless bands below the Fermi level E_F. These initially occupied bands are optically coupled to the $n = 1$ image-potential state that has a parabolic dispersion. This model is clearly not sufficient to provide detailed microscopic understanding, however, as shown below it is able to describe the phase dependence of the photoexcited populations and currents measured at the Cu(100) surface and to analyze their origin.

The considered resonant and nonresonant interband as well as nonresonant intraband transitions result in three qualitatively different excitation pathways that contribute to the population of the excited surface band and that are depicted in Figure 24.4d: (i) A population of the excited band can be generated in second order by two resonant interband excitations of the laser field with frequency ω_a. This is the usual lowest order excitation process that results in a symmetric k-space population in the optically excited surface band. Since this process is induced only by one of the laser fields, the generated population does not depend on the relative phase $\Delta\Phi$ and thus provides a phase-independent background. (ii) Asymmetric k-space distributions, that is, currents, are generated in third order by one excitation with frequency ω_a and two excitations with frequency $\omega_a/2$ where one of the transitions is an intraband excitation [23]. Since the intraband excitation process is proportional to a derivative with respect to the wave vector $\boldsymbol{k}$, this pathway leads to an antisymmetric k-space occupation [23]. The direction of the k-space asymmetry and thus of the photocurrent is proportional to sin $(\Delta\Phi)$.

Considering only processes (i) and (ii), the photoexcited occupations at $+k_{\|}$ and $-k_{\|}$ are 180° out of phase and the total population does not depend on the relative phase $\Delta\Phi$ since process (ii) only redistributes the electrons in k-space. The broken inversion symmetry at the surface, however, requires the inclusion of a third excitation process (iii) that describes nonresonant intersubband transitions between the closely spaced bands below E_F. This excitation process leads to a third-order occupation in the surface band that is symmetric in $k_{\|}$. Therefore, it does not contribute to the photocurrent, but depends on the relative phase as cos $(\Delta\Phi)$ [29].

On the basis of the multiband model described above, the population of the image-potential band has been calculated by solving the optical Bloch equations numerically including all three excitation processes. The results are compared with the phase-dependence of the experimental photoemission intensity for two opposite parallel momenta $k_{\|} = \pm 0.224\,\text{Å}^{-1}$ in Figure 24.4b and c. At this parallel momentum, the fraction of the excited electrons that contribute to the optically induced current reaches about 5% (Figure 24.4c). This fraction can be further increased by increasing the intensity ratio between the visible and the UV pulses [16].

As already mentioned, process (i) provides the phase-independent background. Since the processes (ii) and (iii) are antisymmetric and symmetric in $k_{\|}$, respectively, their superposition results in a population of the image-potential band that oscillates with the relative phase $\Delta\phi$ and has a finite phase shift for opposite parallel momenta. The size of this phase shift is determined by the relative strength of processes (ii) and (iii) that depends on the transition dipoles. By adjusting the transition dipoles and the amplitudes of the two incident pulses, we achieve very good agreement with the experimental data. Clearly, the measured dependence of the current and the population on the relative phase between the two incident beams can only be described by incorporating a surface-specific process (iii) into the model.

24.4 Time-Resolved Observation of Current Decay

Definitely, the most interesting aspect of the current detection using time-resolved photoelectron spectroscopy is the possibility to observe the dynamics of the current decay in the time domain. For a conventionally generated electric current in a metal that is induced by a low-frequency electric field, the current is carried out by electrons with energies close to the Fermi level. Such current decays by means of quasielastic collision processes of the electrons with phonons and defects. In k-space, these scattering processes lead to a fast randomization of any asymmetry of the electron population generated by the applied electric field. For electrons that are excited several electron volts above the Fermi level, such as electrons in image-potential states, an additional decay channel that ultimately limits the current flow is electron–hole-pair decay due to inelastic scattering with bulk electrons [30]. This inelastic scattering leads to a decay of the total population in the excited band, which is typically observed in conventional 2PPE experiments using the most basic excitation scheme in which one laser pulse populates the excited band while a second laser pulse is used for the photoemission [31–33]. The elastic scattering within the band, however, does not affect the population in the excited band as long as the optical excitation leads to a symmetric population. For this reason, optical techniques that are sensitive to elastic scattering typically observe the loss (dephasing) of an optically induced coherence between electronic states, which decays due to inelastic as well as elastic scattering. This includes optical surface spectroscopies such as five-wave mixing [34, 35], interferometric time-resolved 2PPE [16, 36], and linewidth measurements for 2PPE in the frequency domain [37] as well as quantum-beat spectroscopy of image-potential states [38]. Elastic scattering of the excited electrons, however, is often only one of the several contributions to the observed phase decay. In contrast to this, the time-resolved observation of the decay of optically induced currents by photoelectron spectroscopy is able to monitor elastic scattering in a direct way in terms of incoherent population dynamics in momentum space. The coherent optical excitation scheme is only required for the creation of an initially asymmetric population distribution and it is not necessary to establish a fixed phase relationship between the excitation pulses and the time-delayed photoemission probe pulse.

The simultaneous observation of elastic and inelastic scattering is demonstrated in Figure 24.5a–d, for the $n = 1$ image-potential band where the current decay is compared with the decay of the average population. The comparison of both data sets clearly shows that the asymmetry of the excited state population levels off considerably faster than the overall population decays. At $\Delta t = 150$ fs, for example, there is still considerable population left in the image-potential band ($\simeq 20\%$) although the asymmetry between positive and negative values of $k_{\|}$ has vanished below the detection limit. To quantify the difference between elastic and inelastic scattering

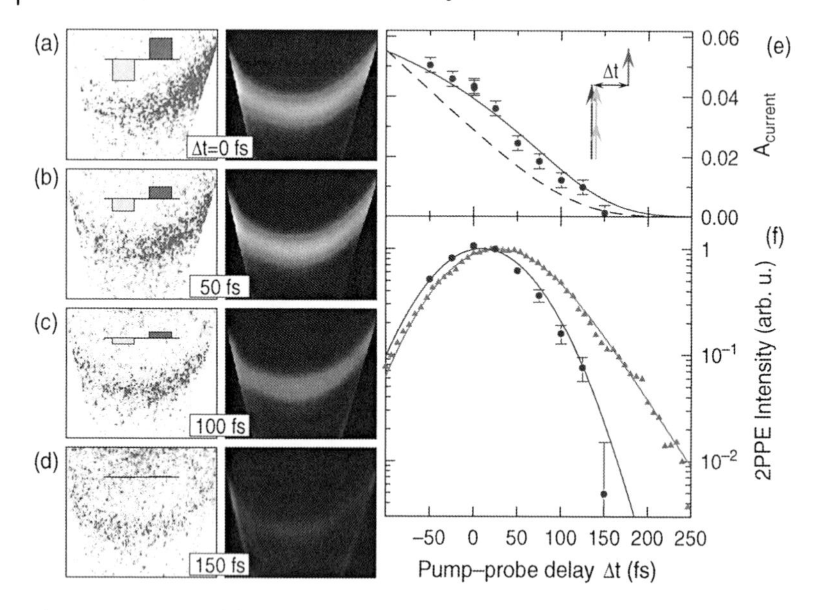

Figure 24.5 Decay of the current in comparison with the average population decay. ((a)–(d)) Difference $E(k_{\|})$ spectra (left) and phase-averaged $E(k_{\|})$ spectra (right) for different time delays Δt of the photoemission laser pulse with respect to the current generation. (e) Relative contribution of the electron population that carries the current with respect to the total population as a function of Δt evaluated at $k_{\|} = \pm 0.15\,\text{Å}^{-1}$. (f) Population decay at the band bottom compared to the current decay as determined from e). Reprinted with permission from [11]. Copyright 2006 by the American Physical Society. (Please find a color version of this figure on the color plates.)

times, Figure 24.5e shows the asymmetry of the photoemission intensity for opposite parallel momenta $A_{\text{current}} = (I^+ - I^-)/(I^+ + I^-)$ evaluated at $k_{\|} = \pm 0.15\,\text{Å}^{-1}$, which corresponds to the relative contribution of the current to the total excited population. The decay of A_{current} with the time delay Δt clearly shows that the current decays faster than the total population, whereas $A_{\text{current}}(\Delta t)$ would be constant for a current decay time that is long compared to the population decay. The dashed curve shows the limit of an instantaneous decay of the current under the assumption of laser pulses with a pure Gaussian shape. Based on a simple rate-equation model, the solid line shows the result for a current decay time of 10 fs that is much faster than the lifetime of the population for inelastic decay of about 35 fs. This short decay time of the current has been attributed to electron scattering with steps and surface defects [6].

24.5 Summary

In summary, we have shown that the combination of a coherent two-color excitation with time- and angle-resolved 2PPE makes it possible to observe elastic scattering of excited electrons in a direct way by monitoring the decay of an initially asymmetric population distribution in momentum space. This method would clearly benefit from the development of electron spectrometers that cover the momentum space in two dimensions [39], which would make it possible to observe the full dynamics of electrons at surfaces or in other two-dimensional systems [24].

Finally, we would like to emphasize that the method is not restricted to the investigation of surface currents. The escape depth of low-kinetic energy photoelectrons from solids amounts to several tens of atomic layers [40]. Therefore, it should thus be possible to generate current pulses of electrons in the bulk conduction band of anorganic or organic semiconductors and to monitor their dynamics in the near-surface region in a similar way as done here for surface-specific states. Since the electron mobility $\mu_e = e\tau/m_e$ of semiconductors is usually larger than that of metals, the experiments should in fact become easier in some respects due to a longer scattering time τ.

Acknowledgments

We acknowledge funding by the Deutsche Forschungsgemeinschaft through grant No. HO 2295/1–4 and by the Center for Optodynamics, Marburg.

References

1. Nitzan, A. and Ratner, M.A. (2003) Electron transport in molecular wire junctions. *Science*, **300**, 1384–1389.
2. Agraït, N., Yeyati, L., and van Ruitenbeek, J.M. (2003) Quantum properties of atomic-sized conductors. *Phys. Rep.*, **377**, 81.
3. Landauer, R. (1957) Spatial variation of currents and fields due to localized scatteres in metallic conduction. *IBM J. Res. Dev.*, **1**, 223.
4. Ashcroft, N.W. and Mermin, N.D. (1976) *Solid State Physics*, Holt, Rinehart and Winston, New York.
5. Drude, P. (1900) Zur elektronentheorie der metalle. *Ann. Phys.*, **306**, 566.
6. Güdde, J., Rohleder, M., Meier, T., Koch, S.W., and Höfer, U. (2007) Time-resolved investigation of coherently controlled electric currents at a metal surface. *Science*, **318**, 1287–1291.
7. Dupont, E., Corkum, P.B., Liu, H.C., Buchanan, M., and Wasilewski, Z.R. (1995) Phase-controlled currents in semiconductors. *Phys. Rev. Lett.*, **74**, 3596–3599.
8. Atanasov, R., Haché, A., Hughes, J.L.P., van Driel, H.M., and Sipe, J.E. (1996) Coherent control of photocurrent generation in bulk semiconductors. *Phys. Rev. Lett.*, **76**, 1703–1706.
9. Haché, A., Kostoulas, Y., Atanasov, R., Hughes, J.L.P., Sipe, J.E., and van Driel, H.M. (1997) Observation of coherently controlled photocurrent in unbiased, bulk GaAs. *Phys. Rev. Lett.*, **78**, 306–309.
10. Stevens, M.J., Smirl, A.L., Bhat, R.D.R., Najmaie, A., Sipe, J.E., and van Driel, H.M. (2003) Quantum interference control of ballistic pure spin currents in semiconductors. *Phys. Rev. Lett.*, **90**, 136603.
11. Hübner, J., Rühle, W.W., Klude, M., Hommel, D., Bhat, R.D.R., Sipe, J.E., and van Driel, H.M. (2003) Direct observation of optically injected spin-polarized currents in semiconductors. *Phys. Rev. Lett.*, **90**, 216601.

12 Cote, D., Fraser, J.M., DeCamp, M., Bucksbaum, P.H., and van Driel, H.M. (1999) Thz Emission from coherently controlled photocurrents in GaAs. *Appl. Phys. Lett.*, **75**, 3959–3961.

13 Brumer, P. and Shapiro, M. (1986) Control of unimolecular reactions using coherent light. *Chem. Phys. Lett.*, **126**, 541–546.

14 Echenique, P.M. and Pendry, J.B. (1978) The existence and detection of Rydberg states at surfaces. *J. Phys. C Solid State*, **11**, 2065–2075.

15 Giesen, K., Hage, F., Himpsel, F.J., Riess, H.J., Steinmann, W., and Smith, N.V. (1987) Effective mass of image-potential states. *Phys. Rev. B*, **35**, 975–978.

16 Güdde, J., Rohleder, M., and Höfer, U. (2006) Time-resolved two-color interferometric photoemission of image-potential states on Cu(100). *Appl. Phys. A*, **85**, 345–350.

17 Güdde, J., Rohleder, M., Meier, T., Koch, S.W., and Höfer, U. (2009) Generation and time-resolved detection of coherently controlled electric currents at surfaces. *Phys. Stat. Sol. (c)*, **6**, 461–465.

18 Reuß, Ch., Shumay, I.L., Thomann, U., Kutschera, M., Weinelt, M., Fauster, Th., and Höfer, U. (1999) Control of dephasing of image-potential states by adsorption of CO on Cu(100). *Phys. Rev. Lett.*, **82**, 153–156.

19 Rohleder, M., Duncker, K., Berthold, W., Güdde, J., and Höfer, U. (2005) Momentum-resolved dynamics of Ar/Cu (100) interface states probed by time-resolved two-photon photoemission. *New J. Phys.*, **7**, 103.

20 Aversa, C. and Sipe, J.E. (1996) Coherent current control in semiconductors: a susceptibility perspective. *IEEE J. Quantum Electron.*, **32**, 1570–1573.

21 Aversa, C. and Sipe, J.E. (1995) Nonlinear-optical susceptibilities of semiconductors: results with a length-gauge analysis. *Phys. Rev. B*, **52**, 14636–14645.

22 Meier, T., Rossi, F., Thomas, P., and Koch, S.W. (1995) Dynamic localization in anisotropic Coulomb systems: field-induced crossover of the exciton dimension. *Phys. Rev. Lett.*, **75**, 2558–2561.

23 Duc, H.T., Vu, Q.T., Meier, T., Haug, H., and Koch, S.W. (2006) Temporal decay of coherently optically injected charge and spin currents due to carrier–LO–phonon and carrier–carrier scattering. *Phys. Rev. B*, **74**, 165328.

24 Duc, H.T., Meier, T., and Koch, S.W. (2005) Microscopic analysis of the coherent optical generation and the decay of charge and spin currents in semiconductor heterostructures. *Phys. Rev. Lett.*, **95**, 086606.

25 Meier, T., Duc, H.T., Vu, Q.T., Pasenow, B., Hübner, J., Chatterjee, S., Rühle, W.W., Haug, H., and Koch, S.W. (2007) Ultrafast dynamics of optically-induced charge and spin currents in semiconductors. *Adv. Sol. State Phys.*, **46**, 199–210.

26 Haug, H. and Koch, S.W. (1990) *Quantum Theory of the Optical and Electronic Properties of Semiconductors*, World Scientific, Singapore.

27 Shumay, I.L., Höfer, U., Thomann, U., Reuß, Ch., Wallauer, W., and Fauster, Th. (1998) Lifetimes of image-potential states on Cu(100) and Ag(100) surfaces measured by femtosecond time-resolved two-photon photoemission. *Phys. Rev. B*, **58**, 13974–13981.

28 Weida, M.J., Ogawa, S., Nagano, H., and Petek, H. (2000) Ultrafast interferometric pump–probe correlation measurements in systems with broadened bands or continua. *J. Opt. Soc. Am. B*, **17**, 1443–1451.

29 Fraser, J.M., Shkrebtii, A.I., Sipe, J.E., and van Driel, H.M. (1999) Quantum interference in electron–hole generation in noncentrosymmetric semiconductors. *Phys. Rev. Lett.*, **83**, 4192–4195.

30 Echenique, P.M., Berndt, R., Chulkov, E.V., Fauster, Th., Goldmann, A., and Höfer, U. (2004) Decay of electronic excitations at metal surfaces. *Surf. Sci. Rep.*, **52**, 219.

31 Williams, R.T., Royt, T.R., Rife, J.C., Long, J.P., and Kabler, M.N. (1982) Picosecond time-resolved photoelectron spectroscopy of ZnTe. *J. Vac. Sci. Technol.*, **21**, 509–513.

32 Bokor, J. (1989) Ultrafast dynamics at semiconductor and metal surfaces. *Science*, **246**, 1130–1134.

33 Haight, R. (1995) Electron dynamics at surfaces. *Surf. Sci. Rep.*, **21**, 277–325.

34 Voelkmann, C., Reichelt, M., Meier, T., Koch, S.W., and Höfer, U. (2004) Five-wave-mixing spectroscopy of ultrafast electron dynamics at a Si(001) surface. *Phys. Rev. Lett.*, **92**, 127405.

35 Meier, T., Reichelt, M., Koch, S.W., and Höfer, U. (2005) Femtosecond time-resolved five-wave mixing at silicon surfaces. *J. Phys. Condens. Matter*, **17**, S221–S244.

36 Petek, H. and Ogawa, S. (1997) Femtosecond time-resolved two-photon photoemission studies of electron dynamics in metals. *Prog. Surf. Sci.*, **56**, 239–310.

37 Boger, K., Roth, M., Weinelt, M., Fauster, T., and Reinhard, P.G. (2002) Linewidths in energy-resolved two-photon photoemission spectroscopy. *Phys. Rev. B*, **65**, 075104.

38 Höfer, U., Shumay, I.L., Reuß, Ch., Thomann, U., Wallauer, W., and Fauster, Th. (1997) Time-resolved coherent photoelectron spectroscopy of quantized electronic states at metal surfaces. *Science*, **277**, 1480–1482.

39 Kirchmann, P.S., Rettig, L., Nandi, D., Lipowski, U., Wolf, M., and Bovensiepen, U. (2008) A time-of-flight spectrometer for angle-resolved detection of low energy electrons in two dimensions. *Appl. Phys. A*, **91**, 211.

40 Hüfner, S. (2003) *Photoelectron Spectroscopy - Principles and Applications*, Springer, Berlin.

25
Ultrabroadband Terahertz Studies of Correlated Electrons

Rupert Huber and Alfred Leitenstorfer

25.1
Introduction

Electromagnetic radiation at terahertz frequencies (1 THz = 10^{12} Hz corresponds to a photon energy of 4.1 meV) is of central importance in nature. A large variety of fundamental low-energy excitations in all phases of matter couple resonantly to THz electromagnetic waves. Among such transitions are, for example, vibrations of the molecular frame or the ionic crystal lattice of solids, collective charge density oscillations or internal excitonic transitions in semiconductors, characteristic energy gaps in strongly correlated electron systems, or magnon resonances in magnetically ordered materials. The microscopic dynamics of these elementary excitations has been primarily studied by rather indirect means involving nonresonant light pulses in the near-infrared (NIR) or visible (VIS) domain with durations on the 10–100 fs scale (1 fs = 10^{-15} s). In contrast, few-cycle THz transients have been increasingly recognized as the most direct probes of low-energy modes, opening up a vast field of applications in fundamental science and technology [1].

Despite its interdisciplinary relevance, the THz window used to be notoriously difficult to access. Only in the past few years, THz optoelectronics has closed the spectral gap by systematically combining electronic and optical concepts: Advantages known from electronics, such as the possibility to synthesize precisely defined waveforms and to record the real-time evolution of the electric field, are now available in the entire infrared regime. Few- and even single-cycle THz pulses have been demonstrated throughout the frequency range from 0.1 THz to beyond 100 THz [2–8]. Concomitantly, field-sensitive electro-optic detection has become sufficiently broadband to resolve even NIR light pulses [8–13]. In a modern definition, *ultrabroadband* THz technology, thus, covers almost the entire band of infrared light (see Figure 25.1).

These advances have set the stage for a new generation of femtosecond experiments studying the ultrafast response of correlated electrons to well-defined *electric fields* rather than light *intensities*. In an optical pump/multi-THz probe scheme, a first laser pulse of a duration of a few femtoseconds prepares a state far from thermal

Dynamics at Solid State Surface and Interfaces Vol.1: Current Developments
Edited by Uwe Bovensiepen, Hrvoje Petek, and Martin Wolf

ISBN: 978-3-527-40937-2

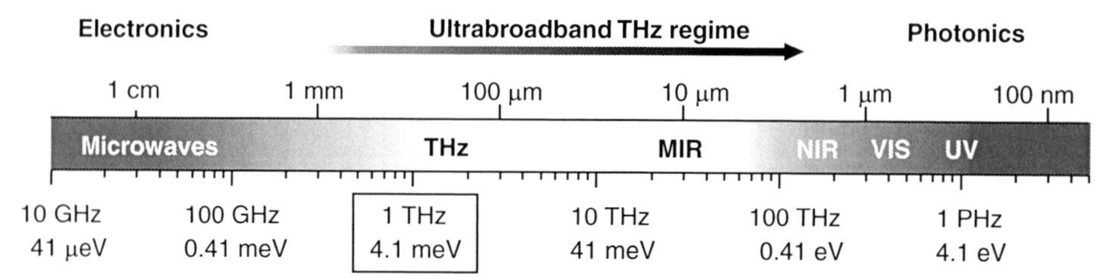

Figure 25.1 The ultrabroadband THz spectral regime. MIR: mid infrared, NIR: near infrared, VIS: visible light, UV: ultraviolet. Vacuum wavelengths, frequencies, and photon energies are indicated.

equilibrium. The subsequent low energy dynamics is traced by an ultrabroadband multi-THz pulse. Measuring both amplitude and phase of the transmitted test wave, we retrieve the complex-valued dielectric function with unprecedented temporal resolution [14]. This technique has matured in recent years to become a practical tool in solid state physics [14–22]. As a first example, we discuss how this technique allows us to shed new light onto a long-standing paradigm question in strongly correlated materials: the femtosecond light-induced insulator–metal transition of vanadium dioxide (VO_2). By monitoring the THz conductivity as one of the central parameters during the transition, on the femtosecond scale, we reveal an intriguing interplay of a coherent structural wave packet motion and electronic conductivity [21]. Our findings suggest a novel model of the phase transition.

Latest THz sources have been pushed to extreme peak electric field strengths in excess of 100 MV/cm [13]. Such transients facilitate the exciting step from THz probing to THz coherent control. In our second application, we test this idea with excitons, Coulomb-bound pairs of one electron and one hole in a semiconductor. THz photons couple directly to internal Rydberg-like quantum transitions. With intense THz fields, we drive Rabi cycles of the 1s–2p transition of a gas of optically dark, dense, and cold *para* excitons in the semiconductor Cu_2O. This way, we coherently control the internal quantum state with an efficiency of more than 80% [22]. The results mark an encouraging route toward systematic THz quantum optics of condensed matter and test recent microscopic theories of the ultrafast dynamics.

This chapter is organized as follows: After a review of state-of-the-art multi-THz technology in Section 25.2, we discuss our studies of MIR conductivity of VO_2 during a femtosecond insulator–metal transition in Section 25.3. THz control of excitons will be the subject of Section 25.4, followed by a summary and an outlook.

25.2
Phase-Locked Few-Cycle THz Pulses: From Ultrabroadband to High Intensity

From the stupendously fast-growing field of femtosecond technology, THz optoelectronics stands out due to three major aspects: (i) THz detectors and emitters cover a hard-to-access spectral region (see Figure 25.1), (ii) THz transients are routinely generated with fixed carrier phase, and (iii) field-sensitive detection has become routinely available in this context. Here we show how to generate THz pulses in an ultrabroadband range, what phase stability means, why THz pulses are intrinsically phase locked, and how the real-time oscillation of the electric field of infrared light may be captured. Finally, we will present a novel high-field source that generates peak electric fields as high as 108 MV/cm and thus paves the route to extremely nonlinear THz optics.

In the past two decades, a variety of FIR and MIR femtosecond sources have been introduced, including THz emission from integrated-circuit Hertzian dipoles [2, 23], optical rectification of ultrashort laser pulses at GaAs and InP surfaces [24], and optically switched high-field transport in semiconductor pin diodes [4]. The central frequency of all these sources is below 15 THz with limited tunability. The most broadband THz transients with widely tunable center frequencies have been obtained by optical rectification in the bulk of nonlinear crystals [5–8, 10, 13]. This technique will be reviewed here in more detail.

A generic multi-THz system is displayed in Figure 25.2. We typically start with NIR pulses of a duration of ≈10 fs, which are routinely available from Ti:sapphire lasers and, recently, also from compact Er:fiber systems [25]. The major part of the laser

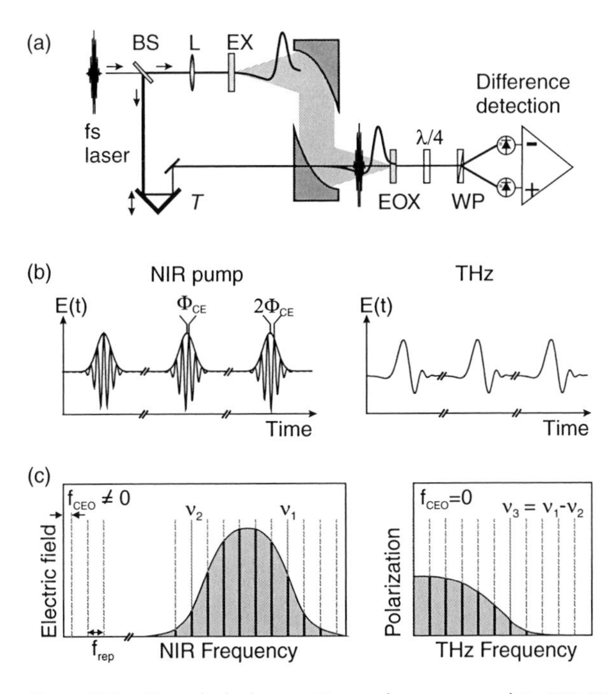

Figure 25.2 Phase-locked generation and field-sensitive detection of THz pulses. (a) Schematic setup for optical rectification of femtosecond near-infrared pulses and electro-optic sensing of the THz field; BS: beam splitter, L: lens or concave mirror, EX: THz emitter crystal, EOX: electro-optic crystal, λ/4: quarter waveplate, WP: Wollaston polarizing beam splitter. (b) Schematic representation of the electric field of the NIR pump pulses (left). The CEP changes from shot to shot. In contrast, the electric waveform of the THz transients is inherently stable (right). (c) Corresponding laser and THz spectra.

intensity is focused into a nonlinear optical crystal (THz emitter) to generate ultrabroadband THz transients by optical rectification, a nonlinear $\chi^{(2)}$ process. Different Fourier components ν_1 and ν_2 of the laser pulse drive a nonlinear polarization at the difference frequency $\nu_3 = \nu_1 - \nu_2$, radiating THz transients into free space [26]. The maximum spectral bandwidth depends on the duration of the pump pulse. Values of ν_3 beyond 100 THz have been obtained with a 10 fs pump [10]. Efficient generation of a difference frequency requires the group velocity of the pump pulse and the THz phase velocity to match. This phase matching condition has been successfully fulfilled over ultrabroad THz bandwidths exploiting thin crystals of birefringent materials such as GaSe or $LiIO_3$ [5–7, 10].

A unique property of these THz transients is their inherent phase stability. Typical femtosecond lasers generate a sequence (repetition rate: f_{rep}) of pulses with identical intensity envelopes. The so-called carrier envelope phase (CEP), that is, the phase of the electric field (carrier wave) with respect to the pulse envelope, however, changes from shot to shot (Figure 25.2b). The rate at which the CEP shifts – the so-called carrier envelope offset frequency f_{CEO} – is determined by the difference of the group and the phase velocity inside the laser oscillator [27]. Without special precautions, such as complex electronic feedback loops, these velocities are different. The absence of a well-defined absolute optical phase is one of the fundamental reasons why femtosecond pump–probe experiments in the NIR and VIS are typically sensitive to intensities only, rather than electric fields. In contrast, the field profile of the THz transients generated from optical rectification is strictly identical in every pulse (right panel in Figure 25.2b).

CEP stability is a consequence of optical rectification as seen most obviously in the Fourier domain. Mathematically, a periodic function transforms into a discrete Fourier series rather than a continuous spectrum. A femtosecond pulse train is thus composed of a comb of spectral lines with a regular spacing given by the repetition rate f_{rep}. A nonvanishing offset frequency manifests itself by a shift of the comb by f_{CEO} [27]. The goal of electronic stabilization schemes for precision metrology is to lock f_{CEO} ideally to zero in order to exploit the regular comb as a stable frequency ruler [27]. Optical rectification reaches this goal elegantly without any electronic servo lock: Let us first note that arbitrary frequency components ν_1 and ν_2 of the laser spectrum may be represented as $\nu_{1,2} = m_{1,2} f_{rep} + f_{CEO}$ with appropriately chosen integers $m_{1,2}$. For a difference frequency component ν_3, we thus obtain $\nu_3 = \nu_1 - \nu_2 = (m_1 - m_2) f_{rep}$. f_{CEO} inherently cancels out; the THz comb is automatically locked to zero (Figure 25.2c). In the time domain, we obtain a well-defined THz waveform with constant CEP (Figure 25.2b), independent of the CEO of the pump pulse.

Another singular advantage of THz optoelectronics comes from the fact that the THz temporal waveform may be directly recorded in the time domain. After the first demonstration of amplitude and phase-sensitive sampling of free space far-infrared transients [2], two technically mature methods have become most popular: photoconductive and free space electro-optic sampling (EOS). Both schemes have been advanced to operate at frequencies ranging from below 0.1 THz to far above 100 THz ($\lambda = 3\,\mu m$) [8, 10, 12].

Figure 25.2a depicts the idea of electro-optic sampling: We use off-axis parabolic mirrors to focus the THz transients into a second nonlinear optical crystal operated as an electro-optic sensor. The THz field induces a momentary birefringence of the crystal via a quasi-instantaneous Pockels nonlinearity. The resulting phase retardation φ between the fast and slow axis is sampled by a copropagating second portion of the laser output, analyzed by subsequent polarization optics, and detected by a pair of balanced photodiodes. By scanning the time delay T between the THz wave and the NIR gate pulse, the temporal shape of the transient may be mapped out. Efficient detection requires the gate pulse to sample a constant THz field strength during its propagation through the sensor. For one, the laser pulse, therefore, has to be substantially shorter than the oscillation period of the THz transient to be detected. Second, both pulses have to copropagate equally fast through the electro-optic crystal. The latter condition may be satisfied either by employing particularly thin crystals (so-called quasiphase matching) [3] or by critical phase matching in birefringent crystals [9, 10], similar to the generation process. Detailed analyses of the spectral response functions of either scheme may be found in Refs [4, 9, 11, 28].

Representative multi-THz transients obtained with the setup of Figure 25.2 are displayed in Figure 25.3 together with corresponding amplitude spectra. The well-defined field profiles as a function of the delay time T provide the ultimate proof of CEP stability (panels (a)–(c)). Signal-to-noise ratios as high as $10^5\sqrt{\mathrm{Hz}}$ may be achieved. Even shorter laser pulses, tunable phase matching schemes, and a variety of crystal materials with high $\chi^{(2)}$ nonlinearities, such as ZnTe, GaP, GaSe, or $LiIO_3$, have enabled field control throughout the far- and mid-infrared (see Figure 25.3d). Phase-matched generation and detection in 30 μm thick GaSe crystals, for example, yield transients covering approximately 8 optical octaves (Figure 25.3c and d) [10]. Seven-femtosecond visible light pulses have been converted in a thin birefringent $LiIO_3$ emitter to support multi-THz spectra with frequency components larger than 140 THz, corresponding to photon energies in excess of 0.58 eV (Figure 25.3d) [7]. Most recently, we introduced a compact, turnkey Er:fiber laser system that pushes field-sensitive optoelectronics to the NIR and may lend itself to real-world applications outside specialized laboratories [8]. Thus, the bandwidth of THz optoelectronics seamlessly covers all low-energy resonances in condensed matter.

With peak electric fields of the order of a few kV/cm, these transients are superb *probes*, for example, of phonons, internal exciton transitions, or electronic conductivity in the MIR as shown below. Yet, the perspective of a novel class of *THz nonlinear optical experiments* and *coherent control* (Section 25.4) has motivated intense efforts to boost the field strength. Free electron lasers have generated THz peak powers of up to 10^6 W [29], but do not generally provide few-cycle transients with a controllable CEP. Tabletop sources exploiting optical rectification of amplified millijoule laser pulses have reached fields of $\approx$1 MV/cm centered below 3 THz [30] or – with some tunability – at up to 35 THz [31]. In the latter case, however, only the wings of the pump spectra contribute to THz generation accounting for low quantum efficiencies of the order of 10^{-4} and maximum THz pulse energies on the nanojoule level.

Our recent concept of a novel Er:fiber/Ti:sapphire hybrid laser overcomes this problem and enhances the maximum THz amplitude by orders of magnitude [13].

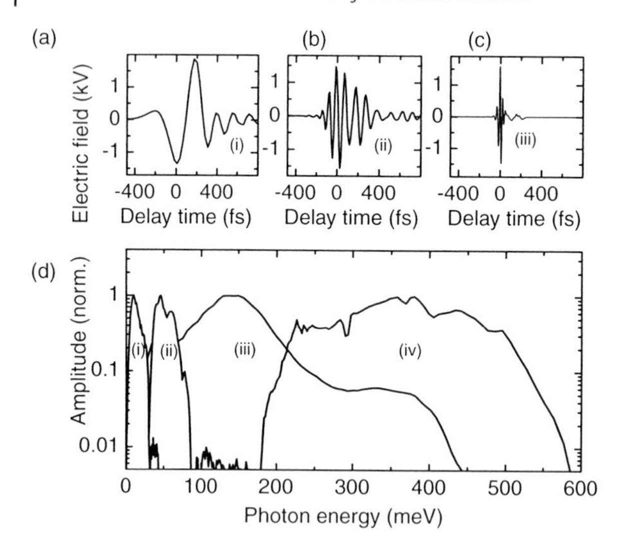

Figure 25.3 Phase-stable multi-THz transients generated via optical rectification and detected by electro-optic sampling in nonlinear optical crystals. (a) Emitter and detector: (110)-GaP ($d = 200\,\mu m$). (b) Emitter: GaSe ($d = 100\,\mu m$), detector: (110)-ZnTe ($d = 15\,\mu m$). (c) Emitter and detector: GaSe ($d = 30\,\mu m$) [10]. (d) Corresponding amplitude spectra (i)–(iii) continuously covering the far- and mid-infrared regime. Spectrum (iv) corresponds to a phase-locked pulse generated by phase-matched optical rectification of a 7 fs near-infrared pulse in a $LiIO_3$ element [7].

A schematic is shown in Figure 25.4. The output of a mode-locked Er:fiber oscillator is seeded into two parallel Er:fiber amplifiers. The first branch is frequency doubled and postamplified in a high-power Ti:sapphire system to a pulse energy of 5 mJ. The beam is split to pump two identical optical parametric amplifiers (OPAs) seeded by the same white light continuum. The independently tunable output pulses are subsequently superimposed for difference frequency generation (DFG) in a nonlinear optical crystal. As both OPAs share the same frequency comb, the CEO is expected to cancel in the DFG process. In contrast to optical rectification, the maximum THz frequency is no longer bound by the bandwidth of the pump spectrum, but may be conveniently set by the relative detuning of both OPAs between 1 THz and more than 100 THz. Since the full spectra of both pump pulses contribute to the DFG process, the quantum efficiency may be enhanced beyond 10%, outperforming optical rectification of a single pulse [31] by three orders of magnitude.

The THz transients are electro-optically sampled by 8 fs gating pulses derived from the second Er:amplifier branch [25]. Figure 25.5a displays a typical waveform that contains only 2 cycles within its FWHM and reaches peak electric fields of 30 MV/cm in a beam waist of 31 μm (FWHM of intensity). In slightly longer pulses, we observed peak fields of up to 108 MV/cm, intensities of 15 TW/cm^2, and energies as high as 19 μJ [13]. These values compare very favorably even to large-scale facilities, such as synchrotrons and latest generation free electron lasers (Figure 25.5b). While the latter sources reach unrivaled average THz powers of up to 20 W [29], our tabletop system

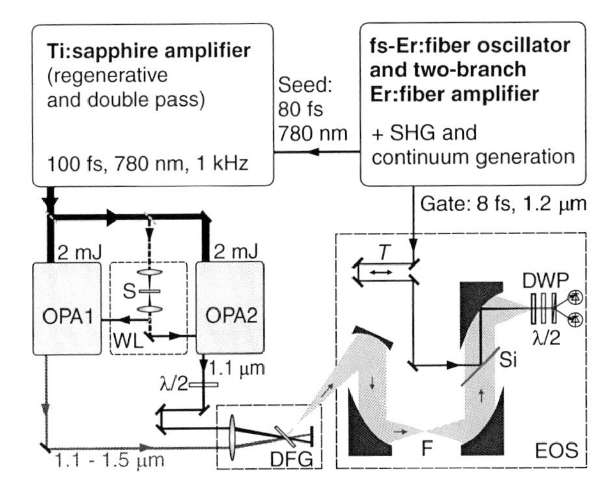

Figure 25.4 Setup of our high-field THz source. SHG: second harmonic generation in periodically poled $LiNbO_3$; S: sapphire window; WL: white-light seed for OPA1 and OPA2; DFG: noncollinear difference frequency mixing of both OPA outputs; EOS: electro-optic sampling with balanced detection; F: intermediate THz focus for transmission experiments; *T*: electro-optic sampling time set via a delay line; Si: silicon beam splitter; D: electro-optic sensor crystal; λ/2: half-wave plate; WP: Wollaston prism.

provides by far the highest peak fields. For nonlinear optical and coherent control experiments discussed below, few-cycle pulses with maximum peak fields are, in fact, most desirable. With this system, we have completed our arsenal of phase-locked sources and field-sensitive detectors of ultrabroadband and high-intensity infrared pulses. We will now demonstrate how this technology may be utilized to monitor and control low-energy femtosecond dynamics in condensed matter.

25.3 Ultrafast Insulator–Metal Transition of VO_2

Insulator–metal transitions in strongly correlated electron systems are among the most intriguing phenomena in solid state physics [32]. A delicate balance of cooperative interactions of crystal structure and electronic degrees of freedom drives the materials into a critical regime. Vanadium dioxide is a classic example: It undergoes a first-order transition from a high-temperature metallic to a low-temperature insulating phase at $T_c = 340\,K$, while the crystal symmetry is reduced from rutile (R) to monoclinic (M1) due to the formation of V dimers (see Figure 25.6). The driving force of this transition has been discussed controversially. Both a structural Peierls instability with a band-like energy gap in the M1 phase [33] and Coulomb repulsion and charge localization typical of a Mott insulator [34, 35] have been proposed.

Time-resolved studies, using femtosecond light pulses to photoinject electron–hole pairs into VO_2 and trigger an insulator–metal transition combined with

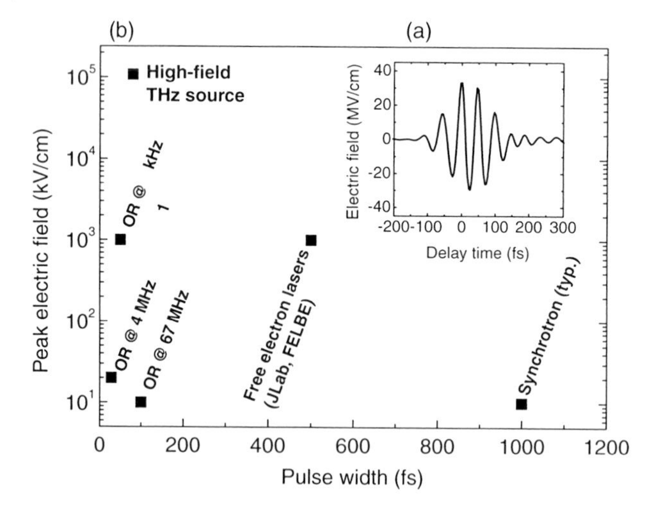

Figure 25.5 The most intense THz fields. (a) Few-cycle THz transient generated and detected via the system of Figure 25.4 (see Ref. [13]). (b) Overview of the peak THz fields as function of the pulse duration for a selection of state-of-the-art high-field sources. OR: optical rectification of femtosecond pulses from Ti:sapphire amplifiers operating at 1 kHz, 4 MHz, or 67 MHz, respectively. JLab: Jefferson Laboratory Free Electron Laser, FELBE: Dresden Free Electron Laser.

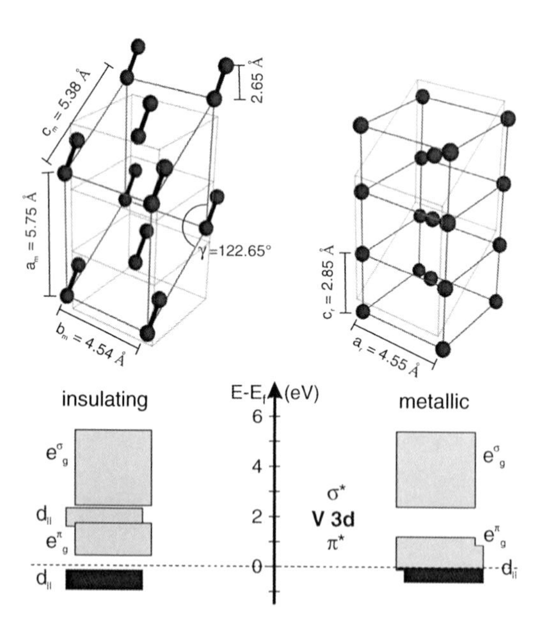

Figure 25.6 Vanadium sublattice (top) and schematic band structure (bottom) of the insulating monoclinic (left) and the metallic rutile (right) phase of VO_2. Labeling, energetic positions, and widths of the electronic bands have been adopted from Ref. [36].

structural rearrangement, have promised insight into the microscopic dynamics. Optical [37, 38] and X-ray [39] spectroscopies as well as – most recently – electron diffraction [40] have been employed to follow ultrafast phase transitions. For VO_2, sufficient temporal resolution to reveal the inherent timescales has been reported, however, only in femtosecond optical reflectivity data. They suggest a 75 fs structural bottleneck for photoswitching [38]. Nevertheless, the microscopic dynamics remains elusive for visible light pulses.

In contrast, ultrabroadband THz pulses provide direct and selective access to the lattice and electronic degrees of freedom on the femtosecond timescale [21], and thus afford important information on the interplay of the microscopic mechanism during the phase transition. Our experiment is based on the NIR pump/multi-THz probe scheme depicted in Figure 25.7. An intense 12 fs laser pulse centered at a photon energy of 1.55 eV excites electrons from the occupied $d_{||}$ band into conduction band states (Figure 25.6) of the insulating phase of a 120 nm thick film of polycrystalline VO_2. After a variable delay time t_D, an ultrashort multi-THz test transient is transmitted. Electro-optic detection yields the profile of the probe electric field as a function of the delay time T. By comparing the THz transients before and after the sample – both with and without optical pump – we may extract the dielectric response in the steady state and far from equilibrium. Most important, both amplitude and phase of the THz fields are known. We may thus exploit straightforward electrodynamical relations to retrieve the full complex-valued THz conductivity without resorting to Kramers–Kronig analysis [14]. Technically, we create a 2D map of the nonequilibrium response by systematic variation of the two independent time axes T and t_D. A careful Fourier analysis of these data then uncovers the dynamics of the complex-valued ultrabroadband conductivity with subcycle resolution [41]. For the present experiment, it is most appropriate to Fourier transform the 2D data along axis T for selected delay times $\tau = T + t_D$, that is, at a well-defined delay between excitation of the VO_2 film and electro-optic sampling of its response. In this way, we retrieve the MIR conductivity $\sigma(\omega)$ and its pump-induced changes $\Delta\sigma(\omega, \tau)$ with femtosecond resolution [14, 41].

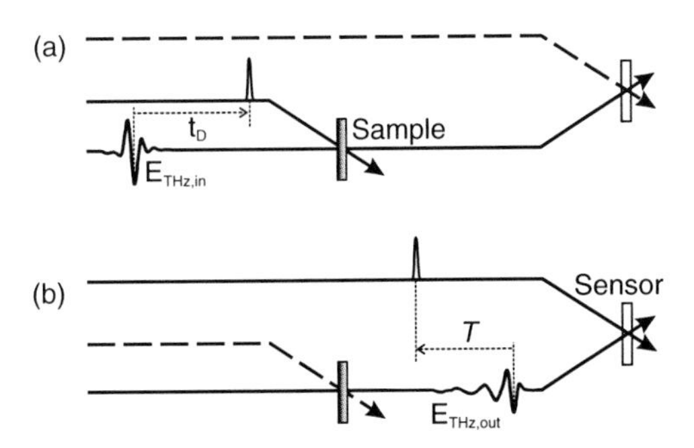

Figure 25.7 The principle of the femtosecond pump/THz probe experiment. (a) An optical pump prepares the sample in a strongly nonequilibrium state. A multi-THz transient $E_{THz,\,in}$ introduced after a delay time t_D probes the subsequent low-energy dynamics. (b) The transmitted field profile $E_{THz,\,out}$ is recorded electro-optically as a function of a second delay T.

Figure 25.8a shows the real part $\sigma_1(\omega)$ of the THz conductivity of the unexcited VO_2 film below and above T_c. The large value and the spectral shape of $\sigma_1(\omega)$ in the metallic phase are consistent with time-integrated mid-IR data of Ref. [42]. Although the electronic conductivity of insulating VO_2 should vanish, $\sigma_1(\omega)$ exhibits three pronounced maxima at $\hbar\omega = 50$, 62, and 74 meV, corresponding to the polarizability of the highest frequency transverse optical phonons associated with vibrations of the oxygen cages surrounding the V atoms [43]. The spectral region above 85 meV is free of IR-active resonances. Hence, the THz conductivity in this window is almost two orders of magnitude lower than in the rutile phase.

The contour plots of Figure 25.8 depict the spectral shape of the pump-induced conductivity changes $\Delta\sigma_1(\omega, \tau)$ as a function of the time delay τ. Panel (b) shows data for an excitation density of $\Phi = 3\ \mathrm{mJ/cm^2}$. We consider two domains (labeled P and E in Figure 25.8) that correspond to different physical processes: Since spectral region E ($\hbar\omega \geq 85$ meV) is free of phonon resonances (see Figure 25.8a), the pump-induced THz conductivity in this region derives solely from *electronic* degrees of freedom.

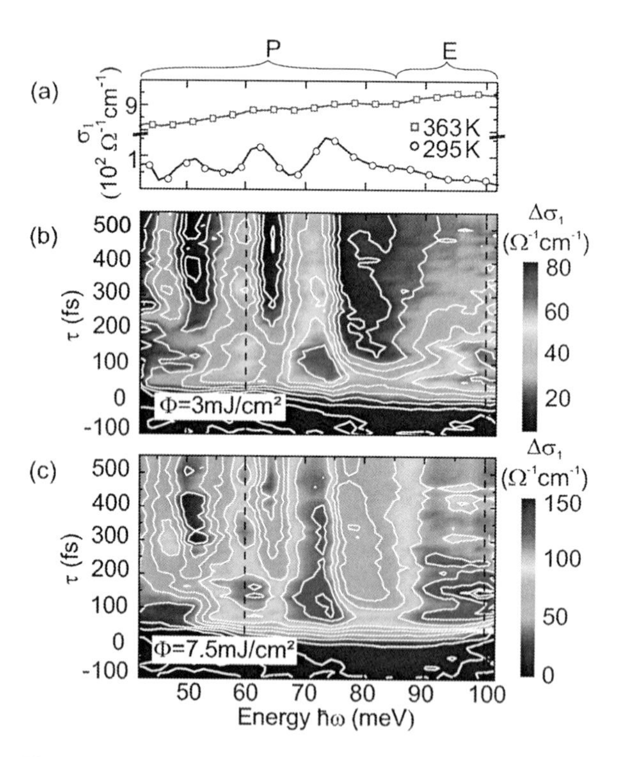

Figure 25.8 2D optical pump–THz probe data. (a) Equilibrium conductivity of metallic (squares) and insulating (circles) VO_2. Color plot of the pump-induced *changes* of the conductivity $\Delta\sigma_1(\omega, \tau)$ for pump fluences (b) $\Phi = 3\ \mathrm{mJ/cm^2}$ and (c) $\Phi = 7.5\ \mathrm{mJ/cm^2}$ at $T_L = 250$ K. The broken vertical lines indicate the frequency positions of cross sections reproduced in Figure 25.9. The spectral domain P comprises predominantly changes of the phonon resonances, while region E reflects the electronic conductivity. (Please find a color version of this figure on the color plates.)

Features in the energy regime P (40 meV $< \hbar\omega <$ 85 meV) relate to the IR-active phonon resonances, exposing the *lattice* dynamics.

The differing origins of the signals in the two spectral windows are underscored by qualitatively different temporal dynamics: Ultrafast photodoping induces a quasi-instantaneous onset of conductivity in region E due to directly injected mobile carriers. $\Delta\sigma_1(\omega, \tau)$ in region E decays promptly within 0.4 ps. In contrast, the phononic contribution (domain P) is more long-lived. Photoexcitation induces an increase of polarizability on the low-frequency side of all three phonon resonances, while a smaller change is seen on the blue wing. The change in frequency is superimposed on a striking coherent modulation of $\Delta\sigma_1(\omega, \tau)$, along the pump–probe delay axis τ. This phenomenon is most notable at a THz photon energy of 60 meV (vertical broken line in Figure 25.8b). The corresponding cross section for the 2D data is reproduced in Figure 25.9a. The Fourier transform of the oscillations along τ is centered at 6 THz (Figure 25.9b). Phonons in this frequency regime are critical to the metal–insulator transition of VO_2. In the ground state, A_g lattice modes associated with stretching and tilting of V–V dimers are found at 5.85 and 6.75 THz [38]. These vibrations map the M1 structure onto the R lattice [35]. As discussed below, the periodic modulation of the IR-active phonons is the first manifestation of anharmonic lattice coupling with such A_g phonons.

Figure 25.8c displays results of the 2D THz experiment at a higher excitation fluence. For early delay times $\tau <$ 500 fs, the spectral features in region P qualitatively resemble the corresponding region in Figure 25.8b, scaled by the excitation density. On the other hand, the dynamics of the conductivity in window E differs profoundly (see also cross sections in Figure 25.9c): After the onset of $\Delta\sigma_1(\omega, \tau)$ due to photodoping, the electronic conductivity shows one cycle of modulation in phase with the coherent lattice motion. A distinct signal maximum occurs at a temporal

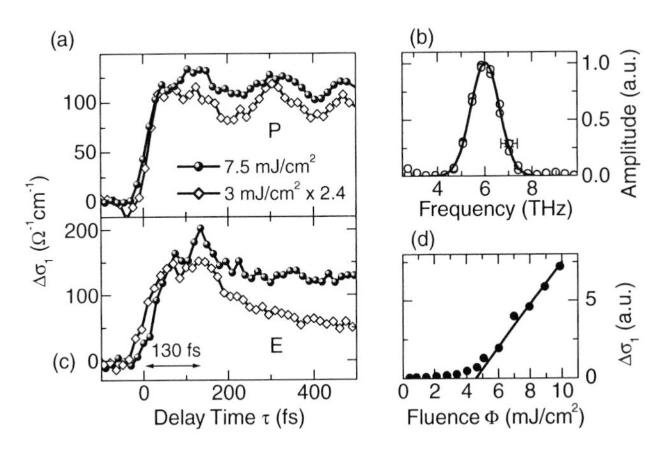

Figure 25.9 Cross sections through the 2D data of Figure 25.8b and c along the time axis τ for a photon energy of (a) $\hbar\omega$ = 60 meV and (c) $\hbar\omega$ = 100 meV. The traces taken at Φ = 3 mJ/cm^2 are scaled by a factor of 2.4 in amplitude. (b) Fourier transform of the periodic modulation of the THz response in (a). (d) Fluence dependence of $\Delta\sigma_1(\tau)$ at τ = 1 ps (T_L = 295 K).

delay of $\tau = 130$ fs. Subsequently, the THz conductivity settles at a constant value for at least 10 ps, indicating the transition of the electronic system into a metallic phase. Remarkably, for $\tau > 130$ fs, no more periodic modulation is imprinted on the electronic conductivity, while the data in region P demonstrate that the crystal lattice is still in a state of large-amplitude coherent oscillations. For late delay times τ beyond 1 ps, the phononic differential transmission completely vanishes for the lower excitation density, indicating a return to the M1 configuration.

A systematic fluence dependence of the (spectrally integrated) conductivity at $\tau = 1$ ps is shown in Figure 25.9d. The THz conductivity vanishes below a threshold fluence Φ_c, while it grows nonlinearly above. This dependence is symptomatic of a cooperative insulator–metal transition and substantiates our assignment of the long-lived electronic conductivity to the ultrafast transition into the metal phase. We find that Φ_c decreases with increasing lattice temperature, for example, $\Phi_c(T_L = 250\text{ K}) = 5.3\text{ mJ/cm}^2$ and $\Phi_c(T_L = 325\text{ K}) = 3\text{ mJ/cm}^2$.

With this rich dynamics described, we propose a qualitative picture of the ultrafast phase transition of VO_2 that captures the essence of the observed dynamics. The idea is inspired by recent cluster dynamical mean field calculations [44]. This theory approximates the dielectric phase as a molecular crystal of V–V dimers embedded in a matrix of oxygen octahedra and shows that the bandgap arises from the strong correlations between two electrons on each singlet modeled in a basis of bonding and antibonding Heitler–London orbitals. The energy dependence of such states on the nuclear coordinate (i.e., V–V separation) is depicted schematically in Figure 25.10. The minimum of the bonding energy surface defines the atomic position in the M1 phase. Absorption of a near-IR photon removes an electron from the bonding orbital, destabilizing the dimer, while the lattice site is left in an excited Franck–Condon state (marker (i) in Figure 25.10). In an isolated molecule, the energy surface of the excited state would lead to dissociation. Due to the repulsion by the nearest neighbors, an energy minimum of the antibonding orbitals will be located near the R configuration by symmetry. Ultrafast photoexcitation thus launches a coherent structural defor-

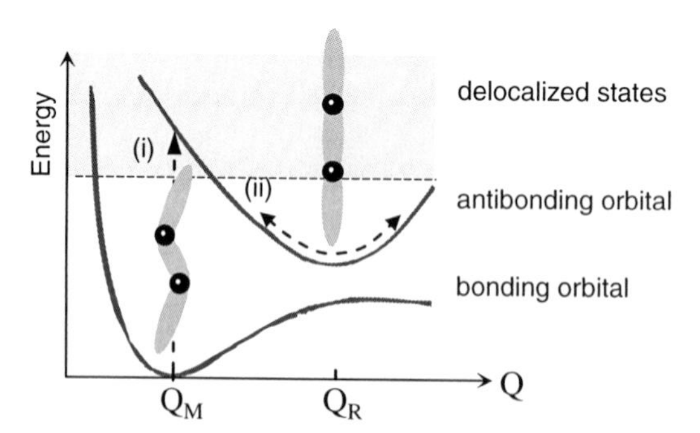

Figure 25.10 Schematic potential surfaces of localized dimers. Q_M and Q_R represent different spatial configurations of the V–V dimers in the structural phases M1 and R, respectively. (i) Photoexcitation of spin singlets into a conductive state. (ii) Structural relaxation and coherent vibrations about the new energy minimum.

mation of excited dimers (marker (ii) in Figure 25.10) followed by oscillations at 6 THz around the new potential minimum [45].

From the known lattice constants of the R and M1 phase, we estimate changes of the V–V distance as large as 0.2 Å [46]. Such extreme deformation imposes strong distortions on the surrounding oxygen octahedra and affect their elastic tensors. Our experiment directly demonstrates the influence of a coherent lattice motion on other phonon resonances: In region P of Figure 25.8, an overall redshift of the oxygen modes attests to a modified average structure of the V–V dimers, while the coherent modulation at 6 THz reflects large-amplitude oscillations about the new potential minimum in the excited state (Figure 25.10).

Our tentative model also provides insight into the nature of the transient electronic conductivity. Photodoping initially leaves the electrons in a highly mobile state in the continuum of delocalized bands (Figure 25.10). The rapid nonexponential decay of the electronic conductivity observed below Φ_c (Figure 25.9c)) indicates that the structural deformation shifts the energy of excited dimers below a mobility edge for extended electronic states – a process reminiscent of self-trapping of excitons [45].

In stark contrast, the electronic conductivity does not decay rapidly at high pump fluences. After an initial regime with one modulation cycle in parallel with the V–V coherent wave packet motion, the conductivity settles to a constant value with no further modulations discernible. Such behavior would be very surprising for a standard solid where the electronic system obeys the Born–Oppenheimer approximation and therefore reacts instantly to the lattice configuration. The decoupling of the electronic conductivity from the coherent lattice motion, which we observe after one V–V oscillation cycle, may be regarded as a direct consequence of the strongly correlated situation in the low-temperature phase of VO_2: Obviously, the local electronic correlations on each dimer that are a hallmark of the dielectric phase cannot be restored fast enough in the presence of a critical density of photoexcited lattice sites and thermal fluctuations preset by the substrate temperature. This retardation effect stabilizes the metallic conductivity at a stage where the lattice is still far from equilibrium. Our model assumption of a reaction time of the local electron correlations that becomes slower than the V–V oscillation period might also explain why the metallic lattice configuration is ultimately adopted after excitation above threshold [39].

Thus, the simultaneous MIR spectral and subcycle temporal resolution allowed us to disentangle the dynamics of both lattice and electronic degrees of freedom substantiating a novel qualitative picture of the photoinduced phase transition. It will be particularly interesting to see how future theoretical studies compare to our model.

25.4
THz Coherent Control of Excitons

Excitons represent a particularly well-defined class of correlated electrons. In contrast to the situation encountered in VO_2, excitons in semiconductors may be treated as a perturbation of the single-particle band structure. Optical interband absorption promoting electrons from the valence into the conduction band prepares a highly complex nonequilibrium scenario of many-body interactions. In the dilute limit, the description of this system may be greatly simplified by regarding the electron–hole ensemble as a gas of excitons, Coulomb-bound electron–hole pairs with a hydrogen-like fine structure in the meV range [47]. The analogy to atoms has motivated an intense search for macroscopically ordered states and Bose–Einstein condensation (BEC) also in excitons [48]. Recent reports of quantum degeneracy in the related system of exciton-polaritons have refueled these hopes [49].

Unfortunately, excitons suffer an important drawback. While sophisticated quantum optical protocols had paved the way to the first observation of BEC of atomic gases, comparable optical control of excitons has been missing because systems relevant for BEC often exhibit weak, if any, radiative interband coupling. Excitons of the yellow series in Cu_2O illustrate this dilemma most drastically [48, 50–53]: An unusually strong exchange interaction splits the 1s ground state (binding energy: 150 meV) by 12 meV into a triply degenerate Γ_5^+ *ortho* and a lower lying nondegenerate Γ_2^+ *para* variety. While interband recombination of *ortho* excitons is dipole forbidden, *para* states are optically dark to all multipole orders due to their spin of unity. This fact ensures long lifetimes on the microsecond scale and thus makes 1s *para* excitons interesting for quasiequilibrium BEC [50, 52]. On the other hand, spin conservation also prohibits the direct observation and control of *para* states via optical interband transitions.

Terahertz pulses couple directly to the internal excitations of quasiparticles, irrespective of interband matrix elements [19, 51–55]. Studies of *intraexcitonic absorption* have provided novel insight into the formation dynamics, fine structure, density, and temperature of excitons [52, 54]. The recent observation of the inverse quantum process of *stimulated* THz *emission* from internal 3p–2s transitions in Cu_2O has raised the hope for future control of dark excitons via coherent THz nonlinearities [17]. In this section, we exploit intense THz transients to control the orbital degree of freedom of a dense gas of 1s *para* excitons in Cu_2O via internal Rabi oscillations.

Our sample is a naturally grown single crystal of Cu_2O (thickness: 264 μm) kept at a temperature of $T_L = 5$ K. In a first step, we prepare a dense and cold gas of 1s *para* excitons. Since these states are strictly dark, they may not be created by resonant interband photoinjection. Instead, we employ near-infrared 12 fs pulses centered at an energy of 1.5 eV to generate unbound electron–hole pairs with a homogeneous density throughout the crystal length by two-photon absorption. Energy relaxation via phonon emission causes a delayed formation of bound exciton states within several picoseconds after excitation. This dynamics is resonantly traced via NIR pump/multi-THz probe spectroscopy following the scheme of Figure 25.7. The complex dielectric response of the electron–hole gas is retrieved analogously to Section 25.3.

Figure 25.11a depicts pump-induced changes of the absorption coefficient as a function of t_D. Starting with an almost Drude-like response of the photogenerated unbound *e–h* pairs, at $t_D = 100$ fs, the spectra change dramatically on a few picosecond scale. At $t_D = 11$ ps, the hallmark 1s–2p lines attest to a population of

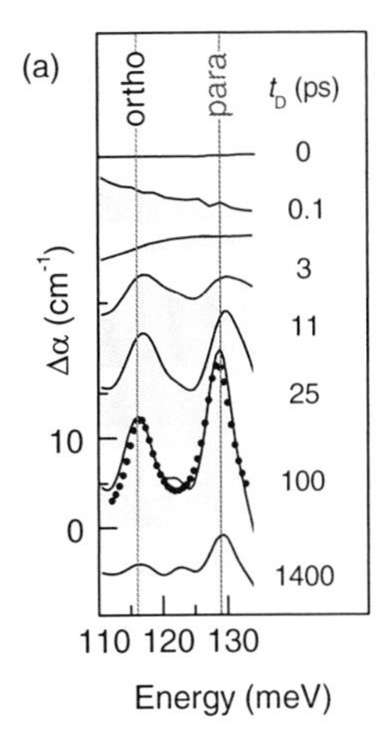

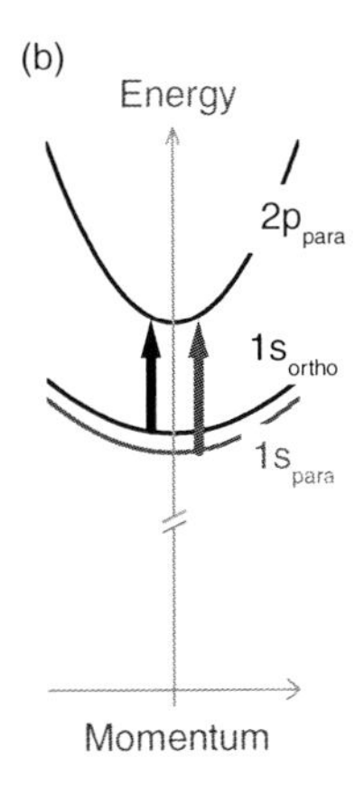

Figure 25.11 (a) Formation and cooling dynamics of 1s excitons in Cu_2O probed by multi-THz spectroscopy: Pump-induced changes of the absorption $\Delta\alpha$ (shaded area) for various delay times t_D after two-photon absorption of 12 fs pulses centered at 1.55 eV. Vertical lines: 1s–2p resonances at vanishing center-of-mass momentum. Dots: Theoretical fits with $n_{1s}^{para} = 1 \times 10^{16}$ cm^{-3}, $n_{1s}^{ortho} = 0.5 \times 10^{16}$ cm^{-3}, and $T_{1s} = 10$ K. (b) Due to different effective masses of 1s and 2p excitons, THz transitions (arrows) depend on the center-of-mass momentum.

1s *para* and *ortho* states [52]. Both intraexcitonic lines narrow and shift to lower frequencies during the first 100 ps, while the ratio of *para* versus *ortho* densities increases. Due to different effective masses of 1s and 2p states (Figure 25.11b), the temperature T_{1s} of the ensemble is encoded in the THz line shape [51, 52]. For $t_D = 100$ ps, line fits provide best agreement with the experiment when we assume densities of $n_{1s}^{para} = (1 \pm 0.2) \times 10^{16}$ cm^{-3} and $n_{1s}^{ortho} = (0.5 \pm 0.1) \times 10^{16}$ cm^{-3} and a temperature of $T_{1s} = 10 \pm 4$ K (Figure 25.11a). n_{1s}^{para} is thus approximately only one order of magnitude below critical values for potential BEC.

In the next step, we actively control the internal quantum state of the dark quasiparticles. To this end, we replace the broadband probe with an intense THz transient E_{THz} generated resonantly to the 1s–2p *para* transition (photon energy: 129 meV, width: 4 meV). Curves (i)–(vi) in Figure 25.12a represent the response of the 1s *para* excitons to the incident waveform at six different peak fields E_{THz}^{peak}. All THz traces are recorded directly in the time domain by ultrabroadband electro-optic detection, resolving both amplitude and phase of the electric field. At low excitation intensity (curves (i) and (ii)), the re-emitted field reaches its maximum toward the end of the driving pulse E_{THz} and then decays with a finite dephasing time of 0.7 ps. Strikingly, for intermediate and high intensities of the driving field (curves (iii)–(vi)], the response is not a linearly increased version of the low-field case. Rather, the re-emitted field rises more rapidly, reaches its maximum before the peak of the pump transient, and decreases within the coherence window. Also, the maximum value of the amplitude saturates for increasing excitation density. This intensity-dependent response, in particular, the temporal shift of the amplitude maximum from

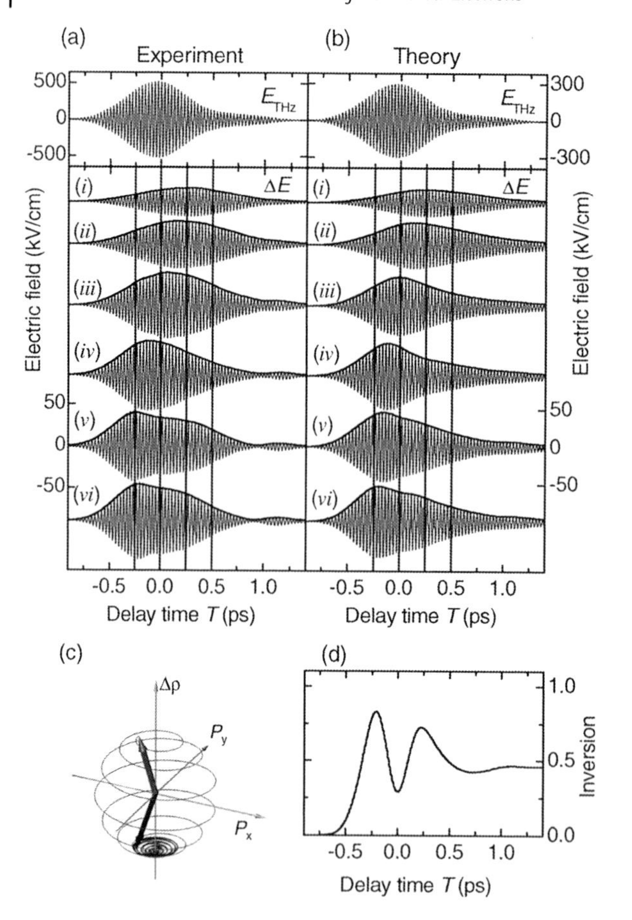

Figure 25.12 Nonlinear THz response of 1s *para* excitons: quantitative comparison of (a) experiment and (b) microscopic theory. Upper panels: Real-time profile of the exciting THz pulse E_{THz}. Lower panels: Re-emitted THz field ΔE (vertically offset) for six values of the peak driving field, with $E_{THz}^{peak} = 0.13, 0.24, 0.45, 0.67, 0.92, 1.0 \times E_0$ for curves (i)–(vi), respectively ($E_0 = 0.5$ MV/cm in experiment and $E_0 = 0.3$ MV/cm in theory). (c) Schematic of the Bloch sphere of the 1s–2p intraexcitonic two-level system. (d) Calculated population inversion of the two-level system consisting of 1s and 2p *para* exciton states at the center of the THz spot.

$T = 0.25$ ps (curve (i)) to $T = -0.25$ ps (curve (vi)), clearly shows that the THz field saturates the 1s–2p transition inducing a coherent nonlinearity. Most remarkably, curves (v) and (vi) exhibit a structured envelope with a first maximum at $T = -0.25$ ps and a second less pronounced side peak at 0.25 ps (curve v) and 0.24 ps (curve vi), respectively. The onset of an oscillatory behavior indicates that the intense THz beam leads to nonlinear dynamics of dark exciton populations well beyond the perturbative regime.

This dynamics is a manifestation of Rabi oscillations qualitatively explained in a Bloch picture of the two-level system (Figure 25.12c). The diagonal (population difference $\Delta\varrho = \varrho_{22} - \varrho_{11}$) and off-diagonal (polarizations P_x and P_y) elements of the density matrix are depicted as the vertical and horizontal coordinates. For low driving fields, the Bloch vector performs a Larmor precession in the vicinity of the south pole of the Bloch sphere, near the ground state of the two-level system. Experimentally, the projection of this trajectory onto the polarization axis is directly mapped out in the time domain (Figure 25.12a). With increasing THz field, the state vector may be driven toward the north pole inducing strong population inversion. During this Rabi cycle, the projection onto the polarization axis reaches a maximum at the equator and decreases from there on. Further rotation of the Bloch vector leads to a periodic increase and decrease of the polarization response. The real-time data of Figure 25.12a display indications of this dynamics and permit reconstruction of the actual motion of the Bloch vector.

To this end, the data may be quantitatively compared with a state-of-the-art microscopic theory by Koch and coworkers [22, 55, 56]. The model starts with a realistic band structure of Cu_2O and includes the THz response of all bound exciton levels, ionization, and ponderomotive terms, as well as the Gaussian-shaped spatial profile of the THz field that leads to an inhomogeneous THz excitation scenario. This approach reproduces both amplitude and phase of the re-emitted field convincingly (Figure 25.12b). The population inversion of the 1s–2p two-level subsystem computed for the center of the THz spot at the maximum field strength exhibits approximately two Rabi oscillation cycles (Figure 25.12d). The 1s–2p population inversion performs strong oscillations and reaches a peak value of $(80 \pm 5)\,\%$ at the spot center. Due to the Gaussian THz field distribution, the total detected signal ΔE averages over various pulse areas and hence displays only residual Rabi oscillations. We note that the novel high-field THz source presented in Section 25.2 will permit experiments under more homogeneous spatial excitation conditions. These studies are currently under way. Our results point out a novel route toward ultrafast nonlinear control of optically dark exciton states. In the near future, advanced protocols known from atomic systems may open new perspectives for preparing ultracold and dense exciton gases via high-field THz transients.

25.5 Conclusions and Perspectives

Ultrabroadband THz technology provides precise access to the electric field of infrared few-cycle pulses. These waveforms couple resonantly to low-energy excitations, with a time resolution better than one oscillation period of the carrier wave. Two classes of experiments are exemplified:

1) Multi-THz transients represent a unique *probe* of the MIR conductivity. We test this idea by studying the interplay of lattice and electronic degrees of freedom during an ultrafast insulator–metal transition of photoexcited VO_2. This case illustrates that the technique is well suited for ultrafast studies of correlated

electrons. A rapidly growing series of applications has exploited this potential: Starting with the first observation of the ultrafast buildup of screening and many-body correlations in a semiconductor [14, 19], studies of electronic and vibrational dynamics in graphite [16], carbon nanotubes [18], and organic semiconductors [57], as well as phase-resolved measurements of stimulated emission in a laser have been reported [20]. We expect a yet more comprehensive insight into correlated electron dynamics from a quantitative comparison with complementary approaches such as ultrafast photoelectron spectroscopy or X-ray and electron diffraction (part II, chapter V,6), once all techniques have reached equally high temporal resolution.

2) High-field THz sources have now set the stage for strong THz *nonlinearities*. As a specific example, we exploit THz fields of the order of MV/cm to control the internal quantum state of *para* excitons in Cu_2O. This work underpins the fact that intense THz pulses open up a very general paradigm: direct amplitude and phase-sensitive control of condensed matter via low-energy resonances. In contrast to intense laser pulses in the visible and near-infrared regimes, high-field THz transients manipulate many-body systems in their electronic ground level, preparing an especially well-defined nonequilibrium initial state. Perspectives of systematic THz quantum optics of solids, field-driven phase transitions of strongly correlated electron systems [58], and coherent nonlinearities in molecules are attracting particular attention. All these studies have exploited exclusively the leading term in the multipole expansion of light–matter interaction – the electric dipole moment – while coupling to magnetic dipoles has been weak enough to be neglected. Recent advances in high-field THz sources may change this situation dramatically: Record THz electric fields of 108 MV/cm [13] are associated with giant magnetic fields of up to 30 T. It will be interesting to see how these fields interact with magnetic dipoles of electron spins. Ultimately, THz control may be combined even with nonoptical probes of the electronic energy distribution or the real-space structure of the ionic lattice. Latest experiments presented throughout this book may have just pushed open the door to this exciting new territory.

References

1 Ferguson, B. and Zhang, X.-C. (2002) Materials for THz science and technology. *Nat. Mater.*, **1**, 26–33; Tonouchi, M. (2007) Cutting-edge THz technology. *Nat. Photonics*, **1**, 97–105.

2 Fattinger, C. and Grischkowsky, D. (1989) THz beams. *Appl. Phys. Lett.*, **54**, 490–492;Smith, P. *et al.* (1988) Subpicosecond photoconducting dipole antennas. *IEEE J. Quantum Electron.*, **24**, 255–260.

3 Wu, Q. and Zhang, X.-C. (1997) 7 THz broadband GaP electro-optic sensor. *Appl. Phys. Lett.*, **70**, 1784–1786.

4 Leitenstorfer, A. *et al.* (1999) Detectors and sources for ultrabroadband electro-optic sampling: experiment and theory. *Appl. Phys. Lett.*, **74**, 1516–1518;Leitenstorfer, A. *et al.* (1999) Femtosecond charge transport in polar semiconductors. *Phys. Rev. Lett.*, **82**, 5140–5143.

5 Kaindl, R.A. *et al.* (1999) Broadband phase-matched difference frequency mixing of femtosecond pulses in GaSe: experiment and theory. *Appl. Phys. Lett.*, **75**, 1060–1062.
6 Huber, R. *et al.* (2000) Generation and field-resolved detection of femtosecond electromagnetic pulses tunable up to 41 THz. *Appl. Phys. Lett.*, **76**, 3191–3193.
7 Zentgraf, T. *et al.* (2007) Ultrabroadband 50–130 THz pulses generated via phase-matched difference frequency mixing in $LiIO_3$. *Opt. Express*, **15**, 5775–5781.
8 Sell, A. *et al.* (2008) Field-resolved detection of phase-locked infrared transients from a compact Er:fiber system tunable between 55 and 107 THz. *Appl. Phys. Lett.*, **93**, 251107.
9 Liu, K., Xu, J., and Zhang, X.-C. (2004) GaSe crystals for broadband THz wave detection. *Appl. Phys. Lett.*, **85**, 863–865.
10 Kübler, C. *et al.* (2004) Ultrabroadband detection of multi-THz field transients with GaSe electro-optic sensors: approaching the near infrared. *Appl. Phys. Lett.*, **85**, 3360–3362.
11 Kampfrath, T., Nötzold, J., and Wolf, M. (2007) Sampling of broadband THz pulses with thick electro-optic crystals. *Appl. Phys. Lett.*, **90**, 231113.
12 Ashida, M. *et al.* Ultrafast Phenomena XVI, Corkum, P., de Silvestri, S., Nelson, K.A., Riedle, E., Schoenlein, R.W. eds., *Springer Verlag, Berlin, Springer Series in Chemical Physics*, **92**, 979–981 (2009).
13 Sell, A., Leitenstorfer, A., and Huber, R. (2008) Phase-locked generation and field-resolved detection of widely tunable THz pulses with amplitudes exceeding 100 MV/cm. *Opt. Lett.*, **23**, 2767–2769.
14 Huber, R. *et al.* (2001) How many-particle interactions develop after ultrafast excitation of an electron–hole plasma. *Nature*, **414**, 286–289.
15 Luo, C.W. *et al.* (2004) Phase-resolved nonlinear response of a two-dimensional electron gas under femtosecond intersubband excitation. *Phys. Rev. Lett.*, **92**, 047402.
16 Kampfrath, T. *et al.* (2005) Strongly coupled optical phonons in the ultrafast dynamics of the electronic energy and current relaxation in graphite. *Phys. Rev. Lett.*, **95**, 187403.
17 Huber, R. *et al.* (2005) Stimulated THz emission from intraexcitonic transitions in Cu_2O. *Phys. Rev. Lett.*, **96**, 017402.
18 Perfetti, L. *et al.* (2006) Ultrafast dynamics of delocalized and localized electrons in carbon nanotubes. *Phys. Rev. Lett.*, **96**, 027401.
19 Huber, R. *et al.* (2005) Femtosecond formation of phonon–plasmon coupled modes in InP: ultrabroadband THz experiment and quantum kinetic theory. *Phys. Rev. Lett.*, **94**, 027401.
20 Kröll, J. *et al.* (2007) Phase-resolved measurements of stimulated emission in a laser. *Nature*, **449**, 698–701.
21 Kübler, C. *et al.* (2007) Coherent structural dynamics and electronic correlations during an ultrafast insulator-to-metal phase transition in VO_2. *Phys. Rev. Lett.*, **99**, 116401.
22 Leinß, S. *et al.* (2008) THz coherent control of optically dark paraexcitons in Cu_2O. *Phys. Rev. Lett.*, **101**, 246401.
23 Dreyhaupt, A. *et al.* (2006) Optimum excitation conditions for the generation of high-electric-field THz radiation from an oscillator-driven photoconductive device. *Opt. Lett.*, **31**, 1546–1548.
24 Rice, A. *et al.* (1994) THz optical rectification from ⟨110⟩ zinc-blende crystals. *Appl. Phys. Lett.*, **64**, 1324–1326.
25 Sell, A. *et al.* (2009) 8-fs pulses from a compact Er:fiber system: quantitative modeling and experimental implementation. *Opt. Express*, **17**, 1070–1077.
26 Boyd, R.W. (1992) *Nonlinear Optics*, Academic Press.
27 Jones, D.J. *et al.* (2000) Carrier-envelope phase control of femtosecond mode-locked lasers and direct optical frequency synthesis. *Science*, **288**, 635–639; Apolonski, A. *et al.* (2000) Controlling the phase evolution of few-cycle light pulses. *Phys. Rev. Lett.*, **85**, 740–743.
28 Gallot, G. and Grischkowsky, D. (1999) Electro-optic detection of THz radiation. *J. Opt. Soc. Am. B*, **16**, 1204–1212.
29 Carr, G.L. *et al.* (2002) High-power THz radiation from relativistic electrons. *Nature*, **420**, 153–156.

30 Hebling, J. *et al.* (2008) Generation of high-power THz pulses by tilted-pulse-front excitation and their application possibilities. *J. Opt. Soc. Am. B*, **25**, B6–B19.

31 Reimann, K. *et al.* (2003) Direct field-resolved detection of THz transients with amplitudes of megavolts per centimeter. *Opt. Lett.*, **28**, 471–473.

32 Imada, M., Fujimori, A., and Tokura, Y. (1998) Metal–insulator transitions. *Rev. Mod. Phys.*, **70**, 1039–1263.

33 Goodenough, J.B. (1971) The two components of the crystallographic transition in VO_2. *J. Solid State Chem.*, **3**, 490; Wentzcovitch, R.M., Schulz, W.W., and Allen, P.B. (1994) VO_2: Peierls or Mott–Hubbard? A view from band theory. *Phys. Rev. Lett.*, **72**, 3389–3392.

34 Zylberstejn, A. and Mott, N. (1975) Metal–insulator transition in VO_2. *Phys. Rev. B*, **11**, 4383–4395.

35 Paquet, D. and Leroux-Hugon, P. (1980) Electron correlations and electron–lattice interactions in the metal–insulator, ferroelastic transition in VO_2: a thermodynamical study. *Phys. Rev. B*, **22**, 5284–5301.

36 Koethe, T.C. *et al.* (2006) Transfer of spectral weight and symmetry across the metal–insulator transition in VO_2. *Phys. Rev. Lett.*, **97**, 116402.

37 Becker, M.F. *et al.* (1994) Femtosecond laser excitation of the semiconductor–metal phase transition in VO_2. *Appl. Phys. Lett.*, **65**, 1507–1509.

38 Cavalleri, A. *et al.* (2004) Evidence for a structurally-driven insulator-to-metal transition in VO_2: a view from the ultrafast timescale. *Phys. Rev. B*, **70**, 161102.

39 Cavalleri, A. *et al.* (2001) Femtosecond structural dynamics in VO_2 during an ultrafast solid–solid phase transition. *Phys. Rev. Lett.*, **87**, 237401;(2005) Band-selective measurements of electron dynamics in VO_2 using femtosecond near-edge X-ray absorption. *Phys. Rev. Lett.*, **95**, 067405.

40 Baum, P., Yang, D.-S., and Zewail, A.H. (2007) 4D visualization of transitional structures in phase transformations by electron diffraction. *Science*, **318**, 788–792.

41 Kindt, J.T. and Schmuttenmaer, C.A. (1999) Theory for determination of the low-frequency time-dependent response function in liquids using time-resolved THz pulse spectroscopy. *J. Chem. Phys.*, **110**, 8589–8596.

42 Choi, H.S. *et al.* (1996) Mid-infrared properties of a VO_2 film near the metal–insulator transition. *Phys. Rev. B*, **54**, 4621–4628.

43 Schilbe, P. (2002) Raman scattering in VO_2. *Physica B*, **316**, 600–602.

44 Biermann, S. *et al.* (2005) Dynamical singlets and correlation-assisted Peierls transition in VO_2. *Phys. Rev. Lett.*, **94**, 026404.

45 Dexheimer, S.L. *et al.* (2000) Femtosecond vibrational dynamics of self-trapping in a quasi-one-dimensional system. *Phys. Rev. Lett.*, **84**, 4425–4428.

46 Marezio, M. *et al.* (1972) Structural aspects of the metal–insulator transitions in Cr-Doped VO_2. *Phys. Rev. B*, **5**, 2541–2551.

47 Zimmermann, R. (1987) *Many-Particle Theory of Highly Excited Semiconductors*, Teubner-Texte zur Physik 18, BSB Teubner;Haug, H. and Koch, S.W. (2004) *Quantum Theory of the Optical and Electronic Properties of Semiconductors*, World Scientific.

48 Moskalenko, S.A. and Snoke, D.W. (2000) *Bose–Einstein Condensation of Excitons and Biexcitons*, Cambridge University Press.

49 Kasprzak, J. *et al.* (2006) Bose–Einstein condensation of exciton polaritons. *Nature*, **443**, 409–414.

50 Snoke, D.W. *et al.* (1990) Evidence for Bose–Einstein condensation of a two-component exciton gas. *Phys. Rev. Lett.*, **64**, 2543–2546;O'Hara, K.E. and Wolfe, J.P. (2000) Relaxation kinetics of excitons in Cu_2O. *Phys. Rev. B*, **62**, 12909–12922.

51 Johnsen, K. and Kavoulakis, G.M. (2001) Probing Bose–Einstein condensation of excitons with electromagnetic radiation. *Phys. Rev. Lett.*, **86**, 858–861.

52 Kubouchi, M. *et al.* (2005) Study of orthoexciton-to-paraexciton conversion in Cu_2O by excitonic Lyman spectroscopy. *Phys. Rev. Lett.*, **94**, 016403.

53 Jörger, M. *et al.* (2005) Midinfrared properties of Cu_2O: high-order lattice vibrations and intraexcitonic transitions of the 1s paraexciton. *Phys. Rev. B*, **71**, 235210.

54 Kaindl, R.A. *et al.* (2003) Ultrafast THz probes of transient conducting and insulating phases in an electron–hole gas. *Nature*, **423**, 734–738.

55 Koch, S.W. *et al.* (2006) Semiconductor excitons in new light. *Nat. Mat.*, **5**, 523–531, and references therein.

56 Steiner, J.T., Kira, M., and Koch, S.W. (2008) Optical nonlinearities and Rabi flopping of an exciton population in a semiconductor interacting with strong THz fields. *Phys. Rev. B*, **77**, 165308.

57 Koeberg, M. *et al.* (2007) Simultaneous ultrafast probing of intramolecular vibrations and photoinduced charge carriers in rubrene using broadband time-domain THz spectroscopy. *Phys. Rev. B*, **75**, 195216.

58 Rini, M. *et al.* (2007) Control of the electronic phase of a manganite by mode-selective vibrational excitation. *Nature*, **449**, 72–74.

Index

Dynamics at Solid State Surface and Interfaces Vol.1: Current Developments
Edited by Uwe Bovensiepen, Hrvoje Petek, and Martin Wolf

ISBN: 978-3-527-40937-2